# Pigment–Protein Complexes in Plastids:
Synthesis and Assembly

# CELL BIOLOGY: A Series of Monographs

## EDITORS

**D. E. BUETOW**
Department of Physiology
and Biophysics
University of Illinois
Urbana, Illinois

**I. L. CAMERON**
Department of Cellular and
Structural Biology
The University of Texas
Health Science Center at San Antonio
San Antonio, Texas

**G. M. PADILLA**
Department of Cell Biology
Duke University Medical Center
Durham, North Carolina

**A. M. ZIMMERMAN**
Department of Zoology
University of Toronto
Toronto, Ontario, Canada

*Volumes published since 1984*

Gary S. Stein and Janet L. Stein (editors). RECOMBINANT DNA AND CELL PROLIFERATION, 1984

Prasad S. Sunkara (editor). NOVEL APPROACHES TO CANCER CHEMOTHERAPY, 1984

B. G. Atkinson and D. B. Walden (editors). CHANGES IN EUKARYOTIC GENE EXPRESSION IN RESPONSE TO ENVIRONMENTAL STRESS, 1985

Reginald M. Gorczynski (editor). RECEPTORS IN CELLULAR RECOGNITION AND DEVELOPMENTAL PROCESSES, 1986

Govindjee, Jan Amesz, and David Charles Fork (editors). LIGHT EMISSION BY PLANTS AND BACTERIA, 1986

Peter B. Moens (editor). MEIOSIS, 1986

Robert A. Schlegel, Margaret S. Halleck, and Potu N. Rao (editors). MOLECULAR REGULATION OF NUCLEAR EVENTS IN MITOSIS AND MEIOSIS, 1987

Monique C. Braude and Arthur M. Zimmerman (editors). GENETIC AND PERINATAL EFFECTS OF ABUSED SUBSTANCES, 1987

E. J. Rauckman and George M. Padilla (editors). THE ISOLATED HEPATOCYTE: USE IN TOXICOLOGY AND XENOBIOTIC BIOTRANSFORMATIONS, 1987

Heide Schatten and Gerald Schatten (editors). THE MOLECULAR BIOLOGY OF FERTILIZATION, 1989

Heide Schatten and Gerald Schatten (editors). THE CELL BIOLOGY OF FERTILIZATION, 1989

Anwar Nasim, Paul Young, and Byron F. Johnson (editors). MOLECULAR BIOLOGY OF THE FISSION YEAST, 1989

Mary P. Moyer and George Poste (editors). COLON CANCER CELLS, 1990

Gary S. Stein and Jane B. Lian (editors). MOLECULAR AND CELLULAR APPROACHES TO THE CONTROL OF PROLIFERATION AND DIFFERENTIATION, 1991

Vitauts I. Kalnins (editor). THE CENTROSOME, 1992

Carl M. Feldherr (editor). NUCLEAR TRAFFICKING, 1992

Christer Sundqvist and Margareta Ryberg (editors). PIGMENT–PROTEIN COMPLEXES IN PLASTIDS: SYNTHESIS AND ASSEMBLY, 1993

David H. Rohrbach and Rupert Timpl (editors). MOLECULAR AND CELLULAR ASPECTS OF BASEMENT MEMBRANES, 1993

Danton O'Day (editor). SIGNAL TRANSDUCTION DURING BIOMEMBRANE FUSION, 1993. *In preparation*

# Pigment–Protein Complexes in Plastids: Synthesis and Assembly

*Edited by*

**Christer Sundqvist**
**Margareta Ryberg**
*Department of Plant Physiology*
*Botanical Institute*
*University of Göteborg*
*Göteborg, Sweden*

ACADEMIC PRESS, INC.
Harcourt Brace Jovanovich, Publishers

San Diego   New York   Boston   London   Sydney   Tokyo   Toronto

This book is printed on acid-free paper. ∞

Copyright © 1993 by ACADEMIC PRESS, INC.
All Rights Reserved.
No part of this publication may be reproduced or transmitted in any form or by any means, electronic or mechanical, including photocopy, recording, or any information storage and retrieval system, without permission in writing from the publisher.

Academic Press, Inc.
1250 Sixth Avenue, San Diego, California 92101-4311

*United Kingdom Edition published by*
Academic Press Limited
24–28 Oval Road, London NW1 7DX

Library of Congress Cataloging-in-Publication Data

Pigment-protein complexes in plastids : synthesis and assembly / edited by Christer Sundqvist and Margareta Ryberg.
   p.  cm. – (Cell biology)
  Includes bibliographical references and index.
  ISBN 0-12-676960-5
  1. Chloroplasts.  2. Plastids.  3. Thylakoids.  4. Plant pigments.
5. Plant proteins.  I. Sundqvist, Christer.  II. Ryberg, Margareta.
III. Series.
QK725.P54   1993
581.87'33–dc20                                                 92-26705
                                                                                                   CIP

PRINTED IN THE UNITED STATES OF AMERICA
   93  94  95  96  97    EB    9  8  7  6  5  4  3  2  1

# Contents

Contributors .................................................... xi
Preface ........................................................ xiii

## 1 The Chloroplast as Site of Chlorophyll Formation and Photosynthesis: A Short History

*Hemming I. Virgin*

I. Introduction ................................................. 1
II. The Chloroplast as an Organelle .............................. 2
III. The Chloroplast as Site for Photosynthetic Pigments .......... 5
IV. The Chloroplast as Site for Photosynthesis ................... 11
V. Concluding Remarks .......................................... 16
References ................................................... 16

## 2 Plastid Ultrastructure and Development

*Hans Ryberg, Margareta Ryberg, and Christer Sundqvist*

I. Introduction ................................................ 25
II. Common Plastid Features ..................................... 26
III. Plastid Types and Development ............................... 32
IV. Concluding Remarks .......................................... 50
References ................................................... 51

## 3 Light and Temperature Regulation of Chloroplast Development

*Kenneth Eskins*

I. Introduction ................................................ 63
II. Chloroplast Pigments ........................................ 65
III. Pigment–Protein Complexes ................................... 67
IV. Light-Regulated Development ................................. 68

|   |   |   |
|---|---|---|
| V. | Light Effects on Transcription and Translation | 78 |
| VI. | Case Study: Regulation of LHCPII | 79 |
| VII. | Temperature Effects on Chloroplast Development | 81 |
| VIII. | Concluding Remarks | 83 |
|   | References | 83 |

## 4   Biosynthesis of the Chlorophyll Chromophore of Pigmented Thylakoid Proteins

*William R. Richards*

|   |   |   |
|---|---|---|
| I. | Introduction | 91 |
| II. | Early Stages of Chlorophyll Synthesis: The Pathway to Protoporphyrin IX | 94 |
| III. | Later Stages of Chlorophyll Synthesis: The Magnesium Branch | 124 |
|   | References | 156 |

## 5   Protochlorophyllide Reductase: A Key Enzyme in the Greening Process

*Rüdiger Schulz and Horst Senger*

|   |   |   |
|---|---|---|
| I. | Introduction | 180 |
| II. | Light Regulation of Protochlorophyllide Reductase | 182 |
| III. | Pigment–Enzyme Complex | 190 |
| IV. | Enzyme Activity of Protochlorophyllide Reductase | 193 |
| V. | Cloning of Protochlorophyllide Reductase Genes and Primary Structure of Enzyme Protein | 196 |
| VI. | Active Site and Membrane Association of Protochlorophyllide Reductase | 203 |
| VII. | Concluding Remarks | 205 |
|   | References | 206 |

## 6   Esterification of Chlorophyllide and Its Implication for Thylakoid Development

*Wolfhart Rüdiger*

|   |   |   |
|---|---|---|
| I. | Introduction | 219 |
| II. | Detection and Characterization of Chlorophyll Synthetase Activity | 220 |
| III. | Hydrogenation of Geranylgeranyl Derivatives | 229 |
| IV. | Translation of Plastid-Encoded Chlorophyll *a* Apoproteins and Chlorophyll Synthetase Reaction | 231 |
| V. | Influence of Metal Chelators and 5-Aminolevulinate on Chlorophyll Synthetase of Developing Plastids | 234 |
| VI. | Concluding Remarks | 236 |
|   | References | 237 |

# Contents

## 7 Chloroplast Lipids and the Assembly of Membranes

*Eva Selstam and Anna Widell Wigge*

|     |     |     |
| --- | --- | --- |
| I.   | Introduction | 241 |
| II.  | Plastid Lipids and Localization of Biosynthesis | 242 |
| III. | Heterogeneity of Lipids in Inner Plastid Membranes | 254 |
| IV.  | Specific Interactions between Thylakoid Lipids and Chlorophyll Protein Complexes in Photosystem II | 258 |
| V.   | Assembly of Prolamellar Body Membranes | 261 |
|      | References | 268 |

## 8 Regulation, Synthesis, and Integration of Chloroplast- and Nuclear-Encoded Proteins

*Wolfgang Hachtel and Andreas Friemann*

|     |     |     |
| --- | --- | --- |
| I.   | Introduction | 279 |
| II.  | Chloroplast-Encoded Chlorophyll Apoproteins | 281 |
| III. | Nuclear-Encoded Chlorophyll-Binding Proteins | 289 |
| IV.  | Coordination of Nuclear and Plastid Gene Expression | 297 |
| V.   | Concluding Remarks | 299 |
|      | References | 301 |

## 9 Import and Routing of Chloroplast Proteins

*Douwe de Boer and Peter Weisbeek*

|     |     |     |
| --- | --- | --- |
| I.   | Introduction | 311 |
| II.  | Chloroplast Targeting Signals | 313 |
| III. | Chloroplast Envelope Translocation | 317 |
| IV.  | Intraorganelle Routing | 323 |
|      | References | 328 |

## 10 Assembly and Reconstitution of Chlorophyll *a/b*-Containing Complexes

*Harald Paulsen*

|     |     |     |
| --- | --- | --- |
| I.   | Introduction | 335 |
| II.  | Chlorophyll *a/b*-Containing Complexes and Apoproteins | 336 |
| III. | Assembly of Light Harvesting Complex II | 341 |
| IV.  | Reconstitutions | 351 |
| V.   | Concluding Remarks | 356 |
|      | References | 356 |

## 11 Photosynthetic Activities during Early Assembly of Thylakoid Membranes

*Fabrice Franck*

|      |                                                                                                                 |     |
| ---- | --------------------------------------------------------------------------------------------------------------- | --- |
| I.   | Introduction                                                                                                    | 365 |
| II.  | Time Course of Development of Photosynthetic Activities after Single Turnover of Protochlorophyllide Reductase  | 366 |
| III. | Spectral Changes of Chlorophyll(ide) in Relation to Formation of Photosystems                                   | 371 |
| IV.  | Synthesis of Chlorophyll *a*-Binding Polypeptides Induced by Short Illumination                                 | 375 |
| V.   | Concluding Remarks                                                                                              | 377 |
|      | References                                                                                                      | 377 |

## 12 Structure, Function, and Assembly of Photosystem I

*Birgitte Andersen and Henrik Vibe Scheller*

|      |                                                              |     |
| ---- | ------------------------------------------------------------ | --- |
| I.   | Introduction                                                 | 383 |
| II.  | Structural and Functional Characterization of Photosystem I  | 387 |
| III. | Assembly of Photosystem I                                    | 404 |
| IV.  | Concluding Remarks                                           | 410 |
|      | References                                                   | 411 |

## 13 Function and Organization of Photosystem II

*Hans-Erik Åkerlund*

|      |                                                           |     |
| ---- | --------------------------------------------------------- | --- |
| I.   | Introduction                                              | 419 |
| II.  | Function of Photosystem II                                | 420 |
| III. | Structural Organization of Photosystem II                 | 424 |
| IV.  | Photoinhibition and Turnover of Photosystem II Components | 433 |
| V.   | Dynamic Changes and Photosystem II Heterogeneity          | 433 |
| VI.  | Concluding Remarks                                        | 436 |
|      | References                                                | 437 |

## 14 Carotenoids in Chloroplast Pigment–Protein Complexes

*George Britton*

|      |                                                              |     |
| ---- | ------------------------------------------------------------ | --- |
| I.   | Introduction                                                 | 448 |
| II.  | Occurrence and Distribution of Carotenoids                   | 448 |
| III. | Location of Carotenoids in Pigment–Protein Complexes         | 453 |
| IV.  | Functions of Carotenoids in Pigment–Protein Complexes        | 455 |
| V.   | Biosynthesis of Carotenoids                                  | 459 |
| VI.  | Carotenoid Biosynthesis as Part of Chloroplast Development   | 467 |

|      |                                              |     |
| ---- | -------------------------------------------- | --- |
| VII. | Concluding Remarks                           | 473 |
|      | Appendix                                     | 473 |
|      | References                                   | 478 |

## 15 Molecular Biology of Chromoplast Development

*Carl A. Price, Miguel Cervantes-Cervantes, Noureddine Hadjeb, Lee A. Newman, and Michal Oren-Shamir*

|       |                                                              |     |
| ----- | ------------------------------------------------------------ | --- |
| I.    | Introduction                                                 | 486 |
| II.   | Chromoplasts: Ultrastructure, Chemistry, and Function        | 486 |
| III.  | Genes Affecting Chromoplasts                                 | 490 |
| IV.   | Chromoplast DNA                                              | 494 |
| V.    | Chromoplast Proteins                                         | 496 |
| VI.   | Chromoplasts and Leucoplasts: Importance of Not Being Green  | 500 |
| VII.  | Conclusions and Prospects                                    | 501 |
|       | References                                                   | 502 |

Index ........................................................... 507

# Contributors

*Numbers in parentheses indicate the pages on which the authors' contributions begin.*

**Hans-Erik Åkerlund** (419), Department of Plant Biochemistry, University of Lund, S-220 07 Lund, Sweden

**Birgitte Andersen** (383), Plant Biochemistry Laboratory, Department of Plant Biology, Royal Veterinary and Agricultural University, 40 Thorvaldsensvej, DK-1871 Frederiksberg C, Copenhagen, Denmark

**George Britton** (448), Department of Biochemistry, University of Liverpool, Liverpool L69 3BX, United Kingdom

**Miguel Cervantes-Cervantes** (486), Department of Biochemistry, Escuela Nacional de Ciencias Biológicas, Instituto Politecnico Nacional, Mexico City, Mexico

**Douwe de Boer** (311), Agrotechnological Research Institute (ATO-DLO), 6700 AA Wageningen, The Netherlands

**Kenneth Eskins** (63), Phytoproducts Research, U.S. Department of Agriculture, Agricultural Research Service, National Center for Agricultural Utilization Research, Peoria, Illinois 61604-3999

**Fabrice Franck** (365), Department of Botany B22, University of Liege, Liege, B-4000 Sart-Tilman, Belgium

**Andreas Friemann** (279), Botanical Institute, University of Bonn, W-5300 Bonn 1, Germany

**Wolfgang Hachtel** (279), Botanical Institute, University of Bonn, W-5300 Bonn 1, Germany

**Noureddine Hadjeb** (486), Waksman Institute, Rutgers University, Piscataway, New Jersey 08855-0758

**Lee A. Newman** (486), Waksman Institute, Rutgers University, Piscataway, New Jersey 08855-0758

**Michal Oren-Shamir** (486), Department of Plant Genetics, Weizmann Institute, Rehovot 76100, Israel

**Harald Paulsen** (335), Botanisches Institut der Universität München, D-8000 München 19, Germany

**Carl A. Price** (486), Waksman Institute, Rutgers University, Piscataway, New Jersey 08855-0758
**William R. Richards** (91), Department of Chemistry, Simon Fraser University, Burnaby, British Columbia, Canada V5A 1S6
**Wolfhart Rüdiger** (219), Botanisches Institut der Universität München, D-8000 München 19, Germany
**Hans Ryberg** (25), Department of Plant Physiology, Botanical Institute, University of Göteborg, S-413 19 Göteborg, Sweden
**Margareta Ryberg** (25), Department of Plant Physiology, Botanical Institute, University of Göteborg, S-413 19 Göteborg, Sweden
**Henrik Vibe Scheller** (383), Plant Biochemistry Laboratory, Department of Plant Biology, Royal Veterinary and Agricultural University, 40 Thorvaldsensvej, DK-1871 Frederiksberg C, Copenhagen, Denmark
**Rüdiger Schulz** (180), Fachbereich Biologie, Botanik, Philipps-Universität, D-3550 Marburg, Germany
**Eva Selstam** (241), Department of Plant Physiology, University of Umeå, S-901 87 Umeå, Sweden
**Horst Senger** (180), Fachbereich Biologie, Botanik, Philipps-Universität, D-3550 Marburg, Germany
**Christer Sundqvist** (25), Department of Plant Physiology, Botanical Institute, University of Göteborg, S-413 19 Göteborg, Sweden
**Hemming I. Virgin** (1), Department of Plant Physiology, Botanical Institute, University of Göteborg, S-413 19 Göteborg, Sweden
**Peter Weisbeek** (311), Department of Molecular Cell Biology, University of Utrecht, 3508 TB Utrecht, The Netherlands
**Anna Widell Wigge** (241), Department of Plant Physiology, University of Umeå, S-901 87 Umeå, Sweden

# Preface

The chloroplast is the most distinguishing feature of the plant cell. It possesses an oxygen-producing and carbon dioxide-reducing system unmatched by organelles in other cells. The chloroplast membranes, the thylakoids, house the energy-regenerating machinery for oxygen production and carbon dioxide reduction. As a result of extensive research the dynamic organization of the thylakoids is just in the process of being revealed. The more knowledge we gain, the more aware we become that thylakoid formation is not just a matter of diffusion or self-assembly, but is a multibiosynthetic process, which proceeds in parallel with a highly ordered transport mechanism and a sophisticated membrane insertion apparatus.

Pigment–protein complexes are among the main building units of the thylakoids. In this book we have tried to cover different aspects of their biosynthesis, assembly, and function, but have avoided detailed descriptions of the many pigment–protein complexes found in the literature. The central theme is the chlorophyll–protein complexes, and these are dealt with from different points of view, covering their molecular biology and physiological relevance in a broad sense.

The regulation and biosynthesis of chlorophyll proteins involve a coordinated expression of nuclear and plastid genes and require communication among the cell organelles to respond properly to changing light and temperature conditions. The very biosynthesis of the carotenoids and the chlorophyll chromophore provide examples of precision and efficiency, not to mention the fascinating topic of transport and recognition of thylakoid protein complexes during plastid development. The pigment–protein complexes are functional units of the two multicomplexes, photosystems I and II. Their main function is to collect light energy for electron transport and carbon dioxide reduction. Most contain chlorophyll, but there are carotenoid-containing complexes which, in addition, have a protective role. The fact that the internal membrane system is so intricately constructed and that one organelle can exhibit so many characteristics, of which only a few are dealt with in this book, makes chloroplast research on the cellular level a fascinating area.

Researchers in the fields of plant molecular biology, genetics, plant physiology, plant biochemistry, and cell biology will find valuable information in the different chapters. We have encouraged the contributors to write for a broad group of readers. It is thus our hope that the book will also be of interest to students and teachers in agriculture, horticulture, biology, and biochemistry. Likewise, the book is an attempt to interest aspiring researchers working with pigment–protein complexes, which are of such fundamental importance, in linking living matter with its physical feeder of energy—the sun.

The sequence of chapters is not arbitrary. The first chapters are to a certain degree introductory and give a brief background for beginners in the field. The following ones describe, in progression, the formation, regulation, incorporation, and function of chlorophyll- and carotenoid-containing proteins.

To facilitate the reading of each chapter some overlap among the contributions has been allowed. When a topic is dealt with in more than one chapter cross-references are used. One and the same problem is sometimes treated from different aspects in different chapters. Slightly divergent opinions and interpretations can be found. Only continued research and additional data will eventually answer the outstanding questions. Although the chapters have been edited the responsibility for emphasis lies with the authors.

We express our sincere thanks to Professor Dennis Buetow, who encouraged us to undertake the editing of this book; to Nancy Illenius, who helped us by carefully reading the manuscripts; and to Clas Dahlin, who made valuable comments on the scientific content. To all the contributors who willingly wrote and responded to our comments on their manuscripts, we wish to express our appreciation.

Christer Sundqvist<br>Margareta Ryberg

# 1

# The Chloroplast as Site of Chlorophyll Formation and Photosynthesis: A Short History

### HEMMING I. VIRGIN

Department of Plant Physiology
Botanical Institute
University of Göteborg
S-413 19 Göteborg, Sweden

 I. Introduction
 II. The Chloroplast as an Organelle
  A. Early Research
  B. Inner Structure of Chloroplast and Etioplast
 III. The Chloroplast as Site for Photosynthetic Pigments
  A. Early History
  B. The Modern Era
 IV. The Chloroplast as Site for Photosynthesis
  A. Early History
  B. The Modern Era of Photosynthesis Research
 V. Concluding Remarks
  References

## I. INTRODUCTION

 This chapter is a short and personally colored account of some highlights in the history of research on the chloroplast, the organelle that characterizes the green plant. The literature on the subject is extensive, but contains definite milestones of decisive importance for further research. Special emphasis is placed on chloroplast structure, the role of the chloroplast as the site of photosynthesis, and its role as carrier of the pigment–protein complexes necessary for this process. Data on the subject are found in many textbooks and reviews treating various aspects of this vast subject, for example, Czygan (1980), Frey-Wyssling (1948), Gibbs (1971), Hillman (1970), Kirk and Tilney-Bassett (1978), Rabinowitch (1945, 1951), Reinert (1980), Scheer (1991), Shlyk (1970), Sironval (1967), Sironval and Brouers (1984), Takamiya (1974), and Trebst and Avron (1977).

The very rapid development of modern empirical science is, to a large extent, the result of the introduction of new methods and of new and more sophisticated equipment, allowing more and more exact measurements. Although these advances have benefitted all science, they are particularly evident in the development of chloroplast research. The introduction of the electron microscope, the molecular biology techniques, and the new spectroscopic methods, in combination with the use of modern computer science, have been of vital importance in acquiring our present knowledge of the molecular structure of this organelle.

As in all research, great achievements by a few outstanding scientists are based on discoveries made by innumerable researchers, all of which have generated our present knowledge. Although the foundation for what we know today was laid by prominent scientists in the 19th century who turned plant physiology into a science, only during the last decades have we witnessed an extreme development in the field of chloroplast research.

## II. THE CHLOROPLAST AS AN ORGANELLE

### A. Early Research

The green organelle in the plant cell, the chloroplast, was first described by von Mohl (1837). Sachs (1862), the founder of modern plant physiology, showed that the light-dependent formation of starch was connected closely to this organelle. Böhm (1883) noticed that starch could disappear from a leaf kept in darkness. Using motile aerobic bacteria, Engelmann (1881), in a classic experiment with *Spirogyra* cells, showed that oxygen was evolved from its spiral chloroplast, but only from irradiated parts. These observations definitely tied the photosynthetic process to the chloroplast.

Among early studies on chloroplasts, the work by Senn (1908) on chloroplast movements, by Blackman (1905) on the separation of photosynthesis into a photochemical and a biochemical temperature-dependent process, and by Willstätter and Stoll (1913) on the structure of chlorophyll pigments should be mentioned. All are milestones in the study of the chloroplast and were fundamental to further research.

In the early 20th century, Molisch (1925) confirmed data obtained earlier (Molisch, 1904) that oxygen was produced from irradiated dried and rewet-

# 1. A Short History

ted leaf powder. Also, dried heated leaves were active and evolved oxygen. Later, Hill (1939) and Hill and Scarisbrick (1940) repeated this experiment and showed that the oxidant 2,6-dichlorophenolindophenol was reduced when added to leaf preparations that were irradiated. Then, $O_2$ evolution was shown to occur using many other oxidants called "Hill reactants." These observations heralded a new era in photosynthesis research.

Chloroplasts always are imbedded in the cytoplasm, but their location in the cells varies. Localization is determined primarily by light intensity and the direction of the irradiation beam, as described in detail by Senn (1908). Chloroplast movements require energy and most likely are governed by filaments attached to the plastids (Schönbohm, 1973). The movement is controlled by phytochrome and by a blue-light receptor as well as by chlorophyll (Chl) itself (Zyrzycka, 1951; Haupt, 1959).

## B. Inner Structure of Chloroplast and Etioplast

### 1. Fundamental Structures

In the 1930s, the chloroplasts were described as microscopically homogeneous. Küster (1935) considered the contents of the chloroplast to constitute a very liquid hydrogel, although some observations spoke against such an interpretation (their flattened shape and autonomic configuration; Senn, 1908). Living chloroplasts indeed also contained indications of a structure that was on the edge of microscopic visibility. Already Meyer (1883) and Schimper (1885) had stated that chloroplasts contained granulae in which the pigments were present, but colloid chemists at that time refuted this view, since they had reached the conclusion that all living components of cells were fluid and optically empty (see Frey-Wyssling, 1948). Therefore, any kind of microstructure made visible by fixation was considered a form of precipitation, structure coagulation, artificial product, or artifact in general. However, Heitz (1932) definitively showed that the chloroplasts had a granular structure. The introduction of the uv-microscope, which allowed magnifications up to 4000×, clearly revealed a lamellar structure for the green subunits.

These units were called "grana" and were thought to contain exclusively the pigments with photosynthetic activity. The medium surrounding the grana was called "stroma." Under the fluorescence microscope, the grana showed red fluorescence, indicating the presence of Chl.

The real breakthrough in structural research came with the introduction of the electron microscope, invented by Ruska, who received the Nobel

Prize for the invention in 1986. Suddenly, magnifications in the range 100,000–200,000× were possible. Although attainable only on fixed materials, these enhanced magnifications provided much new information. Today, when electron micrographs are found in every textbook in botany, it is difficult for the young scientist to realize the excitement felt when a new, once-hidden world opened up before our eyes. The first electron micrographs were rather blurry, but soon improved in quality. Micrographs obtained today do not differ very much in resolution from the first good micrographs obtained.

The first electron micrographs revealed that the grana are composed of stacks of membranes and that the plastid is surrounded by two membranes, the double envelope. The inner membranes were named thylakoids by Menke (1961). Their shape and distribution varied considerably among species. Small spheric bodies, plastoglobuli, were scattered among the grana stacks and later were found to be rich in quinones (Tevini and Lichtenthaler, 1970).

Early electron microscopic studies by, among others, Leyon (1953), von Wettstein (1958), and Menke (1962) revealed that etioplasts, developed in darkness, contain a lattice-like semicrystallinic cubic structure called the prolamellar body (PLB) (Hodge *et al.*, 1956). Flat membranes, now called prothylakoids, extend from this structure. The very regular structure of the PLB has been, and still is, the subject of many speculations. Von Wettstein (1958) proposed that the PLB consisted of tubules arranged in a cubic lattice, a model that was confirmed by Gunning (1965). The proposed model still prevails and is summarized excellently by Gunning and Steer (1975). Eriksson *et al.* (1961) showed that irradiation caused the PLB to dissolve rapidly into small vesicles. After some hours of irradiation, the length of thylakoid membrane increased and grana stacks started to form. Virgin *et al.* (1963) found that the action spectrum for vesicle formation is similar to the absorption spectrum of protochlorophyll (Pchl). The PLBs have become very important since these studies indicated that the precursor of chlorophyll a (Chl a) is present in the PLB.

## 2. Structure and Function

As soon as electron microscopists began to acquire detailed micrographs of ultrastructure, attempts were made to associate the Chl formation site and the various photosynthetic subreactions with certain structures seen in the micrographs. Using the freeze fracture technique (Moor, 1964; Mühlethaler *et al.*, 1965), it became possible to break up the thylakoid membrane into inner and outer membrane surfaces, which could be split further. A typical electron micrograph of freeze-fractured thylakoid mem-

# 1. A Short History

branes reveals that each type of fracture face exhibits a distinct population of intramembrane particles, consistent with the current view that membrane stacking produces both a structural and a functional differentiation of thylakoid membranes. The first particles that could be related to an identified function were a population of particles, with an average size of 100 Å on the outer (matrix) surface, that could be shown to represent coupling factor 1 (CF1).

Improvements in modern electron microscopes have made it possible to scrutinize in detail the membrane system of the plastids. New fixation and preparation techniques, for example, freeze etching and immunolocalization, have increased the usefulness of this instrument further. Using improved separation techniques, much information has been collected about the thylakoid membranes. Now we have a fairly good idea of the composition of the photosynthetic membrane. The thylakoid proteins can be divided into five inner membrane complexes, closely associated with outer membrane proteins that perform light-harvesting, electron transport, proton translocation, and ATP synthesis functions. Thus, it is possible to identify the structures that are related to the two photosystems (Anderson, 1986). A number of histochemical reactions have been used to correlate the structures seen in the specimens with biochemical data. Using serological techniques, it has been possible to identify and localize a number of proteins involved in the photosynthetic pathway, for example, ferredoxin, carboxydismutase, plastocyanin, cytochrome $f$, and many others. Based on this knowledge, the electron flow in the Z-scheme of photosynthesis now is tied definitively to this membrane system. Schematic models of protein positions and electron flow in the membranes have been presented (Fig. 1).

## III. THE CHLOROPLAST AS SITE FOR PHOTOSYNTHETIC PIGMENTS

### A. Early History

The chemistry of Chl formation is comprehensively described in Chapter 4. The book by Willstätter and Stoll (1913) also contains fundamental information about Chl and its derivatives. The total *in vitro* synthesis of Chl was achieved by Woodward (1961). However, only in the last few years was the whole sequence of reactions from 5-aminolevulinic acid (ALA) to Chl a elucidated, thanks to efforts made by many teams (see Bogorad, 1976). Research on the last light-driven step in Chl biosynthesis, that is, the reduction of Mg–2,4-divinylphaeoporphyrin $a_5$ monomethylester, protochlorophyllide (Pchlide), has attracted great interest for a long

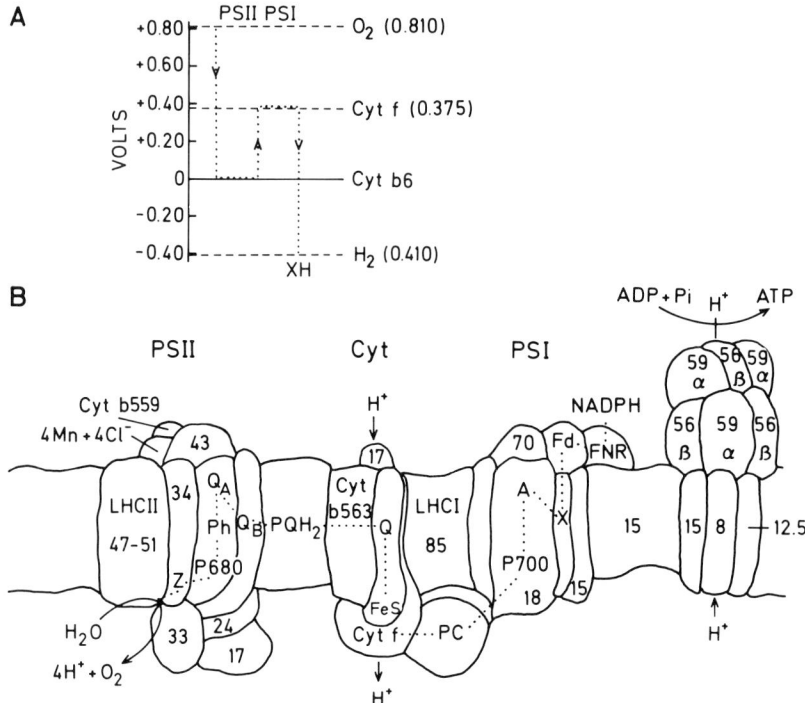

**Fig. 1.** Summary of 25 years of development in photosynthesis research. (A) The Hill–Bendall scheme for the light reactions in photosynthesis. (Adapted from Hill and Bendall, 1960.) (B) Schematic outline of the two photosystems and the ATP-forming complexes incorporated into the photosynthetic membrane. Indication of experimentally identified protein complexes. (Adapted from Anderson, 1986.)

period of time. In a comprehensive textbook, Miller (1938) makes the following remarks about this light-driven step:

> The formation of chlorophyll is a physiological process that occurs only in living cells and under conditions favorable to life. The substance or substances from which chlorophyll arises have never been isolated, and their existence is only inferred.

This statement was essentially true at the time it was written, but considerable early literature on the physiology of Chl formation also reported the fluorescence of leaf pigments. Monteverde (1893) coined the name Pchl for the Chl precursor. In several papers, Lubimenko (1927, 1928) reported experiments on light-induced Chl formation. Palladin (1922) considered that a pigment called "chlorophyllogen" is formed in darkness in the chloroplasts and that this pigment is transformed rapidly into Chl in light.

# 1. A Short History

The pigment showed a distinct fluorescence and had an absorption band between 620 and 640 nm. Eyster (1928) claimed that Pchl is a pigment that is formed without the aid of light and that changes photochemically into Chl on exposure to light.

## B. The Modern Era

### 1. Protochlorophyllide, Protochlorophyllide Holochrome, and NADPH–Protochlorophyllide Oxidoreductase

Modern research on Chl formation can be said to have started when high-resolution absorption and fluorescence spectrophotometers were constructed. With few exceptions, such instruments first came into general use after the end of World War II. The appearance of the now classic Beckman DU ultraviolet spectrophotometer meant a revolution in the technique of measuring absorption spectra. Although qualitative studies of the fluorescence of etiolated plants rather quickly revealed the existence of Pchl, only in the early 1950s were more serious studies on the light-driven transformation process initiated. A very accurate action spectrum for Chl formation, that is, the phototransformation of Pchl to Chl in normal and albino corn, was presented by Koski *et al.* (1951). These spectra showed only one peak at 650 nm in the red region. Current belief is that only the 650-nm form of Pchl is transformed rapidly to Chl. The spectra obtained are also among the best examples of how other inactive pigments can work as screening pigments. In this case, the screening pigments are carotenoids. Absorption measurements of dark-grown plant material finally revealed that the pigment *in vivo* had a distinct absorption peak at 650 nm in the red region and an evident shoulder at 636 nm (Hill *et al.*, 1953). Sometimes a peak at 628 nm was present also. The three spectral forms of Pchl also can be recognized clearly in the fluorescence spectra of live material cooled to liquid nitrogen temperature to prevent phototransformation to chlorophyllide (Chlide) a (Gassman and Bogorad, 1967; Sironval *et al.*, 1968; Granick and Gassman, 1970; Kahn *et al.*, 1970; Schoch, 1978; Cohen and Rebeiz, 1981). As with Chl, pigment extracts in organic solvents were always identical with only one peak, the position of which depended on the solvent. It thus became obvious that Pchl *in vivo* was bound to different protein complexes, forming "holochromes" (Smith *et al.*, 1956; Boardman, 1962a).

Studies by Smith and Benitez (1954) on the temperature dependence of the phototransformation revealed that the transformation process is a mixture of a temperature-independent photochemical process and a temperature-dependent one, indicating the participation of several reactants. However, administration of light energy divided into millisecond flashes gave the same yield as the same energy administered continuously (Virgin, 1955). This experiment showed that there could not be many long-lived intermediates in the reaction chain. Boardman (1962b) suggested that Pchl and the donor molecule were integral parts of the holochrome.

Unfortunately, purified NADPH was not readily available at that time, and the enzymatic experiments performed did not reveal the bound inaccessible NADPH. Disclosure of the role of NADPH did not occur until 12 years later. Isolation and purification of the holochromes was done in several laboratories. The molecular weight could be determined, but values obtained varied among different plants and depended on the method used for determination (see Virgin, 1981). Thus, values between 29 and 600 kDa were reported. Discovery of the holochrome had a great impact on further studies of Chls and photosynthesis, because it became clear that the pigments *in vivo* mostly work in connection with a protein and never as a single chromophore. Pioneering work in the study of Chl–protein complexes was done by Smith, who prepared soluble Chl–protein complexes using digitonin (Smith, 1938). Early research in photosynthesis was held back because the chromophore separates so easily from the protein. A long time passed before Chl–protein complexes were accepted widely as functional units.

The finding by Granick (1959) that Pchlide in etiolated leaves treated with ALA forms large amounts of the nonesterified 636-nm form implied great progress in Chl research and attracted great attention. Some indications were that this Pchlide was not bound to any protein. However, it was shown that, after a short irradiation, the 636-nm form was transformed slowly into the 650-nm form, which then could be transformed by light to Chlide a (Gassman and Bogorad, 1967; Sundqvist, 1969, 1970; Granick and Gassman, 1970). The absorption shift from 636 nm to 650 nm could be due to the formation of a holochrome, indicating the participation of an enzyme in the photoreduction process. Early Pchl holochrome preparations never showed enzymatic activity. An enzyme-dependent transformation of Pchlide to Chlide *in vitro* was demonstrated by Griffiths (1975). Griffiths (1975, 1978) and Brodersen (1976) showed that NADPH was the most likely hydrogen donor for the reduction of Pchlide. The enzyme, called NADPH–protochlorophyllide oxidoreductase (Pchlide-reductase), is a photoenzyme present in relatively large amounts in dark-grown leaves. Work on isolation of pure preparations of the enzyme was soon successful (Apel *et al.*, 1980; Beer and Griffiths, 1981) and the enzyme was shown by Apel (1981) to be encoded by nuclear DNA as a 44-kDa precursor. The amino acid composition (Röper *et al.*, 1987) and the sequence of cDNA

## 1. A Short History

encoding the enzyme was determined (Schulz et al., 1989). This research is treated in detail in Chapter 5.

### 2. Esterified Pigments

As the reader may have noticed, no distinction was made between Pchl and Pchlide in early research in the field. The use of paper and column chromatographic methods during the 1950s revealed the different forms. The main pigment in the dark-grown material was shown to be Pchlide and was transformed by light into Chlide. A varying but mostly small amount of Pchl also was found in the dark-grown plants. More or less immediately after a short irradiation, minor amounts of Chl could be detected. This Chl was always only Chl a. Chlorophyll b (Chl b) appeared at the earliest, 5–15 min after the start of irradiation. The origin of Chl b is Chl a, but probably only newly molecules of the pigment (Shlyk, 1970). According to Oelze-Karow and Mohr (1978), phytochrome in $P_{fr}$ form accelerates the formation of Chl b. The mechanism governing the accurate relationship between the two Chl complexes in photosystem I (PSI) and photosystem II (PSII) is not known.

During the 1960s, lively discussion took place about the degree of esterification of Pchl, the photoconversion of different esterified forms, and their presence in the different spectral forms. The chromophore in the 650-nm form is now well established to be Pchlide (see Ryberg and Sundqvist, 1991). Parts of the other forms are probably esterified, that is, are true Pchls (Liljenberg, 1974).

After irradiation of a dark-grown leaf, Chlide is esterified to Chl a within half an hour at room temperature (Wolff and Price, 1957; Virgin, 1960). The rate of this reaction is regulated by phytochrome (Liljenberg, 1966) and is dependent on seedling age (Sironval et al., 1965; Schoch, 1978). For a long time, the esterifying alcohol was believed to be phytol (Phy) and chlorophyllase (chlase) was believed to have a dual function. Evidence slowly emerged, however, that, at least in dark-grown leaves, after irradiation Pchlide is esterified with geranylgeranyldiphosphate (GGPP), which is then hydrogenated stepwise to Phy in the pigment (Liljenberg, 1974; Rüdiger et al., 1977; Schoch et al., 1977; Schoch, 1978). However, the Chl-synthetase has not been identified and the role of a direct esterification with Phy later during the greening period cannot be excluded.

### 3. Shibata Shift and Lag Phase in Chlorophyll Formation

At the time of early studies on Pchl transformation, recording spectrophotometers had not yet been constructed. Making a carefully plotted spectrum took hours with a Beckman DU absorption spectrophotometer. Recording instruments came into general use in early 1954. Using such instruments, a complete absorption spectrum was obtained within minutes. This provides another example of the importance of instrumental development, which in this case led to the discovery of absorption shifts of newly formed Chl a by Shibata (1957). These findings gave rise to great amounts of work in many laboratories around the world on intermediates in Chl formation, to which Litvin and Belajeva (1971), Dujardin and Sironval (1977), and others have made important contributions. The rate of the Shibata shift was dependent on seedling age, temperature, and phytochrome activation (Jabben and Mohr, 1975; Axelsson, 1977). The Shibata shift has been correlated with the esterification process and the transport of Pchlide-reductase from the PLBs to the prothylakoids. A detailed analysis of seedling age dependence of the Shibata shift and the esterification showed that the shift can precede the esterification. Change in the size of the Chlide–protein molecular complexes is a more probable cause of the shift (Akoyunoglou and Michalopoulos, 1971).

A lag phase in further Chl formation (greening) follows the primary rapid phototransformation of Pchlide in a dark-grown leaf. During this lag, the thylakoids are formed gradually, building up the grana. Withrow *et al.* (1956) found that this lag was shortened by a red light pulse, indicating that the phytochrome system was involved. An action spectrum for the light effect verified the involvement of phytochrome (Virgin, 1961). The description and interpretation of the lag phase engaged several research groups during the 1960s and 1970s. The lag phase was found to be influenced by growth regulators: cytokinins decreased the lag phase and abscissic acid increased it (Beevers *et al.*, 1970). In excised cucumber, the removal of the hypocotyl hook increased the lag phase in the cotyledons (Hardy *et al.*, 1970) but exogenous ALA could partly abolish the delay. Seedling age (Akoyunoglou and Argyroudi-Akoyunoglou, 1969; Nadler and Granick, 1970) and, in more general terms, the physiological state of the tissue seemed to be extremely important for the onset of early Chl biosynthesis leading to Chl accumulation. Evidence indicates that the lag phase is an effect of the slow activation of the protein-synthesizing machinery, probably on the transcriptional level (Mohr, 1977). Nevertheless, the greening period of previously dark-grown plants was studied intensively to find the mechanism behind Chl formation and accumulation, as well as the mechanism behind the development of the photochemical activities associated with Chl–protein complexes (Anderson and Boardman, 1964; Hiller *et al.*, 1973).

# 1. A Short History

## IV. THE CHLOROPLAST AS SITE FOR PHOTOSYNTHESIS

### A. Early History

Miller (1938) gave a summary of the theories popular at that time about the process of photosynthesis. He quoted James (1934), who considered that a diffusion reaction and at least one thermal "dark" reaction were involved in the process. James (1934) proposed the following steps in photosynthesis, ideas that were based largely on the statements of Willstätter and Stoll (1913) and Warburg (1919):

> 1. The carbon dioxide arriving at the surface of the chloroplasts reacts as carbonic acid or some simple derivative with magnesium of the Chl to form a dissociable addition compound.
> 2. Under the influence of light, this compound undergoes internal rearrangements of the molecule giving an unstable compound, Chl–formaldehyde–peroxide, in which the oxygen is only loosely held.
> 3. This compound in the presence of peroxidase readily breaks down by two stages, each releasing half a molecule of oxygen.
> 4. The gaseous oxygen escapes from the system and the formaldehyde eventually polymerizes to sugar.

Literature on the photosynthetic mechanism was already abundant at this time but hardly contributed to any clearer view of the subject. The formaldehyde hypothesis for sugar formation (first proposed by Baeyer, 1870) was more or less generally accepted, yet it was never proved to be a metabolic reaction in the plant.

Miller said

> The student can easily judge from this discussion of the chemistry of photosynthesis that little is known concerning the mechanism of the process. The facts known are that the process takes place under the influence of the visible portion of the spectrum, that Chl, water, oxygen, and carbon dioxide are necessary, that carbohydrates are formed and oxygen eliminated, and that the process proceeds under the general conditions which are suitable to the activities of living organisms. *These facts are not new and have been known in a general way for more than 70 years.* (emphasis by this author)

Miller continues:

> The enormous amount of research upon the process of photosynthesis during the past half century has thrown little or no additional light upon the subject. The problem is evidently too complex for specialists in any one field of science to solve.

In the words of Spoehr (1916):

> The subject of photosynthesis requires for its investigation the masterly application of many branches of science—physics, chemistry, plant physiology—and the frank cooperation of the most skilled workers in these various fields.

Note that the name van Niel is not found in Miller's book. Perhaps at that time van Niel's suggestion (van Niel, 1931) that the oxygen evolved in photosynthesis could come from water appeared so unlikely that it was worthy of note.

## B. The Modern Era of Photosynthesis Research

### 1. Origin of Oxygen, ATP Formation, and the Calvin Cycle

As mentioned earlier, Blackman (1905) proposed that the photosynthetic process was composed of one photochemical and one biochemical part. This proposal hinted at what could be hidden in the "black box." Using the isotope method for which von Hevesy received the Nobel Prize in 1944, Ruben *et al*. (1941) found that the oxygen evolved was derived from water, in strong support of the idea of van Niel (1931). These discoveries, along with that of the Hill reaction (Hill, 1939), indicated that carbon dioxide was not involved in the primary photosynthetic reactions and was not the source of oxygen. With these findings, two of the fundamental ideas about photosynthesis for that period had fallen. Excitement was great among scientists in the field.

In the early studies, described by Arnon (1977), it was not possible to demonstrate the Hill reaction in isolated chloroplasts completely free from surrounding cytoplasm, because

> isolated chloroplasts lack either the appropriate hydrogen carriers or enzymes, or both, required for the reduction of carbon dioxide by the hydrogen derived from photolysis of water, and . . . in the intact cell these factors are found chiefly or wholly outside the chloroplast. (Arnon, 1951)

The chloroplast was thought to be the site only for the light-absorbing and water-splitting reactions. It was shown later (Arnon *et al*., 1954), using a proper isolation medium, that the reduction of carbon dioxide also took place in the chloroplast. Thus, complete photosynthesis was shown to be restricted to the chloroplast, a fact that Sachs and Pfeffer, giants in the plant physiology field, had postulated already.

After these discoveries, it became clear that the reduction of carbon dioxide required a strong reducing power and an energy source that primarily was to be found in the light energy absorbed. Ruben (1943) had postulated—without any experimental evidence—that photosynthetic carbon dioxide assimilation was a process dependent on photochemically generated ATP and reduced pyridine nucleotides. Also, Emerson *et al*. (1944) had postulated, on the basis of experiments with *Chlorella*, that the "sole function of light energy in photosynthesis is the formation of energy-rich phosphate bonds."

However, Arnon and collaborators finally showed experimentally that ATP was formed in the chloroplast, contrary to the ideas of many other scientists who anticipated that the mitochondria delivered the ATP. Photophosphorylation became firmly established as a concept by a series of papers published in 1954 (see Arnon, 1977).

Simultaneously, Calvin, Bassham, and Benson were studying the formation of carbohydrates. In a series of papers (Calvin and Benson, 1948; Bassham and Calvin, 1957; Calvin and Bassham, 1962), they reported on the reduction of carbon dioxide by NADPH and the enzymatic reaction cycle that resulted in the formation of starch using ATP as an energy source. With the discovery of the primary acceptor for carbon dioxide, ribulose 1,5-bisphosphate, these investigators were able to make a complete scheme of the starch formation, now known as the "Calvin cycle." For this work, Calvin received the Nobel Prize in 1961. The work was based on the use of radioactive isotopes and the paper chromatographic method for which Martin and Synge were awarded the Nobel Prize in 1952. It was now unobjectionably proved that light was not necessary for carbon dioxide to be used for carbohydrate formation, provided that the necessary enzymes, an energy source, and reducing power were available. With these discoveries, Blackman's (1905) hypothesis was verified fully—46 years after its presentation!

This concept was substantiated by Trebst *et al.* (1958), who succeeded in physically separating the two reactions by fractionating isolated chloroplasts. Only the green lamellar fraction of the chloroplasts was able to perform the light reaction, whereas the Chl-free fraction could fix carbon dioxide in the dark in the presence of NADPH and ATP.

## 2. The Two Photosystems

The findings that NADPH was involved in photosynthesis (Arnon *et al.*, 1957; San Pietro, 1958) and that the oxygen evolved derived from water suggested that "the black box of photosynthesis" might contain an electron transport system similar to that found earlier for the respiration. Indeed, using modern spectrophotometric techniques, vitamin $K_1$ (Dam *et al.*, 1948; Arnon *et al.*, 1955), cytochromes (cyt) $b6$, $b$-559, and $f$ (Hill and Scarisbrick, 1951; Davenport and Hill, 1952; Lundegårdh, 1962), plastocyanin (Katoh, 1960), and different kinds of plastoquinones (Arnon and Crane, 1965) were discovered in rapid succession as components of the chloroplasts. Thus, when the first indications of redox substances in the chloroplasts were seen, models of the photochemical part of photosynthesis began to be suggested. Exact photosynthetic action spectra for oxygen production were measured also (Haxo and Blinks, 1950) using the polarographic method for which Heyrovsky received the Nobel Prize in 1959.

Real insight in the mechanisms of the light-dependent part of photosynthesis was not obtained until Emerson *et al.* (1957) published their studies on the quantum efficiency of complete photosynthesis in *Chlorella* cells. These researchers found that the action spectrum for oxygen evolution dropped earlier than the curve for light absorption did. "The red drop," as described by Emerson and Lewis (1943), became a famous phenomenon. This drop could be counteracted by the addition of supplementary monochromatic irradiation. According to Emerson and his collaborators, the phenomenon could be explained by the assumption that the full efficiency of photosynthesis at long wavelengths of light, absorbed by Chl a, could be obtained only with a simultaneous absorption of light of shorter wavelengths by accessory pigments, which started a second light reaction. These pigments were Chl b and carotenoids in *Chlorella* and phycobilins in red algae.

In 1960, Hill and Bendall—without any firm experimental evidence—suggested that one light reaction oxidizes Cyt $f$ and reduces $NADP^+$ and a second light reaction reduces Cyt $b6$ and oxidizes water. In schematic form, their model had the shape of a Z. In principle, this model is still valid. Of course many details have been added since its first presentation (cf. Chapters 12 and 13). Experimental proof of how the photoreactions are coupled was given by Duysens *et al.* (1961) and Duysens and Amesz (1962) with studies on the photosynthetic bacterium *Rhodospirillum rubrum*. Duysens introduced the use of absorption difference spectroscopy in photosynthesis research: if a photosynthetic intermediate is a colored redox substance, its absorption spectrum should change with the redox state on illumination. This work is a striking example of the important role of new techniques in scientific discovery, in this case in the discovery of electron transport in photosynthesis. To cite Duysens (1989b), "Before 1947, nothing was known about photosynthetic intermediates and their reactions with exception of largely incorrect or weakly supported speculations." By introducing absorption difference spectrophotometry to photosynthesis research, Duysens and his collaborators were able to see reversible changes in the absorption spectra of purple bacteria when irradiated with alternating short-wavelength red and long-wavelength red light. These observations gradually resulted in the theory of two photosystems, named PSI and PSII, and how they are linked together in a sequence by an electron transport chain in which cyto-

## 1. A Short History

chromes play an important role (Duysens and Amesz, 1962; Fig. 1). The idea of two photosystems had been suggested earlier, as shown previously, but was substantiated experimentally by Duysen's work (Duysens, 1989a). The discovery of the absorption changes at 700 nm (Kok, 1956), which were supposed to arise from PSI, further strengthened the theory. The reaction center for PSII was later identified and has been designated P680 (Butler, 1961; Döring et al., 1969).

Today the two photosystems are considered to be multiprotein complexes in which the chromophores are embedded. The two photosystems consist of reaction centers with several pigment–protein complexes. They are surrounded by several hundred Chl molecules (the antennae) that collect the light energy and transfer it to the centers. The different action spectra characteristic for activation of the two systems require the existence of Chl complexes with different optical properties. These differences are not only caused by different relationships between the amounts of Chl a and Chl b but also by different protein moieties. Anderson and Boardman (1966), using modern protein separation techniques, isolated two pigment–protein complexes with retained activities. Computerized deconvolution of absorption spectra from different species (French et al., 1972) into several absorption species also gave a hint of the existence of several pigment–protein complexes. Fluorescence measurements have played a great role in the study of energy transport between the two photosystems. The use of fluorescence in this connection can be credited to the work by Katz (1949). Studies of low-temperature fluorescence are now important in distinguishing specific Chl–protein complexes belonging to various components of the two photosystems (van Dorssen et al., 1987; Siefermann-Harms, 1988).

The introduction of molecular biology has opened new domains of study and has made it possible to sequence the whole chloroplast genome (Ohyama et al., 1986; Shinozaki et al., 1986). The different light-harvesting complexes have been isolated and characterized (e.g., Melis and Anderson, 1983). The photochemical reactions tied to PSI also will be understood soon (Glazer and Melis, 1987).

The secrets of the oxygen-evolving part of PSII have, until recently, resisted all attempts at discovery. Through intensive work that, during the last decade, has focused on protein separation and on the function of the water oxidation reaction, PSII is now among the best characterized multiunit membrane protein complexes (see Chapter 13). An interesting finding is that the primary acceptor is a pheophytin molecule that transfers an electron to plastoquinone (Klimov et al., 1977, 1980). Highly purified PSII pigment–protein complexes capable of $O_2$ evolution have been isolated by several research teams. Using EPR techniques, the role of manganese is becoming clearer. Thus, hope exists that the "black box" of photosynthesis will open a door to allow our entrance.

> Perhaps the greatest surprise to photosynthesis researchers in the past decade was the finding that PSII can be manipulated like any other enzyme; the surprise of the next decade should be interesting indeed. (Ghanotakis and Yocum, 1990)

## V. CONCLUDING REMARKS

Our current knowledge of the chloroplast is comprehensive. Most of it has been achieved during the last 50 years. We have insight into the principles for its ontogenesis, its ultrastructure, and its role in photosynthesis, for which the Chls it contains are a prerequisite. However, only during the last few years have we begun to understand the molecular basis of the processes involved. This research promises many new discoveries in the years to come. Also, we still do not understand many manifestations tied to the chloroplasts. A few of these, such as chloroplast movements and changes in form, are phenomena that belong to the field of classical plant physiology.

We are now in the era of molecular biology. Predicting the future is difficult but, assuming a continuous development at the rate that we have witnessed during the last 50 years, with construction of new measuring apparatuses and development of new techniques, it is probable that we shall be able to solve the classical problems just mentioned in the half century to come.

Some years ago, when the "black box" was still unbroken, it was believed that the photosynthetic process was less complicated than it has turned out to be. As our knowledge of the complexity of the reactions involved increases, doubt about the development of plants with increased efficiency in photosynthesis also increases. The realization of *in vitro* photosynthesis is possible. On the other hand, with our increasing knowledge, the development of new instruments and methods is proceeding at such a rate that we can measure and study processes today that a few years ago would have been considered science fiction.

## REFERENCES

Akoyunoglou, G., and Argyroudi-Akoyunoglou, J. H. (1969). Effects of intermittent and continuous light on the chlorophyll formation in etiolated plants at various ages. *Physiol. Plant.* **22,** 288–295.

# 1. A Short History

Akoyunoglou, G., and Michalopoulos, G. (1971). The relation between the phytylation and the 682–672 shift *in vivo* of chlorophyll *a*. *Physiol. Plant.* **25**, 324–329.

Anderson, J. M. (1986). Photoregulation of the composition, function, and structure of thylakoid membranes. *Annu. Rev. Plant Physiol.* **37**, 93–136.

Anderson, J. M., and Boardman, N. (1964). Studies on the greening of dark-grown bean plants. II. Development of photochemical activity. *Austr. J. Biol. Sci.* **17**, 93–101.

Anderson, J. M., and Boardman, N. (1966). Fractionation of the photochemical systems of photosynthesis. I. Chlorophyll contents and photochemical activities of particles isolated from spinach chloroplasts. *Biochim. Biophys. Acta* **112**, 403–421.

Apel, K. (1981). The protochlorophyllide holochrome of barley (*Hordeum vulgare* L.). Phytochrome-induced decrease of translatable mRNA coding for the NADPH : protochlorophyllide oxidoreductase. *Eur. J. Biochem.* **120**, 89–93.

Apel, K., Santel, H-J., Redlinger, T. E., and Falk, H. (1980). The protochlorophyllide of barley (*Hordeum vulgare* L.). Isolation and characterization of the NADPH : protochlorophyllide oxidoreductase. *Eur. J. Biochem.* **111**, 251.

Arnon, D. I. (1951). Extracellular photosynthetic reactions. *Nature (London)* **167**, 1008–1010.

Arnon, D. I. (1977). Photosynthesis 1950–75: Changing concepts and perspectives. *In* "Photosynthesis I, Photosynthetic Electron Transport and Photophosphorylation" (A. Trebst and M. Avron, eds.), Encyclopedia of Plant Physiology, Vol. 5. Springer-Verlag, Berlin.

Arnon, D. I., and Crane, F. L. (1965). Role of quinones in photosynthetic reactions. *In* "Biochemistry of Quinones" (R. A. Morton, ed.), pp. 433–458. Academic Press, London.

Arnon, D. I., Whatley, F. R., and Allen, M. B. (1954). Photosynthesis by isolated chloroplasts. II. Photosynthetic phosphorylation, the conversion of light into phosphate bond energy. *J. Am. Chem. Soc.* **76**, 6324–6329.

Arnon, D. I., Whatley, F. R., and Allen, M. B. (1955). Vitamin K as a cofactor of photosynthetic phosphorylation. *Biochim. Biophys. Acta* **16**, 607–708.

Arnon, D. I., Whatley, F. R., and Allen, M. B. (1957). Triphosphopyridine nucleotide as a catalyst of photosynthetic phosphorylation. *Nature (London)* **180**, 182–185.

Axelsson, L. (1977). The photostability of different chlorophyll forms in dark grown leaves of wheat. III. Dependence on age of the plants. *Physiol. Plant.* **41**, 217–222.

Baeyer, A. (1870). Über die Wasserentziehung und ihre Bedeutung für das Pflanzenleben und die Gährung. *Ber. Chem. Ges.* **3**, 63–78.

Bassham, J. A., and Calvin, M. (1957). "The Path of Carbon in Photosynthesis." Prentice-Hall, Englewood Cliffs, New Jersey.

Beer, N. S., and Griffiths, W. I. (1981). Purification of the enzyme NADPH : protochlorophyllide oxidoreductase. *Biochem. J.* **195**, 93–192.

Beevers, L., Loveys, B., Pearson, J. A., and Wareing, P. F. (1970). Phytochrome and hormonal control of expansion and greening of etiolated wheat leaves. *Planta* **90**, 286–294.

Blackman, F. F. (1905). Optima and limiting factors. *Ann. Bot.* **19**, 281–295.

Boardman, N. K. (1962a). Studies on a protochlorophyll–protein complex. I. Purification and molecular weight determination. *Biochim. Biophys. Acta* **62**, 63–79.

Boardman, N. K. (1962b). Studies on a protochlorophyll–protein complex. II. The photoconversion of protochlorophyll to chlorophyll *a* in the isolated complex. *Biochim. Biophys. Acta* **64**, 279–293.

Böhm, J. A. (1883). Über Stärkebildung aus Zucker. *Bot. Ztg.* **41**, 33–38, 49–54.

Bogorad, L. (1976). Chlorophyll biosynthesis. *In* "Chemistry and Biochemistry of Plant Pigments" (T. W. Goodwin, ed.), 2nd Ed., Vol. 1, pp. 64–148. Academic Press, New York.

Brodersen, P. (1976). Factors affecting the photoconversion of protochlorophyllide to chlorophyllide in etioplast membranes isolated from barley. *Photosynthetica* **10**, 33–39.
Butler, W. L. (1961). A far-red absorbing form of chlorophyll *in vivo*. *Arch. Biochem. Biophys.* **93**, 413–422.
Calvin, M., and Bassham, J. A. (1962). "The Photosynthesis of Carbon Compounds." Benjamin, New York.
Calvin, M., and Benson, A. A. (1948). The path of carbon in photosynthesis. *Science* **107**, 476–480.
Cohen, C. E., and Rebeiz, C. A. (1981). Chloroplast biogenesis. 34. Spectrofluorometric characterization *in situ* of the protochlorophyll species in etiolated tissues of higher plants. *Plant Physiol.* **67**, 98–103.
Czygan, F. C. (ed.) (1980). "Pigments in Plants," 2nd Ed. Fischer, Stuttgart.
Dam, H., Hjorth, E., and Kruse, I. (1948). On the determination of vitamin K in chloroplasts. *Physiol. Plant.* **1**, 379–381.
Davenport, H. E., and Hill, R. (1952). Cytochrome components in chloroplasts. *Nature (London)* **179**, 1112–1114.
Döring, G., Renger, G., Vater, J., and Witt, H. T. (1969). Properties the photoactive chlorophyll-*a* in photosynthesis. *Z. Naturforsch.* **24b**, 1139–1143.
Dujardin, E., and Sironval, C. (1977). Transitory pigment–protein complexes similar to photosynthesis active centres during protochlorophyll(ide) photoreduction. *Plant Lett.* **10**, 347–355.
Duysens, L. N. M. (1989a). The discovery of the two photosynthetic systems: A personal account. *Photosynth. Res.* **21**, 61–79.
Duysens, L. N. M. (1989b). The study of reaction centers and of the primary and associated reactions of photosynthesis by means of absorption difference spectrophotometry: A commentary. *Biochim. Biophys. Acta* **1000**, 395–400.
Duysens, L. N. M., and Amesz, J. (1962). Function and identification of two photochemical systems in photosynthesis. *Biochem. Biophys. Acta* **64**, 243–260.
Duysens, L. N. M., Amesz, J., and Kamp, B. M. (1961). Two photochemical systems in photosynthesis. *Nature (London)* **190**, 510–511.
Emerson, R., and Lewis, C. M. (1943). The dependence of the quantum yield of *Chlorella* photosynthesis on wavelength of light. *Am. J. Bot.* **39**, 165–178.
Emerson, R. L., Stauffer, J. F., and Umbreit, W. W. (1944). Relationships between phosphorylation and photosynthesis in *Chlorella*. *Am. J. Bot.* **31**, 107–120.
Emerson, R. L., Chalmers, R. V., and Cederstrand, C. (1957). Some factors influencing the long-wave limit of photosynthesis. *Proc. Natl. Acad. Sci. USA* **43**, 133–144.
Engelmann, T. W. (1881). Neue Methode zur Untersuchung der Sauerstoffausscheidung pflanzlicher und tierischer Organismen. *Bot. Ztg.* **39**, 441–448.
Eriksson, G., Kahn, A., Walles, B., and von Wettstein, D. (1961). Zur makromolecularen Physiologie der Chloroplasten III. *Ber. Deutsch. Bot. Ges.* **74**, 222–232.
Eyster, W. H. (1928). Protochlorophyll. *Science* **68**, 509–570.
French, C. S., Brown, J. S., and Lawrence, M. C. (1972). Four universal forms of chlorophyll *a*. *Plant Physiol.* **49**, 422–429.
Frey-Wyssling, A. (1948). "Submicroscopic Morphology of Protoplasm and Its Derivatives." Elsevier, New York.
Gassman, M.,and Bogorad, L. (1967). Studies on the regeneration of protochlorophyllide after brief illumination of etiolated bean leaves. *Plant Physiol.* **42**, 781–784.
Gibbs, M. (1971). "Structure and Function of Chloroplasts." Springer-Verlag, Berlin.
Ghanotakis, D. F., and Yocum, C. F. (1990). Photosystem II and the oxygen-evolving complex. *Annu. Rev. Plant Physiol. Plant Mol. Biol.* **41**, 255–276.

Glazer, A. N., and Melis, A. (1987). Photochemical, reaction centers: Structure, organization, and function. *Annu. Rev. Plant Physiol.* **38,** 11–45.
Granick, S. (1959). Magnesium porphyrins formed by barley seedlings treated with aminolevulinic acid. *Plant Physiol. (Suppl.)* **34,** XVIII.
Granick, S., and Gassman, M. (1970). Rapid regeneration of protochlorophyllide 650. *Plant Physiol.* **45,** 201–205.
Griffiths, W. T. (1975). Characterization of the terminal stages of chlorophyll(ide) synthesis in etioplast membrane preparations. *Biochem. J.* **152,** 623–635.
Griffiths, W. T. (1978). Reconstitution of chlorophyllide formation by isolated etioplast membranes. *Biochem. J.* **174,** 681–692.
Gunning, B. E. S. (1965). The greening process in plants. I. The structure of the prolamellar body. *Protoplasma* **60,** 111–130.
Gunning, B. E. S., and Steer, M. W. (1975). "Ultrastructure and the Biology of Plant Cells." Arnold, London.
Hardy, S. I., Castelfranco, P. A., and Rebeiz, A. C. (1970). Effect of the hypocotyl hook on greening in etiolated cucumber cotyledons. *Plant Physiol.* **46,** 705–707.
Haupt, W. (1959). Die Chloroplastendrehung bei *Mougeotia*. I. Über den quantitativen und qualitativen Lichtbedarf der Schwachlichtbewegung. *Planta (Berlin)* **53,** 484–501.
Haxo, F. T., and Blinks, L. R. (1950). Photosynthetic action spectra of marine algae. *J. Gen. Physiol.* **33,** 389–422.
Heitz, E. (1932). Die Herkunft der Chromocentren. Dritter Beitrag zur Kenntnis der Beziehung zwischen Kernstruktur und qualitativer Verschiedenheit der Chromosomen in ihrer Langsrichtung. *Planta (Berlin)* **18,** 571–636.
Hill, R. (1939). Oxygen produced by isolated chloroplasts. *Proc. R. Soc. London Ser. B.* **127,** 192–210.
Hill, R., and Bendall, F. (1960). Function of the two cytochrome components in chloroplasts: A working hypothesis. *Nature (London)* **186,** 136–137.
Hill, R., and Scarisbrick, R. (1940). Production of oxygen by illuminated chloroplasts. *Nature (London)* **146,** 61–62.
Hill, R., and Scarisbrick, R. (1951). The haematin compounds of leaves. *New Phytol.* **50,** 88–111.
Hill, R., Smith, J. H. C., and French, C. S. (1953). The absorption and fluorescence spectra of natural protochlorophyll. *Carnegie Inst. Wash. Yearbook* **52,** 153–155.
Hiller, R. G., Pilger, D., and Genge, S. (1973). Photosystem II activity and pigment–protein complexes in flashed bean leaves. *Plant Sci. Lett.* **1,** 81–88.
Hillman, W. S. (ed.) (1970). "Papers in Plant Physiology." Holt, Rinehart, & Winston, New York.
Hodge, A. J., McLean, J. D., and Mercer, F. W. (1956). A possible mechanism for the morphogenesis of lamellar systems in plant cells. *J. Biophys. Biochem. Cytol.* **2,** 597–608.
Jabben, M., and Mohr, H. (1975). Stimulation of the Shibata shift by phytochrome in the cotyledons of the mustard seedling *Sinapis alba* L. *Photochem. Photobiol.* **22,** 55–58.
James, W. O. (1934). The dynamics of photosynthesis. *New Phytol.* **33,** 8–40.
Kahn, A., Boardman, N. K., and Thorne, S. W. (1970). Energy transfer between protochlorophyllide molecules: Evidence for multiple chromophores in the photoactive protochlorophyllide–protein complex *in vivo* and *in vitro*. *J. Mol. Biol.* **48,** 85–101.
Katoh, S. (1960). A new copper protein from *Chlorella ellipsoidea*. *Nature (London)* **186,** 533–534.
Katz, E. (1949). Chlorophyll fluorescence as an energy flowmeter for photosynthesis. *In* "Photosynthesis in Plants" (J. Franck and W. E. Loomis, eds.), pp. 287–292. Iowa State College Press, Ames.

Kirk, J. T. O., and Tilney-Bassett, R. A. E. (1978). "The Plastids, Their Chemistry, Structure, Growth, and Inheritance." Freeman, London.
Klimov, V. V., Klevanik, A. V., and Shuvalov, V. A. (1977). Reduction of pheophytin in the primary light reaction of photosystem II. *FEBS Lett.* **82,** 183–186.
Klimov, V. V., Dolan, E., Shaw, E. R., and Ke, B. (1980). Interaction between the intermediary electron acceptor (pheophytin) and a possible plastoquinone–iron complex in photosystem II reaction centers. *Proc. Natl Acad. Sci. USA* **77,** 7227–7231.
Kok, B. (1956). On the reversible absorption change at 705 m$\mu$ in photosynthetic organisms. *Biochim. Biophys. Acta* **48,** 527–533.
Koski, V. M., French, C. S., and Smith, J. H. C. (1951). The action spectrum for the transformation of protochlorophyll to chlorophyll *a* in normal and albino corn seedlings. *Arch. Biochem. Biophys.* **31,** 1–7.
Küster, E. (1935). "Die Pflanzenzelle." Fischer, Jena.
Leyon, H. (1953). The structure of chloroplasts. II. The first differentiation of the chloroplast structure in *Vallota* and *Taraxacum* studied by means of electron microscopy. *Exp. Cell Res.* **5,** 520–529.
Liljenberg, C. (1966). The effect of light on the phytolization of chlorophyllide *a* and the spectral dependence of the process. *Physiol. Plant.* **19,** 403–410.
Liljenberg, C. (1974). Characterization and properties of a protochlorophyll ester in leaves of dark grown barley with geranylgeraniol as esterifying alcohol. *Physiol. Plant.* **32,** 208–213.
Litvin, F. F., and Belayeva, O. B. (1971). Sequence of photochemical and dark reactions in the terminal stage of chlorophyll biosynthesis. *Photosynthetica* **5,** 200–209.
Lubimenko, W. (1927). Les pigments de plastes et leur transformation dans les tissus vivants de la plante. *Rev. Gén. Bot.* **39,** 547–559, 619–637, 698–710.
Lubimenko, W. (1928). Les pigments de plastes et leur transformation dans les tissus vivants de la plante. *Rev. Bén. Bot.* **40,** 23–29, 88–94, 146–155, 226–243, 303–318, 372–381.
Lundegårdh, H. (1962). Response of chloroplast cytochromes to oxygen. *Physiol. Plant.* **15,** 399–400.
Melis, A., and Anderson, J. M. (1983). Structural and functional organization of the photosystems in *Spinacia* chloroplasts. Antenna size, relative electron transport capacity, and chlorophyll composition. *Biochim. Biophys. Acta* **724,** 473–484.
Menke, W. (1961). Über die Chloroplasten von *Anthoceros punctatus*. *Z. Naturforsch.* **16b,** 334–336.
Menke, W. (1962). Über die Struktur der Heitz–Leyonschen Kristalle. *Z. Naturforsch.* **17b,** 188–190.
Meyer, A. (1883). "Das Chlorophyllkorn in Chemischer, Morphologischer und Biologischer Beziehung." Felix, Leipzig.
Miller, E. C. (1938). "Plant Physiology." McGraw-Hill, New York.
Mohr, H. (1977). Phytochrome and chloroplast development. *Endevour New Sci.* **1,** 107–114.
Molisch, H. (1904). Ueber Kohlensäure-Assimilations-Versuche mittels der Leuchtbacterie Methode. *Botan. Z.* **62,** 1–10.
Molisch, H. (1925). Carbon dioxide in dead leaves. *Z. Bot.* **17,** 577–593.
Monteverde, N. A. (1893). Das Absorbtionsspektrum des Chlorophylls. *Acta Horti Petropolitani* **13,** 121–178.
Moor, H. (1964). Die Gefrier-Fixation lebender Zellen und ihre Anwendung in der Elektronenmikroskopie. *Z. Zellforsch.* **62,** 546–580.
Mühlethaler, K., Moor, H., and Szarkowski, J. M. (1965). The ultrastructure of the chloroplast lamellae. *Planta (Berlin)* **67,** 305–323.

Nadler, K., and Granick, S. (1970). Controls on chlorophyll synthesis in barley. *Plant Physiol.* **46**, 240–246.
Oelze-Karow, H., and Mohr, H. (1978). Control of chlorophyll *b* biosynthesis by phytochrome. *Photochem. Photobiol.* **27**, 189–193.
Ohyama, K., Fukuzawa, H., Kohci, T., Shirai, H., and Sano, T. (1986). Chloroplast gene organization deduced from complete sequence of liverwort *Marchantia polymorpha* chloroplast DNA. *Nature (London)* **322**, 572–574.
Palladin, V. I. (1922). "Plant Physiology" (Engl. transl.). Blakiston's Son, Philadelphia.
Rabinowitch, E. (1945). "Photosynthesis and Related Processes," Vol. I. Interscience, New York.
Rabinowitch, E. (1951). "Photosynthesis and Related Processes," Vol. II. Interscience, New York.
Reinert, J. (ed.) (1980). "Chloroplasts, Results and Problems in Cell Differentiation," Vol. 10. Springer-Verlag, Berlin.
Röper, U., Prinz, H., and Lütz, C. (1987). Amino acid composition of the enzyme NADPH: protochlorophyllide oxidoreductase. *Plant Sci.* **52**, 15–19.
Ruben, S. (1943). Photosynthesis and phosphorylation. *J. Am. Chem. Soc.* **65**, 279–282.
Ruben, S., Randall, M., Kamen, M., and Hyde, J. L. (1941). Heavy oxygen ($^{18}O$) as a tracer in the study of photosynthesis. *J. Am. Chem. Soc.* **63**, 877–879.
Rüdiger, W., Hedden, P., Köst, H.-P., and Chapman, D. J. (1977). Esterification of chlorophyllide by geranylgeranyl pyrophosphate in a cell-free system from maize shoots. *Biochem. Biophys. Res. Commun.* **74**, 1268–1272.
Ryberg, M., and Sundqvist, C. (1991). Structural and functional significance of pigment–protein complexes of chlorophyll precursors. *In* "Chlorophylls" (H. Scheer, ed.), pp. 587–612. CRC Press, Boca Raton, Florida.
Sachs, J. (1862). Über den Einfluss des Lichtes auf die Bildung des Amylums in den Chlorophyllkörnern. *Bot. Ztg.* **20**, 365–373.
San Pietro, A. (1958). Photosynthetic pyridine nucleotide reductase. I. Partial purification and properties of the enzyme from spinach. *J. Biol. Chem.* **231**, 211–229.
Scheer, H. (ed.) (1991). "Chlorophylls." CRC Press, Boca Raton, Florida.
Schimper, A. F. W. (1885). Untersuchungen über die Chlorophyllkörper und die ihnen homologen Gebilde. *Jb. Wiss. Bot.* **16**, 1–246.
Schoch, S. (1978). The esterification of chlorophyllide *a* in greening bean leaves. *Z. Naturforsch.* **33C**, 712.
Schoch, S., Lempert, U., and Rüdiger, W. (1977). Ueber die letzten Stufen der Chlorophyll Biosynthese: Zwischenprodukte zwischen Chlorophyllide und phytolhaltigem Chlorophyll. *Z. Pflanzenphysiol.* **83**, 427–436.
Schönbohm, E. (1973). Kontraktile Fibrillen als aktive Elemente bei der Mechanik der Chloroplasten Verlagerung. *Ber. Deutsch. Bot. Ges.* **86**, 407–422.
Schulz, R., Steinmüller, K., Klaas, M., Forreither, C., Rasmussen, S., Hiller, C., and Apel, K. (1989). Nucleotide sequence of a cDNA coding for the NADPH : protochlorophyllide-oxidoreductase (PCR) of barley (*Hordeum vulgare* L.) and its expression in *Escherichia coli*. *Mol. Gen. Genet.* **217**, 355–361.
Senn, G. (1908). "Die Gestalts- und Lageveränderungen der Pflanzenchromatophoren." Engelmann Verlag, Leipzig.
Shibata, K. (1957). Spectroscopic studies on chlorophyll formation in intact leaves. *J. Biochem. (Tokyo)* **44**, 147–173.
Shinozaki, M., Ohme, M., Tanaka, M., Wakasugi, T., and Hayashida, N. (1986). The

complete nucleotide sequence of the tobacco chloroplast genome: Its gene organization and expression. *EMBO J.* **5**, 2043–2049.
Shlyk, A. A. (1970). "Chlorophyll Metabolism in Green Plants" (T. N. Godnev, ed.). pp. 137–160. Israel Program for Scientific Translations, Jerusalem.
Siefermann-Harms, D. (1988). Fluorescence properties of isolated chlorophyll–protein complexes. *In* "Application of Chlorophyll Fluorescence" (H. K. Lichtenthaler, ed.), pp. 45–54. Kluwer, Dordrecht, The Netherlands.
Sironval, C. (1967). "Le Chloroplast—Croissance et Vieillissement." Masson, Paris.
Sironval, C., and Brouers, M. (eds.) (1984). "Protochlorophyllide Reduction and Greening." Nijhoff/Junk, The Hague, The Netherlands.
Sironval, C., Michel-Wolwertz, M. R., and Madsen, A. (1965). On the nature and possible function of the 673- and 684-nm forms *in vivo* of chlorophyll. *Biochim. Biophys. Acta.* **94**, 344–354.
Sironval, C., Brouers, M., Michel, J.-M., and Kuipper, Y. (1968). The reduction of protochlorophyllide into chlorophyllide. I. The kinetics of the P657–647 to P688–676 phototransformation. *Photosynthetica* **2**, 268–287.
Smith, E. L. (1938). Solution of chlorophyll–protein compounds (phyllochlorins) extracted from spinach. *Science* **88**, 170–171.
Smith, J. H. C., and Benitez, A. (1954). The effect of temperature on the conversion of protochlorophyll to chlorophyll *a* in etiolated barley leaves. *Plant Physiol.* **29**, 135–148.
Smith, J. H. C., Kupke, D. N., and Giese, A. T. (1956). On the preparation, purification, and nature of the protochlorophyll holochrome. *Carnegie Inst. Wash. Yearbook* **55**, 243–248.
Spoehr, H. A. (1916). Theories of photosynthesis in the light of some new facts. *Plant World* **19**, 1–16.
Sundqvist, C. (1969). Transformation of protochlorophyllide formed from exogenous δ-aminolevulinic acid in continuous light and in flash light. *Physiol. Plant.* **22**, 147–156.
Sundqvist, C. (1970). The conversion of protochlorophyllide$_{636}$ to protochlophyllide$_{650}$ in leaves treated with δ-aminolevulinic acid. *Physiol. Plant.* **23**, 412–424.
Takamiya, A. (ed.) (1974). "Selected Papers in Biochemistry," Vol. 12. University Park Press, Baltimore.
Tevini, M., and Lichtenthaler, H. K. (1970). Untersuchungen über die Pigment- und Lipochinonanstattung der zwei photosynthetischen Pigmentsysteme. *Z. Pflanzenphysiol.* **62**, 17–32.
Trebst, A., and Avron, M. (eds.). (1977). "Photosynthesis I, Encyclopedia of Plant Physiology, New Series," Vol. 5. Springer-Verlag, Berlin.
Trebst, A. V., Tsujimoto, H. Y., and Arnon, D. I. (1958). Separation of light and dark phases in the photosynthesis of isolated chloroplasts. *Nature (London)* **182**, 351–355.
van Dorssen, R. J., Plijter, J. J., Dekker, J. P., Den Ouden, A., Amesz, J., and van Gorkom, H. J. (1987). Spectroscopic properties of chloroplast grana membranes and of the core of photosystem II. *Biochim. Biophys. Acta* **890**, 134–143.
van Niel, C. B. (1931). On the morphology and physiology of the purple and green sulphur bacteria. *Arch. Microbiol.* **3**, 1–12.
Virgin, H. I. (1955). The conversion of protochlorophyll to chlorophyll *a* in continuous and intermittent light. *Physiol. Plant.* **8**, 389–403.
Virgin, H. I. (1960). Pigment transformations in leaves of wheat after irradiation. *Physiol. Plant.* **13**, 155–164.
Virgin, H. I. (1961). Action spectrum for the elimination of the lag phase in chlorophyll formation in previously dark-grown leaves of wheat. *Physiol. Plant.* **14**, 439–452.

Virgin, H. I. (1981). The physical state of protochlorophyll(ide) in plants. *Annu. Rev. Plant Physiol.* **32,** 451–463.

Virgin, H. I., Kahn, A., and von Wettstein, D. (1963). The physiology of chlorophyll formation in relation to structural changes in chloroplasts. *Photochem. Photobiol.* **2,** 83–91.

von Mohl, H. (1837). Ph.D. Thesis. Untersuchungen über die anatomischen Vernältnisse des Chlorophylis. University of Tübingen, Germany.

von Wettstein, D. (1958). The formation of plastid structures. The photochemical apparatus. *Brookhaven Symp. Biol.* **11,** 138–159.

Warburg, O. (1919). Über die Geschwindigkeit der photochemischen Kohlensäurezersetzung in lebenden Zellen. I. *Biochem. Z.* **100,** 230–270.

Willstätter, R., and Stoll, A. (1913). "Untersuchungen über Chlorophyll." Springer-Verlag, Berlin.

Withrow, R. B., Wolff, J. B., and Price, L. (1956). Elimination of the lag phase of chlorophyll synthesis in dark-grown leaves by a treatment with low irradiances of monochromatic light. *Plant Physiol. (Suppl.)* **31,** XIII.

Wolff, J. B., and Price, L. (1957). Terminal steps of chlorophyll *a* biosynthesis in higher plants. *Arch. Biochim. Biophys.* **72,** 293–301.

Woodward, R. B. (1961). The total synthesis of chlorophyll. *Pure Appl. Chem.* **2,** 383–404.

Zyrzycka, A. (1951). The influence of the wavelength of light on the movements of chloroplasts in *Lemna trisulca*. *Acta Soc. Bot. Pol.* **21,** 17–37.

# 2

# Plastid Ultrastructure and Development

**HANS RYBERG, MARGARETA RYBERG, AND CHRISTER SUNDQVIST**

Department of Plant Physiology
Botanical Institute
University of Göteborg
S-413 19 Göteborg, Sweden

I. Introduction
II. Common Plastid Features
   A. Plastid Envelope
   B. Nucleoids
   C. Ribosomes
   D. Microtubule-like Structures
   E. Soluble Stromal Compounds
   F. Stromal Inclusions
III. Plastid Types and Development
   A. Proplastids
   B. Plastid Development
   C. Chloroplasts
   D. Etioplasts
   E. Chromoplasts
   F. Storage Plastids
   G. Other Plastids
IV. Concluding Remarks
   References

## I. INTRODUCTION

All plastids originate more or less directly from the undifferentiated proplastids of meristematic cells or tissues. Most plants have genetically identical plastids that are inherited from the mother plant. When plastids are inherited from both parent plants, one plastid type usually is eliminated. In only a few types of mature plant cells have both plastid types survived. The proplastids develop into various types of plastids with different functions, depending on the tissue and on the environmental conditions. Phanerogams produce chloroplasts only in light, whereas many cryptogams can develop chloroplasts in darkness. Different types of storage plastids containing starch, proteins, or lipids are found mainly in seeds, roots, and central parts of stems. In many yellow or red flowers and fruits, the color is created by carotenoids located in special plastids, the chromoplasts. This chapter is meant to inspire extended research on the regulation of plastid development and differentiation. For more information the reader is referred to the excellent reviews on plastid structure, function, and development that have been published over the last 20 years (e.g., Gunning and Steer, 1975; Kirk and Tilney-Bassett, 1978; Schiff, 1980; Staehelin, 1986; Wrischer *et al.*, 1986; Wellburn, 1987).

## II. COMMON PLASTID FEATURES

Despite the profound diversity in plastid appearance, several common features can be recognized. A two-membrane envelope surrounds an amorphous stroma rich in soluble proteins. Ribosomes, nucleoids, microtubule-like structures, plastoglobuli, and membranes occur as defined structures in the stroma. The organization of the inner membranes, the thylakoids, varies depending on the developmental stage and the plastid type, and is treated in Section III.

### A. Plastid Envelope

#### 1. Structural Features

The plastid envelope of higher plants consists of two membranes. A high degree of homology exists between the envelope of nongreen plastids and the chloroplast envelope (Alban *et al.*, 1988). The thickness of each membrane is approximately 6 nm (Gunning and Steer, 1975). The space between the two envelope membranes is electron transparent, and the distance between them can be as great as 20 nm. At some sites, the two envelope membranes are in contact (Cline *et al.*, 1985). These sites may be of significance for protein import (see Chapter 9). Freeze-fracture electron microscopic analyses of chloroplast envelopes revealed differences between the two membranes. The inner membrane showed particles at a much higher density than did the outer membrane (Simpson, 1978; Cline *et al.*, 1985).

## 2. Plastid Ultrastructure and Development

Invaginations from the inner envelope are observed frequently and are the starting points of thylakoid formation (see Section III,B,3). Tubular evaginations from the outer envelope membrane into the cytoplasm have been observed and result in enlarged contact areas between the two compartments (Schötz and Diers, 1975). Also, reports have been made on contacts between the outer membrane and membranes in the cytoplasm, for example, the plasma membrane, tonoplast, endoplasmic reticulum, or outer mitochondrial membrane (Crotty and Ledbetter, 1973). No evidence suggests a physiological function for these contact sites.

The outer surface of the outer envelope membrane is strongly negatively charged (Neuburger *et al.*, 1977). The outer membrane is freely permeable to small molecules up to a molecular mass of ~8 kDa (Heldt, 1976). The inner membrane is impermeable to sucrose and is a highly efficient barrier against movement of most ions. The selective permeability of the inner envelope membrane to a few species of anions is due to specific translocators (Heldt, 1976).

### 2. Polypeptide Composition

As many as 75 polypeptides were identified in the envelope, with molecular sizes ranging from 10 to 140 kDa (Joyard *et al.*, 1982). The predominant polypeptides have molecular sizes of 14, 30, and 54 kDa. The 30-kDa protein has a role in phosphate translocation (Flügge and Heldt, 1991). Two polypeptides with apparent molecular sizes of 10 and 24 kDa were localized to the outside of the outer membrane (Joyard *et al.*, 1983). Immunocytochemical labeling showed the presence of a 20-kDa polypeptide (van Berkel *et al.*, 1986) and of protochlorophyllide (Pchlide) reductase (Joyard *et al.*, 1990) in the outer membrane. Many envelope proteins are encoded by nuclear DNA (Dorne *et al.*, 1982b).

The diaphorase activity of a nitrate reductase was suggested to be localized in the space between the envelope membranes. This suggestion was based on a specific staining reaction requiring ferricyanide as an electron acceptor (Ekés, 1981). This example is one of few demonstrations of enzyme activity within the space between the two membranes. No suitable preparation method has been described to study the biochemistry of this compartment.

### 3. Lipid Composition

The two envelope membranes show evident differences in the ratio of acyl lipids to protein, with a ratio of 2.3–3 for the outer membrane and 0.8–1 for the inner membrane (Block *et al.*, 1983). Both membranes contain monogalactosyl diacylglycerol (MGDG), digalactosyl diacylglycerol (DGDG), and sulfolipids in high amounts and phosphatidylethanolamine in low amounts compared with cytoplasmic membranes. The outer membrane is low in MGDG content but has a high content of phosphatidylcholine compared with other plastid membranes. The lipid composition of the inner membrane resembles that of the thylakoids, with a high MGDG content.

The biosynthesis of galactolipids occurs in the plastid envelope. The inner membrane seems to be the site for the incorporation of galactose into galactolipids (Dorne et al., 1982a) and is also the site for incorporation of fatty acids synthesized in the stroma into MGDG and DGDG (Douce and Joyard, 1984). Chloroplasts are the only site for synthesis of $C_{16}$ and $C_{18}$ fatty acids in leaf cells. These substances thus must be exported to other organelles during membrane growth. At the same time, lipids must be transferred to the thylakoids, inferring a special sorting mechanism. Sulfolipid biosynthesis also might be confined to the envelope, whereas most phospholipid biosynthesis seems to take place in the cytoplasm (Douce and Joyard, 1984).

## 4. Pigment Composition

The plastid envelope contains carotenoids and is especially enriched in violaxanthin. The inner envelope contains about 5 times more carotenoids than the outer envelope, which is enriched in neoxanthin. The ratio of xanthophyll to carotene is higher in the envelope than in the thylakoids (Block et al., 1983). The synthesis of carotenoids seems to occur in the inner envelope, as does the synthesis of the $C_{20}$ prenyl unit of phylloquinone, plastoquinone-9, and $\alpha$-tocopherol (Douce and Joyard, 1984). The prenylquinones present in the envelope are the same as those in the thylakoids. Whereas plastoquinone is enriched in the thylakoids, $\alpha$-tocopherol predominates in the envelope. The function of $\alpha$-tocopherol could be to protect the lipids from harmful oxidations. Some investigations indicate presence of Pchlide in low but significant amounts in the envelope (Pineau et al., 1986).

## B. Nucleoids

Plastid DNA is present in electron-transparent areas, the so-called nucleoids. The DNA molecules are thread-like and circular when viewed with the electron microscope after negative staining. They form fibrils 2.5 nm in diameter and can be made visible in the light microscope by staining with the DNA-specific fluorochrome 4′,6-diamidino-2-phenylindole (DAPI). The plastid DNA probably is supercoiled *in vivo* and can be attached to thylakoids (Yoshida *et al.*, 1978). The size is about 150 kb in higher plants but varies among species. Numerous copies of plastid DNA are present in each plastid. A mature wheat chloroplast can contain as many as 300 copies (Boffey and Leech, 1982). The nucleoids are located in the periphery, as in wheat chloroplasts (Selldén and Leech, 1981), or more centrally, as in the chloroplasts of many dicotyledons. DNA levels and nucleoid location are influenced by plastid development (Lindbeck *et al.*, 1987, and references therein). Lindbeck and associates showed that thylakoid growth affected the location of nucleoids in greening etioplasts of *Phaseolus vulgaris*. In etioplasts, the nucleoids were found close to and surrounding the prolamellar body (PLB). Exposure to light probably caused a spreading of the DNA as thylakoid growth proceeded, since the DNA was found in small nucleoids between the thylakoids throughout the plastid.

## C. Ribosomes

The plastid ribosomes have a diameter of about 17 nm. They are smaller than cytoplasmic ribosomes and are referred to as 70 S ribosomes. The mature ribosome is made of protein to approximately 50% by weight. Plastid ribosomes are composed of two subunits with sedimentation constants of 50 S and 30–35 S, respectively, which correspond to aggregate masses of about $1.7 \times 10^6$ and $1.0 \times 10^6$ Da. The large subunit of the chloroplast ribosome contains rRNA species of 4.5, 5, and 23 S, respectively. The small subunit contains a 16 S rRNA. Chloroplast ribosomes are found throughout the stroma. They can become attached to the thylakoids as polysomes on the stroma-exposed side. Ribosomes or ribosome-like particles also have been found in the stromal part within the PLBs of etioplasts (Wellburn *et al.*, 1977; Wellburn, 1982).

## D. Microtubule-like Structures

Tubular structures frequently are observed in the plastid stroma. Some of them resemble the peripheral reticulum (Harris, 1978), whereas others are more similar to the cytoplasmic microtubules (Vaughn and Wilson, 1981). These microtubule-like structures (MTLS) have been observed in developing and transforming plastids of algae and higher plants (Lawrence and Possingham, 1984). Plastid MTLSs of higher plants have a smaller diameter (13 nm) than microtubules in the cytoplasm (23 nm). They often appear in clusters of 30–40 individual MTLSs. In a longitudinal section, they have a length of at least 0.5 μm. MTLSs often are found around the periphery of PLBs in etioplasts as well as close to developing thylakoids (Artus et al., 1990). They cosediment with thylakoids (Sprey, 1975). Their composition and function are not known, but they may play a role in the reorganization of membrane components in developing plastids. MTLSs are insensitive to microtubule inhibitors and thus, probably are not composed of tubulin (Artus et al., 1990).

## E. Soluble Stromal Compounds

The stroma is composed mainly of soluble proteins. In $C_3$ plants, the major protein component is ribulose bisphosphate carboxylase (Rubisco). Immunoelectron microscopic analyses showed an even distribution of the enzyme in the stroma (Shaw and Henwood, 1985; Dehesh et al., 1986; Vaughn, 1987). The guard cell chloroplasts contain 0.5% of the amount of Rubisco present in mesophyll cells (Vaughn, 1987). In *Euglena,* the majority is localized in the pyrenoid and only a minor portion is found in the stroma (Osafune et al., 1990b,c).

In the red alga *Porphyridium cruentum,* high amounts of carbonic anhydrase were found in the chloroplast when the supply of carbon dioxide was low (Yagawa et al., 1987). Other enzymes participating in the reductive photosynthetic cycles are also present in high amounts. In addition, many enzymes responsible for the formation of proteins, lipids, porphyrins, terpenoids, quinonoids, and aromatic compounds are present. Several enzymes have been localized to both the cytoplasm and the chloroplast stroma. However, distinguishing between the continuous presence of an enzyme and the transient presence of a precursor protein can be difficult (Prunkard et al., 1986). Immunoelectron microscopy studies also have shown the presence of substances such as the growth regulators abscisic acid (Sossountzov et al., 1986) and indole-3-acetic acid (Ohmiya et al., 1990).

## F. Stromal Inclusions

The stroma and the intrathylakoidal space frequently contain crystalloid inclusions. These inclusions often are formed as a result of stress such as water deficit (de Greef and Verbelen, 1973; Wrischer, 1973) or chilling (Wrischer, 1970). They also occur in detached leaves (Wrischer, 1978) and in isolated protoplasts (Takebe et al., 1973).

## 1. Proteins

Protein bodies in the form of whorl-like clusters are common features of plastids and are called stroma-centers (Gunning, 1965; Gunning et al., 1968; Wellburn and Wellburn, 1971). Different proteins can be included. The fiber-like structures are about 8.5 nm in diameter and 200 nm in length. The stroma-center in *Avena*, described by Gunning and associates (Gunning et al., 1968) and tentatively regarded as Rubisco, later was shown by immunoelectron microscopy to be a β-glucosidase that activates the saponins of the vacuole when the cells are damaged (Nisius and Ruppel, 1987; Nisius, 1988).

Another crystalline inclusion, also called a stroma-center, was found in *Kalanchoë* (Lee and Thompson, 1973), a plant with crassulacean acid metabolism (CAM). This inclusion was suggested to be a storage of enzymes of the CAM pathway (Thompson et al., 1977). *Sedum*, another CAM plant, has the same type of inclusions, which accumulate during the day and disappear during the night (Santos and Salema, 1983). The presence of these inclusions appeared not to be governed by the hydration conditions of the plants (Santos and Salema, 1984) and, thus, were not the result of water deficit stress.

Electron-opaque materials arranged in rows or in paracrystalline aggregates are found in many plastids and can be due to an iron-binding protein, phytoferritin. Phytoferritin is present in high amounts in developing plastids and probably acts as iron storage during development (Perrin, 1970; Sprey et al., 1976,1978) or during senescence (Barton, 1970). Membrane-bound amorphous proteinaceous inclusions are found in many plants and are suggested to contain precursors for thylakoid membrane growth (Ames and Pivorun, 1974; Cran and Possingham, 1974).

## 2. Lipids

Electron-dense spherical bodies called plastoglobuli, or osmiophilic globuli, are common in most plastids. They can be seen in close connection with PLBs or thylakoids. During the life span of a plastid, the number and size of the plastoglobuli change. In expanding photosynthetically highly active cells, the number of plastoglobuli is low. When development of the cells is arrested, or during senescence, the number and size of plastoglobuli increase (Tuquet and Newman, 1980; Harris and Schaefer, 1981; Greening et al., 1982; Modrusan and Wrischer, 1987). Chloroplasts in some cacti contain very large plastoglobuli, even in young tissues (Lichtenthaler, 1969; Thomson and Platt, 1973). Plastoglobuli usually have a diameter of ~50 nm; there can be as many as 1000 in a chloroplast. Plastoglobuli can contain large amounts of lipids, carotenoids, or carotenoid precursors, and often a high amount of plastoquinone (Tevini and Lichtenthaler, 1970; Dahlin and Ryberg, 1986).

## 3. Starch

Starch grains appear as electron-transparent structures, occasionally with electron-opaque regions. Photosynthetic products are deposited as starch temporarily in chloroplasts, where they accumulate during the day and are distributed during the night (c.f. Gunning and Steer, 1975; see also Section III,F). In chloroplasts, the starch grains can be as large as 2 $\mu$m in length.

## III. PLASTID TYPES AND DEVELOPMENT

### A. Proplastids

The proplastid, also called eoplast (Whatley, 1977), is the organelle from which all other plastids originate. Proplastids are found in meristematic tissues, where the number of plastids per cell is low and remains more or less constant by division (Thomson *et al.*, 1972; Whatley, 1974). The ultrastructure of proplastids is simple and consists of the envelope, often with invaginations from the inner membrane, and a stroma with some ribosomes, nucleoids, plastoglobuli, and occasionally some inner membranes and starch grains (Fig. 1). Proplastids are more or less spherical and have an average diameter of about 1 $\mu$m (Khandakar and Bradbeer, 1989).

### B. Plastid Development

#### 1. Plastid Division

As the leaves develop, the number of plastids per cell increases from 10–15 in young cells to 100–150 in mature leaf cells. In a thorough study, Whatley (1980) described plastid growth and division in the primary leaf of *Phaseolus vulgaris* grown under a 12-hr light/12-hr dark photoperiod. Plastids divided at all stages of plastid development. Plastid division and cell division were not synchronous; plastids continued to divide during cell expansion. Dividing chloroplasts were found in wheat leaves in a discrete band of cells 15–25 mm from the leaf base and separated from the basal intercalary meristem (Leech and Pyke, 1988). Chloroplast division

## 2. Plastid Ultrastructure and Development

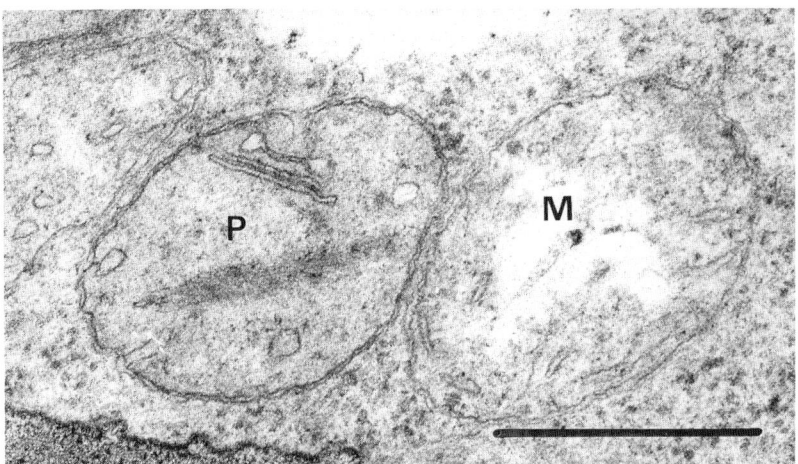

**Fig. 1.** Proplastid (P) in the cotyledon of a 1-day-old radish seedling (*Raphanus sativus*) with few inner membranes and some invaginations from the inner envelope membrane. The size of the proplastid is similar to that of a mitochondrion (M). Bar: 0.5 μm.

comprises a sequence of morphological changes. The plastid changes from a rounded to a dumbbell shape; by continual constriction, the two daughter chloroplasts finally separate. For wheat, the whole process is finished within 1 hr and the daughter chloroplasts double in size within 4 hr (Leech and Pyke, 1988). Microtubules and filamentous structures in the cytoplasm are involved in plastid division (Tewinkel and Volkmann, 1987; Chida and Ueda, 1991).

### 2. Plastid Differentiation

The development of a plastid during ontogenesis involves profound changes in plastid size and internal structure. In light as well as in darkness, the plastids pass through an amyloplast and an amoeboid stage during their further development to chloroplasts or etioplasts (Whatley, 1980). The final appearance of the plastid depends not only on light intensity and quality, but also on other external factors such as temperature, water, and nutrient supply, as well as on the tissue (Sundqvist *et al.*, 1980; see Chapter 3). The differentiation of a proplastid or etioplast to a chloroplast depends on the location of the cell in relationship to a neighboring meristem, as well as on the age of the leaf (Rascio *et al.*, 1976, 1980, 1984). Within and in the immediate vicinity of the meristem, no chloroplast differentiation takes place.

The size of the plastids varies considerably among different tissues of the plant. The diameter of the plastid increases during leaf development from about 1 μm at the proplastid stage to about 6 μm at the mature chloroplast stage in the mesophyll of a leaf. This increase parallels the increase in leaf area and cell enlargement. In similar types of tissues, there is often a variation in the chloroplast size depending on the position in the plant. For example, the leaves of the upper crown part of a tree have larger plastids than those of the lower crown part (Goryshina, 1980).

Most reports on chloroplast ultrastructure development lack detailed biochemical information. In some reports, the biochemistry of proplastid to chloroplast development under photoperiodic growth is analyzed also, for example, chlorophyll a/b ratio (Kirchanski, 1975), chlorophyll, carotenoids, starch, ATP, and 3-phosphoglyceric acid content (Wellburn et al., 1982), chlorophyll, lipid, and protein content, and photosynthetic activity (Guillot-Salomon et al., 1987). That the ultrastructure as well as the biochemistry of the plastids varies during the light–dark cycle is not always taken into account. Profound differences can be seen when sampling is made after the light and after the dark period (Lott, 1970). This fact is of importance, for instance, when analyses are made to correlate structure to function or composition.

## 3. Thylakoid Formation

Thylakoids are generally believed to be formed by continued invaginations from the inner envelope membrane, at least at early stages of chloroplast development (Douce and Joyard, 1984). Such invaginations probably form the perforated thylakoids found in young developing plastids (Whatley et al., 1982). Tubules have been seen in contact with both the thylakoids and the inner envelope membrane in developing plastids (Oross and Possingham, 1991). In later stages, transfer of membrane material may occur by vesicles budding from the envelope and migrating toward the thylakoid membrane (cf. Carde et al., 1982; Whatley et al., 1982; Hoober et al., 1991). At least lipid components are suggested to be transferred in this way from the envelope to the thylakoids (Morré et al., 1991a,b). In that case, the vesicles must be processed before they fuse with the already existing thylakoids. Another possibility is a successive addition of material to existing thylakoids (Anderson, 1981). Lipids, pigments, and proteins are synthesized within the plastid; proteins also are transported into the plastid and incorporated into the growing thylakoids (see Chapters 8, 9, and 10).

As thylakoid growth proceeds, some of the membranes overlap and form stacks called grana. Several models of the events leading to grana formation have been presented (Wehrmeyer and Röbbelen, 1965; Wieckowski, 1967; Diers and Schötz, 1969; Argyroudi-Akoyunoglou *et al.*, 1976). The number of thylakoids per granum is influenced greatly by the irradiation intensity during development. Species growing in the sun usually have fewer thylakoids per granum than species growing in the shade (Guillot-Salomon *et al.*, 1978; Rühle and Wild, 1985). Similar differences can be seen among individuals of the same species grown under different light conditions; the number of thylakoids per granum also increases with decreasing light intensity (Mousseau and Bourdu, 1968). Even within a single leaf a gradient in the size of the grana is seen. Chloroplasts in the palisade parenchyma usually have fewer thylakoids per granum than the chloroplasts in the spongy parenchyma (Terashima and Inoue, 1985). Reversing the direction of the light can result in a complete inversion of the intraleaf gradient of the number of thylakoids per granum (Terashima *et al.*, 1986). Light quality influences the size of the grana; red light increases the number of thylakoids per granum over blue light (Buschmann *et al.*, 1978). Individuals of the same species of mountain plants show variations in grana size depending on the altitudes at which they grow. The higher the altitude, the fewer the number of thylakoids per granum (Miroslavov and Kravkina, 1991). Whether this is a response to different light conditions has not been examined.

## C. Chloroplasts

### 1. Seed Plants

The inner membranes of the chloroplasts, the thylakoids, constitute a system of membranes with chlorophylls, carotenoids, proteins, cofactors, and lipids, all of which are necessary for the primary reactions of photosynthesis. The thylakoids usually are arranged in stacks called grana (Fig. 2). Single stromal thylakoids connect the different grana and, by running helically around the granum, also connect the thylakoids within one granum. The intrathylakoidal space, or thylakoid lumen, is about 10 nm wide. The lumen of the stromal thylakoids is continuous with the lumen of the granal thylakoids. Thus, the intrathylakoidal space within one chloroplast is a continuous, although very complex, compartment (Paolillo, 1970; Mustardy and Janossy, 1979). The appressed region at which two thylakoids make contact is called a partition. The number of thylakoids in one granum can vary between 2 and 100 but is often 10–25. The number of grana within one chloroplast is usually 40–60. Freeze–fracture studies have revealed differences in stromal and granal thylakoids. The density

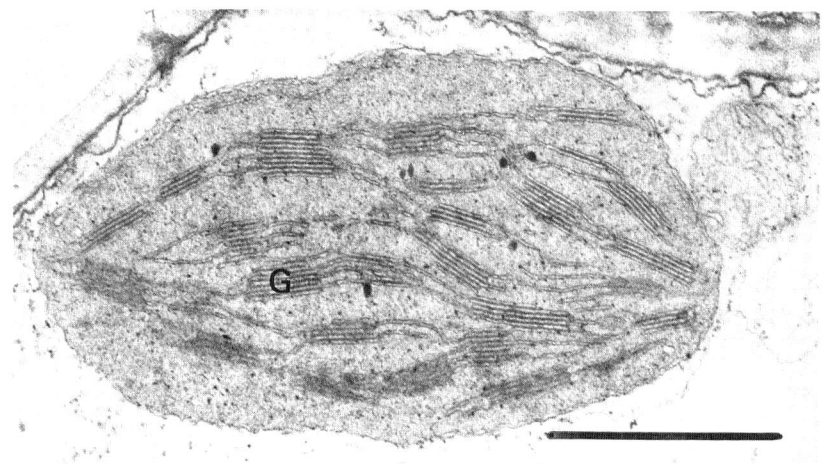

**Fig. 2.** Chloroplast of a 6-day-old dark-grown wheat leaf (*Triticum aestivum*) irradiated for 48 hr. The thylakoids occur in stacks, grana (G), or as single thylakoids. Almost all thylakoids are oriented in the same plane, a feature typical of mature chloroplasts. Bar: 1 µm.

and size of the intramembrane particles differ between stacked (appressed) and unstacked (nonappressed) regions; these differences are due to a lateral heterogeneity of the components in the photosynthetic membranes (Miller and Staehelin, 1976; Staehelin, 1976; Simpson, 1978).

A number of thylakoid proteins have been localized to the grana and/or stroma thylakoids. Immunoelectron microscopy is widely used to localize antigens and is a powerful supplement to fractionation methods for localization of plastid proteins (Table I). Generally, the results of immunolocalization studies *in situ* support the conclusions drawn from fractionation experiments. The distribution of various proteins of the photosystems to appressed or nonappressed regions can be determined, provided the data are treated statistically and the resolution limits borne in mind. Immunoelectron microscopy is a powerful method with which to study the distribution of a protein at various stages of development or under conditions that would make fractionation studies tedious or impossible (cf. Fig. 3). As another example, immunolabeling showed that, during senescence, the lateral heterogeneity of the apoprotein of the light-harvesting complex of photosystem II (LHCPII) is lost (Hilditch *et al.*, 1989). Detailed descriptions of the protein composition of the photosystems are given in Chapters 12 and 13.

An interesting pattern in the distribution of a plastid protein is seen in

## TABLE I
**Localization of Membrane-Associated Plastid Proteins by Immunoelectron Microscopy**

| Protein | Main location | Plant | Reference |
|---|---|---|---|
| Seed plants | | | |
| Chloroplasts | | | |
| CP1 | Nonappressed thylakoids | *Spinacia, Hordeum* | Vallon et al. (1986, 1987); Hoyer-Hansen et al. (1988) |
| LHCI | Nonappressed thylakoids | *Hordeum* | Vallon et al. (1986) |
| CP47 | Grana thylakoids | *Hordeum* | Vallon et al. (1987) |
| CP43 | Grana thylakoids | *Hordeum* | Vallon et al. (1987); Simpson and von Wettstein (1989) |
| CP29 | Appressed thylakoids | *Hordeum* | Vallon et al. (1987); Hoyer-Hansen et al. (1988) |
| LHCPII | Appressed thylakoids | *Spinacia, Hordeum* | Shaw and Henwood (1985); Simpson et al. (1987); Hoyer-Hansen et al. (1988); Vallon et al. (1989) |
| D1 | Appressed thylakoids | *Spinacia, Hordeum* | Vallon et al. (1986, 1987) |
| D2 | Appressed thylakoids | *Spinacia* | Vallon et al. (1986) |
| Cyt $b$-559 | Appressed thylakoids | *Spinacia, Hordeum* | Vallon et al. (1987, 1989) |
| 16 kDa ($O_2$-ev) | Appressed thylakoids | *Spinacia* | Goodchild et al. (1985a) |
| 23 kDa ($O_2$-ev) | Appressed thylakoids | *Spinacia* | Goodchild et al. (1985a) |
| 33 kDa ($O_2$-ev) | Appressed thylakoids | *Spinacia, Hordeum* | Goodchild et al. (1985a); Vallon et al. (1987) |
| Cyt $f$ | Appressed and nonappressed thylakoids | *Spinacia, Hordeum* | Goodchild et al. (1985b); Shaw and Henwood (1985) |
| CF1 | Nonappressed thylakoids | *Spinacia, Hordeum* | Shaw and Henwood (1985); Vallon et al. (1986, 1987, 1989) |

*(continued)*

**TABLE I**—*Continued*

| Protein | Main location | Plant | Reference |
|---|---|---|---|
| Etioplasts | | | |
| Pchlide-reductase | Polamellar bodies (and prothylakoids during greening) | *Phaseolus, Lycopersicon, Hordeum, Zea, Triticum* | Shaw *et al.* (1985); Dehesh *et al.* (1986); Ryberg and Dehesh (1986); Grevby *et al.* (1989a); Forreiter *et al.* (1990) |
| CF1 | Prothylakoids | *Hordeum* | Shaw *et al.* (1985) |
| Cyt *f* | Prothylakoids | *Hordeum* | Shaw *et al.* (1985) |
| Algae | | | |
| Chloroplasts | | | |
| PSI | More in nonappressed than in appressed thylakoids | *Fucus, Cryptomonas* | Lichtlé *et al.* (1992a,b) |
| LHC | Appressed thylakoids | *Chlamydomonas* | Vallon *et al.* (1986) |
| | Thylaoids (and Golgi during greening) | *Euglena* | Osafune *et al.* (1990a,1991a,b) |
| | Even distribution along all thylakoids | *Fucus, Laminaria, Cryptomonas, Rhodomonas* | Grevby *et al.* (1989b); Lichtlé *et al.* (1992a,b); Vesk *et al.* (1992) |
| Phycoerythrin | Thylakoid lumen | *Cryptomonas, Rhodomonas* | Spear-Bernstein and Miller (1989); Lichtlé *et al.* (1992b); Vesk *et al.* (1992) |
| D1 | Appressed thylakoids | *Chlamydomonas* | Vallon *et al.* (1986) |
| D2 | Appressed thylakoids | *Chlamydomonas* | Vallon *et al.* (1986) |
| CF1 | Nonappressed thylakoids | *Chlamydomonas* | Vallon *et al.* (1986) |

*Euglena gracilis*, in which LHCPII is found in both the Golgi apparatus and the plastid thylakoids (Osafune *et al.*, 1990a, 1991a,b; Schiff *et al.*, 1991). Based on the localization pattern of the protein during a light–dark cycle and during greening, Schiff and associates (1991) suggested the following scenario. Light permits translation of the LHCPII precursor on cytoplasmic ribosomes of the rough endoplasmic reticulum (ER). The

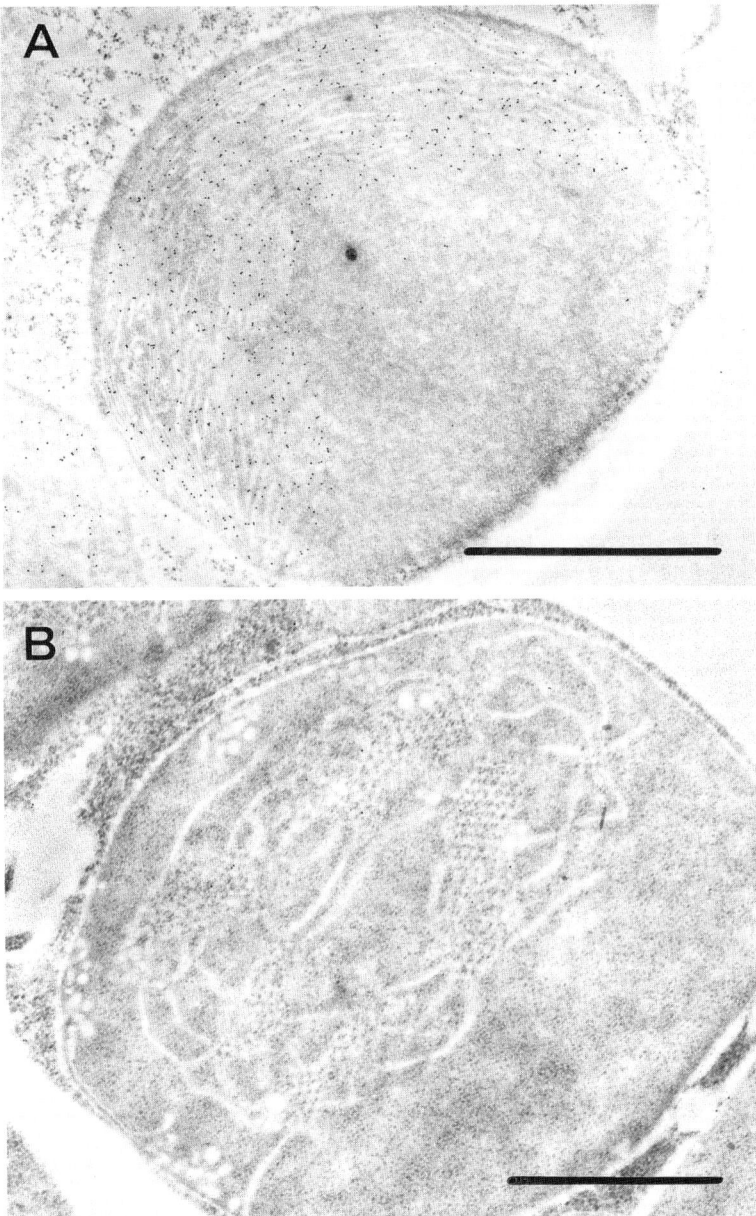

**Fig. 3.** Localization of LHCPII in (A) a mature chloroplast and (B) an etioplast of spinach leaves (*Spinacia oleracea*) by immunoelectron microscopy. The accumulation of the protein in the chloroplast and its absence in the etioplast is obvious. Black dots caused by protein A–gold, 10 nm, indicate the site of the protein. The leaves were embedded in Lowicryl K4M at low temperature. Bars: 1 $\mu$m.

protein passes through the ER to the Golgi, presumably for further modifications. The Golgi vesicles can transport LHCPII to the plastid and assist in the uptake of the protein.

Plastids in roots grown in light also can develop into chloroplasts. These chloroplasts are structurally indistinguishable from leaf chloroplasts. However, one antenna complex of photosystem II (PSII) found in leaf chloroplasts was missing in root chloroplasts (Casadoro *et al.*, 1990).

The chloroplasts of the bundle sheath cells differ from those of the mesophyll cells in $C_4$ plants (Laetsch, 1971; Hatch *et al.*, 1975). In corn and sugarcane, the bundle sheath chloroplasts lack grana and the thylakoids traverse the full length of the chloroplast. The peripheral reticulum frequently observed in $C_4$ plants is a system of anastomosing membranes that is continuous with the inner membrane of the envelope (Gracen *et al.*, 1972; Sprey and Laetsch, 1978). The function of this system is probably to facilitate transport of metabolites. The peripheral reticulum is more frequent in chloroplasts of the mesophyll cells than in chloroplasts of the bundle sheath cells. A peripheral reticulum has been observed also in a $C_3$ plant (van Steveninck *et al.*, 1972) and in a gymnosperm (Whatley, 1975). A chloroplast with an unusual granal configuration is found in the "resurrection" plant *Myrothamnus*, in which the grana are arranged in a staircase fashion (Wellburn and Wellburn, 1976).

## 2. Algae

The ultrastructure and pigment composition of chloroplasts differ widely among the algal groups. The evolutionary classification of the algae made by Lee (1989) is based on the number of membranes surrounding the chloroplast. The algae are divided in four groups: (1) the prokaryotic Cyanophyta and Prochlorophyta that actually lack chloroplasts, (2) algae with chloroplasts surrounded by the two membranes of the "normal" envelope (Glaucophyta, Rhodophyta, and Chlorophyta), (3) algae with chloroplasts surrounded by an additional membrane, the chloroplast endoplasmic reticulum, giving a total of three membranes that separate the stroma from the cytoplasm (Euglenophyta and Dinophyta), and (4) algae with chloroplasts surrounded by two extra membranes of chloroplast ER, giving four membranes between the stroma and the cytoplasm (Cryptophyta, Chrysophyta, Prymnesiophyta, Bacillariophyta, Xanthophyta, Eustigmatophyta, Raphidophyta, and Phaeophyta). Evolutionary schemes also have been proposed by Whatley and Whatley (1981) and by Billard (1985).

The thylakoid system and pigment composition varies significantly among the algal classes (Table II; Figs. 4–6; see Larkum and Barrett,

**TABLE II**

**Characteristics of Algal Chloroplasts**[a]

| Algal class | Chlorophyll | Phycobilin[b] | Number of membranes in envelope | Thylakoids per band |
|---|---|---|---|---|
| Cyanophyta | a | PE, PC, APC | — | 1 |
| Prochlorophyta | a, b | — | — | 1–5 |
| Glaucophyta | a | + | — | 1 |
| Rhodophyta | a, (d) | PE, PC, APC | 2 | 1 |
| Chlorophyta | a, b | — | 2 | 2–6 |
| Euglenophyta | a, b | — | 3 | 3 |
| Dinophyta | a, (b)[c], c | (+)[d] | 3 | 3, (2)[d] |
| Cryptophyta | a, c | PE, PC | 4 | 2 |
| Chrysophyta | a, $c_1$, $c_2$ | — | 4 | 3 |
| Prymnesiophyta | a, $c_1$, $c_2$ | — | 4 | 3 |
| Bacillariophyta | a, $c_1$, $c_2$ | — | 4 | 3 |
| Xanthophyta | a, c | — | 4 | 3 |
| Eustigmatophyta | a | — | 4 | 3 |
| Raphidophyta | a, c | — | 4 | 3 |
| Phaeophyta | a, $c_1$, $c_2$ | — | 4 | 3 |

[a] Adapted from Lee (1989).
[b] PE, phycoerythrin; PC, phycocyanin; APC, allophycocyanin.
[c] Watanabe et al. (1987).
[d] Schnepf and Elbrächter (1988).

1983, for review). In the prokaryotic Prochlorophyta, which have the photosynthesizing membranes in the cytoplasm, the thylakoids are organized in grana-like structures and contain chlorophyll b (Cox, 1986). The chloroplasts of green algae resemble those of higher plants by having thylakoids stacked in grana, although usually with fewer thylakoids per granum (Fig. 4). A comparison of the localization of thylakoid protein complexes in *Chlamydomonas,* spinach, and barley was made by Olive and Vallon (1991). These workers showed that there are obvious similarities in the distribution of most thylakoid proteins in the alga and higher plants. Several major chlorophyll–protein complexes of higher plants were identified in five species of Chlorophyta (Levavasseur, 1989). Another similarity between green algae and higher plants is that both store the photosynthetic products inside the chloroplast. All other algae but the Cryptophyta store photosynthetic products in the cytoplasm. In Cryptophyta, these molecules are stored between the envelope and the chloroplast ER. At the same site is a structure called the nucleomorph, which is suggested to be a remnant of the nucleus of an endosymbiont (Gillott and

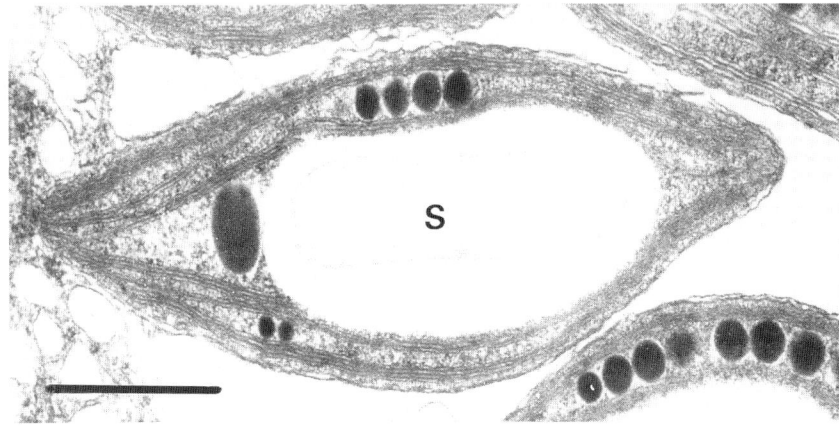

**Fig. 4.** Chloroplast of a green alga (*Codium fragile*). The thylakoids form grana with few thylakoids per granum, a feature typical of Chlorophyta. A large area is occupied by starch (S). Plastoglobuli are frequent between the thylakoid stacks. Bar: 0.5 μm.

Gibbs, 1980). Another peculiarity in the Cryptophyta is the organization of the thylakoids. They are present in pairs with some regions appressed; the accessory pigment, the phycobiliproteins, are located within the lumen of the thylakoid and probably also within the thylakoid membrane (SpearBernstein and Miller, 1989). The other organisms with phycobilins, the Cyanophyta, Glaucophyta, and Rhodophyta, contain these accessory pigments in special disk-shaped or hemispherical bodies, the phycobilisomes, on the surface of the stromal side of the thylakoids (Gantt, 1990).

Ultrastructure and pigment composition is influenced greatly by growth irradiance, as can be seen in the red unicellular alga *Porphyridium* (Cunningham *et al.*, 1989, 1991). The amounts of photosystem I (PSI), PSII, and phycobilisome pigment–proteins decreased at similar rates with increasing irradiance, whereas the content of the carotenoid zeaxanthin increased. Zeaxanthin probably is not an accessory pigment but has photoprotecting properties (see Chapter 14).

The light-harvesting complexes of many algae differ from those of higher plants. For instance, a fucoxanthin–chlorophyll a/c–protein complex from the brown seaweed *Petalonia* has been isolated and characterized (Katoh and Ehara, 1990). From the chrysophyte *Ochromonas*, which also has the fucoxanthin–chlorophyll a/c–protein, a new chlorophyll a–carotenoid–protein complex was characterized (Gibbs and Biggins, 1991). Also, the thylakoid arrangement of species belonging to the Phaeophyta and related groups differs from that of higher plants. The thylakoids

## 2. Plastid Ultrastructure and Development

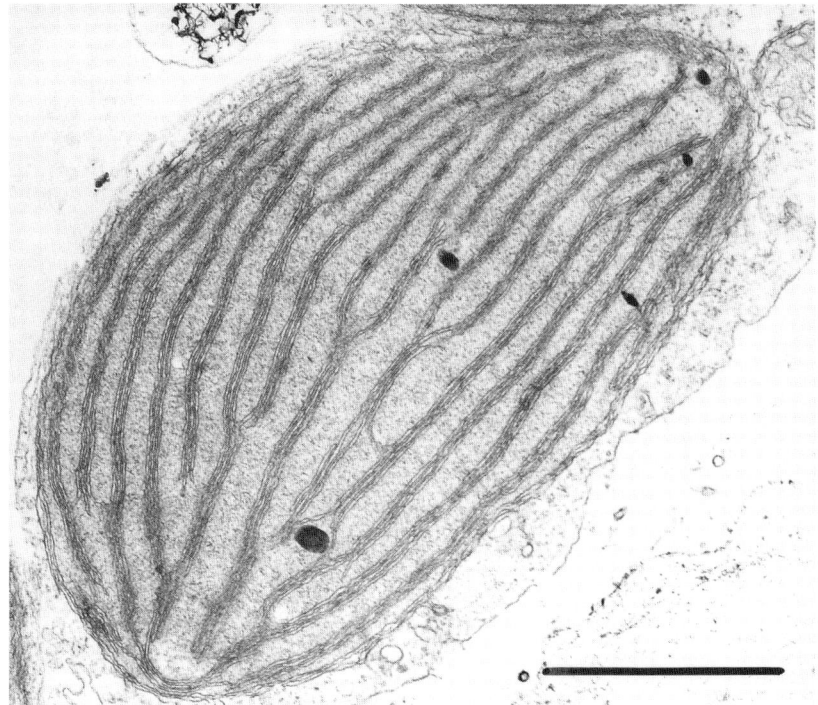

**Fig. 5.** Chloroplast of a brown alga (*Chorda filum*). The thylakoids occur in bands of three, a feature typical of Phaeophyta. Bar: 1 μm.

are grouped in bands of three, which run the full length of the chloroplast (Fig. 5).

A structure that can be found in chloroplasts of all eukaryotic algal groups is the pyrenoid (Fig. 6). It is structurally distinct from the stroma and has a fine granular appearance. It can be located centrally or in an evagination from the chloroplast envelope and, thus, in a pocket separated from the thylakoids. Thylakoids often run through the pyrenoid. Examples of different pyrenoids from the Dinophyta are described by Dodge and Crawford (1971). Since the pyrenoids often are covered by starch products from photosynthetic activity, it was assumed that they contained enzymes for carbohydrate metabolism. It has now been shown by immunoelectron microscopy that the pyrenoid of many algae contains Rubisco, for example, Glaucophyta (Mangeney and Gibbs, 1987), Rhodophyta (McKay and Gibbs, 1990), Chlorophyta (Lacoste-Royal and Gibbs, 1987), and Euglenophyta (Osafune *et al.*, 1990b,c). In the work by McKay and Gibbs (1990),

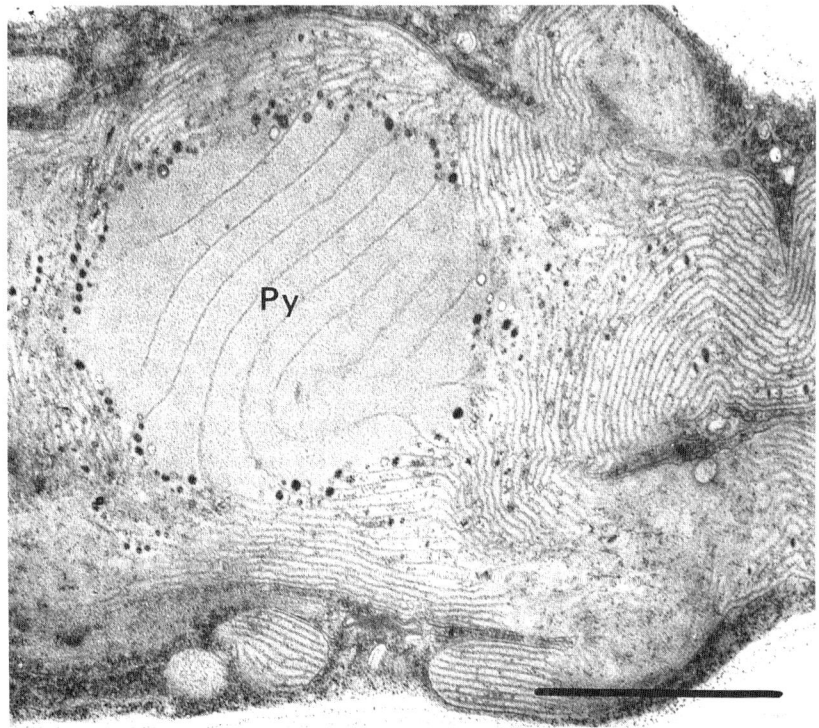

**Fig. 6.** Chloroplast of a red alga (*Porphyra umbilicalis*). The pyrenoid (Py) is located centrally and is traversed by several single thylakoids. The thylakoids also occur as single membranes in the stroma. Plastoglobuli are frequent, especially close to the pyrenoid. Only about half the plastid is shown. Bar: 2 μm.

it also was shown that the thylakoids that traverse the pyrenoid in *Porphyridium* lack phycoerythrin and PSII activity. The Cyanophyta and Prochlorophyta lack pyrenoids but have small polyhedral bodies, carboxysomes, in the cytoplasm with a fine structure similar to that of pyrenoids. Immunogold localization has revealed that the carboxysomes contain Rubisco (Swift and Leser, 1989). Rubisco is also localized in the pyrenoids of several mosses belonging to the Anthocerotae (Vaughn *et al.*, 1990).

### D. Etioplasts

In tissues and organs destined to become green and photosynthesizing in light, the proplastids develop into etioplasts in darkness (Bradbeer *et*

## 2. Plastid Ultrastructure and Development

*al.*, 1974a; Khandakar and Bradbeer, 1988). In early literature, the term proplastid included all stages of development up to the chloroplast, that is, also the etioplast (e.g., Klein and Bogorad, 1964).

The etioplast is characterized by the presence of one or several PLBs, structures consisting of tubular membranes connected into a highly regular three-dimensional lattice (Fig. 7; Gunning and Jagoe, 1967; see also Chapter 7). In a fully developed etioplast with a size of 3–4 $\mu$m, the PLB has a diameter of about 1 $\mu$m. In the most common form, the basic unit of the PLB is built up by four tubular membranes merging to one point; the other ends are directed toward the corners of a regular tetrahedron. These basic units can be combined in many different ways, resulting in different structural appearances of the PLBs (Gunning and Steer, 1975). At the periphery of the PLB, and connected to it, many flat perforated membranes, prothylakoids (PTs), stretch out into the stroma (Weier and Brown, 1970). The membranes of the PLBs and PTs thus divide the interior of the etioplast in two different continuous phases. The stroma and the space between the tubules of the PLBs make up one compartment; the intrathylakoidal space of the PTs and the interior of the PLB tubules form the other.

Irradiation of dark-grown leaves leads to dramatic morphological changes (Henningsen and Boynton, 1970, 1974; Rosinski and Rosen, 1972; Bradbeer *et al.*, 1974b, 1977). The regular structure of the PLBs is lost rapidly, an event that is initiated by the phototransformation of Pchlide to chlorophyllide (Chlide). The pigment change is probably not sufficient to cause the ultrastructural changes. Both *in situ* and isolated, PLBs can keep their regular structure after phototransformation of Pchlide to Chlide (Treffry, 1970; Ryberg and Sundqvist, 1988). Chlide undergoes a spectral shift, the Shibata shift, which seems to be a prerequisite for the dispersal of the PLBs.

Pchlide-reductase is the dominating protein of PLBs, where it is associated to Pchlide in large aggregates (Böddi *et al.*, 1990; Ryberg and Sundqvist, 1991; Chapter 5). During the early phase of PLB dispersal, the Pchlide-reductase aggregates dissociate and relocation of Pchlide-reductase to the PTs occurs (Ryberg and Dehesh, 1986). The Shibata shift probably reflects such a dissociation, which seems to be a prerequisite for esterification of Chlide (Treffry, 1970; Lindsten *et al.*, 1990; Artus *et al.*, 1992). After a complete dispersal, only a few plastoglobuli indicate the place of the former PLB. During continued irradiation, the PTs are transformed into thylakoids and the membranes start to overlap (incipient grana formation). Eventually, typical grana are formed. The time sequence of these changes varies with light intensity and quality, temperature, water supply, and age. For mature etioplasts in 5- to 7-day-old leaves, the follow-

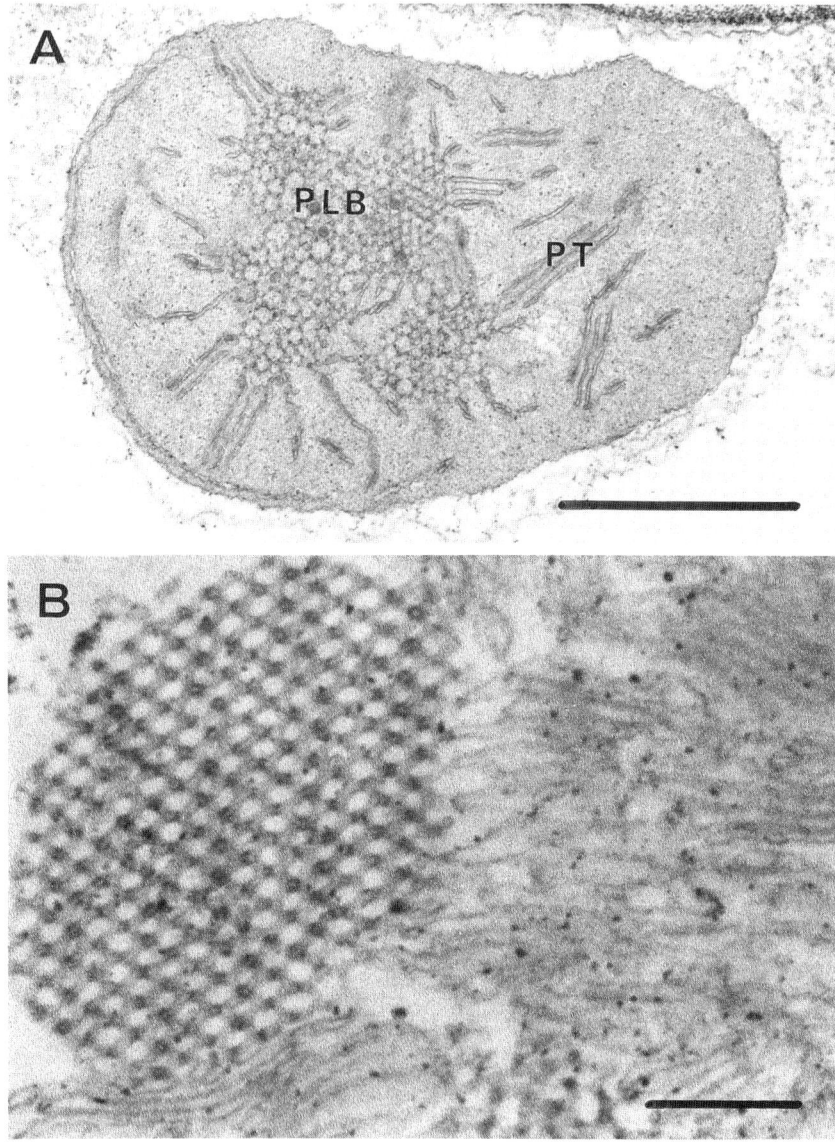

**Fig. 7.** (A) Etioplast of a 6-day-old dark-grown wheat leaf. The internal membranes form a three-dimensional network of tubular membranes, the prolamellar body (PLB), connected with flat perforated membranes, the prothylakoids (PT). Bar: 1 μm. (B) The intimate connection between the PLB and the thylakoids is seen clearly in this micrograph of isolated plastid membranes. The PLBs were formed in darkness in *Zebrina* plastids that were green and contained grana when the leaf was transferred to the dark. Bar: 0.2 μm.

## 2. Plastid Ultrastructure and Development

ing durations of the processes are good approximations. The reduction of Pchlide to Chlide occurs within milliseconds, the Shibata shift is finished within 10–20 min, a relocation of Pchlide-reductase to the PTs is evident after 5 min, and after 30 min it is found along the entire PTs. The first thylakoid overlaps can be seen after a few minutes, grana are recognized after 1–2 hr, and after 4 hr the dispersal of the PLBs is completed. After 12–16 hr of irradiation, the etioplasts have developed into chloroplasts (Selldén, 1977).

An interesting finding is that plastids of dark-grown pea leaves exposed to a daily heat-shock treatment were as large as mature chloroplasts and contained well-developed inner membranes. They also accumulated otherwise light-induced transcripts of certain plastid proteins (Kloppstech et al., 1991). Thus, light is not always required for development beyond the etioplast stage.

PLBs are not restricted to dark-grown leaves. Small PLBs, in addition to grana, are present in developing chloroplasts of light-grown plants. Leaf age, rather than light, seems to affect the formation of PLBs in differentiating chloroplasts (Rascio et al., 1980). PLBs are formed also in chloroplasts of leaves grown under low-light conditions (Wrischer, 1966) and in plastids of greening leaves when returned to darkness or low light, as long as the plastid is not completely mature (Weier et al., 1970; Ikeda, 1971; Minkov et al., 1988). The idea that the PLBs are in a dynamic equilibrium with the growing thylakoids in developing leaves is further strengthened by their appearance during the night under natural conditions (Fig. 8; Rebeiz and Rebeiz, 1986).

The onset of photosynthesis in greening dark-grown leaves is coupled tightly to the structural changes of the etioplasts (e.g., Robertson and Laetsch, 1974). Many proteins needed for the development of the photosystems are present already in darkness, for example, coupling factor 1 (CF1), cytochrome $f$, cytochrome $b$-559, plastocyanin, and the Rieske Fe–S center protein (Plesnicar and Bendall, 1973, Herrmann et al., 1985; Takabe, 1986; Takabe et al., 1986), whereas chlorophyll and light-harvesting complex (LHCP) and P700 apoproteins need light for synthesis (Tanaka and Tsuji, 1983,1985; Mullet, 1988; see Chapters 8, 12, and 13).

In storage organs and roots, plastid differentiation can be arrested for a long time. In darkness, the development stops at the proplastid or amyloplast stage. No differentiation to etioplasts occurs. Pchlide content is extremely low or absent (Björn, 1976; McEwen et al., 1991; Virgin and Sundqvist, 1992). Differentiation of the proplastids or amyloplasts to chloroplasts can be induced by light under certain conditions (Björn, 1967; Thomson and Whatley, 1980).

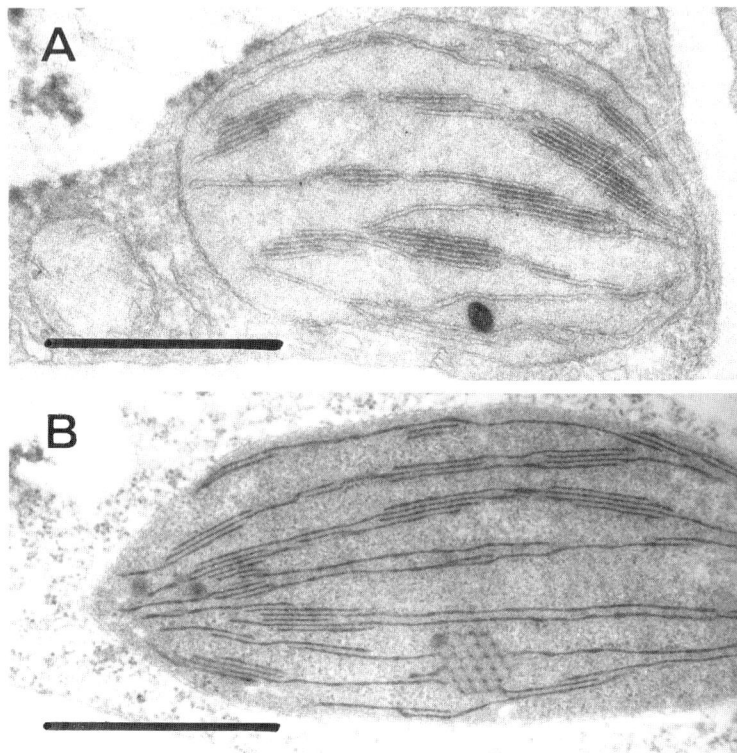

**Fig. 8.** Developing chloroplasts of young beech leaves (*Fagus silvatica*) a few days after budburst. (A) The leaf was fixed after several hours in light. (B) The leaf was fixed after 12 hr in darkness. PLBs were formed among the developing thylakoids during the dark period. Note that the section has become "negatively" stained, which gives the dark appearance of the thylakoid and PLB-membrane lumen. The reason for this phenomenon is not known, but it is observed occasionally, especially in immature plastids. Bars: 1 $\mu$m.

### E. Chromoplasts

Chromoplasts are characterized by their high carotenoid content, which gives a yellow, orange, or red color to tissues. They are found frequently in petals and fruits and serve to attract animals for cross-pollination and seed dispersal. Some roots, for example, carrots, have chromoplasts with crystals of $\beta$-carotene. A special type of chromoplast is the plastid found in aging leaves. As the chlorophyll is broken down, the structure of the senescing chloroplasts resembles chromoplasts with large plastoglobuli (Ikeda, 1979). These plastids are called gerontoplasts (Whatley and What-

## 2. Plastid Ultrastructure and Development

ley, 1987) to distinguish them from true chromoplasts. Sitte (1974) classified different types of chromoplasts based on the structure in which the carotenoids are localized (see Chapter 15).

Chromoplasts can develop from many different plastids, for example, chloroplasts, proplastids, or amyloplasts (Falk, 1976; Thomson and Whatley, 1980; Knoth, 1981; Huyskens et al., 1985; Brett and Sommerard, 1986). Chromoplasts also can revert to chloroplasts as, for example, in pumpkin fruits (Devidé and Ljubesic, 1974). In some petals, chromoplast-like plastids without any carotenoids have been reported (Whatley and Whatley, 1987). The plastids of the seed coat of Cucurbitaceae contain large amounts of protochlorophyll and resemble chromoplasts (Fig. 9; Sundqvist and Ryberg, 1979; Ryberg et al., 1980).

### F. Storage Plastids

All nonpigmented plastids are sometimes referred to as leukoplasts. Large amounts of starch, protein, or lipids often are found in these plastids. Plastids with few thylakoids and large deposits of starch are called amyloplasts. The whole plastid can be filled with starch grains of various forms. Amyloplasts are typical of storage organs and can be as large as 30 μm (Gunning and Steer, 1975). Amyloplasts also are found close to the apical meristems of both roots and shoots, where they act as statoliths for

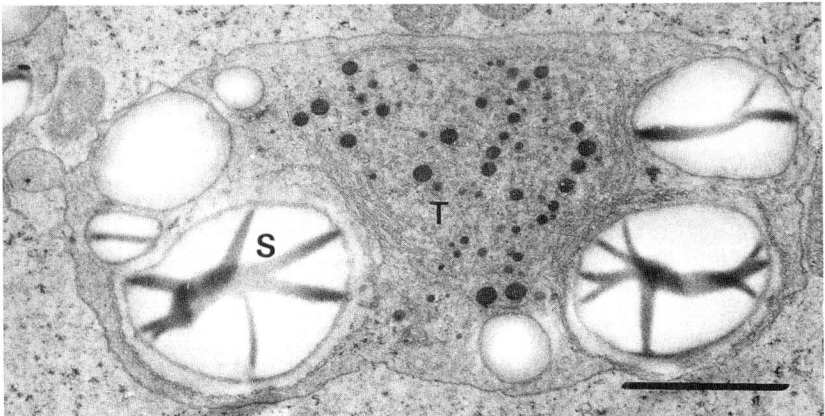

**Fig. 9.** Chromoplast-like plastid of the inner seedcoat of *Cucurbita pepo*. Irregularly arranged tubular membranes (T) and large starch grains (S) occupy most of the plastid space. Plastoglobuli are scattered among the membranes. Bar: 2 μm.

graviperception (Juniper and French, 1973). During chloroplast development, the plastid passes through a short-lived amyloplast-like stage (Whatley, 1980; Wellburn, 1987).

Proteinoplasts, or proteoplasts, have large deposits of protein, either amorphous or as crystalloids (Thomson and Whatley, 1980). They often are observed as a temporary stage during plastid development. Proteinoplasts develop from proplastids in a cell layer between the palisade and spongy mesophyll layers in primary leaves of mung bean (Hurkman and Kennedy, 1976). The plastids pass through an amyloplast stage before they develop an amorphous spherical membrane-bound protein body. In light the proteinoplasts rapidly degenerate, whereas in darkness they stay intact for weeks.

Plastids filled with oil droplets are called elaioplasts (Wellburn, 1987). They resemble the globulous type of chromoplasts, but do not contain carotenoids. These plastids are found mostly in epidermal cells of monocotyledonous plants, and have not been investigated closely.

### G. Other Plastids

Sieve tubes of vascular plants contain spherical plastids with a diameter of about 1 $\mu$m (Behnke, 1972). These plastids usually contain starch or protein crystals. Usually they are not termed amyloplasts or proteinoplasts but simply sieve-tube plastids. It has been suggested that they are involved in the sealing of sieve plates of sieve elements (Eleftheriou, 1984).

Crystalloid inclusions frequently are observed in plastids of epidermal cells, where they occur not only in the stroma but also in the thylakoid lumen (e.g., Williams, 1974; Hoefert and Esau, 1975; Martin and Larbalestier, 1977; Sprey and Lambert, 1977; Platt-Aloia and Thomson, 1979; Mikulska et al., 1981). The epidermal plastids usually do not develop into chloroplasts but stop at a less differentiated stage. Slightly differentiated chloroplasts, often with starch grains, are present in guard cells of most plants (Williams, 1974; Hoefert and Esau, 1975; Miyake and Maeda, 1976; Mikulska et al., 1981). Fully differentiated leukoplasts can be found with chloroplasts in the same cells, as in the variegated leaves of *Hedera* (Salema and Abreu, 1972) and *Solanum* (Toyama, 1972).

### IV. CONCLUDING REMARKS

The progress in chloroplast research at the suborganelle level continuously reveals more and more details about the function and complexity of this organelle. Purification and identification of various substructures have given important information about compartmentalization of the organelle. Ultrastructural studies also have made us aware of the incredible variations in plastid appearance. Immunocytochemistry at the electron microscope level has increased our knowledge of the locations of various proteins. By improved *in situ* techniques, which will allow even higher resolution than the techniques used today, it certainly will become possible to determine, for instance, the orientation of a thylakoid protein within the membrane. The rather slow fixation methods commonly used today have hampered studies on the dynamics and fast changes of the plastids. Freeze fixation techniques, which allow a rapid termination of most chemical reactions and ultrastructural changes, will certainly become more used in the future. The identification of plant genes and the improved understanding of gene regulation makes it even more rewarding to look for correlations between plastid ultrastructure and the presence and function of proteins during plastid development and differentiation.

## REFERENCES

Alban, C., Joyard, J., and Douce, R. (1988). Preparation and characterization of envelope membranes from nongreen plastids. *Plant Physiol.* **88,** 709–717.

Ames, I. H., and Pivorun, J. P. (1974). A cytochemical investigation of a chloroplast inclusion. *Am. J. Bot.* **61,** 794–797.

Anderson, J. M. (1981). Consequences of spacial separation of photosystem 1 and 2 in thylakoid membranes of higher plant chloroplasts. *FEBS Lett.* **124,** 1–10.

Argyroudi-Akoyunoglou, J. H., Kondylaki, S., and Akoyunoglou, G. (1976). Growth of grana from "primary" thylakoids in *Phaseolus vulgaris*. *Plant Cell Physiol.* **17,** 939–954.

Artus, N. N., Ryberg, M., and Sundqvist, C. (1990). Plastid microtubule-like structures in wheat are insensitive to microtubule inhibitors. *Physiol. Plant.* **79,** 641–648.

Artus, N. N., Ryberg, M., Lindsten, A., Ryberg, H., and Sundqvist, C. (1992). The Shibata shift and the transformation of etioplasts to chloroplasts in wheat with clomazone (FMC 57020) and amiprophos-methyl (Tokunol M). *Plant Physiol.* **98,** 253–263.

Barton, R. (1970). The production and behaviour of phytoferritin particles during senescence of *Phaseolus* leaves. *Planta* **94,** 73–77.

Behnke, H.-D. (1972). Sieve-tube plastids in relation to angiosperm systematics—An attempt towards a classification by ultrastructural analysis. *Bot. Rev.* **38,** 155–197.

Billard, C. (1985). Le complexe nucleoplastidial chez les chromophytes: Structure, fonction et intérêt dans une perspective phylogénétique. *Cryptog. Algol.* **6,** 191–211.

Björn, L. O. (1967). The light requirement for different steps in the development of chloroplasts in excised wheat roots. *Physiol. Plant.* **20,** 483–499.

Björn, L. O. (1976). The state of protochlorophyll and chlorophyll in corn roots. *Physiol. Plant.* **37,** 183–184.

Block, M. A., Dorne, A. J., Joyard, J., and Douce, R. (1983). Preparation and characterization of membrane fractions enriched in outer and inner envelope membranes from spinach chloroplasts. II. Biochemical characterization. *J. Biol. Chem.* **258,** 13281–13286.

Böddi, B., Lindsten, A., Ryberg, M., and Sundqvist, C. (1990). Phototransformation of aggregated forms of protochlorophyllide in isolated etioplast inner membranes. *Photochem. Photobiol.* **52,** 83–87.

Boffey, S. A., and Leech, R. M. (1982). Chloroplast DNA levels and the control of chloroplast division in light-grown wheat leaves. *Plant Physiol.* **69**, 1387–1391.

Bradbeer, J. W., Ireland, H. M. M., Smith, J. W., Rest, J., and Edge, H. J. W. (1974a). Plastid development in primary leaves of *Phaseolus vulgaris*. VII. Development during growth in continuous darkness. *New Phytol.* **73**, 263–270.

Bradbeer, J. W., Gyldenholm, A. O., Ireland, H. M. M., Smith, J. W., Rest, J., and Edge, H. J. W. (1974b). Plastid development in primary leaves of *Phaseolus vulgaris*. VIII. The effect of the transfer of dark-grown plants to continuous illumination. *New Phytol.* **73**, 271–279.

Bradbeer, J. W., Arron, G. P., Herrera, A., Kemble, R. J., Montes, G., Sherratt, D., and Wara-Aswapati, O. (1977). The greening of leaves. *In* "Integration of Activity in the Higher Plant" (D. H. Jennings, ed.), pp. 195–219. Cambridge University Press, Cambridge.

Brett, D. W., and Sommerard, A. P. (1986). Ultrastructural development of plastids in the epidermis and starch layer of glossy *Ranunculus* petals. *Ann. Bot.* **58**, 903–910.

Buschmann, C., Meier, D., Kleudgen, H. K., and Lichtenthaler, H. K. (1978). Regulation of chloroplast development by red and blue light. *Photochem. Photobiol.* **27**, 195–198.

Carde, J.-P., Joyard, J., and Douce, R. (1982). Electron microscopic studies of envelope membranes from spinach plastids. *Biol. Cell* **44**, 315–324.

Casadoro, G., Pasqua, G., Vecchia, F. D., and Rascio, N. (1990). Chloroplasts of greened roots. *Cytobios* **64**, 73–79.

Chida, Y., and Ueda, K. (1991). Division of chloroplasts in a green alga, *Trebouxia potteri*. *Ann. Bot.* **67**, 435–442.

Cline, K., Keegstra, K., and Staehelin, L. A. (1985). Freeze-fracture electron microscopic analysis of ultrarapidly frozen envelope membranes on intact chloroplasts and after purification. *Protoplasma* **125**, 111–123.

Cox, G. (1986). Comparison of *Prochloron* from different hosts. I. Structural and ultrastructural characteristics. *New Phytol.* **104**, 429–445.

Cran, D. G., and Possingham, J. V. (1974). Plastid thylakoid formation. *Ann. Bot.* **38**, 843–847.

Crotty, W. J., and Ledbetter, M. C. (1973). Membrane continuities involving chloroplasts and other organelles in plant cells. *Science* **182**, 839–841.

Cunningham, F. X., Jr., Dennenberg, R. J., Mustardy, L., Jursinic, P. A., and Gantt, E. (1989). Stoichiometry of photosystem I, photosystem II, and phycobilisomes in the red alga *Porphyridium cruentum* as a function of growth irradiance. *Plant Physiol.* **91**, 1179–1187.

Cunningham, F. X., Jr., Mustárdy, L., and Gantt, E. (1991). Irradiance effects on thylakoid membranes of the red alga *Porphyridium cruentum*. An immunocytochemical study. *Plant Cell Physiol.* **32**, 419–426.

Dahlin, C., and Ryberg, H. (1986). Accumulation of phytoene in plastoglobuli of SAN-9789 (Norflurazon)-treated dark-grown wheat. *Physiol. Plant.* **68**, 39–45.

de Greef, J. A., and Verbelen, J. P. (1973). Physiological stress and crystallites in leaf plastids of *Phaseolus vulgaris* L. *Ann. Bot.* **37**, 593–596.

Dehesh, K., van Cleve, B., Ryberg, M., and Apel, K. (1986). Light-induced changes in the distribution of the 36,000-$M_r$ polypeptide of NADPH–protochlorophyllide oxidoreductase within different cellular compartments of barley (*Hordeum vulgare* L.). II. Localization by immunogold labelling in ultrathin sections. *Planta* **169**, 172–183.

Devidé, Z., and Ljubešic, N. (1974). The reversion of chromoplasts to chloroplasts in pumpkin fruits. *Z. Pflanzenphysiol.* **73**, 296–306.

Diers, L., and Schötz, F. (1969). Über ring- und schalenförmige Thylakoidbildungen in den Plastiden. *Z. Pflanzenphysiol.* **60**, 187–210.

Dodge, J. D., and Crawford, R. M. (1971). A fine-structural survey of dinoflagellate pyrenoids and food-reserves. *Bot. J. Linn. Soc.* **64**, 105–115.

Dorne, A.-J., Block, M. A., Joyard, J., and Douce, R. (1982a). Studies on the localization of enzymes involved in galactolipid metabolism in chloroplast envelope membranes. *In* "Biochemistry and Metabolism of Plant Lipids" (J. F. G. M. Wintermans and P. J. C. Kuiper, eds.), pp. 153–164. Elsevier/North-Holland, Amsterdam.

Dorne, A.-J., Carde, J.-P., Börner, T., Joyard, J., and Douce, R. (1982b). Polar lipid composition of a plastids ribosome-deficient barley mutant. *Plant Physiol.* **69**, 1467–1470.

Douce, R., and Joyard, J. (1984). The regulatory role of the plastid envelope during development. *In* "Chloroplast Biogenesis" (N. R. Baker and J. Barber, eds.), Vol. 5, pp. 71–132. Elsevier, Amsterdam.

Ekés, M. (1981). Ultrastructural demonstration of ferricyanid reductase (diaphorase) activity in the envelopes of the plastids of etiolated barley (*Hordeum vulgare* L.) leaves. *Planta* **151**, 439–446.

Eleftheriou, E. P. (1984). Sieve-element plastids of *Triticum* and *Aegilops* (Poaceae). *Plant Syst. Evol.* **145**, 119–133.

Falk, H. (1976). Chromoplasts of *Tropaeolum majus* L.: Structure and development. *Planta* **128**, 15–22.

Flügge, U.-I., and Heldt, H. W. (1991). Metabolite translocators of the chloroplast envelope. *Annu. Rev. Plant Physiol. Plant Mol. Biol.* **42**, 129–144.

Forreiter, C., van Cleve, B., Schmidt, A., and Apel, K. (1990). Evidence for a general light-dependent negative control of NADPH–protochlorophyllide oxidoreductase in angiosperms. *Planta* **183**, 126–132.

Gantt, E. (1990). Pigmentation and photoacclimation. *In* "Biology of the Red Algae" (K. M. Cole and R. G. Sheath, eds.), pp. 203–219. Cambridge University Press, Cambridge.

Gibbs, P. B., and Biggins, J. (1991). Thylakoid organization in the chromophyte alga *Ochromonas danica*. *Plant Physiol.* **97**, 381–387.

Gillott, M. A., and Gibbs, S. P. (1980). The cryptomonad nucleomorph: Its ultrastructure and evolutionary significance. *J. Phycol.* **16**, 558–568.

Goodchild, D. J., Anderson, J. M., and Andersson, B. (1985a). Immunocytochemical localization of the cytochrome $b/f$ complex of chloroplast thylakoid membranes. *Cell Biol. Intern. Rep.* **9**, 715–721.

Goodchild, D. J., Andersson, B., and Anderson, J. M. (1985b). Immunocytochemical localization of polypeptides associated with the oxygen evolving system of photosynthesis. *Eur. J. Cell Biol.* **36**, 294–298.

Goryshina, T. K. (1980). Structural and functional features of the leaf assimilatory apparatus in plants in a forest-steppe oakwood. II. Seasonal dynamics of the plastid apparatus in the herbaceous understory. *Acta Oecol. Oecol. Plant.* **1**, 201–208.

Gracen, V. E., Jr., Hilliard, J. H., Brown, R. H., and West, S. H. (1972). Peripheral reticulum in chloroplasts of plants differing in $CO_2$ fixation pathways and photorespiration. *Planta* **107**, 189–204.

Greening, M. T., Butterfield, F. J., and Harris, N. (1982). Chloroplast ultrastructure during senescence and regreening of flax cotyledons. *New Phytol.* **92**, 279–285.

Grevby, C., Axelsson, L., and Sundqvist, C. (1989a). Light-independent plastid differentiation in the brown alga *Laminaria saccharina* (Phaeophyceae). *Phycologia* **28**, 375–384.

Grevby, C., Engdahl, S., Ryberg, M., and Sundqvist, C. (1989b). Binding properties of NADPH–protochlorophyllide oxidoreductase as revealed by detergent and ion treatments of isolated and immobilized prolamellar bodies. *Physiol. Plant.* **77**, 493–503.

Guillot-Salomon, T., Tuquet, C., De Lubac, M., Hallais, M.-F., and Signol, M. (1978). Analyse comparative de l'ultrastructure et de la composition lipidique des chloroplastes de plantes d'ombre et de soleil. *Cytobiology* **17**, 442–452.

Guillot-Salomon, T., Farineau, N., Cantrel, C., Oursel, A., and Tuquet, C. (1987). Isolation and characterisation of developing chloroplasts from light-grown barley leaves. *Physiol. Plant.* **69,** 113–122.
Gunning, B. E. S. (1965). The fine structure of chloroplast stroma following aldehyde osmium-tetroxide fixation. *J. Cell Biol.* **24,** 79–93.
Gunning, B. E. S., and Jagoe, M. P. (1967). The prolamellar body. *In* "Biochemistry of Chloroplasts" (T. W. Goodwin, ed.), Vol. II, pp. 655–676. Academic Press, London.
Gunning, B. E. S., and Steer, M. W. (1975). "Ultrastructure and the Biology of Plant Cells." Arnold, London.
Gunning, B. E. S., Steer, M. W., and Cochrane, M. P. (1968). Occurrence, molecular structure, and induced formation of the "stromacentre" in plastids. *J. Cell Sci.* **3,** 445–456.
Harris, J. B. (1978). Development of a tubular apparatus in chloroplasts of ageing *Cyphomandra* leaves. *Cytobios* **21,** 151–164.
Harris, J. B., and Schaefer, V. G. (1981). Some correlated events in aging leaf tissues of tree tomato and tobacco. *Bot. Gaz.* **142,** 43–54.
Hatch, M. D., Kagawa, T., and Craig, S. (1975). Subdivision of $C_4$-pathway species based on differing $C_4$ acid decarboxylating systems and ultrastructural features. *Aust. J. Plant Physiol.* **2,** 111–128.
Heldt, H. W. (1976). Metabolite transport in intact spinach chloroplasts. *In* "The Intact Chloroplast" (J. Barber, ed.), pp. 215–234. Elsevier, Amsterdam.
Henningsen, K. W., and Boynton, J. E. (1970). Macromolecular physiology of plastids. VIII. Pigment and membrane formation in plastids of barley greening under low light intensity. *J. Cell Biol.* **44,** 290–304.
Henningsen, K. W., and Boynton, J. E. (1974). Macromolecular physiology of plastids. IX. Development of plastid membranes during greening of dark-grown barley seedlings. *J. Cell Sci.* **15,** 31–55.
Herrmann, R. G., Westhoff, P., Alt, J., Tittgen, J., and Nelson, N. (1985). Thylakoid membrane proteins and their genes. *In* "Molecular Form and Function of the Plant Genome" (L. van Vloten-Doting, G. S. P. Groot, and T. C. Hall, eds.), pp. 233–256. Plenum, Amsterdam.
Hilditch, P. I., Thomas, H., Thomas, B. J., and Rogers, L. J. (1989). Leaf senescence in a non-yellowing mutant of *Festuca pratensis:* Proteins of photosystem II. *Planta* **177,** 265–272.
Hoefert, L. L., and Esau, K. (1975). Plastid inclusions in epidermal cells of *Beta. Am. J. Bot.* **62,** 36–40.
Hoober, J. K., Boyd, C. O., and Paavola, L. G. (1991). Origin of thylakoid membranes in *Chlamydomonas reinhardtii* y-1 at 38°C. *Plant Physiol.* **96,** 1321–1328.
Hoyer-Hansen, G., Bassi, R., Honberg, L. S., and Simpson, D. J. (1988). Immunological characterization of chlorophyll $a/b$-binding proteins of barley thylakoids. *Planta* **173,** 12–21.
Hurkman, W. J., and Kennedy, G. S. (1976). Fine structure and development of proteoplasts in primary leaves of mung bean. *Protoplasma* **89,** 171–184.
Huyskens, S., Timberg, R., and Gross, J. (1985). Pigment and plastid ultrastructural changes in kumquat (*Fortunella margarita*) "Nagami" during ripening. *J. Plant Physiol.* **118,** 61–72.
Ikeda, T. (1971). Prolamellar body formation under different light and temperature conditions. *Bot. Mag. (Tokyo)* **84,** 363–375.
Ikeda, T. (1979). Electron microscopic evidence for the reversible transformation of *Eunonymus* plastids. *Bot. Mag. (Tokyo)* **92,** 23–30.

## 2. Plastid Ultrastructure and Development

Joyard, J., Grossman, A. R., Bartlett, S. G., Douce, R., and Chua, N.-H. (1982). Characterization of envelope membrane polypeptides from spinach chloroplasts. *J. Biol. Chem.* **257,** 1095–1101.

Joyard, J., Billecocq, A., Bartlett, S. G., Block, M. A., Chua, N.-H., and Douce, R. (1983). Localization of polypeptides to the cytosolic side of the outer envelope membrane of spinach chloroplasts. *J. Biol. Chem.* **258,** 10000–10006.

Joyard, J., Block, M., Pineau, B., Albrieux, C., and Douce, R. (1990). Envelope membranes from mature spinach chloroplasts contain a NADPH : protochlorophyllide reductase on the cytosolic side of the outer membrane. *J. Biol. Chem.* **265,** 21820–21827.

Juniper, B. E., and French, A. (1973). The distribution and redistribution of endoplasmic reticulum (ER) in geoperceptive cells. *Planta* **109,** 211–224.

Katoh, T., and Ehara, T. (1990). Supramolecular assembly of fucoxanthin–chlorophyll–protein complexes isolated from a brown alga, *Petalonia fascia*. Electron microscopic studies. *Plant Cell Physiol.* **31,** 439–447.

Khandakar, K., and Bradbeer, J. W. (1988). Primary leaf growth in bean (*Phaseolus vulgaris*). II. Cell and plastid development during growth in darkness and after transfer to illumination at various stages of dark growth. *Bangladesh J. Bot.* **17,** 173–188.

Khandakar, K., and Bradbeer, J. W. (1989). Primary leaf growth in bean (*Phaseolus vulgaris*). III. Some qualitative and quantitative studies in cell fine structure with developmental changes of plastids during the first four days of dark development. *Cytologia* **54,** 409–417.

Kirchanski, S. J. (1975). The ultrastructural development of the dimorphic plastids of *Zea mays* L. *Am. J. Bot.* **62,** 695–705.

Kirk, J. T. O., and Tilney-Bassett, R. A. E. (1978). "The Plastids. Their Chemistry, Structure, Growth and Inheritance." Elsevier/North-Holland, Amsterdam.

Klein, S., and Bogorad, L. (1964). Fine structural changes in proplastids during photodestruction of pigments. *J. Cell Biol.* **22,** 443–451.

Kloppstech, K., Otto, B., and Sierralta, W. (1991). Cyclic temperature treatments of dark-grown pea seedlings induce a rise in specific transcript levels of light-regulated genes related to photomorphogenesis. *Mol. Gen. Genet.* **225,** 468–473.

Knoth, R. (1981). Ultrastructure of lycopene-containing chromoplasts in fruits of *Algaonema commutatum* Schott (Araceae). *Protoplasma* **106,** 249–259.

Lacoste-Royal, G., and Gibbs, S. P. (1987). Immunocytochemical localization of ribulose-1,5-bisphosphate carboxylase in the pyrenoid and thylakoid region of the chloroplast of *Chlamydomonas reinhardtii*. *Plant Physiol.* **83,** 602–606.

Laetsch, W. M. (1971). Chloroplast structural relationships in leaves of $C_4$ plants. *In* "Photosynthesis and Photorespiration" (M. D. Hatch, C. B. Osmond, and R. O. Slatyer, eds.), pp. 323–349. Wiley, New York.

Larkum, A. W. D., and Barrett, J. (1983). Light-harvesting processes in algae. *In* "Advances in Botanical Research" (H. W. Woolhouse, ed.), Vol. 10, pp. 1–219. Academic Press, London.

Lawrence, M. E., and Possingham, J. V. (1984). Observations of microtubule-like structures within spinach plastids. *Biol. Cell* **52,** 77–82.

Lee, R. E. (1989). "Phycology." Cambridge University Press, New York.

Lee, R. E., and Thompson, A. (1973). The stromacentre of plastids of *Kalanchoë pinnata* Persoon. *J. Ultrastruct. Res.* **42,** 451–456.

Leech, R. M., and Pyke, K. A. (1988). Chloroplast division in higher plants with particular reference to wheat. *In* "Division and Segregation of Organelles" (S. A. Boffey and D. Lloyd, eds.), pp. 39–62. Cambridge University Press, Cambridge.

Levavasseur, G. (1989). Analyse comparée des complexes pigment–protéines de chlorophycophytes marines benthiques. *Phycologia* **28,** 1–14.
Lichtenthaler, H. K. (1969). Plastoglobuli und Lipochinongehalt der Chloroplasten von *Cereus peruvianus* (L.) Mill. *Planta* **87,** 304–310.
Lichtlé, C., McKay, R. M. L., and Gibbs, S. P. (1992a). Immunogold localization of photosystem-I and of photosystem-II light-harvesting complexes in cryptomonad thylakoids. *Biol. Cell* **74,** 187–194.
Lichtlé, C., Spilar, A., and Duval, J. C. (1992b). Immunogold localization of light-harvesting and photosystem I complexes in the thylakoids of *Fucus serratus* (Phaeophyceae). *Protoplasma* **166,** 99–106.
Lindbeck, A. G. C., Rose, R. J., Lawrence, M. E., and Possingham, J. V. (1987). The role of chloroplast membranes in the location of chloroplast DNA during the greening of *Phaseolus vulgaris* etioplasts. *Protoplasma* **139,** 92–99.
Lindsten, A., Welch, C. J., Schoch, S., Ryberg, M., Rüdiger, W., and Sundqvist, C. (1990). Chlorophyll synthetase is latent in well preserved prolamellar bodies of etiolated wheat. *Physiol. Plant.* **80,** 277–285.
Lott, J. N. A. (1970). Changes in the cotyledons of *Cucurbita maxima* during germination. III. Plastids and chlorophyll. *Can. J. Bot.* **48,** 2259–2265.
McEwen, B., Virgin, H. I., Böddi, B., and Sundqvist, C. (1991). Protochlorophyll forms in roots of dark-grown plants. *Physiol. Plant.* **81,** 455–461.
McKay, R. M. L., and Gibbs, S. P. (1990). Phycoerythrin is absent from the pyrenoid of *Porphyridium cruentum:* Photosynthetic implications. *Planta* **180,** 249–256.
Mangeney, E., and Gibbs, S. P. (1987). Immunocytochemical localization of ribulose-1,5-bisphosphate carboxylase/oxygenase in the cyanelles of *Cyanophora paradoxa* and *Glaucocystis nostochinearum. Eur. J. Cell Biol.* **43,** 65–70.
Martin, E. S., and Larbalestier, G. (1977). A membrane-bound plastid inclusion in the epidermis of leaves of *Taraxacum officinale. Can. J. Bot.* **55,** 222–225.
Mikulska, E., Damsz, B., and Zolnierowicz, H. (1981). Structural and functional polymorphism of plastids in leaves of *Clivia miniata* Rgl. I. Ontogenesis of plastids in epidermis and guard cells. *Acta Soc. Bot. Pol.* **50,** 381–389.
Miller, K. R., and Staehelin, L. A. (1976). Analysis of the thylakoid outer surface. Coupling factor is limited to unstacked membrane regions. *J. Cell Biol.* **68,** 30–47.
Minkov, I. N., Ryberg, M., and Sundqvist, C. (1988). Properties of reformed prolamellar bodies from illuminated and redarkened etiolated wheat plants. *Physiol. Plant.* **72,** 725–732.
Miroslavov, E. A., and Kravkina, I. M. (1991). Comparative analysis of chloroplasts and mitochondria in leaf chlorenchyma from mountain plants grown at different altitudes. *Ann. Bot.* **68,** 195–200.
Miyake, H., and Maeda, E. (1976). The fine structure of plastids in various tissues in the leaf blade of rice. *Ann. Bot.* **40,** 1131–1138.
Modrušan, Z., and Wrischer, M. (1987). Seasonal changes in chloroplasts of blackberry leaves. *Acta Bot. Croat.* **46,** 23–31.
Morré, D. J., Morré, J. T., Morré, S. R., Sundqvist, C., and Sandelius, A. S. (1991a). Chloroplast biogenesis. Cell-free transfer of envelope monogalactosylglycerides to thylakoids. *Biochim. Biophys. Acta* **1070,** 437–445.
Morré, D. J., Selldén, G., Sundqvist, C., and Sandelius, A. S. (1991b). Stromal low temperature compartment derived from the inner membrane of the chloroplast envelope. *Plant Physiol.* **97,** 1558–1564.
Mousseau, M., and Bourdu, R. (1968). Influence des conditions écologiques d'éclairement

pendant la croissance sur la structure et l'activité des chloroplastes de *Teucrium scorodonia* L. *Bull. Soc. Franç. Physiol. Végét.* **14,** 307–315.

Mullet, J. E. (1988). Chloroplast development and gene expression. *Annu. Rev. Plant Physiol. Plant Mol. Biol.* **39,** 475–502.

Mustárdy, L. A., and Jánossy, A. G. S. (1979). Evidence of helical thylakoid arrangement by scanning electron microscopy. *Plant Sci. Lett.* **16,** 281–284.

Neuburger, M., Joyard, J., and Douce, R. (1977). Strong binding of cytochrome $c$ on the envelope of spinach chloroplasts. *Plant Physiol.* **59,** 1178–1181.

Nisius, A. (1988). The stromacentre in *Avena* plastids: An aggregation of $\beta$-glucosidase responsible for the activation of oat-leaf saponins. *Planta* **173,** 474–481.

Nisius, A., and Ruppel, H. G. (1987). Immunocytological and chemical studies on the stromacentre-forming protein from *Avena* plastids. *Planta* **171,** 443–452.

Ohmiya, A., Hayashi, T., and Kakiuchi, N. (1990). Immuno-gold localization of indole-3-acetic acid in peach seedlings. *Plant Cell Physiol.* **31,** 711–715.

Olive, J., and Vallon, O. (1991). Structural organization of the thylakoid membrane: Freeze-fracture and immunocytochemical analysis. *J. Electr. Microsc. Tech.* **18,** 360–374.

Oross, J. W., and Possingham, J. V. (1991). Tubular structures in developing plastids of three dicotyledonous species. *Can. J. Bot.* **69,** 136–139.

Osafune, T., Schiff, J. A., and Hase, E. (1990a). Immunogold localization of LHCP II apoprotein in the Golgi of *Euglena*. *Cell Struct. Funct.* **15,** 99–105.

Osafune, T., Sumida, S., Ehara, T., Kodama, M., and Hase, E. (1990b). Immunelectron microscopic evidence for the occurrence of ribulose-1,5-bisphosphate carboxylase/oxygenase in "propyrenoids" of *Euglena gracilis*. *J. Electr. Microsc.* **39,** 101–104.

Osafune, T., Yokota, A., Sumida, S., and Hase, E. (1990c). Immunogold localization of ribulose-1,5-bisphosphate carboxylase with reference to pyrenoid morphology in chloroplasts of synchronized *Euglena gracilis* cells. *Plant Physiol.* **92,** 802–808.

Osafune, T., Schiff, J. A., and Hase, E. (1991a). Stage-dependent localization of LHCP II apoprotein in the Golgi of synchronized cells of *Euglena gracilis* by immunogold electron microscopy *Exp. Cell Res.* **193,** 320–330.

Osafune, T., Sumida, S., Schiff, J. A., and Hase, E. (1991b). Immunolocalization of LHCP II apoprotein in the Golgi during light-induced chloroplast development in non-dividing *Euglena* cells. *J. Electr. Microsc.* **40,** 41–47.

Paolillo, D. J., Jr. (1970). The three-dimensional arrangement of intergranal lamellae in chloroplasts. *J. Cell Sci.* **6,** 243–255.

Perrin, A. (1970). Diversité des formes d'accumulation de la phytoferritine dans les cellules constituant l'épithème des hydathodes de *Taraxacum officinale* Weber et *Saxifraga aizoon* Jacq. *Planta* **93,** 71–81.

Pineau, B., Dubertret, G., Joyard, J., and Douce, R. (1986). Fluorescence properties of the envelope membranes from spinach chloroplasts. Detection of protochlorophyllide. *J. Biol. Chem.* **261,** 9210–9215.

Platt-Aloia, K. A., and Thomson, W. W. (1979). Membrane bound inclusions in epidermal plastids of developing sesame leaves and cotyledons. *New Phytol.* **83,** 793–799.

Plesnicar, M., and Bendall, D. S. (1973). The photochemical activities and electron carriers of developing barley leaves. *Biochem. J.* **136,** 803–812.

Prunkard, D. E., Bascomb, N. F., Robinson, R. W., and Schmidt, R. R. (1986). Evidence for chloroplastic localization of an ammonium-inducible glutamate dehydrogenase and synthesis of its subunit from cytosolic precursor-protein in *Chlorella sorokiniana*. *Plant Physiol.* **81,** 349–355.

Rascio, N., Orsenigo, M., and Arboit, D. (1976). Prolamellar body transformation with increasing cell age in the maize leaf. *Protoplasma* **90**, 253–263.

Rascio, N., Colombo, P. M., and Orsenigo, M. (1980). The ultrastructural development of plastids in leaves of maize plants exposed to continuous illumination. *Protoplasma* **102**, 131–139.

Rascio, N., Mariani, P., and Casadoro, G. (1984). Etioplast–chloroplast transformation in maize leaves: Effects of tissue age and light intensity. *Protoplasma* **119**, 110–120.

Rebeiz, C. C., and Rebeiz, C. A. (1986). Chloroplast biogenesis. 53. Ultrastructural study of chloroplast development during photoperiodic greening. *In* "Regulation of Chloroplast Differentiation" (G. Akoyunoglou and H. Senger, eds.), pp. 389–396. A. R. Liss, New York.

Robertson, D., and Laetsch, W. M. (1974). Structure and function of developing barley plastids. *Plant Physiol.* **54**, 148–159.

Rosinski, J., and Rosen, W. G. (1972). Chloroplast development: Fine structure and chlorophyll synthesis. *Q. Rev. Biol.* **47**, 160–191.

Rühle, W., and Wild, A. (1985). Die Anpassung des Photosyntheseapparates höherer Pflanzen an die Lichtbedingungen. *Naturwissensch.* **72**, 10–16.

Ryberg, H., Liljenberg, C., and Sundqvist, C. (1980). Crystalloid formation in protochlorophyll-accumulating plastids from the inner seed coat of *Cyclanthera explodens*. *Physiol. Plant.* **50**, 333–339.

Ryberg, M., and Dehesh, K. (1986). Localization of NADPH–protochlorophyllide oxidoreductase in dark-grown wheat (*Triticum aestivum*) by immuno-electron microscopy before and after transformation of the prolamellar bodies. *Physiol. Plant.* **66**, 616–624.

Ryberg, M., and Sundqvist, C. (1988). The regular ultrastructure of isolated prolamellar bodies depends on the presence of membrane-bound NADPH–protochlorophyllide oxidoreductase. *Physiol. Plant.* **73**, 218–226.

Ryberg, M., and Sundqvist, C. (1991). Structural and functional significance of pigment–protein complexes of chlorophyll precursors. *In* "Chlorophylls" (H. Scheer, ed.), pp. 587–612. CRC Press, Boca Raton, Florida.

Salema, R., and Abreu, I. (1972). Fine structure of chloroplasts and leukoplasts of the leaves of *Hedera helix* L. cv. *argenteo-variegata*. *Bol. Soc. Broteriana* **46**, 259–271.

Santos, I., and Salema, R. (1983). Stereological study of the variation of chloroplast tubules and volume in the CAM plant *Sedum telephium*. *Z. Pflanzenphysiol.* **113**, 29–37.

Santos, I., and Salema, R. (1984). Effects of hydration conditions on the stroma inclusion of some CAM chloroplasts. *Plant Cell Environ.* **7**, 541–544.

Schiff, J. A. (1980). Development, inheritance, and evolution of plastids and mitochondria. *In* "The Biochemistry of Plants" (N. E. Tolbert, ed.), Vol. 1, pp. 209–272. Academic Press, New York.

Schiff, J. A., Schwartzbach, S. D., Osafune, T., and Hase, E. (1991). Photocontrol and processing of LHCP-II apoprotein in *Euglena:* Possible role of Golgi and other cytoplasmic sites. *J. Photochem. Photobiol. B* **11**, 219–236.

Schnepf, E., and Elbrächter, M. (1988). Cryptophycean-like double membrane-bound chloroplast in the dinoflagellate, *Dinophysis* Ehrenb.: Evolutionary, phylogenetic and toxicological implications. *Bot. Acta* **101**, 196–203.

Schötz, F., and Diers, L. (1975). Vergrößerung der Kontaktfläche zwischen Chloroplasten und ihrer cytoplasmatischen Umgebung durch tubuläre Ausstülpungen der Plastidenhülle. *Planta* **124**, 277–285.

Selldén, G. (1977). "The Development of the Photosynthetic Apparatus in Greening Barley Seedlings." Ph.D. Thesis. University of Göteborg, Sweden.

Selldén, G., and Leech, R. M. (1981). Localization of DNA in mature and young wheat

chloroplasts using the fluorescent probe 4',6-diamidino-2-phenylindole. *Plant Physiol.* **68,** 731–734.
Shaw, P. J., and Henwood, J. A. (1985). Immuno-gold localization of cytochrome $f$, light-harvesting complex, ATP synthase and ribulose 1,5-bisphosphate carboxylase/oxygenase. *Planta* **165,** 333–339.
Shaw, P., Henwood, J., Oliver, R., and Griffiths, T. (1985). Immunogold localisation of protochlorophyllide oxidoreductase in barley etioplasts. *Eur. J. Cell Biol.* **39,** 50–55.
Simpson, D. J. (1978). Freeze-fracture studies on barley plastid membranes II. Wildtype chloroplast. *Carlsberg Res. Commun.* **43,** 365–389.
Simpson, D. J., and von Wettstein, D. (1989). The structure and function of the thylakoid membrane. *Carlsberg Res. Commun.* **54,** 55–65.
Simpson, D. J., Bassi, R., and Hinz, U. G. (1987). Cell-specific expression of LHCII and the organisation of the photosynthetic reaction centres in chloroplast thylakoids. *In* "Plant Molecular Biology" (D. von Wettstein and N.-H. Chua, eds.), pp. 93–104. Plenum, New York.
Sitte, P. (1974). Plastiden-Metamorphose und Chromoplasten bei *Chrysosplenium. Z. Pflanzenphysiol.* **73,** 243–265.
Sossountzov, L., Sotta, B., Maldiney, R., Sabbagh, I., and Miginiac, E. (1986). Immunoelectron-microscopy localization of abscisic acid with colloidal gold on Lowicryl-embedded tissues of *Chenopodium polyspermum* L. *Planta* **168,** 471–481.
Spear-Bernstein, L., and Miller, K. R. (1989). Unique location of the phycobiliprotein light-harvesting pigment in the Cryptophyceae. *J. Phycol.* **25,** 412–419.
Sprey, B. (1975). Membranassoziierte Tubuli während der Chloroplastengenese von *Hordeum vulgare* L. *Protoplasma* **84,** 197–203.
Sprey, B., and Laetsch, W. M. (1978). Structural studies of peripheral reticulum in $C_4$ plant chloroplasts of *Portulaca oleracea* L. *Z. Pflanzenphysiol.* **87,** 37–53.
Sprey, B., and Lambert, C. (1977). Lamellae-bound inclusions in isolated spinach chloroplasts. II. Identification and composition. *Z. Pflanzenphysiol.* **83,** 227–247.
Sprey, B., Gliem, G., and Jánossy, A. G. S. (1976). Iron containing inclusions in chloroplasts of *Nicotiana clevelandii* X *Nicotiana glutinosa*. I. X-Ray microanalysis and ultrastructure. *Z. Pflanzenhysiol.* **79,** 165–176.
Sprey, B., Gliem, G., and Jánossy, A. G. S. (1978). Iron and phosphorus containing inclusions in chloroplasts of *Nicotiana clevelandii* X *Nicotiana glutinosa*. II. Development of etioplasts to chloroplasts in cotyledons. *Z. Pflanzenphysiol.* **88,** 69–82.
Staehelin, L. A. (1976). Reversible particle movements associated with unstacking and restacking of chloroplast membranes *in vitro. J. Cell Biol.* **71,** 136–158.
Staehelin, L. A. (1986). Chloroplast structure and supramolecular organization of photosynthetic membranes. *In* "Encyclopedia of Plant Physiology" (L. A. Staehelin and C. J. Arntzen, eds.), Vol. 19, pp. 1–84. Springer-Verlag, Berlin.
Sundqvist, C., and Ryberg, H. (1979). Structure of protochlorophyll-containing plastids in the inner seed coat of pumpkin seeds (*Cucurbita pepo*). *Physiol. Plant.* **47,** 124–128.
Sundqvist, C., Björn, L. O., and Virgin, H. I. (1980). Factors in chloroplast differentiation. *In* "Results and Problems in Cell Differentiation" (J. Reinert, ed.), Vol. 10, pp. 201–224. Springer-Verlag, Berlin.
Swift, H., and Leser, G. P. (1989). Cytochemical studies on prochlorophytes: Localization of DNA and ribulose 1,5-bisphosphate carboxylase-oxygenase. *J. Phycol.* **25,** 751–761.
Takabe, T. (1986). Accumulation of chloroplast proteins during greening of rice leaves. *Plant Science* **43,** 193–199.
Takabe, T., Takabe, T., and Akazawa, T. (1986). Biosynthesis of P700-chlorophyll $a$ protein complex, plastocyanin, and cytochrome $b_6/f$ complex. *Plant Physiol.* **81,** 60–66.

Takebe, I., Otsuki, Y., Honda, Y., Nishio, T., and Matsui, C. (1973). Fine structure of isolated mesophyll protoplasts of tobacco. *Planta* **113**, 21–27.
Tanaka, A., and Tsuji, H. (1983). Formation of chlorophyll–protein complexes in greening cucumber cotyledons in light and then in darkness. *Plant Cell Physiol.* **24**, 101–108.
Tanaka, A., and Tsuji, H. (1985). Appearance of chlorophyll–protein complexes in greening barley seedlings. *Plant Cell Physiol.* **26**, 893–902.
Terashima, I., and Inoue, Y. (1985). Palisade tissue chloroplasts and spongy tissue chloroplasts in spinach: Biochemical and ultrastructural differences. *Plant Cell Physiol.* **26**, 63–75.
Terashima, I., Sakaguchi, S., and Hara, N. (1986). Intra-leaf and intracellular gradients in chloroplast ultrastructure of dorsiventral leaves illuminated from the adaxial or abaxial side during their development. *Plant Cell Physiol.* **27**, 1023–1031.
Tevini, M., and Lichtenthaler, H. K. (1970). Untersuchungen über die Pigment- und Lipochinonausstattung der zwei photosynthetischen Pigmentsysteme. *Z. Pflanzenphysiol.* **62**, 17–32.
Tewinkel, M., and Volkmann, D. (1987). Observations on dividing plastids in the protonema of the moss *Funaria hygrometrica* Sibth. *Planta* **172**, 309–320.
Thompson, A., Vogel, J., and Lee, R. E. (1977). Carbon dioxide uptake in relation to a plastid inclusion body in the succulent *Kalanchoë pinnata* Persoon. *J. Exp. Bot.* **28**, 1037–1041.
Thomson, W. W., and Platt, K. (1973). Plastid ultrastructure in the barrel cactus, *Echinocactus acanthodes*. *New Phytol.* **72**, 791–797.
Thomson, W. W., and Whatley, J. M. (1980). Development of nongreen plastids. *Annu. Rev. Plant Physiol.* **31**, 375–394.
Thomson, W. W., Foster, P., and Leech, R. M. (1972). The isolation of proplastids from roots of *Vicia faba*. *Plant Physiol.* **49**, 270–272.
Tôyama, S. (1972). Electron microscope studies on the morphogenesis of plastids. VI. Plastid development and fine structure in variegated leaves of tomato. *Bot. Mag. (Tokyo)* **85**, 1–10.
Treffry, T. (1970). Phytylation of chlorophyllide and prolamellar-body transformation in etiolated peas. *Planta* **91**, 279–284.
Tuquet, C., and Newman, D. W. (1980). Aging and regreening in soybean cotyledons. 1. Ultrastructural changes in plastids and plastoglobuli. *Cytobios* **29**, 43–59.
Vallon, O., Wollman, F. A., and Olive, J. (1986). Lateral distribution of the main protein complexes of the photosynthetic apparatus in *Chlamydomonas reinhardtii* and in spinach: An immunocytochemical study using intact thylakoid membranes and a PS II enriched membrane preparation. *Photobiochem. Photobiophys.* **12**, 203–220.
Vallon, O., Hoyer-Hansen, G., and Simpson, D. J. (1987). Photosystem II and cytochrome *b*-559 in the stroma lamellae of barley chloroplasts. *Carlsberg Res. Commun.* **52**, 405–421.
Vallon, O., Tae, G.-S., Cramer, W. A., Simpson, D., Hoyer-Hansen, G., and Bogorad, L. (1989). Visualization of antibody binding to the photosynthetic membrane: The transmembrane orientation of cytochrome *b*-559. *Biochim. Biophys. Acta* **975**, 132–141.
van Berkel, J., Steup, M., Völker, W., Robenek, H., and Flügge, U. I. (1986). Polypeptides of the chloroplast envelope membranes as visualized by immunochemical techniques. *J. Histochem. Cytochem.* **34**, 577–583.
van Steveninck, M. E., Goldney, D. C., and van Steveninck, R. F. M. (1972). Chloroplast peripheral reticulum in *Nymphoides indica*. *Z. Pflanzenphysiol.* **67**, 155–160.
Vaughn, K. C. (1987). Two immunological approaches to the detection of ribulose-1,5-bisphosphate carboxylase in guard cell chloroplasts. *Plant Physiol.* **84**, 188–196.

Vaughn, K. C., and Wilson, K. G. (1981). Improved visualization of plastid fine structure: Plastid microtubules. *Protoplasma* **108**, 21–27.
Vaughn, K. C., Campbell, E. O., Hasegawa, J., Owen, H. A., and Renzaglia, K. S. (1990). The pyrenoid is the site of ribulose 1,5-bisphosphate carboxylase/oxygenase accumulation in the hornwort (Bryophyta: Anthocerotae) chloroplast. *Protoplasma* **156**, 117–129.
Vesk, M., Dwarte, D., Fowler, S., and Hiller, R. G. (1992). Immunocytochemical localization of light-harvesting pigment complexes in a cryptophyte. *Protoplasma* (in press).
Virgin, H. I., and Sundqvist, C. (1992). Pigment formation in potato tubers (*Solanum tuberosum*) exposed to light followed by darkness. *Physiol. Plant.* (in press).
Watanabe, M. M., Takeda, Y., Sasa, T., Inouye, I., Suda, S., Sawaguchi, T., and Chihara, M. (1987). A green dinoflagellate with chlorophylls *a* and *b*: Morphology, fine structure of the chloroplast, and chlorophyll composition. *J. Phycol.* **23**, 382–389.
Wehrmeyer, W., and Röbbelen, G. (1965). Räumliche Aspekte zur Membranschichtung in den Chloroplasten einer *Arabidopsis*-Mutante unter Auswertung von Serienschnitten. III. Über Membranbildungsprozesse im Chloroplasten. *Planta* **64**, 312–329.
Weier, T. E., and Brown, D. L. (1970). Formation of the prolamellar body in 8-day, dark-grown seedlings. *Am. J. Bot.* **57**, 267–275.
Weier, T. E., Sjoland, R. D., and Brown, D. L. (1970). Changes induced by low light intensities on the prolamellar body of 8-day, dark-grown seedlings. *Am. J. Bot.* **57**, 276–284.
Wellburn, A. R. (1982). Bioenergetic and ultrastructural changes associated with chloroplast development. *Int. Rev. Cyt.* **80**, 133–191.
Wellburn, A. R. (1987). Plastids. *Int. Rev. Cytol. (Suppl.)* **17**, 149–210.
Wellburn, A. R., Quail, P. H., and Gunning, B. E. S. (1977). Examination of ribosome-like particles in isolated prolamellar bodies. *Planta* **134**, 45–52.
Wellburn, A. R., Robinson, D. C., and Wellburn, F. A. M. (1982). Chloroplast development in low light-grown barley seedlings. *Planta* **154**, 259–265.
Wellburn, F. A. M., and Wellburn, A. R. (1971). Developmental changes occurring in isolated intact etioplasts. *J. Cell Sci.* **9**, 271–287.
Wellburn, F. A. M., and Wellburn, A. R. (1976). Novel chloroplasts and unusual cellular ultrastructure in the "resurrection" plant *Myrothamnus flabellifolia* Welw. (Myrothamnaceae). *Bot. J. Linn. Soc.* **72**, 51–54.
Whatley, J. M. (1974). Chloroplast development in primary leaves of *Phaseolus vulgaris*. *New Phytol.* **73**, 1097–1110.
Whatley, J. M. (1975). The occurrence of a peripheral reticulum in plastids of the gymnosperm, *Welwitschia mirabilis*. *New Phytol.* **74**, 215–220.
Whatley, J. M. (1977). Variations in the basic pathway of chloroplast development. *New Phytol.* **78**, 407–420.
Whatley, J. M. (1980). Plastid growth and division in *Phaseolus vulgaris*. *New Phytol.* **86**, 1–16.
Whatley, J. M., and Whatley, F. R. (1981). Chloroplast evolution. *New Phytol.* **87**, 233–247.
Whatley, J. M., and Whatley, F. R. (1987). When is a chromoplast? *New Phytol.* **106**, 667–678.
Whatley, J. M., Hawes, C. R., Horne, J. C., and Kerr, J. D. A. (1982). The establishment of the plastid thylakoid system. *New Phytol.* **90**, 619–629.
Wieckowski, S. (1967). Chloroplasts in growing bean leaf. *Acta Soc. Bot. Pol.* **36**, 161–169.
Williams, E. (1974). Fine structure of vascular and epidermal plastids of the mature maize leaf. *Protoplasma* **79**, 395–400.
Wrischer, M. (1966). Neubildung von Prolamellarkörpern in Chloroplasten. *Z. Pflanzenphysiol.* **55**, 296–299.

Wrischer, M. (1970). Intrathylakoidal protein crystalloids in spinach plastids. *Acta Bot. Croat.* **29,** 39–42.

Wrischer, M. (1973). Protein crystalloids in the stroma of bean plastids. *Protoplasma* **77,** 141–150.

Wrischer, M. (1978). Ultrastructural changes in plastids of detached spinach leaves. *Z. Pflanzenphysiol.* **86,** 95–106.

Wrischer, M., Ljubešic, N., Marcenko, E., Kunst L., and Hloušek-Radojcic, A. (1986). Fine structural studies of plastids during their differentiation and dedifferentiation. *Acta Bot. Croat.* **45,** 43–54.

Yagawa, Y., Muto, S., and Miyachi, S. (1987). Carbonic anhydrase of a unicellular red alga *Porphyridium cruentum* R-1. II. Distribution and role in photosynthesis. *Plant Cell Physiol.* **28,** 1509–1516.

Yoshida, Y., Laulhere, J.-P., Rozier, C., and Mache, R. (1978). Visualization of folded chloroplast DNA from spinach. *Biol. Cell.* **32,** 187–190.

# 3

# Light and Temperature Regulation of Chloroplast Development

## KENNETH ESKINS

Phytoproducts Research
U.S. Department of Agriculture
Agricultural Research Service
National Center for Agricultural Utilization Research
Peoria, Illinois

    I. Introduction
   II. Chloroplast Pigments
  III. Pigment–Protein Complexes
  IV. Light-Regulated Development
      A. Long-Term Low Irradiance Light
      B. Periodic Light
      C. Light Effects in Mutants
      D. Light Quality Studies
   V. Light Effects on Transcription and Translation
  VI. Case Study: Regulation of LHCPII
 VII. Temperature Effects on Chloroplast Development
VIII. Concluding Remarks
      References

## I. INTRODUCTION

Chloroplast development is a complex process controlled by a genetic program that involves interaction between the nuclear and chloroplast genomes (Taylor, 1989). Multiple genetic pathways in this program are subject to environmental regulation. Since plants lack mobility, evolution has provided pathways that are specific for many possible environmental scenarios and are activated by the appropriate cues and signals, that is, the expression of a particular gene or group of genes can be modified to adapt to a particular environment. The genetic system, or the plants genetic makeup, is reasonably stable under changing environmental conditions, but the expression of those genes is responsive to alterations in the surroundings (Lee, 1988).

Environmental factors that influence chloroplast development include temperature, light, water (Duysen and Freeman, 1974), and nutrition (Guikema and Sherman, 1983). Among these, we will consider the effects of light and temperature. These factors are especially important for the development of chloroplasts, which are responsible for the conversion of light into chemical energy via pigment–protein complexes associated with thylakoid membranes.

Response to the light environment is mediated through photoreceptors that are present at the earliest stages of development. Activation of these photoreceptors initiates and continually regulates the form and function of the chloroplasts. Three photoreceptors, protochlorophyllide (Pchlide), phytochrome, and blue-light receptor, have been characterized. Of these, the blue-light receptor is the least known and has not been isolated or chemically identified. It is characterized mainly by the regulatory processes that are activated only by light of blue or ultraviolet wavelengths. A fourth receptor, responsive to the ultraviolet B wavelengths, is known primarily for the regulation of anthocyanin synthesis (Drumm-Herrel and Mohr, 1982) but also may be involved in chloroplast regulation.

Phytochrome is well characterized; both the chromophore and the apoprotein have been isolated (Smith, 1983). Phytochrome functions as a switching device activated by a far-red absorbing form ($P_{fr}$) and deactivated by a red light absorbing form ($P_r$). The switching function of phytochrome occurs during very early stages of development when etioplasts or proplastids are being converted to chloroplasts (Nagy et al., 1988). In green plants, and after long-term irradiation of dark-grown plants, a second form of phytochrome (type II) predominates. At present, it is not known whether or not this form of phytochrome is involved in plant responses that are regulated by the $P_{fr}/P_{tot}$ ratio and the irradiance (Smith and Morgan, 1983).

Pchlide is the initial receptor of light energy for pigment synthesis and can be activated by a broad range of wavelengths (Kamiya et al., 1981). The light dependent conversion of Pchlide to chlorophyllide (Chlide) initiates the synthesis of chlorophylls (Chls) and signals the beginning of greening and chloroplast development (Kasemir, 1983). The synthesis of Pchlide via 5-aminolevulinic acid is controlled by phytochrome (Kasemir and Mohr, 1981).

Chloroplast development is coordinated with other plant regulatory programs (Butterfass, 1980) that control cell and whole plant growth. The process involves continuous feedback and modification of responses among many systems. Thus, a light signal received at an inappropriate stage of development or an incorrect temperature range will not be effective. Conversely, temperature and limited or spectrally modified light can be used to attenuate growth or selectively initiate only parts of the overall developmental program.

In contrast to light receptors, the systems that sense temperature and are responsible for the coordination of light and temperature regulation have been explored only slightly. However, the association and movement of pigment–protein complexes (Sundby and Andersson, 1985), the diurnal expression pattern of Chl a/b proteins (Piechulla and Riesselmann, 1990), and the response of pigment deficient mutants to light intensity (Yang *et al.*, 1990) are known to be modified by temperature. This effect is likely to involve temperature-mediated changes in membrane fluidity (Laval-Martin and Troton, 1990) or conformational changes of receptors. Thus, the tremendous range of responsiveness of plants, which is so essential to their survival, is to us a study in environmental regulation of gene expression.

The mechanics of chloroplast development involve gene transcription, synthesis of proteins and enzymes, assembly of multiple-component complexes into lipid bilayers, and light use via photosynthesis. A complete description of these processes is beyond the scope of this chapter, which is limited to the effects of light and temperature on the regulation of synthesis and assembly of reaction center and light-harvesting pigment–protein complexes of the chloroplast. Photoreceptors, which are also pigment–protein complexes, will be considered only as agents of regulation. The environmental factor light will be subdivided further into light-quality effects, irradiance effects, and the interaction of the two. Interaction between light and temperature will be considered also.

## II. CHLOROPLAST PIGMENTS

The chloroplast pigments associated with greening and development are the Chls and carotenoids. Flavonoids are associated with the chloroplast also and may have significant but unknown regulatory or structural functions. However, only the major Chls and carotenoids will be considered here, since they are sufficient to detail the course of chloroplast development. Chl a accumulation is a measure of the degree of development of either the whole plant or the chloroplasts. The accumulation of Chl b is a measure of the degree of formation of light-harvesting pigment–proteins. In addition, the major xanthophylls neoxanthin and violaxanthin are markers for light-harvesting complexes. β-Carotene is a marker for reaction center complexes (Rawyler *et al.*, 1980; Lichtenthaler *et al.*, 1982, Eskins *et al.*, 1983b).

The course of pigment accumulation during normal development can be followed in several ways. One is the obvious analysis performed during greening. In this procedure, etiolated tissue must be developed by the plant so there is sufficient tissue to analyze. The disadvantage is that early developmental responses mediated by light are now affected by the stage of development. Nevertheless, many studies using etiolated tissue of various ages have provided general information about the synthesis of pigments during greening and chloroplast formation. These data show that, after light triggering, there is a short lag period in Chl a formation (Erdos *et al.*, 1987) followed by a steady increase in the accumulation of Chl a until a plateau is reached at maturity. During the lag period, Pchlide is converted to Chlide and further into Chl a by steps involving geranylgeraniol esters (Schoch, 1978; Hoober *et al.*, 1990). Under continuous light, the formation of Chl a is followed by the formation of Chl b and the increased synthesis of oxygenated carotenoids, neoxanthin, and violaxanthin. Studies that limit the amount of light during development can slow or inhibit the formation of Chl b and the subsequent formation of mature thylakoid structure. Importantly, these studies showed the relationship between irradiance and pigment synthesis and led to information about the pigments associated with chloroplast pigment–protein complexes.

A second method of connecting pigment accumulation to development uses mutant plants that are deficient in pigments and express this deficiency as a function of light intensity (Allen *et al.*, 1988) and stage of development (Eskins *et al.*, 1981,1982). The connection between accessory pigments and Chl a during chloroplast development can be shown directly by experiments that plot accessory pigments against Chl a content in a number of pigment-deficient mutants analyzed at various stages of development. Such studies help determine how pigments accumulate as pigment–protein complexes in mature tissue grown under natural growth conditions. In addition, extrapolition of the data to zero Chl a provides a model of chloroplast development that correlates well with data derived from greening of etiolated tissue.

The realization that all chloroplast pigments are associated with protein complexes (Markwell *et al.*, 1979) makes our interpretation of developmentally regulated changes in pigments easier. Further discussion in this chapter considers the effects of light, temperature, and developmental stage on the accumulation of these pigment–protein complexes.

## III. PIGMENT–PROTEIN COMPLEXES

A large number of pigment–protein complexes exist in the chloroplast that differ from one another in molecular weight, function, and the amount and kinds of associated pigment. Some of these complexes are associated primarily with stacked or unstacked chloroplast membranes or with photosystem I (PSI) or photosystem II (PSII) particles isolated from broken chloroplasts (Allen et al., 1990). These complexes have been identified by means of relative molecular weights, relative hydrophobicity (Siegenthaler and Dumont, 1990), component analysis of photosystem particles, photochemical analysis, and comparison of mutant and wild-type plants (Simpson, 1990). The function of every complex is not known, but specific assignments of pigment–protein and function have been made for reaction centers I and II and for some of the associated light-harvesting complexes.

Identification of the major pigment–protein complexes and association of them with specific functions in the chloroplast will facilitate this discussion. The list will not be exhaustive since extensive reviews already exist in the literature (Thornber, 1986; Bassi et al., 1990). The pigment–proteins are members of the two photosystems and can be subdivided further into reaction-center and light-harvesting complexes. The literature contains multiple designations for these complexes; the number and character of each is a direct result of the system used to separate or analyze the complexes. For this discussion, the pigment–protein complexes of PSI will be CP1a, CP1, and LHCPI.

CP1a is the largest complex and is composed of the PSI reaction center (CP1) and associated PSI light-harvesting complex (LHCPI) (Mullet et al., 1980). CP1, like other reaction-center complexes, contains Chl a and $\beta$-carotene whereas LHCPI contains Chl a, Chl b, xanthophylls, and lutein. The reaction center complex (CP1) is composed of two proteins, each with apparent molecular masses calculated from SDS–PAGE of 60–68 kDa (Vierling and Alberte, 1983). In certain systems, LHCPI can be separated into two distinct complexes (LHCPI$_{730}$ and LHCPI$_{680}$), which differ in apparent molecular mass and in fluorescence (Bassi and Simpson, 1987). The apoproteins of LHCPI run between 20 and 25 kDa in SDS–PAGE systems (Haworth et al., 1983; Chapter 12).

The pigment–protein complexes of PSII are the reaction-center complex composed of D1, D2, and cytochrome (Cyt) $b$-559 proteins (Namba and Satoh, 1987), the connecting Chl a LHCPII antennae (CP47 and CP43) (Bricker, 1990) and the major light-harvesting complex of PSII, LHCPII (CP64). The minor complexes, CP29 (Camm and Green, 1989), CP26, and CP24 (Dainese et al., 1990), function to connect the major LHCPII to the inner Chl a antennae. Additional information on these complexes is readily available in the literature (Bassi *et al.*, 1990).

## IV. LIGHT-REGULATED DEVELOPMENT

Studies with etioplasts isolated from dark-grown plants have been very useful in mapping the course of development in the chloroplasts (Thorne and Boardman, 1971). In particular, these studies have given us information about the dark formation of prolamellar bodies (PLBs) and their conversion by light to functioning thylakoid membranes.

During dark growth, leaves build up Pchlide and form a three-dimensional membrane structure, the PLB, in the etioplasts. The light-regulated conversion of etioplasts to chloroplasts is mediated by phytochrome and by protochlorophyllide oxidoreductase (Pchlide-reductase; Batschauer and Apel, 1984) in the PLB (Lindsten *et al.*, 1988). Absorption of light, primarily of red and blue wavelengths, converts Pchlide to Chlide. Continued irradiation converts Chlide to Chl a via a series of geranylgeraniol esters.

Figure 1A shows the pigment constitution of etiolated tissue to be mainly Pchlide, neoxanthin, violaxanthin, lutein epoxide, lutein, protochlorophyll, and $\beta$-carotene. After 5 min of light treatment, Pchlide is converted to Chlide; after 30 min light (Fig. 1B), Chl a and geranylgeraniol precursors are present (Eskins and Harris, 1981). Small amounts of Chl b also are present.

Changes in pigment content are paralleled by changes in protein patterns and by structural changes in membranes. PLBs dissolve and are replaced by vesicles; prothylakoids are converted to thylakoids with the formation of lipid bilayer membranes (Henningsen and Boynton, 1974; Sundqvist and Ryberg, 1989). These membranes act as structure for the assembly of large complexes, which are made of proteins imbedded in the lipid membranes. The proteins are threaded through the lipid bilayer (Cramer *et al.*, 1985) and exposed to both sides of the bilayer (Pichersky and Green, 1990).

Proteins that are encoded by the chloroplast are translated on chloroplast ribosomes, then inserted into membranes. Nuclear-encoded proteins are translated on cytoplasmic ribosomes and transported through the chloroplast membranes to be inserted into thylakoid membranes (Chitnis *et al.*, 1987). Processing and the attachment of pigments to stabilize LHCPI and LHCPII can follow (see also Chapters 9 and 10). The presence of Chl b is essential for complete stability of these complexes, but other stability factors are also at work (Terao and Katoh, 1990).

## 3. Regulation of Chloroplast Development

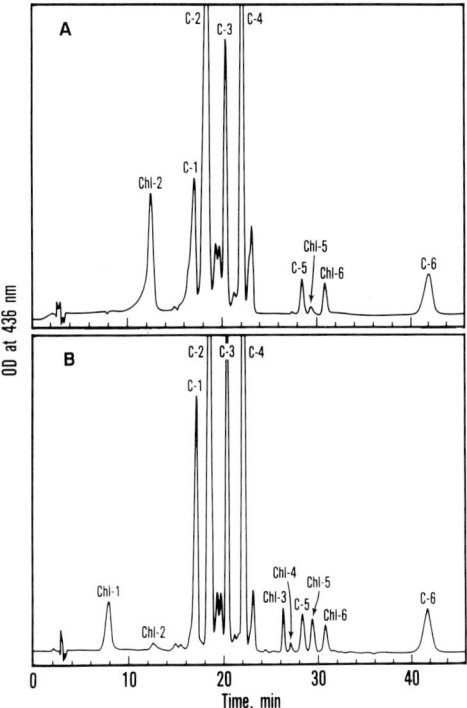

**Fig. 1.** (A) Separation of pigments from etiolated leaves of red kidney bean. Chl-2, protochlorophyllide; C-1, neoxanthin; C-2, violaxanthin; C-3, lutein epoxide; C-4, lutein; C-5, unknown; Chl-5, Chl-a; Chl-6, protochlorophyll a; C-6, $\beta$-carotene. (B) HPLC separation of pigment from white light illuminated (30 min) leaves of red kidney bean. Chl-1, chlorophyllide; Chl-2, protochlorophyllide; C-1, neoxanthin; C-2, violaxanthin; C-3, lutein epoxide; C-4, lutein; Chl-3 and Chl-4, geranylgeranyl chlorophylls; C-5, unknown; Chl-5, chlorophyll a; Chl-6, protochlorophyll a; C-6, $\beta$-carotene.

Continued development organizes some of these thylakoids into appressed and nonappressed regions (Cline, 1988). In these membranes, pigment–protein complexes are assembled in stepwise fashion and segregated into groups with similar functions. The large PSI and PSII particles are examples. Each is composed of many pigment–protein complexes that serve either a light-harvesting or a reaction-center function. The amounts and relationships of these complexes are responsive to the quality and irradiance levels of light in a regulated process that maintains maximum photosynthetic efficiency (Chow et al., 1990). These complexes also are regulated by plant age and are disassembled in a stepwise fashion during senescence (Eskins and McCarthy, 1987a).

Studies of the light regulation of this assembly process can proceed in several ways. We limit ourselves to discussing the effects of long-term light on the synthesis of pigments and the assembly of pigment–protein complexes. We use the assembly of the light-harvesting complex of PSII (LHCPII) as an example of the coordination of many regulatory signals. We look at methods that slow down this process by using very low levels of continuous light or by using short bursts of higher intensity intermittent light in long dark periods. Alternatively, mutant plants whose development is retarded by high light intensities can be used to document the early stages of chloroplast development (Lemoine *et al.*, 1987). We also examine the use of different qualities of light (red, far-red, blue) and the interaction of light quality and irradiance.

## A. Long-Term Low Irradiance Light

First, a general description of changes in ultrastructure and polypeptide patterns of plants grown under normal light–dark cycles at various irradiance levels is given. Figure 2 shows the effect of irradiance level on chloroplast development in 10-day-old mesophyll cells grown under 0.02, 0.08, and 0.45 W m$^{-2}$ of red light.

At 0.02 W m$^{-2}$, Chl a levels are 300 nmol g$^{-1}$ fwt, Chl a/b ratios are 8.0, and crystalline PLBs are still prominent. Single stroma thylakoids are present that stream out of the PLBs and extend the length of the chloroplast. At 0.08 W m$^{-2}$, Chl a levels have increased to 500 nmol g$^{-1}$ fwt, Chl a/b ratios have dropped to 5.0, and PLBs are still present but less extensive. Thylakoids have formed many stacked regions (grana). At 0.45 W m$^{-2}$, Chl a has increased to 800 nmol g$^{-1}$ fwt, Chl a/b ratios are 3.7, PLBs have disappeared, and the number of grana has increased.

Typical polypeptide patterns that correspond to these ultrastructural changes and irradiance levels are shown in Fig. 3. Etioplasts typically have polypeptide bands at 36 kDa (Pchlide-reductase), 50–58 kDa (coupling factor and large subunit of Rubisco), and at 10–20 kDa, all of which decrease with development. Formation of the PSI reaction center (66 kDa), Chl a antennae of PSII (47 and 43 kDa, also known as CP47 and CP43), the PSII reaction center, and multiple forms of the PSII light-harvesting complex (32–36 kDa) is detectable even at low irradiance levels (0.02 W m$^{-2}$). Coupling factor, Pchlide-reductase, and low molecular

---

**Fig. 2.** Electron micrographs of corn mesophyll cells grown under various irradiances of red light. Bar: 0.34 μm.

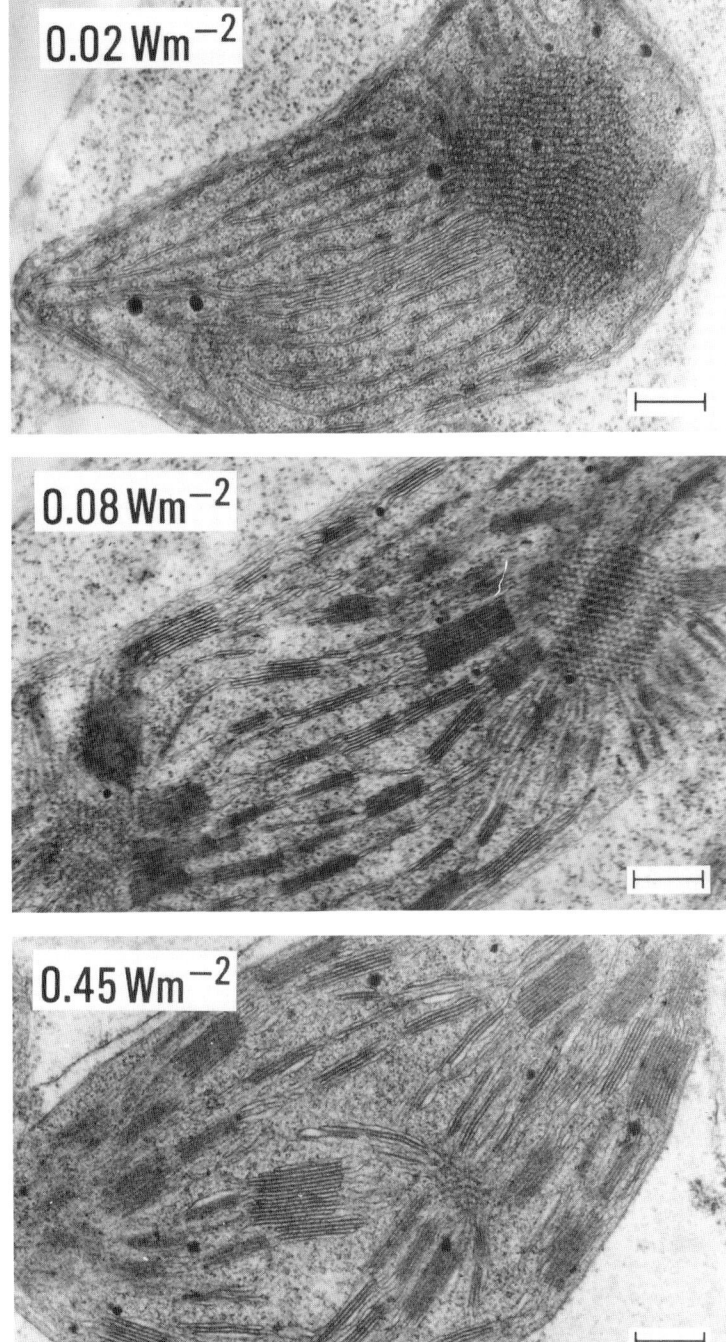

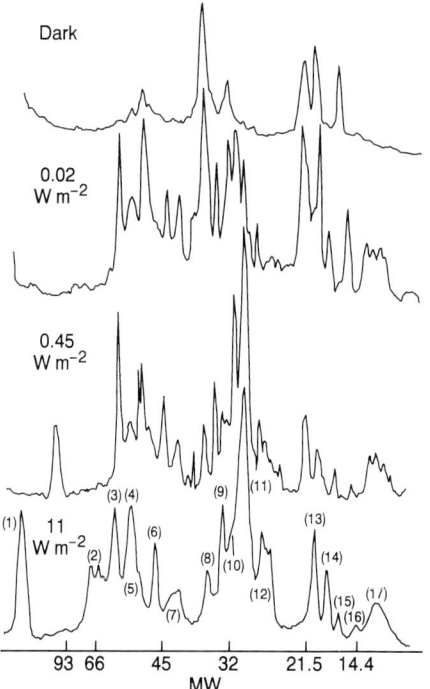

**Fig. 3.** Scans of electrophoretic gels of corn mesophyll membrane polypeptides grown in the dark and under various irradiances of red light. Peaks are (1) CP1 complex, 110–120 kDa; (2) PS1 reaction center protein, 66 kDa; (3,4) coupling factors $\alpha$ and $\beta$; (5,6) CP47, CP43; (7) 40–43-kDa protein; (8) 36.5-kDa protochlorophyllide oxidoreductase; (9) 34-kDa PS2; (10) 32-kDa protein; (11) LHCPII, 31–27 kDa; (12) LHCPI, 21–26 kDa; (13) 21-kDa PSI; (14) 18-kDa PSI; (15) 15-kDa protein; (16) 12-kDa protein; (17) LMW.

weight proteins are still prominent, however. With increasing irradiance levels (Bennett *et al.*, 1987), LHCPI binds to the PSI reaction center to form a 110-kDa complex, and LHCPII shifts from major bands at 32 kDa to bands at 29 and 27 kDa. This shift corresponds to an increase in Chl b and in the degree of stacking.

### B. Periodic Light

Studies with periodic light also have been useful in determining the effects of light on chloroplast development. In one such study (Akoyunoglou *et al.*, 1978), plants were grown in the dark for various periods (2–9 days), then subjected to light–dark cycles (LDC) of 2 min light and 98 min

## 3. Regulation of Chloroplast Development

dark. Under these low energy conditions, the developmental process was completed only partially and the degree of development was related directly to the number of LDCs. When the number of LDCs is small (14), the chloroplasts formed are agranal and are reduced in pigment content. They are devoid of Chl b and lack light-harvesting pigment–protein complexes (Argyroudi-Akoyunoglou et al., 1984). Plants grown under these conditions do, however, transcribe and translate the mRNA for the light-harvesting complexes. The PSI and PSII reaction centers are present and very active, but the ratio of PSII to PSI is higher than in normal green chloroplasts. With greater numbers of LDCs, small amounts of Chl b are produced, and elongated thylakoid stacks extend the length of the chloroplast. This energy-arrested developmental state is quite similar to that produced by very low light given continuously. Curiously, a very similar state of chloroplast development can be produced by very high light intensity in pigment-deficient mutant plants of many species (Faludi-Daniel et al., 1986; Allen et al., 1987). This issue is addressed in more detail in Section IV,C.

If the amount of energy available to the plant is increased by increasing the ratio of light to dark (2 min light/28 min dark), the rate of Chl a and Chl b synthesis is increased, light-harvesting complexes appear, and normal chloroplast development proceeds, although it does not reach the levels seen in plants grown in continuous light. These results suggest that the rate of Chl synthesis is controlled by irradiance level and that the availability of Chl determines the stability of light-harvesting complexes (Tzinas et al., 1986).

Perhaps the most interesting result of this study was the discovery that the assembly process operates during developmental windows, that is, chloroplasts formed under relatively few LDCs continue normal development when exposed to continuous light. Those that were exposed to many LDCs, however, were not able to develop normally when exposed to continuous light. This evidence argues for possible environmental conditioning during very early stages of development (Jackson and Lyndon, 1990). Additional evidence for this hypothesis was produced in experiments using different qualities of light for the LDCs (Akoyunoglou et al., 1980).

### C. Light Effects in Mutants

Mutant plants among many different species have altered pigment formation and chloroplast development (Freeman et al., 1982). Early studies of some of these pigment-deficient mutants gave the first evidence of the connection between Chl b and the formation of grana in the chloroplasts.

Usefully, in some of these mutants, the degree of expression of the mutation is directly responsive to increasing irradiance levels of light. In normal plants, the amount of pigments and the number of grana and light-harvesting complexes increase with the availability of light up to a certain point, beyond which increasing irradiance causes a decrease in pigments and light-harvesting complexes. This response is a mechanism by which the plant prevents the photosystems from taking in more quanta than can be processed. The mutants appear to have an exaggerated response to irradiance that results in plants with low pigment content, altered photosystem stoichiometry (Eskins et al., 1983a; Ghirardi and Melis, 1988), and arrested development under high light. A genetic isoline series of these mutants, each of which expresses different degrees of arrested development, has been used to construct the course of normal chloroplast development (Eskins and Banks, 1979).

Figure 4 shows such a construct for pigment accumulation, in which the amount of Chl a is used as a measure of development. When the accessory pigments are plotted against Chl a and the linear relationship is extrapolated to zero Chl a (etioplast), the construct predicts that lutein and carotene exist prior to Chl a and that neoxanthin, violaxanthin, and Chl b are generated at approximately the same time (Chl a content between 30 and 50 nmol $g^{-1}$ fwt). Although neoxanthin and violaxanthin are present in the etioplast, their rate of synthesis is increased greatly, concurrently with that of Chl b, during the genesis of light-harvesting complexes. Neoxanthin

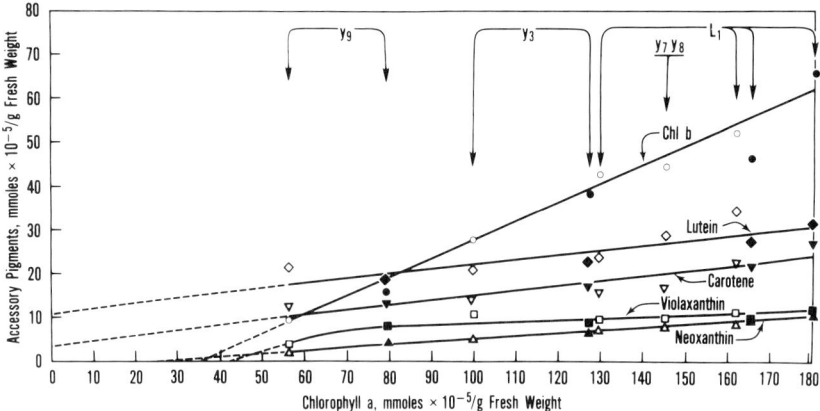

**Fig. 4.** Chlorophyll a compared with accessory pigments for soybean normal and mutant genotypes. Open symbols, field grown. Closed symbols, growth chamber. $y_3y_3$, $y_7y_8$, and $y_9y_9$ are chlorophyll-deficient mutants exhibiting different patterns of expression of the mutation.

directly increases with Chl a; violaxanthin first increases rapidly, then levels off. Before Chl b synthesis, Chl a and carotene are assembled into reaction-center complexes for PSI and PSII.

Polypeptide patterns that correspond to these arrested stages of chloroplast development are shown in Fig. 5. With increasing development relatively fewer coupling factor, PSI reaction center, and 33–32 kDa proteins and more LHCPI proteins are bound to PSI reaction center and PSII LHCPII. As noted previously for plants grown under low light irradiances, increasing development results in a shift of PSII proteins from 31–32 kDa to Chl b-containing LHCPII pigment–proteins at 29 and 27 kDa. Further information about LHCPII development in these mutants is discussed in Section VI.

## D. Light Quality Studies

Red and blue light are both effective for photosynthesis; both are capable of converting Pchlide to Chlide. Red and blue light both activate phytochrome, although the $P_{fr}/P_{tot}$ ratio established in blue light (0.4–0.5) is lower than that established in red light (0.8; Sponga *et al.*, 1986), implying that blue and red light have many common regulatory points. However, responses that are solely blue-light driven are found also.

It is quite common to think of certain qualities of light as connected. The red/far-red couple operates an on/off switch associated with early developmental responses. When discussing long-term irradiance-connected responses, blue-light receptor and phytochrome receptor are paired as cooperative regulators (Mohr, 1986). Some responses to blue light are mediated through phytochrome by establishing a permissible ratio of $P_{fr}/P_{tot}$. Among these responses may be those "high-irradiance" responses (Mancinelli and Robino, 1978) in which blue and far-red light are seen to favor similar developmental responses and, perhaps, to be separate excitors of the same photoreceptor. Other responses are mediated through the blue-light receptor. Thus, the interaction between red and blue light is seen as the interaction between separate blue-light and phytochrome photoreceptors.

Possibly, red light and blue light may interact through a common photoreceptor that measures red and blue light as a function of total energy, similar in mechanism to Pchlide, that is, the photoreceptor is used up as part of the response. In comparisons of red and blue responses, red light appears to interact with blue light in some manner during development. To see this interaction, the action of each individual light source as a function of energy (irradiance level) must be known. A comparison of

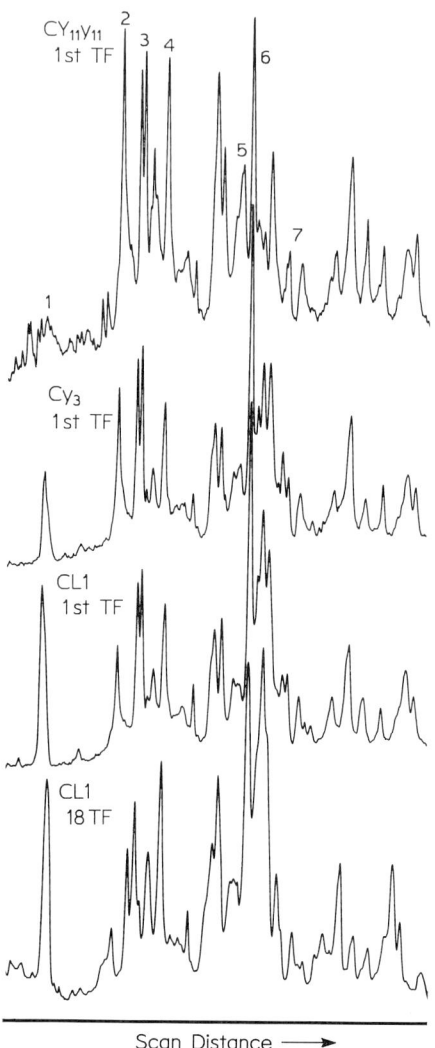

**Fig. 5.** Scans of electrophoretic gels of membrane polypeptides isolated from soybean first trifoliolate leaves. From top to bottom: Mutant Clark $Cy_{11}y_{11}$, mutant Clark $y_3y_3$, wild-type Clark L1, and wild-type Clark L1, mature plant. Peaks are (1) CP1 complex, (2) PSI reaction center, (3) coupling factor $\alpha$, $\beta$, (4) PSII 47- and 43-kDa proteins, (5) 34–33-kDa PSII proteins, LHCPII 33–26-kDa proteins, and (7) PSI LHCPI 21–24 kDa proteins.

## 3. Regulation of Chloroplast Development

these light quality–irradiance responses suggests that their behavior may be coupled.

Long-term growth of plants and algae under various parts of the light spectrum produces distinctive types of chloroplasts (Senger and Bauer, 1987). A typical pattern found in barley (*Hordeum vulgare*) is that blue-light grown plants have chloroplasts whose thylakoids are unstacked and contain high ratios of reaction centers to light-harvesting proteins. Compared with chloroplasts from plants grown in red light, blue-light grown chloroplasts are lower in pigment content and Chl a/b ratios are relatively higher. Red-light grown plants have chloroplasts that contain more Chl, more Chl b, and more light-harvesting proteins. Their chloroplasts also contain many regions in which thylakoids are organized into grana (Lichtenthaler *et al.*, 1980). In pigment content and thylakoid structure, blue-light grown plants are similar to pigment-deficient mutants grown at high intensity and red-light grown plants are similar to those mutants grown under low light intensity.

Plants do not green under long wavelength far-red light, but do green under far-red light that contains shorter wavelength red light (Eskins *et al.*, 1986). Many factors of chloroplast development, such as pigment and pigment–protein accumulation, appear to be independent of the amount of far-red light and are controlled instead by the amount of red light in the far-red source, despite the fact that the far-red/red ratio of the light source controls leaf growth and other whole plant responses. This result may correspond to a long-term very low fluence (VLF) response (Kaufman *et al.*, 1985).

The patterns of red and blue light-mediated chloroplast development just described for barley plants do not apply to all species. For instance, in *Sinapis alba* (Wild and Holzapfel, 1980) and in the unicellular green alga *Scenedesmus obliquus*, adaption to blue and red light is opposite to that seen in barley (Humbeck *et al.*, 1988). Compared with red-light grown cells, blue-light adapted *Scenedesmus* has higher Chl content, higher ratio of Chl to carotenoids and Chl to Cyt $f$, smaller Chl a/b ratios, faster half-rise time of fluorescence induction, and increased amounts of light-harvesting complexes.

The response to light quality also seems to be related to irradiance level, that is, the response to blue and red is opposite in plants that are habituated to high or low light. In low light habituated ferns, blue light produces high pigment, stacked thylakoids, and increased light-harvesting proteins (Leong *et al.*, 1985). This contrasts with the response found in high light habituated maize (Eskins and McCarthy, 1987b). In many species, such as soybean and pine, the accumulation of Chl b and light-harvesting proteins is enhanced by blue light over red at very low irradiance. However, increasing irradiance causes a switch to enhancement by red over blue (Milivojevic and Eskins, 1991).

The response to blue and red is also dependent on the stage of development and the tissue type of the plant. This effect is shown by the fact that root tissue and tissue cultures of plant sections require blue light to initiate greening (Richter and Wessel, 1985). These results suggest the presence of a photoreceptor or combination of photoreceptors that measures blue and red light as a function of irradiance. This light quality–irradiance switch also is connected to the stage of development and, perhaps, to temperature. The connection between blue-light receptor and phytochrome may be irradiance and age of tissue regulated (Tavladoraki et al., 1986).

## V. LIGHT EFFECTS ON TRANSCRIPTION AND TRANSLATION

Several reviews have examined the light regulation of chloroplast development (Tobin and Silverthorne, 1985; Rodermel et al., 1987; Gruissem, 1989). Some specific examples of such regulation are presented here. The light regulation of LHCPII is used as an example of the interaction of many levels of control. For a general description, we look at the regulation of plastid PSII proteins during light-initiated greening of etiolated seedlings.

In spinach (Westhoff et al., 1990), the inner antennae Chl proteins, CP47 and CP43, and the reaction-center proteins, D1 and D2, are not found in etiolated seedlings. Light is a requirement for the synthesis of these chloroplast-encoded proteins. However, nuclear-encoded proteins associated with the water splitting apparatus (33, 23, and 16 kDa) are present; light has little effect on their accumulation. The third constituent of the reaction center, Cyt $b$-559, is present in dark-grown seedlings but increases in light. Since analysis of spinach etioplasts shows that the mRNA transcripts for CP47, CP43, D1, and D2 are present in the dark, the light control of expression of these genes does not occur at the level of transcription. Evidence suggests, instead, that translational control is important in the accumulation of PSII (Klein and Mullet, 1987) and also PSI reaction-center proteins (Kreuz et al., 1986). Work with cell-free translation products of dark- and light-grown *Euglena* also suggests that the synthesis of many chloroplast-encoded proteins is regulated by light at the level of translation (Rikin et al., 1987).

The importance of plant type in determining the mode of gene regulation is shown by a study of greening in sorghum (Westhoff et al., 1990), in which leaf development itself is strongly dependent on light. In dark-grown

## 3. Regulation of Chloroplast Development

sorghum, the Chl proteins CP47, CP43, D1, and D2 are not present, as in spinach. In contrast to spinach, however, polypeptides of the water splitting apparatus and Cyt $b$-559 also are not present in etiolated sorghum. Further, transcripts for these proteins are absent or present in very low amounts in the dark and increase dramatically in the light. Thus, regulation of these plastid-encoded proteins in sorghum appears to be transcriptional.

A third type of plastid gene regulation operates during the differentiation of maize bundle sheath cells. The gene for CP47 is part of a large polycistronic unit that is transcribed during development of bundle sheath cells. Genes for CP43 and D2 are part of another separate large polycistronic unit. Transcription of these large units is followed by processing into dicistronic and monocistronic units. Analysis of these transcripts reveals differences in the RNA stability of the individual pieces of the polycistronic units, suggesting that differential RNA stability is also an important regulatory process in plastid gene expression (Mullet and Klein, 1987; Oswald et al., 1990).

High rates of transcription and translation occur during the initial phases of chloroplast development (Erdos et al., 1990). Long-term irradiation, however, produces equilibrium levels of RNA and protein that reflect the sum of synthesis and degradation. Most mRNA levels decline with development, but transcripts for D1 and D2 remain high because of the instability of these proteins in the light. Evidence suggests that blue light in particular enhances both the transcription and synthesis of these proteins (Gamble and Mullet, 1989; Eskins and Beremand, 1990; see also Chapter 8).

### VI. CASE STUDY: REGULATION OF LHCPII

The major PSII light-harvesting Chl a/b-binding proteins of the chloroplast (LHCPII) are encoded by a multigene family located in the nucleus. In many species, LHCPII proteins and mRNAs are present at low levels in dark-green plants but are increased substantially by light (Nelson et al., 1984). Reports indicate that a substantial portion of the dark RNA is a Cab-1 transcript that is weakly responsive to light (Sullivan et al., 1989). During the initial stages of greening, light regulation of the transcription of these genes occurs via phytochrome. This signal can be mediated by white, red, or blue light. Translation of LHCPII mRNA occurs on cytoplasmic ribosomes; the product is a precursor protein containing a transit peptide sequence necessary for proper translocation of the protein into the chloroplast.

The fully mature LHCPII is a complex of protein and pigment; additional light regulation of the complex operates via pigment synthesis (Plumley and Schmidt, 1987). Early studies showed that a barley mutant (chlorina f2) that contained no Chl b also contained no LHCPII protein. Since the mutant did contain the LHCPII transcript, synthesis of Chl b is not required for transcription of LHCPII mRNA but is required for stability of the mature protein in the thylakoids (Apel and Kloppstech, 1978). Work with mustard (*Sinapis alba*) using red and far-red light has confirmed that there is no correlation between the capacity for Chl synthesis and accumulation of LHCPII transcripts. However, experiments in which the integrity of the chloroplast itself was impaired suggest that some positive factor originating in the chloroplast is necessary for transcription of the LHCPII genes (Burgess and Taylor, 1988; Oelmüller, 1989).

In contrast to these data, previous work with pigment-deficient mutants that are also deficient in LHCPII showed that blockage of chloroplast translation with chloramphenicol (Duysen *et al.*, 1985) increased the synthesis of LHCPII proteins but had no effect on transcription of LHCPII. This result also argues for a chloroplast regulatory factor in the mutant, but the factor is a negative regulator and operates post-transcriptionally (Mogen *et al.*, 1990).

A number of studies have investigated the effects of light irradiance and light quality on the regulation of LHCPII (Anderson, 1986). Most of this work has been concerned with regulation that takes place during the conversion of etioplasts to chloroplasts. Typical results using red light pulses on dark-grown seedlings (Horwitz *et al.*, 1988) show that activation of LHCPII transcription (Mösinger *et al.*, 1988) occurs at very low fluences but increases with irradiance (Kaufman *et al.*, 1984). Over a fluence range of $10^{-4}$ to $10^4$ $\mu$mol m$^{-2}$ sec$^{-1}$, the response of LHCPII mRNA is biphasic and suggests at least two different threshold levels of phytochrome response. These levels have been called very low fluence (VLF) and low fluence (LF) responses (Kaufman *et al.*, 1985). VLF responses also are triggered by far-red light. Thus, they are not far-red reversible. Studies with plants subjected to long-term irradiation with low red or low red plus far-red light give results that are consistent with the VLF response (Eskins *et al.*, 1986). LF responses are far-red reversible but lose reversibility with time.

Similar studies that explored the effect of changes in blue-light irradiance on the accumulation of LHCPII mRNA in red-light grown pea seedlings (Warpeha and Kaufman, 1990) have indicated the presence of an LF and high fluence (HF) blue-light response. The LF blue response ($10^{-1}$ $\mu$mol m$^{-2}$ sec$^{-1}$) causes an increase in LHCPII mRNA. The HF response returns levels to those seen in red-light grown plants. In dark-grown plants, only a general increase in LHCPII mRNA occurs with increasing irradiance of blue light. Therefore, the HF response is a result of blue/red interaction or red light-mediated changes in the state of development.

In a comparison of increasing irradiances of long-term blue and red light on accumulation of maize LHCPII mRNA, red-light was shown to be more effective than blue light (Eskins and McCarthy, 1987b), in contrast to the blue-light stimulation of LHCPII mRNA in red-light grown peas, but the effectiveness of red light over blue light was infuenced strongly by irradiance and also showed a maximum at low fluences (Eskins et al., 1989). In addition, mixtures of red and blue light were shown to have the same effect as blue light alone (Eskins and Beremand, 1990), indicating that, in maize grown at 10 $\mu$ml m$^{-2}$ sec$^{-1}$, a low fluence of long-term blue light in the presence of red light has a repressive effect and that the effectiveness of blue light for decreasing LHCPII mRNA is independent of phytochrome photoequilibrium. Similar patterns are expressed for far-red light in mustard, that is, a short period of far-red light is stimulatory but long-term irradiation is repressive (Oelmuller and Schuster, 1987). In this laboratory, it has been shown blue light is stimulatory to LHCPII during early stages of development but repressive at latter stages (K. Eskins and L. Smith, 1991, unpublished observations).

The varied responses to blue and red light in different species (Senger and Bauer, 1987) and the change in response to blue or red light within a single species as a function of irradiance or development (Milivojevic and Eskins, 1991) indicate that gene switching in chloroplast and whole plant development is regulated by an interaction between irradiance and quality at the receptor(s) (Voskresenskaya, 1984). This receptor system must operate via a quantum-counting mechanism that couples and uncouples blue/red receptors and may involve multiple sites on the genes to which repressor and activator proteins are bound.

## VII. TEMPERATURE EFFECTS ON CHLOROPLAST DEVELOPMENT

Temperature-mediated regulation of chloroplast development is not as well characterized as light-mediated regulation. A number of studies, however, have investigated chloroplast pigment–protein complexes and other constituents of the thylakoid lipid membranes as a function of growth temperature (Berry and Björkman, 1980; Casadoro et al., 1983; Steffen and Palta, 1987).

In one study, differences in the development of winter rye (*Secale cereale* L. cv *Puma*) at 5°C and 20°C were reflected in the amounts of lipids and in the ratios of oligomeric to monomeric LHCPII (Williams et al., 1987). Specifically, growth at the lower temperature results in a large decrease in 3-transhexadecanoic acid and a decrease in the ratio of oligomeric to monomeric LHCPII. In addition, greening at low temperature causes a significant lag in formation of Chl b and produces smaller grana stacks.

In moss protonemata (*Ceratodon purpureus*), growth at low temperatures increases the ratio of lipid to protein and increases the unsaturation of lipids. Chl a/b ratios are unaffected, but thylakoid stacking is decreased. Also, low temperature grown protonemata recover better and more quickly from high-light photoinhibition (Aro et al., 1987). In winter rye, Chl accumulation under continuous light is reduced at 5°C compared with 20°C but Chl accumulation under intermittent light is enhanced at 5°C (Krol and Huner, 1990). In plants, PSII functions of electron transport are inhibited by low or high temperature stress and are more susceptible to such stress than corresponding PSI functions (Gounaris et al., 1984). High temperatures block PSII reaction centers, increase the ratio of PSII $\alpha$ to PSII $\beta$ centers (Andersson et al., 1987), and dissociate LHCPII from the PSII core. This response also may involve movement of the photosystem II core and the tightly bound antennae from the grana to the stroma (Sundby and Andersson, 1985; Öquist, 1987; Ovaska et al., 1990).

An interesting correlation between high light-induced pigment deficiency and temperature has been noted in mutants from a number of species. In chlorina mutants that were characterized by a high ratio of F680/F740 (low temperature fluorescence) and depleted Chl b and light-harvesting complexes, temperature regulated the characteristics of the mutation, that is, low temperature enhanced and higher temperature decreased the expression of the mutation. Switching from low to high temperature also switched the expression of the mutation in a period of days (Knoetzel and Simpson, 1990). Similar temperature-dependent data are reported for pigment-deficient mutants of sweet clover (*Melilotus alba*) with the notable exceptions of the ch5 (Chl b-less) and ch12 mutants, which show little temperature dependence of mutant expression (Yang et al., 1990). In contrast to chlorina mutants, leaves of sweet clover mutants that developed and expanded at permissive or nonpermissive temperatures are not altered by transition to the other temperature (Markwell et al., 1986). This result suggests a developmental regulation of temperature sensitivity that corresponds with a similar developmental regulation of light sensitivity. Whole-plant systems and the chloroplast also have a mechanism that responds to temperatures that are too high. Plants that have been shocked with high temperatures produce, in response, a series of protective proteins called heat-shock proteins (Lin et al., 1984; Linquist, 1986). The number and nature of these proteins is determined by the biological system and by genetic variation within a species (Vierling and Nguyen, 1990).

## 3. Regulation of Chloroplast Development

Specific heat-shock proteins are associated with the chloroplast and appear to bind to chloroplast membranes at elevated temperatures (Restivo *et al.*, 1986; Vierling *et al.*, 1986; Glaczinski and Kloppstech, 1988). The heat-shock proteins also induce tolerance to subsequent heat treatments (Key and Nobel, 1986) and to high light-induced photoinhibition (Schuster *et al.*, 1988). This effect may help explain why photoinhibition effects are enhanced at lower temperatures (Farage and Long, 1991). The mechanism by which heat-shock proteins protect against high temperature stress is unknown, but is likely to involve the stabilization of membranes and the preservation of protein tertiary structure.

### VIII. CONCLUDING REMARKS

Exploration of the interdependence of light, temperature, and stage of development in regulating expression of nuclear and chloroplast genes is expanding our knowledge of whole-plant development. Transfer of genes into new systems will require knowledge of switching and regulatory pathways. Light and temperature signals delivered at early stages and reinforced at critical transition points will allow us to program and control such development. Especially important will be those repressor and activator proteins that bind to regulatory genes. These proteins must respond to light quality via photoreceptors, but also must be sensitive to the irradiance level. More information is needed about the quantum counting mechanisms of currently known receptors. We also must bear in mind the possibility of new receptors and new interactions among light quality, irradiance level, and temperature as a function of developmental stage.

### REFERENCES

Akoyunoglou, G., Argyoudi-Akoyunoglou, J. H., Christias, C., Tsakiris, S., and Tsimili-Michael, M. (1978). Thylakoid growth and differentiation in continuous light as controlled by the duration of preexposure to periodic light. *In* "Chloroplast Development" (G. Akoyunoglou and J. H. Argyroudi-Akoyunoglou, eds.), pp. 843–856. Elsevier/North Holland, Amsterdam.

Akoyunoglou, G., Anni, H., and Kalosakas, K. (1980). The effect of light quality and the mode of illumination on chloroplast development in etiolated bean leaves. *In* "The Blue Light Syndrome" (H. Senger, ed.), pp. 473–484. Springer-Verlag, Berlin.

Allen, K. D., Duysen, M. E., and Staehelin, L. A. (1987). A chlorophyll *b* deficient mutant of wheat with an altered photoadaption response. *In* "Progress in Photosynthesis Research" (J. Biggens, ed.), Vol. IV, pp. 601–604. Nijhoff, Dordrecht, The Netherlands.

Allen, K. D., Duysen, M. E., and Staehelin, L. A. (1988). Biogenesis of thylakoid membranes is controlled by light intensity in the conditional chlorophyll *b*-deficient CD3 mutant of wheat. *J. Cell Biol.* **107**, 907–919.

Allen, K. D., Falbel, T. G., Shaw, S. L., Bennett, A., and Staehelin, L. A. (1990). Resolution of up to eighteen chlorophyll–protein complexes from vascular plant thylakoids using a new green gel system. *In* "Current Research in Photosynthesis" (M. Baltscheffsky, ed.), Vol. II, pp. 264–272. Kluwer, Dordrecht, The Netherlands.

Anderson, J. M. (1986). Photoregulation of the composition, function, and structure of thylakoid membranes. *Annu. Rev. Plant Physiol.* **37**, 93–136.

Andersson, B., Sundby, C., Larsson, U. K., Maenpaa, P., and Melis, A. (1987). Dynamic aspects of the organization of the thylakoid membrane. *In* "Progress in Photosynthetic Research" (J. Biggens, ed.), Vol. II, pp. 669–676. Nijhoff, Dordrecht, The Netherlands.

Apel, K., and Kloppstech, K. (1978). The plastid membranes of barley (*Hordeum vulgare*). Light-induced appearance of mRNA coding for the apoprotein of the light-harvesting chlorophyll *a/b* protein. *Eur. J. Biochem.* **85**, 581–588.

Argyroudi-Akoyunoglou, J. H., Castorinis, A., and Akoyunoglou, G. (1984). Biogenesis and organization of the pigment–protein complexes. Relation to the low temperature fluorescence characteristics of developing thylakoids. *Israel J. Bot.* **33**, 65–82.

Aro, E-M., Somersalo, S., and Karunen, P. (1987). Thylakoid membrane composition and photoinactivation of $CO_2$ fixation in moss protonemata as influenced by the growth temperature. *In* "Progress in Photosynthesis Research" (J. Biggens, ed.), Vol. IV, pp. 115–118. Nijhoff, Dordrecht, The Netherlands.

Bassi, R., and Simpson, D. J. (1987). Chlorophyll-proteins of barley photosystem I. *Eur. J. Biochem.* **163**, 221–230.

Bassi, R., Rigoni, F., and Giacometti, G. M. (1990). Chlorophyll binding proteins with antenna function in higher plants and green algae. *Photochem. Photobiol.* **52**, 1187–1206.

Batschauer, A., and Apel, K. (1984). An inverse control by phytochrome of the expression of two nuclear genes in barley (*Hordeum vulgare* L). *Eur. J. Biochem.* **143**, 593–597.

Bennett, J., Schwender, J. R., Shaw, E. K., Tempel, N., Ledbetter, M., and Williams, K. S. (1987). Failure of corn leaves to acclimate to low irradiance. Role of protochlorophyllide reductase in regulating levels of five chlorophyll-binding proteins. *Biochim. Biophys. Acta* **892**, 118–129.

Berry, J., and Björkman, O. (1980). Photosynthetic response and adaption to temperature in higher plants. *Annu. Rev. Plant Physiol.* **31**, 491–543.

Bricker, T. M. (1990). The structure and function of CPa-1 and CPa-2 in photosystem II. *Photosyn. Res.* **24**, 1–13.

Burgess, D. G., and Taylor, W. C. (1988). The chloroplast affects the transcription of a nuclear gene family. *Mol. Gen. Genet.* **214**, 89–96.

Butterfass, T. H. (1980). The continuity of plastids and the differentiation of plastid populations. *In* "Results and Problems in Cell Differentiation: Chloroplast" (J. Reinert, ed.), Vol. 10, pp. 29–44. Springer-Verlag, Berlin.

Camm, E. L., and Green, B. R. (1989). The chlorophyll *a/b* complex, CP92, is associated with the photosystem II reaction center core. *Biochim. Biophys. Acta* **974**, 180–184.

Casadoro, G., Hoyer-Hansen, G., Kannangara, C. G., and Gough, S. P. (1983). An analysis of temperature and light sensitivity in tigrina mutants of barley. *Carlsberg Res. Commun.* **48**, 95–129.

Chitnis, P. R., Nechustai, R., Harel, E., and Thornber, J. P. (1987). Some requirements for the insertion of the precursor of apoproteins of *Lemna* light harvesting complex II into barley thylakoids. *In* "Progress in Photosynthesis Research" (J. Biggens, ed.), Vol. IV, pp. 573–576. Nijhoff, Dordrecht, The Netherlands.

Chow, W. S., Melis, A., and Anderson, J. M. (1990). Adjustments of photosystem stoichiom-

## 3. Regulation of Chloroplast Development

etry in chloroplasts improve the quantum efficiency of photosynthesis. *Proc. Natl. Acad. Sci. USA* **87,** 7502–7506.

Cline, K. (1988). Light harvesting chlorophyll *a/b* protein membrane insertion, proleolytic processing, assembly into LHC II, and localization to appressed membranes occurs in chloroplast lysates. *Plant Physiol.* **86,** 1120–1126.

Cramer, W. A., Widger, W. R., Herrmann, R. G., and Trebst, A. (1985). Topography and function of thylakoid membrane proteins. *Trends Biochem. Sci.* **10,** 125–129.

Dainese, P., DiPaolo, M. L., Silvestri, M., and Bassi, R. (1990). Properties of the minor chlorophyll *a/b* proteins CP29, CP26, and CP24 from *Zea mays* photosystem II membranes. *In* "Current Research in Photosynthesis" (M. Baltscheffsky, ed.), Vol. II, pp. 249–252. Kluwer, Dordrecht, The Netherlands.

Drumm-Herrel, H., and Mohr, H. (1982). The effect of prolonged light exposure on the effectiveness of phytochrome in anthocyanin synthesis in tomato seedlings. *Photochem. Photobiol.* **35,** 233–236.

Duysen, M. E., and Freeman, T. P. (1974). Effects of moderate water deficit (stress) on wheat seedling growth and plastid pigment development. *Plant Physiol.* **31,** 262–266.

Duysen, M. E., Freeman, T. P., Williams, N. D., and Huckle, L. L. (1985). Chloramphenicol stimulation of light-harvesting chlorophyll protein complex accumulation in a chlorophyll *b* deficient wheat mutant. *Plant Physiol.* **78,** 531–536.

Erdos, G., Shinohara, K., Chen, H-Q., Lee, S., Gillott, M., and Buetow, D. E. (1987). Chloroplast development and regulation of LHCP-gene expression in greening cultured soybean cells. *In* "Progress in Photosynthesis Research" (J. Biggens, ed.), Vol. IV, pp. 539–542. Nijhoff, Dordrecht, The Netherlands.

Erdos, G., Chen, H-Q., and Buetow, D. E. (1990). Chloroplast developmental stage specific controls on LHCP II genes in cultured soybean cells. *In* "Current Research on Photosynthesis" (M. Baltscheffsky, ed.), Vol. III, pp. 549–552. Kluwer, Dordrecht, The Netherlands.

Eskins, K., and Banks, D. E. (1979). The relationship of accessory pigments to chlorophyll *a* content in chlorophyll-deficient peanut and soybean varieties. *Photochem. Photobiol.* **30,** 585–588.

Eskins, K., and Beremand, P. (1990). Light quality and irradiance level control of light-harvesting complex of photosystem II in maize mesophyll cells. Evidence for a low fluence-rate threshold in blue-light reduction of mRNA and protein. *Physiol. Plant.* **78,** 435–440.

Eskins, K., and Harris, L. (1981). High-performance liquid chromatography of etioplast pigments in red kidney bean leaves. *Photochem. Photobiol.* **33,** 131–133.

Eskins, K., and McCarthy, S. A. (1987a). Comparison of soybean pigment protein complexes during development and senescence. *In* "Plant Senescence." Its Biochemistry and Physiology" (W. W. Thompson, E. A. Nothnagel, and R. C. Huffaker, eds.), pp. 108–113. American Society of Plant Physiology, Rockville, Maryland.

Eskins, K., and McCarthy, S. A. (1987b). Blue, red, and blue plus red light control of chloroplast pigment and pigment–proteins in corn mesophyll cells: Irradiance level–quality interaction. *Physiol. Plant.* **71,** 100–104.

Eskins, K., Harris, L., and Bernard, R. L. (1981). Genetic control of chloroplast pigment development in soybeans as a function of leaf and plant maturity. *Plant Physiol.* **67,** 759–762.

Eskins, K., Kwolek, W. F., and Harris, L. (1982). The accumulation of accessory pigments as a function of chlorophyll *a*. A comparison of development and genetic control. *Physiol. Plant.* **54,** 409–413.

Eskins, K., Delmastro, D., and Harris, L. (1983a). A comparison of pigment–protein com-

plexes among normal, chlorophyll-deficient, and senescent soybean genotypes. *Plant Physiol.* **73**, 51–55.
Eskins, K., Duysen, M. E., and Olson, L. (1983b). Pigment analysis of chloroplast pigment–protein complexes in wheat. *Plant Physiol.* **71**, 777–779.
Eskins, K., McCarthy, S. A., Dybas, L., and Duysen, M. (1986). Corn chloroplast development in low fluence red light and in low fluence red light plus far-red light. *Physiol. Plant.* **67**, 242–246.
Eskins, K., Westhoff, P., and Beremand, P. (1989). Light quality and irradiance level interaction in the control of expression of LHCP2. Pigments, pigment–proteins, and mRNA accumulation. *Plant Physiol.* **91**, 163–169.
Faludi-Daniel, A., Mustardy, L. A., Vass, L., and Kiss, I. G. (1986). Energization and ultrastructural pattern of thylakoids formed under periodic illumination followed by continuous light. *Photosyn. Res.* **9**, 229–238.
Farage, P. K., and Long, S. P. (1991). The occurrence of photoinhibition in an over-wintering crop of oil-seed rape (*Brassica napus L.*) and its correlation with changes in crop growth. *Planta* **185**, 279–286.
Freeman, T. P., Duysen, M. E., Olson, N. H., and Williams, N. D. (1982). Electron transport and chloroplast ultrastructure of a chlorophyll deficient mutant of wheat. *Photosyn. Res.* **3**, 179–189.
Gamble, P. E., and Mullet, J. E. (1989). Blue light regulates the accumulation of two *psb*D–*psb*C transcripts in barley chloroplast. *EMBO J.* **8**, 2785–2794.
Ghirardi, M. L., and Melis, A. (1988). Chlorophyll *b* deficiency in soybean mutants. Effects on photosystem stoichiometry and chlorophyll antenna size. *Biochim. Biophys. Acta* **932**, 130–137.
Glaczinski, H., and Kloppstech, K. (1988). Temperature-dependent binding to the thylakoid membranes of nuclear-coded chloroplast heat-shock proteins. *Eur. J. Biochem.* **173**, 579–583.
Gounaris, K., Brain, A. R. R., Quinn, P. J., and Williams, W. P. (1984). Structural reorganization of chlorophyll thylakoid membranes in response to heat stress. *Biochim. Biophys. Acta* **766**, 198–208.
Gruissem, W. (1989). Chloroplast gene expression. How plants turn their plastids on. *Cell* **56**, 161–170.
Guikema, J. A., and Sherman, L. A. (1983). Organization and function of chlorophyll in membranes of cyanobacteria during iron starvation. *Plant Physiol.* **73**, 250–256.
Haworth, P., Warson, J. L., and Arntzen, C. J. (1983). The detection, isolation, and characterization of a light harvesting complex which is specifically associated with photosystem I. *Biochim. Biophys. Acta* **724**, 151–158.
Henningsen, K. W., and Boynton, J. E. (1974). Macromolecular physiology of plastids. IX. Development of plastid membrane during greening of dark grown barley seedlings. *J. Cell Sci.* **15**, 31–55.
Hoober, J. K., Maloney, M. A., and Marks, B. D. (1990). Aspects of assembly of chlorophyll *a/b* protein complexes. *In* "Current Research in Photosynthesis" (M. Baltscheffsky, ed.), Vol. III, pp. 723–726. Kluwer, Dordrecht, The Netherlands.
Horwitz, B. A., Thompson, W. F., and Briggs, W. R. (1988). Phytochrome regulation of greening in *Pisum*. *Plant Physiol.* **86**, 299–305.
Humbeck, K., Hoffmann, B., and Senger, H. (1988). Influence of energy flux and quality of light on the molecular organization of the photosynthetic apparatus in *Scenedesmus*. *Planta* **173**, 205–212.
Jackson, J. A., and Lynden, R. F. (1990). Habituation: Cultural curiosity or developmental determinant. *Physiol. Plant.* **79**, 579–583.
Kamiya, A., Ikegami, I., and Hase, E. (1981). Effects of light on chlorophyll formation in

cultured tobacco cells. I. Chlorophyll accumulation and photochlorophyll(ide) in callus cells under blue and red light. *Plant Cell Physiol.* **22,** 1385–1396.
Kasemir, H. (1983). Light control of chlorophyll accumulation in higher plants. *In* "Encyclopedia of Plant Physiology" (W. Shropshire, Jr., and H. Mohr, eds.), Vol. 16B, pp. 662–686. Springer-Verlag, Berlin.
Kasemir, H., and Mohr, H. (1981). The involvement of phytochrome in controlling chlorophyll and 5-aminolevulinate formation in a gymnosperm seedling (*Pinus sylvestris*). *Planta* **132,** 291–295.
Kaufman, L. S., Thompson, W. F., and Briggs, W. R. (1984). Different red light requirements for phytochrome-induced accumulation of *Cab* RNA and the *rbc*S RNA. *Science* **226,** 1447–1449.
Kaufman, L. S., Briggs, W. R., and Thompson, W. F. (1985). Phytochrome control of specific RNA levels in developing pea buds: The presence of very low and low fluence responses. *Plant Physiol.* **75,** 388–393.
Key, S. C., and Nobel, P. S. (1986). Concomitant changes in high temperature tolerance and heat-shock proteins in desert succulents. *Plant Physiol.* **80,** 596–598.
Kirsch, W., Seyer, P., and Herrmann, R. G. (1986). Nucleotide sequence of the clustered genes for two P700 chlorophyll *a* apoproteins of the photosystem I reaction center and the ribosomal protein S14 of the spinach plastid chromosome. *Curr. Genet.* **10,** 843–855.
Klein, R., and Mullet, J. E. (1987). Control of gene expression during higher plant chloroplast biogenesis. Protein synthesis and transcript levels of *psb*A, *psa*A–*psa*B, and *rbc*L in dark grown and illuminated barley seedlings. *J. Biol. Chem.* **262,** 4341–4348.
Knoetzel, J., and Simpson, D. (1990). Characterisation of a temperature-sensitive mutant of barley. *In* "Current Research in Photosynthesis" (M. Baltscheffsky, ed.), Vol. II, pp. 867–870. Kluwer, Dordrecht, The Netherlands.
Kreuz, K., Dehesh, K., and Apel, K. (1986). The light-dependent accumulation of the P700 chlorophyll *a* protein of the photosystem I reaction center in barley. Evidence for translational control. *Eur. J. Biochem.* **159,** 459–467.
Krol, M., and Huner, N. P. A. (1990). Low temperature effects on PCR and chlorophyll accumulation in rye seedlings. *In* "Current Research in Photosynthesis" (M. Baltscheffsky, ed.), Vol. III, pp. 761–763. Kluwer, Dordrecht, The Netherlands.
Laval-Martin, D., and Troton, D. (1990). Chilling resistance of photosynthetic performance in diuron adapted *Euglena gracilis*. *Plant Sci.* **72,** 213–222.
Lee, M. (1988). The chromosomal basis of somaclonal variation. *Annu. Rev. Plant Physiol. Plant Mol. Biol.* **39,** 413–437.
Lemoine, Y., Zabulon, G., and Cornu, A. (1987). Chlorophyll protein complexes changes associated with chloroplast development in a virescent *Petunia hybrida* mutant. *In* "Progress in Photosynthesis Research" (J. Biggens, ed.), Vol. II, pp. 371–374. Nijhoff, Dordrecht, The Netherlands.
Leong, T-Y., Goodchild, D. J., and Anderson, J. M. (1985). Effect of light quality on the composition, function, and structure of photosynthetic thylakoid membranes of *Asplenium australasicum* (Sm.) Hook. *Plant Physiol.* **78,** 561–567.
Lichtenthaler, H. K., Buschmann, C., and Rahmdorf, U. (1980). The importance of blue light for the development of sun-type chloroplast. *In* "The Blue Light Syndrome" (H. Senger, ed.), pp. 485–494. Springer-Verlag, Berlin.
Lichtenthaler, H. K., Prenzel, U., and Kuhn, G. (1982). Carotenoid composition of chlorophyll–carotenoid proteins from radish chloroplast. *Z. Naturforsch.* **37C,** 10–12.
Lin, C. Y., Roberts, J. K., and Key, J. L. (1984). Acquisition of thermotolerance in soybean seedlings: Synthesis and accumulation of heat-shock proteins and their cellular localization. *Plant Physiol.* **74,** 152–160.
Lindquist, S. (1986). The heat-shock response. *Annu. Rev. Biochem.* **55,** 1151–1191.

Lindsten, A., Ryberg, M., and Sundqvist, C. (1988). The polypeptide composition of highly purified prolamellar bodies and prothylakoids from wheat (*Triticum aestivum*) as revealed by silver staining. *Physiol. Plant.* **72**, 167–176.

Mancinelli, A. L., and Robino, E. (1978). The "high irradiance responses" of plant photomorphogenesis. *Bot. Rev.* **44**, 129–180.

Markwell, J. P., Thornber, J. P., and Boggs, R. T. (1979). Higher plant chloroplast: Evidence that all the chlorophyll exists as chlorophyll–protein complexes. *Proc. Natl. Acad. Sci. USA* **76**, 1233–1235.

Markwell, J. P., Danko, S. J., Bauwe, H., Osterman, J., Gorz, H. J., and Haskins, F. A. (1986). A temperature-sensitive chlorophyll $b$-deficient mutant of sweetclover (*Melilotus alba*). *Plant Physiol.* **81**, 329–334.

Milivojevic, D., and Eskins, K. (1991). Effects of irradiance quality (blue, red) interaction on the synthesis of pigments and pigment–proteins in maize and black pine mesophyll chloroplast. *Physiol. Plant.* **80**, 624–628.

Mogen, K., Eide, J., Duysen, M., and Eskins, K. (1990). Chloramphenicol stimulates the accumulation of light-harvesting chlorophyll $a/b$ protein II by affecting posttranscriptional events in the chlorina CD3 mutant of wheat. *Plant Physiol.* **92**, 1233–1240.

Mohr, H. (1986). Coaction between pigment systems. *In* "Photomorphogenesis in Plants" (R. E. Kendrick and G. H. M. Kronenberg, eds.), pp. 547–564. Nijhoff/Junk, Dordrecht, The Netherlands.

Mösinger, E., Batschauer, A., Apel, K., Schäfer, E., and Briggs, W. L. (1988). Phytochrome regulation of greening in barley. Effects on mRNA abundance and on transcriptional activity of isolated nuclei. *Plant Physiol.* **86**, 706–710.

Mullet, J., and Klein, R. R. (1987). Transcription and RNA stability are important determinants of higher plant chloroplast RNA levels. *EMBO J.* **6**, 1571–1579.

Mullet, J. E., Burke, J. J., and Arntzen, C. J. (1980). Chlorophyll proteins of photosystem I. *Plant Physiol.* **65**, 814–822.

Nagy, F., Kay, S. A., and Chua, N-H. (1988). Gene regulation by phytochrome. *Trends Gen.* **4**, 37–42.

Namba, O., and Satoh, K. (1987). Isolation of a photosystem II reaction center consisting of D1 and D2 polypeptides and cytochrome $b$-559. *Proc. Natl. Acad. Sci. USA* **84**, 109–112.

Nelson, T., Harpster, M. H., Mayfield, S. P., and Taylor, W. C. (1984). Light regulated gene expression during maize leaf development. *J. Cell Biol.* **98**, 558–564.

Oelmüller, R. (1989). Photooxidative destruction of chloroplast and its effect on nuclear gene expression and extraplastidic enzyme levels. *Photochem. Photobiol.* **49**, 229–239.

Oelmüller, R., and Schuster, C. (1987). Inhibition and promotion by light of the accumulation of translatable mRNA of the light-harvesting chlorophyll $a/b$-binding protein of photosystem II. *Planta* **172**, 60–70.

Öquist, G. (1987). Environmental stress and photosynthesis. *In* "Progress in Photosynthesis Research" (J. Biggens, ed.), Vol. IV, pp. 1–9. Nijhoff, Dordrecht, The Netherlands.

Oswald, A., Streubel, M., Ljungberg, U., Hermans, J., Eskins, K., and Westhoff, P. (1990). Differential biogenesis of photosystem II in mesophyll and bundle sheath cells of malic enzyme NADP+ type C4 plants. A comparative protein and RNA analysis. *Eur. J. Biochem.* **190**, 185–194.

Ovaska, J., Maenpaa, P., Nurmi, A., and Aro, E-M. (1990). Distribution of chlorophyll–protein complexes during chilling in the light compared with heat-induced modification. *Plant Physiol.* **93**, 48–54.

Pichersky, E., and Green, B. R. (1990). The extended family of chlorophyll $a/b$-binding proteins of PSI and PSII. *In* "Current Research in Photosynthesis" (M. Baltscheffsky, ed.), Vol. III, pp. 553–556. Kluwer, Dordrecht, The Netherlands.

Piechulla, B., and Riesselmann, S. (1990). Effect of temperature alterations on the diurnal expression pattern of the chlorophyll $a/b$-binding proteins in tomato seedlings. *Plant Physiol.* **94**, 1903–1906.
Plumley, F. G., and Schmidt, G. W. (1987). Reconstitution of chlorophyll $a/b$ light-harvesting complexes. Xanthophyll-dependent assembly and energy transfer. *Proc. Natl. Acad. Sci. USA* **84**, 146–150.
Rawyler, A., Henry, L. E. A., and Siegenthaler, P-A. (1980). Acyl and pigment lipid composition of two chlorophyll proteins. *Carlsberg Res. Commun.* **45**, 443–451.
Restivo, F. M., Tassi, F., Maestri, E., Lorenzone, C., Puglisi, P. P., and Marmiroli, N. (1986). Identification of chloroplast associated heat-shock proteins in *Nicotiana plumbaginifolia* protoplast. *Curr. Genet.* **11**, 145–149.
Richter, G., and Wessel, K. (1985). Red light inhibits blue light induced chloroplast development in cultured plant cells at the mRNA level. *Plant Mol. Biol.* **5**, 175–182.
Rikin, A., Monroy, A., and Schwartzbach, S. D. (1987). Evidence for translational regulation of *Euglena* protein synthesis by light. *In* "Progress in Photosynthesis Research" (J. Biggens, ed.), Vol. IV, pp. 581–584. Nijhoff, Dordrecht, The Netherlands.
Rodermel, S. R., Crossland, L. D., Lukens, J. H., Muskavitch, K. M., Russel, D. R., Shen, J. Y., Zhu, Y.-S., and Bogorad, L. (1987). Photoregulation of maize plastid genes during light-induced development. *In* "Progress in Photosynthesis Research" (J. Biggens, ed.), Vol. IV, pp. 519–526. Nijhoff, Dordrecht, The Netherlands.
Schoch, S. (1978). The esterification of chlorophyllide $a$ in greening bean leaves. *Z. Naturforsch.* **33C**, 712–714.
Senger, H., and Bauer, B. (1987). The influence of light quality on adaption and function of the photosynthetic apparatus. *Photochem. Photobiol.* **45**, 939–946.
Shuster, G., Even, D., Kloppstech, K., and Ohad, I. (1988). Evidence for protection by heat-shock proteins against photoinhibition during heat-shock. *EMBO J.* **7**, 1–6.
Siegenthaler, P-A., and Dumont, N. (1990). Characteristics of spinach chloroplast envelope, thylakoid, and stroma polypeptides as revealed by Triton X-114 phase partition. *Plant Cell Physiol.* **31**, 1101–1108.
Simpson, D. J. (1990). The structure of photosystem I and II. *In* "Current Research in Photosynthesis" (M. Baltscheffsky, ed.), Vol. II, pp. 725–732. Kluwer, Dordrecht, The Netherlands.
Smith, H., and Morgan, D. C. (1983). The function of phytochrome in nature. *In* "Encyclopedia of Plant Physiology" (W. Shropshire, Jr., and H. Mohr, eds.), Vol. 16B, pp. 491–513. Springer-Verlag, Berlin.
Smith, W. O. (1983). Phytochrome as a molecule. *In* "Encyclopedia of Plant Physiology" (W. Shropshire, Jr., and H. Mohr, eds.), Vol. 16B, pp. 96–115. Springer-Verlag, Berlin.
Sponga, F., Deitzer, G. F., and Mancinelli, A. L. (1986). Cryptochrome, phytochrome, and the photoregulation of anthocyanin production under blue light. *Plant Physiol.* **82**, 952–955.
Steffen, K. L., and Palta, J. P. (1987). Acclimation of light harvesting and light utilization capacity in response to growth temperatures. *In* "Progress in Photosynthetic Research" (J. Biggens, ed.), Vol. IV, pp. 111–114. Nijhoff, Dordrecht, The Netherlands.
Sullivan, T. O., Christensen, A. H., and Quail, P. H. (1989). Isolation and characterization of a maize chlorophyll $a/b$ binding protein gene that produces high levels of mRNA in the dark. *Mol. Gen. Genet.* **215**, 431–440.
Sundby, C., and Andersson, B. (1985). Temperature-induced reversible migration along the thylakoid membrane of photosystem II regulates its association with LHCP-II. *FEBS Lett.* **191**, 24–28.
Sundqvist, C., and Ryberg, M. (1989). The distribution and structural role of NADPH–pro-

tochlorophyllide oxidoreductase in isolated etioplast inner membranes. *Photosynthetica* **23**(3), 427–438.

Tavladoraki, P., Akoyunoglou, G., Bitsch, A., Meyer, G., and Kloppstech, K. (1986). Age and phytochrome induced changes at the level of the translatable mRNA coding for the LHC-II apoprotein of *Phaseolus vulgaris* leaves. *In* "Regulation of Chloroplast Differentiation" (G. Akoyunoglou and H. Senger, eds.), pp. 559–564. Liss, New York.

Taylor, W. C. (1989). Regulatory interaction between nuclear and plastid genomes. *Annu. Rev. Plant Physiol. Mol. Biol.* **40**, 211–233.

Terao, T., and Katoh, S. (1990). Turnover of LHC-I and LHC-II apoproteins in chl *b* deficient mutants of rice and barley. *In* "Current Research in Photosynthesis" (M. Baltscheffsky, ed.), Vol. II, pp. 859–862. Kluwer, The Netherlands.

Thornber, J. P. (1986). Biochemical characterization and structure of pigment–proteins of photosynthetic organisms. *In* "Encyclopedia of Plant Physiology (L. A. Staehelin and C. J. Arntzen, eds.), Vol. 19, pp. 98–142. Springer-Verlag, Berlin.

Thorne, S. W., and Boardman, N. K. (1971). Formation of chlorophyll *b* and the fluorescence properties and photochemical activities of isolated plastids from greening pea seedlings. *Plant Physiol.* **47**, 252–261.

Tobin, E. M., and Silverthorne, J. (1985). Light regulation of gene expression in higher plants. *Annu. Rev. Plant Physiol.* **36**, 569–593.

Tzinas, G., Akoyunoglou, G., and Akoyunoglou, A. (1986). Effects of the rate of chlorophyll *a* formation in thylakoid development in higher plants. *In* "Regulation of Chloroplast Differentation" (G. Akoyunoglou and H. Senger, eds.), pp. 697–702. Liss, New York.

Vierling, E., Mishkind, M. L., Schmidt, G. W., and Key, J. L. (1986). Specific heat-shock proteins are transported into chloroplast. *Proc. Natl. Acad. Sci. USA* **83**, 361–365.

Vierling, E., and Alberte, R. S. (1983). P700 chlorophyll *a* protein. Purification, characterization, and antibody preparation. *Plant Physiol.* **72**, 625–633.

Virling, R. A., and Nguyen, H. T. (1990). Heat-shock protein synthesis and accumulation in diploid wheat. *Crop Sci.* **30**, 1337–1342.

Voskresenskaya, N. P. (1984). Control of the activity of photosynthetic apparatus in higher plants. *In* "Blue Light Effects in Biological Systems" (H. Senger, ed.), pp. 407–418. Springer-Verlag, Berlin.

Warpeha, K. M. F., and Kaufman, L. S. (1990). Two distinct blue light responses regulate the levels of transcripts of specific nuclear coded genes in pea. *Planta* **182**, 553–558.

Westhoff, P., Schrubar, H., Oswald, A., Streubel, M., and Offermann, K. (1990). Biogenesis of photosystem II in C3 and C4 plants. A model system to study developmentally regulated and cell-specific expression of plastid genes. *In* "Current Research in Photosynthesis" (M. Baltscheffsky, ed.), Vol. III, pp. 483–490. Kluwer, Dordrecht, The Netherlands.

Wild, A., and Holzapfel, A. (1980). The effect of blue and red light on the content of chlorophyll, cytochrome *f*, soluble reducing sugars, soluble proteins, and the nitrate reductase activity during growth of the primary leaves of *Sinapis alba*. *In* "The Blue Light Syndrome" (H. Senger, ed.), pp. 444–451. Springer-Verlag, Berlin.

Williams, J. P., Huner, N. P. A., Krol, M., Maissan, E., Low, P. S., Roberts, D., and Thompson, J. E. (1987). *In vivo* low temperature induced decreases 3-transhexadecanoic acid influences oligomerization of LHCII. *In* "Progress in Photosynthetic Research" (J. Biggens, ed.), Vol. IV, pp. 127–130. Nijhoff, Dordrecht, The Netherlands.

Yang, C-M., Osterman, J. C., and Markwell, J. (1990). Temperature sensitivity as a general phenomenon in a collection of chlorophyll deficient mutants of sweetclover (*Melilotus alba*). *Biochem. Gen.* **28**, 31–40.

# 4

# Biosynthesis of the Chlorophyll Chromophore of Pigmented Thylakoid Proteins

**WILLIAM R. RICHARDS**

Department of Chemistry
Simon Fraser University
Burnaby, British Columbia, Canada V5A 1S6

I. Introduction
II. Early Stages of Chlorophyll Synthesis: The Pathway to Protoporphyrin IX
   A. Localization of Enzymes of Protoporphyrin IX Synthesis
   B. $C_5$ Pathway of 5-Aminolevulinic Acid Synthesis
   C. Conversion of 5-Aminolevulinic Acid to Porphobilinogen
   D. Conversion of Porphobilinogen to Uroporphyrinogen III
   E. Conversion of Uroporphyrinogen III to Protoporphyrin IX
III. Later Stages of Chlorophyll Synthesis: The Magnesium Branch
   A. Magnesium Chelatase
   B. S-Adenosyl-L-Methionine : Magnesium Protoporphyrin IX Methyltransferase
   C. Oxidative Cyclase Enzyme System
   D. Monovinyl and Divinyl Pathways of Chlorophyll Synthesis: Reduction of the 4-Vinyl Group
   E. Protochlorophyllide Reductase
   F. Chlorophyll Synthetase
   References

## I. INTRODUCTION

Chlorophyll *a* (Chl *a*) is found universally in all species of oxygenic plants and cyanobacteria, and functions both as an antenna and as a phototrap. Other chromophores[1] that have purely antenna functions in-

---

[1] This chapter will not discuss certain minor chromophores reported to occur in some photosynthetic organisms, including Chl *d, a'*, and RC I (Scheer, 1991), and pigments related to Chl $c_1$ and $c_2$ (Fawley, 1989; Fookes and Jeffery, 1989; Jeffery, 1989).

clude carotenoids (found in most photosynthetic organisms), chlorophyll b (Chl b; found in terrestrial plants, green algae, *Euglena*, and prochlorophytes), and various chlorophylls of the c-type (found in brown algae and the chromophytes, including diatoms and dinoflagellates). Another distinctively different group of antenna complexes, the phycobiliproteins of red algae, cryptophytes, and cyanobacteria, contain as chromophores the linear tetrapyrroles phycoerythrobilin, phycocyanobilin, phycobiliviolin, and phycourobilin. The biosynthesis of linear tetrapyrroles has been reviewed by Brown *et al.* (1990). The pigment–protein complexes of anoxygenic photosynthetic bacteria contain two types of chromophores, carotenoids and bacteriochlorophylls (Bchls), a variety of which are now known (Scheer, 1991). However, their biosynthetic pathways will not be addressed specifically in this review, except for work on enzymes of Bchl a synthesis, which may be related to equivalent enzymes involved in the biosynthesis of plant chlorophylls.

A wide variety of other tetrapyrroles occurs in nature, among which are several hemes (the prosthetic groups of cytochromes and other hemoproteins), vitamin $B_{12}$ (precursor of $B_{12}$-coenzyme), "factor $F_{430}$" (the nickel tetrapyrrole of methanogenic bacteria), and phytochromobilin (the linear tetrapyrrole chromophore of higher plant phytochrome). Given such a diverse collection of tetrapyrroles, which are distributed virtually throughout the biosphere, it is indeed remarkable that the biosynthetic pathways of *all* the known tetrapyrroles proceed through the same monopyrrole precursor, porphobilinogen (PBG), as well as its immediate precursor, 5-aminolevulinic acid (ALA), and the first stable tetrapyrrole product to be formed from PBG, uroporphyrinogen III (urogen III). From this initial linear sequence of intermediates, the various tetrapyrroles are formed via a series of branches from the central pathway (Warren and Scott, 1990; Fig. 1). Urogen III is the first branch point intermediate; each of two branches[2] leads from urogen III to one of two secondary branch point intermediates: precorrin 2 or protoporphyrin IX (proto IX). Three different branches stem from precorrin 2; each involves the eventual incorporation of a different metal. One (involving iron) branches to siroheme, another (involving cobalt) to vitamin $B_{12}$, and a third (involving nickel) to factor $F_{430}$. Similarly, two different branches stem from proto IX, each again incorporating a different metal. Incorporation of iron yields protoheme, the precursor of most hemes (except siroheme) and the metal-free linear tetrapyrroles, whereas incorporation of magnesium yields Mg proto

[2] A minor branch also leads to copper uroporphyrin III (found in bird feathers) and, from several other tetrapyrroles, to such specialized natural products as the tetrapyrrole-containing coloring matter found in certain mollusk shells and avian egg shells (cf. Hendry and Jones, 1980).

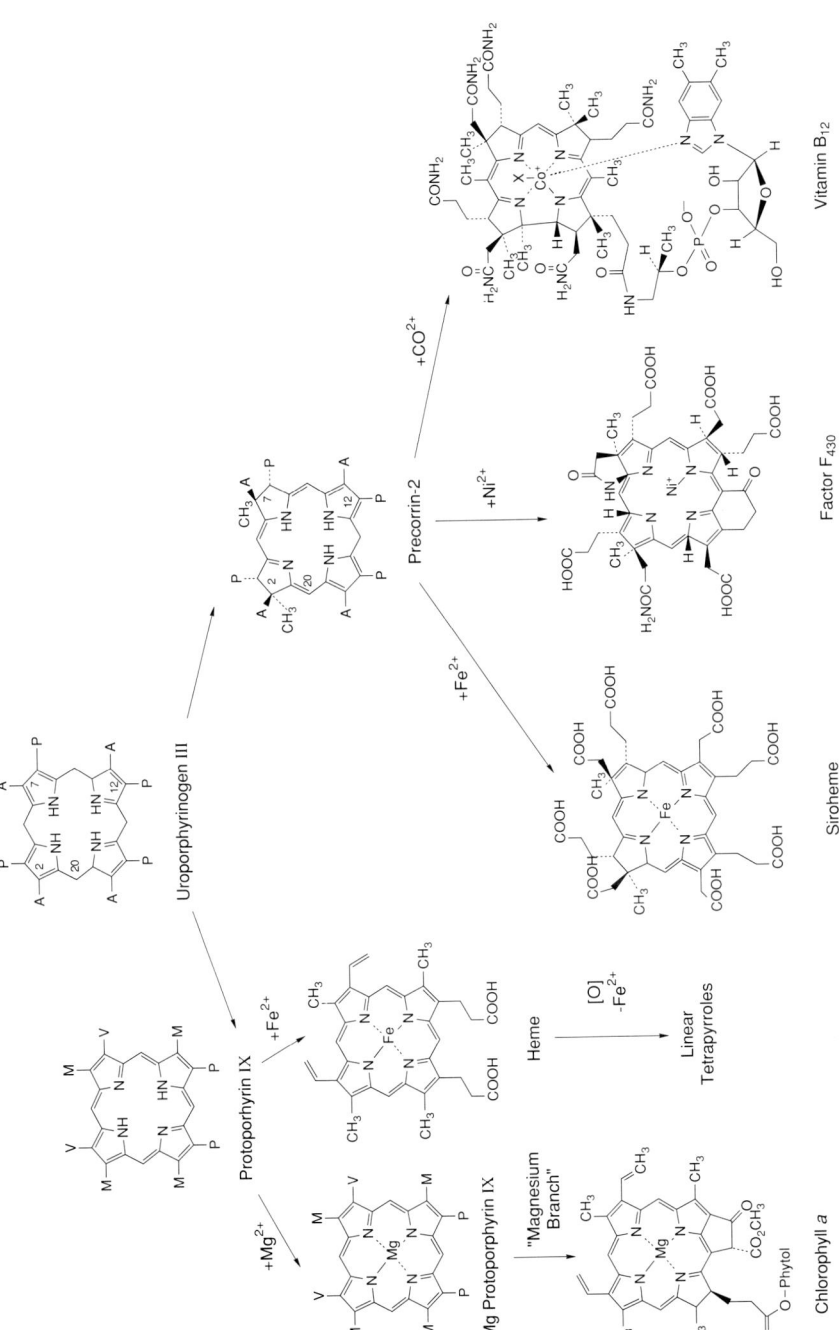

**Fig. 1.** The major branches of tetrapyrrole synthesis leading from uroporphyrinogen III. Adapted from Warren and Scott (1990).

IX, the precursor of all the chlorophylls. This "magnesium branch" is essentially a linear pathway with subsequent branches leading from the central pathway at several points to yield the various plant and bacterial chlorophylls. As discussed in subsequent sections, however, parallel pathways (or "metabolic grids") are likely to be present in the biosynthetic pathways of plant and bacterial chlorophylls.

An important variation exists at the beginning of the tetraphyrrole biosynthetic pathway. Whereas life has chosen only *one* monopyrrole, PBG, as the precursor of all of its tetrapyrroles, it has chosen two distinctly different pathways for the formation of ALA, the immediate precursor of PBG. One pathway, known as the "Shemin pathway" after its discoverer David Shemin (cf. Shemin, 1989), is used by animal mitochondria for the formation of hemes and by purple nonsulfur bacteria for the formation of hemes, Bchl *a*, and vitamin $B_{12}$. This pathway involves condensation of succinyl coenzyme A (succinyl-CoA) with glycine (with concurrent loss of $CO_2$) to form ALA directly. The second pathway (called the "$C_{-5}$ pathway" because L-glutamate supplies all five carbons for the formation of ALA) is used by plant chloroplasts for the formation of ALA for hemes and chlorophylls and by most other photosynthetic bacteria for the formation of all their tetrapyrroles (Section II,B).

Review articles about various aspects of the biosynthesis of plant chlorophylls include those written by Rebeiz and Lascelles (1982), Castelfranco and Beale (1983), Rebeiz *et al.* (1983), Beale (1984, 1990), Battersby (1985, 1986a,b, 1987), Leeper (1985a,b, 1987, 1989, 1991), Rüdiger and Schoch (1988, 1991), Beale and Weinstein (1990, 1991), Dailey (1990), Jordan (1990), Warren and Scott (1990), and Griffiths (1991).

## II. EARLY STAGES OF CHLOROPHYLL SYNTHESIS: THE PATHWAY TO PROTOPORPHYRIN IX

### A. Localization of Enzymes of Protoporphyrin IX Synthesis

The enzymes involved in the early stages of tetrapyrrole synthesis were the first to be studied in detail, perhaps because most were found to be soluble and easily isolated. Proto IX synthesis has been studied in plants and animals, as well as in microorganisms of all types. The intermediates (shown in Fig. 2) in this part of the pathway were assumed to be common for the biosynthesis of hemes and chlorophylls. The enzymes of heme synthesis in animals were found to be localized in two cellular compartments: the mitochondria and the cytoplasm (Fig. 3). The synthesis begins in the mitochondria with the condensation of glycine with succinyl-Co A

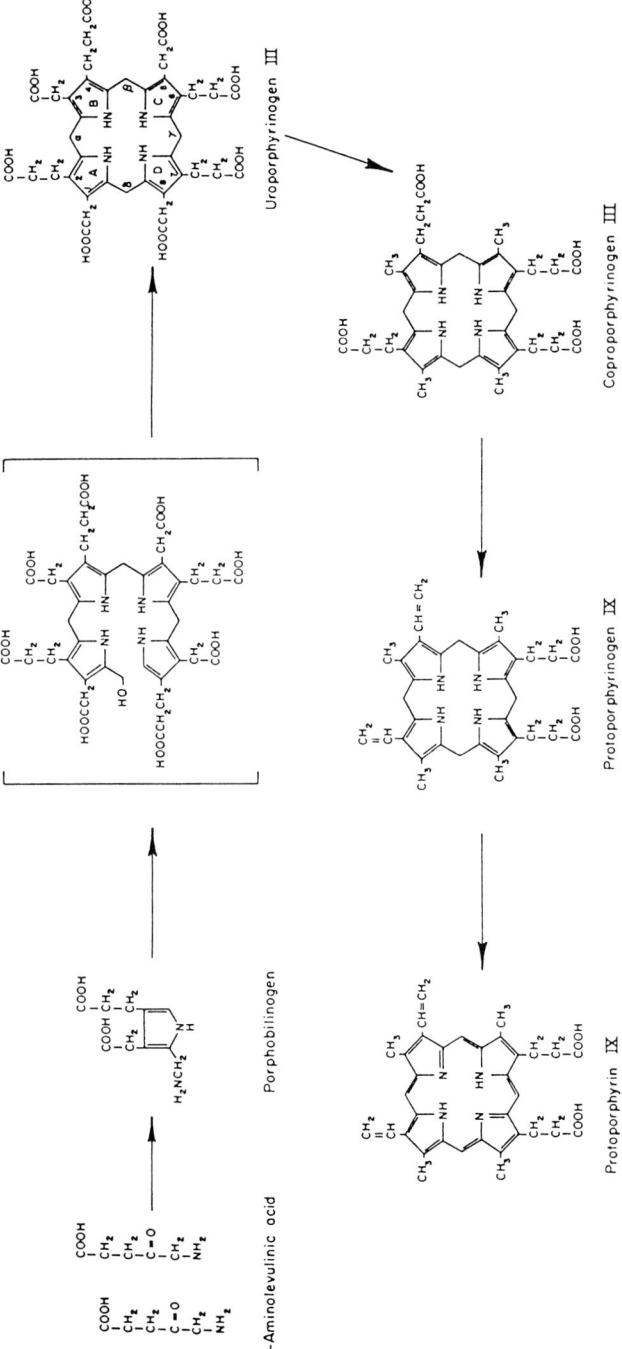

**Fig. 2.** Early stages of chlorophyll synthesis. Conversion of 5-aminolevulinic acid to protoporphyrin IX. Reprinted with permission from Beale (1984).

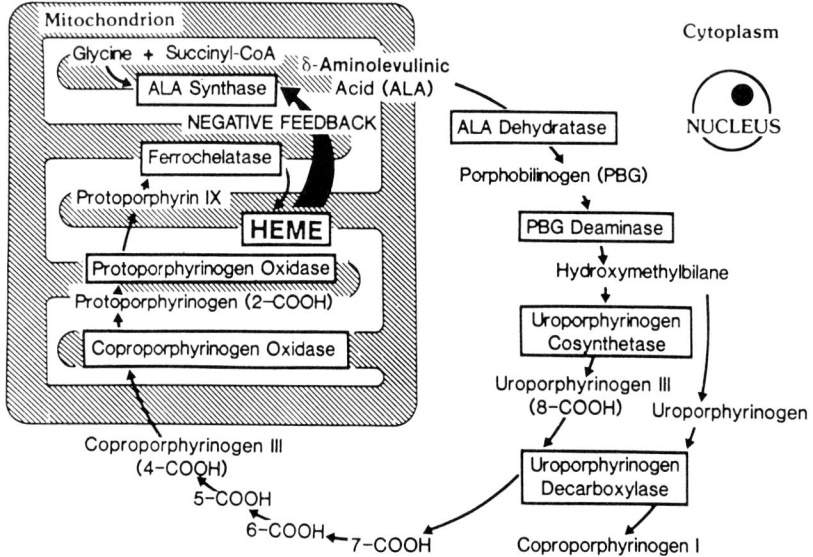

**Fig. 3.** Localization of heme synthesis in animals. Reprinted with permission from Moore (1990). Copyright © 1990 by McGraw-Hill.

by ALA synthase (Jordan, 1990). The product, ALA, passes out of the mitochondria and into the cytoplasm where the next four enzymes are found as soluble proteins. The product of these enzymatic steps, coproporphyrinogen III (coprogen III), returns to the mitochondria where the remaining three enzymes convert it to protoheme. Protoheme is the prosthetic group of cytochrome $b$ and (after addition of cysteine residues of the apoprotein across the double bonds of the two vinyl groups) cytochromes $c$ and $c_1$. Protoheme is also very likely to be the precursor of heme A of cytochrome $aa_3$, although very little is known of its biosynthesis. Some protoheme also must be returned to the cytoplasm where it is incorporated into cytochromes and other hemoproteins of the cytoplasm, endoplasmic reticulum, and peroxisomes.

This particular localization of enzymes provides several distinct advantages to the cell. Mitochondria are the site for the generation of succinyl-CoA by the citric acid cycle and of glycine by serine hydroxymethyltransferase or glycine synthase. As well, the final product, protoheme, acts as a negative feedback inhibitor on the committed-step enzyme ALA synthase (Andrew et al., 1990). The early intermediates, ALA, PBG, urogen III, and coprogen III, are water soluble and are ideal substrates and products for the soluble cytoplasmic enzymes. However, the last two intermediates

### 4. Biosynthesis of the Chlorophyll Chromophore

of the pathway, protoporphyrinogen IX (protogen IX) and proto IX, are dicarboxylates with reduced water solubility; these compounds would not be as suitable for water-soluble enzymes. Hence, in animals at least, the last three enzymes form a complex in the inner mitochondrial membrane. Coprogen oxidase is an easily dissociated extrinsic protein, whereas protogen oxidase and ferrochelatase are firmly bound intrinsic proteins (Dailey, 1990).

The first enzyme of the heme pathway in animals, ALA synthase (EC 2.3.1.37), was demonstrated at quite an early date in the purple nonsulfur bacterium, *Rhodobacter sphaeroides* (previously called *Rhodopseudomonas spheroides*). The early stages of the biosynthesis of hemes and Bchl in this organism were thought to involve common enzymes as well as common intermediates, with the exception that this bacterium can form Bchl under completely anaerobic conditions whereas heme synthesis requires molecular oxygen. Despite the inability of researchers to demonstrate ALA synthase in photosynthetic organisms other than purple nonsulfur bacteria, the idea of common enzymes as well as common intermediates for the early stages of the biosynthesis of hemes and chlorophylls was, nevertheless, still prevalent. Most of the other early stage enzymes had been isolated from higher plants in soluble form. It was generally assumed that, at some point, a common intermediate entered the chloroplast from the cytoplasm to provide a precursor for Chl *a* synthesis.

Two discoveries clarified the confusion about the early stages of Chl *a* synthesis in higher plants. Beale and Castelfranco (1974) demonstrated that L-glutamate (and other 5-carbon compounds such as 2-ketoglutarate and L-glutamine) were precursors of ALA in greening cucumber cotyledons, while Fuesler *et al.* (1984a) demonstrated that chloroplasts *alone* were able to convert glutamate to Chl *a*. Therefore, the chloroplasts must contain all the necessary enzymes for this process. Glutamate was shown to provide all five carbon atoms of ALA, C-1 of glutamate becoming C-5 of ALA (Beale *et al.*, 1975). Etioplasts and chloroplasts of greening barley also incorporated labeled glutamate efficiently into protoheme as well as into Chl *a* (Castelfranco and Jones, 1975). Later, glutamate and other 5-carbon compounds rather than glycine were demonstrated to be used exclusively to form the heme moieties of mitochondrial cytochrome oxidase in the red alga *Cyanidium caldarium* (Weinstein and Beale, 1984) and etiolated maize (Schneegurt and Beale, 1986), as well as the protoheme of peroxidase excreted from peanut cell cultures (Chibbar and van Huystee, 1983). Also, plant microsomal cytochrome P450 is inhibited by gabaculine (an inhibitor of the $C_5$ pathway; Section II,B,2), indicating that its heme is derived from glutamate also (Werck-Reichhart *et al.*, 1988). Therefore, rather than "At what point does a common intermediate enter the chloroplast from the cytoplasm to provide a precursor for Chl *a* synthesis," the question is, "At what point does a common intermediate leave the chloroplast to provide a precursor for nonplastid heme synthesis?"

The last two enzymes of heme synthesis, protogen IX oxidase (Jacobs and Jacobs, 1984, 1987) and ferrochelatase (Porra and Lascelles, 1968; Little and Jones, 1976), have been detected in mitochondria and chloroplasts; each enzyme was found to be associated with membranes in the organelles. Ferrochelatase was analyzed after purification of the two organelles by sucrose density gradient centrifugation; however, cross-contamination by membranes of the alternative organelle had not been excluded completely (Little and Jones, 1976). The soluble enzymes that convert ALA to protogen IX have proved extremely difficult to localize to the chloroplast and/or the cytoplasm. Smith (1988) has studied the distribution of two of these enzymes, ALA dehydratase and PBG deaminase, in pea and in the spadices of *Arum*, where the synthesis of mitochondrial heme is predominant. In both plants, the distribution of these enzymes into various subcellular fractions paralleled the distribution of a soluble chloroplast stromal marker enzyme, but not marker enzymes for the cytoplasm or the mitochondria. These results were consistent with an exclusive plastid location for these two enzymes. Final proof could not be obtained, however, since soluble chloroplast marker enzymes are always released during isolation procedures by breakage of the fragile plastids, and smaller amounts of cytoplasm-specific enzymes could not have been distinguished under these conditions. Hence, although the possibility cannot be excluded that ALA itself, or some other intermediate before protogen IX, is transported from the chloroplast into the cytoplasm, it is more likely that protogen IX is transported from the chloroplast to the mitochondria to complete the synthesis of the protoheme required for nonplastid hemoproteins. Alternatively, heme may be transported from the chloroplast for the synthesis of cytoplasmic hemoproteins. Thomas and Weinstein (1990) have demonstrated that heme efflux can occur from a "free heme" pool inside the chloroplast.

*Euglena gracilis* is the only organism conclusively demonstrated to have *both* the $C_5$ pathway and ALA synthase to date. It may, therefore, have completely separate enzymes and pools of intermediates for the biosynthesis of chloroplast and extrachloroplast tetrapyrroles. The heme A of mitochondrial cytochrome oxidase was found to be formed exclusively from glycine, whereas the tetrapyrroles of chloroplasts were formed exclusively from glutamate (Beale *et al.*, 1981; Weinstein and Beale, 1983). Nevertheless, Shashidhara and Smith (1991) have shown that PBG deaminase is localized in the chloroplast but not the cytoplasm of light-grown *Euglena*, casting some doubt on the complete separation of the heme and chlorophyll biosynthetic pathways in this organism.

### 4. Biosynthesis of the Chlorophyll Chromophore

The localization of enzymes *in* the chloroplast also has been very difficult to study. Smith and Rebeiz (1979) had concluded earlier that all the enzymes for the conversion of ALA to proto IX were soluble stromal enzymes in cucumber, whereas enzymes of the magnesium branch were membrane bound. Castelfranco *et al.* (1988) supported a stromal location for PBG deaminase in cucumber, whereas Nasri *et al.* (1988) found that about two-thirds of the activity of ALA dehydratase was soluble and the remainder was membrane bound in the etiochloroplasts of radish. On the other hand, Carell and Kahn (1964) concluded that, whereas urogen III decarboxylase was soluble, ALA dehydratase was membrane bound in *Euglena*. Lee *et al.* (1991) carried out the osmotic lysis of carefully purified etiochloroplasts of cucumber and reported that nearly 90% of the activity of enzymes converting ALA to proto IX remained with the membrane fraction. Virtually all the activity was released into the supernatant fraction by a high-speed homogenization, however, indicating that these enzymes were associated only loosely with the membrane, perhaps as an extrinsic enzyme complex (Lee *et al.*, 1991).

### B. $C_5$ Pathway of 5-Aminolevulinic Acid Synthesis

The so-called "$C_5$ pathway" subsequently was demonstrated conclusively using $^{13}C$ NMR analysis of Chl *a* formed by *Scenedesmus obliquus* from $^{13}C$-labeled precursors (either glycine or glutamate) by Oh-hama *et al.* (1982). This analysis also was carried out in maize (Porra *et al.*, 1983) and in several species of purple and green sulfur bacteria (Oh-hama *et al.*, 1986a,b; Smith and Huster, 1987). The $C_5$ pathway now is known to be the more common pathway by far, and has been found in all photosynthetic organisms examined to date except the $\alpha$-subgroup of purple nonsulfur bacteria and *Protaminobacter ruber* (Sato *et al.*, 1985a,b), which contain only ALA synthase. The $C_5$ pathway has been found in all higher plants, green and red algae, cyanobacteria, prochlorophytes, purple and green sulfur bacteria, green nonsulfur bacteria, and heliobacteria (Avissar *et al.*, 1989; Beale, 1990). Only in the case of the phytoflagellate *Euglena gracilis* has it been demonstrated unambiguously that both the $C_5$ pathway and ALA synthase are in operation (Section II,A).

The $C_5$ pathway requires four macromolecular components: three enzymes and a small molecular weight RNA molecule. All four components have been separated and shown to be necessary for ALA synthesis from glutamate during reconstitution experiments in barley (Bruyant and Kannangara, 1987) and *Chlorella* (Weinstein *et al.*, 1987). The involvement of a species of RNA was inferred first from studies on the inhibition of *in vitro* ALA synthesis by preincubation of extracts with RNase A (Huang et al., 1984; Kannangara et al., 1984; Weinstein and Beale, 1985b). The RNA species from barley subsequently was isolated, purified, and sequenced by Schön et al., 1986). It proved to be a tRNA$^{Glu}$ with a UUC anticodon in which the first U had been modified to 5-methylaminomethyl-2-thiouridine and the second U to pseudouridine (Schön et al., 1986). This is the only species of tRNA$^{Glu}$ encoded by the chloroplast genome of barley (Kannangara et al., 1984; Schön et al., 1988), which only uses a single codon, GAA, for glutamate (Brown et al., 1990). The tRNA$^{Glu}$ contains a normal 3'-terminal CCA sequence and, when charged with glutamate, must be used for *both* protein and ALA synthesis. In the case of *Synechocystis*, the tRNAs containing UUC anticodons were purified by affinity chromatography directed against UUC (Schneegurt and Beale, 1988) and subsequently separated into two components by high performance liquid chromatography (HPLC; Schneegurt et al., 1988). Both RNA species (which had the same sequences but differed in the nature of their base modifications) were found to participate in protein *and* ALA synthesis (O'Neill et al., 1988). Results by O'Neill and Söll (1990) demonstrated that both tRNA$^{Glu}$ species are formed from a single gene that transcribes a single tRNA precursor.

The $C_5$ pathway, therefore, consists of three enzyme-catalyzed steps (Fig. 4): (1) synthesis of glutamyl-tRNA$^{Glu}$, (2) reduction of the glutamyl residue of the latter to glutamate 1-semialdehyde (GSA) or its equivalent, and (3) net isomerization of GSA to ALA.

## *1. Glutamyl-tRNA Synthetase*

Glutamyl-tRNA synthetases involved in ALA synthesis have been isolated and purified from barley (Bruyant and Kannagara, 1987), *Chlorella* (Weinstein et al., 1987), *Synechocystis* (Rieble and Beale, 1989), and *Chlamydomonas* (Chen et al., 1990a). Most were similar, if not identical, to synthetases participating in protein synthesis, exhibiting $M_r$ values of 54, 73, 63, and 62 kDa, respectively. In *Chlamydomonas*, the authors reported that the synthetase was a monomer; however, Chang et al. (1990) reported that they had isolated a synthetase from *Chlamydomonas* that was a 60-kDa dimer composed of two 32.5-kDa subunits. These authors also reported that their synthetase was inhibited by heme, opening the possibility that its regulation was specific to ALA synthesis. In barley, only one synthetase was isolated that was able to charge with glutamate the only tRNA$^{Glu}$ and both tRNA$^{Gln}$ species present in this organism. Only the product of tRNA$^{Glu}$ activation was used for ALA synthesis, whereas the resulting glutamyl-tRNA$^{Gln}$ products subsequently were γ-amidated to their corresponding glutaminyl derivatives (Bruyant and Kannagara,

## 4. Biosynthesis of the Chlorophyll Chromophore

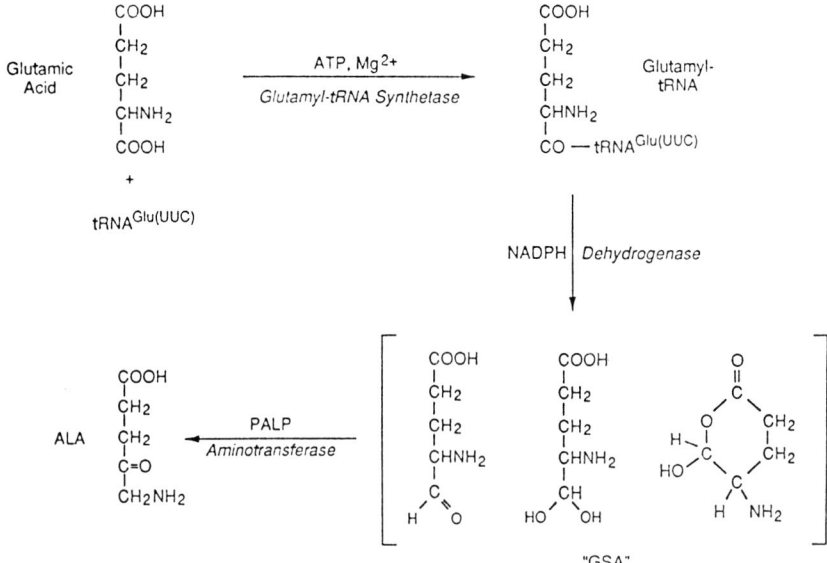

**Fig. 4.** The $C_5$ pathway of 5-aminolevulinic acid (ALA) synthesis in plants. Reprinted with permission from Beale (1990).

1987). In *Chlorella,* several synthetases were present, but only one produced a glutamyl-tRNA used in ALA synthesis (Avissar and Beale, 1988). Using a reconstituted system lacking the synthetase, these authors demonstrated that glutamyl-tRNA and not free glutamate was the true precursor of ALA.

All aminoacyl-tRNA synthetases require ATP, which is converted to AMP plus $PP_i$ during the reaction. However, unlike most other synthetase reactions, the participation of a glutamyl-AMP intermediate in the glutamyl-tRNA synthetase reaction has not been demonstrated. For rat liver glutamyl-tRNA synthetase, failure to observe catalysis of ATP-$PP_i$ exchange by the enzyme in the absence of $tRNA^{Glu}$ suggests that the enzyme does not involve glutamyl-adenylate as an intermediate (Deutscher, 1967). Glutamyl-adenylate may be unstable because of the proximity of the $\gamma$-carboxyl group (Avissar and Beale, 1988), whereas the glutamyl-tRNA ester somehow is protected by the rest of the tRNA molecule (a necessary requirement if glutamate is to be incorporated into proteins). This result suggests that glutamyl-tRNA may be involved in ALA synthesis because it is the only stable $\alpha$-activated glutamate derivative available.

## 2. Glutamyl-tRNA Reductase

Glutamyl-tRNA reductases (sometimes called "dehydrogenases") also have been isolated and purified from barley (Bruyant and Kannagara, 1987), *Chlorella* (Weinstein *et al.*, 1987), and *Chlamydomonas* (Chen *et al.*, 1990b; Krishnasamy and Wang, 1990). In *Chlamydomonas*, the enzyme had an $M_r$ of 130 kDa and was stimulated by the readdition of purified glutamyl-tRNA synthetase, suggesting the possibility of complex formation between it and the synthetase *in vivo*. The coenzyme requirement was found to be quite variable between species: in *Chlorobium*, NADPH was used nearly exclusively (Rieble *et al.*, 1989); in *Chlorella*, NADPH was twice as effective as NADH (Weinstein and Beale, 1985a); in *Euglena*, the two coenzymes were equally effective (Mayer *et al.*, 1987); and in *Synechocystis*, NADPH was more effective than NADH at low concentration whereas the reverse was true at high concentration (Rieble and Beale, 1988). The substrate has been shown to be glutamyl-tRNA$^{Glu}$ in reconstitution experiments (Avissar and Beale, 1988); however, the product of the reduction is less certain. Wang *et al.* (1981) had shown earlier that the product of the reductase co-separated chromatographically with a substance initially thought to be GSA (Kannangara and Gough, 1978). The same substance was shown to be generated by dilute acid hydrolysis of the diethyl acetal of GSA synthesized by Houen *et al.* (1983). However, this substance did not have the physical properties of free GSA, which would be expected to be extremely unstable on chemical grounds. Gough *et al.* (1989) have characterized the major product of the ozonolysis of 4-amino-5-hexenoic acid as the hydrate of GSA and shown it to be convertible to ALA by the next enzyme in the pathway. On the other hand, a cyclic form of GSA, a lactone formed between the γ-carboxyl and the hydrated aldehyde (Fig. 4) that has been named 2-hydroxy-3-aminotetrahydropyran-1-one (HAT), has been proposed to be the actual intermediate in ALA synthesis (Jordan, 1990). Derivatives of GSA with the aldehyde protected in some other way are still possible, for example, in the form of an aziridinolpropionate (Kannangara *et al.*, 1988) or still bound to the tRNA as a 2',3'-acetal or a 3'-hemiacetal (Rüdiger and Schoch, 1988). If, for example, a 3'-hemiacetal of tRNA was formed as the first product of the reduction reaction, it subsequently could be broken down by attack of the γ-carboxyl to form the lactone HAT.

## 3. Glutamate 1-Semialdehyde Aminotransferase

GSA aminotransferases (EC 5.4.3.8) have been isolated and purified from barley and *Synechococcus* (Grimm *et al.*, 1989), *Synechocystis* (Rieble and Beale, 1989), *Chlorella* (Avissar and Beale, 1989), and *Chlamy-*

## 4. Biosynthesis of the Chlorophyll Chromophore

*domonas* (Mau and Wang, 1990; Jahn *et al.*, 1991). These enzymes exhibited $M_r$ values of 80, 46, 99, 60, and 51 or 43 kDa, respectively. In barley, the gene for the aminotransferase has been sequenced from a cDNA clone and the inferred mature protein was found to have a molecular mass of 46,172 Da (Grimm, 1990); hence, the active barley enzyme is a dimer, whereas the *Synechococcus* enzyme is a monomer (Grimm *et al.*, 1989). The aminotransferase requires no substrates other than GSA; hence, the reaction represents a net isomerization of GSA to ALA rather than a transamination involving other amino donors or acceptors. The enzyme has, in fact, been classified as an isomerase (EC Class 5). Although first reported to require no added cofactors, the enzyme from *Chlorella* has been shown to be stimulated by the addition of low concentrations (0.1–3.0 $\mu M$) of pyridoxal 5'-phosphate (PALP) (Avissar and Beale, 1989). In barley, however, PALP was reported to be inhibitory whereas pyridoxamine 5'-phosphate (PAMP) was stimulatory (Kannangara and Gough, 1978). In *Cyanidium*, both PALP and PAMP were stimulatory at low concentrations (< 100 $\mu M$), whereas PAMP was stimulatory and PALP was inhibitory at higher concentrations (Houghton *et al.*, 1989). The aminotransferase also is inhibited by gabaculine (3-amino-2,3-dihydrobenzoic acid), a neurotoxin known to inhibit γ-aminobutyric acid aminotransferase by forming a stable adduct with its coenzyme PALP (Rando, 1977). Houghton *et al.* (1989) have found gabaculine to be a more effective inhibitor of the *Cyanidium* aminotransferase than its proposed PALP adduct *N*-[*m*-carboxyphenyl]pyridoxamine 5'-phosphate; however, the inhibition by gabaculine could be overcome by the addition of excess PAMP.

The results just described suggest that PAMP is the actual form of the coenzyme at the start of the catalytic cycle and that the enzyme may act by a mechanism shown in Fig. 5. In this mechanism (Hoober *et al.*, 1988),

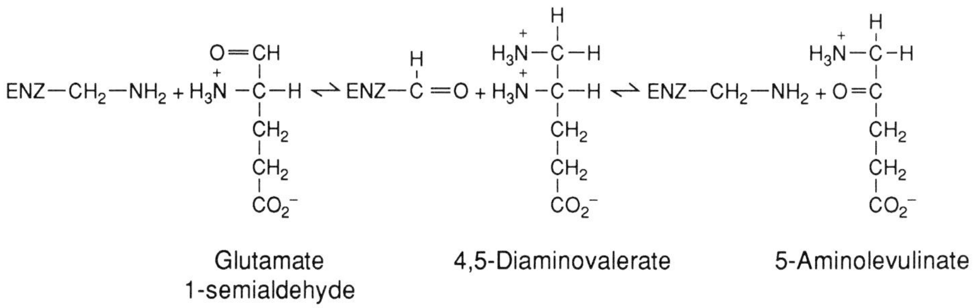

$$\text{ENZ-CH}_2\text{-NH}_2 + \text{H}_3\overset{+}{\text{N}}-\underset{\underset{\underset{\text{CO}_2^-}{|}}{\underset{\text{CH}_2}{|}}}{\overset{\overset{\text{O=CH}}{|}}{\text{C}}}-\text{H} \rightleftarrows \text{ENZ}-\overset{\overset{\text{H}}{|}}{\text{C}}=\text{O} + \text{H}_3\overset{+}{\text{N}}-\underset{\underset{\underset{\text{CO}_2^-}{|}}{\underset{\text{CH}_2}{|}}}{\overset{\overset{\text{H}_3\overset{+}{\text{N}}-\overset{|}{\text{C}}-\text{H}}{|}}{\text{C}}}-\text{H} \rightleftarrows \text{ENZ-CH}_2\text{-NH}_2 + \text{O}=\underset{\underset{\underset{\text{CO}_2^-}{|}}{\underset{\text{CH}_2}{|}}}{\overset{\overset{\text{H}_3\overset{+}{\text{N}}-\overset{|}{\text{C}}-\text{H}}{|}}{\text{C}}}$$

  Glutamate          4,5-Diaminovalerate         5-Aminolevulinate
  1-semialdehyde

**Fig. 5.** Proposed mechanism for the formation of 5-aminolevulinic acid by glutamate 1-semialdehyde aminotransferase. The mechanism predicts that pyridoxamine 5'-phosphate (-CH$_2$-NH$_2$) is the form of the coenzyme at the start of the catalytic cyle. Reprinted with permission from Hoober *et al.* (1988).

PALP is shown as an intermediate in the catalytic cycle, which might be sensitive to inhibition by gabaculine at this point. Other inhibitors (e.g., 4-amino-5-hexynoic acid) resemble the open-chain form of GSA and may inhibit the cycle by becoming covalently bound to the active site of the protein (Beale, 1990). The mechanism in Fig. 5 predicts 4,5-diaminovaleric acid (rather than 4,5-dioxovaleric acid) as an intermediate, as well as *intermolecular* (rather than an intramolecular) transfer of amino groups. This mechanism would be consistent with the results of a study in which a mixture of glutamate molecules separately labeled with either $^{13}C$ or $^{15}N$ yielded a significant proportion of ALA molecules labeled with both isotopes (Mau and Wang, 1988). In this study, however, the isotopic stability of the product, ALA, was not measured, leading to some uncertainty in the interpretation of results that otherwise clearly would indicate an intermolecular amino transfer (Beale, 1990).

## C. Conversion of 5-Aminolevulinic Acid to Porphobilinogen

ALA dehydratase (EC 4.2.1.24) or, more informatively, PBG synthase converts two molecules of ALA to the monopyrrole PBG with the elimination of two molecules of water. The enzyme first was isolated from animal sources in 1955 and from *Rhodobacter sphaeroides* several years later. The animal enzymes are zinc metalloenzymes that exist as octamers with $M_r$ values of about 280 kDa. Although it was first thought that four $Zn^{2+}$ were bound per octamer (Hasnain *et al.*, 1985), evidence has accumulated that eight $Zn^{2+}$ are bound per octamer. Three cysteines and one histidine are involved in binding the zinc in a region of the monomer that contains four conserved cysteine and two conserved histidine residues in the three species examined (summarized by Jordan, 1990). The zinc ions do not participate in binding the substrate ALA molecules (Hasnain *et al.*, 1985); however, all eight are required for maximum activity (Jordan, 1990), although there seem to be only four catalytic units per octamer (Jaffe and Hanes, 1986). The enzyme from *R. sphaeroides* first was reported to be a hexamer with an $M_r$ of 240 kDa that required $K^+$ rather than $Zn^{2+}$ for activity (van Heyningen and Shemin, 1971). However, evidence obtained by Gurne *et al.* (1977) suggested that it, too, was an octamer. The *hemB* gene of *R. sphaeroides* was cloned and sequenced by Delaunay *et al.* (1991) and found to encode a monomer of 39 kDa. The expressed gene product was immunoreactive with antibodies raised against the ALA dehydratases of higher plants. Several of these enzymes have been isolated from various plant and algal sources, including wheat (Nandi and Waygood, 1967), tobacco (Shetty and Miller, 1969), radish (Shibata and Ochiai,

1977), *Chlorella* (Tamai *et al.*, 1979), and spinach (Liedgens *et al.*, 1980,1983). The spinach enzyme has been purified using monoclonal antibodies and appears to be a hexamer with an $M_r$ of 300 kDa that requires $Mg^{2+}$ rather than $Zn^{2+}$ or $K^+$ for activity. The enzyme was localized to the plastids of both pea (Smith, 1988) and radish (Nasri *et al.*, 1988). The nuclear gene for the ALA dehydratase of pea has been isolated and sequenced; it reportedly lacks the $Zn^{2+}$-binding domains characteristic of the animal enzymes (Li *et al.*, 1991).

A mechanism for the action of ALA dehydratase was proposed first by Shemin and co-workers (Nandi and Shemin, 1968; Shemin, 1976; Barnard *et al.*, 1977). They determined that ALA formed a Schiff base with the enzyme and that certain inhibitors (e.g., levulinic acid) competed with ALA for Schiff base formation. Treatment of the enzyme (plus ALA) with sodium borohydride led to its irreversible inactivation. The site of Schiff base formation subsequently has been identified as a lysine residue in the *R. sphaeroides* enzyme (Nandi, 1978) and in animal enzymes (Gibbs and Jordan, 1986). After cDNA sequences became known, the residue was identified further as lysine 252 in human and rat and lysine 247 in *Escherichia coli* (summarized by Jordan, 1990). In the mechanism proposed by Shemin and co-workers, the ALA molecule that becomes the acetic acid side (or "A-side") of the PBG rather than the one that becomes the propionic acid side (or "P-side") forms the Schiff base with the enzyme. This deduction was based on the fact that levulinic acid also reacts with ALA to form a "mixed" PBG derivative (desamino-PBG), in which the levulinic acid has formed the A-side and ALA has formed the P-side. However, Jordan and co-workers, in an elegant series of experiments, used $^{13}C$- or $^{14}C$-labeled ALA to demonstrate that the opposite is true: the ALA molecule that becomes the P-side of PBG forms the Schiff base with the enzyme (Jordan and Seehra, 1980a,b; Jordan and Gibbs, 1985; Jordan, 1990). Their mechanism is shown in Fig. 6. In this mechanism, the A-side ALA molecule forms a Schiff base with the free amino group of the covalently bound P-side ALA, thus activating the former for an aldol condensation into the Schiff base of the latter. Work with ALA labeled with $^3H$ at $C_5$ has demonstrated that the *pro* $_R$ hydrogen is removed selectively by the enzyme (Abboud and Akhtar, 1976; Chaudhry and Jordan, 1976). In addition to the lysine residue, Fig. 6 indicates at least two additional residues in the active site that participate in general base catalysis. Two histidine residues have been implicated as essential in the animal enzyme by photoinactivation studies (Tsukamoto *et al.*, 1979). In addition, the animal and *R. sphaeroides* enzymes contain cysteine residues (not involved in zinc binding) that are essential for activity (Hasnain *et al.*, 1985) and may participate as proton donors or acceptors. The spinach

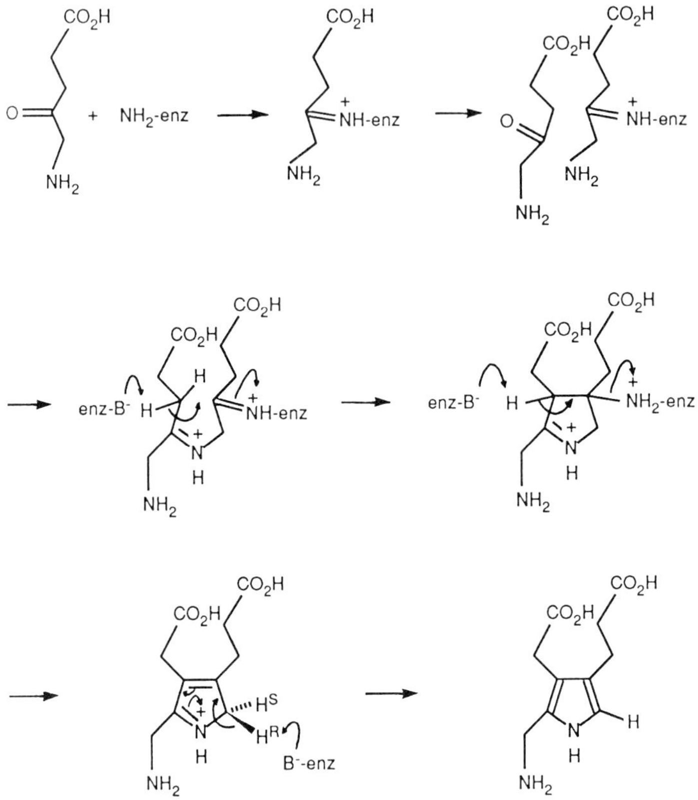

Fig. 6. Proposed mechanism for the formation of porphobilinogen by 5-aminolevulinic acid dehydratase. Reprinted with permission from Jordan (1990). Copyright © 1990 by McGraw-Hill.

enzyme is not inhibited by iodoacetamide and, therefore, has no essential cysteines in its active site; however, one arginine residue per ALA appears to be involved in binding ALA via ion-pair formation with the charged carboxyl group of ALA (Liedgens *et al.*, 1983).

### D. Conversion of Porphobilinogen to Uroporphyrinogen III

Two enzymes work in concert to convert four molecules of PBG to urogen III with the elimination of four molecules of ammonia. PBG can undergo a nonenzymatic acid-catalyzed polymerization to form all four possible urogen isomers of types I, II, III, and IV in the statistically

## 4. Biosynthesis of the Chlorophyll Chromophore

expected ratio of $1:1:4:2$. The statistical predominance of the type III isomer of abiotically generated urogen may have caused early life forms to use it as the precursor of protoheme and, eventually, all other important tetrapyrroles, including chlorophylls, vitamin $B_{12}$, siroheme, and factor $F_{430}$. Other factors also may have been an advantage to early life, for example, the fact that the protoheme derived from the type III isomer has its two charged carboxylates close together, making them able to bind in a more polar region of a hemoprotein (leaving the rest of the macrocycle able to exist in more hydrophobic regions). In any case, once it became advantageous to use the type III isomer of urogen, early organisms required an enzyme system capable of synthesizing such an isomer when the naturally occurring supply began to run out. The discovery of exactly how living organisms accomplish this feat has occupied the interest of a large number of biochemists for more than 30 years. Current knowledge of this mechanism has been the result of a lot of hard work, tremendous insight, and beautifully conceived and executed research. Nevertheless, the literature is strewn with numerous incorrect hypotheses and experimental dead ends. Although a good many details of the mechanism still remain to be elucidated, including the enigmatic "switch" mechanism by which pyrrole ring D in a type I isomer is turned around to form a type III isomer, the discoveries made to date about this remarkable process have been awe inspiring!

Before it was known that two enzyme were involved in the conversion of PBG to urogen III, the corresponding catalytic activity was referred to as "porphobilinogenase." Then it was discovered that heated extracts of spinach converted PBG to urogen I instead of urogen III, indicating that a heat-stable enzyme was responsible for "urogen I synthase" activity whereas a heat-labile enzyme was responsible for "urogen isomerase" activity. The two proteins subsequently were separated by Bogorad (1958a,b), who demonstrated that the "isomerase" could not actually convert urogen I to urogen III and must, therefore, use some other (probably very unstable) intermediate produced by the heat-stable enzyme to form urogen III. In the absence of the heat-labile enzyme, the unstable intermediate must be converted spontaneously to urogen I. The heat-stable enzyme became known as PBG deaminase (EC 4.3.1.8) and the heat-labile enzyme as urogen III cosynthase (EC 4.2.1.75) because it was thought to have activity only in the presence of the deaminase.

PBG deaminases have been isolated from *Rhodobacter sphaeroides* (Davies and Neuberger, 1973; Jordan and Shemin, 1973), wheat germ, spinach (Higuchi and Bogorad, 1975), *Chlorella* (Shioi et al., 1980), *Euglena* (Williams et al., 1981), and pea (Castelfranco et al., 1988). The enzymes all exhibited similar $M_r$ values of 36, 40, 40, 35, 41, and 43 kDa, respectively. The PBG deaminase from pea was purified to homogeneity on Reactive Red 120 Sepharose and found to be a monomer of 36–45 kDa (Spano and Timko, 1991). The deaminase has been localized in the chloroplasts of developing pea leaves (Castelfranco et al., 1988; Smith, 1988) and in the chloroplast but not the cytoplasm of light-grown *Euglena* (Shashidhara and Smith, 1991). A similar molecular mass (33,857 Da) was deduced from sequence studies of the cloned *Escherichia coli hemC* gene for PBG deaminase (Thomas and Jordan, 1986), which showed a high degree of homology with the cDNA clone for human PBG deaminase (Raich et al., 1986). Kotler et al. (1987), however, have isolated a deaminase from *Rhodopseudomonas palustris* that appeared to have an $M_r$ (74 kDa) that was twice that for previously isolated enzymes.

Urogen III cosynthase was isolated first from wheat germ by Bogorad (1958b) and found to be extremely unstable. The enzyme subsequently was purified from wheat germ by Higuchi and Bogorad (1975) and from *Euglena* by Hart and Battersby (1985). In *Euglena*, it was found to have an $M_r$ of 31 kDa, similar to the molecular mass (27,766 Da) deduced for the gene product of the cloned *hemD* gene (urogen III cosynthase) of *E. coli* (Jordan et al., 1988a). In contrast to the results for the deaminase, however, the sequence of the cDNA clone for the human cosynthase (Tsai et al., 1988) has very little homology with the sequence of the cloned *E. coli* gene (Sasarman et al., 1987; Jordan et al., 1987). The human cosynthase has an essential histidine residue (Frydman and Feinstein, 1974) that is missing in the *Euglena* enzyme.

The most likely candidates for the putative intermediate between the two enzymes were linear oligopyrroles. Such compounds, with two, three, or four pyrrole rings (termed dipyrrylmethanes, tripyrranes, and bilanes, respectively), were found to accumulate during incubations of PBG deaminase with PBG in the presence of $NH_3$, $NH_2OH$, or $NH_2OCH_3$ (Pluscec and Bogorad, 1970; Stella et al., 1971; Radmer and Bogorad, 1972; Davies and Neuberger, 1973). These observations resulted in the testing of a number of dipyrrylmethanes (Frydman et al., 1971,1978a; Osgerby et al., 1972), tripyrranes (Frydman et al., 1978b), and bilanes (Batlle and Rossetti, 1977; Battersby et al., 1977, 1978a,b; Diaz et al., 1979) for activity in enzyme extracts. Most of the oligopyrroles tested had aminomethyl groups in one of the $\alpha$ positions; a shorthand version of their structures is given in Tables I and II. None of the aminomethyldipyrrylmethanes or aminomethyltripyrranes was a substrate of the combined deaminase/cosynthase system or either enzyme alone. When incubated with PBG, some of the pyrrole compounds were incorporated into urogens (as indicated in Table I), but all were much poorer substrates than PBG alone and many acted as inhibitors of the incorporation of PBG into urogen III. Studies with synthetic aminomethylbilanes, in which each of the four pyrrole rings

**TABLE I**

**Formation of Uroporphyrinogens from Dipyrrylmethanes and Tripyrranes**

| Compound tested[a] | Incubation with PBG | Results of incubation with deaminase and cosynthase |
|---|---|---|
| Aminomethyldipyrrylmethanes[b] | | |
| $H_2NCH_2$—AP—AP | + | Formed all urogen I |
| $H_2NCH_2$—**PA**—AP | + | Formed 7% yield of urogen III |
| $H_2NCH_2$—AP—**PA** | + | Inhibited PBG incorporation |
| $H_2NCH_2$—**PA**—**PA** | + | Inhibited PBG incorporation |
| Aminomethyltripyrranes[c] | | |
| $H_2NCH_2$—AP—AP—AP | + | Formed all urogen I |
| $H_2NCH_2$—**PA**—AP—AP | + | Inhibited PBG incorporation |
| $H_2NCH_2$—AP—**PA**—AP | + | Formed small amount of urogen III |
| $H_2NCH_2$—**PA**—**PA**—AP | + | Formed 16% yield of urogen III |

[a] A and P refer to acetic and propionic acid groups in the $\beta$ positions of pyrrole rings; for rings in which the normal order of these groups is reversed, the letters are boldfaced.
[b] Frydman et al. (1971, 1978a).
[c] Frydman et al. (1978b).

was reversed in turn (Table II), demonstrated that all but the bilane in which the first pyrrole ring was reversed could act as substrates, but only of the combined enzyme system and not of the cosynthase alone (Battersby et al., 1977, 1978a, 1982). Analysis of the product formed from labeled aminomethylbilanes by $^{13}$C NMR confirmed that the fourth pyrrole ring was reversed in urogen III formation (Battersby et al., 1978b). Also, in pulse-labeling experiments with either $^{14}$C-labeled PBG (Jordan and Seehra, 1979; Seehra and Jordan, 1980) or $^{13}$C-labeled PBG (Battersby et al., 1979b), it was determined that the order of addition of pyrrole rings was A > B > C > D, rather than the reverse.

The detection and identification of the actual intermediate (sometimes called preuroporphyrinogen) between the deaminase and the cosynthase occurred nearly simultaneously in two laboratories (Battersby et al., 1979a; Burton et al., 1979a,b; Jordan et al., 1979). Several detailed accounts of the identification of this intermediate as a hydroxymethylbilane (HMB)—HOCH$_2$-AP-AP-AP-AP—have appeared (Battersby, 1985,1986b; Leeper, 1985a; Jordan, 1990; Warren and Scott, 1990). HMB was demonstrated to be the true substrate of the cosynthase (Battersby et al., 1979a; Jordan et al., 1979). Hence, the prefix "co" has been dropped from the name of the enzyme, which is now known as urogen III synthase. Further, PBG deaminase now may be referred to more accurately as HMB synthase. Also the reason aminomethylbilane required both enzymes for its conversion to urogen III was demonstrated to be that the deaminase

**TABLE II**

**Formation of Uroporphyrinogens from Bilanes**

| Compound tested[a] | Incubation with PBG | Percentage of Urogen isomer with ring D | | Reversal[b] (%) | Relative rate |
|---|---|---|---|---|---|
| | | Unreversed | Reversed | | |
| Aminomethylbilanes[c] | | | | | |
| None | + | 0 (I) | 100 (III) | — | 1562 |
| $H_2NCH_2$—AP—AP—AP—AP | — | 8 (I) | 90 (III) | — | 100 |
| $H_2NCH_2$—**PA**—AP—AP—AP | — | — | — | — | — |
| $H_2NCH_2$—AP—**PA**—AP—AP | — | 90 (III) | 8.5 (II) | — | 39 |
| $H_2NCH_2$—AP—AP—**PA**—AP | — | 90 (III) | 7.5 (IV) | — | 51.+ |
| $H_2NCH_2$—AP—AP—AP—**PA** | — | 76 (III) | 22 (I) | — | 12 |
| $H_2NCH_2$—**PA**—**PA**—**PA**—AP | — | — | — | — | — |
| Hydroxymethylbilanes[d] | | | | | |
| $HOCH_2$—AP—AP—AP—AP | — | 8 (I) | 92 (III) | >98 | 100 |
| $HOCH_2$—AP—**PA**—AP—AP | — | — | — | — | <0.5 |
| $HOCH_2$—AP—AP—**PA**—AP | — | 49 (III) | 51 (IV) | >95 | 5 |
| $HOCH_2$—AP—AP—AP—**PA** | — | 64 (III) | 35 (I) | 45 | 13 |
| $HOCH_2$—**PA**—**PA**—**PA**—AP | — | — | — | — | <0.5 |

[a] A and P refer to acetic and propionic acid groups in the β positions of pyrrole rings; for rings in which the normal order of these groups is reversed, the letters are boldfaced.
[b] Corrected for chemical cyclization.
[c] Incubated with both deaminase and cosynthase; Battersby et al. (1977, 1978a, 1982).
[d] Incubated with cosynthase only; Battersby et al. (1979, 1981b).

### 4. Biosynthesis of the Chlorophyll Chromophore

was necessary to convert it to HMB. Battersby *et al.* (1981b) have prepared several synthetic analogs of HMB (in which each of the four pyrrole rings was reversed) to test as substrates for the synthase. The results were similar to those observed for the synthetic aminomethylbilanes (Table II), except that ring B also must have the correct orientation. However, it was observed that the hydroxymethylbilanes HOCH$_2$-AP-AP-**PA**-AP and HOCH$_2$-AP-AP-AP-**PA** actually yielded the "reversed ring D" isomers urogen IV and urogen I, respectively, in fairly high yield (Table II). Therefore, reversal of ring D occurs regardless of its initial orientation!

Attention then turned to the mechanism of synthesis of HMB. Since the early experiments of Pluscec and Bogorad (1970) and Davies and Neuberger (1973), it had been recognized that intermediates formed from the successive condensation of four PBG molecules probably were bound covalently to the enzyme. Subsequently it was demonstrated that, during incubation of HMB synthase with PBG, intermediate forms of the enzyme with 1–4 bound substrate molecules (referred to as ES$_1$, ES$_2$, ES$_3$, and ES$_4$) could be separated by anion-exchange chromatography or gel electrophoresis and shown to be catalytically active (Anderson and Desnick, 1980; Berry *et al.*, 1981). The substrate molecules were not released by treatment with SDS, indicating that they were bound covalently (Jordan and Berry, 1981). The substrate molecules were not released by treatment with SDS, indicating that they were bound covalently (Jordan and Berry, 1981). Several proposals for the nature of the amino acid attachment site were made, involving lysine (Hart *et al.*, 1984), arginine (Russell *et al.*, 1984), and cysteine (Evans *et al.*, 1986). None of these amino acids proved to be involved in the *direct* covalent linkage to the substrate; however, an indirect linkage to cysteine through a unique cofactor was present. The other amino acids may be required for noncovalent interactions with the substrate.

Three laboratories worked independently and simultaneously on the nature of the covalent linkage. All provided key information for the eventual elucidation of its structure (Hart *et al.*, 1987; Jordan and Warren, 1987; Scott *et al.*, 1988a). The work summarized here relied on large quantities of HMB synthase available from strains of *E. coli* genetically engineered to overproduce the *E. coli hemC* gene (Thomas and Jordan, 1986; Jordan *et al.*, 1988b). A covalently bound cofactor was detected that gave rise to urogen I under the experimental conditions of peptide mapping and gave color reactions typical of a dipyrrylmethane. A study of the ES$_1$ complex derived from [11-$^{13}$C]PBG by NMR indicated a pyrrole–[$^{13}$CH$_2$]-pyrrole linkage. As well, the ES$_2$ complex gave a color reaction indicative of a bilane. Incubation of the enzyme in the presence of labeled PBG resulted in no incorporation of label into the cofactor. Conversely, growth of the recombinant *E. coli* strain in labeled ALA resulted in an enzyme with a labeled cofactor that was not incorporated into product when subsequently incubated with unlabeled PBG.

Further $^{13}$C NMR work established that the dipyrrylmethane cofactor was linked through a sulfur atom (Beifuss *et al.*, 1988; Hart *et al.*, 1988; Jordan *et al.*, 1988c) that was identified as cysteine 242 of the *E. coli* enzyme (Jordan *et al.*, 1988c; Miller *et al.*, 1988; Scott *et al.*, 1988a,b). The structure of the cysteine-linked dipyrrylmethane cofactor is shown in Fig. 7; further details on the determination of its structure are reviewed by Hart *et al.* (1990) and Jordan (1990). The dipyrrylmethane cofactor has been confirmed to be present in purified HMB synthase isolated from *Euglena* (Hart *et al.*, 1987), spinach, and barley (Warren and Jordan, 1988). A "two-site" mechanism for the assembly and release of HMB by the synthase is shown in Fig. 7 (Jordan, 1990).

Studies on the mechanism of urogen III synthase have been hampered because of the instability of the enzyme. This problem may have been alleviated because of the availability of overproducing strains of *E. coli* containing the cloned *hemD* gene. Some evidence suggests that PBG deaminase and urogen III synthase form a complex *in vivo* that may help stabilize the second enzyme. Erythrocyte PBG deaminase immobilized on Sepharose columns associates with the synthase in the absence of PBG, but forms urogen III when PBG is added (Frydman and Feinstein, 1974). The deaminase is an extremely slow enzyme with a turnover rate of one HMB produced from four PBGs every 2 sec. The synthase is much faster, with a turnover rate of about 200 sec$^{-1}$. Hence, the deaminase would be rate limiting; the extremely unstable HMB could be passed from the deaminase to the synthase to be used as fast as it was generated without ever appearing free in the medium (Beale, 1984).

Of the many mechanisms that have been proposed for urogen III synthase over the years, the one involving a spiro intermediate proposed 30 years ago by Mathewson and Corwin (1961) may yet prove to be essentially correct. The difficult task of synthesizing such an intermediate to test its biological activity was undertaken by Battersby and co-workers (Stark *et al.*, 1986; Battersby, 1987; Leeper, 1989). The original spiro intermediate proposed by Mathewson and Corwin (1961) showed the three pyrrole rings in the macrocycle to be in their alternative tautomeric pyrrolenine forms

---

**Fig. 7.** Proposed mechanism for the formation of hydroxymethylbilane by porphobilinogen deaminase. Successive condensations of substrate (S) in the S-site yields first the dipyrrylmethane cofactor and then the intermediates ES through ES$_4$. The reaction terminates by elimination of hydroxymethylbilane and regeneration of the cofactor in the C-site. Reprinted with permission from Jordan (1990). Copyright © 1990 by McGraw-Hill.

(**I,** Fig. 8). These authors thought that the increased number of $sp^3$ carbons would give the maximum amount of flexibility to the macrocycle. During the course of the synthetic work, a spiro-dicyanide (**III**) was formed; X-ray crystal structure analysis of it (Battersby, 1987) revealed that each of the three pyrrole rings was coplanar with its adjacent methylene carbons (cf. Fig. 8). Thus, the spiro intermediate probably existed in its aromatic pyrrole form (**II**) rather than as the pyrrolenine (**I**). Finally, two isomers of the spiro-lactam (**IV**) were synthesized and separated (Stark *et al.*, 1986). One of the isomers proved to be a strong competitive inhibitor of urogen III synthase, whereas the other had no effect at all. Because of restriction in the rotation of the puckered rings of the spiro intermediate, molecules of a given chirality of the quaternary spiro-carbon can have adjacent pyrrole rings both pointed up and the opposite one pointed down or vice versa. Therefore, the isomers separated by Stark *et al.* (1986) were, in fact, atropisomers and each was a racemic mixture of enantiomers. The strongly inhibiting racemic mixture has been resolved into its two enantiomers by Cassidy *et al.* (1991); one enantiomer ($K_i = 1.8\ \mu M$) was found to be a 20-fold more effective competitive inhibitor than the other ($K_i = 38\ \mu M$), providing strong evidence for the presence of a spiro intermediate (**II**) in the transition state. A possible mechanism of urogen III synthase involving the spiro intermediate is shown in Fig. 9.

Scott (1990) has been unable to obtain NMR evidence for the spiro intermediate and has suggested another possibility for the mechanism of ring D reversal, that of a lactone intermediate formed by addition of the acetic acid residue of ring D to the double bond of an azafulvene intermediate. Although no direct evidence for the lactone intermediate has yet been obtained, indirect evidence is provided by the fact that an HMB analog in which the ring D acetic acid was replaced by a methyl group was inactive in the enzyme-catalyzed cyclization, whereas the analog in which the ring D propionic acid was replaced by an ethyl group was active (Battersby *et al.*, 1983). Scott (1990) also pointed out that the "reversed ring D" HMB isomer, although yielding some urogen I, also formed an equal amount of urogen IIII (cf. Table II), indicating that a propionic acid in the acetic acid position was less efficient for ring reversal.

The steric course of the combined HMB synthase and urogen III synthase reactions has been shown to proceed with net retention of the configuration of all methylene carbons (Jones *et al.*, 1984; Neidhart *et al.*, 1985; Schauder *et al.*, 1987). Hence, if the mechanism of either enzyme proceeds with the intermediate formation of azafulvenes (cf. Fig. 9), they must remain bound to the enzyme to allow stereoselective attack of the incoming group, resulting in net retention of configuration.

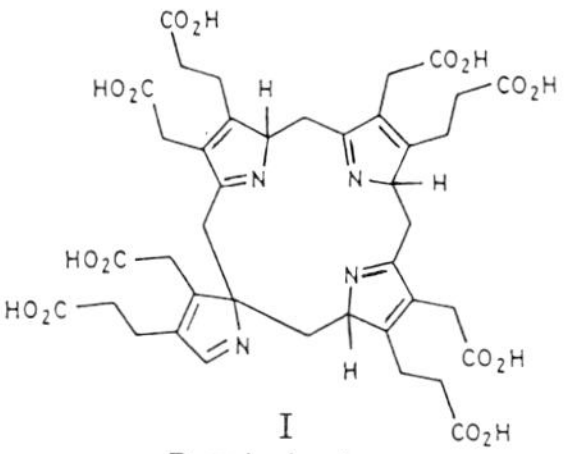

## 4. Biosynthesis of the Chlorophyll Chromophore

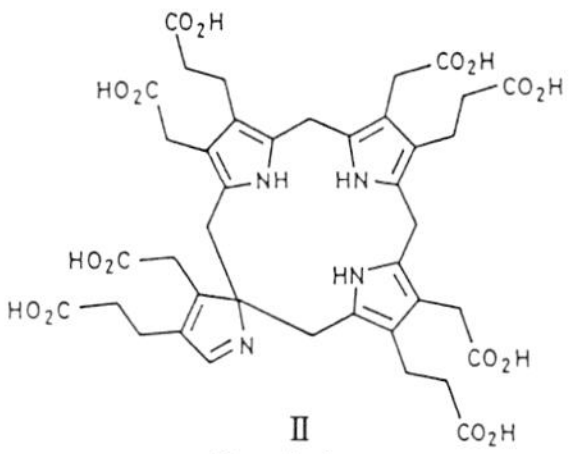

**Fig. 8.** A possible spiro intermediate in the cyclization of hydroxymethylbilane by uroporphyrinogen III synthase, shown as its pyrrolenine tautomeric isomer (I) initially proposed by Mathewson and Corwin (1961) and as its pyrrole tautomeric isomer (II). III and IV are synthetic spiro compounds prepared by Battersby and co-workers (see text for details). Adapted from Leeper (1987). The X-ray structure of III is reprinted with permission from Battersby (1987).

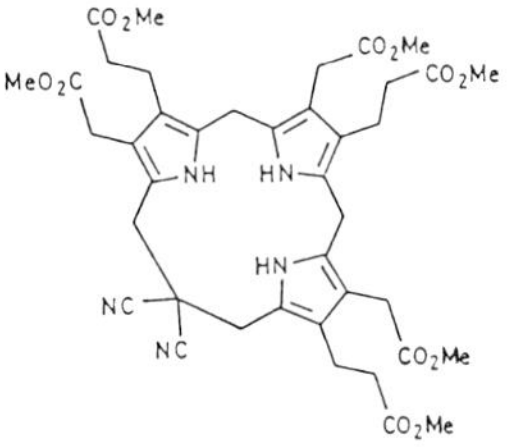

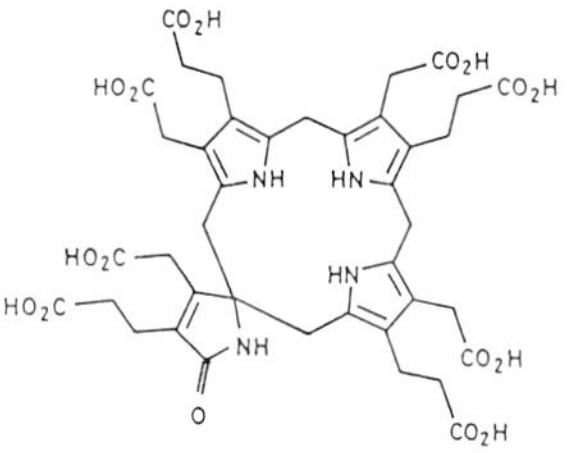

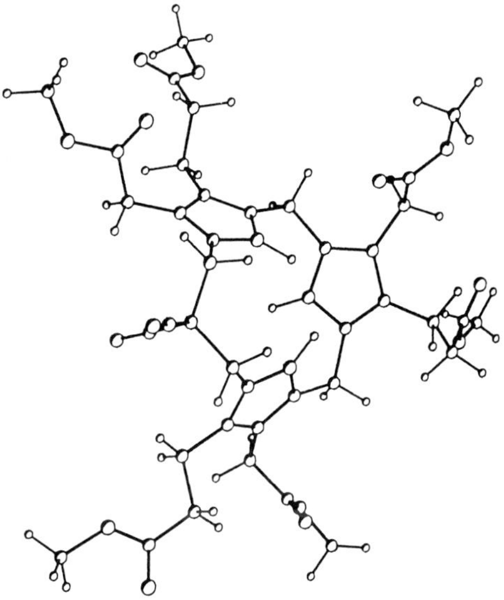

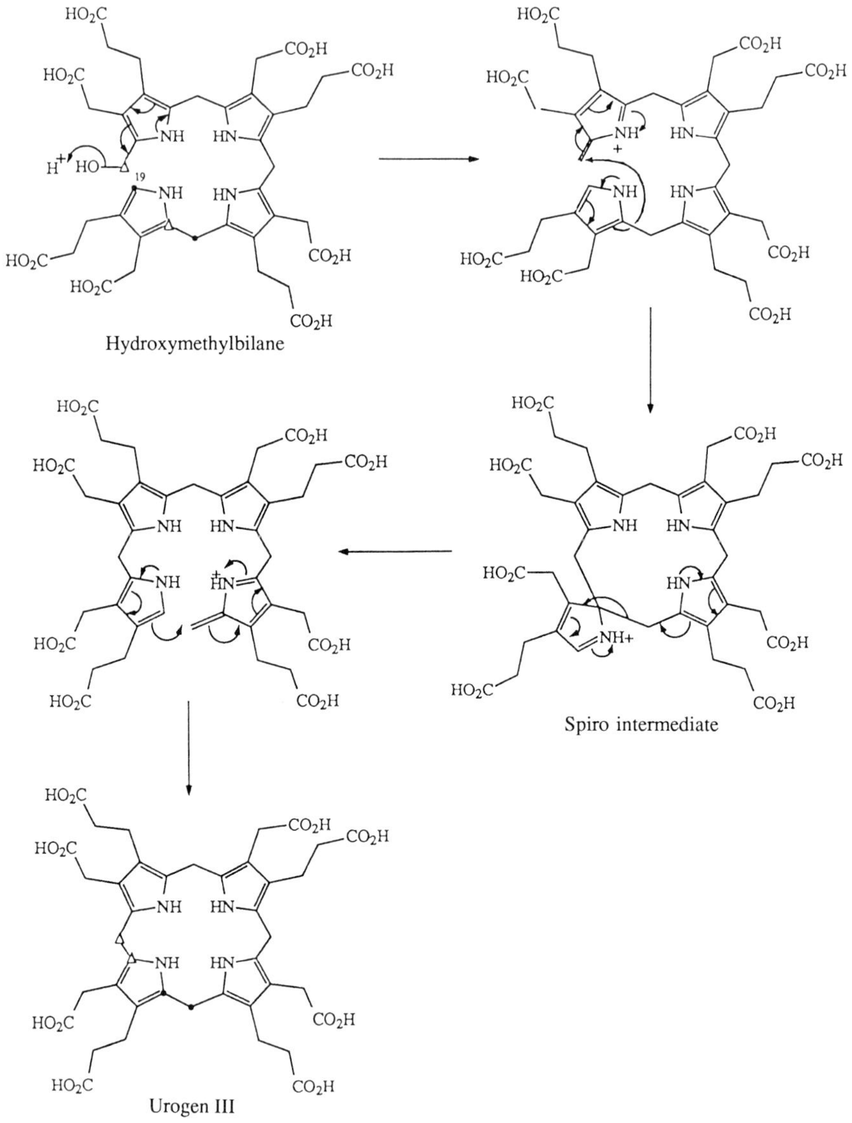

**Fig. 9.** Proposed mechanism for the cyclization of hydroxymethylbilane by uroporphyrinogen III synthase. Reprinted with permission from Leeper (1991). Copyright © 1991 by CRC Press.

## E. Conversion of Uroporphyrinogen III to Protoporphyrin IX

### 1. *Uroporphyrinogen III Decarboxylase*

It has been known since the mid 1950s that reduced (hexahydro) porphyrins, or porphyrinogens, were the true precursors of heme. The enzyme responsible for the decarboxylation of urogen III to coprogen III, uroporphyrinogen III decarboxylase (EC 4.1.1.37), first was studied in erythrocytes, where it was determined that all four isomers of urogen were decarboxylated but that urogen III was a much better substrate than urogen I. A 7-COOH intermediate (named phriaporphyrinogen) was found to accumulate during the reaction; subsequently, 6- and 5-COOH intermediates also were isolated from patients suffering from porphyria cutanea tarda. Urogen III decarboxylase has been detected in several plant species and purified from tobacco by Chen and Miller (1974). The enzyme has no metal requirements and is inhibited by sulfhydryl-binding reagents and metals such as $Cu^{2+}$, $Zn^{2+}$, and $Hg^{2+}$. The enzyme is soluble in the plastids of *Euglena* (Carell and Kahn, 1964) and cucumber (Smith and Rebeiz, 1979). It also has been studied in *Rhodopseudomonas palustris* (Koopman *et al.*, 1986; Koopman and Batlle, 1987) and found to contain essential histidine and lysine residues. The cDNA for the rat and human urogen III decarboxylases have been cloned, sequenced, and found to be 95% homologous (Romeo *et al.*, 1986; Romana *et al.*, 1987). The human gene encodes a protein with a molecular mass of 40,831 Da.

Several groups in 1983 showed that a single enzyme from human erythrocytes was responsible for the decarboxylation of each acetic acid residue of urogen III to a methyl residue, but that the first decarboxylation proceeded much more rapidly than the succeeding ones (reviewed by Jordan, 1990). Earlier indications were that decarboxylation began with the "reversed" pyrrole ring D, and proceeded clockwise around the macrocyclic ring (D > A > B > C) (Battersby *et al.*, 1976a; Jackson *et al.*, 1976, 1980). However, Luo and Lim (1990) have analyzed by HPLC the 7-COOH porphyrins formed from incubation of urogen III with the red blood cells of a normal individual or a patient suffering from porphyria cutanea tarda. They found an almost equal mixture of the four possible isomers in both cases, indicating that the first decarboxylation is, in fact, random. The decarboxylase may have a much higher affinity for urogen III than for any of the 7-COOH intermediates (or urogen I); hence, subsequent decarboxylations proceed more slowly until higher concentrations of the 7-COOH intermediates can be built up. It had been determined earlier that the order of the decarboxylations of urogen I was not specific; both possible 6-COOH isomers were observed (Jackson *et al.*, 1977). The mechanism of the decarboxylation proceeds with overall retention of configuration of the methylene carbons of the acetic acid residues (Fig. 10). The stereochemical course of the reaction was determined by Barnard and Akhtar (1975,1979), who incorporated $(R)$-[2-$^2$H$_1$,$^3$H$_1$]succinate and isolated $(S)$-[$^2$H$_1$,$^3$H$_1$]acetic acid after oxidative degradation of the resulting labeled heme. Similar results were also obtained by Battersby *et al.* (1981a) with *Rhodobacter sphaeroides*.

## 2. Coproporphyrinogen III Oxidase

Coprogen III oxidase (EC 1.3.3.3) oxidatively decarboxylates the propionic acid side chains on pyrrole rings A and B of coprogen III, yielding the vinyl groups of protogen IX. Enzyme extracts that converted coprogen III to proto IX were isolated first from bovine liver and were shown to require molecular oxygen for activity. Subsequently protogen IX,

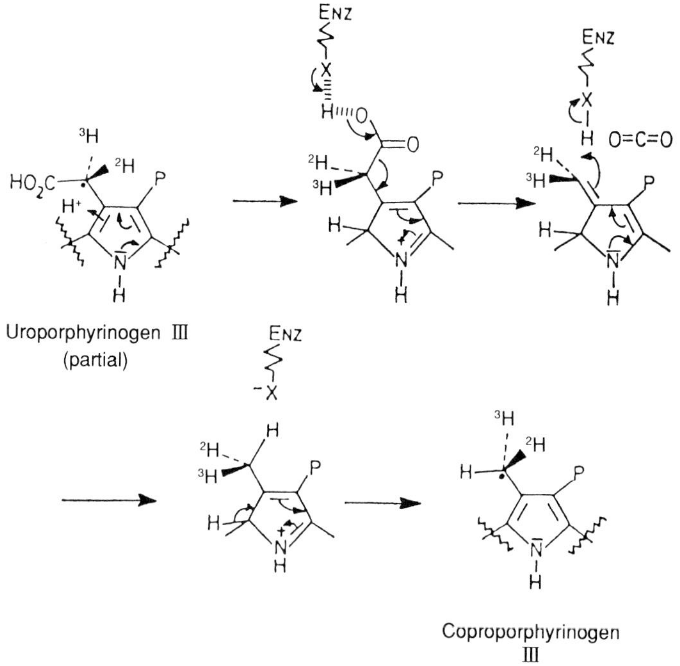

**Fig. 10.** A proposed mechanism for the decarboxylation of the four acetic acid residues of uroporphyrinogen III by uroporphyrinogen III decarboxylase that permits the retention of the configuration of the carbon atoms of the resulting methyl groups. Adapted from Barnard and Akhtar (1975).

## 4. Biosynthesis of the Chlorophyll Chromophore

not proto IX, was demonstrated to be the true product of the enzyme. Bovine liver coprogen III oxidase has been purified ~30,000-fold by Yoshinaga and Sano (1980a) and was reported not to be affected by metals or sulfhydryl-directed reagents. The enzyme from animals was found to be associated with mitochondria and localized in the intramembrane space (Elder and Evans, 1978; Grandchamp *et al.*, 1978), probably as an extrinsic protein loosely associated with the inner membrane (Dailey, 1990). However, in a yeast mutant, the enzyme was found to be a soluble cytoplasmic protein (Camadro *et al.*, 1986). The purified enzyme was a 70-kDa dimer containing two very tightly bound iron atoms per dimer and was inhibited by sulfhydryl-directed reagents. Coprogen III oxidase also was studied in *Euglena* (Battersby *et al.*, 1972; Cavaleiro *et al.*, 1974) and has been isolated from tobacco leaves by Hsu and Miller (1970). The enzyme was stimulated by metal ions such as $Fe^{2+}$, $Mn^{2+}$, and $Co^{2+}$, and inhibited by chelators such as ethylenediaminetetraacetic acid (EDTA).

Coprogen III oxidase also has been isolated from the prokaryotes *Chromatium* (Mori and Sano, 1968), *Rhodobacter sphaeroides* (Tait, 1969,1972; Seehra *et al.*, 1983), *Rhizobium japonicum* (Keithly and Nadler, 1983), and a variety of other microorganisms (Jacobs *et al.*, 1971). In facultative bacteria, the enzyme from aerobic cultures requires oxygen, whereas the enzyme from anaerobic cultures requires $NAD(P)^+$ as an electron acceptor. In *R. sphaeroides*, the anaerobic conversion requires both a soluble and a membrane fraction, whereas the aerobic conversion requires only the soluble fraction (Tait, 1972). The anaerobic enzyme in *R. sphaeroides* also requires ATP plus methionine for activity (Tait, 1972), a requirement also found for the anaerobic coprogen III oxidase of yeast (Poulson and Polglase, 1974). Hence, S-adenosyl-L-methionine (SAM) may act as an intracellular mediator during the light-activated stimulation of Bchl synthesis, either by allosteric activation (Tait, 1972) or by covalent modification (methylation) of the anaerobic enzyme. Now it is known that there are two separate coprogen III oxidases in *R. sphaeroides;* a mutant that lacks only the anaerobic one has been isolated (C. N. Hunter, 1990, personal communication).

Cavaleiro *et al.* (1974) identified a porphyrin with one vinyl and three carboxyl groups that was formed during the course of oxidative decarboxylation as 2-vinyl-4-propionate-deuteroporphyrin IX. This intermediate had been shown earlier to accumulate in the Harderian glands of rats; hence, it was known as harderoporphyrin. Harderoporphyrinogen was demonstrated to be a better substrate than its 4-vinyl-2-propionate isomer, isoharderoporphyrinogen, indicating that the decarboxylation of ring A preceded that of ring B (Cavaleiro *et al.*, 1974; Jackson *et al.*, 1974). Also it has been shown that the oxidase cannot decarboxylate coprogens I and

II, but can convert coprogen IV to proto(gen) XIII (Elder et al., 1978; Jackson et al., 1978; Al-Hazimi et al., 1987). For coprogen IV, however, a 3-COOH intermediate accumulated to a greater extent than for coprogen III. Therefore it was proposed that the enzyme must encounter substituents on adjacent pyrrole rings with the sequence Me(Vi)-Me-Pr-Me before recognizing the propionate for oxidative decarboxylation; this sequence is encountered only in coprogens III and IV and harderoporphyrinogen (Jackson et al., 1978; Yoshinaga and Sano, 1980b).

Two possible mechanisms for coprogen III oxidase are shown in Fig. 11 (Leeper, 1985a). Decarboxylation occurs with retention of both $\alpha$-hydrogens but only one of the $\beta$-hydrogens (Battersby et al., 1972; Zaman et al., 1972) and proceeds by *trans*-elimination of the *pro* S hydrogen and $CO_2$ (Zaman and Akhtar, 1976; Battersby, 1978; Yoshinaga and Sano, 1980b; Seehra et al., 1983). Hence, $\beta$-ketopropionate- or acrylate-containing intermediates are not possible. Sano (1966) had determined earlier that synthetic 2,4-bis($\beta$-hydroxyproprionate)-deuteroporphyrinogen IX was converted to proto IX by a beef liver mitochondrial extract under *anaerobic* conditions. Hence, most of the mechanisms that have been proposed for coprogen III oxidase involve $\beta$-hydroxypropionate side chains as intermediates (Sano, 1966; Chaudhry et al., 1977; Jackson et al., 1978; Yoshinaga and Sano, 1980b), which may be formed by some kind of hydroxylation reaction. Such a mechanism cannot be in effect in anaerobic organisms, however, because molecular oxygen is not involved. Seehra et al. (1983) proposed a mechanism for anaerobic organisms involving hydride removal, which would yield an intermediate in equilibrium with the $\beta$-hydroxpropionate intermediate by hydration of the resulting double bond (Fig. 11). Since the enzymes from aerobic organisms do not appear to have the properties of known hydroxylases, these investigators suggested that the hydride-removal mechanism might be universal. However, the study of bovine liver coprogen III oxidase by Yoshinaga and Sano (1980b) revealed that modification of tyrosine residues inactivated the enzyme to coprogen III but not to 2,4-bis($\beta$-hydroxypropionate)-deuteroporphyrinogen, indicating that one or more tyrosine residues may be involved in a hydroxylation step of the reaction. In addition, results of Camadro et al. (1986) indicate that the aerobic yeast enzyme contains firmly bound iron, a common property of hydroxylases. Hence, there may be two different mechanisms in operation for aerobic and anaerobic coprogen III oxidases.

### 3. *Protoporphyrinogen IX Oxidase*

Although it has been known since 1964 that protogen IX is the true product of coprogen III oxidase, it was not certain that the oxidation of

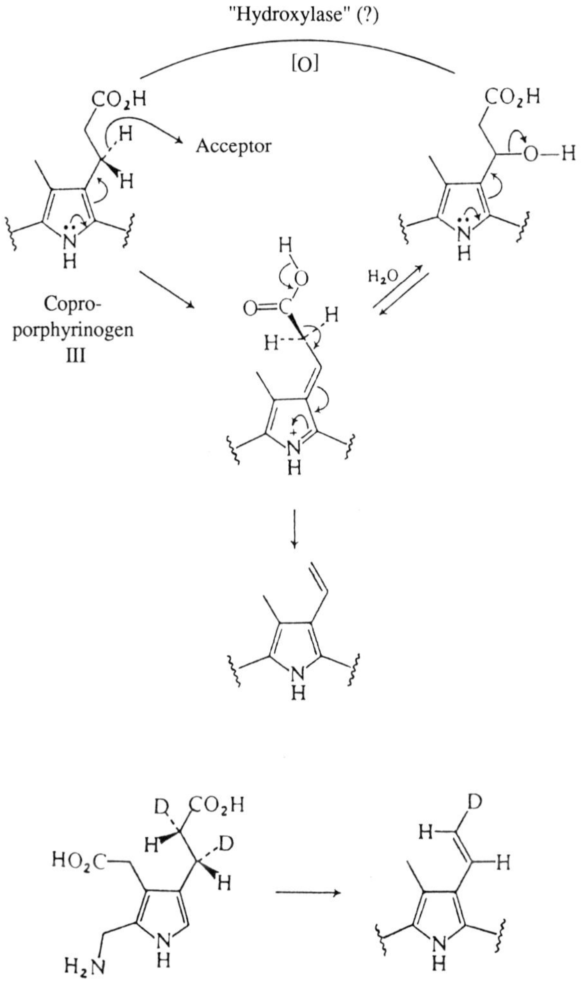

# 4. Biosynthesis of the Chlorophyll Chromophore

**Fig. 11.** Proposed mechanisms for the oxidative decarboxylation of the 2- and 4-propionic acid residues of coproporphyrinogen III by coproporphyrinogen III oxidase. In aerobic organisms, there may be a direct incorporation of one of the atoms of molecular oxygen by a "hydroxylase" step to form a β-hydroxypropionic acid intermediate. Alternatively, molecular oxygen may act only as an electron acceptor for an oxidase step; the β-hydroxypropionic acid would be formed as the side product of a reversible hydration reaction. In anaerobic organisms, the "hydroxylase" mechanism is not possible; thus, the electron acceptor must be a molecule other than molecular oxygen. The geometry of the resulting vinyl groups formed using deuterium-labeled propionic acid residues is shown at the bottom of the figure. Adapted from Leeper (1985a).

protogen IX to proto IX was catalyzed by an enzyme because of the ease of porphyrinogen autoxidation. Jackson *et al.* (1974) and Battersby *et al.* (1976b) demonstrated that the oxidation must be enzymatic when they showed that randomly *meso*-tritiated protogen lost 50% of its activity on incubation in cell-free extracts; it would have been expected to lose only 4% through chemical oxidation due to the tritium isotope effect. The enzyme responsible, protogen IX oxidase (EC 1.3.3.4), first was purified from yeast by Poulson and Polglase (1975) and later from the livers of various mammals (Poulson, 1976; Camadro *et al.*, 1985; Dailey and Karr, 1987; Siepker *et al.*, 1987). All the enzymes required oxygen for activity, exhibited $M_r$ values of 32–35 or 57–65 kDa, and had $K_m$ values in the low $\mu M$ range for protogen IX. A $K_m$ value of 125 $\mu M$ for oxygen was measured for the enzyme from mouse (Ferreira and Dailey, 1988). The mammalian enzymes were all fairly specific for protogen IX, exhibiting <10% of their activity for protogen XIII and meso-, hardero-, isohardero-, and 2(4)-vinyl-4(2)-hydroxyethyl-deuteroporphyrinogen IX (Poulson and Polglase, 1975; Jackson *et al.*, 1978).

Protogen IX oxidase is an intrinsic protein of the inner mitochondrial membrane of animals. Deybach *et al.* (1985) presented evidence that substrate could approach the enzyme from either side of the membrane. However Ferreira *et al.* (1988), using membrane-impermeable inhibitors, concluded that the active site of protogen IX oxidase was on the cytoplasmic side whereas that of ferrochelatase was on the matrix side of the inner membrane. These researchers also suggested that the two enzymes exist as an enzyme complex (probably with the extrinsic enzyme coprogen III oxidase), although they were unable to obtain any direct kinetic evidence for such a complex (Ferreira *et al.*, 1988; Dailey, 1990).

The only reported purification of protogen IX oxidase from a higher plant has been by Jacobs and Jacobs (1984, 1987) from barley leaves. A similar, if not identical, enzyme was found in plastids and mitochondria. The enzyme from both organelles exhibited an $M_r$ value of 36 kDa on SDS–PAGE and had a $K_m$ value for protogen of 5 $\mu M$ (similar to values reported for the mammalian enzymes). Unlike the mammalian enzymes, however, the barley enzyme was equally active with protogen IX and mesoporphyrinogen IX (Jacobs and Jacobs, 1987).

In the case of prokaryotes, the protogen IX oxidases of heterotrophic bacteria (Jacobs and Jacobs, 1979) and *Rhodobacter sphaeroides* (Jacobs and Jacobs, 1981) are also membrane bound but are obligately coupled to the respiratory chain of the cell. Thus, in *Escherichia coli*, if the cells have been growing aerobically, oxygen functions as the terminal electron acceptor, whereas if they have been growing anaerobically, some other electron acceptor such as fumarate or nitrate can substitute; quinones are required for the electron transfer reaction also (summarized in Jacobs and

## 4. Biosynthesis of the Chlorophyll Chromophore

Jacobs, 1979). Probably because of the difficulty in supplying a terminal electron acceptor, an active enzyme has not been isolated from *R. sphaeroides* membranes (Jacobs and Jacobs, 1981); however, Klemm and Barton (1987) have solubilized protogen IX oxidase from membranes of *Desulfovibrio gigas*, using 2,6-dichlorophenolindophenol (DCIP) as an electron acceptor. The enzyme appears to be a trimer of three nonidentical subunits with an overall $M_r$ of 148 kDa.

The mechanism of action of protogen IX oxidase has been inferred from a tritium labeling study carried out by Jones *et al.* (1984). They incorporated (*S*)-[11-$^3$H$_1$]PBG into broken chicken erythrocytes and found that only one-fourth of the activity remained in the isolated heme. All was found at the β-*meso* position, having been completely lost from the α-, γ-, and δ-*meso* positions. Since the formation of urogen III is known to occur with retention of configuration and no loss of hydrogens from the four methylene carbons, Jones *et al.* (1984) proposed that tritium atoms on the α-, γ-, and δ-methylene carbons were lost during hydride ion extraction from one side of the macrocyclic ring (in addition to three protons from the pyrrole nitrogens) whereas a hydrogen on the β-methylene carbon was lost from the other side of the macrocycle via a tautomerization, thereby retaining a tritium label at the β-*meso* position (Fig. 12). The nature of the hydride acceptor is not yet certain. Siepker *et al.* (1987) reported that the

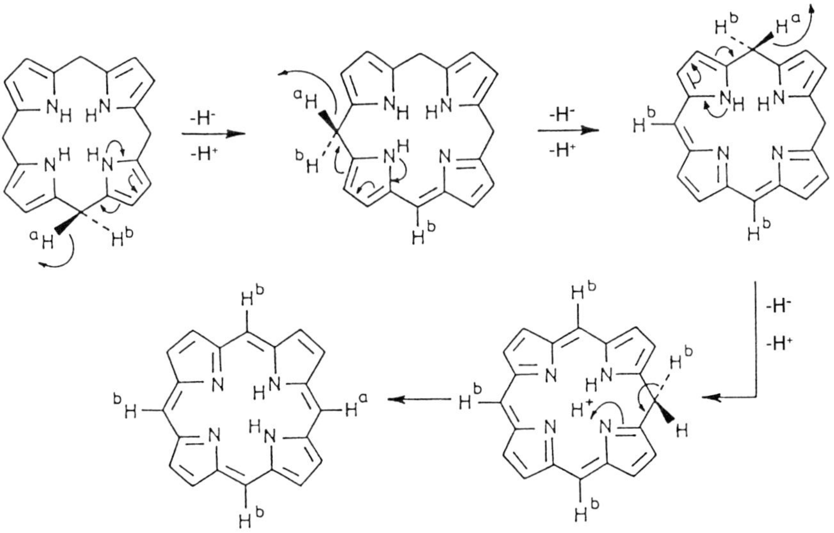

**Fig. 12.** Proposed mechanism for the loss of hydrogen atoms from the *meso* positions of a porphyrinogen by the action of protoporphyrinogen IX oxidase. Reprinted with permission from Jones *et al.* (1984).

oxidase from bovine liver contains a firmly bound reduced flavin ($FADH_2$), while Proulx and Dailey (reported in Dailey, 1990) presented unpublished data that the mouse liver enzyme contains an oxidized flavin (FMN). The stoichiometry of oxygen use indicates the consumption of 3 $O_2$ per mol protogen (Ferreira and Dailey, 1988), presumably yielding 3 $H_2O_2$, as do many other flavoprotein oxidases. However, a kinetic study by Ferreira (1986) was inconsistent with this mechanism. Sulfhydryl groups also seem to be involved since the rat, bovine, and *D. gigas* enzymes (but not the mouse enzyme) were inhibited by sulfhydryl-directed reagents. However, there is no indication of the nature of sulfhydryl involvement, whether through oxidation to disulfides, substrate binding, or participation in the catalytic mechanism (Dailey, 1990).

## III. LATER STAGES OF CHLOROPHYLL SYNTHESIS: THE MAGNESIUM BRANCH

The magnesium branch of Chl *a* synthesis starts with the insertion of magnesium into proto IX (Fig. 13). Earlier versions of the pathway of Chl *a* synthesis (cf. Jones, 1978; Castelfranco and Beale, 1981,1983) depicted the magnesium branch as a linear pathway in which Mg proto was esterified to Mg proto monomethyl ester (MgPME), followed by oxidative cyclization of the 6-methyl propionate group of the latter to yield divinylprotochlorophyllide. This intermediate must be reduced twice. Reduction of the 4-vinyl group was thought to proceed before reduction of pyrrole ring D, yielding a linear sequence involving protochlorophyllide (Pchlide), chlorophyllide (Chlide), and, following esterification of the latter, Chl *a*. As will be discussed in Section III,D, uncertainty about the exact intermediate to undergo the first reduction (the reduction of the 4-vinyl group to ethyl) has led to proposals for the existence of parallel pathways of Chl *a* synthesis involving both divinyl and monovinyl intermediates. In addition to variations among plant species, the reduction of the 4-vinyl substituent also seems to have variations under different physiological growth conditions within the same species. An understanding of the situation is by no means complete; several possible locations for 4-vinyl reduction are indicated in Fig. 13. Along with the parallel pathways indicated in Fig. 13, other parallel pathways have been proposed that involve fully esterified intermediates or monoesters of the 7-propionate (rather than the 6-propionate) group. The intermediates will be discussed in Section III,D, but have not been included in Fig. 13.

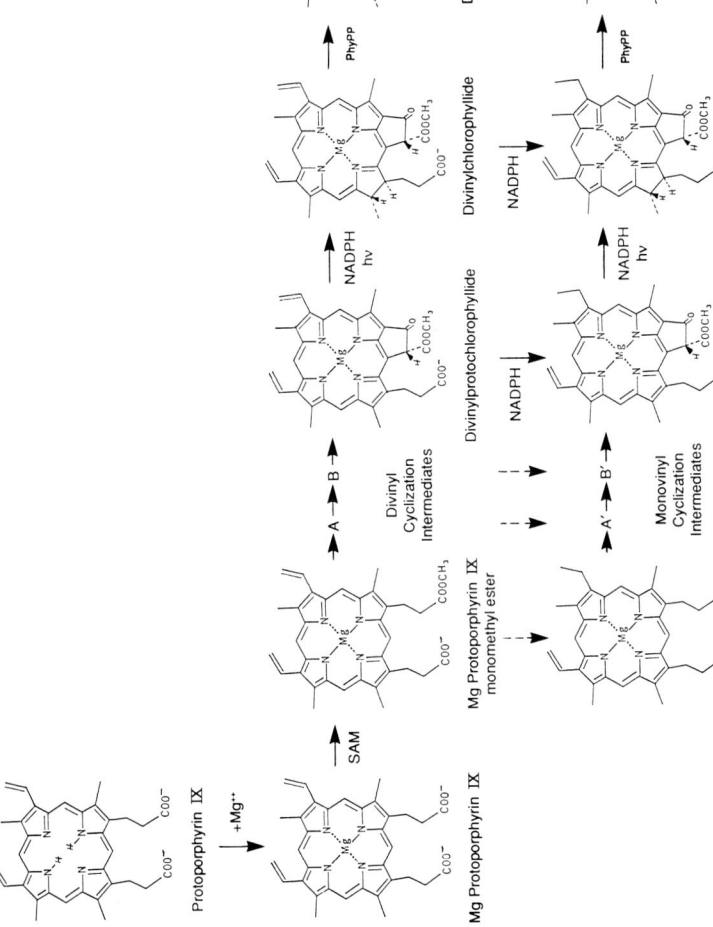

**Fig. 13.** Magnesium branch of chlorophyll synthesis. Conversion of protoporphyrin IX to chlorophyll *a*.

## A. Magnesium Chelatase

For many years, the demonstration of magnesium chelatase activity in plants and bacteria proved elusive. Early attempts to demonstrate the chelation of magnesium by proto IX in *Rhodobacter sphaeroides* (Johnson and Jones, 1964; Neuberger and Tait, 1964) or barley (Little and Kelsey, 1964) led to the formation of zinc proto instead. Zinc chelatase activity was shown to be due to the activity of ferrochelatase in both *R. sphaeroides* (Mazanowska *et al.*, 1966) and barley (Goldin and Little, 1969). Ferrochelatase is an intrinsic membrane protein in the chloroplasts and mitochondria of plants (Porra and Lascelles, 1968; Little and Jones, 1976) and in the chromatophores of *R. sphaeroides* (Jones and Jones, 1970; Dailey, 1982) and is able to insert $Fe^{2+}$, $Zn^{2+}$, or $Co^{2+}$ into a variety of porphyrins (reviewed by Dailey, 1990).

Gorchein (1972,1973) demonstrated the conversion of proto to MgPME in whole cells or spheroplasts of *R. sphaeroides*, although no successful demonstration of magnesium chelatase activity in cell-free extracts of *R. sphaeroides* was reported. The product was identified as the magnesium chelate by atomic absorption spectrometry. The conversion required a metal chelator such as EDTA or ethylene glycol-bis($\beta$-aminoethyl ether)-$N,N'$-tetraacetic acid (EGTA), a phospholipid sol (into which the proto was suspended by ultrasonication), and a source for the generation of energy (most likely ATP), either anaerobically in the light or under low oxygen partial pressures in the dark. The conversion was inhibited at >15% $O_2$. Because the product was the methyl ester, Gorchein (1972) proposed that the magnesium chelatase was coupled very tightly to the subsequent enzyme in the biosynthetic pathway, SAM:Mg proto methyltransferase. Gorchein also suggested that ATP may have had some function other than to supply SAM for the methyltransferase reaction (Gorchein, 1973).

Rebeiz and Castelfranco (1971) developed a cell-free system from etiolated cucumber cotyledons capable of converting [$^{14}$C]ALA into [$^{14}$C]Pchl(ide). Other metalloporphyrins were found to accumulate during the incubation, including one identified as MgPME by thin-layer chromatography (Rebeiz *et al.*, 1970; Rebeiz and Castelfranco, 1971). In an attempt to duplicate these results with cell-free extracts of etiolated wheat, Ellsworth and Lawrence (1973) incubated ALA, PBG, protogen IX, or proto IX in the presence of $^{28}Mg^{2+}$. They concluded that only very small amounts of $^{28}$Mg proto(ME) were formed from ALA, PBG, or protogen (and none at all from proto ), and that the major product was Zn proto(ME). Rebeiz *et al.* (1975) subsequently analyzed the metalloporphyrin produced in cucumber extracts by spectrofluorimetry and reconfirmed its identity as MgPME, which can be distinguished from Zn proto(ME) easily by this method (Smith and Rebeiz, 1977a). The investigators stated, however, that, under certain incubation conditions, massive amounts of Zn proto(ME) could be made to accumulate (Rebeiz et al., 1975).

Although the foregoing studies demonstrated that magnesium chelatase was present and working in cell-free preparations, no one had demonstrated the direct incorporation of magnesium into proto, the supposed substrate of the chelatase, although Shieh et al. (1974) had demonstrated previously the incorporation of [$^{14}$C]proto (but not [$^{14}$C]proto methyl ester) into Chl a in tobacco leaf homogenates. Finally, Smith and Rebeiz (1977b) reported the conversion of proto to Mg proto(ME) (plus other metalloporphyrins) when either etioplasts or developing chloroplasts of cucumber were suplemented with $MgCl_2$, glutathione (GSH), $NAD^+$, ATP, coenzyme A, and methanol. The Mg proto(ME) constituted about 20% of the metalloporphyrins synthesized, the remainder being mostly Zn proto(ME) and "longer wavelength" metalloporphyrins. Smith and Rebeiz (1979) later determined that lysed cucumber plastids, and a membrane fraction sedimented from lysed plastids were capable of converting very small amounts of proto to Mg proto(ME) also.

At about the same time, Castelfranco et al. (1979) reported the conversion of proto to Mg proto(ME) by a cell-free system from etiolated cucumber in the presence of $MgCl_2$, EDTA, GSH, $NAD^+$, ATP, and glutamate. Glutamate was shown to be acting as an ATP-generating system in the mitochondria that contaminated the etioplast preparation; glutamate could be replaced by increasing the ATP concentration (Pardo et al., 1980). It was also found that GSH and $NAD^+$ could be omitted from magnesium chelatase assays without loss of activity (Fuesler et al., 1984a). The product was confirmed to be the magnesium (and not the zinc) chelate by low temperature spectrofluorometry and HPLC; zinc chelation was observed in place of magnesium chelation when ATP was omitted and $ZnCl_2$ was added to the incubation mixture (Fuesler et al., 1981). The product was shown to be Mg proto (not MgPME) by HPLC; MgPME is formed only in the presence of SAM (Fuesler et al., 1982). From a study using a membrane-impermeable inhibitor of magnesium chelatase (p-chloromercuribenzene sulfonate), it was proposed that the magnesium chelatase was located on the chloroplast envelope (Fuesler et al., 1984b). In the same study, the authors also reported that intact plastids were required for magnesium chelatase activity.

Richter and Rienits (1982) also reported the conversion of [$^{14}$C]proto to Mg [$^{14}$C]proto by intact etioplasts and isolated etioplast membranes of cucumber in the presence of $MgCl_2$, EDTA, GSH, $NADP^+$, and ATP. Small amounts of Zn proto(ME) were formed also. When [$^{14}$C]ALA or [$^{14}$C]glutamate was used as a precursor, some MgPME was formed in addition to Mg proto, but little Zn proto(ME) was formed, as determined by thin-layer chromatography and HPLC (Richter and Rienits, 1980). This study also confirmed the results of Ellsworth and Lawrence (1973) that wheat etioplasts formed only Zn proto(ME) whereas cucumber etioplasts formed Mg proto(ME).

A note should be added here on the effects of some of the cofactors required for the demonstration of magnesium chelatase activity. Richter and Rienits (1982) found that Zn proto(ME) almost completely replaced Mg proto if EDTA was omitted from the incubation mixture. This observation is consistent with the results of Smith and Rebeiz (1977b), whose incubation mixture contained no EDTA, that the product was only ~20% MgPME, the majority being Zn proto(ME). Also, Rebeiz and Castelfranco (1971) included methanol in their incubation mixture and found that it greatly stimulated MgPME synthesis as well as Pchl(ide) synthesis. Methanol could not be replaced by other low molecular weight alcohols. When [$^{14}$C]methanol was used, it was found to be incorporated into both MgPME and Pchlide, indicating that it may have been converted to SAM. Interestingly, a much larger incorporation of $^{14}$C was observed into Pchlide ester, raising the possibility that the latter is a methyl ester rather than a phytyl ester of Pchlide (Rebeiz and Castelfranco, 1971). Neither Castelfranco *et al.* (1979) nor Richter and Rienits (1982) included methanol in their incubation mixtures. In both cases, the major product was Mg proto, not MgPME. Coenzyme A also was omitted by these groups, since Rebeiz and Castelfranco (1971) had found it to be required only for Pchlide (and not MgPME) formation.

Magnesium ion forms a very stable hexaaqua complex [$Mg(H_2O)_6^{2+}$] in water. It is very difficult to replace the water molecules with most nitrogen-containing ligands. This property of $Mg^{2+}$ causes serious problems in the chemical synthesis of magnesium tetrapyrroles, which is usually only accomplished under completely anhydrous conditions, for example, with $Mg^{2+}$ in the form of Grignard reagents, $Mg(pyridine)_6(ClO_4)_2$, $Mg(pyridine)_6(I)_2$, or "magnesium viologen" (Wei *et al.*, 1962; Baum *et al.*, 1964). Therefore, it is not surprising that hydrated $Mg^{2+}$ was not the form of magnesium employed for the chelation reaction. Fuesler *et al.* (1981) found that, when ATP is omitted, the product is Zn proto, not Mg proto. Therefore, these investigators assumed that the Mg–ATP complex was the real substrate for the chelatase. Also, they found a sigmoidal dependence of chelatase activity on the added concentrations of ATP, $MgCl_2$, and proto, indicating that the chelatase may be an allosteric regulatory protein (Fuesler et al., 1981). However, these plots were constructed without considering the formation of the Mg–ATP complex. Richter and Rienits (1982), taking the calculated concentration of the Mg–ATP complex into account, demonstrated a classical Michaelis–Menten response of the enzyme to the concentration of the Mg–ATP complex. However, an additional enzyme activation due to $Mg^{2+}$ ion concentration, in excess of that required for Mg–ATP complex formation, was seen (Richter and Rienits, 1982). None of these studies sufficiently demonstrated the actual form of magnesium involved, however. It is still possible that some other organic form of magnesium is required for the chelation process (cf. Duval and Duranton, 1972).

Walker and Weinstein (1991a,b) have begun a reinvestigation of the magnesium chelatase of higher plants. They have determined the optimum conditions for the assay procedure originally developed for greening cucumber etioplasts by Fuesler et al. (1981, 1984a). They have shown that ATP hydrolysis most likely is required for activity, and also have demonstrated that possible feedback inhibitors such as protoheme, Pchlide, and Chlide inhibit no more than the substrate, proto IX, whereas the product, Mg proto, causes almost no inhibition of the enzyme (Walker and Weinstein, 1991a). Also, magnesium chelatase activity has been shown to increase with greening time, indicating that there may be a light-dependent developmental regulation of its activity rather than allosteric inhibition by metabolic effectors. Using developing pea seedlings, an active magnesium chelatase has been obtained for the first time in an organelle-free preparation (Walker and Weinstein, 1991b). The activity has been fractionated into membrane-bound and soluble fractions that can be combined to reconstitute magnesium chelatase activity. It also has been demonstrated that the membrane-bound, rather than the soluble, fraction is inactivated in cucumber during chloroplast lysis.

## B. S-Adenosyl-L-Methionine:Magnesium Protoporphyrin IX Methyltransferase

The enzyme that catalyzes the converison of Mg proto to MgPME, SAM:Mg proto methyltransferase (EC 2.1.1.11), was demonstrated first in the chromatophores of the purple nonsulfur bacterium, *Rhodobacter sphaeroides* (Tait and Gibson, 1961; Gibson et al., 1963; Neuberger et al., 1970). Subsequently the enzyme was reported to be present in the green sulfur bacterium *Chlorobium* (Holt, 1966; Jones, 1968), the alga *Euglena gracilis* (Ebbon and Tait, 1969; Neuberger *et al.*, 1970), and several higher plants, including corn (Radmer and Bogorad, 1967), wheat (Ellsworth and Dullaghan, 1972), barley, cucumber, pea, and soybean (Shieh *et al.*, 1978). The methyltransferase of *R. sphaeroides* was found to be repressed, but not inhibited, by oxygen and induced when cells were incubated subsequently under semianaerobic conditions in the light (Gorchein *et al.*, 1968). Lascelles and Wertlieb (1971) isolated several mutants of *R. sphaeroides* that had lost the ability to regulate the methyltansferase in this manner. The enzyme is located in the cell membrane of *R. sphaeroides* mutants unable to form Bchl (W. R. Richards, 1990, unpublished observations). Also, the product of the enzyme, MgPME, has been found to accumulate in a cell membrane fraction known to be a precursor of the intracytoplasmic membranes[3] (Oelze, 1988). Hence, the methyltransferase very likely is incorporated into the intracytoplasmic membranes via the cytoplasmic membrane. The enzyme has been dissociated from chromatophores in an active form using sodium cholate (Hinchigeri *et al.*, 1984).

Whether the methyltransferase is also membrane bound in higher plants and *Euglena* is less certain. The product, MgPME, was found to accumulate in small membrane particles, characterized as centers of Chl biosynthesis by Shlyk *et al.* (1982), which were isolated from immature wheat or barley leaves under certain developmental conditions. Soluble forms of the enzyme were isolated from etiolated wheat by homogenization with 0.5 $M$ sucrose (Ellsworth and Dullaghan, 1972) or with 0.5 $M$ sucrose plus 1 $M$ NaCl (Hinchigeri *et al.*, 1981). The specific activity of the enzyme was found to decrease somewhat during greening (Ellsworth and St. Pierre, 1976). The methyltransferase was found to be associated with sedimentable chloroplast membranes of greening barley, but could be solubilized from the membrane fraction by 0.5% Tween 80 (Shieh *et al.*, 1978). The soluble form of the enzyme has been localized to the cytoplasmic fraction of etiolated wheat (L. Gibson, C. Stevenson, and W. R. Richards, 1991, unpublished observations). Little or no additional soluble enzyme was released after disruption of isolated etioplasts, in which the methyltransferase activity was all membrane bound. Hence, an active methyltransferase soluble precursor (probably still containing a transit sequence) may accumulate in the cytoplasm of etiolated wheat and may later be transported into etioplasts during development to become the mature membrane-bound enzyme. In *Euglena,* a soluble form of the enzyme was obtained during disruption of the cells in a buffer containing 0.5 $M$ mannitol. However, the extraction of additional enzyme activity

---

[3] The intracytoplasmic membranes contain the photosynthetic apparatus of this bacterium and yield chromatophores on cellular disruption.

from chloroplasts required 0.5% Tween 80 (Ebbon and Tait, 1969). Both forms of the enzyme were present when the alga was grown under heterotrophic conditions in the dark. However, the specific activity of the enzyme increased 2- to 3-fold on incubation in the light, due mostly to an increase in the membrane-bound form (Ebbon and Tait, 1969).

Ebbon and Tait (1969) were able to purify partially the Tween 80-dissociated form (but not the soluble form) of the *Euglena* methyltransferase by gel filtration on Sephadex G-200. However, the enzyme subsequently lost all its activity on storage at 4°C. Attempts by Ellsworth *et al.* (1974) to purify the wheat methyltransferase by gel filtration were also unsuccessful because of the instability of the enzyme. Richards and co-workers found both enzymes to be stabilized by the addition of 0.1 m$M$ sodium ascorbate and 4 m$M$ 2-mercaptoethanol to all buffers during isolation, purification, and storage. Using affinity chromatography, they were able to obtain a 460- to 900-fold purification of both the soluble and Tween 80-dissociated forms of the *Euglena* enzyme on Mg proto- or hemin-linked Sepharose 4B (Hinchigeri *et al.*, 1981; Richards *et al.*, 1981), or the soluble form of the *Euglena* enzyme on $S$-adenosyl-L-homocysteine (SAH)-linked Sepharose 4B by elution with SAM (Hinchigeri and Richards, 1982). The etiolated wheat methyltransferase was purified between 1000- and 2000-fold by affinity chromatography on hemin-linked Sepharose 4B (Hinchigeri *et al.*, 1981).

The ability of methyltransferase from *Euglena* to be purified by chromatography on Mg proto- or SAH-linked affinity columns indicated that it could bind either substrate in either order. Subsequently the enzyme was shown to obey a random bi-bi mechanism with the formation of two dead-end ternary complexes (Equation B, Fig. 14) by a detailed kinetic analysis (Hinchigeri and Richards, 1982). The *R. sphaeroides* methyltransferase was shown by Hinchigeri *et al.* (1984) to exhibit an equilibrium-ordered bi-bi mechanism with Mg proto as the obligatory first substrate (Equation C, Fig. 14). Also, a partial purification of the cholate-solublized enzyme was achieved by elution with SAM during affinity chromatography on SAH–agarose. This behavior was not expected of an enzyme exhibiting the kinetic mechanism just described and demonstrated that care should be taken when using affinity chromatographic behavior of an enzyme as the sole criteria for the prediction of its reaction mechanism. However, such a method uncovered that the enzyme evidently was able to bind SAM also, indicating that the presence of bound SAM in some way hindered the subsequent binding of Mg proto, and was, therefore, a useful supplement to the kinetic data (Hinchigeri *et al.*, 1984).

Ellsworth *et al.* (1974) previously had carried out a detailed kinetic study on the etiolated wheat methyltransferase. They determined that its kinetic

A. ETIOLATED WHEAT
Reaction Mechanism: Ping-Pong

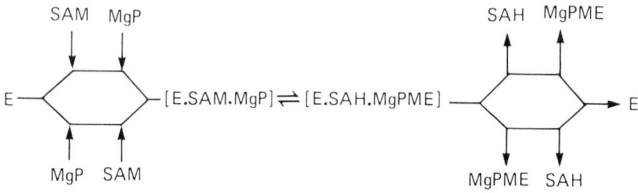

B. EUGLENA GRACILIS
Reaction Mechanism: Random

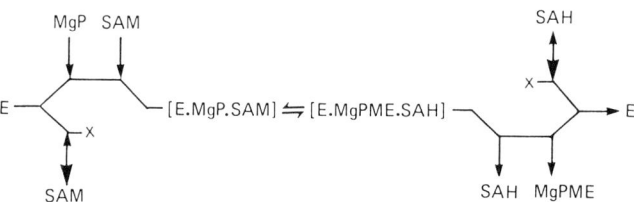

C. RHODOBACTER SPHAEROIDES
Reaction Mechanism: Ordered (magnesium protoporphyrin first)

**Fig. 14.** Summary of the kinetic reaction mechanisms observed for $S$-adenosyl-L-methionine:magnesium protoporphyrin IX methyltransferase in three different photosynthetic organisms. E, enzyme; E-CH$_3$, methylated enzyme; MgP, Mg proto IX. Reprinted with permission from Richards *et al.* (1987a). Copyright © 1987 by Kluwer Academic Publishers.

behavior and the pattern of inhibition by its products were consistent with a ping-pong mechanism (Equation A, Fig. 14) and that there was no indication that it was a regulatory enzyme (Ellsworth and St. Pierre, 1976). Hinchigeri *et al.* (1981) found that the behavior of the wheat methyltransferase during affinity chromatographic purification was also consistent with a ping-pong mechanism, since it bound to hemin–Sepharose 4B columns only in the presence of SAM. Attempts to confirm such a mechanism by detecting the formation of a methylated enzyme during incubation with [*methyl*-$^{14}$C]SAM, however, were unsuccesful (Hinchigeri *et al.*, 1981). A ping-pong mechanism was established firmly by carrying out an exchange reaction, incubating the methyltransferase (purified by hemin–Sepharose

4B affinity chromatography) in the presence of [$^{14}$C]SAH and unlabeled SAM; activity was exchanged into the SAM pool to a level of 70% of the theoretical maximum value (Yee et al., 1989). Such an exchange is only possible through a methylated enzyme intermediate (Equation A, Fig. 14).

The methyltransferases from three different photosynthetic organisms—a higher plant, a unicellular alga, and a photosynthetic bacterium—have been shown to obey three distinctly different kinetic reaction mechanisms (reviewed by Richards et al., 1987a). Discovering whether the three enzymes exhibit a significant amount of homology when the amino acid sequences become known or whether they have evolved separately from different ancestor proteins to fill the same metabolic function will be interesting. If they do prove to be homologous, it would provide an interesting system in which to study the evolution of three different kinetic reaction mechanisms at the protein level.

## C. Oxidative Cyclase Enzyme System

The formation of the isocyclic (or cyclopentenone) ring (ring E) of Chl *a* is no doubt a multistep process requiring the participation of more than one enzyme. Granick (1950) was first to suggest that the modification of the 6-propionic acid group of Mg proto might occur via a mechanism similar to the β-oxidation of fatty acids. Subsequently it was found that MgPME accumulated in a *Chlorella* mutant and in barley treated with ALA and α,α'-dipyridyl (Granick, 1961), indicating that MgPME was likely to be the true precursor during isocyclic ring formation. The intermediates would, therefore, correspond to those shown in Fig. 15. In the case of the methyl β-ketopropionate derivative, the methyl ester would be a necessary intermediate to prevent decarboxylation of the otherwise unstable β-keto acid (Bogorad, 1966).

It is well known that, under basic conditions, β-keto esters generate resonance-stabilized enolates that readily attack electron-deficient carbon atoms via nucleophilic addition or substitution reactions. The *meso* positions. Kenner and co-workers (Cox et al., 1969; Kenner et al., 1972, 1974) known to react in electrophilic rather than nucleophilic substitution reactions. Kenner and co-workers (Cox et al., 1969; Kenner et al., 1972, 1974) have synthesized the 6-β-ketopropionate derivatives of mesoporphyrin IX dimethyl ester and its 2-vinyl analog. They demonstrated that the cyclization reaction is an *oxidative* cyclization. Cyclization of the magnesium chelate of the former porphyrin occurred only after addition of a methanolic solution of sodium carbonate and a mild oxidizing agent (iodine); however, the product was the 10-methoxy derivative of pheoporphyrin $a_5$ dimethyl ester (Cox et al., 1969). When the thallium(III) chelates

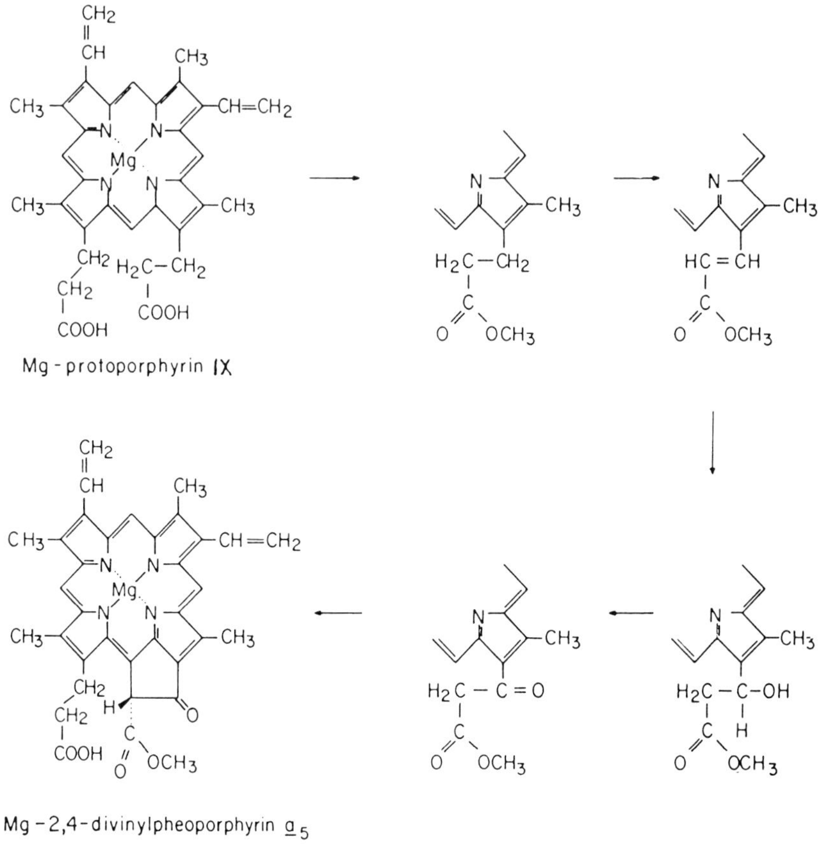

**Fig. 15.** Intermediates in the conversion of magnesium protoporphyrin IX to magnesium 2,4-divinylpheoporphyrin $a_5$ (divinylprotochlorophyllide), proposed by analogy with the β-oxidation of fatty acids. Reprinted with permission from Beale (1984).

of these porphyrins were prepared, however, cyclization (facilitated by a photochemical oxidation of the macrocycle by $Tl^{3+}$) yielded the corresponding pheoporphyrins without 10-methoxy groups (Kenner et al., 1972, 1974).

Since the proposed precursor MgPME contains two vinyl groups, the product of the oxidative cyclization logically should be magnesium 2,4-divinylpheoporphyrin $a_5$ monomethyl ester (also referred to as divinylprotochlorophyllide or DV-Pchlide). DV-Pchlide first was detected to occur naturally in cultures of *Rhodobacter sphaeroides* blocked in Bchl synthesis

## 4. Biosynthesis of the Chlorophyll Chromophore

either by mutation (Lascelles, 1966; Richards and Lascelles, 1969) or by inhibition (Jones, 1963,1967). It also was detected in its fully esterified form (DV-Pchl) in the inner seed coat of marrow (Jones, 1966) and in its monocarboxylate form (DV-Pchlide) in certain marine algae (Ricketts, 1966). Therefore DV-Pchlide was a likely intermediate in the biosynthesis of Bchl and Chl $a$, and reduction of the 4-vinyl group to yield protochlorophyllide (referred to as MV-Pchlide to emphasize its monovinyl nature) was likely to follow isocyclic ring formation in both pathways (Jones, 1963, 1967).

Aronoff and Ellsworth (1968) isolated several mutants of *Chlorella* that were blocked at different points in the latter stages of Chl synthesis. Mutants of the A-type were found to accumulate MgPME and several of its derivatives thought to be involved in isocyclic ring formation (Ellsworth and Aronoff, 1968b,1969). These mutants were, therefore, thought to be defective in the enzyme catalyzing the final condensation step with the $\gamma$-*meso* position (Aronoff *et al.*, 1971; cf. Fig. 16). The MgPME derivatives and their monovinyl (2-vinyl-4-ethyl) and desvinyl (2,4-diethyl) analogs were characterized by analysis of the magnesium-free dimethyl esters by mass spectrometry, degradation, and other chemical analyses, and were found to contain methyl acrylate, methyl $\beta$-hydroxypropionate, and methyl $\beta$-ketoproprionate groups in the 6 position (Ellsworth and Aronoff, 1968b,1969). Although the methyl acrylate derivative could not be isolated directly, it was estimated to be present to the extent of 15% in the MgPME fraction (Ellsworth and Aronoff, 1969). Submutants that accumulated only MgPME (and its monovinyl derivative) were isolated also, and were thought to be defective in an enzyme catalyzing oxidation of the $\beta$-carbon of the 6-propionate group (Fig. 16). Mutants of the B-type were found to accumulate MV-Pchlide and DV-Pchlide in the dark, and were thought to be defective in an enzyme catalyzing a dark (glucose-dependent) reduction of MV-Pchlide to Chlide (Fig. 16). When algal cultures were grown for long periods of time (5–7 days), they accumulated only MV-Pchlide, but when they were grown for shorter periods (3 days), both DV-Pchlide and MV-Pchlide were formed. In general, the longer the incubation, the more monovinyl intermediates accumulated (even intermediates with no vinyl groups, e.g., magnesium mesoporphyrin IX monomethyl ester, although only in very small amounts). Submutants that accumulated only DV-Pchlide in the dark were isolated also, and were thought to be defective in the enzyme catalyzing reduction of the 4-vinyl group; hence, a pathway was proposed in which 4-vinyl reduction could occur at any point between MgPME and DV-Pchlide, but at a slower rate than isocyclic ring formation (Aronoff *et al.*, 1971; cf. Fig. 16). Finally, mutants of the C-type accumu-

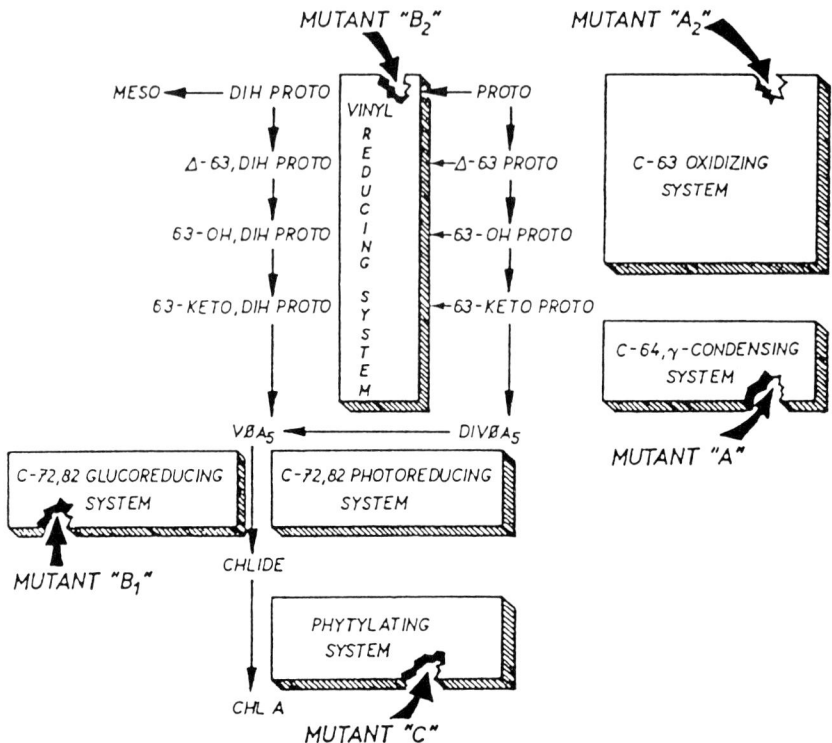

Fig. 16. Parallel pathways for the magnesium branch proposed by Aronoff and coworkers, based on the structures of tetrapyrroles accumulated by mutants of *Chlorella*. The enzyme system inferred to be defective in each mutant type is indicated. PROTO, Mg PME IX; DIH PROTO, 4-ethyl-(4-desvinyl)-MgPME; DIV$\phi$A$_5$, DV-Pchlide; V$\phi$A$_5$, MV-Pchlide. Reprinted with permission from Aronoff *et al.* (1971). Copyright © 1971 by Kluwer Academic Publishers.

lated Chlide $a$ (and pheophorbide $a$), and were thought to be defective in an enzyme involved in phytylation, perhaps in the biosynthesis of phytol itself (Ellsworth and Aronoff 1968a; Aronoff *et al.*, 1971).

Since all the foregoing studies were carried out with mutants, investigators could not be certain that any or all of the compounds detected represented actual biosynthetic intermediates until confirmed by enzymatic studies. For example, it had never been determined that DV-Pchlide actually could be converted to Chl(ide) $a$ until this was demonstrated in etioplasts of barley by Griffiths and Jones (1975) using DV-Pchlide derived from a mutant of *R. sphaeroides*. Then it was demonstrated that both Mg proto (Ellsworth and Hervish, 1975) and MgPME (Mattheis and Rebeiz,

1977; Ellsworth and Murphy, 1979) could be converted to Pchlide by cell-free extracts of etiolated wheat (Ellsworth and Hervish, 1975; Ellsworth and Murphy, 1979) or developing cucumber chloroplasts (Mattheis and Rebeiz, 1977). Both groups found the conversions to require methanol, $NAD^+$, ATP, coenzyme A, and phosphate. However, the relative proportions of MV-Pchlide and DV-Pchlide in the Pchlide produced were not determined. Belanger and Rebeiz (1979,1980a) later reported that the Pchlide pools of four etiolated plants (cucumber, red kidney bean, corn, and barley) consisted of both MV-Pchlide and DV-Pchlide. Further, DV-Pchlide became the major constituent under several different incubation conditions. Rebeiz and co-workers had described "long wavelength metalloporphyrins" earlier (characterized only by their fluorescence spectra) that accumulated during incubations in the presence of ALA (Rebeiz et al., 1975). Although not identified further, these metalloporphyrins were thought to be compatible with the proposed intermediates of isocyclic ring formation (Rebeiz et al., 1975). Ellsworth and Murphy (1979) also observed the formation of possible intermediates between MgPME and Pchlide. Two were isolated, purified by column chromatography, and found to have fluorescence excitation and emission maxima similar to those of synthetic 6-methyl-$\beta$-hydroxypropionate and 6-methyl-$\beta$-ketopropionate derivatives of MgPME. Material was insufficient to allow determination of whether the compounds were monovinyl or divinyl (Ellsworth and Murphy, 1979).

Meanwhile, Castelfranco and co-workers reported the conversion of Mg proto and MgPME by developing cucumber chloroplasts to a product tentatively identified as DV-Pchlide (Chereskin and Castelfranco, 1982; Chereskin et al., 1982). SAM was absolutely required for the conversion of Mg proto and also stimulated the conversion of MgPME. Both $NADP^+$ and NADPH were required, in addition to oxygen and iron, which had been implicated previously in the conversion (Spiller et al., 1982). Based on these requirements, an iron-containing monooxygenase might be required for the formation of the $\beta$-hydroxypropionate side chain, rather than hydration of an acrylate group (Chereskin and Castelfranco, 1982). However, no direct evidence for the involvement of an iron protein could be obtained (Chereskin et al., 1982). Subsequently, the product of the condensation reaction was established definitively as DV-Pchlide by NMR and mass spectrometry, indicating that the reduction of the 4-vinyl group must be much slower than isocyclic ring formation (Chereskin et al., 1983). Also, an HPLC system for the separation of MV-Pchlide and DV-Pchlide was developed by Hanamoto and Castelfranco (1983), allowing for much easier analysis of the products. The enzyme system was termed MgPME (oxidative) cyclase by Wong and Castelfranco (1984), who were able to resolve the enzyme system from developing plastids into two components, a high-speed supernatant and a membrane pellet. The enzyme system could be reconstituted successfully from these two fractions and was shown to require NADPH (or NADH) but not NAD(P)$^+$. ATP, although not required for the cyclization, stimulated the reaction in the absence of NADPH. A requirement for coenzyme A and methanol was not demonstrated. The stimulation of the incorporation of MgPME by SAM was shown to be caused by the presence of a 6-esterase acting on MgPME; SAM was required for the resynthesis of MgPME by the methyltransferase (Wong and Castelfranco, 1984). The MgPME (oxidative) cyclase enzyme system was localized inside the chloroplast envelope (Fuesler et al., 1984b) and was found to be sensitive to sulfhydryl-directed reagents (Fuesler et al., 1984b; Wong and Castelfranco, 1985). Work on the enzyme system by Walker et al. (1991) indicated that the active component(s) in the soluble fraction bound MgPME but not NADPH, since activity was retained by affinity columns containing Mg proto dimethyl ester but not Blue Sepharose. The soluble fraction was purified partially by chromatography on phenyl-Sepharose, indicating that the active component(s) was hydrophobic. The activity of the membrane-associated component(s) was inhibited by pretreatment with the metal chelators 8-hydroxyquinoline and desferal mesylate. Earlier $^{18}$O was shown to be incorporated into the 9-ketone of Pchlide during incubation of MgPME with $^{18}$O$_2$ (Walker et al., 1989). Hence, it seems highly likely that the previously postulated iron-containing monooxygenase (presumably also requiring NADPH and O$_2$) may be located in this fraction.

Studies on the mechanism of the cyclization reaction were carried out also by Castelfranco, Smith, and co-workers (Wong and Castelfranco, 1985; Wong et al., 1985; Walker et al., 1988). Several synthetic derivatives of MgPME, including the 6-methyl-$\beta$-hydroxypropionate and 6-methyl-$\beta$-ketoproprionate (but not the 6-methyl-*trans*-acrylate) derivatives, were found to be effective substrates for isocyclic ring formation. Interestingly, only one stereoisomer of the 6-methyl-$\beta$-hydroxypropionate derivative was found to be active, although its absolute configuration could not be ascertained. A compound identical to this derivative was isolated from incubation with MgPME (but not with the 6-methyl-$\beta$-ketopropionate derivative). Also, since only the *trans*-acrylate isomer was tested, activity for the *cis*-acrylate isomer could not be excluded; however, it would be expected to be quite unstable (Walker et al., 1988). The isolation of the 6-methylacrylate intermediates from the *Chlorella* mutants by Ellsworth and Aronoff (1968b,1969) could be explained by elimination of water from the 6-methyl-$\beta$-hydroxypropionate derivatives. As a result of these experiments, Chereskin and Castelfranco (1982) concluded that the 6-

methylacrylate intermediate originally proposed by Granick (1950) almost certainly could be removed from the pathway, resulting in the pathway shown in Fig. 17.

During the study, a *monovinyl* compound, the 2-vinyl-4-ethyl analog of the 6-methyl-$\beta$-ketopropionate derivative of MgPME, was found to be approximately 4 times more active than the corresponding divinyl derivative, whereas there was no difference between MgPME and its 2-vinyl-4-ethyl analog. (Both of these substances were much better substrates than either the 2-ethyl-4-vinyl or 2,4-diethyl analog.) Therefore, 4-vinyl reduction was suggested to occur somewhere between MgPME and its 6-methyl-$\beta$-ketopropionate derivative (Walker *et al.*, 1988). However, since the final product of the cyclization of MgPME was found to be predominantly DV-Pchlide, the relative rates of monovinyl and divinyl analogs in the final step of the cyclization seem unimportant if the rate of 4-vinyl reduction is much slower than the overall cyclization process. Chereskin *et al.* (1983) noted earlier that MV-Pchlide only accumulated under conditions of etiolation (i.e., when there was more than enough time for 4-vinyl reduction to occur) and that 4-vinyl reduction seemed to occur much more rapidly *after* photoreduction of DV-Pchlide to the corresponding divinylchlorophyllide (DV-Chlide; cf. Section III,D).

Nasrulhaq-Boyce *et al.* (1987) also developed an *in vitro* assay system from etiolated wheat for the conversion of MgPME to Pchlide, by monitoring its subsequent conversion to Chlide. (Visible spectra indicated that the products were the divinyl derivatives in both cases, but these were not assayed specifically.) The conversion required NADPH, oxygen, and intact etioplasts. Proto monomethyl ester and its $Ni^{2+}$ and $Cu^{2+}$ chelates were completely inactive, but ZnPME was as active as MgPME. The conversion was inhibited by lipid-soluble (but not water-soluble) metal-chelating agents. Interestingly, Mg proto dimethyl ester was not a substrate in the wheat system, but Walker *et al.* (1988) had found that dimethyl esters were effective substrates because of the presence of a 7-esterase in their cucumber preparations.

## D. Monovinyl and Divinyl Pathways of Chlorophyll Synthesis: Reduction of the 4-Vinyl Group

As was discussed in Section III,C, Aronoff *et al.* (1971) postulated the existence of parallel pathways for the formation of Chl *a* many years ago (cf. Fig. 16). These parallel pathways were based on the detection of both MV and DV intermediates between MgPME and Pchlide that accumulated in mutants of the green alga *Chlorella*. Hence, the investigators postulated

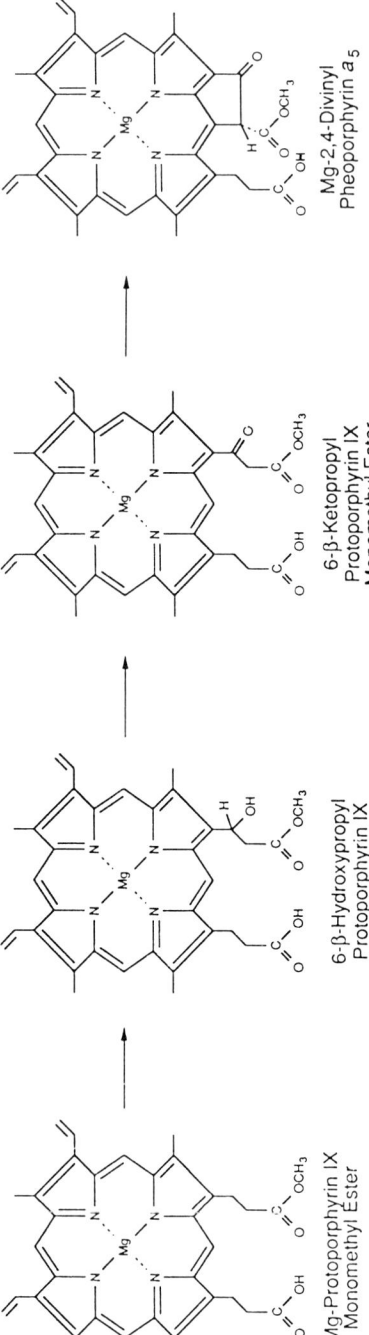

**Fig. 17.** Intermediates proposed by Castelfranco and co-workers in the conversion of magnesium protoporphyrin IX monomethyl ester to divinylprotochlorophyllide by the (oxidative) cyclase enzyme system. Reprinted with permission from Beale and Weinstein (1990). Copyright © 1990 by McGraw-Hill.

## 4. Biosynthesis of the Chlorophyll Chromophore

that reduction of the 4-vinyl group could occur at any point between these two intermediates and that the pathway was a metabolic grid. Shortly thereafter, Ellsworth and Hsing (1973,1974) reported activity due to a 4-vinyl reductase that acted on MgPME in etiolated wheat. Although first reported to use NADPH (Ellsworth, 1972), the enzyme was later demonstrated to use $(R)$-[4-$^3$H$_1$]NADH (generated from [1-$^3$H]ethanol and NAD$^+$ by yeast alcohol dehydrogenase) rather than chemically reduced [4-$^3$H]NADPH. The product was shown to be the 4-ethyl derivative of MgPME. Since the enzyme failed to reduce Mg proto, proto IX, or DV-Pchlide in the presence of chemically reduced [4-$^3$H]NADH (Ellsworth and Hsing, 1973), it was named Mg 4-ethyl-(4-desvinyl)-protoporphyrin IX monomethyl ester:NAD$^+$ oxidoreductase by Ellsworth and Hsing (1974).

Despite the foregoing work, the pathway of Chl *a* biosynthesis usually was represented as a linear pathway with the reduction of the 4-vinyl group occurring during the conversion of DV-Pchlide to MV-Pchlide (cf. Jones, 1978; Castelfranco and Beale, 1981, 1983). Then, in the late 1970s and early 1980s, two separate fractions in each of the intracellular pools of Pchlide, Chlide, Chl *a*, and Chl *b* were detected in several plant species and were characterized by spectrofluorometry as MV and DV analogs (Belanger and Rebeiz, 1979, 1980a,b). In addition, a corn mutant that contained *only* DV-Chl *a* (plus DV-Chl *b*) was isolated by Bazzaz (1981a,b). This mutant's apparent genetic deficiency in a 4-vinyl reductase was indirect evidence for the enzyme's existence, and also proof that all other enzymes in the pathway of Chl *a* and *b* biosynthesis could act on divinyl derivatives. The DV nature of DV-Pchlide, DV-Chlide (Belanger *et al.*, 1982; Wu and Rebeiz, 1984), DV-Chl *a* (Bazzaz and Brereton, 1982; Bazzaz *et al.*, 1982), and DV-Chl *b* (Wu and Rebeiz, 1985) was confirmed subsequently by a variety of chemical and physical methods, including low temperature spectrofluorometry, NMR, and mass spectrometry. Belanger and Rebeiz (1982) also detected the MV-analogs of MgPME and Mg proto in etiolated cucumber and dark-grown *Euglena;* the presence of the MV-analogs of proto IX and protogen IX was inferred also. Rebeiz and co-workers, therefore, proposed parallel MV and DV pathways of Chl biosynthesis from protogen IX to Chls *a* and *b* (Rebeiz *et al.*, 1981; for reviews, cf. Rebeiz, 1982; Rebeiz and Lascelles, 1982). The parallel pathways first were proposed with four branches, including 6- (or 10-) methylester-7-carboxylate and 6- (or 10-) methylester-7-alkylester MV and DV intermediates. The only point of connection (i.e., 4-vinyl reduction) between the DV and MV branches in the *monocarboxylate* pathways was at MgPME, based on the work of Ellsworth and Hsing (1973, 1974). Later, the scheme was expanded to a six-branched pathway by the inclusion of 6- (or 10-) carboxylate-7-alkylester MV and DV intermediates as well

(Rebeiz et al., 1983; cf. Fig. 18). In both cases, however, the 6- (or 10-) methylester-7-carboxylate pathways were described as the major pathways.

In the six-branched pathway (Fig. 18), how the three MV pathways would arise is not altogether clear. Presumably, rather than by reduction of the 4-vinyl group of protogen IX, MV-protogen IX could arise by simple decarboxylation (rather than *oxidative* decarboxylation) of the 4-propionate residue of coprogen III. (No such "coprogen III decarboxylase" activity has been demonstrated, however.) Reduction of the 4-vinyl group was shown to occur only between DV-Chlide and MV-Chlide. The action of a 4-vinyl reductase on DV-Chlide was based on the work of Duggan and Rebeiz (1982a,b) who carried out studies with etiolated cucumber cotyledons induced to accumulate DV-Pchlide devoid of MV-Pchlide by a light–dark pretreatment. The researchers were able to demonstrate conversion of DV-Pchlide to DV-Chlide immediately after a further light flash, and to demonstrate the reduction of DV-Chlide to MV-Chlide during a subsequent dark incubation. Radioactivity from [$^{14}$C]DV-Pchlide synthesized from [$^{14}$C]ALA was traced through both DV-Chlide and MV-Chlide, but no reduction of DV-Pchlide to MV-Pchlide was observed (Duggan and Rebeiz, 1982b). The direct *in vitro* conversion if DV-Pchlide to DV-Chlide by the enzyme NADPH:Pchlide oxidoreductase (Pchlide-reductase), was demonstrated by Bazzaz and Griffiths (1984). Whyte (1989) has shown that the Pchlide-reductase of etiolated wheat has a higher affinity (lower $K_m$) and higher $V_{max}$ for DV-Pchlide than for MV-Pchlide.

Carey and Rebeiz (1985) have classified higher plants as MV or DV based on whether they accumulate predominantly MV-Pchlide or DV-Pchlide when grown in the dark. This classification was expanded to four classes based on the effect of light on MV-Pchlide or DV-Pchlide synthesis induction following dark growth (Carey et al., 1985; Rebeiz et al., 1986). Thus, some 20 different plants were classified as D-DV/L-DV (e.g., cucumber and mustard), D-MV/L-DV (e.g., most commonly studied plants, including wheat, barley, oat, kidney bean, soybean, and tomato), D-DV/L-MV, and D-MV/L-MV; plants in the latter two classes are rare (Rebeiz et al., 1986). In barley in the MV mode, exogenously added DV intermediates (proto IX, Mg proto, and MgPME) were shown to be converted to MV-Pchlide at a point (or points) from proto up to (but not including) DV-Pchlide, whereas little or no such conversion occurred in

**Fig. 18.** Parallel pathways for chlorophyll *a* synthesis proposed by Rebeiz and co-workers. Reprinted with permission from Rebeiz *et al.* (1983). Copyright © 1983 by Kluwer Academic Publishers. Refer to this reference for an explanation of the various intermediates shown.

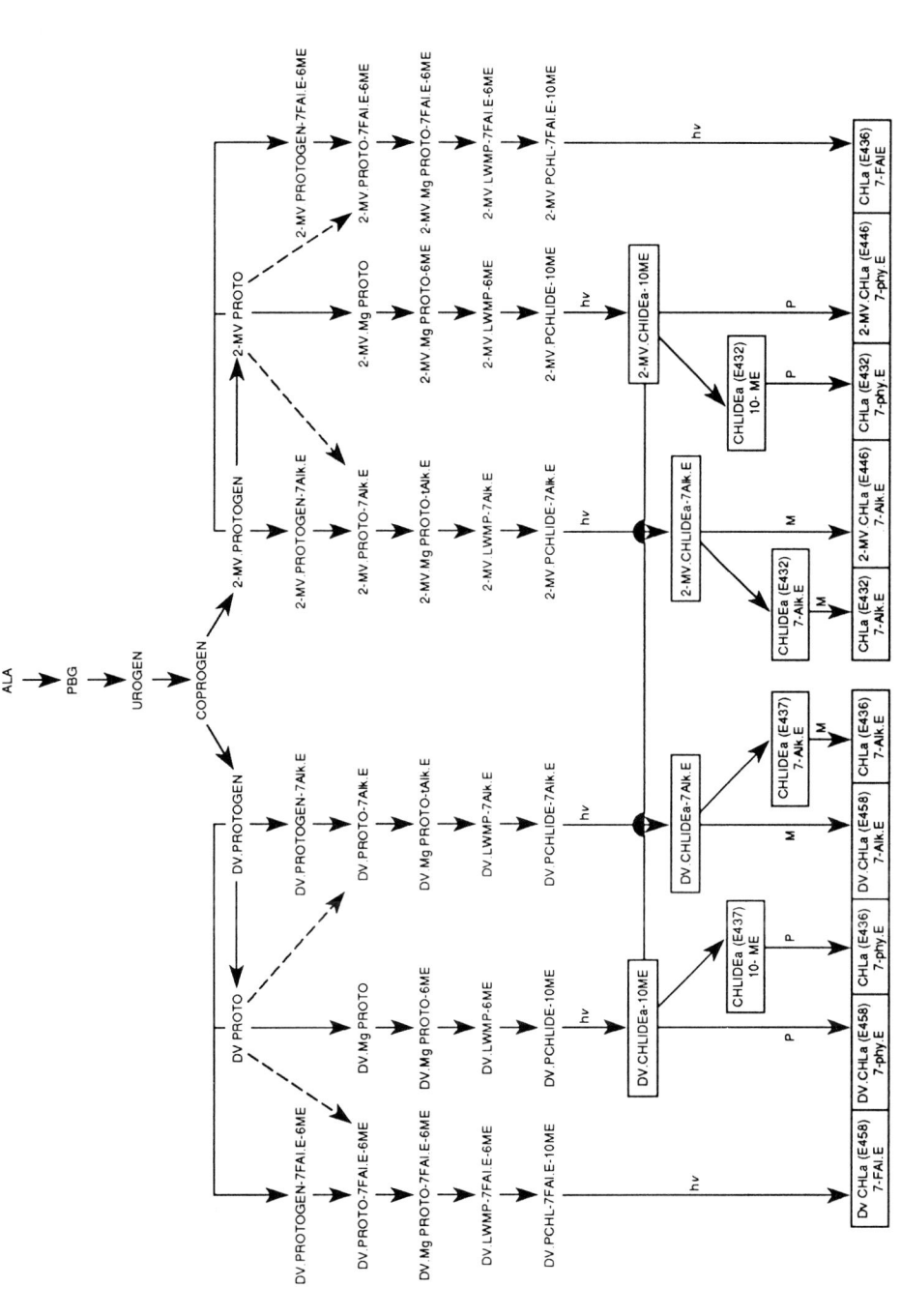

cucumber (Rebeiz *et al.*, 1986; Tripathy and Rebeiz, 1986). Tripathy and Rebeiz (1988) subsequently demonstrated, however, that DV-Pchlide could be reduced to MV-Pchlide in barley (but not in cucumber, at least not on the same time scale) during the transition from the DV to the MV mode of Pchlide synthesis.

Whyte and Griffiths (Whyte, 1989; Griffiths, 1991) have demonstrated a precursor–product relationship between DV-Pchlide and MV-Pchlide in dark-grown 5-day-old wheat seedlings during a 3-hr dark period following 1 hr in the light. The DV-Pchlide and MV-Pchlide were analyzed by HPLC in powdered polyethylene columns by a variation of the method of Shioi and Beale (1987). Thus, the measurements were not subject to inaccuracies inherent in spectrofluorometric determinations due to the presence of decomposition products in the extracts. The rate of the reduction of DV-Pchlide in wheat was estimated to be about 4 times faster than its reduction in etiolated cucumber cotyledons, for which a precursor–product relationship was not evident until well after 8 hr darkness following 1 hr light. In both cases, the $NADPH/NADP^+$ ratio was reduced greatly (10-fold in wheat, 3-fold in cucumber) by the light treatment. Hence, if the reduction of DV-Pchlide by the 4-vinyl reductase requires NADPH and the reaction is near equilibrium (i.e., has a $\Delta G^{0'}$ near 0), then the $NADPH/NADP^+$ ratio will be proportional to the MV-Pchlide/DV-Pchlide ratio. In wheat, Mapleston and Griffiths (1978) showed a rapid recovery of this ratio during a redarkening period (perhaps due to a high activity of the phosphogluconate pathway), which correlated with a corresponding increase in the MV-Pchlide/DV-Pchlide ratio. In cucumber, however, the recovery of the $NADPH/NADP^+$ ratio was only ~30% complete by 6 hr, corresponding to a much slower increase in the MV-Pchlide/DV-Pchlide ratio in this species (Whyte, 1989; Griffiths, 1991).

The reported preference of the 4-vinyl reductase of wheat for NADH over NADPH (Ellsworth and Hsing, 1973) is now seen as curious because of the very low ratios of $NADH/NAD^+$ in wheat (~0.16) and cucumber (~0.30), both of which were little affected by light (Whyte, 1989). Using MgPME and NADH as an assay, Kwan *et al.* (1986) reported a partial purification of a soluble 4-vinyl reductase from etiolated wheat by affinity chromatography on ZnPME-linked Sepharose 4B. However, in preliminary experiments, Richards *et al.* (1987a) were unable to demonstrate a preference of the wheat enzyme for MgPME and NADH. Higher activities were obtained with DV-Pchlide than with MgPME as a substrate and with $(R)$-$[4-^3H_1]$NADPH than with $(S)$-$[4-^3H_1]$NADPH or either stereoisomer of $[4-^3H_1]$NADH as the coenzyme (although no products were isolated to confirm the results). The 4-vinyl reductase of etiolated wheat may have a much higher affinity for MgPME and DV-Pchlide than the 4-vinyl reductase of etiolated cucumber, which may have little or no affinity for MgPME, cyclization intermediates, or DV-Pchlide, but rapidly can convert DV-Chlide to MV-Chlide in whole etioplasts (Duggan and Rebeiz, 1982b) or isolated cotyledons (Hanamoto and Castelfranco, 1983). Results from this laboratory have indicated that broken cucumber etioplasts enriched in DV-Pchlide also are able to convert DV-Chlide to MV-Chlide (but not DV-Pchlide to MV-Pchlide) rapidly in the presence of NADPH during a dark period following a flash that converts some of the DV-Pchlide to DV-Chlide (Richards *et al.*, 1991). The DV-Chlide and MV-Chlide were separated by HPLC on powdered polyethylene by a modification of the method of Shioi and Beale (1987); the separated bands were characterized by spectrofluorometry.

In summary, what can be said about Chl formation *in vivo* as a result of all of the laboratory experiments? The light–dark cycles of Rebeiz *et al.* (1986) were designed to approximate a day–night cycle. During the night, much of the Pchlide synthesized (either DV-Pchlide or MV-Pchlide) would accumulate in the form of a stable ternary complex with NADPH and Pchlide-reductase (cf. Section III,E). However, most of the Pchlide synthesized probably would be in the form of DV-Pchlide since the oxidative cyclase enzyme system of most plants may work much more rapidly than the 4-vinyl reductase (Chereskin *et al.*, 1983). D-MV/L-DV plants (e.g., wheat and barley) have been shown to be able to convert DV-Pchlide to MV-Pchlide in the dark. The concentration of free DV-Pchlide would be expected to be quite low ($<1$ $\mu M$) since wheat Pchlide-reductase has a higher affinity for DV-Pchlide than for MV-Pchlide, exhibiting $K_m$ values of 0.96 and 2.60 $\mu M$, respectively (Whyte, 1989). Hence, DV-Pchlide would be reduced only slowly to MV-Pchlide as it dissociated from the Pchlide-reductase complex and was replaced by the MV-Pchlide produced. Overproduction of DV-Pchlide (as occurs in the presence of ALA), beyond the capacity of the Pchlide-reductase present to bind it, would allow a more rapid reduction of DV-Pchlide to MV-Pchlide. During the day, these plants may synthesize Chl *a* entirely by the DV pathway up to the formation of DV-Chlide.

In D-DV/L-DV plants (e.g., cucumber), the activity of the 4-vinyl reductase may be too low to allow all (or even most) of the DV-Pchlide to be reduced to MV-Pchlide. Light treatments followed by 1 hr darkness increase the content of DV-Pchlide in the regenerated Pchlide pool because there is not enough time for its reduction to MV-Pchlide, which requires a prolonged period of darkness. At the first light of dawn, all DV-Pchlide in a photoconvertible Pchlide-reductase complex would be reduced to DV-Chlide, which would then dissociate from its Pchlide-reductase complex, be converted rapidly (within minutes) to MV-Chlide by 4-vinyl reductase, and, finally, esterified to Chl *a* (Section III,F). Plants that have the D-MV/L-MV mode of Pchlide synthesis may have a 4-vinyl reductase with higher affinities for earlier intermediates; however, D-DV/L-MV plants cannot be explained by a similar argument.

Tripathy and Rebeiz (1986) have shown that exogenously added MV-proto can be converted to MV-Pchlide; hence, the Mg chelatase, methyltransferase, and oxidative cyclase enzyme systems all can work with MV intermediates. Does the 4-vinyl reductase convert an early intermediate to its MV analog in MV plants? Although etiolated wheat 4-vinyl reductase has been shown to be active with exogenous MgPME, it is not known whether it can act on Mg proto and proto IX also. The detection of MV analogs of these intermediates may be explained by the sequential action on MV-MgPME of a 6-esterase (Wong and Castelfranco, 1984) and an Mg dechelatase (Shioi *et al.*, 1991), rather than by the generation of the MV analogs from coprogen III by a nonoxidative decarboxylation of the 4-propionate residue or by the direct action of the 4-vinyl reductase. Also, whenever an intermediate is added as an exogenous substrate or is induced to accumulate, it may become a substrate for 4-vinyl reductase because of its presence in much higher concentrations than are present *in vivo*, where equivalent reductions may not occur. The use of early DV intermediates as substrates (Tripathy and Rebeiz, 1986) may have been an example of such a phenomenon. Other examples include the accumulation of cyclization intermediates in the *Chlorella* mutants (Aronoff *et al.*, 1971) and the accumulation of MgPME by treatment with $\alpha,\alpha'$-dipyridyl and ALA (Belanger and Rebeiz, 1982). Hence, the (monocarboxylate) pathway of Chl *a* actually may be very close to a linear pathway *in vivo*, with 4-vinyl reduction occurring primarily at DV-Pchlide (in the night) in D-MV plants and primarily at DV-Chlide (in the daytime) in plants with D-DV or L-DV modes of synthesis. However, more work must be done before the actual *in vivo* pathways can be known with certainty.

Work on the mechanism of 4-vinyl reductase has been done with the Bchl-synthesizing photosynthetic bacterium *Rhodobacter sphaeroides*. As was discussed in Section II,E,2, vinyl group formation proceeds with retention of both $\alpha$-hydrogens, but only one $\beta$-hydrogen, of the propionic acid residues during the synthesis of protogen IX from coprogen III (Fig. 11). Using this information, Battersby *et al.* (1981a) used $(2S,3S)$-$[2$-$^2H_1,^3H_1,3$-$^2H_1]$ALA (Fig. 19A) and Emery and Akhtar (1985b) used $(2R,3R)$-$[2,3$-$^3H_2]$ALA (Fig. 19B) as precursors to determine that there was *trans*-addition of two hydrogen atoms to the 4-vinyl group from the directions indicated in Fig. 19 during its reduction to an ethyl group by 4-vinyl reductase. These investigators determined the absolute configuration of the methyl and methylene carbons, respectively, of the propionic acid generated after degradation of the resulting Bchl. For methyl carbons,

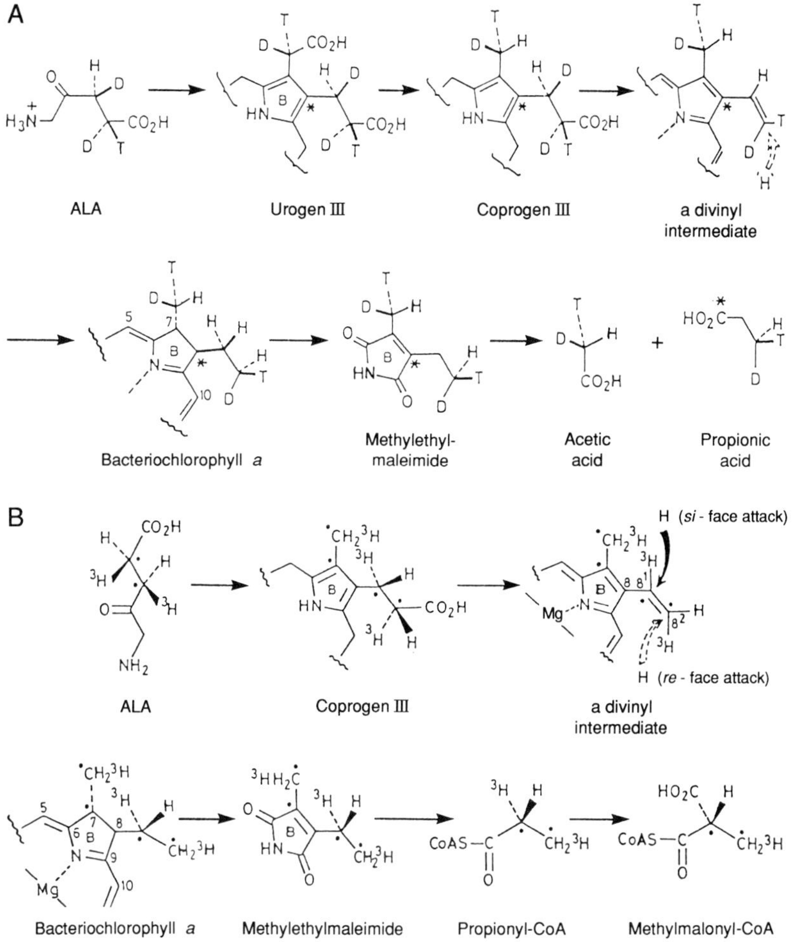

# 4. Biosynthesis of the Chlorophyll Chromophore

**Fig. 19.** Stereochemistry of 4-vinyl reduction in bacteriochlorophyll biosynthesis. Labeling studies were done to determine the stereochemistry of (A) the methyl carbon and (B) the methylene carbon of the resulting 4-ethyl group. Adapted from (A) Battersby et al. (1981a) and (B) Emery and Akhtar (1985b).

the propionic acid was degraded to $(R)$-$[^2H_1, ^3H_1]$acetic acid, whereas for methylene carbons it was converted to its CoA-derivative and carboxylated by propionyl-CoA carboxylase, an enzyme known to be specific for removal of the *pro R* hydrogen of C-2.

Ellsworth and Hsing (1973) had found up to 50% of the maximum theoretical radioisotopic incorporation when [4-$^3$H]NADH was incubated with MgPME in crude extracts of etiolated wheat. This result indicates a direct transfer of a hydride from NADH to one of the carbons of the 4-vinyl group. As described earlier, preliminary work in this laboratory has been carried out on the conversion of DV-Chlide to MV-Chlide in broken etioplasts of cucumber cotyledons in a brief dark period following the photoconversion of accumulated DV-Pchlide to DV-Chlide (Richards et al., 1991). The dark incubation was carried out in the presence of either the ($R$)- or ($S$)-isomer of [4-$^3$H$_1$]NADPH; only in the presence of the ($R$)-isomer was there any significant incorporation of radioactivity into the resulting tetrapyrrole fraction. When the tetrapyrroles were separated by HPLC on powdered polyethylene, however, radioactivity appeared in an unidentified band immediately preceding the MV-Chlide band, which should have contained $^3$H had there been a direct transfer to one of the two carbons of the 4-vinyl group of DV-Chlide (I. Snajdorova and W. R. Richards, 1991, unpublished observations). Therefore, perhaps the hydride was transferred instead to the 3 position of the pyrrole ring which, with protonation of the $\beta$-carbon of the vinyl group, would yield a $^3$H-labeled ethylidine intermediate. The structure of this hypothetical intermediate is identical to that of Bchlide $g$. Michalski et al. (1987) demonstrated that Bchl $g$ can be photoisomerized readily to Chl $a$. However, such an isomerization of the proposed labeled intermediate would lead to loss of the $^3$H atom, yielding unlabeled MV-Chlide. The identity of the labeled band is currently under investigation.

Finally, Pudek and Richards (1975) have raised the possibility of parallel 4-vinyl and 4-ethyl pathways of Bchl synthesis in *R. sphaeroides*. Currently no evidence supports the presence of any 4-vinyl intermediates beyond DV-Pchlide in any of the known mutants. In mutants deficient in either the *bchB* or *bchL* gene, both DV-Pchlide and MV-Pchlide were found to accumulate (Yang and Bauer, 1990). These genes initially were thought to encode Pchlide-reductase and 4-vinyl reductase, respectively, but are now thought to encode components of a dark bacterial Pchlide-reductase (Burke et al., 1991). Hence, 4-vinyl reductase is one of the few enzymes in the Mg branch of Bchl synthesis for which no mutant has been assigned. Both DV-Pchlide and MV-Pchlide also accumulate when nicotinamide, an apparent inhibitor of Pchlide-reductase, is added to cultures of *R. sphaeroides* (Shioi et al., 1988). Perhaps potential mutants of 4-vinyl reductase cannot be detected because 4-vinyl intermediates can be converted to a functional 4-vinyl Bchl.

### E. Protochlorophyllide Reductase

NADPH:Pchlide oxidoreductase (EC 1.6.99.1) is discussed in much more detail by Schulz and Senger in Chapter 5 of this volume. Studies on the molecular nature of Pchlide-reductase began when Horton and Leach

## 4. Biosynthesis of the Chlorophyll Chromophore

(1972) reported that an etioplast preparation of corn retained its photoconvertible Pchlide$_{638/650}$ when supplemented with ATP, and Henningsen and Kahn (1971) reported that the Pchlide holochrome could be dissociated by saponin to an active form with an $M_r$ of 63 kDa. Griffiths (1974a,b) later demonstrated that isolated barley etioplast membranes were also photoactive when supplemented with NADPH. Enzyme activity could be reconstituted with Pchlide (but not Pchl) added exogenously in a cholate sol (Griffiths, 1974b, 1978, 1980). Chlide was labeled by [4-$^3$H]NADPH in the light, but did not label the enzyme in the dark, giving rise to the notion of a dark-stable ternary Enzyme–Pchlide–NADPH complex (Griffiths, 1978, 1981). Mapleston and Griffiths (1978) demonstrated that adequate amounts of NADPH were present in the dark. The various spectral forms of Pchlide and Chlide subsequently were explained in terms of various ternary complexes formed between Pchlide-reductase, Pchlide, Chlide, NADPH, and NADP$^+$ (El Hamouri and Sironval, 1980; El Hamouri et al., 1981, 1983; Oliver and Griffiths, 1982). These spectroscopic forms (cf. Fig. 20) have been reviewed by Griffiths and Oliver (1984), Beale and Weinstein (1990), and Griffiths (1991).

Pchlide-reductase is a major protein constituent of angiosperm etioplast membranes, localized primarily in prolamellar bodies (cf. Sundqvist and Ryberg, 1989). Active enzyme is associated either with a single polypeptide (barley, rye, squash, and wheat) or with a closely spaced doublet (oats and beans) with $M_r$ values between 34 and 38 kDa. Joyard et al. (1990)

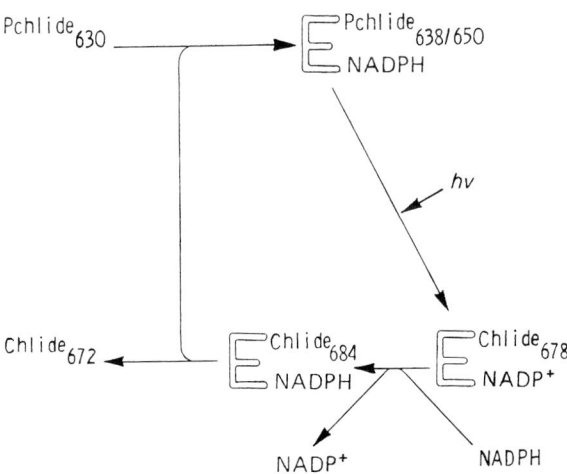

**Fig. 20.** Proposed ternary complexes formed between NADPH:protochlorophyllide oxidoreductase (E) and its substrates and products. Reprinted with permission from Griffiths and Oliver (1984). Copyright © 1984 by Cambridge University Press.

have found, however, that Pchlide-reductase is located on the cytoplasmic side of the outer envelope membrane of *mature* spinach chloroplasts. The molecular properties of Pchlide-reductase and its fate during the development of etioplasts to chloroplasts is discussed in more detail in Chapter 5.

Pchlide-reductase has been purified partially from various plants by a number of workers after its solubilization with Triton X-100 (Apel *et al.*, 1980; Beer and Griffiths, 1981; Ikeuchi and Murakami, 1982; Stadnichuk and Walter, 1984). Richards *et al.* (1987b) found that solubilization with 4 m$M$ 3-[(3-cholamidopropyl)-dimethylammonio]-1-propane-sulfonate (CHAPS) resulted in a photoconvertible Pchlide$_{638/650}$ complex without the readdition of Pchlide. The solubilized complex was irradiated in the presence of an NADPH-generating system just as it entered an affinity chromatography column containing either DV-Pchlide or its zinc analog covalently linked to Sepharose CL-4B. The affinity column was developed with buffers containing NADP$^+$. Some of the Pchlide-reductase was not retained by the column; however, the retained enzyme was not eluted by removal of NADP$^+$ from the buffer, indicating that the latter was not required for firm binding of Pchlide to the enzyme. Active Pchlide-reductase was eluted by the addition of DV-Pchlide (or its zinc analog) to the elution buffer. Although the specific activity of the purified enzyme was reduced by this procedure (perhaps because of partial inactivation of the enzyme), most of the contaminating polypeptides visible by Coomassie blue staining after SDS–PAGE had been removed.

The mechanism of Pchlide-reductase has been studied from several different points of view. The involvement of a cysteine residue, either in the reaction mechanism or in substrate binding, has been inferred from the potent inhibition of Pchlide-reductase by thiol-specific reagents (Griffiths, 1975). Using one of these reagents, [$^3$H]$N$-phenylmaleimide, Oliver and Griffiths (1981) demonstrated that both NADPH and Pchlide protected against labeling (and hence inactivation) of the enzyme; however, NADPH was somewhat more effective in this regard, indicating a closer proximity of the cysteine to the NADPH-binding site. A putative NADPH-binding region was identified in the first 31 amino acids of the N terminus of the mature oat Pchlide-reductase (Darrah *et al.*, 1990). Positively charged amino acid residues also were inferred to be required to bind the 7-propionate group of Pchlide, its substrate analogs (Griffiths, 1980), and the 2′-phospho group of NADPH (Oliver and Griffiths, 1981) and to act as a proton donor during the reduction reaction (Oliver, 1982). Oat Pchlide-reductase has an abundance of basic residues (12% Arg + Lys), although specific residues involved in the functions just discussed have not yet been identified.

## 4. Biosynthesis of the Chlorophyll Chromophore

Griffiths (1981) was the first to demonstrate that one equivalent of hydrogen was transferred to Pchlide during its photoreduction to Chlide with chemically reduced [4-$^3$H]NADPH. (Since this substance is achiral, labeling with 0.5 equivalents of tritium was considered equal to the transfer of one equivalent of hydrogen.) As mentioned in Section III,D, Pchlide-reductase can convert either DV-Pchlide or MV-Pchlide to its corresponding Chlide (Bazzaz and Griffiths, 1984; Whyte, 1989). When DV-Pchlide was incubated with [4-$^3$H]NADPH in place of MV-Pchlide, once again only one equivalent of hydrogen was found to be transferred rather than two, if reduction of the 4-vinyl group were also occurring (Griffiths, 1981). Using stereospecifically labeled (R)- and (S)-isomers of [4-$^3$H$_1$]NADPH, Valera et al. (1987) determined that the pro S hydrogen of NADPH was transferred to Pchlide during the photoreduction. In an elegant labeling study using the two stereospecifically labeled isomers of 4-[$^2$H$_1$]NADPH, Begley and Young (1989) confirmed the results of Valera et al. (1987) by analyzing the products by $^2$H NMR. These investigators further demonstrated that the pro S hydride of NADPH was added to $C_7$ of pyrrole ring D, whereas $C_8$ was protonated by a hydrogen ion in equilibrium with water. Therefore, they proposed the model shown in Fig. 21 for the active site of Pchlide-reductase. Walker and Griffiths (1988) have presented evidence that Pchlide-reductase may be a flavoprotein. The enzyme was inhibited by micromolar concentrations of the flavin antagonists quinacrine and trifluoperazine, and FAD was found to copurify with fractions containing the active Pchlide-reductase. Although the involvement of a flavin in the reduction of Pchlide by NADPH has not yet been confirmed, the results of these studies imply that, if a hydride is first transferred to a bound flavin, the same hydride subsequently must be transferred to the Pchlide.

It has been known since the early 1950s that gymnosperms, algae, and cyanobacteria can form Chl in complete darkness and that Pchlide is an intermediate in the process. Organisms of this type must, therefore, possess a light-independent enzyme for Pchlide reduction that is referred to as dark Pchlide-reductase. Work with the cyanobacterium *Anacystis nidulans* (*Synechococcus* PCC 6301) has indicated that the dark Pchlide-reductase is localized in distinct fractions of the plasma membrane in which both Pchlide and Chlide (but no Chl *a*) accumulate (Peschek et al., 1989a,b). Incubation in the dark with NADPH leads to the conversion of all accumulated Pchlide to Chlide. NADPH was several times more effective as a reductant than NADH. The reaction was inhibited by EGTA, indicating a $Ca^{2+}$ requirement for the reduction. In addition, the reaction was reversible; incubation with $NADP^+$ (but not $NAD^+$) led to conversion of all accumulated Chlide to Pchlide.

Many of these same organisms also have a light-dependent pho-

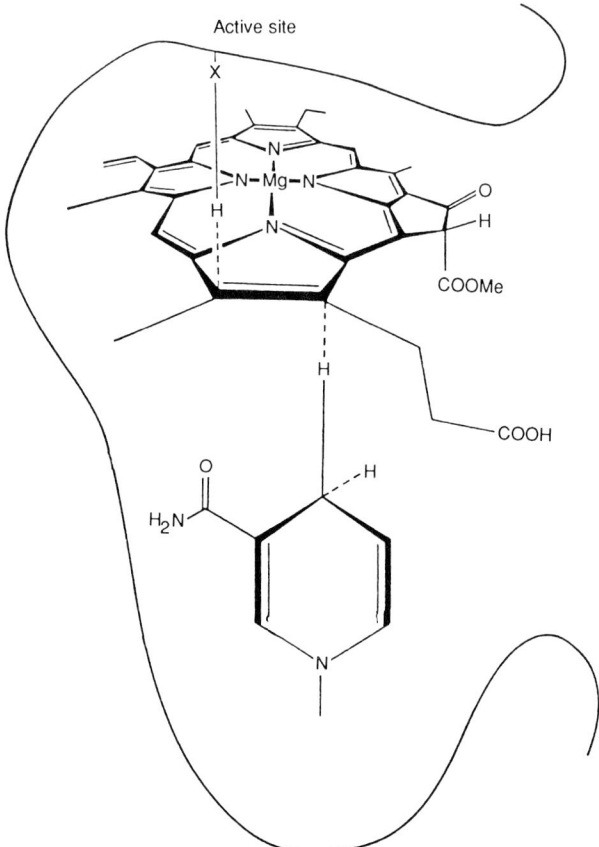

**Fig. 21.** Possible juxtaposition of the substrates in the active site of NADPH:protochlorophyllide oxidoreductase that would result in the pattern of hydrogen atom addition observed for pyrrole ring D. Reprinted with permission from Begley and Young (1989). Copyright © 1989 by the American Chemical Society.

toenzyme. For example, mutants of the green alga *Chlamydomonas reinhardtii* have been isolated that lack one or both of these enzymes (Ford et al., 1981,1983). Whether angiosperms also have a dark Pchlide-reductase has been the subject of a number of investigations. In greening barley incubated in [$^{14}$C]ALA in the dark, Chl(ide) was found to contain only about 1.2 or 3.6%, respectively, of the label incorporated into Pchlide when excised (Apel et al., 1984) or whole (Packer and Adamson, 1986; Packer et al., 1987) seedlings were used. At the same time, although the whole seedlings formed 6.5 times as much Chl *a* in the light as in the dark, approximately 800 times more radioactivity was incorporated into Chl *a* in the light than in the dark (Packer *et al.*, 1987). The authors speculated that the exogenously added [$^{14}$C]ALA was able to enter the light-dependent Pchlide-reductase pathway but not the dark Pchlide-reductase pathway. Perhaps any Pchlide that accumulated under these conditions might have been only in the form of a (light-dependent) Pchlide-reductase complex and therefore protected from the action of the dark Pchlide-reductase. Also, no analysis of the relative amounts of MV- and DV-Pchlides was carried out in these studies. Tripathy and Rebeiz (1987) used [$^{14}$C]ALA and [$^{14}$C]glutamate during Chl *a* formation in the dark in greening barley leaves and isolated barley etiochloroplasts. They found that, in greening leaves, [$^{14}$C]glutamate formed entirely MV-Pchlide and very little radioactivity was found to be passed into the Chl *a* pool, whereas [$^{14}$C]ALA yielded Pchlide with an MV/DV ratio of 1.2, with a much higher incorporation of activity into the Chl *a* pool. In etiochloroplasts, however, activity from both precursors was used almost equally well for Chl *a* synthesis.

Phototrophic purple bacteria also are able to synthesize Bchl in the dark. At least two genes (*bchB* and *bchL*) are thought to be involved in the step that converts Pchlide to Chlide (whether at the MV or DV level is not known). The gene product of one of these genes (*bchL*) has been found to be homologous to an iron–sulfur protein (the gene product of the *nifH* gene) of the nitrogenase enzyme system that exists as a dimer with a single $Fe_4S_4$ center shared between the two monomers, each of which has an ATP binding site (Burke *et al.*, 1991). The gene products of two other genes homologous to the *bchL* gene product have been detected in the chloroplast genomes of *Marchantia polymorphia* and *Chlamydomonas reinhardtii* (C. E. Bauer, 1991, personal communication). These genes, which are referred to as the *frxC* and *gidB* genes, respectively, may be part of a dark Pchlide-reductase enzyme complex in these organisms. DNA complementary to *bchL* also has been found in two cyanobacteria, in the chloroplast genomes of six nonflowering vascular plants (including four ferns), and in five species of gymnosperms, but not in the chloroplast genome of several angiosperms, including corn, tobacco, rice, *Arabidopsis thaliana*, or *Bougainvillea glabra* (C. E. Bauer, 1991, personal communication). On the other hand, Spano *et al.* (1991) have obtained evidence for the existence of a nuclear multigene family for the Pchlide-reductase of dark-grown white pine needles. Two distinctly different expressed forms of the enzyme were found to have sequences similar to the photoconvertible Pchlide-reductases of angiosperms. Whether these expressed forms represented photoconvertible or dark Pchlide-reductases is still unknown.

## F. Chlorophyll Synthetase

The last step of Chl $a$ synthesis occurs by transfer of a 20-carbon isoprenyl group from geranylgranyl diphosphate (GGPP) or phytyl diphosphate (PhyPP) to the 7-carboxylate of Chlide $a$. The enzyme is called Chl synthetase and can be classified as an alkyltransferase (EC 2.5). This enzyme will be discussed in greater detail by Rüdiger in Chapter 6 of this volume. The enzyme also has been reviewed by Rüdiger and Schoch (1988, 1991); hence, only a brief introduction will be given here.

At one time, it was considered possible that Chlide could be esterified by reversal of the hydrolytic enzyme chlorophyllase (EC 3.1.1.14) (Ellsworth et al., 1976). Esterification of Chlide in the presence of very high concentrations of phytol had been known since the early part of the century; however, such a high concentration of phytol has never been observed in plants under physiological conditions (Steffens et al., 1976). Chlorophyllase also can catalyze a transesterification between an alkyl Chlide ester and a free alcohol (Ellsworth, 1971). Hence, a reaction between methyl Chlide and free phytol was considered a possibility; again, however, no evidence for the formation of methyl Chlide was ever obtained (Ellsworth et al., 1976). Rebeiz et al. (1983) have found evidence for the presence of nonisoprenoid long-chain alkyl esters of Chlide (as well as other intermediates). It is possible that these could be converted to (or arise from?) phytol esters by transesterification with chlorophyllase. However, the major pathway of Chl synthesis is still considered to be through intermediates with a free 7-carboxylate. Both chlorophyllase reactions would require attack of the oxygen of free phytol on the carboxyl carbon with elimination of one of the oxygen atoms of the 7-carboxylate of Chlide. Akhtar and co-workers (Akhtar et al., 1984; Ajaz et al., 1985; Emery and Akhtar, 1985a,1987) carried out an $^{18}O$-labeling study with Bchl $a_P$ (of *Rhodobacter sphaeroides*) and Bchl $a_{GG}$ (of *Rhodospirillum rubrum*) and determined that *both* oxygens of the 7-carboxylate were retained during the esterification reaction. These results prove that, in these organisms at least, esterification proceeds by an isoprenyltransferase step; the oxygen of the isoprenyl alcohol is incorporated into a leaving group such as pyrophosphate. Hence, although equivalent studies have not been carried out in plants, an identical mechanism is likely to apply to Chl $a$ biosynthesis. Chlorophyllase may, therefore, be relegated to the function of a Chl degradative enzyme.

Chl synthetase was described first by Rüdiger et al. (1977, 1980). Both Chlide $a$ and Chlide $b$ were esterified equally well (about twice as well as pyrochlorophyllide $a$). Chl synthetase also must be able to esterify DV-Chlide, since both DV-Chl $a$ and DV-Chl $b$ accumulated in the corn

## 4. Biosynthesis of the Chlorophyll Chromophore

mutant of Bazzaz (1981a,b). Pheophorbide *a*, Pchlide, and Bchlide *a* were essentially inactive and showed no inhibition of the esterification of Chlide, indicating that they were not bound by the enzyme (Benz and Rüdiger, 1981). Vezitskii and Shcherbakov (1987) have determined that both the $Zn^{2+}$ and $Cd^{2+}$ chelates of pheophorbide *a* were esterified, whereas the $Cu^{2+}$ and $Ni^{2+}$ chelates were not, in contrast to the action of chlorophyllase, which hydrolyzed the phytol ester of pheophytin *a* and its chelates with all these above metals.

GGPP and PhyPP were found to be approximately equally effective in esterification when present in saturating amounts (20–30 molar excess with respect to Chlide) (Rüdiger *et al.*, 1980); both were much better substrates than farnesyl diphosphate. The enzyme seems to use GGPP preferentially in etioplasts (Rüdiger *et al.*, 1980) incorporating PhyPP in chloroplasts (Soll *et al.*, 1983). The relative reaction rates of GGPP and PhyPP in these two organelles were 2:1 and 1:4, respectively (Rüdiger and Schoch, 1991). A labeling study carried out *in vivo* indicated that only Chl $a_P$ was formed from endogenous isoprenyl alcohols in green plants, whereas the Chl $a_{GG}$ pool was unlabeled (Rüdiger, 1987; Rüdiger and

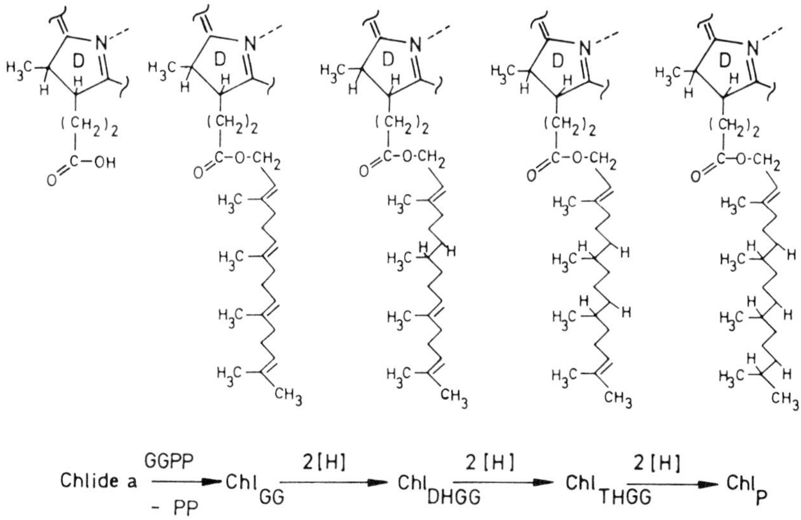

$$\text{Chlide } a \xrightarrow[-\text{PP}]{\text{GGPP}} \text{Chl}_{GG} \xrightarrow{2[H]} \text{Chl}_{DHGG} \xrightarrow{2[H]} \text{Chl}_{THGG} \xrightarrow{2[H]} \text{Chl}_P$$

**Fig. 22.** Esterification of chlorophyllide *a* with geranylgeranyl diphosphate, followed by a three-step reduction of the geranylgeranyl (GG) ester to a phytol (P) ester through dihydrogeranylgeranyl (DHGG) and tetrahydrogeranylgeranyl (THGG) ester intermediates. Reprinted with permission from Rüdiger and Benz (1984). Copyright © 1984 by Cambridge University Press.

Schoch, 1991). Chl synthetase has been localized to the prothylakoid and prolamellar body fractions of etioplasts (Lütz et al., 1981) and the thylakoids of chloroplasts (Soll et al., 1983). In etioplasts of wheat, Lindsten et al. (1990) have shown that Chl synthetase activity in the prolamellar body fraction is latent but becomes fully active in the prothylakoid fraction.

The reduction of GGPP to PhyPP by NADPH has been localized to the plastid envelope membrane (Soll and Schultz, 1981; Soll et al., 1983). GGPP is fairly soluble in the stroma, whereas PhyPP remains in the envelope membrane fraction (Rüdiger, 1987). In etioplasts, Chl $a_{GG}$, the ester formed primarily, may be reduced through a series of intermediates to Chl $a_P$ (Fig. 22; Schoch et al., 1977; Rüdiger and Benz, 1984). Reducing equivalents are provided by NADPH (Benz et al., 1980; Soll and Schultz, 1981). The reduction, but not the esterification, is inhibited by anaerobiosis (Schoch et al., 1980). However, a mutant of *Scenedesmus* unable to carry out the reduction can use Chl $a_{GG}$ as a completely functional Chl (Henry et al., 1986). In chloroplasts, it is not altogether clear how PhyPP reaches the site of esterification in the thylakoids, nor whether, in fact, the Chl synthetase of chloroplasts is the same enzyme as that found in etioplasts (Rüdiger, 1987). Attempts to purify solubilized Chl synthetase by affinity chromatography, using the zinc analogs of either Chlide or pyrochlorophyllide as ligands, have been unsuccessful (P. Lauterbach, W. R. Richards, and W. Rüdiger, 1989, unpublished observations).

## REFERENCES

Abboud, M. M., and Akhtar, M. (1976). Stereochemistry of hydrogen elimination of the enzymic formation of the C2–C3 double bond of porphobilinogen. *J. Chem. Soc. Chem. Commun.* 1007–1008.

Ajaz, A. A., Corina, D. L., and Akhtar, M. (1985). The mechanism of the C-$13^3$ esterification step in the biosynthesis of bacteriochlorophyll *a*. *Eur. J. Biochem.* **150**, 309–312.

Akhtar, M., Ajaz, A. A., and Corina, D. L. (1984). The mechanism of the attachment of esterifying alcohol in bacteriochlorophyll *a* biosynthesis. *Biochem. J.* **224**, 187–194.

Al-Hazimi, H. M. G., Jackson, A. H., Knight, D. W., and Lash, T. D. (1987). Synthetic and biosynthetic studies of porphyrins. 7. The action of coproporphyrinogen oxidase on coproporphyrinogen IV: Syntheses of protoporphyrin XIII, mesoporphyrin XIII, and related tricarboxylic porphyrins. *J. Chem. Soc. Perkin Trans. 1*, 265–276.

Anderson, P. M., and Desnick, R. J. (1980). Purification and properties of uroporphyrinogen I synthase from human erythrocytes. *J. Biol. Chem.* **255**, 1993–1999.

Andrew, T. L., Riley, P. G., and Dailey, H. A. (1990). Regulation of heme biosynthesis in higher animals. *In* "Biosynthesis of Heme and Chlorophylls" (H. A. Dailey, ed.), pp. 163–200. McGraw-Hill, New York.

## 4. Biosynthesis of the Chlorophyll Chromophore

Apel, K., Santel, H.-J., Redlinger, T. E., and Falk, K. (1980). The protochlorophyllide holochrome of barley (*Hordeum vulgare* L.). Isolation and characterization of the NADPH : protochlorophyllide oxidoreductase. *Eur. J. Biochem.* **111**, 251-258.

Apel, K., Motzkus, M., and Dehesh, K. (1984). The biosynthesis of chlorophyll in greening barley (*Hordeum vulgare*): Is there a light-independent protochlorophyllide reductase? *Planta* **161**, 550-554.

Aronoff, S., and Ellsworth, R. K. (1968). The biogenesis of chlorophyll *a*. *Photosynthetica* **2**, 288-297.

Aronoff, S., Houlson, P. R., and Ellsworth, R. K. (1971). Investigations on the biogenesis of chlorophyll *a*. V. Ordering of submutants of ultraviolet chlorophyll mutants of *Chlorella*. *Photosynthetica* **5**, 166-169.

Avissar, Y. J., and Beale, S. I. (1988). Biosynthesis of tetrapyrrole pigment precursors: Formation and utilization of glutamyl-tRNA for δ-aminolevulinic acid synthesis by isolated enzyme fractions from *Chlorella vulgaris*. *Plant Physiol.* **88**, 879-886.

Avissar, Y. J., and Beale, S. I. (1989). Biosynthesis of tetrapyrrole pigment precursors: Pyridoxal requirement of the aminotransferase step in the formation of δ-aminolevulinate from glutamate in extracts of *Chlorella vulgaris*. *Plant Physiol.* **89**, 852-859.

Avissar, Y. J., Ormerod, J. G., and Beale, S. I. (1989). Distribution of δ-aminolevulinic acid biosynthetic pathways among photosynthetic bacteria and related organisms. *Arch. Microbiol.* **151**, 513-519.

Barnard, G. F., and Akhtar, M. (1975). Stereochemistry of porphyrinogen carboxylase reaction in heme biosynthesis. *J. Chem. Soc. Chem. Commun.* 494-496.

Barnard, G. F., and Akhtar, M. (1979). Stereochemical and mechanistic studies on the decarboxylation of uroporphyrinogen III in heme biosynthesis. *J. Chem. Soc. Perkin Trans. 1*, 2354-2360.

Barnard, G. F., Itoh, R., Hohberger, L. H., and Shemin, D. (1977). Mechanism of porphobilinogen synthase. Possible role of essential thiol groups. *J. Biol. Chem.* **252**, 8965-8974.

Batlle, A. M. C., and Rossetti, M. V. (1977). Review. Enzymic polymerization of porphobilinogen into uroporphyrinogens. *Int. J. Biochem.* **8**, 251-267.

Battersby, A. R. (1978). The discovery of nature's biosynthetic pathways. *Experientia* **34**, 1-13.

Battersby, A. R. (1985). Biosynthesis of the pigments of life. *Proc. R. Soc. London B* **225**, 1-26.

Battersby, A. R. (1986a). Biosynthesis of the pigments of life. *Ann. N. Y. Acad. Sci.* **471**, 138-154.

Battersby, A. R. (1986b). Stereochemical and biosynthetic studies on the pigments of life. *In* "Workshop Conferences Hoechst" (W. Bartmann and K. B. Sharpless, eds.), Vol. 17, pp. 283-302. VCH, New York.

Battersby, A. R. (1987). Nature's pathways to the pigments of life. *Natl. Prod. Rep.* **4**, 77-87.

Battersby, A. R., Baldas, J., Collins, J., Grayson, D. H., James, R. J., and McDonald, E. (1972). Mechanism of biosynthesis of the vinyl groups of protoporphyrin IX. *J. Chem. Soc. Chem. Commun.* 1265-1266.

Battersby, A. R., Hunt, E., McDonald, E., Paine, J. B., III, and Saunders, J. (1976a). Biosynthesis of porphyrins and related macrocycles. 8. Enzymic decarboxylation of uroporphyrinogen III: Structure of an intermediate phyriaporphyrinogen III and synthesis of the corresponding porphyrin and of two isomeric porphyrins. *J. Chem. Soc. Perkin Trans. 1*, 1008-1018.

Battersby, A. R., McDonald, E., Redfern, J. R., Staunton, J., and Wightman, R. H. (1976b). Biosynthesis of porphyrins and related macrocycles. V. Structural integrity of the type III porphyrinogen macrocycle in an active biological system; studies on the aromatisation of protophyrinogen IX. *J. Chem. Soc. Perkin Trans. 1*, 266-273.

Battersby, A. R., McDonald, E., William, D. C., and Wurziger, H. K. W. (1977). Biosynthesis of the natural (type-III) porphyrins: Proof that rearrangement occurs after head-to-tail bilane formation. *J. Chem. Soc. Chem. Commun.* 113–115.
Battersby, A. R., Fookes, C. J. R., Matcham, G. W. J., and McDonald, E. (1978a). Biosynthesis of natural porphyrins: Enzymic experiments on isomeric bilanes. *J. Chem. Soc. Chem. Commun.* 1064–1066.
Battersby, A. R., Fookes, C. J. R., McDonald, E., and Meegan, M. (1978b). Biosynthesis of type-III porphyrins: Proof of intact enzymic conversion of head-to-tail bilane into uroporphyrinogen III by intramolecular rearrangement. *J. Chem. Soc. Chem. Commun.* 185–186.
Battersby, A. R., Fookes, C. J. R., Gustafson-Potter, K. E., Matcham, G. W. J., and McDonald, E. (1979a). Proof by synthesis that unrearranged hydroxymethylbilane is the product from deaminase and the substrate for cosynthetase in the biosynthesis of uroporphyrinogen III. *J. Chem. Soc. Chem. Commun.* 1155–1158.
Battersby, A. R., Fookes, C. J. R., Matcham, G. W. J., and McDonald, E. (1979b). Order of assembly of the four pyrrole rings during the biosynthesis of natural porphyrins. *J. Chem. Soc. Chem. Commun.* 539–541.
Battersby, A. R., Gutmann, A. L., Fookes, C. J. R., Gunther, H., and Simon, H. (1981a). Stereochemistry of formation of methyl and ethyl groups in bacteriochlorophyll *a*. *J. Chem. Soc. Chem. Commun.* 645–647.
Battersby, A. R., Fookes, C. J. R., Matcham, G. W. J., and Pandey, P. S. (1981b). Biosynthesis of natural porphyrins: Studies with isomeric hydroxymethylbilanes on the specificity and action of cosynthetase. *Angew. Chem. Int. Ed. Engl.* **20**, 293–295.
Battersby, A. R., Fookes, C. J. R., Gustafson-Potter, K. E., McDonald, E., and Matcham, G. W. J. (1982). Biosynthesis of porphyrins and related macrocycles. 17. Chemical and enzymic transformation of isomeric aminomethylbilanes into uroporphyrinogens: Proof that unrearranged bilane is the preferred enzymic substrate and detection of a transient intermediate. *J. Chem. Soc. Perkin Trans. 1*, 2413–2426.
Battersby, A. R., Fookes, C. J. R., and Pandey, P. S. (1983). Linear tetrapyrrolic intermediates for biosynthesis of the natural porphyrins: Experiments with modified substrates. *Tetrahedron* **39**, 1919–1926.
Baum, S. J., Burnham, B. F., and Plane, R. A. (1964). Studies on the biosynthesis of chlorophyll: Chemical incorporation of magnesium into porphyrins. *Proc. Natl. Acad. Sci. USA* **52**, 1439–1442.
Bazzaz, M. B. (1981a). New chlorophyll *a* and *b* chromophores isolated from a mutant of *Zea mays* L. *Naturwissensch.* **68**, 94.
Bazzaz, M. B. (1981b). New chlorophyll chromophores isolated from a chlorophyll-deficient mutant of maize. *Photobiochem. Photobiophys.* **2**, 199–207.
Bazzaz, M. B., and Brereton, R. G. (1982). 4-Vinyl-4-desethyl chlorophyll *a*: A new naturally occurring chlorophyll. *FEBS Lett.* **138**, 104–108.
Bazzaz, M. B., and Griffiths, W. T. (1984). *In vitro* synthesis of 4-vinyl-4-desethyl chlorophyllide *a* and chlorophyllide *a* in etioplast membranes of maize. *Plant Physiol.* **75**, S-172.
Bazzaz, M. B., Bradley, C. V., and Brereton, R. G. (1982). 4-Vinyl-4-desethyl chlorophyll *a*: Characterization of a new naturally occurring chlorophyll using fast atom bombardment, field desorption, and "in beam" electron impact mass spectroscopy. *Tet. Lett.* **23**, 1211–1214.
Beale, S. I. (1984). Biosynthesis of photosynthetic pigments. *In* "Chloroplast Biogenesis" (N. R. Baker and J. Barber, eds.), pp. 133–205. Elsevier, Amsterdam.
Beale, S. I. (1990). Biosynthesis of the tetrapyrrole pigment precursor, δ-aminolevulinic acid, from glutamate. *Plant Physiol.* **93**, 1273–1279.

## 4. Biosynthesis of the Chlorophyll Chromophore

Beale, S. I., and Castelfranco, P. A. (1974). The biosynthesis of δ-aminolevulinic acid in higher plants. II. Formation of $^{14}$C-δ-aminolevulinic acid from labeled precursors in greening plant tissues. *Plant Physiol.* **53**, 297–303.
Beale, S. I., and Weinstein, J. D. (1990). Tetrapyrrole metabolism in photosynthetic organisms. *In* "Biosynthesis of Heme and Chlorophylls" (H. A. Dailey, ed.), pp. 287–391. McGraw-Hill, New York.
Beale, S. I., and Weinstein, J. D. (1991). Biosynthesis of 5-aminolevulinic acid in phototropic organisms. *In* "Chlorophylls" (H. Scheer, ed.), pp. 385–406. CRC Press, Boca Raton, Florida.
Beale, S. I., Gough, S. P., and Granick, S. (1975). The biosynthesis of δ-aminolevulinic acid from the intact carbon skeleton of glutamic acid in greening barley. *Proc. Natl. Acad. Sci. USA* **72**, 2719–2723.
Beale, S. I., Foley, T., and Dzelzkalns, V. (1981). δ-Aminolevulinic acid synthase from *Euglena gracilis*. *Proc. Natl. Acad. Sci. U.S.A.* **78**, 1666–1669.
Beer, N. S., and Griffiths, W. T. (1981). Purification of the enzyme NADPH : protochlorophyllide oxidoreductase. *Biochem. J.* **195**, 83–92.
Begley, T. P., and Young, H. (1989). Protochlorophyllide reductase. 1. Determination of the regiochemistry and the stereochemistry of the reduction of protochlorophyllide to chlorophyllide. *J. Am. Chem. Soc.* **111**, 3095–3096.
Beifuss, U., Hart, G. J., Miller, A. D., and Battersby, A. R. (1988). $^{13}$C-NMR studies on the pyrromethane cofactor of hydroxymethylbilane synthase. *Tet. Lett.* **29**, 2591–2594.
Belanger, F. C., and Rebeiz, C. A. (1979). Chloroplast biogenesis. XXVII. Detection of novel chlorophyll and chlorophyll precursors in higher plants. *Biochem. Biophys. Res. Commun.* **88**, 365–372.
Belanger, F. C., and Rebeiz, C. A. (1980a). Chloroplast biogenesis: Detection of divinyl protochlorophyllide in higher plants. *J. Biol. Chem.* **255**, 1266–1272.
Belanger, F. C., and Rebeiz, C. A. (1980b). Chloroplast biogenesis. 30. Chlorophyll(ide) (E459 F675) and chlorophyll(ide) (E449 F675): The first detectable products of divinyl and monovinyl protochlorphyll photoreduction. *Plant Sci. Lett.* **18**, 343–350.
Belanger, F. C., and Rebeiz, C. A. (1982). Chloroplast biogenesis: Detection of monovinyl magnesium protoporphyrin monoester and other monovinyl magnesium porphyrins in higher plants. *J. Biol. Chem.* **257**, 1360–1371.
Belanger, F. C., Duggan, J. X., and Rebeiz, C. A. (1982). Chloroplast biogenesis: Identification of chlorophyllide *a* (E458 F674) as a divinyl chlorophyllide *a*. *J. Biol. Chem.* **9**, 4849–4858.
Benz, J., and Rüdiger, W. (1981). Chlorophyll biosynthesis: Various chlorophyllides as exogenous substrates for chlorophyll synthetase. *Z. Naturforsch. C Biosci.* **36**, 51–57.
Benz, J., Wolf, C., and Rüdiger, W. (1980). Chlorophyll biosynthesis: Hydrogenation of geranylgeraniol. *Plant Sci. Lett.* **19**, 225–230.
Berry, A., Jordan, P. M., and Seehra, J. S. (1981). The isolation and characterization of catalytically competent porphobilinogen deaminase–intermediate complexes. *FEBS Lett.* **129**, 220–224.
Bogorad, L. (1958a). The enzymic synthesis of porphyrins from porphobilinogen. I. Uroporphyrin I. *J. Biol. Chem.* **233**, 501–509.
Bogorad, L. (1958b). The enzymic synthesis of porphyrins from porphobilinogen. II. Uroporphyrin III. *J. Biol. Chem.* **233**, 510–515.
Bogorad, L. (1966). The biosynthesis of chlorophylls. *In* "The Chlorophylls" (L. P. Vernon and G. R. Seeley, eds.), pp. 481–510. Academic Press, New York.
Brown, S. B., Houghton, J. D., and Vernon, D. I. (1990). New trends in photobiology. Biosynthesis of phycobilins. Formation of the chromophore of phytochrome, phycocyanin, and phycoerythrin. *J. Photochem. Photobiol. B* **5**, 3–23.

Bruyant, P., and Kannangara, C. G. (1987). Biosynthesis of δ-aminolevulinate in greening barley leaves. VIII. Purification and characterization of the glutamate–tRNA ligase. *Carlsberg Res. Commun.* **52,** 99–109.

Burke, D., Alberti, M., Stein, D., and Hearst, J. (1991). Chlorophyll Fe proteins and other chlorophyll biosynthesis genes from *Rhodobacter capsulatus* to higher plants. *Photochem. Photobiol. Suppl.* **53,** 85S.

Burton, G., Fagerness, P. E., Hosozawa, S., Jordan, P. M., and Scott, A. I. (1979a). $^{13}$C NMR evidence of a new intermediate, preuroporphyrinogen, in the enzymic transformation of porphobilinogen into uroporphyrinogen III. *J. Chem. Soc. Chem. Commun.* 202–204.

Burton, G., Nordlov, H., Hosozawa, S., Matsumoto, H., Jordan, P. M., Fagerness, P. E., Pryde, L. M., and Scott, A. I. (1979b). Structure of preuroporphyrinogen: Exploration of an enzyme mechanism by $^{13}$C and $^{15}$N NMR spectroscopy. *J. Am. Chem. Soc.* **101,** 3114–3116.

Camadro, J.-M., Ibraham, N. G., and Levere, R. D. (1985). Kinetic properties of the membrane-bound human liver mitochondrial protoporphyrinogen oxidase. *Arch. Biochem. Biophys.* **242,** 206–212.

Camadro, J.-M., Chambon, H., Jolles, J., and Labbe, P. (1986). Purification and properties of coproporphyrinogen oxidase from the yeast *Saccharomyces cerevisiae. Eur. J. Biochem.* **156,** 579–588.

Carell, E. F., and Kahn, J. S. (1964). Synthesis of porphyrins by isolated chloroplasts of *Euglena. Arch. Biochem. Biophys.* **108,** 1–6.

Carey, E. E., and Rebeiz, C. A. (1985). Chloroplast biogenesis. 49. Differences among angiosperms in the biosynthesis and accumulation of monovinyl and divinyl protochlorophyllide during photoperiodic greening. *Plant Physiol.* **79,** 1–6.

Carey, E. E., Tripathy, B. C., and Rebiez, C. A. (1985). Chloroplast biogenesis. 51. Modulation of monovinyl and divinyl protochlorophyllide biosynthesis by light and darkness *in vitro. Plant Physiol.* **79,** 1059–1063.

Cassidy, M. A., Crockett, N., Leeper, F. J., and Battersby, A. R. (1991). Synthetic studies on the proposed spiro intermediate for biosynthesis of the natural porphyrins: The stereochemical probe. *J. Chem. Soc. Chem. Commun.* 384–386.

Castelfranco, P. A., and Beale, S. I. (1981). Chlorophyll biosynthesis. *In* "The Biochemistry of Plants" (M. D. Hatch, ed.), Vol. 8, 375–421. Academic Press, New York.

Castelfranco, P. A., and Beale, S. I. (1983). Chlorophyll biosynthesis: Recent advances and areas of current interest. *Annu. Rev. Plant Physiol.* **34,** 241–278.

Castelfranco, P. A., and Jones, O. T. G. (1975). Protoheme turnover and chlorophyll synthesis in greening barley tissue. *Plant Physiol.* **55,** 485–490.

Castelfranco, P. A., Weinstein, J. D., Schwarcz, S., Pardo, A. D., and Wezelman, B. E. (1979). The Mg insertion step in chlorophyll biosynthesis. *Arch. Biochem. Biophys.* **192,** 592–598.

Castelfranco, P. A., Thayer, S. S., Wilkinson, J. Q., and Bonner, B. A. (1988). Labeling of porphobilinogen deaminase by radioactive 5-aminolevulinic acid in isolated developing pea chloroplasts. *Arch. Biochem. Biophys.* **266,** 219–226.

Cavaleiro, J. A. S., Kenner, G. W., and Smith, K. M. (1974). Pyrroles and related compounds. XXXII. Biosynthesis of protoporphyrin-IX from coproporphyrinogen-III. *J. Chem. Soc. Perkin Trans. 1,* 1188–1194.

Chang, T.-E., Wegmann, B., and Wang, W.-Y. (1990). Purification and characterization of glutamyl–tRNA synthetase: An enzyme involved in chlorophyll biosynthesis. *Plant Physiol.* **93,** 1641–1649.

Chaudhry, A. G., and Jordan, P. M. (1976). Stereochemical studies on the formation of porphobilinogen. *Biochem. Soc. Trans.* **4,** 760–761.

Chaudhry, I. A., Clezy, P. S., and Diakiw, V. (1977). Chemistry of pyrrolic compounds XXXVI. Some aspects of the chemistry of 2-hydroxypropionate porphyrins: New synthesis of harderoporphyrin trimethyl ester, "S-411 porphyrin" tetramethyl ester, and related compounds. *Aust. J. Chem.* **30**, 879–895.

Chen, M.-W., Jahn, D., Schön, A., O'Neill, G. P., and Söll, D. (1990a). Purification and characterization of *Chlamydomonas reinhardtii* chloroplast glutamyl–tRNA synthetase, a natural misacylating enzyme. *J. Biol. Chem.* **265**, 4054–4057.

Chen, M.-W., Jahn, D., O'Neill, G. P., and Söll, D. (1990b). Purification of the glutamyl–tRNA reductase from *Chlamydomonas reinhardtii* involved in δ-aminolevulinic acid formation during chlorophyll biosynthesis. *J. Biol. Chem.* **265**, 4058–4063.

Chen, T. C., and Miller, G. W. (1974). Purification and characterization of uroporphyrinogen decarboxylase from tobacco leaves. *Plant Cell Physiol.* **15**, 993–1005.

Chereskin, B. A., and Castelfranco, P. A. (1982). Effects of iron and oxygen on chlorophyll biosynthesis. II. Observations on the biosynthetic pathway in isolated etiochloroplasts. *Plant Physiol.* **69**, 112–116.

Chereskin, B. A., Wong, Y.-S., and Castelfranco, P. A. (1982). *In vitro* synthesis of the chlorophyll isocyclic ring: Transformation of magnesium–protoporphyrin IX and magnesium–protoporphyrin IX monomethyl ester into magnesium-2,4-divinyl pheoporphyrin $a_5$. *Plant Physiol.* **70**, 987–993.

Chereskin, B. A., Castelfranco, P. A., Dallas, J. L., and Straub, K. M. (1983). Mg-2,4-divinyl pheoporphyrin $a_5$: The product of a reaction catalyzed *in vitro* by developing chloroplasts. *Arch. Biochem. Biophys.* **226**, 10–18.

Chibbar, R. N., and van Huystee, R. B. (1983). Glutamic acid is the haem precursor for peroxidase synthesized by peanut cells in suspension culture. *Phytochemistry* **22**, 1721–1723.

Cox, M. T., Howarth, T. T., Jackson, A. H., and Kenner, G. W. (1969). Formation of the isocyclic ring in chlorophyll. *J. Am. Chem. Soc.* **91**, 1232–1233.

Dailey, H. A. (1982). Purification and characterization of membrane-bound ferrochelatase from *Rhodopseudomonas sphaeroides*. *J. Biol. Chem.* **257**, 14714–14718.

Dailey, H. A. (1990). Conversion of coproporphyrinogen to protoheme in higher eukaryotes and bacteria: Terminal three enzymes. *In* "Biosynthesis of Heme and Chlorophylls" (H. A. Dailey, ed.), pp. 123–161. McGraw-Hill, New York.

Dailey, H. A., and Karr, S. W. (1987). Purification and characterization of murine protoporphyrinogen oxidase. *Biochemistry* **26**, 2697–2701.

Darrah, P. M., Kay, S. A., Teakle, G. R., and Griffiths, W. T. (1990). Cloning and sequencing of protochlorophyllide reductase. *Biochem. J.* **265**, 789–798.

Davies, R. C., and Neuberger, A. (1973). Polypyrroles formed from porphobilinogen and amines by uroporphyrinogen synthase of *Rhodopseudomonas spheroides*. *Biochem. J.* **133**, 471–492.

Delaunay, A.-M., Huault, C., and Balange, A. P. (1991). Molecular cloning of the 5-aminolevulinic acid dehydratase gene from *Rhodobacter sphaeroides*. *J. Bacteriol.* **173**, 2712–2715.

Deutscher, M. P. (1967). Rat liver glutamyl ribonucleic acid synthetase. II. Further properties and anomalous pyrophosphate exchange. *J. Biol. Chem.* **242**, 1132–1139.

Deybach, J. C., da Silva, V., Grandchamp, B., and Nordmann, Y. (1985). The mitochondrial location of protoporphyrinogen oxidase. *Eur. J. Biochem.* **149**, 431–435.

Diaz, L., Frydman, R. B., Valasinas, A., and Frydman, B. (1979). Biosynthesis of uroporphyrinogens. Synthesis of alpha aminomethyl bilanes and their interaction with the enzymatic system. *J. Am. Chem. Soc.* **101**, 2710–2716.

Duggan, J. X., and Rebeiz, C. A. (1982a). Chloroplast biogenesis. 37. Induction of chlorophyllide *a* (E459 F675) accumulation in higher plants. *Plant Sci. Lett.* **24,** 27–37.

Duggan, J. X., and Rebeiz, C. A. (1982b). Chloroplast biogenesis. 42. Conversion of divinyl chlorophyllide *a* to monovinyl chlorophyllide *a in vivo* and *in vitro. Plant. Sci. Lett.* **27,** 137–145.

Duval, D., and Duranton, J. (1972). On a non-chlorophyllic magnesium fraction bound to plastidial lamellar proteins from *Zea mays* L. *Biochim. Biophys. Acta* **274,** 240–245.

Ebbon, J. G., and Tait, G. H. (1969). Studies on S-adenosylmethionine: magnesium protoporphyrin methyltransferase in *Euglena gracilis* strain Z. *Biochem. J.* **111,** 573–582.

Elder, G. H., and Evans, J. O. (1978). Evidence that the coproporphyrinogen oxidase activity of rat liver is situated in the intermembrane space of mitochondria. *Biochem. J.* **172,** 345–347.

Elder, G. H., Evans, J. O., Jackson, J. R., and Jackson, A. H. (1978). Factors determining the sequence of oxidative decarboxylation of the 2- and 4-propionate substituents of coproporphyrinogen III by coproporphyrinogen oxidase in rat liver. *Biochem. J.* **169,** 215–223.

El Hamouri, B., and Sironval, C. (1980). $NADP^+/NADPH$ control of the protochlorophyllide–chlorophyllide proteins in cucumber etioplasts. *Photobiochem. Photobiophys.* **1,** 219–233.

El Hamouri, B., Brouers, M., and Sironval, C. (1981). Pathway from photoinactive $P_{633-628}$ protochlorophyllide to the $P_{696-682}$ chlorophyllide in cucumber etioplast suspensions. *Plant Sci. Lett.* **21,** 375–379.

El Hamouri, B., Oliver, R. P., and Griffiths, W. T. (1983). Complexes of NADPH:protochlorophyllide oxidoreductase. *Photobiochem. Photobiophys.* **6,** 305–315.

Ellsworth, R. K. (1971). Studies on chlorophyllase. I. Hydrolytic and esterification activities of chlorophyllase from wheat seedlings. *Photosynthetica* **5,** 226–232.

Ellsworth, R. K. (1972). Chlorophyll biosynthesis. *In* "The Chemistry of Plant Pigments" (C. O. Chichester, ed.), Vol. 3, pp. 85–102. Academic Press, New York.

Ellsworth, R. K., and Aronoff, S. (1968a). Investigations on the biogenesis of chlorophyll *a*. II. Chlorophyllide *a* accumulation by a *Chlorella* mutant. *Arch. Biochem. Biophys.* **125,** 35–39.

Ellsworth, R. K., and Aronoff, S. (1968b). Investigation on the biogenesis of chlorophyll. III. Biosynthesis of Mg–vinylphaeoporphine $a_5$ methylester from Mg–protoporphine IX monomethyl ester as observed in *Chlorella* mutants. *Arch. Biochem. Biophys.* **125,** 269–277.

Ellsworth, R. K., and Aronoff, S. (1969). Investigation on the biogenesis of chlorophyll *a*. IV. Isolation and partial characterization of some biosynthetic intermediates between Mg–protoporphine IX monomethyl ester and Mg–vinylphaeoporphine $a_5$ methylester, obtained from *Chlorella* mutants. *Arch. Biochem. Biophys.* **130,** 374–383.

Ellsworth, R. K., and Dullaghan, J. P. (1972). Activity and properties of (−)-S-adenosyl-L-methionine:magnesium protoporphyrin IX methyltransferase in crude homogenates from wheat seedlings. *Biochim. Biophys. Acta* **268,** 327–333.

Ellsworth, R. K., and Hervish, P. V. (1975). Biosynthesis of protochlorophyllide *a* from Mg–protoporphyrin IX *in vitro. Photosynthetica* **9,** 125–139.

Ellsworth, R. K., and Hsing, A. S. (1973). The reduction of vinyl side-chains of Mg–protoporphyrin IX monomethyl ester *in vitro. Biochim. Biophys. Acta* **313,** 119–129.

Ellsworth, R. K., and Hsing, A. S. (1974). Activity and some properties of Mg-4-ethyl-(4-desvinyl)-protoporphyrin IX monomethyl ester:$NAD^+$ oxidoreductase in crude homogenates from etiolated wheat seedlings. *Photosynthetica* **8,** 228–234.

Ellsworth, R. K., and Lawrence, G. D. (1973). Synthesis of magnesium protoporphyrin IX *in vitro. Photosynthetica* **7,** 73–86.

## 4. Biosynthesis of the Chlorophyll Chromophore

Ellsworth, R. K., and Murphy, S. J. (1979). Biosynthesis of protochlorophyllide *a* from Mg–protoporphyrin IX monomethyl ester *in vitro*. *Photosynthetica* **13**, 392–400.
Ellsworth, R. K., and St. Pierre, M. E. (1976). Biosynthesis and inhibition of (−)-S-adenosyl-L-methionine:magnesium protoporphyrin methyltransferase of wheat. *Photosynthetica* **10**, 291–301.
Ellsworth, R. K., Dullaghan, J. P., and St. Pierre, M. E. (1974). The reaction of S-adenosyl-L-methionine:magnesium protoporphyrin IX methyltransferase of wheat. *Photosynthetica* **8**, 375–384.
Ellsworth, R. K., Tsuk, R. M., and St. Pierre, L. A. (1976). Studies on chlorophyllase. IV. Attribution of hydrolytic and esterifying "chlorophyllase" activities observed *in vitro* to two enzymes. *Photosynthetica* **10**, 312–323.
Emery, V. C., and Akhtar, M. (1985a). Mechanistic studies on the phytylation step in bacteriochlorophyll *a* biosynthesis: An application of the $^{18}$O-induced isotope effect in $^{13}$C NMR spectroscopy. *J. Chem. Soc. Chem. Commun.* 600–601.
Emery, V. C., and Akhtar, M. (1985b). Stereochemistry of the generation of the ethyl group in bacteriochlorophyll *a* biosynthesis. *J. Chem. Soc. Chem. Commun.* 1646–1647.
Emery, V. C., and Akhtar, M. (1987). Mechanistic studies on the phytylation and methylation steps in bacteriochlorophyll *a* biosynthesis: An application of the $^{18}$O-induced isotope effect in $^{13}$C NMR. *Biochemistry* **26**, 1200–1208.
Evans, J. N. S., Burton, G., Fagerness, P. E., Mackenzie, N. E., and Scott, A. I. (1986). Biosynthesis of porphyrins and corrins. 2. Isolation, purification, and NMR investigations of the porphobilinogen–deaminase covalent complex. *Biochemistry* **25**, 905–912.
Fawley, M. W. (1989). A new form of chlorophyll *c* involved in light harvesting. *Plant Physiol.* **91**, 727–732.
Ferreira, G. M. A. D. C. (1986). "Heme Biosynthesis: Characterization of the Two Terminal Membrane-Bound Enzymes." Ph.D. Thesis. University of Georgia, Athens, Georgia.
Ferreira, G. C., and Dailey, H. A. (1988). Mouse protoporphyrinogen oxidase: Kinetic parameters and demonstration of inhibition by bilirubin. *Biochem. J.* **250**, 597–603.
Ferreira, G. C., Andrew, T. L., Karr, S. W., and Dailey, H. A. (1988). Organization of the terminal two enzymes of the heme biosynthetic pathway: Orientation of protoporphyrinogen oxidase and evidence for a membrane complex. *J. Biol. Chem.* **263**, 3835–3839.
Fookes, C. J. R., and Jeffery, S. (1989). The structure of chlorophyll $c_3$: A novel marine photosynthetic pigment. *J. Chem. Soc. Chem. Commun.* 1827–1828.
Ford, C., Mitchell, S., and Wang, W.-Y. (1981). Protochlorophyllide photoconversion mutants of *Chlamydomonas reinhardtii*. *Mol. Gen. Genet.* **184**, 460–464.
Ford, C., Mitchell, S., and Wang, W.-Y. (1983). Characterization of NADPH:protochlorophyllide oxidoreductase in the Y-7 and PC-1 Y-7 mutants of *Chlamydomonas reinhardtii*. *Mol. Gen. Genet.* **192**, 290–292.
Frydman, B., Reil, S., Valasinas, A., Frydman, R. B., and Rapoport, H. (1971). Synthesis of 2-aminomethyldipyrrylmethanes of biosynthetic interest. *J. Am. Chem. Soc.* **93**, 2738–2745.
Frydman, R. B., and Feinstein, G. (1974). Studies on porphobilinogen deaminase and uroporphyrinogen III cosynthase from human erythrocytes. *Biochim. Biophys. Acta* **350**, 358–373.
Frydman, R. B., Levy, E. S., Valasinas, A., and Frydman, B. (1978a). Biosynthesis of uroporphyrinogens: Interaction among 2-aminomethyldipyrrylmethanes and the enzymic systems. *Biochemistry* **17**, 110–115.
Frydman, R. B., Levy, E. S., Valasinas, A., and Frydman, B. (1978b). Biosynthesis of uroporphyrinogens: Interaction among 2-aminomethyltripyrranes and the enzymatic systems. *Biochemistry* **17**, 115–120.

Fuesler, T. P., Wright, L. A., Jr., and Castelfranco, P. A. (1981). Properties of magnesium chelatase in greening etioplasts: Metal ion specificity and effect of substrate concentrations. *Plant Physiol.* **67**, 246–249.

Fuesler, T. P., Hanamoto, C. M., and Castelfranco, P. A. (1982). Separation of Mg–protoporphyrin IX and Mg–protoporphyrin IX monomethyl ester synthesized *de novo* by developing cucumber etioplasts. *Plant Physiol.* **69**, 421–423.

Fuesler, T. P., Castelfranco, P. A., and Wong, Y.-S. (1984a). Formation of Mg-containing chlorophyll precursors from protoporphyrin IX, δ-aminolevulinic acid, and glutamate in isolated, photosynthetically competent, developing chloroplasts. *Plant Physiol.* **74**, 928–933.

Fuesler, T. P., Wong, Y.-S., and Castelfranco, P. A. (1984b). Localization of Mg-chelatase and Mg–protoporphyrin IX monomethyl ester (oxidative) cyclase activities within isolated, developing cucumber chloroplasts. *Plant Physiol.* **75**, 662–664.

Gibbs, P. N. B., and Jordan, P. M. (1986). Identification of lysine at the active site of human 5-aminolevulinic acid dehydratase. *Biochem. J.* **236**, 447–451.

Gibson, K. D., Neuberger, A., and Tait, G. H. (1963). Studies on the biosynthesis of porphyrin and bacteriochlorophyll by *Rhodopseudomonas spheroides*. 4. S-Adenosylmethionine: magnesium protoporphyrin methyltransferase. *Biochem. J.* **88**, 325–334.

Goldin, B. R., and Little, H. N. (1969). Metalloporphyrin chelatase activity from barley. *Biochim. Biophys. Acta* **171**, 321–332.

Gorchein, A. (1972). Magnesium protoporphyrin chelatase activity in *Rhodopseudomonas spheroides:* Studies with whole cells. *Biochem. J.* **127**, 97–106.

Gorchein, A. (1973). Control of magnesium–protoporphyrin chelatase activity in *Rhodopseudomonas spheroides:* Role of light, oxygen, and electron and energy transfer. *Biochem. J.* **134**, 833–845.

Gorchein, A., Neuberger, A., and Tait, G. H. (1968). Adaptation of *Rhodopseudomonas spheroides*. *Proc. R. Soc. London B* **171**, 111–125.

Gough, S. P., Kannangara, C. G., and Bock, K. (1989). A new method for the synthesis of glutamate 1-semialdehyde: Characterization of its structure in solution by NMR spectroscopy. *Carlsberg Res. Commun.* **54**, 99–108.

Grandchamp, B., Phung, N., and Nordmann, Y. (1978). The mitochondrial localization of coproporphyrinogen III oxidase. *Biochem. J.* **176**, 97–102.

Granick, S. (1950). The structural and functional relationships between heme and chlorophyll. *Harvey Lectures* **44**, 220–245.

Granick, S. (1961). Magnesium protoporphyrin monoester and protoporphyrin monomethyl ester in chlorophyll biosynthesis. *J. Biol. Chem.* **236**, 1168–1172.

Griffiths, W. T. (1974a). Source of reducing equivalents for the *in vitro* synthesis of chlorophyll from protochlorophyll. *FEBS Lett.* **46**, 301–304.

Griffiths, W. T. (1974b). Protochlorophyll and protochlorophyllide as precursors for chlorophyll synthesis *in vitro*. *FEBS Lett.* **49**, 196–200.

Griffiths, W. T. (1975). Characterization of the terminal stages of chlorophyll(ide) synthesis in etioplast membrane preparations. *Biochem. J.* **152**, 623–635.

Griffiths, W. T. (1978). Reconstitution of chlorophyllide formation by isolated etioplast membranes. *Biochem. J.* **174**, 681–692.

Griffiths, W. T. (1980). Substrate-specificity studies on protochlorophyllide reductase in barley (*Hordeum vulgare*) etioplast membranes. *Biochem. J.* **186**, 267–278.

Griffiths, W. T. (1981). Role of NADPH in chlorophyll synthesis by protochlorophyllide reductase. *In* "Photosynthesis" (G. Akoyunoglou, ed.), Vol. V, pp. 65–71. Balaban, Philadelphia.

Griffiths, W. T. (1991). Protochlorophyllide photoreduction. *In* "Chlorophylls" (H. Scheer, ed.), pp. 433–449. CRC Press, Boca Raton, Florida.

Griffiths, W. T., and Jones, O. T. G. (1975). Magnesium 2,4-divinyl phaeoporphyrin $a_5$ as a substrate for chlorpohyll biosynthesis *in vitro*. *FEBS Lett.* **50,** 355–358.

Griffiths, W. T., and Oliver, R. P. (1984). Protochlorophyllide reductase—Structure, function, and regulation. *In* "Chloroplast Biogenesis" (R. J. Ellis, ed.), pp. 245–258. Cambridge University Press, Cambridge.

Grimm, B. (1990). Primary structure of a key enzyme in plant tetrapyrrole synthesis: Glutamate 1-semialdehyde aminotransferase. *Proc. Natl. Acad. Sci. U.S.A.* **87,** 4169–4173.

Grimm, B., Bull, A., Welinder, K. G., Gough, S. P., and Kannangara, C. G. (1989). Purification and partial amino acid sequence of the glutamate 1-semialdehyde aminotransferase of barley and *Synechococcus*. *Carlsberg Res. Commun.* **54,** 67–79.

Gurne, D., Chen, J., and Shemin, D. (1977). Dissociation and reassociation of immobilized porphobilinogen synthase: Use of immobilized subunits for enzyme isolation. *Proc. Natl. Acad. Sci. U.S.A.* **74,** 1383–1387.

Hanamoto, C. M., and Castelfranco, P. A. (1983). Separation of monovinyl and divinyl protochlorophyllides and chlorophyllides from etiolated and phototransformed cucumber cotyledons. *Plant Physiol.* **73,** 79–81.

Hart, G. J., and Battersby, A. R. (1985). Purification and properties of uroporphyrinogen III synthase (cosynthase) from *Euglena gracilis*. *Biochem. J.* **232,** 151–160.

Hart, G. J., Leeper, F. J., and Battersby, A. R. (1984). Modification of hydroxymethylbilane synthase (porphobilinogen deaminase) by pyridoxal phosphate: Demonstration of an essential lysine residue. *Biochem. J.* **222,** 93–102.

Hart, G. J., Miller, A. D., Leeper, F. J., and Battersby, A. R. (1987). Biosynthesis of the natural porphyrins: Proof that hydroxymethylbilane synthase (porphobilinogen deaminase) uses a novel binding group in its catalytic action. *J. Chem. Soc. Chem. Commun.* 1762–1765.

Hart, G. J., Miller, A. D., and Battersby, A. R. (1988). Evidence that the pyrromethane cofactor of hydroxymethylbilane synthase (porphobilinogen deaminase) is bound through the sulfur atom of a cysteine residue. *Biochem. J.* **252,** 909–912.

Hart, G. J., Miller, A. D., Beifuss, U., Leeper, F. J., and Battersby, A. R. (1990). Biosynthesis of porphyrins and related macrocylces. 35. Discovery of a novel dipyrrolic cofactor essential for the catalytic action of hydroxymethylbilane synthase (porphobilinogen deaminase). *J. Chem. Soc. Perkin Trans. 1,* 1979–1993.

Hasnain, S. S., Wardell, E. M., Garner, C. D., Schlosser, M., and Beyersmann, D. (1985). Extended X-ray absorption–fine structure investigations of zinc in 5-aminolevulinic acid dehydratase. *Biochem. J.* **230,** 625–633.

Hendry, G. A. F., and Jones, O. T. G. (1980). Haems and chlorophylls: Comparison of function and formation. *J. Med. Genet.* **17,** 1–14.

Henningsen, K. W., and Kahn, A. (1971). Photoactive subunits of protochlorophyll(ide) holochrome. *Plant Physiol.* **47,** 685–690.

Henry, A., Powls, R., and Pennock, J. F. (1986). *Scenedesmus obliquus* PS28: A tocopherol-free mutant which cannot form phytol. *Biochem. Soc. Trans.* **14,** 958–959.

Higuchi, M., and Bogorad, L. (1975). Purification and properties of uroporphyrinogen I synthase and uroporphyrinogen III cosynthetase: Interaction between the enzymes. *Ann. N.Y. Acad. Sci.* **244,** 401–418.

Hinchigeri, S. B., and Richards, W. R. (1982). The reaction mechanism of S-adenosyl-L-methionine:magnesium protoporphyrin methyltransferase from *Euglena gracilis*. *Photosynthetica* **16,** 554–560.

Hinchigeri, S. B., Chan, J. C.-S., and Richards, W. R. (1981). Purification of S-adenosyl-L-methionine:magnesium protoporphyrin methyltransferase by affinity chromatography. *Photosynthetica* **15,** 351–359.

Hinchigeri, S. B., Nelson, D. W., and Richards, W. R. (1984). The purification and reaction mechanism of S-adenosyl-L-methionine:magnesium protoporphyrin methyltransferase from Rhodopseudomonas sphaeroides. Photosynthetica **18**, 168–178.

Holt, A. S. (1966). Recently characterized chlorophylls. In "The Chlorophylls" (L. P. Vernon and G. R. Seely, eds.), pp. 111–118. Academic Press, New York.

Hoober, J. K., Kahn, A., Ash, D. E., Gough, S., and Kannangara, C. G. (1988). Biosynthesis of δ-aminolevulinate in greening barley leaves. IX. Structure of the substrate, mode of gabaculine inhibition, and the catalytic mechanism of glutamate 1-semialdehyde aminotransferase. Carlsberg Res. Commun. **53**, 11–25.

Horton, P., and Leech, R. M. (1972). The effect of ATP on photoconversion of protochlorophyllide in isolated etioplasts. FEBS Lett. **26**, 277–280.

Houen, G., Gough, S. P., and Kannangara, C. G. (1983). δ-Aminolevulinate synthesis in greening barley. V. The structure of glutamate 1-semialdehyde. Carlsberg Res. Commun. **48**, 567–572.

Houghton, J. D., Brown, S. B., Gough, S. P., and Kannangara, C. G. (1989). Biosynthesis of δ-aminolevulinate in Cyanidium caldarium: Characterization of tRNA$^{Glu}$, ligase, dehydrogenase, and glutamate 1-semialdehyde aminotransferase. Carlsberg Res. Commun. **54**, 131–143.

Hsu, W. P., and Miller, G. W. (1970). Coproporphyrinogenase in tobacco (Nicotiana tabacum L.). Biochem. J. **117**, 215–220.

Huang, D.-D., Wang, W.-Y., Gough, S. P., and Kannangara, C. G. (1984). δ-Aminolevulinic acid-synthesizing enzymes need an RNA moiety for activity. Science **225**, 1482–1484.

Ikeuchi, M., and Murakami, S. (1982). Measurement and identification of NADPH:protochlorophyllide oxidoreductase solubilized with Triton X-100 from etioplast membranes of squash cotyledons. Plant Cell Physiol. **23**, 1089–1099.

Jackson, A. H., Games, D. E., Couch, P. W., Jackson, J. R., Belcher, R. V., and Smith, S. G. (1974). Conversion of coproporphyrinogen III to protoporphyrin IX. Enzyme **17**, 81–87.

Jackson, A. H., Sancovich, H. A., Ferramola, A. M., Evans, N., Games, D. E., and Matlin, S. A. (1976). Macrocyclic intermediates in the biosynthesis of porphyrins. Phil. Trans. R. Soc. London B **273**, 191–206.

Jackson, A. H., Rao, K. R. N., Supphagen, D. M., and Smith, S. G. (1977). Intermediates between uroporphyrinogen I and coproporphyrinogen I. J. Chem. Soc. Chem. Commun. 696–698.

Jackson, A. H., Elder, G. H., and Smith, S. G. (1978). The metabolism of coproporphyrinogen III into protoporphyrinogen IX. Int. J. Biochem. **9**, 877–882.

Jackson, A. H., Sancovich, H. A., and Ferramola, A. M. (1980). Synthetic and biosynthetic studies of porphyrins. III. Structures of the intermediates between uroporphyrinogen III and coproporphyrinogen III: Synthesis of fourteen heptacarboxylic, hexacarboxylic, and pentacarboxylic porphyrins related to uroporphyrin III. Bioorg. Chem. **9**, 71–120.

Jacobs, J. M., and Jacobs, N. J. (1987). Oxidation of protoporphyrinogen to protoporphyrin, a step in chlorophyll and heme biosynthesis: Purification and partial characterization of the enzyme from barley organelles. Biochem. J. **244**, 219–224.

Jacobs, N. J., and Jacobs, J. M. (1979). Microbial oxidation of protoporphyrinogen, an intermediate in heme and chlorophyll biosynthesis. Arch. Biochem. Biophys. **197**, 396–403.

Jacobs, N. J., and Jacobs, J. M. (1981). Protoporphyrinogen oxidation in Rhodopseudomonas sphaeroides. A step in heme and bacteriochlorophyll synthesis. Arch. Biochem. Biophys. **211**, 305–311.

## 4. Biosynthesis of the Chlorophyll Chromophore

Jacobs, N. J., and Jacobs, J. M. (1984). Protoporphyrinogen oxidation, an enzymatic step in heme and chlorophyll synthesis: Partial characterization of the reaction in plant organelles and comparison with mammalian and bacterial systems. *Arch. Biochem. Biophys.* **229**, 312–319.

Jacobs, N. J., Jacobs, J. M., and Brent, P. (1971). Characterization of the late steps of microbial heme synthesis: Conversion of coproporphyrinogen to protoporphyrin. *J. Bacteriol.* **107**, 203–209.

Jaffe, E. K., and Hanes, D. (1986). Dissection of the early steps in the porphobilinogen synthase catalysed reaction: Requirements for Schiff base formation. *J. Biol. Chem.* **261**, 9348–9353.

Jahn, D., Chen, M.-W., and Söll, D. (1991). Purification and functional characterization of glutamate- 1-semialdehyde aminotransferase from *Chlamydomonas reinhardtii*. *J. Biol. Chem.* **266**, 161–167.

Jeffrey, S. W. (1989). Chlorophyll *c* pigments and their distribution in the chromophyte algae. In "The Chromophyte Algae: Problems and Perspectives" (J. C. Green, B. S. C. Leadbeater, and W. L. Diver, eds.), Vol. 38, p. 13. Clarendon Press, Oxford.

Johnson, A., and Jones, O. T. G. (1964). Enzymic formation of haems and other metalloporphyrins. *Biochim. Biophys. Acta* **93**, 171–173.

Jones, C., Jordan, P. M., and Akhtar, M. A. (1984). Mechanism and stereochemistry of the porphobilinogen deaminase and protoporphyrinogen IX oxidase reactions: Stereospecific manipulation of hydrogen atoms at the four methylene bridges during the biosynthesis of heme. *J. Chem. Soc. Perkin Trans. 1*, 2625–2633.

Jones, M. S., and Jones, O. T. G. (1970). Ferrochelatase of *Rhodopseudomonas spheroides*. *Biochem. J.* **119**, 453–462.

Jones, O. T. G. (1963). Magnesium 2,4-divinylphaeoporphyrin $a_5$ monomethyl ester, a protochlorophyll-like pigment produced by *Rhodopseudomonas spheroides*. *Biochem. J.* **89**, 182–189.

Jones, O. T. G. (1966). A protein–protochlorophyll complex obtained from inner seed coats of *Cucurbita pepo*. *Biochem. J.* **101**, 153–160.

Jones, O. T. G. (1967). Intermediates in chlorophyll biosynthesis in *Rhodopseudomonas spheroides:* Effects of substrates and inhibitors. *Phytochemistry* **6**, 1355–1362.

Jones, O. T. G. (1968). Biosynthesis of chlorophylls. In "Porphyrins and Related Compounds" (T. W. Goodwin, ed.), pp. 131–145. Academic Press, New York.

Jones, O. T. G. (1978). Chlorophyll biosynthesis. In "The Porphyrins" (D. Dolphin, ed.), Vol. VI A, pp. 179–232. Academic Press, New York.

Jordan, P. M. (1990). The biosynthesis of 5-aminolevulinic acid and its transformation into coproporphyrinogen in animals and bacteria. In "Biosynthesis of Heme and Chlorophylls" (H. A. Dailey, ed.), pp. 55–121. McGraw-Hill, New York.

Jordan, P. M., and Berry, A. (1981). Mechanism of action of porphobilinogen deaminase: The participation of stable enzyme substrate covalent intermediates between porphobilinogen and the porphobilinogen deaminase from *Rhodopseudomonas sphaeroides*. *Biochem. J.* **195**, 177–181.

Jordan, P. M., and Gibbs, P. N. B. (1985). Mechanism of action of 5-aminolevulinic acid dehydratase from human erythrocytes. *Biochem. J.* **227**, 1015–1020.

Jordan, P. M., and Seehra, J. S. (1979). The biosynthesis of uroporphyrinogen III: Order of assembly of the four porphobilinogen molecules in the formation of the tetrapyrrole ring. *FEBS Lett.* **104**, 364–366.

Jordan, P. M., and Seehra, J. S. (1980a). $^{13}$C NMR as probe for the study of enzyme catalysed reactions: Mechanism of action of 5-aminolevulinic acid dehydratase. *FEBS Lett.* **114**, 283–286.

Jordan, P. M., and Seehra, J. S. (1980b). Mechanism of action of 5-aminolevulinic acid dehydratase: Stepwise order of addition of the two molecules of 5-aminolevulinic acid in the enzymic synthesis of porphobilinogen. *J. Chem. Soc. Chem. Commun.* 240–242.

Jordan, P. M., and Shemin, D. (1973). Purification and properties of uroporphyrinogen I synthase from *Rhodopseudomonas spheroides*. *J. Biol. Chem.* **248**, 1019–1024.

Jordan, P. M., and Warren, M. J. (1987). Evidence for a dipyrromethane cofactor at the catalytic site of *E. coli* porphobilinogen deaminase. *FEBS Lett.* **225**, 87–92.

Jordan, P. M., Burton, G., Nordlov, H., Schneider, M., Pryde, L., and Scott, A. I. (1979). Preuroporphyrinogen, a substrate for uroporphyrinogen III cosynthase. *J. Chem. Soc. Chem. Commun.* 204–205.

Jordan, P. M., Mgbeje, I. A. B., Alwan, A. F., and Thomas, S. D. (1987). Nucleotide sequence of *hem*D, the second gene in the *hem* operon of *Escherichia coli* K-12. *Nucleic Acids Res.* **15**, 10583.

Jordan, P. M., Mgbeje, I. A., Thomas, S. D., and Alwan, A. F. (1988a). Nucleotide sequence of the *hem*D gene of *Escherichia coli* encoding uroporphyrinogen III synthase and initial evidence for a *hem* operon. *Biochem. J.* **249**, 613–616.

Jordan, P. M., Thomas, S. D., and Warren, M. J. (1988b). Purification, crystallization, and properties of porphobilinogen deaminase from a recombinant strain of *Escherichia coli* K12, *Biochem. J.* **254**, 427–435.

Jordan, P. M., Warren, M. J., Williams, H. J., Stolowich, N. J., Roessner, C. A., Grant, S. K., and Scott, A. I. (1988c). Identification of a cysteine-242 residue as the binding site for the dipyrromethane cofactor at the active site of *Escherichia coli* porphobilinogen deaminase. *FEBS Lett.* **235**, 189–193.

Joyard, J., Block, M., Pineau, B., Albrieux, C., and Douce, R. (1990). Envelope membranes from mature spinach chloroplasts contain a NADPH:protochlorophyllide reductase on the cystolic side of the outer membrane. *J. Biol. Chem.* **265**, 21820–21827.

Kannangara, C. G., and Gough, S. P. (1978). Biosynthesis of δ-aminolevulinate in greening barley leaves: Glutamate 1-semialdehyde aminotransferase. *Carlsberg Res. Commun.* **43**, 185–194.

Kannangara, C. G., Gough, S. P., Oliver, R. P., and Rasmussen, S. K. (1984). Biosynthesis of δ-aminolevulinate in greening barley leaves. VI. Activation of glutamate by ligation to RNA. *Carlsberg Res. Commun.* **49**, 417–437.

Kannangara, C. G., Gough, S. P., Bruyant, P., Hoober, J. K., Kahn, A., and von Wettstein, D. (1988). tRNA$^{Glu}$ as a cofactor in δ-aminolevulinate biosynthesis: Steps that regulate chlorphyll synthesis. *Trends Biochem. Sci.* **13**, 139–143.

Keithly, J. H., and Nadler, K. D. (1983). Protoporphyrin formation in *Rhizobium japonicum*. *J. Bacteriol.* **154**, 838–845.

Kenner, G. W., McCombie, S. W., and Smith, K. M. (1972). Porphyrin β-keto-esters and their cyclisation to phaeoporphyrins. *J. Chem. Soc. Chem. Commun.* 844–845.

Kenner, G. W., McCombie, S. W., and Smith, K. M. (1974). Pyrroles and related compounds. XXX. Cyclisation of porphyrin β-keto-esters to phaeoporphyrins. *J. Chem. Soc. Perkin Trans. 1*, 572–530.

Klemm, D. J., and Barton, L. L. (1987). Purification and properties of protoporphyrinogen oxidase from an anaerobic bacterium. *Desulfovibrio gigas. J. Bacteriol.* **169**, 5209–5215.

Koopman, G. E., and Batlle, A. M. C. (1987). Biosynthesis of porphyrins in *Rhodopseudomonas palustris*. VI. The effect of metals, thiols, and other reagents on the activity of uroporphyrinogen decarboxylase. *Int. J. Biochem.* **19**, 373–378.

Koopman, G. E., De Geralnik, A. A. J., and Batlle, A. M. C. (1986). Porphyrin biosynthesis in *Rhodopseudomonas palustris*. V. Purification of uroporphyrinogen decarboxylase and some unusal properties. *Int. J. Biochem.* **18**, 935–944.

Kotler, M. L., Fumagalli, S. A., Juknat, A. A., and Batlle, A. M. C. (1987). Porphyrin biosynthesis in *Rhodopseudomonas palustris*. VIII. Purification and properties of deaminase. *Comp. Biochem. Physiol.* **87B**, 601–606.

Krishnasamy, S., and Wang, W.-Y. (1990). Purification of the second enzyme of chlorophyll biosynthesis from *Chlamydomonas reinhardtii*. *Plant. Physiol.* **93**, S-62.

Kwan, L. Y.-M., Darling, D. L., and Richards, W. R. (1986). Affinity chromatography of two enzymes of the latter stages of chlorophyll synthesis. *In* "Regulation of Chloroplast Differentiation" (G. Akoyunoglou and H. Senger, eds.), pp. 57–62. A. R. Liss, New York.

Lascelles, J. (1966). The accumulation of bacteriochlorophyll precursors by mutant and wild type strains of *Rhodopseudomonas spheroides*. *Biochem. J.* **100**, 175–183.

Lascelles, J., and Wertlieb, D. (1971). Mutant strains of *Rhodopseudomonas spheroides* which form photosynthetic pigments aerobically in the dark: Growth characteristics and enzymic activities. *Biochim. Biophys. Acta* **266**, 328–340.

Lee, H. J., Ball, M. D., and Rebeiz, C. A. (1991). Intraplastidic localization of the enzymes that convert δ-aminolevulinic acid to protoporphyrin IX in etiolated cucumber cotyledons. *Plant Physiol.* **96**, 910–915.

Leeper, F. J. (1985a). The biosynthesis of porphyrins, chlorophylls, and vitamin $B_{12}$. *Natl. Prod. Rep.* **2**, 19–47.

Leeper, F. J. (1985b). The biosynthesis of porphyrins, chlorophylls, and vitamin $B_{12}$. *Natl. Prod. Rep.* **2**, 561–580.

Leeper, F. J. (1987). The biosynthesis of porphyrins, chlorophylls, and vitamin $B_{12}$. *Natl. Prod. Rep.* **4**, 441–469.

Leeper, F. J. (1989). The biosynthesis of porphyrins, chlorophylls, and vitamin $B_{12}$. *Natl. Prod. Rep.* **6**, 171–203.

Leeper, F. J. (1991). Intermediate steps in the biosynthesis of chlorophylls. *In* "Chlorophylls" (H. Scheer, ed.), pp. 407–431. CRC Press, Boca Raton, Florida.

Li, J., Spano, A. J., and Timko, M. P. (1991). Isolation and characterization of nuclear genes encoding the ALA dehydratase of pea (*Pisum sativum*). *Plant Physiol.* **96**, 5–127.

Liedgens, W., Grützmann, R., and Schneider, H. A. W. (1980). Highly efficient purification of the labile plant enzyme 5-aminolevulinate dehydratase (EC 4.2.1.24) by means of monoclonal antibodies. *Z. Naturforsch. C Biosci.* **35**, 958–962.

Liedgens, W., Lutz, C., and Schneider, H. A. W. (1983). Molecular properties of 5-aminolevulinic acid dehydratase from *Spinacia olereiea*. *Eur. J. Biochem.* **135**, 75–79.

Lindsten, A., Welch, C. J., Schoch, S., Ryberg, M., Rüdiger, W., and Sundqvist, C. (1990). Chlorophyll synthetase is latent in well preserved prolamellar bodies of etiolated wheat. *Physiol. Plant.* **80**, 277–285.

Little, H. N., and Jones, O. T. G. (1976). The subcellular localization and properties of the ferrochelatase of etiolated barley. *Biochem. J.* **156**, 309–314.

Little, H. N., and Kelsey, M. I. (1964). Incorporation of $Zn^{II}$ into protoporphyrin by extracts from barley. *Fed. Proc.* **23**, 223.

Luo, J., and Lim, C. K. (1990). Decarboxylation of uroporphyrinogen III by erythrocyte uroporphyrinogen decarboxylase. *Biochem. J.* **268**, 513–515.

Lütz, C., Benz, J., and Rüdiger, W. (1981). Esterification of chlorophyllide in prolamellar body (PLB) and prothylakoid (PT) fractions from *Avena sativa* etioplasts. *Z. Naturforsch. C Biosci.* **36**, 58–61.

Mapleston, R. E., and Griffiths, W. T. (1978). Effects of illumination of etiolated leaves on the redox state of NADP in the plastids. *FEBS. Lett.* **92**, 168–172.

Mathewson, J. H., and Corwin, A. H. (1961). Biosynthesis of pyrrole pigments: A mechanism for porphobilinogen polymerization. *J. Am. Chem. Soc.* **83**, 135–137.

Mattheis, J. R., and Rebeiz, C. A. (1977). Chloroplast biogenesis: Net synthesis of protochlorophyllide from magnesium protoporphyrin monoester by developing chloroplasts. *J. Biol. Chem.* **252,** 4022–4024.

Mau, Y.-H. L., and Wang, W.-Y. (1988). Biosynthesis of δ-aminolevulinic acid in *Chlamydomonas reinhardtii:* Study of the transamination mechanism using specifically labeled glutamate. *Plant Physiol.* **86,** 793–797.

Mau, Y.-H. L., and Wang, W.-Y. (1990). Purification of glutamate 1-seminaldehyde aminotransferase, the third enzyme of chlorophyll biosynthesis, from *Chlamydomonas reinhardtii. Plant Physiol.* **93,** S-62.

Mayer, S. M., Weinstein, J. D., and Beale, S. I. (1987). Enzymatic conversion of glutamate to δ-aminolevulinate in soluble extracts of *Euglena gracilis. J. Biol. Chem.* **262,** 12541–12549.

Mazanowska, A. M., Nueberger, A., and Tait, G. H. (1966). Effect of lipids and organic solvents on the enzymic formation of zinc protoporphyrin and haem. *Biochem. J.* **98,** 117–127.

Michalski, T. J., Hunt, J. E., Bowman, M. K., Smith, U., Bardeen, K., Gest, H., Norris, J. R., and Katz, J. J. (1987). Bacteriopheophytin *g:* Properties and some speculations on a possible role for bacteriochlorophylls *b* and *g* in the biosynthesis of chlorophylls. *Proc. Natl. Acad. Sci. USA* **84,** 2570–2574.

Miller, A. D., Hart, G. J., Packman, L. C., and Battersby, A. R. (1988). Evidence that the pyrromethane cofactor of hydroxymethylbilane synthase (porphobilinogen deaminase) is bound to the protein through the sulphur atom of cysteine-242. *Biochem. J.* **254,** 915–918.

Moore, M. R. (1990). Historical introduction to porphyrins and porphyrias. *In* "Biosynthesis of Heme and Chlorophylls" (H. A. Dailey, ed.), pp. 1–54. McGraw-Hill, New York.

Mori, M., and Sano, S. (1968). Protoporphyrin formation from coproporphyrinogen III by *Chromatium* cell extracts. *Biochem. Biophys. Res. Commun.* **32,** 610–615.

Nandi, D. L., (1978). Lysine as the substrate binding site of porphobilinogen synthase of *Rhodopseudomonas sphaeroides. Z. Naturforsch. C Biosci.* **33,** 799–800.

Nandi, D. L., and Shemin, D. (1968). 5-Aminolevulinic acid dehydratase of *Rhodopseudomonas spheroides.* III. Mechanism of porphobilinogen synthesis. *J. Biol. Chem.* **243,** 1236–1242.

Nandi, D. L., and Waygood, E. R. (1967). Biosynthesis of porphyrins in wheat leaves. II. 5-Aminolevulinate hydrolyase. *Can. J. Biochem.* **45,** 327–336.

Nasri, F., Huault, C., and Belangé, A. P. (1988). 5-Aminolevulinate dehydratase activity in thylakoid-related structures of etiochloroplasts from radish cotyledons. *Phytochemistry* **27,** 1289–1295.

Nasrulhaq-Boyce, A., Griffiths, W. T., and Jones, O. T. G. (1987). The use of continuous assays to characterize the oxidative cyclase that synthesizes the chlorophyll isocyclic ring. *Biochem. J.* **243,** 23–29.

Neidhart, W., Anderson, P. C., Hart, G. J., and Battersby, A. R. (1985). Synthesis of (11$S$)- and (11$R$)-[11-$^{2}$H$_{1}$] porphobilinogen; Stereochemical studies on hydroxymethylbilane synthase (PBG deaminase). *J. Chem. Soc. Chem. Commun.* 924–927.

Neuberger, A., and Tait, G. H. (1964). Studies on the biosynthesis of porphyrin and bacteriochlorophyll by *Rhodopseudomonas spheroides. Biochem. J.* **90,** 607–616.

Neuberger, A., Ebbon, J. G., and Tait, G. H. (1970). S-Adenosylmethionine:magnesium protoporphyrin methyltransferase (*Rhodopseudomonas spheroides* and *Euglena gracilis*). *In* "Methods in Enzymology" (S. P. Colowick, and N. O. Kaplan, eds.), Vol. 17, pp. 222–226. Academic Press, New York.

Oelze, J. (1988). Regulation of tetrapyrrole synthesis by light in chemostat cultures of *Rhodobacter sphaeroides. J. Bacteriol.* **170,** 4652–4657.

Oh-hama, T., Seto, H., Otake, N., and Miyachi, S. (1982). $^{13}$C-NMR evidence for the pathway of chlorophyll biosynthesis in green algae. *Biochem. Biophys. Res. Commun.* **105**, 647–652.

Oh-hama, T., Seto, H., and Miyachi, S. (1986a). $^{13}$C-NMR evidence of bacteriochlorophyll *a* formation by the $C_5$ pathway in *Chromatium*. *Arch. Biochem. Biophys.* **246**, 192–198.

Oh-hama, T., Seto, H., and Miyachi, S. (1986b). $^{13}$C-NMR evidence for bacteriochlorophyll *c* formation by the $C_5$ pathway in green sulfur bacterium, *Prosthecochloris*. *Eur. J. Biochem.* **159**, 189–194.

Oliver, R. P. (1982). "Protochlorophyllide Reductase." Ph.D. Thesis, University of Bristol, England.

Oliver, R. P., and Griffiths, W. T. (1981). Covalent labelling of the NADPH:protochlorophyllide oxidoreductase from etioplast membranes with [$^3$H]*N*-phenylmaleimide. *Biochem. J.* **195**, 93–101.

Oliver, R. P., and Griffiths, W. T. (1982). Pigment–protein complexes of illuminated etiolated leaves. *Plant Physiol.* **70**, 1019–1025.

O'Neill, G. P., and Söll, D. (1990). Expression of the *Synechocystis* sp. strain PCC 6803 glutamic acid transfer RNA gene provides transfer RNA for protein and chlorophyll biosynthesis. *J. Bacteriol.* **172**, 6363–6371.

O'Neill, G. P., Peterson, D. M., Schön, A., Chen, M.-W., and Söll, D. (1988). Formation of the chlorophyll precursor δ-aminolevulinic acid in cyanobacteria requires aminoacylation of a tRNA$^{Glu}$ species. *J. Bacteriol.* **170**, 3810–3816.

Osgerby, J. M., Pluscec, J., Kim, Y. C., Boyer, F., Stojanac, N., Mah, H. D., and MacDonald, S. F. (1972). Possible intermediates in biosynthesis of porphyrins. *Can. J. Chem.* **50**, 2652–2660.

Packer, N., and Adamson, H. (1986). Incorporation of 5-aminolevulinic acid into chlorophyll in darkness in barley. *Physiol. Plant.* **68**, 220–230.

Packer, N., Adamson, H., and Walmsley, J. (1987). Comparison of chlorophyll accumulation and $^{14}$C-ALA incorporation into chlorophyll in dark and light in green barley. *In* "Progress in Photosynthesis Research" (J. Biggins, ed.), Vol. IV, pp. 487–490. Nijhoff, Boston.

Pardo, A. D., Chereskin, B. M., Castelfranco, P. A., Franceschi, V. R., and Wezelman, B. E. (1980). ATP requirement for Mg chelatase in developing chloroplasts. *Plant Physiol.* **65**, 956–960.

Peschek, G. A., Hinterstoisser, B., Pineau, B., and Missbichler, A. (1989a). Light-independent NADPH:protochlorophyllide oxidoreductase activity in purified plasma membrane from the cyanobacterium *Anacystis nidulans*. *Biochem. Biophys. Res. Commun.* **162**, 71–78.

Peschek, G. A., Hinterstoisser, B., Wastyn, M., Kuntner, O., Pineau, B., Missbichler, A., and Lang, J. (1989b). Chlorophyll precursors in the plasma membrane of a cyanobacterium, *Anacystis nidulans*. *J. Biol. Chem.* **264**, 11827–11832.

Pluscec, J., and Bogorad, L. (1970). A dipyrrylmethane intermediate in the enzymic synthesis of uroporphyrinogen. *Biochemistry* **9**, 4736–4743.

Porra, R. J., and Lascelles, J. (1968). Studies on ferrochelatase: The enzymic formation of haem in proplastids, chloroplasts, and plant mitochondria. *Biochem. J.* **108**, 343–348.

Porra, R. J., Klein, O., and Wright, P. E. (1983). The proof by $^{13}$C-NMR spectroscopy of the predominance of the $C_5$ pathway over the Shemin pathway in chlorophyll biosynthesis in higher plants and the formation of the methyl ester group of chlorophyll from glycine. *Eur. J. Biochem.* **130**, 509–516.

Poulson, R. (1976). The enzymic conversion of protoporphyrinogen IX to protoporphyrin IX in mammalian mitochondria. *J. Biol. Chem.* **251**, 3730–3733.

Poulson, R., and Polglase, W. J. (1974). Aerobic and anaerobic coproporphyrinogenase activities in extracts from *Saccharomyces cerevisiae*. *J. Biol. Chem.* **249,** 6367–6371.

Poulson, R., and Polglase, W. J. (1975). The enzymic conversion of protoporphyrinogen IX to protoporphyrin IX: Protoporphyrinogen oxidase activity in mitochondrial extracts of *Saccharomyces cerevisiae*. *J. Biol. Chem.* **250,** 1269–1274.

Pudek, M. R., and Richards, W. R. (1975). A possible alternate pathway of bacteriochlorophyll biosynthesis in a mutant of *Rhodopseudomonas sphaeroides*. *Biochemistry* **14,** 3132–3137.

Radmer, R. J., and Bogorad, L. (1967). (−)$S$-adenosyl-L-methionine: magnesium protoporphyrin methyltransferase, an enzyme in the biosynthetic pathway of chlorophyll in *Zea mays*. *Plant. Physiol.* **42,** 463–465.

Radmer, R., and Bogorad, L. (1972). A tetrapyrrole intermediate in the enzymic synthesis of uroporphyrinogen. *Biochemistry* **11,** 904–910.

Raich, N., Romeo, P.-H., Dubart, A., Beaupain, D., Cohen-Sohal, M., and Goossens, M. (1986). Molecular cloning and complete primary sequence of human erythrocyte porphobilinogen deaminase. *Nucleic Acids Res.* **14,** 5955–5968.

Rando, R. R. (1977). Mechanism of the irreversible inhibition of 2-aminobutyric acid:$\alpha$-ketoglutaric acid transaminase by the neurotoxin gabaculine. *Biochemistry* **16,** 4604–4610.

Rebeiz, C. A. (1982). Chlorophyll: Anatomy of a discovery. *Chemtech.* **12,** 52–63.

Rebeiz, C. A., and Castelfranco, P. A. (1971). Protochlorophyll biosynthesis in a cell-free system from higher plants. *Plant Physiol.* **47,** 24–32.

Rebeiz, C. A., and Lascelles, J. (1982). Biosynthesis of pigments in plants and bacteria. *In* "Photosynthesis: Energy Conversion by Plants and Bacteria" (Govindjee, ed.), Vol. I, pp. 699–780. Academic Press, New York.

Rebeiz, C. A., Abou-Haïdar, M., Yaghi, M., and Castelfranco, P. (1970). Porphyrin biosynthesis in cell-free homogenates from higher plants. *Plant Physiol.* **46,** 543–549.

Rebeiz, C. A., Smith, B. B., Mattheis, J. R., Rebeiz, C. C., and Dayton, D. F. (1975). Chloroplast biogenesis: Biosynthesis and accumulation of Mg–protoporphyrin-IX monoester and other metalloporphyrins by isolated etioplasts and developing chloroplasts. *Arch. Biochem. Biophys.* **167,** 351–365.

Rebeiz, C. A., Belanger, F. C., McCarthy, S. A., Freyssinet, G., Duggan, J. X., Wu, S.-M., and Mattheis, J. R. (1981). Biosynthesis and accumulation of novel chlorophyll *a* and *b* chromophoric species in green plants. *In* "Photosynthesis" (G. Akoyunoglou, ed.), Vol. V, pp. 197–212. Balaban, Philadelphia.

Rebeiz, C. A., Wu, S.-M., Kuhadja, M., Daniell, H., and Perkins, E. J. (1983). Chlorophyll *a* biosynthetic routes and chlorophyll *a* chemical heterogeneity in plants. *Mol. Cell. Biochem.* **57,** 97–125.

Rebeiz, C. A., Tripathy, B. C., Wu, S.-W., Montazer-Zouhoor, A., and Carey, E. E. (1986). Chloroplast biogenesis 52: Demonstration *in toto* of monovinyl and divinyl monocarboxylic chlorophyll biosynthetic routes in higher plants. *In* "Regulation of Chloroplast Differentiation" (G. Akoyunoglou and H. Senger, eds.), pp. 13–24. A. R. Liss, New York.

Richards, W. R., and Lascelles, J. (1969). The biosynthesis of bacteriochlorophyll: The characterization of the latter stage intermediates from mutants of *Rhodopseudomonas spheroides*. *Biochemistry* **8,** 3473–3482.

Richards, W. R., Chan, J. C.-S., and Hinchigeri, S. B. (1981). Affinity chromatographic purification of an enzyme of chlorophyll synthesis. *In* "Photosynthesis" (G. Akoyunoglou, ed.), Vol. V, pp. 243–252. Balaban, Philadelphia.

## 4. Biosynthesis of the Chlorophyll Chromophore

Richards, W. R., Fung, M., Wessler, A. N., and Hinchigeri, S. B. (1987a). The purification and properties of three later-stage enzymes of chlorophyll synthesis. In "Progress in Photosynthesis Research" (J. Biggins, ed.), Vol. IV, pp. 475–482. Nijhoff, Boston.

Richards, W. R., Walker, C. J., and Griffiths, W. T. (1987b). The affinity chromatographic purification of NADPH: protochlorophyllide oxidoreductase from etiolated wheat. *Photosynthetica* **21**, 462–471.

Richards, W. R., Fidai, S., Gibson, L., Lauterbach, P., Snajdarova, I., Valera, V., Wieler, J. S., and Yee, W. C. (1991). Enzymology of the magnesium branch of chlorophyll and bacteriochlorophyll biosynthesis. *Photochem. Photobiol. Suppl.* **53**, 84S–85S.

Richter, M. L., and Rienits, K. G. (1980). The synthesis of magnesium and zinc protoporphyrin and their methyl esters in etioplast preparations studied by high pressure liquid chromatography. *FEBS Lett.* **116**, 211–216.

Richter, M. L., and Rienits, K. G. (1982). The synthesis of magnesium protoporphyrin IX by etiochloroplast membrane preparations. *Biochim. Biophys. Acta* **717**, 255–264.

Ricketts, T. R. (1966). Magnesium 2,4-divinylphaeoporphyrin $a_5$ monomethyl ester, a protochlorophyll-like pigment present in some unicellular flagellates. *Phytochemistry* **5**, 223–229.

Rieble, S., and Beale, S. I. (1988). Enzymatic transformation of glutamate to δ-aminolevulinic acid by soluble extracts of *Synechocystis* sp. 6803 and other oxygenic prokaryotes. *J. Biol. Chem.* **263**, 8864–8871.

Rieble, S., and Beale, S. I. (1989). Separation of the enzymes required for transformation of glutamate to δ-aminolevulinic acid in extracts of *Synechocystis* sp. PCC 6803. *Plant Physiol.* **89**, S-51.

Rieble, S., Ormerod, J. G., and Beale, S. I. (1989). Transformation of glutamate to δ-aminolevulinic acid by soluble extracts of *Chlorobium vibrioforme*. *J. Bacteriol.* **171**, 3782–3787.

Romana, M., le Boulch, P., and Romeo, P.-H. (1987). Rat uroporphyrinogen decarboxylase cDNA: Nucleotide sequence and comparison to human uroporphyrinogen decarboxylase. *Nucleic Acids Res.* **15**, 7211.

Romeo, P.-H., Raich, N., Dubart, A., Beaupain, D., Pryor, M., Kushner, J. P., Cohen-Sohal, M., and Goossens, M. (1986). Molecular cloning and nucleotide sequence of a complete human uroporphyrinogen decarboxylase cDNA. *J. Biol. Chem.* **261**, 9825–9831.

Rüdiger, W. (1987). Chlorophyll synthetase and its implication for regulation of chlorophyll biosynthesis. In "Progress in Photosynthesis Research" (J. Biggins, ed.), Vol. IV, pp. 461–467. Nijhoff, Boston.

Rüdiger, W., and Benz, J. (1984). Synthesis of chloroplast pigments. In "Chloroplast Biogenesis" (R. J. Ellis, ed.), pp. 225–244. Cambridge University Press, Cambridge.

Rüdiger, W., and Schoch, S. (1988). Chlorophylls. In "Plant Pigments" (T. W. Goodwin, ed.), pp. 1–59. Academic Press, New York.

Rüdiger, W., and Schoch, S. (1991). The last steps of chlorophyll biosynthesis. In "Chlorophylls" (H. Scheer, ed.), pp. 451–464. CRC Press, Boca Raton, Florida.

Rüdiger, W., Hedden, P., Köst, H.-P., and Chapman, D. J. (1977). Esterification of chlorophyllide by geranylgeranyl pyrophosphate in a cell-free system from maize shoots. *Biochem. Biophys. Res. Commun.* **74**, 1268–1272.

Rüdiger, W., Benz, J., and Guthoff, C. (1980). Detection and partial characteriztion of activity of chlorophyll synthetase in etioplast membranes. *Eur. J. Biochem.* **109**, 193–200.

Russell, C. S., Polack, S., and James, J. (1984). Studies on the active site of wheat germ porphobilinogen deaminase. *Ann. N.Y. Acad. Sci.* **435**, 202–204.

Sano, S. (1966). 2,4-Bis-(β-hydroxypropionic acid) deuteroporphyrinogen IX, a possible intermediate between coproporphyrinogen III and protoporphyrin IX. *J. Biol. Chem.* **241,** 5276–5283.

Sasarman, A., Nepveu, A., Echelard, Y., Dymetryszyn, J., Drolet, M., and Goyer, C. (1987). Molecular cloning and sequencing of the *hem*D gene of *Escherichia coli* and preliminary data on the *uro* operon. *J. Bacteriol.* **169,** 4257–4262.

Sato, K., Ishida, K., Mutsushika, O., and Shimizu, S. (1985a). Purification and some properties of δ-aminolevulinic acid synthases from *Protaminobacter ruber* and *Rhodopseudomonas spheroides. Agric. Biol. Chem.* **49,** 3415–3422.

Sato, K., Ishida, K., Shirai, M., and Shimizu, S. (1985b). Occurrence and some properties of two types of δ-aminolevulinic acid synthase in a facultative methylotroph *Protaminobacter ruber. Agric. Biol. Chem.* **49,** 3423–3428.

Schauder, J. R., Jendrezejewski, S., Abell, A., Hart, G. J., and Battersby, A. R. (1987). Stereochemistry of formation of the hydroxymethyl group of hydroxymethylbilane, the precursor for uroporphyrinogen III. *J. Chem. Soc. Chem. Commun.* 436–439.

Scheer, H. (1991). Structure and occurrence of chlorophylls. *In* "Chlorophylls" (H. Scheer, ed.), pp. 3–30. CRC press, Boca Raton, Florida.

Schneegurt, M. A., and Beale, S. I. (1986). Biosynthesis of protoheme and heme *a* from glutamate in maize. *Plant Physiol.* **81,** 965–971.

Schneegurt, M. A., and Beale, S. I. (1988). Characterization of the RNA required for biosynthesis of δ-aminolevulinic acid from glutamate: Purification by anticodon-based affinity chromatography and determination that the UUC glutamate anticodon is a general requirement for function in ALA biosynthesis. *Plant Physiol.* **86,** 497–504.

Schneegurt, M. A., Rieble, S., and Beale, S. I. (1988). tRNA$^{Glu(UUC)}$ of plants and algae: HPLC separation into two subfractions, and determination of their relative effectiveness in homologous *in vitro* translation and ALA-forming systems. *Plant Physiol.* **86,** S-61.

Schoch, S., Lempert, U., and Rüdiger, W. (1977). On the last steps of chlorophyll biosynthesis. Intermediates between chlorophyllide and phytol-containing chlorophyll. *Z. Pflanzenphysiol.* **83,** 427–436.

Schoch, S., Hehlein, C., and Rüdiger, W. (1980). Influence of anaerobiosis on chlorophyll biosynthesis in greening oat seedlings *Avena sativa* L. *Plant Physiol.* **66,** 576–579.

Schön, A., Krupp, G., Gough, S., Berry-Lowe, S., Kannangara, C. G., and Söll, D. (1986). The RNA required in the first step of chlorophyll biosynthesis is a chloroplast glutamate tRNA. *Nature (London)* **322,** 281–284.

Schön, A., Kannangara, C. G., Gough, S., and Söll, D. (1988). Protein biosynthesis in organelles requires misaminoacylation of tRNA. *Nature (London)* **331,** 187–190.

Scott, A. I. (1990). Mechanistic and evolutionary aspects of vitamin $B_{12}$ biosynthesis. *Acc. Chem. Res.* **23,** 308–317.

Scott, A. I., Roessner, C. A., Stolowich, N. J., Karuso, P., Williams, H. J., Grant, S. K., Gonzalez, M. D., and Hoshino, T. (1988a). Site-directed mutagenesis and high-resolution NMR spectroscopy of the active site of porphobilinogen deaminase. *Biochemistry* **27,** 7984–7990.

Scott, A. I., Stolowich, N. J., Williams, H. J., Gonzalez, M. D., Roessner, C. A., Grant, S. K., and Pichon, C. (1988b). Concerning the catalytic site of porphobilinogen deaminase. *J. Am. Chem. Soc.* **110,** 5898–5900.

Seehra, J. S., and Jordan, P. M. (1980). Mechanisms of action of porphobilinogen deaminase: Ordered addition of the four porphobilinogen molecules in the formation of preuroporphyrinogen. *J. Am. Chem. Soc.* **102,** 6841–6846.

Seehra, J. S., Jordan, P. M., and Akhtar, M. (1983). Anaerobic and aerobic coproporphyrinogen III oxidases of *Rhodopseudomonas sphaeroides. Biochem. J.* **209,** 709–718.

Shashidhara, L. S., and Smith, A. G. (1991). Expression and subcellular location of the tetrapyrrole synthesis enzyme porphobilinogen deaminase in light-grown *Euglena gracilis* and three nonchlorophyllous cell lines. *Proc. Natl. Acad. Sci. USA* **88,** 63–67.
Shemin, D. (1976). 5-Aminolevulinic acid dehydratase: Structure, function, and mechanism. *Philos. Trans. R. Soc. London B* **273,** 109–115.
Shemin, D. (1989). An illustration of the use of isotopes: The biosynthesis of porphyrins. *BioEssays* **10,** 30–35.
Shetty, A. S., and Miller, G. W. (1969). Purification and general properties of δ-aminolaevulinate dehydratase from *Nicotiana tabacum* L. *Biochem. J.* **114,** 331–337.
Shibata, H., and Ochiai, H. (1977). Purification and properties of δ-aminolevulinic acid dehydratase from radish cotyledons. *Plant Cell Physiol.* **18,** 421–429.
Shieh, J., Miller, G. W., and Psenak, M. (1974). The incorporation of protoporphyrin IX into chlorophyll *a* in higher plants and iron effects. *Biochem. Physiol. Pflanzen* **165,** 100–103.
Shieh, J., Miller, G. W., and Psenak, M. (1978). Properties of *S*-adenosyl-L-methionine: magnesium protoporphyrin IX methyltransferase from barley. *Plant Cell Physiol.* **19,** 1051–1059.
Shioi, Y., and Beale, S. I. (1987). Polyethylene-based high performance liquid chromatography of chloroplast pigments: Resolution of mono- and divinyl chlorophyllides and other pigment mixtures. *Anal. Biochem.* **162,** 493–499.
Shioi, Y., Nagamine, M., Kuroki, M., and Sasa, T. (1980). Purification by affinity chromatography and properties of uroporphyrinogen I synthetase from *Chlorella regularis*. *Biochim. Biophys. Acta* **616,** 300–309.
Shioi, Y., Doi, M., and Böddi, B. (1988). Selective inhibition of chlorophyll biosynthesis by nicotinamide. *Arch. Biochem. Biophys.* **267,** 69–74.
Shioi, Y., Tatsumi, Y., and Shimokawa, K. (1991). Enzymatic degradation of chlorophyll in *Chenopodium album*. *Plant Cell Physiol.* **32,** 87–94.
Shlyk, A. A., Averina, N. G., and Shalygo, N. V. (1982). Metabolism and inermembrane location of magnesium protoporphyrin IX monomethyl ester in centers of chlorophyll biosynthesis. *Photobiochem. Photobiophys.* **3,** 197–224.
Siepker, L. J., Ford, M., de Kock, R., and Kramer, S. (1987). Purification of bovine protoporphyrinogen oxidase: Immunological cross-reactivity and structural relationship to ferrochelatase. *Biochim. Biophys. Acta* **913,** 349–358.
Smith, A. G. (1988). Subcellular localization of two porphyrin-synthesis enzymes in *Pisum sativum* (pea) and *Arum* (cuckoo-pint) species. *Biochem. J.* **249,** 423–428.
Smith, B. B., and Rebeiz, C. A. (1977a). Spectrofluorometric determination of Mg–protoporphyrin monoester and longer wavelength metalloporphyrins in the presence of Zn–protoporphyrin. *Photochem. Photobiol.* **26,** 527–532.
Smith, B. B., and Rebeiz, C. A. (1977b). Chloroplast biogenesis: Detection of Mg–protoporphyrin chelatase *in vitro*. *Arch. Biochem. Biophys.* **180,** 178–185.
Smith, B. B., and Rebeiz, C. A. (1979). Chloroplast biogenesis. XXIV. Intrachloroplastic localization of the biosynthesis and accumulation of protoporphyrin IX, magnesium protoporphyrin monoester, and longer wavelength metalloporphyrins during greening. *Plant Physiol.* **63,** 227–231.
Smith, K. M., and Huster, M. S. (1987). Bacteriochlorophyll-*c* formation via the glutamate $C_5$ pathway in *Chlorobium* bacteria. *J. Chem. Soc. Chem. Commun.* 14–16.
Soll, J., and Schultz, G. (1981). Phytol synthesis from geranylgeraniol in spinach chloroplasts. *Biochem. Biophys. Res. Commun.* **99,** 907–912.
Soll, J., Schultz, G., Rüdiger, W., and Benz, J. (1983). Hydrogenation of geranylgeraniol. Two pathways exist in spinach chloroplasts. *Plant Physiol.* **71,** 849–854.

Spano, A. J., and Timko, M. P. (1991). Isolation, characterization, and partial amino acid sequence of a chloroplast-localized porphobilinogen deaminase from pea *Pisum sativum* L. *Biochim. Biophys. Acta* **1076,** 29–36.

Spano, A. J., He, Z., and Timko, M. P. (1991). The NADPH-protochlorophyllide oxidoreductase in pine: Evidence for conservation in protein structure between angiosperms and gymnosperms enzymes. *Plant Physiol.* **96,** S-127.

Spiller, S. C., Castelfranco, A. M., and Castelfranco, P. A. (1982). Effects of iron and oxygen on chlorophyll biosynthesis. I. *In vivo* observations on iron and oxygen-deficient plants. *Plant Physiol.* **69,** 107–111.

Stadnichuk, I. N., and Walter, G. (1984). Isolation and polypeptide composition of the protochlorophyllide holochrome of etiolated wheat leaves. *Sov. Plant Physiol. Engl. Transl.* **31,** 671–677.

Stark, W. M., Hart, G. J., and Battersby, A. R. (1986). Synthetic studies on the proposed spiro intermediate for the biosynthesis of the natural porphyrins: Inhibition of cosynthase, *J. Chem. Soc. Chem. Commun.* 465–467.

Steffens, D., Blos, I., Schoch, S., and Rüdiger, W. (1976). Lichtabhängigkeit der Phytolakkumulation. *Planta* **130,** 151–158.

Stella, A. M., Parera, V. E., Llambias, E. B. C., and Batlle, A. M. C. (1971). Pyrryl methane intermediates in the synthesis of uroporphyrinogen III by soybean-D callus porphobilinogenase. *Biochim. Biophys. Acta* **252,** 481–488.

Sundqvist, C., and Ryberg, M. (1989). The distribution and structural role of NADPH:protochlorophyllide oxidoreductase in isolated etioplast inner membranes. *Photosynthetica* **23,** 427–438.

Tait, G. H. (1969). Coproporphyrinogenase activity in extracts from *Rhodopseudomonas spheroides*. *Biochem. Biophys. Res. Commun.* **37,** 166–172.

Tait, G. H. (1972). Coprophyrinogenase activity in extracts of *Rhodopseudomonas spheroides* and *Chromatium* D. *Biochem. J.* **128,** 1159–1169.

Tait, G. H., and Gibson, K. D. (1961). The enzymic formation of magnesium protoporphyrin monomethyl ester. *Biochim. Biophys. Acta* **52,** 614–616.

Tamai, H., Shioi, Y., and Sasa, T. (1979). Purification and characterization of δ-aminolevulinic acid dehydratase from *Chlorella regularis*. *Plant Cell Physiol.* **20,** 435–444.

Thomas, J., and Weinstein, J. D. (1990). Measurement of heme efflux and heme content in isolated developing chloroplasts. *Plant Physiol.* **94,** 1414–1423.

Thomas, S. D., and Jordan, P. M. (1986). Nucleotide sequence of the *hemC* locus encoding porphobilinogen deaminase of *Escherichia coli* K12. *Nucleic Acids Res.* **14,** 6215–6226.

Tripathy, B. C., and Rebeiz, C. A. (1986). Chloroplast biogenesis: Demonstration of the monovinyl and divinyl monocarboxylic routes of chlorophyll biosynthesis in higher plants. *J. Biol. Chem.* **261,** 13556–13564.

Tripathy, B. C., and Rebeiz, C. A. (1987). Non-equivalence of glutamic and δ-aminolevulinic acids as substrates for protochlorophyllide and chlorophyll biosynthesis in darkness. In "Progress in Photosynthesis Research" (J. Biggins, ed.), Vol. IV, pp. 439–443. Nijhoff, Boston.

Tripathy, B. C., and Rebeiz, C. A. (1988). Chloroplast biogenesis. 60. Conversion of divinyl protochlorophyllide to monovinyl protochlorophyllide in green(ing) barley, a dark monovinyl/light divinyl plant species. *Plant Physiol.* **87,** 89–94.

Tsai, S.-F., Bishop, D. F., and Desnick, R. J. (1988). Human uroporphyrinogen III synthase: Molecular cloning, nucleotide sequence, and expression of full-length cDNA. *Proc. Natl. Acad. Sci. USA* **85,** 7049–7053.

Tsukamoto, I., Yoshinaga, T., and Sano, S. (1979). The role of zinc with special reference to the essential thiol groups in 5-aminolevulinic acid dehydratase of bovine liver. *Biochim. Biophys. Acta* **570,** 167–178.

Valera, V., Fung, M., Wessler, A. N., and Richards, W. R. (1987). Synthesis of 4$R$ and 4$S$ tritium-labeled NADPH for the determination of the coenzyme stereospecificity of NADPH:protochlorophyllide oxidoreductase. *Biochem. Biophys. Res. Commun.* **148**, 515–520.
van Heyningen, S., and Shemin, S. (1971). Quarternary structure of 5-aminolevulinic acid dehydratase from *Rhodopseudomonas spheroides*. *Biochemistry* **10**, 4676–4682.
Vezitskii, A. Y., and Shcherbakov, R. A. (1987). Metal analogues of chlorophyll pigments as substrates of enzymes that catalyze the esterification of chlorophyllide and the hydrolysis of chlorophyll in plants. *Biochem. Engl. Transl.* **52**, 677–681.
Walker, C. J., and Griffiths, W. T. (1988). Protochlorophyllide reductase: A flavoprotein? *FEBS Lett.* **239**, 259–262.
Walker, C. J., and Weinstein, J. D. (1991a). Further characterization of the magnesium chelatase in isolated developing cucumber chloroplasts. *Plant Physiol.* **95**, 1189–1196.
Walker, C. J., and Weinstein, J. D. (1991b). *In vitro* assay of the chlorophyll biosynthetic enzyme Mg-chelatase: Resolution of the activity into soluble and membrane-bound fractions. *Proc. Natl. Sci. USA* **88**, 5789–5793.
Walker, C. J., Mansfield, K. E., Rezzano, I. N., Hanamoto, C. H., Smith, K. M., and Castelfranco, P. A. (1988). The magnesium protoporphyrin IX (oxidative) cyclase system. Studies on the mechanism and specificity of the reaction sequence. *Biochem. J.* **255**, 685–692.
Walker, C. J., Mansfield, K. E., Smith, K. M., and Castelfranco, P. A. (1989). Incorporation of atmospheric oxygen into the carbonyl functionality of the protochlorophyllide isocyclic ring. *Biochem. J.* **257**, 599–602.
Walker, C. J., Castelfranco, P. A., and Whyte, B. J. (1991). Synthesis of divinyl protochlorophyllide. *Biochem. J.* **276**, 691–697.
Wang, W.-Y., Gough, S. P., and Kannangara, C. G. (1981). Biosynthesis of δ-aminolevulinate in greening barley leaves. IV. Isolation of three soluble enzymes required for the conversion of glutamate to δ-aminolevulinate. *Carlsberg. Res. Commun.* **46**, 243–257.
Warren, M. J., and Jordan, P. M. (1988). Further evidence for the involvement of a dipyrromethane cofactor at the active site of porphobilinogen deaminases. *Biochem. Soc. Trans.* **16**, 963–965.
Warren, M. J., and Scott, A. I. (1990). Tetrapyrrole assembly and modification into the ligands of biologically functional cofactors. *Trends Biochem. Sci.* **15**, 486–491.
Wei, P. E., Corwin, A. H., and Arellano, R. (1962). Preparation of chelates of porphyrins and phthalocyanine with "Magnesium Viologen." *J. Org. Chem.* **27**, 3344–3346.
Weinstein, J. D., and Beale, S. I. (1983). Separate physiological roles and subcellular compartments for two tetrapyrrole biosynthetic pathways in *Euglena gracilis*. *J. Biol. Chem.* **258**, 6799–6807.
Weinstein, J. D., and Beale, S. I. (1984). Biosynthesis of protoheme and heme $a$ precursors solely from glutamate in the unicellular red alga, *Cyanidium caldarium*. *Plant Physiol.* **74**, 146–151.
Weinstein, J. D., and Beale, S. I. (1985a). Enzymatic conversion of glutamate to δ-aminolevulinate in soluble extracts of the unicellular green alga, *Chlorella vulgaris*. *Arch. Biochem. Biophys.* **237**, 454–464.
Weinstein, J. D., and Beale, S. I. (1985b). RNA is required for enzymatic conversion of glutamate to δ-aminolevulinic acid by extracts of *Chlorella vulgaris*. *Arch. Biochem. Biophys.* **239**, 87–93.
Weinstein, J. D., Mayer, S. M., and Beale, S. I. (1987). Formation of δ-aminolevulinic acid from glutamic acid in algal extracts: Separation into an RNA and three required enzyme components by serial affinity chromatography. *Plant Physiol.* **84**, 244–250.

Werck-Reichhart, D., Jones, O. T. G., and Durst, F. (1988). Haem synthesis during cytochrome P-450 induction in higher plants: 5-Aminolaevulinic acid synthesis through a five-carbon pathway in *Helianthus tuberosus* tuber tissues aged in the dark. *Biochem. J.* **249**, 473–480.

Whyte, B. J. (1989). "Characterization of the Terminal Stages of Chlorophyll Biosynthesis." Ph.D. Thesis. University of Bristol, England.

Williams, D. C., Morgan, G. S., McDonald, E., and Battersby, A. R. (1981). Purification of porphobilinogen deaminase from *Euglena gracilis* and studies of its kinetics. *Biochem. J.* **193**, 301–310.

Wong, Y.-S., and Castelfranco, P. A. (1984). Resolution and reconstitution of the Mg–protoporphyrin IX monomethyl ester (oxidative) cyclase, the enzyme system responsible for the formation of the chlorophyll isocyclic ring. *Plant Physiol.* **75**, 658–661.

Wong, Y.-S., and Castelfranco, P. A. (1985). Properties of the Mg–protoporphyrin IX monomethyl ester (oxidative) cyclase system. *Plant Physiol.* **79**, 730–733.

Wong, Y.-S., Castelfranco, P. A., Goff, D. A., and Smith, K. M. (1985). Intermediates in the formation of the chlorophyll isocyclic ring. *Plant Physiol.* **79**, 725–729.

Wu, S.-M., and Rebeiz, C. A. (1984). Chloroplast biogenesis 45: Molecular structure of protochlorophyllide (E443 F625) and of chlorophyllide $a$ (E458 F674). *Tetrahedron* **40**, 659–664.

Wu, S.-M., and Rebeiz, C. A. (1985). Chlorophyll biogenesis. Molecular structure of chlorophyll $b$ (E489 F666). *J. Biol. Chem.* **260**, 3632–3634.

Yang, Z., and Bauer, C. E. (1990). *Rhodobacter capsulatus* genes involved in early steps of the bacteriochlorophyll biosynthetic pathway. *J. Bacteriol.* **172**, 5001–5010.

Yee, W. C., Eglsaer, S. J., and Richards, W. R. (1989). Confirmation of a ping-pong mechanism for $S$-adenosyl-L-methionine:magnesium protoporphyrin methyltransferase of etiolated wheat by an exchange reaction. *Biochem. Biophys. Res. Commun.* **162**, 483–490.

Yoshinaga, T., and Sano, S. (1980a). Coproporphyrinogen oxidase. I. Purification, properties, and activation by phospholipids. *J. Biol. Chem.* **255**, 4722–4726.

Yoshinaga, T., and Sano, S. (1980b). Coproporphyrinogen oxidase. II. Reaction mechanism and role of tyrosine residues on the activity. *J. Biol. Chem.* **255**, 4727–4731.

Zaman, Z., and Akhtar, M. (1976). Mechanism and stereochemistry of vinyl-group formation in haem biosynthesis. *Eur. J. Biochem.* **61**, 215–223.

Zaman, Z., Abboud, M. M., and Akhtar, M. (1972). Mechanism and stereochemistry of vinyl group formation in haem biosynthesis. *J. Chem. Soc. Chem. Commun.* 1263–1264.

# 5

# Protochlorophyllide Reductase: A Key Enzyme in the Greening Process

## RÜDIGER SCHULZ AND HORST SENGER

Fachbereich Biologie/Botanik
Philipps-Universität
W-3550 Marburg, Germany

I. Introduction
  A. Role of Protochlorophyllide Reduction in Photomorphogenesis
  B. Isolation and Characterization of Enzyme Protein
II. Light Regulation of Protochlorophyllide Reductase
  A. Protochlorophyllide Reduction in Angiosperms, Gymnosperms, Ferns, Mosses, Algae, and Cyanobacteria
  B. Light Regulation of Protochlorophyllide Reductase at the Molecular Level
III. Pigment–Enzyme Complex
  A. Spectral Properties
  B. Localization
IV. Enzyme Activity of Protochlorophyllide Reductase
  A. Substrate Specificity
  B. Influence of Internal and External Factors
V. Cloning Protochlorophyllide Reductase Genes and Primary Structure of Enzyme Protein
  A. Isolation and Characterization of cDNA Clones Encoding Protochlorophyllide Reductase of Barley
  B. Comparison of Different Protochlorophyllide Reductase Protein Sequences
VI. Active Site and Membrane Association of Protochlorophyllide Reductase
  A. Active Site
  B. Hydrophobicity and Membrane Association
VII. Concluding Remarks
  A. Existence of a Light-Dependent and a Light-Independent Protochlorophyllide Reductase
  B. Existence of a Plastid Factor
References

## I. INTRODUCTION

### A. Role of Protochlorophyllide Reduction in Photomorphogenesis

Light is the most important environmental factor in the development of plants. Processes of differentiation that are regulated by light are called "photomorphogenesis" (Mohr and Shropshire, 1983; Schopfer and Apel, 1983). At least three different photoreceptor systems mediate responses to light in plants: phytochrome (Furuya, 1989), the near-UV/blue-light photoreceptors (Senger, 1980, 1984, 1987a; Galland and Senger, 1991), and protochlorophyllide (Pchlide) (Tobin and Silverthorne, 1985). In addition to direct enzyme regulation and translational control (Anderson, 1979; Ruyters, 1987), a major function of these photoreceptors is the light-dependent regulation of a series of genes involved in photomorphogenesis (Harpster and Apel, 1985; Jenkins, 1988). The most spectacular light-regulated process is the greening of angiosperms (Castelfranco and Beale, 1983; Griffiths et al., 1984). This process includes the transformation of etioplasts, the plastid type present in dark-grown plants, into photosynthetically active chloroplasts, including the breakdown of the prolamellar body and the formation of thylakoid membranes (Kahn, 1968b; Boardman et al., 1978; Virgin and Egneus, 1983; Griffiths and Walker, 1987). The first detectable light-dependent step of plastid photomorphogenesis is the photoreduction of Pchlide to chlorophyllide (Chlide) catalyzed by the enzyme NADPH:protochlorophyllide oxidoreductase (Pchlide reductase; EC 1.3.1.33; Griffiths, 1978; Griffiths and Oliver, 1984; Fig. 1).

During photoreduction, the pigment–protein complex serves as its own light receptor. The action spectrum for photoreduction is nearly identical to the absorption spectrum of phototransformable Pchlide (Frank, 1946; Koski et al., 1951; Björn, 1969; Egneus and Sundqvist, 1970; Ogawa et al., 1973; Egan and Schiff, 1974; Brinkmann and Senger, 1980). This process plays a crucial regulatory role in the transformation of etioplasts to chloroplasts, as shown by the strong similarities in the action spectra for these two processes (Virgin et al., 1963). Current discussion focuses on whether the transformation of Pchlide to Chlide also plays a regulatory role in the synthesis of plastid proteins. For example, Laing et al. (1988) showed that the P700-chlorophyll (Chl) $a$ protein was regulated at the translational or post-translational level by the Pchlide reduction process. This control, however, may also involve cytoplasmic signals, since the synthesis of the P700 apoprotein is not turned on in illuminated isolated etioplasts.

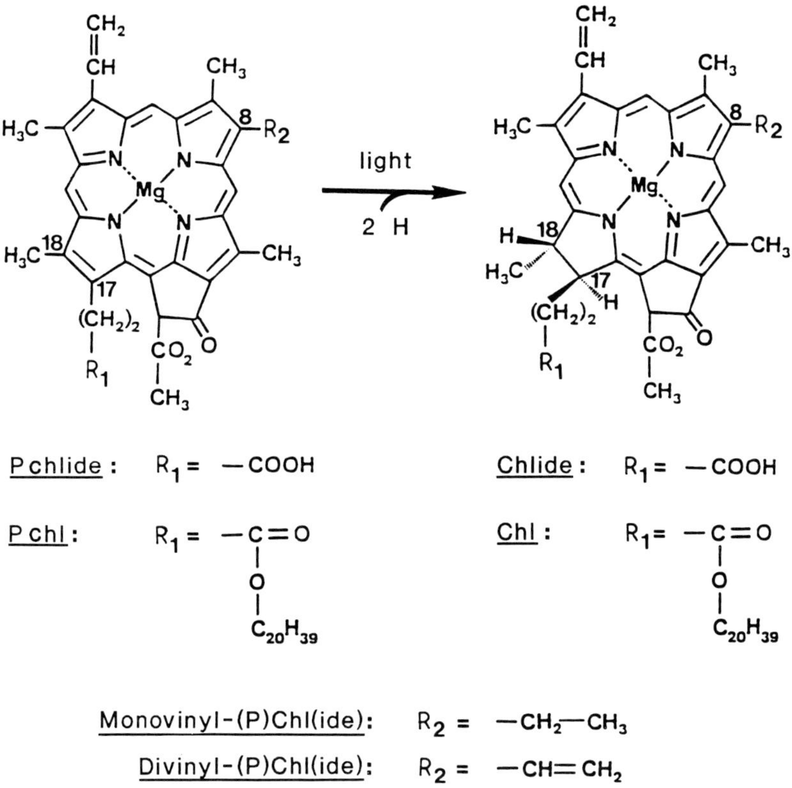

## 5. Protochlorophyllide Reductase

Pchlide : $R_1 = -COOH$

Pchl : $R_1 = -C=O$
       $\quad\quad\quad |$
       $\quad\quad\quad O$
       $\quad\quad\quad |$
       $\quad\quad\quad C_{20}H_{39}$

Chlide : $R_1 = -COOH$

Chl : $R_1 = -C=O$
      $\quad\quad\quad |$
      $\quad\quad\quad O$
      $\quad\quad\quad |$
      $\quad\quad\quad C_{20}H_{39}$

Monovinyl-(P)Chl(ide): $R_2 = -CH_2-CH_3$

Divinyl-(P)Chl(ide): $R_2 = -CH=CH_2$

**Fig. 1.** Photoreduction of Pchl(ide) to Chl(ide) catalyzed by the enzyme Pchlide reductase. Monovinyl (MV) and divinyl (DV) forms of the pigments are shown.

### B. Isolation and Characterization of Enzyme Protein

The protein of the originally isolated holochrome that mediates Pchlide reduction was identified as NADPH : protochlorophyllide oxidoreductase by Griffiths (1978). The Pchlide reductase of angiosperms is the most abundant protein in etioplast membranes (Ikeuchi and Murakami, 1983; Dehesh and Ryberg, 1985). Isolation procedures for the enzymatically active protein have been published by Apel *et al.* (1980), Beer and Griffiths (1981), Ikeuchi and Murakami (1982b), and Richards *et al.* (1987), among others. The enzyme is synthesized outside the plastids in the cytoplasm on 80 S ribosomes, as was shown by high temperature-induced deficiency of plastid ribosomes (Batschauer *et al.*, 1982; Griffiths and Beer, 1982).

Using a polyclonal antiserum against the enzyme protein, Apel (1981) showed that, in barley, the enzyme was synthesized as a precursor protein of about 44 kDa, approximately 8 kDa larger than the mature protein inside the plastids of barley. The size reported for the Pchlide reductase of different plant species varies between 33 and 38 kDa (Apel, 1981; Oliver and Griffiths, 1981; Ikeuchi and Murakami, 1982b; Ryberg and Sundqvist, 1982a; Röper *et al.*, 1987; Selstam *et al.*, 1987; Benli *et al.*, 1991). Isoelectric focusing indicated the presence of at least four different isoelectric species of Pchlide reductase (Ikeuchi and Murakami, 1982a; Dehesh *et al.*, 1986a).

## II. LIGHT REGULATION OF PROTOCHLOROPHYLLIDE REDUCTASE

### A. Protochlorophyllide Reduction in Angiosperms, Gymnosperms, Ferns, Mosses, Algae, and Cyanobacteria

Algae, ferns, mosses, and the cotyledons of most gymnosperms are able to synthesize Chl in the absence of light, whereas angiosperms generally are unable to photoreduce Pchl(ide) and develop photosynthetically active chloroplasts in darkness (Bogorad, 1976; Kirk and Tilney-Bassett, 1978; Castelfranco and Beale, 1983; Kasemir, 1983; Beale, 1984). For green redarkened angiosperms, however, a light-independent Pchl(ide) reduction was postulated. This phenomenon was found by investigating *Hordeum vulgare* (Adamson, 1982; Adamson *et al.*, 1985; Walmsley and Adamson, 1989), *Pisum sativum* (Adamson and Packer, 1984; Adamson *et al.*, 1984), *Tradescantia albiflora* (Adamson, 1978; Adamson *et al.*, 1980), *Zostera capricorni*, and *Posidonia australis* (Adamson and Hiller, 1981; Packer *et al.*, 1984). Chl synthesis in angiosperms grown in complete darkness was reported twice by investigating a mutant of *Arabidopsis thaliana* (Röbbelen, 1956) and seedlings of wheat (Adamson *et al.*, 1990). In contrast, Apel *et al.* (1984) reported that the reduction of radioactively labeled Pchlide to Chlide in excised barley leaves was strictly light dependent. This result was not consistent with the existence of a light-independent Pchlide reductase. In response to these results, Packer and Adamson (1986) confirmed that excised leaves from greened etiolated barley plants did not incorporate radioactively labeled 5-aminolevulinic acid (ALA) into Chl in darkness. Intact barley seedlings, however, were able to synthesize Chl under the same conditions from internal precursors. As a result of these observations, Adamson *et al.* (1987) proposed two enzymes that might be involved in Chl biosynthesis in angiosperms: the well-known light-dependent Pchlide reductase and an uncharacterized light-independent enzyme.

In contrast to angiosperms, the cotyledons of most gymnosperms turn green in complete darkness. Burgerstein (1900) investigated about 60 species of gymnosperms and found that most of them were capable of greening in the dark. Corresponding to the phylogenetic separation of different gymnosperm species from the angiosperms, however, are large variations in the ability to form Chl without illumination and to build a photosynthetically active chloroplast. The cotyledons of the primitive gymnosperm *Metasequoia glyptostroboides* turn green, even when kept in darkness during sprouting (Laudi and Manzini, 1975). The Chl content of leaves returned to darkness, although only half the content observed in light-grown leaves, was remarkably high in comparison with that usually found in etiolated leaves of other plants. The plastids of darkened leaves lacked starch and contained prolamellar bodies of small size that apparently were still differentiating and were surrounded by structures typical of the photosynthetic apparatus. This plastid type was called an etiochloroplast (Brouers and Wolwertz, 1980). Laudi and Manzini (1975) pointed out that leaves of adult *Metasequoia* plants also synthesized Chl in darkness, a typical archaic feature of this plant. *Ginkgo biloba*, however, a true living phylogenetic link between the angiosperms and the gymnosperms, is not able to synthesize Chl in darkness (Rascio *et al.*, 1984). Like angiosperms, *Ginkgo* accumulates Pchlide and builds small prolamellar bodies and a few prothylakoid membranes. This behavior is very similar to that of *Larix decidua*, a plant that is of great evolutionary interest because of its annual leaf shedding (Mariani *et al.*, 1990). In the absence of light, larch shows only a limited ability to synthesize Chl and forms voluminous prolamellar bodies and only a few thylakoid membranes. After illumination, the rate of phototransformation of prolamellar bodies is slow. By investigating the gnetales, an old taxonomic group associated with the gymnosperms but having several characteristics of the angiosperms, Jeske and Senger (1981) found, when comparing the synthesis of Chl and the development of photosynthetic activity in dark-grown seedlings of *Ephedra distachya*, *Welwitschia mirabilis*, and *Gnetum ula*, that synthesis of Chl in the dark occurred only in *Ephedra*. Photosynthetic activity was present after a short time of illumination, which was observed also in other gymnosperms undergoing greening without light (Mariani *et al.*, 1984). *Welwitschia* and *Gnetum*, in contrast, showed the behavior of angiosperms by synthesizing Chl and building a photosynthetically active apparatus only in the light. Because of these observed differences, it was questioned whether *Ephedra* belongs in the same taxonomic group with *Gnetum* and

*Welwitschia* (Jeske and Senger, 1981). Since Chl formation and development of the photosynthetic apparatus of *Gnetum* and *Welwitschia* are angiosperm-like, one must assume that these two species have common ancestors with the angiosperms. Thus, the light requirement for Pchl(ide) reduction was established at a very early phylogenetic stage in the evolution of plants. In the genera *Pinus* (Bogdanovic, 1973a; Jeske and Senger, 1979) and *Picea* (Mariani *et al.*, 1990), the formation of Chl in darkness apparently was restricted to germinating seedlings. Selstam *et al.* (1987) observed that, in pine, in spite of the light-independent Pchlide reductase activity, Pchlide was present in dark-grown seedlings in the same amount seen in dark-grown wheat. The light-dependent activity of the Pchlide reductase consistently was localized in the prolamellar body fraction of the etiochloroplasts of pine cotyledons. Bogorad (1950) reported that, in *Pinus jeffreyi*, the Chls formed in light and darkness are spectroscopically similar. *Pinus* cotyledons were found to green as long as they were immersed in the megagametophyte and, therefore, in complete darkness. To complete Chl synthesis in the dark, pine embryos need contact with living megagametophyte tissue (Bogorad, 1950). The existence of a specific diffusible factor that is necessary for dark Chl formation is indicated by the results of Bogdanovic (1973b), who found that wheat embryos, when grafted onto megagametophyte tissue of black pine, developed Chl in darkness.

As mentioned by Kirk and Tilney-Bassett (1978), among the pteridophytes, ferns can produce Chl in the dark, as can *Selaginella* and *Isoetes*, whereas members of the *Equisetaceae* do not form Chl in the dark (Bittner, 1905; Stahl, 1909). Bryophytes (mosses) are able to synthesize Chl in the dark (Bittner, 1905).

The green algae (*Chlamydomonas, Chlorella,* and *Scenedesmus*) show the ability to perform complete Chl biosynthesis in darkness if an organic carbon source is provided for heterotrophic growth (Kirk and Tilney-Bassett, 1978). The exceptional inability of *Chlorella variegata* and *Chlorella luteo-viridis* to form Chl in darkness (Kirk and Tilney-Bassett, 1978) may be due to naturally induced mutations in Chl biosynthesis or to substrate dependency. The phytoflagellate *Euglena* and the red alga *Cyanidium* generally require light for Chl production (Bogorad, 1976). The pattern of Pchlide synthesis, proplastid organization (with prolamellar body-like structures), and the processes of phototransformation to develop photosynthetically active chloroplasts are highly similar in dark-grown *Euglena* and young etiolated higher plants (Cohen and Schiff, 1976; Schiff, 1978; Schwartzbach, 1990). A very interesting group of induced algae mutants are unable to synthesize Chl in darkness, in contrast to the wild-type strains. These mutants synthesize Chl only on transfer to light and, like angiosperms, accumulate Pchlide in the dark. Examples of this group of induced algae mutants are listed in Table I. Roitgrund and Mets (1990), analyzing chloroplast DNA deletions of a yellow-in-the-dark mutant of *Chlamydomonas reinhardtii*, showed that in this organism the light-independent reduction of Pchlide is a chloroplast gene function. In contrast to the typical chloroplast developed during dark growth of *Scenedesmus obliquus*, cells of the C-2A' mutant formed prolamellar body-like structures comparable to the tubular membrane system of angiosperm etioplasts (Senger *et al.*, 1974; Wellburn *et al.*, 1980; Fig. 2).

In addition, in the cyanobacterium *Anabaena*, presence of a light-independent Pchilde reduction other than the light-dependent pathway was shown (Adamson *et al.*, 1987). For *Anacystis nidulans*, an enzyme activity closely associated with the plasma membrane was reported that transformed Pchlide into Chlide *in vitro* and in the dark when the membrane preparation was incubated with NADPH (Hinterstoisser *et al.*, 1988; Peschek *et al.*, 1989a,b).

TABLE I

**Green Algae Mutants that Lack the Ability to Accumulate Chl when Growing in the Dark but Turn Green in the Light**[a]

| Organism | Mutant | Reference |
| --- | --- | --- |
| *Chlamydomonas reinhardtii* | y-1 | Ohad *et al.* (1967a,b); Matsuda *et al.* (1971); Wang *et al.* (1977); Maloney *et al.* (1989) |
| *Chlorella fusca* | G10 | Bauer and Wild (1976) |
| *C. pyrenoidosa* | g-1 | Galling (1978) |
|  | g-2 | Galling (1981) |
| *C. regularis* | YG-1 | Sasa and Sugahara (1976) |
|  | YG-6 | Shioi and Sasa (1984) |
| *C. vulgaris* | y-1 | Beale (1971) |
| *C. vulgaris* | 610-y | Herron and Mauzerall (1972) |
| *C. vulgaris* | 5/520 | Dubertret and Joliot (1974) |
| *C. vulgaris* | 69 | Wild and Fuldner (1977) |
| *Scenedesmus obliquus* | C-2A' | Senger and Bishop (1972); Bishop and Senger (1972); Oh-hama and Hase (1980); Senger and Brinkmann (1986); Senger and Kotzabasis (1991) |

[a] Adapted from Senger (1987b).

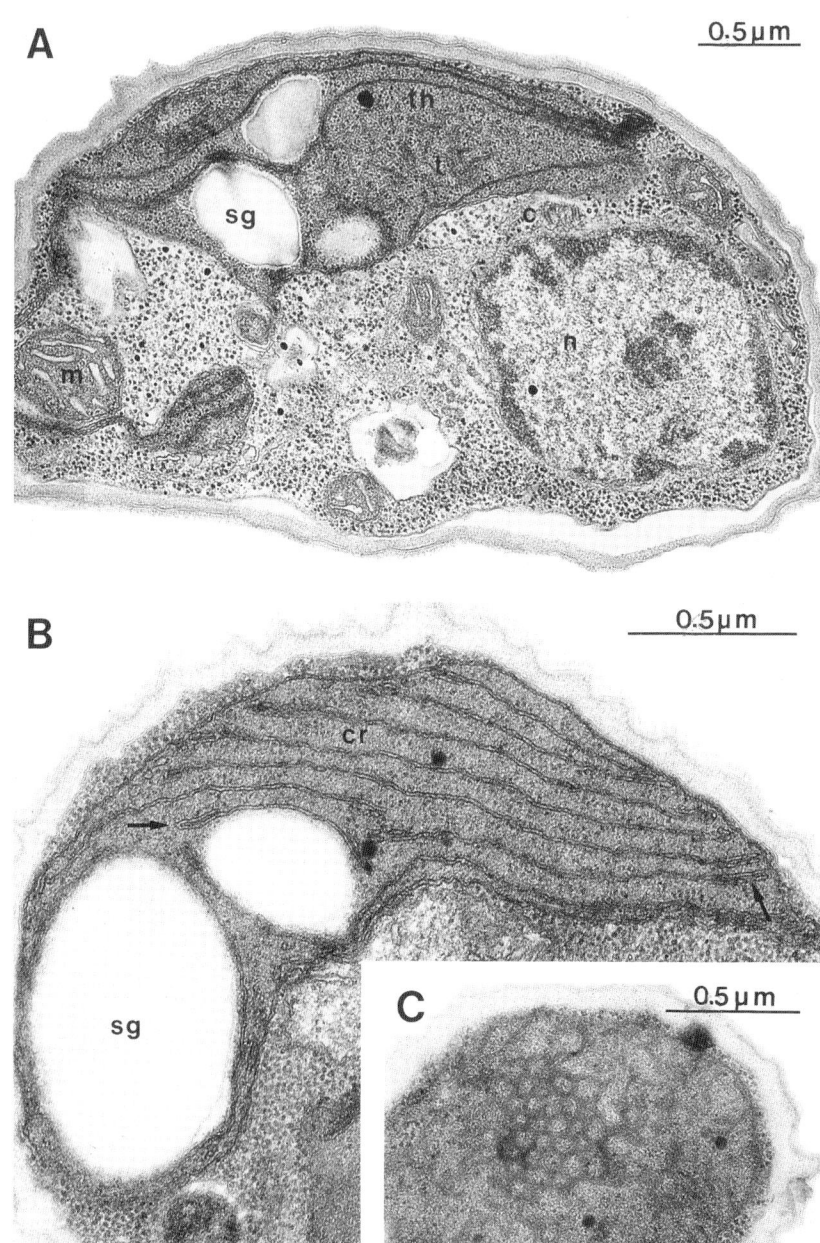

## B. Light Regulation of Protochlorophyllide Reductase at the Molecular Level

Different levels of light regulation of Pchlide reductase during the greening process must be distinguished: (1) the regulation of transcription of mRNA encoding Pchlide reductase and the stability of the transcripts, (2) the regulation of translation of the enzyme and the turnover of the enzyme protein, and (3) the regulation of enzyme activity. The results obtained on light regulation of Pchlide reductase in different plants are inconsistent. Although most authors reported a negative effect of light on the amount of mRNA, the enzyme protein level, and the enzyme activity of Pchlide reductase, others could not find any light regulation (Table II). The report of a positive influence of light on the concentration of mRNA encoding Pchlide reductase and on the protein level (Meyer et al., 1983) as well as the speculation of reciprocal light regulation of Pchlide reductase in mono- and dicotyledonous plants could not be confirmed (Forreiter et al., 1990). Obviously, in all plant species studied, enzyme activity declines within a few minutes on illumination. Mostly a reduction of enzyme protein level was induced as well (Table II). However, Western blot investigations with *Lolium temulentum* (Davies et al., 1989), *Pinus pinea* (Ou et al., 1990), and 5-day-old seedlings of *Sorghum bicolor* (Schrubar et al., 1990) revealed constant levels of Pchlide reductase protein after several hours illumination. Reports on levels of the mRNA transcript encoding Pchlide reductase varied strongly among species. In many cereals, such as barley (Batschauer and Apel, 1984), oat (Häuser et al., 1987; Darrah et al., 1990), wheat (Meyer et al., 1983), and rice (Kay et al., 1989), light induced a drastic decline in the level of Pchlide reductase mRNA whereas in others, such as maize (Forreiter et al., 1990), and in several dicotyledonous plants (Forreiter et al., 1990; Kittsteiner et al., 1990; Benli et al., 1991) only small or no changes in the mRNA level were induced over an extended time of illumination (Table II). As an example of the light regulation of Pchlide-reductase, the amount of mRNA, protein, and enzyme activity and the formation of Chl during the greening process of barley is shown (Fig. 3). On illumination, the amount of mRNA encoding Pchlide

---

**Fig. 2.** Electron micrographs of heterotrophically grown cells of the pigment mutant C-2A' of the unicellular green alga *Scenedesmus obliquus*. (A) Dark-grown cell. (B) After 2 hr of illumination with white light of 10,000 lux. The arrows mark the beginning of thylakoid doubling at the ends of the single-layered thylakoids. (C) The prolamellar-body-like structures inside an etioplast of a dark-grown cell extend into tubules occasionally linked with prothylakoids. c, centriole; m, mitochondrion; n, nucleus; sg, starch granule; t, tubules; th, thylakoids; cr, chloroplast ribosomes. Adapted from Senger et al. (1974).

**TABLE II**

**Light Effect on the Enzyme Activity, Protein Content, and Steady State Concentration of mRNA Transcripts Encoding the Pchlide Reductase of Different Plant Species**[a]

| Organism | A | P | R | Reference |
|---|---|---|---|---|
| *Arabidopsis* | − | − | − | Benli et al. (1991) |
| Barley | − |   |   | Mapleston and Griffiths (1980) |
|   |   | − | − | Apel (1981) |
|   |   | − | − | Santel and Apel (1981) |
|   |   | − | − | Häuser et al. (1984) |
|   |   | − |   | Meyer et al. (1983) |
|   |   |   | − | Batschauer and Apel (1984) |
|   | − |   |   | Dehesh et al. (1986a,b) |
|   |   | − | − | Häuser et al. (1987) |
|   |   |   | − | Schulz et al. (1989) |
|   | − | − | − | Forreiter et al. (1990) |
| Bean |   | + | + | Meyer et al. (1983) |
|   | − | − | − | Forreiter et al. (1990) |
| Cress |   | − | +/− | Kittsteiner et al. (1990) |
| Cucumber | +/− |   |   | El Hamouri (1984) |
| *Lolium* | +/− |   |   | Davies et al. (1989) |
| Maize | − | − | − | Forreiter et al. (1990) |
| Mustard | − | − | − | Forreiter et al. (1990) |
| Oat | − |   |   | Kay and Griffiths (1983) |
|   |   | − |   | Meyer et al. (1983) |
|   |   |   | − | Häuser et al. (1987) |
|   | − |   |   | Schrubar et al. (1990) |
| Pea |   | + | + | Meyer et al. (1983) |
|   | − | − | − | Forreiter et al. (1990) |
| Pine | +/− |   |   | Ou et al. (1990) |
| Rice |   |   | − | Kay et al. (1989) |
| Rye | − |   |   | Kay and Griffiths (1983) |
|   |   | − |   | Meyer et al. (1983) |
| *Sorghum* |   |   |   |   |
| 5 days | +/− |   |   | Schrubar et al. (1990) |
| 11 days | − |   |   | Schrubar et al. (1990) |
| Squash |   | − |   | Ikeuchi and Murakami (1982a) |
| Sunflower | − | − | − | Forreiter et al. (1990) |
| Tomato |   | +/− | +/− | Meyer et al. (1983) |
|   | − | − | − | Forreiter et al. (1990) |
| Wheat | − |   |   | Mapleston and Griffiths (1980) |
|   |   | − |   | Meyer et al. (1983) |

[a] A, enzyme activity; P, protein content; R, steady state mRNA concentration. Negative effect, −; negligible or no effect, +/−; positive effect, +.

## 5. Protochlorophyllide Reductase

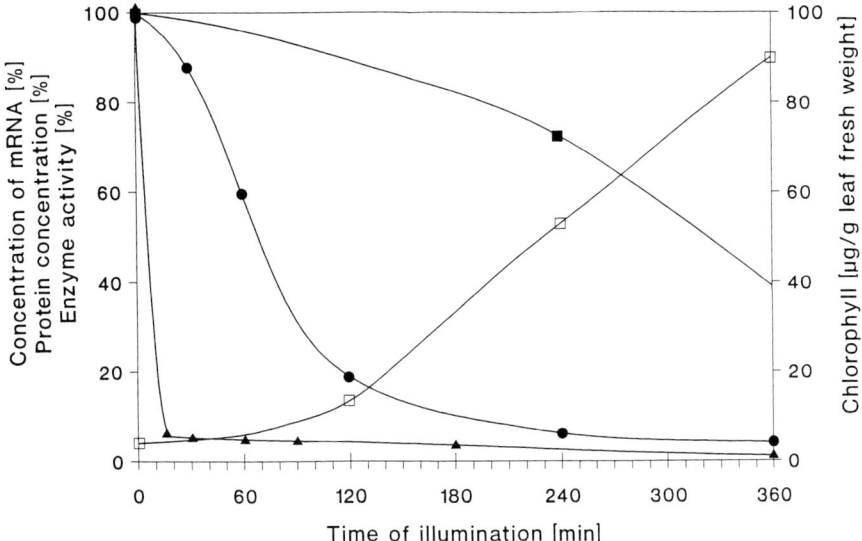

**Fig. 3.** The effect of light on the relative amount of mRNA encoding Pchlide reductase (●), the relative concentration of the enzyme protein (■), the specific enzyme activity of the Pchlide reductase (▲), and the relative Chl content (□) in barley during the transformation of etioplasts to chloroplasts. Adapted from Santel and Apel (1981), Batschauer and Apel (1984), and Dehesh et al. (1986a,b).

reductase of barley decreases drastically. This light effect is mediated by phytochrome (Apel, 1981). Using a short cDNA clone, isolated by differential screening (Apel et al., 1983), that encodes the carboxy terminus of Pchlide reductase, Mösinger et al. (1985) found that phytochrome activation decreases the transcription rate of the nuclear gene encoding Pchlide reductase enzyme in barley. The gene encoding Pchlide reductase, as well as the genes encoding phytochrome itself (Colbert et al., 1983), belong to a small group of negatively light-regulated plant genes (Thompson and White, 1991). In addition to the decrease in the mRNA encoding Pchlide-reductase during illumination of dark-grown angiosperms, a specific proteolytic degradation of the enzyme protein takes place. This degradation appears to be the reason for the decline of the enzyme protein level (Dehesh and Apel, 1983). Proteolytic activities already present in the membranes (Hampp and De Filippis, 1980) are influenced by the loss of binding of the enzyme to the substrate Pchlide and the cosubstrate NADPH. Maintenance of this binding apparently stabilizes the enzyme against proteolysis (Kay and Griffiths, 1983; Häuser et al.,

1984). This result was confirmed by Röper and Lütz (1984) using proteases with known cleavage specificities.

Why Pchlide reductase occurs in such relative abundance in etiolated plants, whereas after illumination and during light-dependent Chl accumulation the enzyme level and activity level decline so dramatically (Dehesh et al., 1983), is not known. To show that the high amount of Pchlide reductase is not an artifact of an extended dark period during the germination of the seeds, as Griffiths et al. (1985) expected, Häuser et al. (1987) investigated this problem by varying the growth conditions of barley. These authors confirmed that even in very young seedlings, which were not grown for an artificially long time in darkness, high amounts of Pchlide-reductase were detectable. Ryberg and Sundqvist (1988) discussed whether the regular ultrastructure of prolamellar bodies of etioplasts depends on the presence of membrane-associated Pchlide reductase as a stabilizing protein in dark-grown plants. For illuminated barley, Santel and Apel (1981) suggested that the function of the enzyme may be restricted to the initial phase of greening and that, later on, the conversion of Pchlide to Chlide is mediated by another, as yet unknown, mechanism. Other authors, however, pointed out that the activities of the Pchlide reductase assayed *in vitro* are more than adequate to account for the rates of Chl formation measured *in vivo* during greening (Griffiths, 1978; Griffiths et al., 1985; Bennett et al., 1987). A photoactive Pchlide–protein complex in the light during the greening of barley was shown by Franck and Strzalka (1992).

## III. PIGMENT–ENZYME COMPLEX

### A. Spectral Properties

Because of the investigations by Shibata (1957) on Pchlide photoconversion in etiolated bean and maize leaves, it is known that photoconversion *in vivo* is accompanied by a series of spectral changes in addition to the spectral shift caused by the disappearance of Pchlide and the formation of Chlide. It is generally agreed that etioplasts of dark-grown leaves contain three spectroscopically distinguishable Pchl(ide) forms: $Pchl(ide)_{628-632}$, $Pchl(ide)_{636-657}$, and $Pchl(ide)_{650-657}$ (Fig. 4). Although $Pchl(ide)_{636-657}$ and $Pchl(ide)_{650-657}$ are photoreduced easily to Chl(ide), $Pchl(ide)_{628-632}$ as the free pigment is not (Bovey et al., 1974; Schneider, 1980; Brouers and Wolwertz, 1981; Virgin, 1981; Ryberg and Sundqvist, 1982b). Oliver and Griffiths (1982) suggested that the occurrence of the different spectral forms of Pchlide depends on their binding to Pchlide-reductase and to NADPH. Griffiths (1978,1981) showed that Pchlide-reductase functions

## 5. Protochlorophyllide Reductase

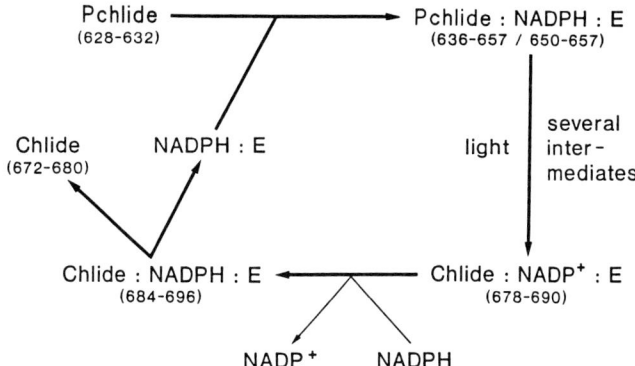

**Fig. 4.** Scheme for the photoconversion of Pchlide (E, Pchlide reductase). Wavelengths refer to the different spectral forms of the pigments (see text). Adapted from Oliver and Griffiths (1982).

by forming a photoactive ternary complex (Pchlide$_{650-657}$), binding the two substrates Pchlide and NADPH which, when exposed to light, cause hydrogen transfer from the coenzyme to the porphyrin to produce Chlide (Dujardin and Sironval, 1978). Kotzabasis *et al.* (1990b) presented evidence for the hypothesis that Pchl(ide)$_{650-657}$ *in vivo* may correspond with the aggregation form of DV-Pchlide *in vitro* and that Pchl(ide)$_{636-657}$ could be an aggregate of MV-Pchlide *in vivo*. The photoreduction of Pchlide to Chlide involves several short-lived intermediates (Franck and Mathis, 1980; Dobek *et al.*, 1981; Inoue *et al.*, 1981; Iwai *et al.*, 1984). The first long-lived form is Chlide$_{678-690}$. Within 30 sec of irradiation at 20°C, this intermediate shifts to Chlide$_{684-696}$. Completion of a further shift to Chlide$_{672-680}$, known as the Shibata shift, is reached after approximately 30 min (Shibata, 1957; Thorne, 1971; Oliver and Griffiths, 1982; Sundqvist and Ryberg, 1989). Shifts in absorption and fluorescence maxima of Chl(ide) in spectra of dark-grown leaves after illumination vary depending on the amount of Pchlide accumulated through treatment with ALA (Klockare and Sundqvist, 1977). The esterified Chlide is incorporated into the growing thylakoid membranes as Chl (Sundqvist and Ryberg, 1989).

### B. Localization

When sucrose gradient centrifugation techniques were used to separate the various fractions of etioplast membrane preparations, contradictory results were obtained for the localization of the Pchlide reductase–pigment complex. For oat and bean, the Pchlide reductase complex was reported to be located mainly in the prothylakoids (Lütz et al., 1981; Gerday et al., 1984). Using purified etioplast membrane fractions of wheat, however, most of the Pchlide reductase was found in the prolamellar body (Ryberg and Sundqvist, 1982a; Böddi et al., 1989). Ikeuchi and Murakami (1983), working on squash cotyledons, also claimed that the photoactive Pchlide NADPH–protein complex in etioplasts was concentrated in the prolamellar body. Investigating the spectral forms of Pchlide in prolamellar bodies and prothylakoids fractionated from wheat etioplasts, Ryberg and Sundqvist (1982b) found that the dominating Pchlide–protein complex in the prolamellar bodies was the phototransformable complex $Pchlide_{650-657}$, whereas the dominant form in the prothylakoids was $Pchlide_{628-632}$. The presence of Pchlide reductase was confirmed, primarily for the prolamellar body, by immunogold labeling with specific antisera, whereas only small amounts of labeling were observed in prothylakoids (Shaw et al., 1985; Dehesh et al., 1986b; Ryberg and Dehesh, 1986; Benli et al., 1991). Lindsten et al. (1988) confirmed this result by analyzing the polypeptide composition of highly purified prolamellar bodies and prothylakoids from wheat. Mainly based on immunogold labeling of ultrathin sections, Ryberg and Dehesh (1986) discussed a translocation of Pchlide reductase from the prolamellar bodies to the prothylakoids that is initiated by the light-dependent transformation of Pchlide to Chlide. Artus et al. (1992) supposed that the occurrence of the Shibata shift may be a prerequisite of the relocation of Pchlide reductase from prolamellar bodies to thylakoids.

A protein reacting with an antiserum against the Pchlide reductase of barley was found outside the plastids of barley in the area of the plasmalemma (Dehesh et al., 1986b). This finding could not be reproduced using the same antibody against the Pchlide reductase isolated from barley (Apel, 1981) or an antibody against the Pchlide reductase expressed in *Escherichia coli* (Schulz, 1990).

The occurrence of Pchlide and Chlide (Pineau et al., 1986) and of a photoactive and Pchlide reductase-immunoreactive protein was reported on the outer surface of the outer envelope membrane of mature spinach chloroplasts (Joyard et al., 1990).

Investigating different organs of plants for the Pchlide reductase–pigment complex, Dehesh et al. (1983) showed that, along the barley leaf gradient, the amount of enzyme was highest in the tip of the leaf and declined rapidly with decreasing age of the leaf tissue, being almost undetectable in the leaf base. McEwen et al. (1991) found Pchl in roots of dark-grown plants of seven species. The highest pigment content was observed in young root tips and decreased upward along the roots. The authors pointed out, however, that it is not known whether the Pchl accumulated in dark-grown roots is the precursor of Chl found in illuminated roots.

## IV. ENZYME ACTIVITY OF PROTOCHLOROPHYLLIDE REDUCTASE

### A. Substrate Specificity

Although the Pchlide accumulated in etiolated plant tissues is generally accepted to be the main substrate used for the phototransformation to Chlide (Griffiths, 1974), controversy about the substrate specificity of Pchlide reductase still continues.

The first subject of discussion, protochlorophyll (Pchl) (the Pchlide ester; Fig. 1), is always present in small amounts in etiolated tissues (Jones, 1966; Boardman, 1966; Sundqvist *et al.*, 1980; Virgin, 1984; Kotzabasis *et al.*, 1989a). The ratio of Pchlide to Pchl depends on the age of the seedling. The juvenile form of the photopigment is Pchl; Pchlide becomes predominant under conditions of prolonged etiolation (Lancer *et al.*, 1976). The ratio of Pchlide to Pchl also can be influenced by application of ALA (Kotzabasis and Senger, 1990). A series of investigations was carried out to determine whether Pchlide reductase accepts Pchl as substrate (Table III). In addition to these investigations of Pchl phototransformation in plant material, *in vitro* tests were done using a hybrid protein of Pchlide reductase expressed in *Escherichia coli* from a complete cDNA clone encoding Pchlide-reductase of barley (Schulz *et al.*, 1989). Pchl extracted

**TABLE III**

**Phototransformation of Esterified Pchlide *in Vivo* and *in Vitro***

| Species | *In vivo* | *In vitro* | Reference |
|---|---|---|---|
| *Chlorella regularis* | | | |
| YG-1 | Yes | | Sasa and Sugahara (1976) |
| YG-6 | No | | Shioi and Sasa (1984) |
| *Cucumis sativus* | Yes | | Rebeiz and Castelfranco (1973) |
| | Yes | | Belanger and Rebeiz (1980b) |
| | No | | Shioi and Sasa (1983) |
| *Cucurbita pepo* | | No | Jones (1966) |
| *Euglena gracilis* | Yes | | Cohen and Schiff (1976) |
| *Hordeum vulgare* | Yes | | Liljenberg (1974) |
| | | No | Griffiths (1974) |
| | | No | Griffiths (1980) |
| *Phaseolus vulgaris* | Yes | | Lancer *et al.* (1976) |
| | No | | Bovey *et al.* (1974) |
| *Scenedesmus obliquus* | | | |
| WT D3 | | Yes | Kotzabasis *et al.* (1989a) |
| C-2A' | Yes | Yes | Kotzabasis *et al.* (1989a) |

from *Scenedesmus obliquus* (C-2A' mutant cells) and NADPH were added. After illumination with dim white light, a photoreduction of Pchl to Chl could be detected by a partial shift of fluorescence emission from 634 nm to 675 nm (R. Knaust, R. Schulz, and H. Senger, unpublished results). Results with this heterologous system support the conclusion that Pchl is certainly feasible as a substrate of Pchlide reductase. Kotzabasis *et al.* (1991) supposed that, in mutant C-2A' of *Scenedesmus*, Chl, which is rapidly formed on illumination from Pchl, is incorporated directly into the reaction centers of the photosynthetic apparatus, thus avoiding the slow esterification step after the onset of light.

The second subject of discussion is the fact that Chl and the precursors Pchlide and Pchl were identified as monovinyl (MV) and divinyl (DV) forms (Houssier and Sauer, 1969; Belanger and Rebeiz, 1980a,b; Kotzabasis *et al.*, 1990b; Fig. 1). On the basis of two parallel MV- or DV-monocarboxylic biosynthesis routes (Rebeiz *et al.*, 1986) that predominate at night or in daylight, higher plants were observed to fall into one of four greening groups; dark DV/light DV, dark MV/light MV, dark MV/light DV, and dark DV/light MV (Carey and Rebeiz, 1985; Carey *et al.*, 1985; Rebeiz *et al.*, 1986; Tripathy and Rebeiz, 1988). No information is available about the function of these (P)Chl(ide) derivatives. According to the model of different Chl biosynthetic pathways published by Rebeiz *et al.* (1986), the photoreduction of MV- and DV-Pchl(ide) was shown using the hybrid protein of Pchlide reductase expressed in *E. coli* (R. Knaust, R. Schulz, and H. Senger, unpublished results).

## B. Influence of Internal and External Factors

In addition to light, several other internal and external factors were found to play a regulatory role in enzyme activity of Pchlide reductase. First, NADPH must be mentioned. Griffiths (1974) reported evidence suggesting that NADPH is the hydrogen donor for Pchlide reduction and that the binding of NADPH to the enzyme probably occurs through the 2'-phosphate group of the cosubstrate (Griffiths, 1980). The free pigment Pchlide, neither bound to NADPH nor to the active site of Pchlide reductase, is not phototransformable (Oliver and Griffiths, 1982). Addition of NADPH to etioplast membrane preparations was shown to result in a stabilization of the binding of Pchlide to the Pchlide reductase enzyme and in a conversion of inactive Pchl(ide) into a pigment–protein complex that is phototransformable and disappears as Chl(ide) is formed (Griffiths, 1975; El Hamouri *et al.*, 1981; Ryberg and Sundqvist, 1982b; Gerday *et al.*, 1984; Böddi *et al.*, 1989).

## 5. Protochlorophyllide Reductase

Synthesis of Pchl(ide) only in small amounts in darkness, compared with the amount of Chl in mature plants (Akoyunoglou and Siegelmann, 1968), and no overproduction of Chl in light demands very effective regulating mechanisms. The reduced level of Pchlide probably reflects the maximum amount of pigment with which Pchlide reductase can be loaded (Manetas and Akoyunoglou, 1978). Accumulated Pchlide was demonstrated to serve as a negative feed-back inhibitor of tRNA$^{glu}$-ligase, an enzyme that catalyses one of the first steps in ALA biosynthesis. Further synthesis of ALA and subsequent Pchlide formation was blocked by this mechanism (Dörnemann et al., 1989). The increase of Pchl(ide) content after incubation of dark-grown plants with ALA (Granick and Gassman, 1970; Gassman, 1973; Kotzabasis and Senger, 1990) results in the accumulation of a 632-nm inactive Pchl(ide) form (Gassman and Bogorad, 1967; Gassman, 1973). Excess Pchl(ide) generated after addition of ALA is photodegraded very easily (Sundqvist, 1969). The transformation of this photoinactive form to Chl(ide) occurs via the 650-nm Pchl(ide) complex (Sundqvist, 1970; Walter, 1974).

Calcium was shown to have an effect on the amount of Pchl(ide) formed and on the activity of Pchlide reductase in the pigment mutant C-2A' of *Scenedesmus obliquus* (Kotzabasis et al., 1990c). Depletion of $Ca^{2+}$ caused an increase in Pchl(ide), but a decrease in Pchlide reductase activity, whereas addition of $Ca^{2+}$ had the opposite effect. Stimulation of Pchlide reductase by $Ca^{2+}$ also was reported for *Anacystis nidulans* (Peschek et al., 1989b).

Horton and Leech (1975) found that surplus ATP prevented the conversion of Pchlide$_{650-657}$ to Pchlide$_{628-632}$ and stimulated the phototransformation of Pchlide to Chlide (Horton and Leech, 1972). These investigators proposed that ATP participates in organizing the pigment–protein complex in the proper configuration for photoconversion. However, Griffiths (1975) could not find any increase in activity of Pchlide reductase by adding ATP to etioplast membranes. Grevby et al. (1987) showed that low pH results in an increased transformation of photoinactive Pchlide$_{628-632}$ to phototransformable Pchlide$_{650-657}$ in darkness. This effect was enhanced in the presence of ATP. Their interpretation suggests that ATP was operative on the prolamellar bodies, which might explain why no effect of ATP was obtained during investigation of preparations of etioplast membrane systems (Griffiths, 1975).

Temperature should influence the photoreduction of Pchl(ide) to Chl(ide) the way it does other light-catalyzed processes (Smith and Benitez, 1954; van Huystee and Hodgins, 1989). By lowering the growth temperature from 33°C to 20°C, however, the pigment mutant C-2A' of the unicellular green alga *S. obliquus*, which reduces only traces of Pchlide under optimal growth conditions at 33°C in darkness, surprisingly recovers the ability to reduce Pchlide in darkness (Oh-hama et al., 1987; Kotzabasis et al., 1990a). At 20°C, this process is about 10 times more active than at 33°C, but reaches only about 13% of the light-dependent rate of Chl biosynthesis. Apparently, the rate of the light-independent reaction at 20°C is not limited by the amount and activity of Pchlide reductase but by the availability of ALA. A similar behavior was seen in temperature-sensitive yellow mutants of *Chlamydomonas reinhardtii* (Ford and Wang, 1980).

Bereza et al. (1984a) found that the cysteine/cystine redox couple had an activating effect, specifically on the Pchlide reductase of bean etioplasts. The redox potential of the microenvironment was suggested to regulate the activity of the enzyme and its affinity to substrates, reaching the optimal value near $-400$ mV (Bereza et al., 1984b).

Reduced plastid thioredoxin $f$ of *S. obliquus*, which is formed in the C-2A' mutant during greening, stimulates Pchlide reductase activity *in vitro* in light and darkness (Kotzabasis et al., 1989b). So far this is a single observation that thioredoxin is a new factor in the regulation of Chl biosynthesis.

## V. CLONING PROTOCHLOROPHYLLIDE REDUCTASE GENES AND PRIMARY STRUCTURE OF ENZYME PROTEIN

### A. Isolation and Characterization of cDNA Clones Encoding Protochlorophyllide Reductase of Barley

On illumination of dark-grown barley seedlings, drastic changes in the composition of the poly(A)-containing RNA fraction are induced. One of the mRNA species whose concentration drops rapidly during illumination encodes the 44-kDa precursor polypeptide of Pchilde reductase (Apel, 1981). To obtain more information about the Pchlide reductase enzyme and the corresponding RNA transcript, Apel et al. (1983) isolated a cDNA fragment called pHvDF1 by differential screening of dark- and light-specific cDNA clones. Investigations using this cDNA insert supported the idea that a single nuclear gene encodes the Pchlide reductase of barley (Steinmüller et al., 1985) and that the gene is only expressed in leaf tissue (Steinmüller et al., 1986). In Northern blots, the isolated cDNA hybridized efficiently with RNA of dark-grown barley seedlings (Schulz et al., 1989; Fig. 5). With RNA isolated from light-grown barley seedlings, only a very weak reaction is detectable by overexposing the blot (R. Schulz, unpublished results).

A comparison of the size of the clone F1, which is 538 bp in length, to the size of the hybridizing transcript of 1.7 kb revealed that the isolated

## 5. Protochlorophyllide Reductase

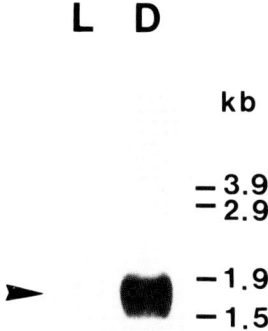

**Fig. 5.** The effect of white light on the appearance of transcripts encoding the Pchlide reductase of barley. Poly(A)-containing RNA was isolated from dark-grown (D) and 18 hr-illuminated (L) barley leaves, denatured, and separated electrophoretically on an agarose gel. The fractionated RNA was transferred to nitrocellulose and hybridized with the $^{32}$P-labeled cDNA of clone F1 (see text). The numerals indicate the size (in kb) of RNA molecules from barley and *E. coli* used as size markers. Reprinted with permission from Schulz *et al.* (1989).

cDNA insert is not complete and contains not more than one-third of the total sequence of the transcript encoding Pchlide reductase. To isolate a full-length cDNA, a polyclonal antiserum directed against the purified Pchlide reductase of barley (Apel, 1981) was used to screen a cDNA-expression library made from poly(A)-containing RNA of dark-grown barley seedlings. The longest cDNA clone isolated from the library was named A7 and was used for all further studies (Schulz *et al.*, 1989). To test the completeness of the reading frame of cDNA A7, it was transcribed *in vitro*. The transcripts were analyzed by *in vitro* translation followed by immunoprecipitation. A single polypeptide was detected that corresponded in size with the immunoprecipitated precursor polypeptide of Pchlide reductase. The result of this experiment indicates that the cDNA of clone A7 carries the entire protein-coding region for the precursor polypeptide of the Pchlide reductase of barley.

DNA sequencing of cDNA A7 revealed a length of 1509 bp and 100% homology to the shorter cDNA F1. The derived amino acid sequence consists of 388 residues with a calculated molecular mass of about 41.2 kDa, which corresponds to the size of the precursor polypeptide (Apel, 1981; Fig. 6). This predicted amino acid sequence of the precursor

```
                                    1               5              10
Hordeum vulgare                   Met Ala Leu Gln --- --- --- Leu Leu Pro ---
Avena sativa
Arabidopsis thaliana              Met Ala Leu Gln Ala Ala Ser --- Leu Val Ser
Pisum sativum                     Met Ala Leu Gln Thr Ala Ser Met Leu Pro Ala

                 15              20              25              30
Ser Thr Leu Ser Val Pro Lys --- --- Lys Gly Ser --- Ser Met Gly Ala Val Ala Val

Ser Ala Phe Ser Val Arg Lys Asp Ala Lys Leu Asn Ala Ser Ser Ser Ser Phe --- ---
Ser --- Phe Ser Ile Pro Lys Glu Gly Lys Ile Gly Ala Ser --- --- --- --- --- Leu

             35              40              45              50
Lys Asp Thr Ala Ala Phe Leu Gly Val Ser Ser --- --- --- Lys Ala Lys --- --- Lys

Lys Asp Ser Ser Leu Phe --- Gly Ala Ser --- Ile Thr Asp Gln Ile Lys Ser Glu His
Lys Asp Ser Thr Leu Phe --- Gly Val Ser Ser Leu Ser Asp Ser Leu Lys Gly Asp Phe

             55              60              65              70
Ala Ser Leu Ala Val Arg Thr Gln Val --- Ala Thr Ala --- Pro Ser Pro Val --- ---

Gly Ser Ser Ser Leu Arg Phe Lys Arg Glu Gln Ser Leu Arg Asn Leu Ala Ile Arg Ala
Thr Ser Ser Ala Leu Arg Cys Lys Arg Glu Leu Arg Gln Lys Val Gly Ala Val Arg Ala

             75              80              85              90
Thr Thr Ser Pro Gly Ser Thr Ala --- --- Ser --- Ser Pro Ser --- Gly Lys Lys Thr

Gln Thr Ala Ala Thr Ser Ser Pro Thr Val Thr Lys Ser Val --- Asp Gly Lys Lys Thr
Glu Thr Ala Ala Pro Ala Thr Pro Ala Val Asn Lys Ser Ser Ser Glu Gly Lys Lys Thr

             95             100             105             110
Leu Arg Gln Gly --- Val Val Ile Thr Gly Ala Ser Ser Gly Leu Gly Leu Ala Ala
                   Val Val Val Ile Thr Gly Ala Ser Ser Gly Leu Gly Leu Ala Ala
Leu Arg Lys Gly Asn Val Val --- Thr Gly Ala Ser Ser Gly Leu Gly Leu Ala Thr
Leu Arg Lys Gly Asn --- Val Val Ile Thr Gly Ala Ser Ser Gly Leu Gly Leu Ala Thr

             115             120             125             130
Ala Lys Ala Leu Ala Glu Thr Gly Lys Trp His Val Val Met Ala Cys Arg Asp Phe Leu
Ala Lys Ala Leu Ala Glu Thr Gly Lys Trp His Val Val Met Ala Cys Arg Asp Phe Leu
Ala Lys Ala Leu Ala Glu Thr Gly Lys Trp Asn Val Ile Met Ala Cys Arg Asp Phe Leu
Ala Lys Ala Leu Ala Glu Ser Gly Lys Trp Asn Val Ile Met Ala Cys Arg Asp Tyr Leu

             135             140             145             150
Lys Ala Ser Lys Ala Ala Lys Ala Ala Gly Met Ala --- Asp Gly Ser Tyr Thr Val Met
Lys Ala Ser Lys Ala Ala Lys Ala Ala Gly Met Ala --- Asp Gly Ser Tyr Thr Val Met
Lys Ala Glu Arg Ala Ala Lys Ser Val Gly Met Pro Lys Asp Ser Tyr Thr Val Met
Lys Ala Ala Arg Ala Ala Lys Ser Ala Gly Leu Ala Lys Glu --- Asn Tyr Thr Ile Met

             155             160             165             170
His Leu Asp Leu Ala Ser Leu Asp Ser Val Arg Gln Phe Val Asp Ala Phe Arg Arg Ala
His Leu Asp Leu Ala Ser Leu Asp Ser Val Arg Gln Phe Val Asp Ala Phe Arg Arg Ala
His Leu Asp Leu Ala Ser Leu Asp Ser Val Arg Gln Phe Val Asp Asn Phe Arg Arg Thr
His Leu Asp Leu Ala Ser Leu Asp Ser Val Arg Gln Phe Val Asp Asn Phe Arg Arg Ser

             175             180             185             190
Glu Met Pro Leu Asp Val Leu Val Cys Asn Ala Ala Ile Tyr Arg Pro Thr Ala Arg Thr
Glu Met Pro Leu Asp Val Leu Val Cys Asn Ala Ala Ile Tyr Arg Pro Thr Ala Arg Lys
Glu Thr Pro Leu Asp Val Leu Val Cys Asn Ala Ala Val Tyr Phe Pro Thr Ala Lys Glu
Glu Met Pro Leu Asp Val Leu Ile Asn Asn Ala Ala Val Tyr Phe Pro Thr Ala Lys Glu

             195             200             205             210
Pro Thr Phe Thr Ala Asp Gly His Glu Met Ser Val Gly Val Asn His Leu Gly His Phe
Pro Thr Phe Thr Ala Glu Gly Val Glu Met Ser Val Gly Val Asn His Leu Gly His Phe
Pro Thr Tyr Ser Ala Glu Gly Phe Glu Leu Ser Val Ala Thr Asn His Leu Gly His Phe
Pro Ser Phe Thr Ala Asp Gly Phe Glu Ile Ser Val Gly Thr Asn His Leu Gly His Phe

             215             220             225             230
Leu Leu Ala Arg Leu Leu Met Glu Asp Leu Gln Lys Ser Asp Tyr Pro Ser Arg Arg Met
Leu Leu Ala Arg Leu Leu Leu Glu Asp Leu Gln Lys Ser Asp Tyr Pro Ser Arg Arg Leu
Leu Leu Ala Arg Leu Leu Leu Asp Asp Leu Lys Lys Ser Asp Tyr Pro Ser Lys Arg Leu
Leu Leu Ser Arg Leu Leu Leu Glu Asp Leu Lys Lys Ser Asp Tyr Pro Ser Lys Arg Leu
```

|     | 235 |     |     |     |     | 240 |     |     |     |     | 245 |     |     |     |     | 250 |
| --- | --- | --- | --- | --- | --- | --- | --- | --- | --- | --- | --- | --- | --- | --- | --- | --- |
| Val | Ile | Val | Gly | Ser | Ile | Thr | Gly | Asn | Ser | Asn | Thr | Leu | Ala | Gly | Asn | Val | Pro | Pro | Lys |
| Val | Ile | Val | Gly | Ser | Ile | Thr | Gly | Asn | Asp | Asn | Thr | Leu | Ala | Gly | Asn | Val | Pro | Pro | Lys |
| Ile | Ile | Val | Gly | Ser | Ile | Thr | Gly | Asn | Thr | Asn | Thr | Leu | Ala | Gly | Asn | Val | Pro | Pro | Lys |
| Ile | Ile | Val | Gly | Ser | Ile | Thr | Gly | Asn | Thr | Asn | Thr | Leu | Ala | Gly | Asn | Val | Pro | Pro | Lys |

|     |     | 255 |     |     |     |     | 260 |     |     |     |     | 265 |     |     |     |     | 270 |
| --- | --- | --- | --- | --- | --- | --- | --- | --- | --- | --- | --- | --- | --- | --- | --- | --- | --- |
| Ala | Ser | Leu | Gly | Asp | Leu | Arg | Gly | Leu | Ala | Gly | Gly | Leu | Ser | Gly | Ala | Ser | Gly | Ala |
| Ala | Asn | Leu | Gly | Asp | Leu | Arg | Gly | Leu | Ala | Gly | Gly | Leu | Thr | Gly | Ala | Ser | Gly | Ser | Ala |
| Ala | Asn | Leu | Gly | Asp | Leu | Arg | Gly | Leu | Ala | Gly | Gly | Leu | Asn | Gly | Leu | Asn | Ser | Ser | Ala |
| Ala | Asn | Leu | Gly | Asp | Leu | Arg | Gly | Leu | Ala | Gly | Gly | Leu | Thr | Gly | Leu | Asn | Ser | Ser | Ala |

|     |     |     | 275 |     |     |     | 280 |     |     |     |     | 285 |     |     |     |     | 290 |
| --- | --- | --- | --- | --- | --- | --- | --- | --- | --- | --- | --- | --- | --- | --- | --- | --- | --- |
| Met | Ile | Asp | --- | Gly | Asp | Glu | Ser | Phe | Asp | Gly | Ala | Lys | Ala | Tyr | Lys | Asp | Ser | Lys | Val |
| Met | Ile | Asp | --- | Gly | Asp | Glu | Ser | Phe | Asp | Gly | Ala | Lys | Ala | Tyr | Lys | Asp | Ser | Lys | Val |
| Met | Ile | Asp | Gly | Gly | Asp | --- | --- | Phe | Asp | Gly | Ala | Lys | Ala | Tyr | Lys | Asp | Ser | Lys | Val |
| Met | Ile | Asp | Gly | Gly | Asp | --- | --- | Phe | Asp | Gly | Ala | Lys | Ala | Tyr | Lys | Asp | Ser | Lys | Val |

|     | 295 |     |     |     |     | 300 |     |     |     |     | 305 |     |     |     |     | 310 |
| --- | --- | --- | --- | --- | --- | --- | --- | --- | --- | --- | --- | --- | --- | --- | --- | --- |
| Cys | Asn | Met | Leu | Thr | Met | Gln | Glu | Phe | His | Arg | Arg | Tyr | His | Glu | Glu | Thr | Gly | Ile | Thr |
| Cys | Asn | Met | Leu | Thr | Met | Gln | Glu | Phe | His | Arg | Arg | Tyr | His | Glu | Asp | Thr | Gly | Ile | Thr |
| Cys | Asn | Met | Leu | Thr | Met | Gln | Glu | Phe | His | Arg | Arg | Phe | His | Glu | Glu | Thr | Gly | Val | Thr |
| Cys | Asn | Met | Leu | Thr | Met | Gln | Glu | Phe | His | Arg | Arg | Try | His | Glu | Glu | Thr | Gly | Ile | Thr |

|     | 315 |     |     |     |     | 320 |     |     |     | 325 |     |     |     |     | 330 |
| --- | --- | --- | --- | --- | --- | --- | --- | --- | --- | --- | --- | --- | --- | --- | --- |
| Phe | Ser | Ser | Leu | Tyr | Pro | Gly | Cys | Ile | Ala | Thr | Thr | Gly | Leu | Phe | Arg | Glu | His | Ile | Pro |
| Phe | Ser | Ser | Leu | Tyr | Pro | Gly | Cys | Ile | Ala | Thr | Thr | Gly | Leu | Phe | Arg | Glu | His | Ile | Pro |
| Phe | Ala | Ser | Leu | Tyr | Pro | Gly | Cys | Ile | Ala | Ser | Thr | Gly | Leu | Phe | Arg | Glu | His | Ile | Pro |
| Phe | Ala | Ser | Leu | Tyr | Pro | Gly | Cys | Ile | Ala | Thr | Thr | Gly | Leu | Phe | Arg | Glu | His | Ile | Pro |

|     | 335 |     |     |     |     | 340 |     |     |     |     | 345 |     |     |     |     | 350 |
| --- | --- | --- | --- | --- | --- | --- | --- | --- | --- | --- | --- | --- | --- | --- | --- | --- |
| Leu | Phe | Arg | Thr | Leu | Phe | Pro | Pro | Phe | Gln | Lys | Phe | Val | Thr | Lys | Gly | Phe | Val | Ser | Glu |
| Leu | Phe | Arg | Thr | Leu | Phe | Pro | Pro | Phe | Gln | Lys | Phe | Val | Thr | Lys | Gly | Phe | Val | Ser | Glu |
| Leu | Phe | Arg | Ala | Leu | Phe | Pro | Pro | Phe | Gln | Lys | Tyr | Ile | Thr | Lys | Gly | Tyr | Val | Ser | Glu |
| Leu | Phe | Arg | Thr | Leu | Phe | Pro | Pro | Phe | Gln | Lys | Tyr | Ile | Thr | Lys | Gly | Tyr | Val | Ser | Glu |

|     | 355 |     |     |     |     | 360 |     |     |     |     | 365 |     |     |     |     | 370 |
| --- | --- | --- | --- | --- | --- | --- | --- | --- | --- | --- | --- | --- | --- | --- | --- | --- |
| Ala | Glu | Ser | Gly | Lys | Arg | Leu | Ala | Gln | Val | Val | Ala | Glu | Pro | Val | Leu | Thr | Lys | Ser | Gly |
| Ala | Glu | Ser | Gly | Lys | Arg | Leu | Ala | Gln | Val | Val | Gly | Glu | Pro | Ser | Leu | Thr | Lys | Ser | Gly |
| Thr | Glu | Ser | Gly | Lys | Arg | Leu | Ala | Gln | Val | Val | Ser | Asp | Pro | Ser | Leu | Thr | Lys | Ser | Gly |
| Glu | Glu | Ser | Gly | Lys | Arg | Leu | Ala | Gln | Val | Val | Ser | Asp | Pro | Ser | Leu | Thr | Lys | Ser | Gly |

|     | 375 |     |     |     |     | 380 |     |     |     |     | 385 |     |     |     |     | 390 |
| --- | --- | --- | --- | --- | --- | --- | --- | --- | --- | --- | --- | --- | --- | --- | --- | --- |
| Val | Tyr | Trp | Ser | Trp | Asn | Lys | Asp | Ser | Ala | Ser | Phe | Glu | Asn | Gln | Leu | Ser | Gln | Glu | Ala |
| Val | Tyr | Trp | Ser | Trp | Asn | Lys | Asp | Ser | Ala | Ser | Phe | Glu | Asn | Gln | Leu | Ser | Gln | Glu | Ala |
| Val | Tyr | Trp | Ser | Trp | Asn | Asn | Ala | Ser | Ala | Ser | Phe | Glu | Asn | Gln | Leu | Ser | Glu | Glu | Ala |
| Val | Tyr | Trp | Ser | Trp | Asn | Asn | Ala | Ser | Ala | Ser | Phe | Glu | Asn | Gln | Leu | Ser | Glu | Glu | Ala |

|     |     | 395 |     |     |     |     | 400 |     |     |     |     | 405 |     |     |     |     | 420 |
| --- | --- | --- | --- | --- | --- | --- | --- | --- | --- | --- | --- | --- | --- | --- | --- | --- | --- |
| Ser | Asp | Pro | Glu | Lys | Ala | Arg | Lys | Val | Trp | Glu | Leu | Ser | Glu | Lys | Leu | Val | Gly | Leu | Ala |
| Ser | Asp | Pro | Glu | Lys | Ala | Arg | Lys | Val | Trp | Glu | Leu | Ser | Glu | Lys | Leu | Val | Gly | Leu | Ala |
| Ser | Asp | Pro | Val | Glu | Lys | Ala | Arg | Lys | Val | Trp | Glu | Ile | Ser | Asp | Lys | Leu | Val | Gly | Leu | Ala |
| Ser | Asp | Ala | Glu | Lys | Ala | Arg | Lys | Val | Trp | Glu | Val | Ser | Glu | Lys | Leu | Val | Gly | Leu | Ala |

**Fig. 6.** Comparison of the derived amino acid sequences of DNA clones encoding the Pchlide reductase of barley (*Hordeum vulgare*), oat (*Avena sativa*), *Arabidopsis thaliana* and pea (*Pisum sativum*). Boxes indicate amino acids that are identical in all sequences. Positions of cysteine residues are shaded for emphasis. The putative cleavage sites between the transit peptides and the mature proteins are indicated by arrowheads. Amino acids of the derived amino acid sequences that are confirmed by N-terminal sequencing of the mature protein or of isolated tryptic fragments of purified Pchlide reductase of barley, oat, and pea are underlined. Adapted from Schulz *et al.* (1989), Darrah *et al.* (1990), Benli *et al.* (1991), and Spano *et al.* (1992).

molecule of Pchlide reductase of barley had been confirmed by partial sequencing of tryptic peptides derived from the mature protein (Benli *et al.*, 1991; Fig. 6). For this purpose, Pchlide reductase was isolated from etiolated barley seedlings and was digested with trypsin. The polypeptide fragments were separated by high performance liquid chromatography (HPLC) and the N-terminal amino acid sequences of these peptide fractions and of the undigested mature enzyme protein were determined. It was possible to confirm more than one-third of the predicted amino acid sequence derived from the cDNA clone A7. The putative cleavage site of the transit peptide was identified by determining the N-terminal region of the undigested protein and by alignment and sequence comparison with the deduced amino acid sequence reported for *Arabidopsis thaliana* (Benli *et al.*, 1991; Fig. 6). The predicted molecular mass of the derived amino acid sequence of the mature protein of barley was 34.7 kDa.

As further evidence that cDNA clone A7 encodes the precursor molecule of the Pchlide reductase of barley, a polyclonal antiserum was raised against the A7 protein expressed in *Escherichia coli* cells. As expected, this antiserum cross-reacted with a 36-kDa protein that decreased rapidly on illumination, as tested by Western blots (Forreiter *et al.*, 1990). By immunogold labeling, the area of the prolamellar body of etiolated barley leaves was found to be the predominant location of the enzyme protein, which disappeared on illumination (Schulz, 1990).

The strongest evidence for the identification of cDNA clone A7 as the sequence encoding the Pchlide reductase of barley, however, was the expression of this cDNA in *Escherichia coli,* which led to the synthesis of an active enzyme molecule (Schulz *et al.* 1989). Lambda gt11-phage lysogens were constructed that carried the cDNA insert A7. In a control experiment, wild-type DNA of lambda phage gt11 without an insert was used. Synthesis of $\beta$-galactosidase in the control experiment or synthesis of the fusion protein of Pchlide reductase and $\beta$-galactosidase was induced, bacteria were lysed, NADPH and Pchlide were added, and the reduction of Pchlide to Chlide was measured by fluorescence spectroscopy (Fig. 7). Enzymatic activity could be detected only in the construct that carried cDNA A7, but not in the control lysogen. This reduction of Pchlide to Chlide took place only in light. When these bacterial lysates were exposed to light for various lengths of time, increasing amounts of Chlide were formed. This photoreduction was completely dependent on the presence of NADPH and Pchlide. Without either of these two substrates, Chlide synthesis could not be observed. The results of this experiment have two important implications. First, in the etiolated barley seedling, Pchlide reductase is part of the highly organized structure of the prolamellar body (Ikeuchi and Murakami, 1983; Dehesh and Ryberg, 1985;

## 5. Protochlorophyllide Reductase

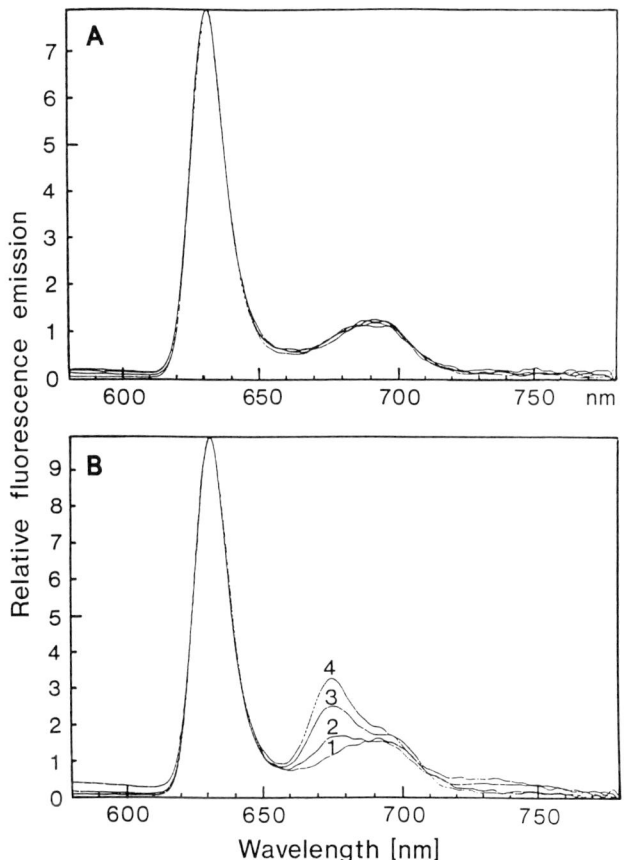

**Fig. 7.** The measurement of light-dependent reduction of Pchlide to Chlide in extracts of induced lysogenic bacteria. Extracts of induced lysogenic *Escherichia coli* cells carrying (A) wild-type DNA of the phage lambda gt11 or (B) recombinant DNA of the lambda gt11 vector and the cDNA A7 encoding the Pchlide of barley were prepared and supplied with Pchlide and NADPH. Fluorescence emission spectra were measured at $-196°C$ (1) in the dark sample, (2) after 20 light flashes, (3) after 20 light flashes and 3 hr dim white light, and (4) after 20 light flashes and 16 hr dim white light with an excitation wavelength of 440 nm. The spectra are normalized at the fluorescence emission maximum of Pchlide. Photoreduction of Pchlide (629 nm) to Chlide (675 nm) occurred only in extracts of lysogenic bacteria carrying the cDNA A7. Adapted from Schulz *et al.* (1989).

Shaw et al., 1985; Ryberg and Dehesh, 1986). On illumination, the prolamellar body rapidly disintegrates and, at the same time, the activity of Pchlide reductase drops instantly to a small percentage of its dark level (Mapleston and Griffiths, 1980; Santel and Apel, 1981). The result of the expression experiment in *Escherichia coli* demonstrated that the activity of Pchlide reductase does not depend exclusively on the integration of the enzyme protein in the prolamellar body structure. Even the precursor polypeptide of Pchlide reductase as part of a large fusion protein showed enzymatic activity. Second, measurement of enzymatic activity in bacterial lysates should allow a detailed analysis of the functional importance of different protein domains of Pchlide reductase in further investigations.

## B. Comparison of Different Protochlorophyllide Reductase Protein Sequences

As shown in Fig. 6, the amino acid sequence of Pchlide reductase of barley as derived from the nucleotide sequence of clone A7 (Schulz et al., 1989) displayed 96% similarity to the incomplete open reading frame of cDNA p127 encoding Pchlide reductase of oat (Darrah et al., 1990). A comparison of the deduced amino acid sequence of barley with those of a full-length cDNA of *Arabidopsis thaliana* (Benli et al., 1991) and of the nuclear gene *lpcr* of pea (Spano et al. 1992) revealed sequence similarities of 81 and 83%, respectively, in the region of the mature protein. Obviously, all the deduced amino acid sequences of the Pchlide reductase enzyme proteins contain dominant highly conserved regions. The amino acid sequences have a relatively high basic amino acid content and also are characterized by a relatively high proportion of hydrophobic amino acids.

In contrast, the similarity between the transit peptides of the precursor molecules for the Pchlide reductases of barley, *Arabidopsis*, and pea is very low (Fig. 6). However, these small similarities and the specific amino acid composition are in good agreement with the general structure of transit peptides, which mediate the transport of proteins into the plastids (von Heijne et al., 1989).

In addition, the total amino acid composition of all the deduced amino acid sequences is in good agreement with results reported earlier for the Pchlide reductase of oat by Röper et al. (1987). However, after hydrolysis of the isolated protein of oat, only one cysteine residue has been reported in the Pchlide reductase polypeptide, whereas the presently known deduced amino acid sequences revealed three or four cysteine residues surrounded by well-conserved amino acid sequence regions (Fig. 6).

## VI. ACTIVE SITE AND MEMBRANE ASSOCIATION OF PROTOCHLOROPHYLLIDE REDUCTASE

### A. Active Site

NMR studies using purified Pchlide reductase of oat seedlings indicated that the reaction catalyzed by the enzyme is a *trans* reduction (Begley and Young, 1989). Regarding the origin of the hydrogen atoms, the NMR studies demonstrated that a hydride is delivered to the C-17 position of Pchlide from NADPH and that the C-18 position is protonated by water or an active site acid of the enzyme itself (Fig. 1). By light-dependent covalent labeling of Pchlide reductase with radioactive *N*-phenylmaleimide, it was shown that cysteine residues are involved in binding Pchlide and catalyzing its light-dependent reduction to Chlide (Griffiths, 1978; Oliver and Griffiths, 1980; Oliver and Griffiths, 1981; Dehesh *et al.*, 1986a). Currently it is not clear whether all cysteine residues found in the deduced amino acid sequences of Pchlide reductases (Fig. 6) are engaged in the binding of Pchlide or whether some of them might be involved in maintaining the three-dimensional structure of the enzyme protein by the formation of disulfide bridges. In previous studies, two to three Pchlide molecules have been found to be bound to one enzyme molecule (Apel *et al.*, 1980). Based on these results, more than one cysteine residue might be involved in substrate binding.

### B. Hydrophobicity and Membrane Association

Different approaches were undertaken to obtain information about the hydrophobicity and the membrane-binding properties of Pchlide reductase, including electron microscopy of the tubular membranes in prolamellar bodies (Kahn, 1968a), measurement of Pchlide reductase activity in etioplast membrane fractions (Griffiths, 1975), analysis of etioplast membrane phosphoproteins (Covello *et al.*, 1987), and treatment of isolated and immobilized prolamellar bodies with different detergents and ions followed by analysis of the structure of the membrane system and the content of Pchlide reductase and its activity (Grevby *et al.*, 1989). All these investigations showed that Pchlide reductase is a constituent of etioplast membranes. By Triton X-114 partitioning (Selstam and Widell-Wigge, 1989) and by salt treatment of isolated etioplast membrane fractions (Widell-Wigge and Selstam, 1990), it was possible to show that Pchlide reductase, as a protein with amphiphilic properties, is associated intimately with the membrane lipids.

To acquire more information about the membrane-binding properties of Pchlide reductase of barley, the derived amino acid sequence was used to perform a hydropathy plot according to Kyte and Doolittle (1982; Fig. 8). As indicated before, several hydrophobic regions are found in the deduced amino acid sequence of the Pchlide reductase of barley. According to Kyte and Doolittle (1982), however, none of them is long enough to span a membrane. In agreement with the analysis of the hydrophobicity of the deduced amino acid sequences of Pchlide reductase of oat (Darrah *et al.*, 1990) and of *Arabidopsis* (Benli *et al.*, 1991), the enzyme is believed not to be membrane spanning but rather to be closely associated to the membrane according to the relatively high number of hydrophobic amino acids (See also Chapter 7).

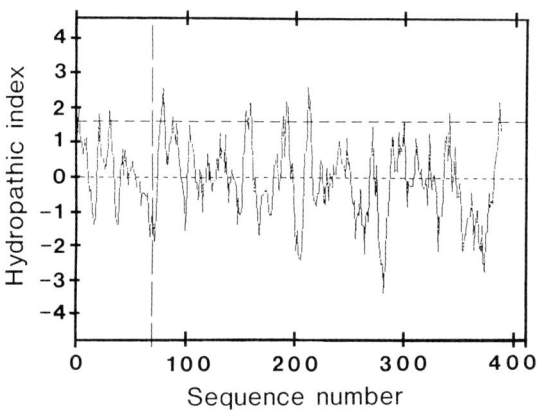

**Fig. 8.** Hydropathy plot of the deduced amino acid sequence of the Pchlide reductase of barley. The hydrophobicity curve represents the average of the specific hydrophobicity index over a window of seven amino acid residues according to Kyte and Doolittle (1982). The putative cleavage site between the transit peptide and the mature protein is indicated by a vertical broken line. The horizontal line marks a hydrophobicity index of 1.6, which must be exceeded by the index sum of at least 19 amino acids to point to a membrane spanning region of the protein. Adapted from Benli *et al.* (1991).

## VII. CONCLUDING REMARKS

### A. Existence of a Light-Dependent and a Light-Independent Protochlorophyllide Reductase

Despite the fact that Scots pine has a light-independent biosynthetic pathway for Chl, light-dependent Pchlide reductase and phototransformable Pchlide nevertheless accumulate in its prolamellar bodies (Selstam and Widell, 1986). Induced mutants of a number of green algae, which unlike the wild-type synthesize Chl only in the light (Table I), demonstrate that there may be two separate pathways of Chl biosynthesis in these organisms. In addition, for the cyanobacteria *Anabaena* (Adamson *et al.*, 1987) and *Anacystis* (Peschek *et al.*, 1989b) and for green redarkened angiosperms, light-independent Pchl(ide) reduction was postulated (Röbbelen, 1956; Adamson, 1978, 1982; Adamson *et al.*, 1980, 1984, 1985; Adamson and Hiller, 1981; Adamson and Packer, 1984; Packer *et al.*, 1984; Walmsley and Adamson, 1989).

Homologs of the chloroplast gene *frxC*, originally described from liverwort (Ohyama *et al.*, 1988), were shown to be involved in light-independent reduction of Pchlide (Fujita *et al.*, 1992; Suzuki and Bauer, 1992). Cross-hybridizing bands were observed only with DNA prepared from organisms that show light-independent Pchlide reduction (Suzuki and Bauer, 1992).

### B. Existence of a Plastid Factor

Interactions between the nuclear and chloroplastic genomes play key roles in leaf development and function. Several reports are available showing that, in addition to the functioning phytochrome and the near-UV/blue-light photoreceptors, the integrity of the plastid is essential for the expression of pertinent nuclear genes involved in chloroplast formation (Bradbeer *et al.*, 1979; Reiß *et al.*, 1983; Harpster *et al.*, 1984; Mayfield and Taylor, 1984; Batschauer *et al.*, 1986; Oelmüller and Mohr, 1986; Oelmüller *et al.*, 1986; Schmidt *et al.*, 1987; Oelmüller, 1989; Stockhaus *et al.*, 1989; Taylor, 1989; Rapp and Mullet, 1991; Susek and Chory, 1992). It was concluded that a signal from intact plastids is needed to permit phytochrome-mediated appearance of translatable mRNAs for those genes whose protein products become integral constituents of the plastid. The basis of this apparent coordinate gene expression is unknown. By analyzing the transcriptional control of the nuclear gene encoding the apoproteins of the light harvesting complex (LHCP) in carotenoid-deficient albina mutants of barley and in barley plants treated with the herbicide Norflurazon, Batschauer et al. (1986) supposed that, if any of the reactions leading to Chl formation in higher plants is involved in the control of LHCP mRNA accumulation, it should be one between the formation of Pchlide and the esterification of Chlide. Häuser et al. (1987) speculated that the function of Pchlide reductase may not be restricted to that of an "ordinary" enzyme, equivalent to others that take part only in the biosynthesis of Chl, but also may be involved more directly in the light-dependent control of chloroplast development (Harpster and Apel, 1985; Benli et al., 1991). Pchlide as the photoreceptor of the plastids may play a regulatory role in the expression of plastid proteins (Laing et al., 1988). Also, the photoreduction of Pchlide to Chlide, as the main step in the development of the chloroplast, also may have some influence on the activity of the nucleus and the cytoplasm by mediation of a still unknown plastid factor.

## REFERENCES

Adamson, H. (1978). Evidence for the accumulation of both chlorophyll $a$ and $b$ in darkness in an angiosperm. In "Chloroplast Development" (G. Akoyunoglou and J. H. Argyroudi-Akoyunoglou, eds.), pp. 135–140. Elsevier/North-Holland, Amsterdam.

Adamson, H. (1982). Evidence for a light-independent protochlorophyllide reductase in green barley leaves. In "Cell Function and Differentiation" (G. Akoyunoglou, A. E. Evangelopoilos, J. Georgtsos, G. Palaiologos, A. Trakatellis, and C. P. Tsiganos, eds.), pp. 33–41. A. R. Liss, New York.

Adamson, H., and Hiller, R. G. (1981). Chlorophyll synthesis in the dark in angiosperms. In "Photosynthesis. V. Chloroplast Development" (G. Akoyunoglou, ed.), pp. 213–221. Balaban, Philadelphia.

Adamson, H., and Packer, N. (1984). Dark synthesis of chlorophyll in vivo and dark reduction of protochlorophyllide in vitro by pea chloroplasts. In "Protochlorophyllide Reduction and Greening" (C. Sironval and M. Brouers, eds.), pp. 353–363. Nijhoff/Junk, The Hague, The Netherlands.

Adamson, H. Y., Hiller, R. G., and Vesk, M. (1980). Chloroplast development and the synthesis of chlorophyll $a$ and $b$ and chlorophyll protein complexes I and II in the dark in Tradescantia albiflora (Kunth). Planta 150, 269–274.

Adamson, H., Packer, N., and Sanders, B. (1984). Chlorophyll synthesis in the dark in peas. In "Advances in Photosynthesis Research" (C. Sybesma, ed.), Vol. IV, pp. 705–708. Nijhoff/Junk, The Hague, The Netherlands.

Adamson, H., Griffiths, T., Packer, N., and Sutherland, M. (1985). Light-independent accumulation of chlorophyll $a$ and $b$ and protochlorophyllide in green barley (Hordeum vulgare). Physiol. Plant. 64, 345–352.

Adamson, H., Walker, C., Bees, A., and Griffiths, T. (1987). Protochlorophyllide reduction in Anabaena. In "Progress in Photosynthesis Research" (J. Biggins, ed.), Vol. IV, pp. 483–486. Nijhoff, Dordrecht, The Netherlands.

Adamson, H., Lennon, M., Ou, K., Packer, N., and Walmsley, J. (1990). Evidence for a light-independent chlorophyll biosynthetic pathway in angiosperm seeds germinated in darkness. In "Current Research in Photosynthesis" (M. Baltscheffsky, ed.), Vol. III, pp. 687–690. Kluwer, Dordrecht, The Netherlands.

## 5. Protochlorophyllide Reductase

Akoyunoglou, G. A., and Siegelman, H. W. (1968). Protochlorophyllide resynthesis in dark-grown bean leaves. *Plant Physiol.* **43**, 66–68.

Anderson, L. E. (1979). Interaction between photochemistry and activity of enzymes. *In* "Photosynthesis II. Photosynthetic Carbon Metabolism and Related Processes" (M. Gibbs and E. Latzko, eds.), pp. 271–281. Springer-Verlag, Berlin.

Apel, K. (1981). The protochlorophyllide holochrome of barley (*Hordeum vulgare* L.). Phytochrome-induced decrease of the translatable mRNA coding for the NADPH : protochlorophyllide oxidoreductase. *Eur. J. Biochem.* **120**, 89–93.

Apel, K., Santel, H.-J., Redlinger, T. E., and Falk, H. (1980). The protochlorophyllide holochrome of barley (*Hordeum vulgare* L.). Isolation and characterization of the NADPH : protochlorophyllide oxidoreductase. *Eur. J. Biochem.* **111**, 251–258.

Apel, K., Gollmer, I., and Batschauer, A. (1983). The light-dependent control of chloroplast development in barley (*Hordeum vulgare* L.). *J. Cell Biochem.* **23**, 181–189.

Apel, K., Motzkus, M., and Dehesh, K. (1984). The biosynthesis of chlorophyll in greening barley (*Hordeum vulgare*). Is there a light-independent protochlorophyllide reductase? *Planta* **161**, 550–554.

Artus, N. N., Ryberg, M., Lindsten, A., Ryberg, H., and Sundqvist, C. (1992). The Shibata shift and the transformation of etioplasts to chloroplasts in wheat with Clomazone (FMC 57020) and Amiprophos-Methyl (Tokunol M). *Plant. Physiol.* **98**, 253–263.

Batschauer, A., and Apel, K. (1984). The inverse control by phytochrome of the expression of two nuclear genes in barley (*Hordeum vulgare* L.). *Eur. J. Biochem.* **143**, 593–597.

Batschauer, A., Santel, H.-J., and Apel, K. (1982). The presence and synthesis of the NADPH-protochlorophyllide oxidoreductase in barley leaves with a high temperature-induced deficiency of plastid ribosomes. *Planta* **154**, 459–464.

Bauer, K., and Wild, A. (1976). Die Wirkung von Blaulicht auf den photosynthetischen Elektronentransport bei gelbgrünen Mutanten von *Chlorella fusca*. *Z. Pflanzenphysiol.* **80**, 443–456.

Beale, S. I. (1971). Studies on the biosynthesis and metabolism of δ-aminolevulinic acid in *Chlorella*. *Plant Physiol.* **48**, 316–319.

Beale, S. I. (1984). Biosynthesis of photosynthetic pigments. *In* "Chloroplast Biogenesis" (N.R. Baker and J. Barber, eds.), pp. 133–205. Elsevier, Amsterdam.

Beer, N. S., and Griffiths, W. T. (1981). Purification of the enzyme NADPH : protochlorophyllide oxidoreductase. *Biochem. J.* **195**, 83–92.

Begley, T. P., and Young, H. (1989). Protochlorophyllide reductase. I. Determination of the regiochemistry and the stereochemistry of the reduction of protochlorophyllide to chlorophyllide. *J. Am. Chem. Soc.* **111**, 3095–3096.

Belanger, F. C., and Rebeiz, C. A. (1980a). Chloroplast biogenesis: Detection of divinyl protochlorophyllide in higher plants. *J. Biol. Chem.* **255**, 1266–1272.

Belanger, F. C., and Rebeiz, C. A. (1980b). Chloroplast biogenesis: Detection of divinylprotochlorophyllide ester in higher plants. *Biochemistry* **19**, 4875–4883.

Benli, M., Schulz, R., and Apel, K. (1991). Effect of light on the NADPH-protochlorophyllide oxidoreductase of *Arabidopsis thaliana*. *Plant Mol. Biol.* **16**, 615–625.

Bennett, J., Schwender, J. R., Shaw, E. K., Tempel, N., Ledbetter, M., and Williams, R. S. (1987). Failure of corn leaves to acclimate to low irradiance. Role of protochlorophyllide reductase in regulating levels of five chlorophyll-binding proteins. *Biochim. Biophys. Acta* **892**, 118–129.

Bereza, B., Laskowski, M., and Hendrich, W. (1984a). The regulatory effect of the cysteine/cystine redox couple on protochlorophyllide photoreduction in etioplasts. *In* "Protochlorophyllide Reduction and Greening" (C. Sironval and M. Brouers, eds.), pp. 149–159. Nijhoff/Junk, The Hague, The Netherlands.

Bereza, B., Laskowski, M., and Hendrich, W. (1984b). The influence of the cysteine-cystine redox couple on the photoreduction of protochlorophyllide–protein complexes in etioplasts. In "Advances in Photosynthesis Research" (C. Sybesma, ed.), Vol. IV, pp. 749–752. Nijhoff/Junk, The Hague, The Netherlands.
Bishop, N. I., and Senger, H. (1972). The development of structure and function in chloroplasts of greening mutants of *Scenedesmus*. II. Development of the photosynthetic apparatus. *Plant Cell Physiol.* **13**, 937–953.
Bittner, K. (1905). Über Chlorophyllbildung im Finstern bei Kryptogamen. *Östereich. Bot. Zeitschrift.* **55**, 302–312.
Björn, L. O. (1969). Action spectra for transformation and fluorescence of protochlorophyll holochrome from bean leaves. *Physiol. Plant.* **22**, 1–17.
Boardman, N. K. (1966). Protochlorophyll. In "The Chlorophylls" (L. P. Vernon and G. R. Seely, eds.), pp. 437–479. Academic Press, New York.
Boardman, N. K., Anderson, J. M., and Goodchild, D. J. (1978). Chlorophyll–protein complexes and structure of mature and developing chloroplasts. In "Current Topics in Bioenergetics" (D. R. Sanadi and L. P. Vernon, eds.), Vol. 8, pp. 35–109. Academic Press, New York.
Böddi, B., Lindsten, A., Ryberg, M., and Sundqvist, C. (1989). On the aggregational states of protochlorophyllide and its protein complexes in wheat etioplasts. *Physiol. Plant.* **76**, 135–143.
Bogdanovic, M. (1973a). Chlorophyll formation in the dark. I. Chlorophyll in pine seedlings. *Physiol. Plant.* **29**, 17–18.
Bogdanovic, M. (1973b). Chlorophyll formation in the dark. II. Chlorophyll in wheat leaves transplanted to pine megagametophytes. *Physiol. Plant.* **29**, 19–21.
Bogorad, L. (1950). Factors associated with the synthesis of chlorophyll in the dark in seedlings of *Pinus jeffreyi*. *Bot. Gaz.* **111**, 221–241.
Bogorad, L. (1976). Chlorophyll biosynthesis. In "Chemistry and Biochemistry of Plant Pigments" (T. H. Goodwin, ed.), Vol. 1, pp. 64–148. Academic Press, London.
Bovey, F., Ogawa, T., and Shibata, K. (1974). Photoconvertible and non-photoconvertible forms of protochlorophyll(ide) in etiolated bean leaves. *Plant Cell Physiol.* **15**, 1133–1137.
Bradbeer, J. W., Atkinson, Y. E., Börner, T., and Hagemann, R. (1979). Cytoplasmatic synthesis of plastid polypeptides may be controlled by plastid-synthesised RNA. *Nature (London)* **279**, 816–817.
Brinkmann, G., and Senger, H. (1980). Is there a regulatory effect of red light during greening of *Scendesmus* mutant C-2A'? In "Photoreceptors and Plant Development" (J. De Greef, ed.), pp. 209–218. University Press, Antwerp, Belgium.
Brouers, M., and Wolwertz, M. R. (1980). Stability of etiochloroplasts isolated from pine cotyledons as studied by means of low temperature absorption and fluorescence spectroscopy. *Photosynth. Res.* **1**, 93–104.
Brouers, M., and Wolwertz, M. R. (1981). Biosynthesis of chlorophyllide from ALA–protochlorophyllide *in vitro*: Role of NADPH and $NADP^+$. In "Photosynthesis. V. Chloroplast Development" (G. Akoyunoglou, ed.), pp. 185–196. Balaban, Philadelphia.
Burgerstein, A. (1900). Über das Verhalten der Gymnospermen-Keimlinge im Lichte und im Dunkeln. *Ber. Dtsch. Bot. Ges.* **18**, 168–184.
Carey, E. E., and Rebeiz, C. A. (1985). Chloroplast biogenesis 49. Differences among angiosperms in the biosynthesis and accumulation of monovinyl and divinyl protochlorophyllide during photoperiodic greening. *Plant Physiol.* **79**, 1–6.
Carey, E. E., Tripathy, B. C., and Rebeiz, C. A. (1985). Chloroplast biogenesis 51. Modulation of monovinyl and divinyl protochlorophyllide biosynthesis by light and darkness *in vitro*. *Plant Physiol.* **79**, 1059–1063.

Castelfranco, P. A., and Beale, S. I. (1983). Chlorophyll biosynthesis: Recent advances and areas of current interest. *Ann. Rev. Plant Physiol.* **34**, 241–278.
Cohen, C. E., and Schiff, J. A. (1976). Events surrounding the early development of *Euglena* chloroplasts. XI. Protochlorophyll(ide) and its photoconversion. *Photochem. Photobiol.* **24**, 555–566.
Colbert, J. T., Hershey, H. P., and Quail, P. H. (1983). Autoregulatory control of translatable phytochrome mRNA levels. *Proc. Natl. Acad. Sci. USA* **80**, 2248–2252.
Covello, P. S., Webber, A. N., Danko, S. J., Markwell, J. P., and Baker, N. R. (1987). Phosphorylation of thylakoid proteins during chloroplast biogenesis in greening etiolated and light-grown wheat leaves. *Photosynth. Res.* **12**, 243–254.
Darrah, P. M., Kay, S. A., Teakle, G. R., and Griffiths, W. T. (1990). Cloning and sequencing of protochlorophyllide reductase. *Biochem. J.* **265**, 789–798.
Davies, T. G. E., Ougham, H. J., Thomas, H., and Rogers, L. J. (1989). Leaf development in *Lolium temulentum:* Plastid membrane polypeptides in relation to assembly of the photosynthetic apparatus and leaf growth. *Physiol. Plant.* **75**, 47–54.
Dehesh, K., and Apel, K. (1983). The function of proteases during the light-dependent transformation of etioplasts to chloroplasts in barley (*Hordeum vulgare* L.). *Planta* **157**, 381–383.
Dehesh, K., and Ryberg, M. (1985). The NADPH-protochlorophyllide oxidoreductase is the major protein constituent of prolamellar bodies in wheat (*Triticum aestivum* L.). *Planta* **164**, 396–399.
Dehesh, K., Häuser, I., Apel, K., and Kloppstech, K. (1983). The distribution of NADPH-protochlorophyllide oxidoreductase in relation to chlorophyll accumulation along the barley leaf gradient. *Planta* **158**, 134–139.
Dehesh, K., Klaas, M., Häuser, I., and Apel, K. (1986a). Light-induced changes in the distribution of the 36,000 $M_r$ polypeptide of NADPH-protochlorophyllide oxidoreductase within different cellular compartments of barley (*Hordeum vulgare* L.). I. Localization by immunoblotting in isolated plastids and total leaf extracts. *Planta* **169**, 162–171.
Dehesh, K., van Cleve, B., Ryberg, M., and Apel, K. (1986b). Light-induced changes in the distribution of the 36,000 $M_r$ polypeptide of NADPH-protochlorophyllide oxidoreductase within different cellular compartments of barley (*Hordeum vulgare* L.). II. Localization by immunogold labeling in ultrathin sections. *Planta* **169**, 172–183.
Dobek, A., Dujardin, E., Franck, F., Sironval, C., Breton, J., and Roux, E. (1981). The first events of protochlorophyll(ide) photoreduction investigated in etiolated leaves by means of the fluorescence excited by short, 610 nm laser flashes at room temperature. *Photobiochem. Photobiophys.* **2**, 35–44.
Dörnemann, D., Kotzabasis, K., Richter, P., Breu, V., and Senger, H. (1989). The regulation of chlorophyll biosynthesis by the action of protochlorophyllide on $^{glu}$tRNA–ligase. *Bot. Acta* **102**, 112–115.
Dubertret, G., and Joliot, P. (1974). Structure and organization of system II photosynthetic units during the greening of a dark-grown *Chlorella* mutant. *Biochim. Biophys. Acta* **357**, 399–411.
Dujardin, E., and Sironval, C. (1978). The mechanism of photoreduction of protochlorophyll(ide). *In* "Chloroplast Development" (G. Akoyunoglou and J. H. Argyroudi-Akoyunoglou, eds.), pp. 83–98. Elsevier/North Holland, Amsterdam.
Egan, J. M., Jr., and Schiff, J. A. (1974). A reexamination of the action spectrum for chlorophyll synthesis in *Euglena gracilis*. *Plant Sci. Lett.* **3**, 101–105.
Egneus, H., and Sundqvist, C. (1970). An action spectrum for the transformation of ALA–protochlorophyllide to ALA-chlorophyllide in the wavelength region 605–675 nm. *Photosynthetica* **4**, 81–83.
El Hamouri, B. (1984). Protochlorophyllide and chlorophyllide–protein complexes during

greening. In "Protochlorophyllide Reduction and Greening" (C. Sironval and M. Brouers, eds.), pp. 129–138. Nijhoff/Junk, The Hague, The Netherlands.

El Hamouri, B., Brouers, M., and Sironval, C. (1981). Pathway from photoinactive $P_{633-628}$ protochlorophyllide to the $P_{696-682}$ chlorophyllide in cucumber etioplasts. Plant Sci. Lett. **24,** 375–379.

Ford, C., and Wang, W. (1980). Temperature-sensitive yellow mutants of Chlamydomonas reinhardtii. Mol. Gen. Genet. **180,** 5–10.

Forreiter, C., van Cleve, B., Schmidt, A., and Apel, K. (1990). Evidence for a general light-dependent negative control of NADPH-protochlorophyllide oxidoreductase in angiosperms. Planta **183,** 126–132.

Franck, F., and Mathis, P. (1980). A short-lived intermediate in the photo-enzymatic reduction of protochlorophyll(ide) into chlorophyll(ide) at a physiological temperature. Photochem. Photobiol. **32,** 799–803.

Franck, F., and Strzalka, K. (1992). Detection of the protochlorophyllide-protein complex in the light during the greening of barley. FEBS Lett. **309,** 73–77.

Frank, S. R. (1946). The effectiveness of the spectrum in chlorophyll formation. J. Gen. Physiol. **29,** 157–179.

Fujita, Y., Takahashi, Y., Chuganji, M., and Matsubara, H. (1992). The nifH-like (frxC) gene is involved in the biosynthesis of chlorophyll in the filamentous cyanobacterium Plectonema boryanum. Plant Cell Physiol. **33,** 81–92.

Furuya, M. (1989). Molecular properties and biogenesis of phytochrome I and II. Adv. Biophys. **25,** 133–167.

Galland, P., and Senger, H. (1991). Flavins as possible blue light photoreceptors. In "Photoreceptor Evolution and Function" (M. G. Holmes, ed.), pp. 65–124. Academic Press, London.

Galling, G. (1978). Development of chloroplast structure and function in a temperature-sensitive mutant of Chlorella. In "Chloroplast Development" (G. Akoyunoglou and J. H. Argyroudi-Akoyunoglou, eds.), pp. 439–444. Elsevier/North-Holland, Amsterdam.

Galling, G. (1981). Development of thylakoids and photosynthetic activity in thermosensitive and light-dependent mutants of Chlorella pyrenoidosa. In "Photosynthesis. V. Chloroplast Development" (G. Akoyunoglou, ed.), pp. 465–472. Balaban, Philadelphia.

Gassman, M. L. (1973). The conversion of photoinactive protochlorophyllide$_{633}$ to phototransformable protochlorophyllide$_{650}$ in etiolated bean leaves treated with δ-aminolevulinic acid. Plant Physiol. **52,** 590–594.

Gassman, M., and Bogorad, L. (1967). Studies on the regeneration of protochlorophyllide after brief illumination of etiolated bean leaves. Plant Physiol. **42,** 781–784.

Gerday, C., Michel-Wolwertz, M.-R., and Brouers, M. (1984). Characterization of PChlide protein complexes in PT and PLB enriched fractions of bean etioplasts. Effect of NADPH. In "Protochlorophyllide Reduction and Greening" (C. Sironval and M. Brouers, eds.), pp. 53–67. Nijhoff/Junk, The Hague, The Netherlands.

Granick, S., and Gassman, M. (1970). Rapid regeneration of protochlorophyllide$_{650}$. Plant Physiol. **45,** 201–205.

Grevby, C., Ryberg, M., and Sundqvist, C. (1987). Transformation of photoinactive to photoactive protochlorophyllide in isolated prolamellar bodies of wheat (Triticum aestivum) exposed to low pH and ATP. Physiol. Plant. **70,** 155–162.

Grevby, C., Engdahl, S., Ryberg, M., and Sundqvist, C. (1989). Binding properties of NADPH-protochlorophyllide oxidoreductase as revealed by detergent and ion treatments of isolated and immobilized prolamellar bodies. Physiol. Plant. **77,** 493–503.

Griffiths, W. T. (1974). Protochlorophyll and protochlorophyllide as precursors for chlorophyll synthesis *in vitro*. *FEBS Lett.* **49,** 196–200.
Griffiths, W. T. (1975). Characterization of the terminal stages of chlorophyll(ide) synthesis in etioplast membrane preparations. *Biochem. J.* **152,** 623–635.
Griffiths, W. T. (1978). Reconstitution of chlorophyllide formation by isolated etioplast membranes. *Biochem. J.* **174,** 681–692.
Griffiths, W. T. (1980). Substrate-specificity studies on protochlorophyllide reductase in barley (*Hordeum vulgare*) etioplast membranes. *Biochem. J.* **186,** 267–278.
Griffiths, W. T. (1981). The role of NADPH in chlorophyll synthesis by protochlorophyllide reductase. *In* "Photosynthesis. V. Chloroplast Development" (G. Akoyunoglou, ed.), pp. 65–71. Balaban, Philadelphia.
Griffiths, W. T., and Beer, N. S. (1982). Site of synthesis of NADPH. Protochlorophyllide oxidoreductase in rye (*Secale cereale*). *Plant Physiol.* **70,** 1014–1018.
Griffiths, W. T., and Oliver, R. P. (1984). Protochlorophyllide reductase—Structure, function and regulation. *In* "Chloroplast Biogenesis" (R. J. Ellis, ed.), pp. 245–258. Cambridge University Press, Cambridge.
Griffiths, W. T., and Walker, C. J. (1987). Photoreactions in chloroplast development. *In* "Progress in Photosynthesis Research" (J. Biggens, ed.), Vol. IV, pp. 469–474. Nijhoff, Dordrecht, The Netherlands.
Griffiths, W. T., Oliver, R. P., and Kay, S. A. (1984). A critical appraisal of the role and regulation of NADPH-protochlorophyllide oxidoreductase in greening plants. *In* "Protochlorophyllide Reduction and Greening" (C. Sironval and M. Brouers, eds.), pp. 19–29. Nijhoff/Junk, The Hague, The Netherlands.
Griffiths, W. T., Kay, S. A., and Oliver, R. P. (1985). The presence and photoregulation of protochlorophyllide reductase in green tissues. *Plant Mol. Biol.* **4,** 13–22.
Häuser, I., Dehesh, K., and Apel, K. (1984). The proteolytic degradation *in vitro* of the NADPH-protochlorophyllide oxidoreductase of barley (*Hordeum vulgare* L.). *Arch. Biochem. Biophys.* **228,** 577–586.
Häuser, I., Dehesh, K., and Apel, K. (1987). Light-induced changes in the amounts of the 36000-$M_r$ polypeptide of NADPH-protochlorophyllide oxidoreductase and its mRNA in barley plants grown under a diurnal light/dark cycle. *Planta* **170,** 453–460.
Hampp, R., and De Filippis, L. (1980). Plastid protease activity and prolamellar body transformation during greening. *Plant Physiol.* **65,** 663–668.
Harpster, M., and Apel, K. (1985). The light-dependent regulation of gene expression during plastid development in higher plants. *Physiol. Plant.* **64,** 147–152.
Harpster, M. H., Mayfield, S. P., and Taylor, W. C. (1984). Effects of pigment-deficient mutants on the accumulation of photosynthetic proteins in maize. *Plant Mol. Biol.* **3,** 59–71.
Herron, H. A., and Mauzerall, D. (1972). The development of photosynthesis in a greening mutant of *Chlorella* and an analysis of the light saturation curve. *Plant Physiol.* **50,** 141–148.
Hinterstoisser, B., Missbichler, A., Pineau, B., and Peschek, G. A. (1988). Detection of chlorophyllide in chlorophyll-free plasma membrane preparations from *Anacystis nidulans*. *Biochem. Biophys. Res. Commun.* **154,** 839–846.
Horton, P., and Leech, R. M. (1972). The effect of ATP on photoconversion of protochlorophyllide into chlorophyllide in isolated etioplasts. *FEBS Lett.* **26,** 277–280.
Horton, P., and Leech, R. M. (1975). The effect of ATP on the photoconversion of protochlorophyllide in isolated etioplasts of *Zea mays*. *Plant Physiol.* **56,** 113–120.
Houssier, C., and Sauer, K. (1969). Optical properties of the protochlorophyll pigments. 1. Isolation, characterization, and infrared spectra. *Biochim. Biophys. Acta* **172,** 476–491.

Ikeuchi, M., and Murakami, S. (1982a). Behavior of the 36,000-dalton protein in the internal membranes of squash etioplasts during greening. *Plant Cell Physiol.* **23,** 575–583.
Ikeuchi, M., and Murakami, S. (1982b). Measurement and identification of NADPH : protochlorophyllide oxidoreductase solubilized with Triton X-100 from etioplast membranes of squash cotyledons. *Plant Cell Physiol.* **23,** 1089–1099.
Ikeuchi, M., and Murakami, S. (1983). Separation and characterization of prolamellar bodies and prothylakoids from squash etioplasts. *Plant Cell Physiol.* **24,** 71–80.
Inoue, Y., Kobayashi, T., Ogawa, T., and Shibata, K. (1981). A short lived intermediate in the photoconversion of protochlorophyllide to chlorophyllide *a*. *Plant Cell Physiol.* **22,** 197–204.
Iwai, J., Ikeuchi, M., Inoue, Y., and Kobayashi, T. (1984). Early processes of protochlorophyllide photoreduction as measured by nanosecond and picosecond spectrophotometry. *In* "Protochlorophyllide Reduction and Greening" (C. Sironval and M. Brouers, eds.), pp. 99–112. Nijhoff/Junk, The Hague, The Netherlands.
Jenkins, G. I. (1988). Photoregulation of gene expression in plants. *Photochem. Photobiol.* **48,** 821–832.
Jeske, C., and Senger, H. (1979). Synthese von Pigmenten und Entwicklung des Photosyntheseapparates in einigen Höheren Pflanzen. *Ber. Dtsch. Bot. Ges.* **92,** 609–617.
Jeske, C., and Senger, H. (1981). The development of the photosynthetic apparatus of *Gnetum:* Angiosperm type? *In* "Photosynthesis. V. Chloroplast Development" (G. Akoyunoglou, ed.), pp. 387–396. Balaban, Philadelphia.
Jones, O. T. G. (1966). A protein–protochlorophyll complex obtained from inner seed coats of *Cucurbita pepo. Biochem. J.* **101,** 153–160.
Joyard, J., Block, M., Pineau, B., Albrieux, C., and Douce, R. (1990). Envelope membranes from mature spinach chloroplasts contain a NADPH : protochlorophyllide reductase on the cytosolic side of the outer membrane. *J. Biol. Chem.* **265,** 21820–21827.
Kahn, A. (1968a). Developmental physiology of bean leaf plastids. II. Negative contrast electron microscopy of tubular membranes in prolamellar bodies. *Plant Physiol.* **43,** 1769–1780.
Kahn, A. (1968b). Developmental physiology of bean leaf plastids. III. Tube transformation and protochlorophyll(ide) photoconversion by a flash irradiation. *Plant Physiol.* **43,** 1781–1785.
Kasemir, H. (1983). Light control of chlorophyll accumulation in higher plants. *In* "Encyclopedia of Plant Physiology" (W. Shropshire and H. Mohr, eds.), Vol. 16B, pp. 662–686. Springer-Verlag, Berlin.
Kay, S. A., and Griffiths, W. T. (1983). Light-induced breakdown of NADPH-protochlorophyllide oxidoreductase *in vitro. Plant Physiol.* **72,** 229–236.
Kay, S. A., Keith, B., Shinozaki, K., Chye, M.-L., and Chua, N.-H. (1989). The rice phytochrome gene: Structure, autoregulated expression, and binding of GT-1 to a conserved site in the 5' upstream region. *Plant Cell* **1,** 351–360.
Kirk, J. T. O., and Tilney-Bassett, R. A. E. (1978). "The Plastids: Their Chemistry, Structure, Growth, and Inheritance." Elsevier/North-Holland, Amsterdam.
Kittsteiner, U., Paulsen, H., Schendel, R., and Rüdiger, W. (1990). Lack of light regulation of NADPH : protochlorophyllide oxidoreductase mRNA in cress seedlings (*Lepidium sativum* L.). *Z. Naturforsch.* **45,** 1077–1079.
Klockare, B., and Sundqvist, C. (1977). Shifts in absorption and fluorescence maxima of chlorophyll(ide) in spectra of dark grown wheat leaves after irradiation. *Photosynthetica* **11,** 189–199.
Koski, V. M., French, C. S., and Smith, J. H. C. (1951). The action spectrum for the transformation of protochlorophyll to chlorophyll *a* in normal and albino corn seedlings. *Arch. Biochem. Biophys.* **31,** 1–17.

Kotzabasis, K., and Senger, H. (1990). The influence of 5-aminolevulinic acid on protochlorophyllide and protochlorophyll accumulation in dark-grown *Scenedesmus*. *Z. Naturforsch.* **45**, 71–73.

Kotzabasis, K., Schüring, M.-P., and Senger, H. (1989a). Occurrence of protochlorophyll and its phototransformation to chlorophyll in mutant C-2A′ of *Scenedesmus obliquus*. *Physiol. Plant.* **75**, 221–226.

Kotzabasis, K., Senger, H., Langlotz, P., and Follmann, H. (1989b). Stimulation of protochlorophyllide oxidoreductase by thioredoxin. *J. Photochem. Photobiol. B Biol.* **3**, 333–339.

Kotzabasis, K., Römer, S., and Senger, H. (1990a). Temperature dependent reduction of protochlorophyllide in darkness followed by the assembly of active photosystems in pigment mutant C-2A′ of *Scenedesmus obliquus*. *Physiol. Plant.* **78**, 635–639.

Kotzabasis, K., Senge, M., Seyfried, B., and Senger, H. (1990b). Aggregation of monovinyl- and divinyl-protochlorophyllide in organic solvents. *Photochem. Photobiol.* **52**, 95–101.

Kotzabasis, K., Miyachi, S., and Senger, H. (1990c). Influence of calcium on formation and reduction of protochlorophyllide in the pigment mutant C-2A′ of *Scenedesmus obliquus*. *Plant Cell Physiol.* **31**, 419–422.

Kotzabasis, K., Humbeck, K., and Senger, H. (1991). Incorporation of photoreduced protochlorophyll into reaction centres. *J. Photochem. Photobiol. B Biol.* **8**, 255–262.

Kyte, J., and Doolittle, R. F. (1982). A simple method for displaying hydropathic character of a protein. *J. Mol. Biol.* **157**, 105–132.

Laing, W., Kreuz, K., and Apel, K. (1988). Light-dependent, but phytochrome-independent, translation control of the accumulation of the P700 chlorophyll-*a* protein of photosystem I in barley (*Hordeum vulgare* L.). *Planta* **176**, 269–276.

Lancer, H. A., Cohen, C. E., and Schiff, J. A. (1976). Changing ratios of phototransformable protochlorophyll and protochlorophyllide of bean seedlings developing in the dark. *Plant Physiol.* **57**, 369–374.

Laudi, G., and Manzini, M. L. (1975). Chlorophyll content and plastid ultrastructure in leaflets of *Metasequoia glyptostroboides*. *Protoplasma* **84**, 185–190.

Liljenberg, C. (1974). Characterization and properties of a protochlorophyllide ester in leaves of dark grown barley with geranylgeraniol as esterifying alcohol. *Physiol. Plant.* **32**, 208–213.

Lindsten, A., Ryberg, M., and Sundqvist, C. (1988). The polypeptide composition of highly purified prolamellar bodies and prothylakoids from wheat (*Triticum aestivum*) as revealed by silver staining. *Physiol. Plant.* **72**, 167–176.

Lütz, C., Röper, U., Beer, N. S., and Griffiths, T. (1981). Sub-etioplast localization of the enzyme NADPH: protochlorophyllide oxidoreductase. *Eur. J. Biochem.* **118**, 347–353.

McEwen, B., Virgin, H. I., Böddi, B., and Sundqvist, C. (1991). Protochlorophyll forms in roots of dark-grown plants. *Physiol. Plant.* **81**, 455–461.

Maloney, M. A., Hoober, J. K., and Marks, D. B. (1989). Kinetics of chlorophyll accumulation and formation of chlorophyll–protein complexes during greening of *Chlamydomonas reinhardtii* y-1 at 38°C. *Plant Physiol.* **91**, 1100–1106.

Manetas, Y., and Akoyunoglou, G. (1978). Studies on the fate of δ-aminolevulinic acid induced-protochlorophyllide. *In* "Chloroplast Development" (G. Akoyunoglou and J. H. Argyroudi-Akoyunoglou, eds.), pp. 105–110. Elsevier/North Holland, Amsterdam.

Mapleston, R. E., and Griffiths, W. T. (1980). Light modulation of the activity of protochlorophyllide reductase. *Biochem. J.* **189**, 125–133.

Mariani, P., Rascio, N., Orsenigo, M., and Tiveron, D. (1984). Photosynthetic apparatus differentiation in *Ephedra distachya* L. *In* "Advances in Photosynthesis Research" (C. Sybesma, ed.), Vol. IV, pp. 657–660. Nijhoff/Junk, The Hague, The Netherlands.

Mariani, P., de Carli, M. E., Rascio, N., Baldan, B., Casadoro, G., Gennari, G., Bodner, M., and Larcher, W. (1990). Synthesis of chlorophyll and photosynthetic competence in etiolated and greening seedlings of *Larix decidua* as compared with *Picea abies*. *J. Plant Physiol.* **137,** 5–14.

Matsuda, Y., Kikuchi, T., and Ishida, M. R. (1971). Studies on chloroplast development in *Chlamydomonas reinhardtii*. I. Effect of brief illumination on chlorophyll synthesis. *Plant Cell Physiol.* **12,** 127–135.

Mayfield, S. P., and Taylor, W. C. (1984). Carotenoid-deficient maize seedlings fail to accumulate light-harvesting chlorophyll *a/b* binding protein (LHCP) mRNA. *Eur. J. Biochem.* **144,** 79–84.

Meyer, G., Bliedung, H., and Kloppstech, K. (1983). NADPH-protochlorophyllide oxidoreductase: Reciprocal regulation in mono- and dicotyledonean plants. *Plant Cell Rep.* **2,** 26–29.

Mösinger, E., Batschauer, A., Schäfer, E., and Apel, K. (1985). Phytochrome control of *in vitro* transcription of specific genes in isolated nuclei from barley (*Hordeum vulgare*). *Eur. J. Biochem.* **147,** 137–142.

Mohr, H., and Shropshire, W., Jr. (1983). An introduction to photomorphogenesis for the general reader. *In* "Encyclopedia of Plant Physiology" (W. Shropshire and H. Mohr, eds.), Vol. 16A, pp. 25–38. Springer-Verlag, Berlin.

Oelmüller, R. (1989). Photooxidative destruction of chloroplasts and its effect on nuclear gene expression and extraplastidic enzyme levels. *Photochem. Photobiol.* **49,** 229–239.

Oelmüller, R., and Mohr, H. (1986). Photooxidative destruction of chloroplasts and its consequence for expression of nuclear genes. *Planta* **167,** 106–113.

Oelmüller, R., Levitan, I., Bergfeld, R., Rajasekhar, V. K., and Mohr, H. (1986). Expression of nuclear genes as affected by treatments acting on the plastids. *Planta* **168,** 482–492.

Ogawa, T., Inoue, Y., Kitajima, M., and Shibata, K. (1973). Action spectra for biosynthesis of chlorophylls *a* and *b* and β-carotene. *Photochem. Photobiol.* **18,** 229–235.

Ohad, I., Siekevitz, P., and Palade, G. E. (1967a). Biogenesis of chloroplast membranes. I. Plastid dedifferentiation in a dark-grown algal mutant (*Chlamydomonas reinhardi*). *J. Cell Biol.* **35,** 521–552.

Ohad, I., Siekevitz, P., and Palade, G. E. (1967b). Biogenesis of chloroplast membranes. II. Plastid differentiation during greening of a dark-grown algal mutant (*Chlamydomonas reinhardi*). *J. Cell Biol.* **35,** 553–584.

Oh-hama, T., and Hase, E. (1980). Formation of protochlorophyll(ide) in wild-type and mutant C-2A' cells of *Scenedesmus obliquus*. *Plant Cell Physiol.* **21,** 1263–1272.

Oh-hama, T., Kotzabasis, K., and Senger, H. (1987). Temperature inducible protochlorophyllide reduction in darkness in a pigment mutant of *Scenedesmus obliquus*. *Physiol. Plant.* **69,** 29–34.

Ohyama, K., Kohchi, T., Sano, T., and Yamada, Y. (1988). Newly identified groups of genes in chloroplasts. *TIBS 13*, 19–22.

Oliver, R. P., and Griffiths, W. T. (1980). Identification of the polypeptides of NADPH-protochlorophyllide oxidoreductase. *Biochem. J.* **191,** 277–280.

Oliver, R. P., and Griffiths, W. T. (1981). Covalent labeling of the NADPH : protochlorophyllide oxidoreductase from etioplast membranes with ($^3$H)*N*-phenylmaleimide. *Biochem. J.* **195,** 93–101.

Oliver, R. P., and Griffiths, W. T. (1982). Pigment–protein complexes of illuminated etiolated leaves. *Plant Physiol.* **70,** 1019–1025.

Ou, K., Packer, N., and Adamson, H. (1990). Immunodetection and photostability of NADPH-protochlorophyllide oxidoreductase in *Pinus pinea* L. *Photosynth. Res.* **23,** 89–94.

## 5. Protochlorophyllide Reductase

Packer, N., and Adamson, H. (1986). Incorporation of 5-aminolevulinic acid into chlorophyll in darkness in barley. *Physiol. Plant.* **68**, 222–230.

Packer, N., Adamson, H., and Gregory, J. (1984). Chloroplast development and associated changes in protochlorophyll(ide) and chlorophyll levels in seagrasses transferred to darkness. *In* "Advances in Photosynthesis Research" (C. Sybesma, ed.), Vol. IV, pp. 745–748. Nijhoff/Junk, The Hague, The Netherlands.

Peschek, G. A., Hinterstoisser, B., Wastyn, M., Kuntner, O., Pineau, B., Missbichler, A., and Lang, J. (1989a). Chlorophyll precursors in the plasma membrane of a cyanobacterium, *Anacystis nidulans*. Characterization of protochlorophyllide and chlorophyllide by spectrophotometry, spectrofluorimetry, solvent partition, and high performance liquid chromatography. *J. Biol. Chem.* **264**, 11827–11832.

Peschek, G. A., Hinterstoisser, B., Pineau, B., and Missbichler, A. (1989b). Light-independent NADPH-protochlorophyllide oxidoreductase activity in purified plasma membrane from the cyanobacterium *Anacystis nidulans*. *Biochem. Biophys. Res. Commun.* **162**, 71–78.

Pineau, B., Dubertret, G., Joyard, J., and Douce, R. (1986). Fluorescence properties of the envelope membranes from spinach chloroplasts: Detection of protochlorophyllide. *J. Biol. Chem.* **261**, 9210–9215.

Rapp, J. C., and Mullet, J. E. (1991). Chloroplast transcription is required to express the nuclear genes *rbcS* and *cab*. Plastid DNA copy number is regulated independently. *Plant Mol. Biol.* **17**, 813–823.

Rascio, N., Mariani, P., and Orsenigo, M. (1984). Photosynthetic apparatus differentiation in *Ginkgo biloba* L. *In* "Advances in Photosynthesis Research" (C. Sybesma, ed.), Vol. IV, pp. 661–664. Nijhoff/Junk, The Hague, The Netherlands.

Rebeiz, C. A., and Castelfranco, P. A. (1973). Protochlorophyll and chlorophyll biosynthesis in cell-free systems from higher plants. *Annu. Rev. Plant Physiol.* **24**, 129–172.

Rebeiz, C. A., Tripathy, B. C., Wu, S.-M., Montazer-Zouhoor, A., and Carey, E. E. (1986). Chloroplast biogenesis 52: Demonstration *in toto* of monovinyl and divinyl monocarboxylic chlorophyll biosynthetic routes in higher plants. *In* "Regulation of Chloroplast Differentiation" (G. Akoyunoglou and H. Senger, eds.), pp. 13–24. A. R. Liss, New York.

Reiß, T., Bergfeld, R., Link, G., Thien, W., and Mohr, H. (1983). Photooxidative destruction of chloroplasts and its consequences for cytosolic enzyme levels and plant development. *Planta* **159**, 518–528.

Richards, W. R., Walker, C. J., and Griffiths, W. T. (1987). The affinity chromatographic purification of NADPH : protochlorophyllide oxidoreductase from etiolated wheat. *Photosynthetica* **21**, 462–471.

Röbbelen, G. (1956). Über die Protochlorophyllreduktion in einer Mutante von *Arabidopsis thaliana* (L) Heynh. *Planta* **47**, 532–546.

Röper, U., and Lütz, C. (1984). Proteolytic effects on the protochlorophyllide-oxidoreductase in oat. *In* "Protochlorophyllide Reduction and Greening" (C. Sironval and M. Brouers, eds.), pp. 43–52. Nijhoff/Junk, The Hague, The Netherlands.

Röper, U., Prinz, H., and Lütz, C. (1987). Amino acid composition of the enzyme NADPH : protochlorophyllide oxidoreductase. *Plant Sci.* **52**, 15–19.

Roitgrund, C., and Mets, L. J. (1990). Localization of two novel chloroplast genome functions: *trans*-Splicing of RNA and protochlorophyllide reduction. *Curr. Genet.* **17**, 147–153.

Ruyters, G. (1987). Control of enzyme capacity and enzyme activity. *In* "Blue Light Responses: Phenomena and Occurrence in Plants and Microorganisms" (H. Senger, ed.), Vol. II, pp. 71–88. CRC Press, Boca Raton, Florida.

Ryberg, M., and Dehesh, K. (1986). Localization of NADPH-protochlorophyllide oxidoreductase in dark-grown wheat (*Triticum aestivum*) by immuno-electron microscopy before and after transformation of the prolamellar bodies. *Physiol. Plant.* **66**, 616–624.

Ryberg, M., and Sundqvist, C. (1982a). Characterization of prolamellar bodies and prothylakoids fractionated from wheat etioplasts. *Physiol. Plant.* **56**, 125–132.

Ryberg, M., and Sundqvist, C. (1982b). Spectral forms of protochlorophyllide in prolamellar bodies and prothylakoids fractionated from wheat etioplasts. *Physiol. Plant.* **56**, 133–138.

Ryberg, M., and Sundqvist, C. (1988). The regular ultrastructure of isolated prolamellar bodies depends on the presence of membrane-bound NADPH-protochlorophyllide oxidoreductase. *Physiol. Plant.* **73**, 218–226.

Santel, H.-J., and Apel, K. (1981). The protochlorophyllide holochrome of barley (*Hordeum vulgare* L.). The effect of light on the NADPH:protochlorophyllide oxidoreductase. *Eur. J. Biochem.* **120**, 95–103.

Sasa, T., and Sugahara, K. (1976). Photoconversion of protochlorophyll to chlorophyll *a* in a mutant of *Chlorella regularis*: *Plant Cell Physiol.* **17**, 273–279.

Schiff, J. A. (1978). Photocontrol of chloroplast development in *Euglena*. *In* "Chloroplast Development" (G. Akoyunoglou, and J. H. Argyroudi-Akoyunoglou, eds.), pp. 747–767. Elsevier/North-Holland, Amsterdam.

Schmidt, S., Drumm-Herrel, H., Oelmüller, R., and Mohr, H. (1987). Time course of competence in phytochrome-controlled appearance of nuclear encoded plastidic proteins and messenger RNAs. *Planta* **170**, 400–407.

Schneider, H. (1980). Chlorophyll biosynthesis. Enzymes and regulation of enzyme activities. *In* "Pigments in Plants" (F.-C. Czygan, ed.), 2d Ed., pp. 237–307. Fischer, Stuttgart.

Schopfer, P., and Apel, K. (1983). Intracellular photomorphogenesis. *In* "Encyclopedia of Plant Physiology" (W. Shropshire, Jr., and H. Mohr, eds.), Vol. 16A, pp. 258–288. Springer-Verlag, Berlin.

Schrubar, H., Wanner, G., and Westhoff, P. (1990). Transcriptional control of plastid gene expression in greening *Sorghum* seedlings. *Planta* **183**, 101–111.

Schulz, R. (1990). Isolierung und Charakterisierung von cDNAs der Gerste (*Hordeum vulgare* L.), die für die NADPH-Protochlorophyllid Oxidoreduktase und die Ferredoxin-NADP Oxidoreduktase kodieren. Thesis, Christian-Albrechts-Universität, Kiel, Germany.

Schulz, R., Steinmüller, K., Klaas, M., Forreiter, C., Rasmussen, S., Hiller, C., and Apel, K. (1989). Nucleotide sequence of a cDNA coding for the NADPH-protochlorophyllide oxidoreductase (PCR) of barley (*Hordeum vulgare* L.) and its expression in *Escherichia coli*. *Mol. Gen. Genet.* **217**, 355–361.

Schwartzbach, S. D. (1990). Photocontrol of organelle biogenesis in *Euglena*. *Photochem. Photobiol.* **51**, 231–254.

Selstam, E., and Widell, A. (1986). Characterization of prolamellar bodies, from dark-grown seedlings of Scots pine, containing light- and NADPH-dependent protochlorophyllide oxidoreductase. *Physiol. Plant.* **67**, 345–352.

Selstam, E., and Widell-Wigge, A. (1989). Hydrophobicity of protochlorophyllide oxidoreductase, characterized by means of Triton X-114 partitioning of isolated etioplast membrane fractions. *Physiol. Plant.* **77**, 401–406.

Selstam, E., Widell, A., and Johansson, L. B.-A. (1987). A comparison of prolamellar bodies form wheat, Scots pine, and Jeffrey pine. Pigment spectra and properties of protochlorophyllide oxidoreductase. *Physiol. Plant.* **70**, 209–214.

Senger, H. (1980). "The Blue Light Syndrome." Springer-Verlag, Berlin.

Senger, H. (1984). "Blue Light Effects in Biological Systems." Springer-Verlag, Berlin.
Senger, H. (1987a). "Blue Light Responses: Phenomena and Occurrence in Plants and Microorganisms." CRC Press, Boca Raton, Florida.
Senger, H. (1987b). Blue light control of pigment biosynthesis—Chlorophyll biosynthesis. *In* "Blue Light Responses: Phenomena and Occurrence in Plants and Microorganisms" (H. Senger, ed.), Vol. I, pp. 75–85. CRC Press, Boca Raton, Florida.
Senger, H., and Bishop, N. I. (1972). The development of structure and function in chloroplasts of greening mutants of *Scenedesmus*. I. Formation of chlorophyll. *Plant Cell Physiol.* **13**, 633–649.
Senger, H., and Brinkmann, G. (1986). Protochlorophyll(ide) accumulation and degradation in the dark and photoconversion to chlorophyll in the light in pigment mutant C-2A' of *Scenedesmus obliquus. Physiol. Plant.* **68**, 119–124.
Senger, H., and Kotzabasis, K. (1991). New apsects of biosynthesis of chlorophylls from protochlorophyllides in *Scenedesmus*. *In* "Light in Biology and Medicine" (R. H. Douglas, J. Moan, and G. Ronto, eds.), Vol. 2, 147–152. Plenum, London.
Senger, H., Bishop, N. I., Wehrmeyer, W., and Kulandaivelu, G. (1974). Development of structure and function of the photosynthetic apparatus during light-dependent greening of a mutant of *Scenedesmus obliquus. In* "Proceedings of the Third International Congress on Photosynthesis" (M. Avron, ed.), pp. 1913–1923. Elsevier, Amsterdam.
Shaw, P., Henwood, J., Oliver, R., and Griffiths, T. (1985). Immunogold localisation of protochlorophyllide oxidoreductase in barley etioplasts. *Eur. J. Cell Biol.* **39**, 50–55.
Shibata, K. (1957). Spectroscopic studies on chlorophyll formation in intact leaves. *J. Biochem.* **44**, 147–173.
Shioi, Y., and Sasa, T. (1983). Formation and degradation of protochlorophylls in etiolated and greening cotyledons of cucumber. *Plant Cell Physiol.* **24**, 835–840.
Shioi, Y., and Sasa, T. (1984). Chlorophyll formation in the YG-6 mutant of *Chlorella regularis*: Accumulation of protochlorophyllide and protochlorophyll esterified with geranylgeraniol. *Plant Cell Physiol.* **25**, 131–137.
Smith, J. H. C., and Benitez, A. (1954). The effect of temperature on the conversion of protochlorophyll to chlorophyll *a* in etiolated barley leaves. *Plant Physiol.* **29**, 135–143.
Spano, A. J., He, Z., Michel, H., Hunt, D. F., and Timko, M. P. (1992). Molecular cloning, nuclear gene structure, and developmental expression of NADPH:protochlorophyllide oxidoreductase in pea (*Pisum sativum* L.). *Plant Mol. Biol. 18*, 967–972.
Stahl, E. (1909). "Zur Biologie des Chlorophylls." Fischer, Jena, Germany.
Steinmüller, K., Batschauer, A., Mösinger, E., Schäfer, E., Rasmussen, S. K., and Apel, K. (1985). The light-induced greening of barley. *In* "Molecular Form and Function of the Plant Genome" (L. van Vloten-Doting, G. S. P. Groot, and T. C. Hall, eds.), 277–290. Plenum, New York.
Steinmüller, K., Batschauer, A., and Apel, K. (1986). Tissue-specific and light-dependent changes of chromatin organization in barley (*Hordeum vulgare*). *Eur. J. Biochem.* **158**, 519–525.
Stockhaus, J., Schell, J., and Willmitzer, L. (1989). Correlation of the expression of the nuclear photosynthetic gene ST-LS1 with the presence of chloroplasts. *EMBO J.* **8**, 2445–2451.
Sundqvist, C. (1969). Transformation of protochlorophyllide, formed from exogenous δ-aminolevulinic acid, in continuous light and in flashlight. *Physiol. Plant.* **22**, 147–156.
Sundqvist, C. (1970). The conversion of protochlorophyllide$_{636}$ to protochlorophyllide$_{650}$ in leaves treated with δ-aminolevulinic acid. *Physiol. Plant.* **23**, 412–424.
Sundqvist, C., and Ryberg, M. (1989). The distribution and structural role of NADPH-protochlorophyllide oxidoreductase in isolated etioplast inner membranes. *Photosynthetica* **23**, 427–438.

Sundqvist, C., Ryberg, H., Böddi, B., and Lang, F. (1980). Spectral properties of a long-wavelength absorbing form of protochlorophyll in seeds of *Cyclanthera explodens*. *Physiol. Plant.* **48**, 297–301.

Susek, R. E., and Chory, J. (1992). A tale of two genomes: Role of a chloroplast signal in coordinating nuclear and plastid genome expression. *Aust. J. Plant Physiol. 19*, 387–399.

Suzuki, J. Y., and Bauer, C. E. (1992). Light-independent chlorophyll biosynthesis: Involvement of the chloroplast gene *chlL* (*frxC*). *Plant Cell* **4**, 929–940.

Taylor, W. C. (1989). Regulatory interactions between nuclear and plastid genomes. *Annu. Rev. Plant Physiol. Plant Mol. Biol.* **40**, 211–233.

Thompson, W. F., and White, M. J. (1991). Physiological and molecular studies of light-regulated nuclear genes in higher plants. *Annu. Rev. Plant Physiol. Plant Mol. Biol.* **42**, 423–466.

Thorne, S. W. (1971). The greening of etiolated bean leaves. I. The initial photoconversion process. *Biochim. Biophys. Acta* **226**, 113–127.

Tobin, E. M., and Silverthorne, J. (1985). Light regulation of gene expression in higher plants. *Annu. Rev. Plant Physiol.* **36**, 569–593.

Tripathy, B. C., and Rebeiz, C. A. (1988). Chloroplast biogenesis 60. Conversion of divinyl protochlorophyllide to monovinyl protochlorophyllide in green(ing) barley, a dark monovinyl/light divinyl plant species. *Plant Physiol.* **87**, 89–94.

van Huystee, R. B., and Hodgins, R. R. W. (1989). Chlorophyll synthesis from protochlorophyll(ide) in chill-stressed maize (*Zea mays* L.). *J. Exp. Bot.* **40**, 431–435.

Virgin, H. I. (1981). The physical state of protochlorophyll(ide) in plants. *Annu. Rev. Plant Physiol.* **32**, 451–463.

Virgin, H. I. (1984). The initial inhibition of protochlorophyll regeneration in dark-grown leaves of barley after a short light pulse, and in greening leaves subjected to dark periods followed by a short light pulse. *Physiol. Plant.* **61**, 303–307.

Virgin, H. I., and Egneus, H. S. (1983). Control of plastid development in higher plants. *In* Encyclopedia of Plant Physiology (W. Shropshire and H. Mohr, eds.), Vol. 16A, pp. 289–311. Springer-Verlag, Berlin.

Virgin, H. I., Kahn, A., and von Wettstein, D. (1963). The physiology of chlorophyll formation in relation to structural changes in chloroplast. *Photochem. Photobiol. (Chlor. Metabol. Sym.)* **2**, 83–91.

von Heijne, G., Steppuhn, J., and Herrmann, R. G. (1989). Domain structure of mitochondrial and chloroplast targeting peptides. *Eur. J. Biochem.* **180**, 535–545.

Walmsley, J., and Adamson, H. (1989). Chlorophyll accumulation and breakdown in light-grown barley transferred to darkness: Effect of seeding age. *Physiol. Plant.* **77**, 312–319.

Walter, G. (1974). Spektrale Änderungen *in vivo* während der Chlorophyll-Bildung etiolierter Weizenkeimpflanzen nach experimentell variiertem Anteil der Protochlorophyll(id)-Holochrome $P_{650}$ und $P_{635}$ durch Behandlung mit δ-Aminolävulinsäure. *Photosynthetica* **8**, 40–46.

Wang, W., Boynton, J. E., and Gillham, N. W. (1977). Genetic control of chlorophyll biosynthesis: Effect of increased δ-aminolevulinic acid synthesis on the phenotype of y-1 mutant of *Chlamydomonas*. *Mol. Gen. Genet.* **152**, 7–12.

Wellburn, F. A. M., Wellburn, A. R., and Senger, H. (1980). Changes in ultrastructure and photosynthetic capacity within *Scenedesmus obliquus* mutants C-2A′, C-6D, and C-6E on transfer from dark grown to illuminated conditions. *Protoplasma* **103**, 35–54.

Widell-Wigge, A., and Selstam, E. (1990). Effects of salt wash on the structure of the prolamellar body membrane and the membrane binding of NADPH-protochlorophyllide oxidoreductase. *Physiol. Plant.* **78**, 315–323.

Wild, A., and Fuldner, K.-H. (1977). The concentration of cytochrome *f* and P700 in chlorophyll-deficient mutants of *Chlorella fusca*. *Planta* **136**, 281–282.

# 6

# Esterification of Chlorophyllide and Its Implication for Thylakoid Development

## WOLFHART RÜDIGER

Botanisches Institut der Universität München
W-8000 München 19, Germany

I. Introduction
II. Detection and Characterization of Chlorophyll Synthetase Activity
   A. Specificity for Substrates
   B. Localization of Chlorophyll Synthetase and Substrates
   C. Temperature-Dependent Esterification
III. Hydrogenation of Geranylgeranyl Derivatives
IV. Translation of Plastid-Encoded Chlorophyll $a$ Apoproteins and Chlorophyll Synthetase Reaction
V. Influence of Metal Chelators and 5-Aminolevulinate on Chlorophyll Synthetase of Developing Plastids
VI. Concluding Remarks
   References

## I. INTRODUCTION

Thylakoid development requires biosynthesis of a large number of compounds, for example, proteins, lipids, pigments, and metal ions. The pigment–protein complexes that are functional in photosynthesis contain such compounds in precisely defined stoichiometric amounts. Stoichiometry can arise through a selection process during assembly of the complexes (see Chapter 10) and through coordination of biosynthesis of the individual compounds. This coordination implies interaction and linkage of distinct biosynthetic pathways. Esterification of chlorophyllide (Chlide) is an example of such linkage. The substrates for this reaction are derived from the tetrapyrrole pathway and from the isoprenoid pathway. Since the final isoprenoid moiety in chlorophyll (Chl) is phytol (Phy), this process often has been called phytylation. As will be shown here, this reaction is more complicated than a simple reaction between the carboxylic acid compound Chlide and the allylic alcohol compound Phy.

Esterification is the last step in Chl biosynthesis. The reaction increases the hydrophobicity of the pigment. The Phy moiety once was considered to be the anchor for Chl in the lipid phase of the thylakoid membrane. Growing knowledge about the protein compounds of photosynthetic membranes, however, revealed that most if not all Chl molecules are protein bound. We must assume, therefore, that the hydrophobic Phy moiety of Chl plays an important role in Chl–protein interaction. In this chapter, our knowledge about the esterification reaction will be summarized first. Then interactions of (esterified) Chl with proteins and implications of such interactions on thylakoid development will be considered.

## II. DETECTION AND CHARACTERIZATION OF CHLOROPHYLL SYNTHETASE ACTIVITY

The first enzyme of Chl metabolism to be detected was *chlorophyllase*, which catalyzes cleavage of Phy from Chl (Willstätter and Stoll, 1913). Since chlorophyllase can catalyze the reverse reaction if a great excess of substrate (i.e., Phy) is supplied, such a reaction often was discussed as a possible mechanism of Chl biosynthesis (Bogorad, 1976). Granick pointed out as early as 1960 that enzymatic esterification reactions generally make use of activated substrates, either carboxylic acids (usually by thioester formation) or alcohols. Experiments with $^{18}$O-labeled precursors (Akhtar *et al.*, 1984; Emery and Akhtar, 1987) excluded the activated carboxy group of Chlide but were compatible with activated alcohol groups of Phy precursors in bacteriochlorophyll biosynthesis (Fig. 1). The isoprenoid alcohols are biosynthesized as diphosphates, that is, in an activated form.

Experiments studying activated alcohols started with $^{14}$C-labeled geranylgeranyldiphosphate (GGPP) and crude homogenates of etiolated maize seedlings (Rüdiger *et al.*, 1977). The second substrate, Chlide, was formed from Pchlide by short irradiation of the seedlings or the homogenate. Incorporation of GGPP into newly formed Chl was detected after extraction and purification of pigments by thin-layer chromatography. The labeled compound, which later turned out to be geranylgeranylchlorophyllide ($Chl_{GG}$), migrated slightly behind authentic chlorophyll (phytylchlorophyllide, $Chl_P$). Identification of Chl derivatives was facilitated by the characteristic shift in $R_F$ value by acidification; this led to removal of magnesium and formation of the corresponding pheophytin, which comi-

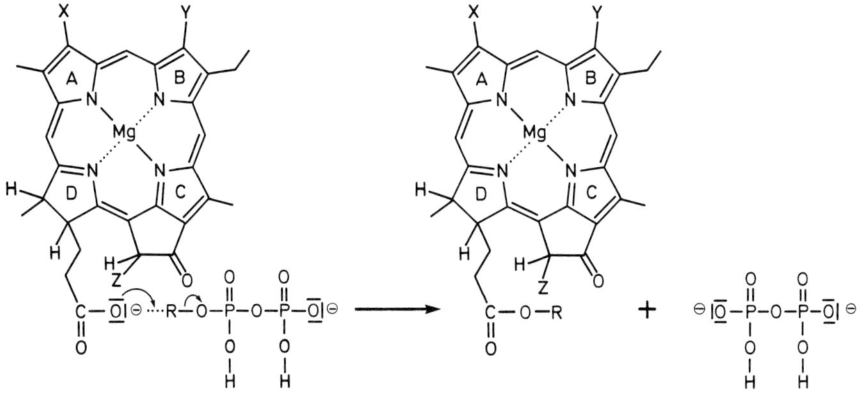

# 6. Esterification of Chlorophyllide

**Fig. 1.** The esterification reaction and its substrate specificity. Tetrapyrroles: X = $-CH=CH_2$ or $-COCH_3$, Y = $-CH_3$ or CHO, Z = $-COOCH_3$ (or $-H$). Alcohols: R = phytyl, geranylgeranyl, or farnesyl. Poor substrates are Z = H and R = farnesyl.

grated with authentic pheophytin$_{GG}$ (pheophorbide geranylgeranyl ester) (Rüdiger et al., 1976).

## A. Specificity for Substrates

The enzyme responsible for esterification proved to be membrane bound. The original system was refined using purified etioplasts that were broken osmotically for the enzyme test (Rüdiger et al., 1980). Incorporation of GGPP into Chl was 4–5 times higher than in the previously used homogenate of etiolated maize seedlings. The enzyme present in this membrane preparation was named Chl synthetase. Substrate specificity was tested with this membrane system (Table I). The proper substrates proved to be the diphosphate of isoprenoid alcohols; the free alcohols or their monophosphates were incorporated into Chl only in the presence of ATP, apparently because of an isoprenoid kinase. The kinase probably is located in the stroma (Soll et al., 1983). The $C_{20}$ compounds Phy and geranylgeraniol (GG) were much better substrates than the $C_{15}$ compound farnesol; no reaction was found with the $C_{10}$ compound geraniol. In the presence of isopentenyldiphosphate (IPP), however, geraniol and farnesol were incorporated into Chl, but only after chain elongation to the $C_{20}$ compound GG (Benz and Rüdiger, 1981a). The preparation must, therefore, contain the prenyltransferase activity leading to GGPP also (Block et al., 1980).

## TABLE I
### Substrate Specificity of Membrane-Bound Chlorophyll Synthetase of Etioplasts

| Substrate | Expected product | Esterification (%) | Conditions[a] |
|---|---|---|---|
| Protochlorophyllide | Protochlorophyll | 0.9 | A |
| Pheophorbide a | Pheophytin | 6 | A |
| Bacteriochlorophyllide a | Bacteriochlorophyll | 4 | A |
| Chlorophyllide a | Chlorophyll a | 52 | A |
| Chlorophyllide b | Chlorophyll b | 53 | A |
| 3-[Acetyl]chlorophyllide a | 3-[Acetyl]chlorophyll a | 52 | A |
| Pyrochlorophyllide a | Pyrochlorophyll | 26 | A |
| Phytyldiphosphate | Chlorophyll a (P) | 70 | B |
| Geranylgeranyldiphosphate | Chlorophyll a (GG) | 65 | B |
| Farnesyldiphosphate | Chlorophyll a (F) | 25 | B |
| Geranylgeraniol | | 0 | B |
| Geranylgeraniol + ATP | | 45–50 | B |
| Phytol | | 0 | B |
| Phytol + ATP | | 48–53 | B |
| Farnesol | | 0 | B |
| Farnesol + ATP | | 7–12 | B |
| Geraniol + ATP | | 0 | B |
| Pheophorbide a | | 0 | C |
| Zn-Pheophorbide a | Zn-Pheophytin | 20–30 | C |
| Ni-Pheophorbide a | Ni-Pheophytin | 1 | C |
| Cu-Pheophorbide a | Cu-Pheophytin | 1 | C |
| Co-Pheophorbide a | Co-Pheophytin | 2 | C |

[a] A, the tetrapyrrole substrate (10–40 nmol) and GGPP (200–800 nmol) were dissolved in 0.05 $M$ Hepes buffer, pH 7.2, containing 0.1% cholate and incubated for 60 min with the membrane fraction containing 8 nmol Pchlide (after Benz and Rüdiger, 1981b). B, samples of broken etioplasts containing about 6 nmol Pchlide were incubated with the indicated isoprenoid substrates (about 20 nmol/nmol pigment). Reaction was started by flash irradiation which produced Chlide (80% of Pchlide). Esterification is increased over control without exogenous isoprenoid substrate (after Rüdiger et al., 1980). C, incubation with the etioplast membrane fraction, containing 1.5 nmol Pchlide per sample, was similar to Condition A, but the buffer contained 0.05% Brij-W1 instead of cholate (M. Helfrich and W. Rüdiger, 1992, unpublished data).

The first experiments had been performed after photoconversion of Pchlide that was contained in the membrane. Whether Chlide had to be released from Pchlide-reductase for esterification or whether Chl synthetase could react with the pigment while it was still attached to the reductase was not clear from these first experiments. It was then discovered that the esterification proceeded equally well independent of the presence or absence of Pchlide. Therefore Chlide was applied in the dark, that is, without removal of Pchlide by phototransformation (Benz and Rüdiger, 1981b). In a membrane sample containing only 8 nmol Pchlide, up to

# 6. Esterification of Chlorophyllide

40 nmol Chl were produced by adding sufficient Chlide and GGPP (Benz and Rüdiger, 1981b). This result cannot be explained if one assumes that only Pchlide-reductase-bound Chlide can be esterified. Most likely, Chlide binds to Chl synthetase directly and the enzyme releases Chl into the lipid phase of the membrane. Chl-binding proteins are not present in sufficient quantity in the etioplast membrane.

These experiments permitted variation of Chlide structure to test the specificity of the enzyme for this substrate (Table I). Chlide $a$, Chlide $b$, and 3-(acetyl)Chlide $a$ proved to be good substrates; pyrochlorophyllide is a poor substrate, but pheophorbide, bacteriochlorophyllide, and Pchlide are not accepted as substrates by the enzyme (Benz and Rüdiger, 1981b). The results suggest that Chl synthetase requires a chlorin derivative as a substrate that contains magnesium as the central metal ion (Fig. 1). Zinc can substitute for magnesium to a certain extent, but complexes with other metal ions such as nickel, copper, or cobalt are not accepted (Table I). The $\beta$-substituents must be those of Chl $a$ or $b$ but some variation is allowed at ring A or—to a certain extent—at the isocylic ring. A hydrogenated ring D seems to be a prerequisite for substrate structure, at least for the enzyme of etiolated oat seedlings. We cannot exclude the possibility that the enzyme is somewhat "leaky" with respect to this prerequisite so traces of Pchl will be formed over time. Whether esterification of Pchlide contributes at all to thylakoid development in oat seedlings is doubtful. Careful analysis showed that the small amount of esterified Pchl (about 5% of Pchlide) occurring in etiolated oat seedlings is not phototransformed into Chl (Schoch et al., 1977). Reports on other plants differ, however (see Chapter 5).

## B. Localization of Chlorophyll Synthetase and Substrates

Pchlide photoreduction occurs in etiolated plants mainly in the prolamellar body (PLB). Therefore, it is logical to assume that esterification of Chlide also takes place in the PLB or in its proximity. Membrane fractionation revealed the presence of Chl synthetase activity in prothylakoid (PT) and PLB fractions of etiolated oat and wheat seedlings (Lütz et al., 1981; Lindsten et al., 1990). Interestingly, enzyme activity is latent in intact PLBs as long as they are preserved in isolation buffer containing NADPH and 50% (w/w) sucrose (Lindsten et al., 1990). Activity appears after dilution, which also leads to dissociation of the highly regular PLB structure. Disaggregation of Chlide–Pchlide-reductase complexes (determined by the Shibata shift) and esterification of Chlide seem to be closely related. The herbicides clomazone and amiprophos-methyl always inhibit the Shibata shift and esterification of Chlide to the same extent in vivo (Artus *et al.*, 1992). No inhibition of Chl synthetase was observed with these herbicides *in vitro*.

Heat treatment of developing grass seedling leads to deficiency in plastid ribosomes and, hence, to bleaching of leaves (inhibition of greening; Feierabend, 1977). Investigation of such heat-bleached primary leaves of rye and oat seedlings revealed nearly normal or only partially reduced activity of Chl synthetase (Hess *et al.*, 1992). Thus, the enzyme cannot be formed on plastid ribosomes but must be, like other enzymes of Chl biosynthesis, a product of cytoplasmic ribosomes and, hence, encoded by a nuclear gene. Expression of the Chl synthetase gene does not depend on the postulated plastid factor (Hess *et al.* 1992), which is not present in heat-treated seedlings and is essential for expression of nuclear genes encoding certain plastid proteins (e.g. *Cab* and *RbcS* genes; Oelmüller, 1989; Taylor, 1989).

Irradiation of etiolated oat seedlings with white light leads to a slow decrease in Chl synthetase activity (Fig. 2). This decrease is much slower than the decrease in Pchlide reductase activity under identical conditions (Forreiter *et al.*, 1991). Nevertheless, Chl synthetase activity can be found in mature chloroplasts. Careful fractionation of spinach chloroplasts revealed the presence of this enzyme activity in the thylakoid membrane,

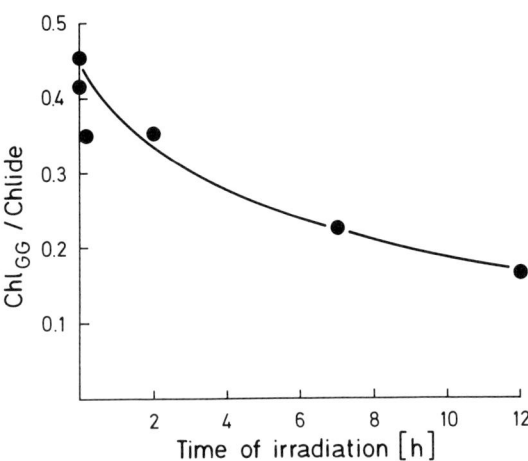

**Fig. 2.** Chl synthetase activity in greening oat seedlings. Seedlings were irradiated with white light for various time periods. Chl synthetase then was solubilized with 50% glycerol. The relative activity was determined by incubation with Chlide and GGPP for 20 min under otherwise standard conditions (Rüdiger *et al.*, 1980). Esterification was determined as $Chl_{GG}$ in addition to endogenous $Chl_P$. Adapted from Schindler (1983).

but its absence in envelope membrane and stroma (Soll et al., 1983). Synthetase activity also was found in membrane fractions of chloroplasts and chromoplasts from *Capsicum annuum* (Dogbo et al., 1984) and in the stroma fraction of chromoplasts from daffodil (Kreuz and Kleinig, 1981). The enzyme seems to persist during plastid development, although whether the observed activity is always due to the same enzyme protein is not yet clear. In oat etioplasts, the relative substrate specificities for exogenous GGPP, PhyPP, and farnesyldiphosphate (FPP) were 6, 3, and 1, respectively, as determined by mixing the substrates (Rüdiger et al., 1980). In spinach chloroplasts, the relative substrate specificities for exogenous GGPP and PhyPP were 1 and 4 for esterification of Chlide *a* (Soll et al., 1983). FPP was somewhat less effective than GGPP in these experiments. GGPP and PhyPP were incorporated at the same rate into Chlide *b*, whereas FPP was not accepted for esterification of Chlide *b*. No such comparison of substrates was described in other systems containing Chl synthetase. Whether the reported differences reflect properties of different plant species, of different membrane environments for the enzymes, or of different enzymes in etioplasts and chloroplasts is unknown. Differences also were found in intact plants. Chl formed within the first 15–20 min of the esterification reaction is mainly $Chl_{GG}$ in etiolated oat and bean seedlings (Schoch et al., 1977; Schoch, 1978) but exclusively $Chl_P$ in green barley seedlings (Rüdiger, 1987).

Chlide that is formed in the PLB can migrate with Pchlide-reductase into PTs during light-dependent dissociation of the PLB (Dehesh et al., 1986; Ryberg and Dehesh, 1986). For esterification, Chlide must be transferred to Chl synthetase. This transfer from one membrane-bound enzyme to another probably takes place in the liquid phase of the membrane. Because of the partition coefficient, some Chlide theoretically can be transferred into the aqueous (stromal) phase also, but there is no indication of a soluble Chlide pool to date.

The situation differs for the second substrate, which has been identified indirectly as GGPP with some PhyPP in etiolated oat and bean seedlings (Schoch et al., 1977; Schoch, 1978). Quantitative analysis of etiolated oat seedlings demonstrated that GGPP is mainly soluble, whereas PhyPP is mainly contained in membrane fractions. The soluble fraction contains 8-fold more GG-phosphates than Phy-phosphates (Table II) whereas the membrane fraction contains about 7-fold more Phy-phosphates than GG-phosphates (Benz et al., 1983b). GGPP is formed in the stroma of plastids (Block et al., 1980) but does not remain in the plastid compartment. We must assume that it leaches out into the cytoplasm because the plastid envelope is permeable to GGPP. This molecule is lost during isolation of intact plastids and can be incorporated into Chl when applied from outside

TABLE II

Quantitative Analysis of GGPP and PhyPP in Etiolated Oat Seedlings[a, b]

| Compound | Amount (nmol · g fresh wt$^{-1}$) |
|---|---|
| GGPP | 13.7 |
| GGMP | 2.3 |
| PhyPP | 1.7 |
| PhyMP | 0.3 |

[a] (After Benz et al., 1983a).

[b] The total amount of GGPP was determined by isolation; the yield was checked by the isotope dilution method. The ratio of diphosphates to monophosphates was calculated from densitometry after staining of thin-layer chromatograms; the ratio GG/Phy was determined by gas–liquid chromatography.

to intact plastids (Benz et al., 1981). GGPP is even incorporated into Chl in growing tobacco cell suspension cultures (Benz et al. 1984). This property differs from that of Chlide, which cannot be metabolized when applied from the outside to intact plastids (U. Röper and W. Rüdiger, 1987, unpublished observations). The solubility and penetration properties of GGPP make it a good candidate for regulatory functions in the cell, for example, for feedback inhibition of the isoprenoid pathway (Steiger et al., 1985).

## C. Temperature-Dependent Esterification

In contrast to the photoconversion to Pchlide, esterification of Chlide is strongly temperature dependent in intact plants. This property already has been described for *Phaseolus vulgaris* (Wolff and Price, 1957) and investigated in detail in *Avena sativa* (Keckeis, 1983). Esterification kinetics in primary leaves of 8-day-old etiolated oat seedlings at 0°C, 8°C, and 28°C are presented as an example in Fig. 3. The reaction was started by pulse irradiation (1 min white light) of the leaves, which photoconverted 90% of the Pchlide independent of the temperature. Esterification of Chlide was nearly completed (Chl yield, 91–96% of initial Chlide) within 30–45 min at 28°C but was not completed within 90 min at lower temperatures.

## 6. Esterification of Chlorophyllide

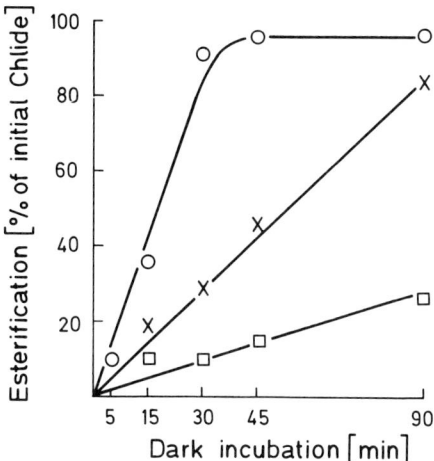

**Fig. 3.** Kinetics of esterification of Chlide in 8-day-old etiolated oat seedlings. Primary leaves were kept at 0°C (□), 8°C (x), or 28°C (○). The time course of esterification was determined by pigment analysis at 5, 15, 30, 45, and 90 min after photoconversion of Pchlide. For experimental details, see Table III. Adapted from Keckeis (1983).

The initial esterification rate was nearly 4-fold at 8°C and 8-fold at 28°C than the initial rate at 0°C. These ratios vary somewhat with the age of the seedlings, as shown in Table III. Because of the increase in Pchlide with the age of the seedlings, the amount of Chlide produced by photoconversion also increased with the age of seedlings. Whereas the initial rates of esterification (Chl formed per minute) were nearly constant with respect to seedling age at a given temperature when presented as a percentage of initial Chlide, the rates increased steadily with age when presented in pmol per g fresh weight (Table III).

The strong temperature dependency of the esterification reaction is postulated to be related to some transport process that in turn depends on the membrane fluidity (Rüdiger, 1987). However, whether only substrates or also protein compounds (e.g., enzymes) must be transported is not clear. Since GGPP leaches out into the cytoplasm (see Section II,B), its transport back into etioplasts might depend on the fluidity of the inner envelope membrane. Chl synthetase activity is latent in highly regular PLBs (see Section II,B). It appears only after a spectral shift that probably reflects transformation of aggregated forms of Chlide–Pchlide-reductase normally present in PLBs (Böddi et al., 1989) into nonaggregated forms normally present in PTs. This disaggregation is also strongly temperature dependent (see Chapters 2 and 7).

**TABLE III**

**Initial Esterification Rates of Chlide in Etiolated Oat Seedlings *in Vivo*[a,b]**

| Temperature (°C) | Age of seedlings (days) | | | |
|---|---|---|---|---|
| | 4 | 5 | 6 | 8 |
| | Initial esterification rates (pmol Chl · g fresh wt$^{-1}$ · min$^{-1}$) | | | |
| 0 | 9 | 10 | 12 | 17 |
| 8 | 17 | 23 | 36 | 63 |
| 28 | 59 | 80 | 128 | 143 |
| | Initial esterification rates (Chl, % of initial Chlide · min$^{-1}$) | | | |
| 0 | 0.5 | 0.4 | 0.3 | 0.3 |
| 8 | 1.0 | 0.9 | 0.9 | 1.1 |
| 28 | 3.4 | 3.2 | 3.3 | 2.5 |
| | Amount of Chlide produced by photoconversion (nmol · g fresh wt$^{-1}$)[c] | | | |
| | 1.7 ± 0.2 | 2.5 ± 0.2 | 4.0 ± 0.3 | 5.7 ± 0.7 |

[a] Calculated from data of Keckeis (1983).

[b] Seedlings were grown in the dark for the indicated time at 28°C. Primary leaves were harvested under dim green safelight. Some samples were cooled within 15 min to the indicated temperature. All samples were illuminated with white light (3000 lux) for 1 min, leading to 90–93% photoconversion of Pchlide irrespective of the temperature. The samples then were kept in the dark at the indicated temperature for another 90 min. Pigments were extracted and analyzed as pheophytin and pheophorbide (Rüdiger *et al.*, 1980).

[c] The values are mean values from the three temperatures (± SD, $n = 12$).

The temperature range that was used in these experiments (Table III) can occur in the natural environment of plants in temperate zones. Differences and fluctuations in the esterification rate of Chlide therefore might be a natural phenomenon. A low esterification rate should result in a corresponding low accumulation of Chl. Temperature adaptation of plants growing in extreme environments is expected, however.

Examples of plants growing in extreme environments are alpine plants. An interesting observation is the occurrence of pale leaves of several alpine plants growing in or near snow fields or in contact with melting snow. These plants possess typical characteristics of "etiolated" leaves, although they receive full sunlight: they not only lack bulk Chl but also contain typical etioplasts with PLBs (Bergweiler, 1987; C. Lütz, 1991, personal communication). Chl synthetase activity of several of these "etiolated" plants was investigated (Blank-Huber, 1986). A temperature dependency of the esterification reaction *in vitro* can be seen clearly (Table IV). The presumed adaptation of the plants to low temperature apparently does not exclude a faster esterification at elevated temperatures. The

## 6. Esterification of Chlorophyllide

**TABLE IV**

**Chl Synthetase Activity in Alpine Plants**[a,b]

| Plant material | Newly formed chlorophyll (nmol per sample) at | | |
|---|---|---|---|
| | 25°C | 12°C | 6°C |
| *Taraxacum alpinum* | | | |
| Grown under snow | 0.78 | 0.46 | 0.12 |
| Grown beside snow | 1.86 | 0.38 | 0.16 |
| Green control leaves | 5.11 | 0.79 | 2.42 |
| *Eriophorum angustifolium* | 1.12 | 0.35 | 0.24 |
| *Luzula lutea* | n.d. | 0.39 | 0.10 |

[a] Reprinted with permission from Blank-Huber (1986).

[b] Plastids were prepared from 5 g fresh leaves and incubated under standard condition with 6 nmol Chlide and 150 μg [$^{14}$C]GGPP for 45 min at the indicated temperature. Pigments were extracted and analyzed for newly formed Chl (Rüdiger et al., 1980).

investigation demonstrates further, that—at least for *Taraxacum officinale*—the green tissues contain more enzyme activity than the "etiolated" tissues. A possible reason for the lack of greening in these plants grown in natural daylight could be the lack of active Chl synthetase. However, it cannot be excluded that other factors necessary for greening are absent also.

## III. HYDROGENATION OF GERANYLGERANYL DERIVATIVES

Hydrogenation of GG to Phy can occur at two different stages, either at the stage of diphosphates (GGPP → PhyPP) or at the stage of the pigments (Chl$_{GG}$ → Chl$_P$; see Section II,B). The first reaction takes place in the envelope membrane of chloroplasts (Soll et al., 1983), the second reaction at the site of Chlide formation, that is, in the thylakoid membrane of chloroplasts (Soll et al., 1983) or in PTs and PLBs of etioplasts (Lindsten et al., 1990). The hydrogen donor in both cases is NADPH (Benz et al., 1980). Slowing of esterification at low temperature (see Section II,C) always resulted in slowing of hydrogenation also (Keckeis, 1983; Blank-Huber, 1986). Esterification and hydrogenation could, nevertheless, be separated from each other.

The hydrogenase seems to be a more sensitive enzyme than the synthetase. Benz (1980) found that extensive homogenization of etioplast membranes inactivated the hydrogenase much more quickly than the synthetase. Treatment of plants with nonlethal concentrations of the herbicides

3-amino-1,2,4-triazol (Rüdiger et al., 1976; Rüdiger and Benz, 1979) or S-ethyl-dipropylthiocarbamate (Wilkinson, 1985) yielded reduced amounts of $Chl_P$ and increased amounts of $Chl_{GG}$ and the intermediates ($Chl_{H2GG}$, $Chl_{H4GG}$). Anaerobic pretreatment of intact seedlings proved to be most effective in inhibiting this hydrogenase. As shown in Fig. 4, production of $Chl_P$ can be inhibited completely without inhibition of the esterification rate of total Chl after prolonged anaerobiosis (3–10 hr). Even after shorter periods of anaerobiosis (0.5–2 hr), formation of $Chl_P$ is reduced (Schoch et al., 1980).

The photoconversion of Pchlide to Chlide can be used as a sensitive

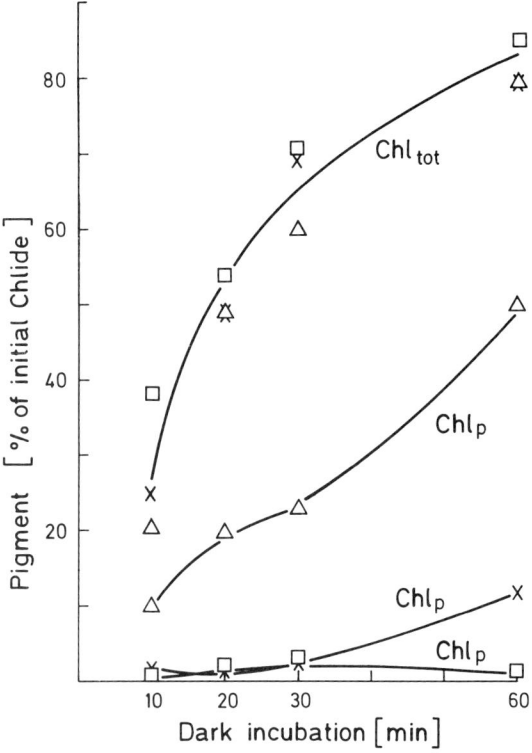

**Fig. 4.** Kinetics of esterification of Chlide and hydrogenation of GG in 7-day-old etiolated seedlings. The seedlings were kept anaerobically in the dark for 3 hr (x) or 10 hr (□); controls (△) were kept under normal air in the dark for the same time. Pigments then were extracted and analyzed by HPLC. Esterification (total esterified chlorophyll, $Chl_{tot}$) occurred with nearly identical kinetics after all treatments. Formation of $Chl_P$ was reduced after anaerobiosis. Adapted from Hehlein (1979).

indicator of the NADPH level in plastids (Mapleston and Griffiths, 1978). Since this photoconversion was not decreased under our experimental conditions (Schoch *et al.*, 1980), decreased hydrogenation of $Chl_{GG}$ probably is not caused by lack of NADPH. We conclude that the activity of the hydrogenase is affected by anaerobic pretreatment. Etiolated oat seedlings treated anaerobically as described can recover in the light. After illumination with white light for 10 hr in anaerobic environment, considerable greening and formation of $Chl_P$ was seen (Schoch *et al.*, 1980).

## IV. TRANSLATION OF PLASTID-ENCODED CHLOROPHYLL *a* APOPROTEINS AND CHLOROPHYLL SYNTHETASE REACTION

In etioplasts, neither Chl nor Chl-binding proteins are accumulated in the dark. On illumination, both Chl and Chl-binding proteins accumulate as a part of the light-dependent etioplast–chloroplast transformation. This transformation is mediated by at least three photoreceptors: phytochrome, a blue-light receptor, and Pchlide bound to the reductase (Buetow, 1986). Phytochrome regulates transcription of several nuclear genes encoding Chl-binding proteins, for example, the *Cab* gene family encoding the light-harvesting Chl *a/b* binding proteins of photosystem II (Apel, 1979; Apel and Kloppstech, 1980; Bennett, 1981). The apoproteins are presumed not to be stabilized until Chl is produced. Accumulation of these proteins therefore should depend not only on phytochrome but also on the light-dependent Pchlide-reductase. The situation differs for plastid encoded Chl-proteins. Transcripts are already present in etioplasts in the dark but no translation of most transcripts can be detected (Herrmann *et al.*, 1985; Klein and Mullet, 1986; Kreuz *et al.*, 1986). Radiolabel from [$^{35}$S]methionine is incorporated in the dark only into the D2 protein (Klein *et al.*, 1988); incorporation into the other apoproteins (D1, CP64, CP43, CP47) requires light. Illumination of intact etiolated barley seedlings with white light (100 $\mu$E · m$^{-2}$ · sec$^{-1}$) for 1 min is sufficient for measurable accumulation of apoproteins (Klein *et al.*, 1988); such illumination does not change the transcript levels (Klein and Mullet, 1986). Synthesis of Chl *a* apoproteins (e.g., CP64) could not be detected after illumination of isolated intact etioplasts (Laing *et al.*, 1988). Therefore, a cytoplasmic factor was postulated to be essential for translation in the plastids. Eichacker *et al.* (1990), however, demonstrated incorporation of labeled methionine into Chl *a* apoproteins (D1, CP64, CP43, CP47) in isolated etioplasts after irradiation (photoconversion of Pchlide to Chlide) and incubation with PhyPP (Fig. 5). Light was not necessary when Chlide with PhyPP was added to lysed etioplasts. This result demonstrated for the first time the direct connection

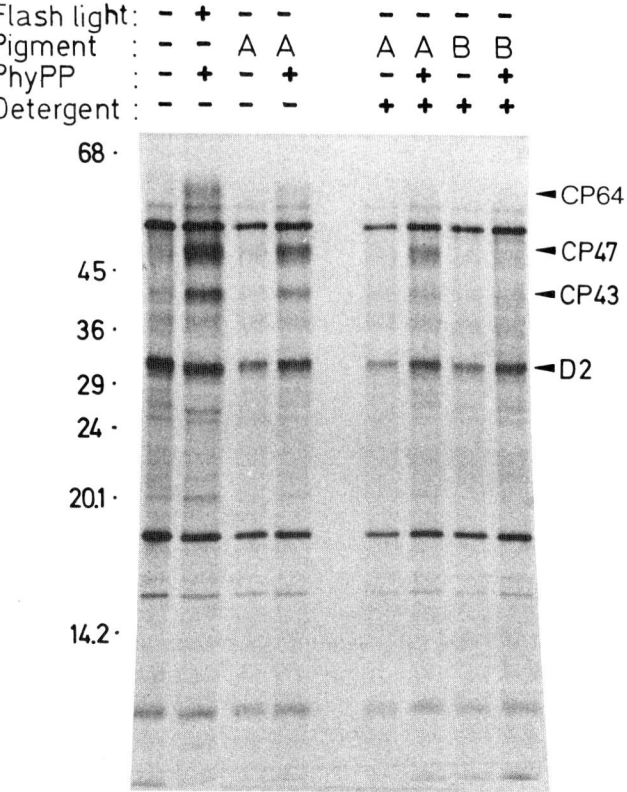

**Fig. 5.** *In vitro* induction of Chl *a* apoprotein synthesis in barley etioplasts (L. Eichacker, unpublished data). Lysed etioplasts were incubated with or without PhyPP with Chlide *a* (A) or Chl *a* (B); detergent, 0.0013% Brij-W1. Incorporation of [$^{35}$S]methionine was detected by fluorography after separation of the membrane fraction on a 12.5% SDS–PAGE gel containing 4 *M* urea. (For further experimental details, see Eichacker *et al.*, 1990.) In addition to the protein bands of the noninduced control, labeled bands of CP47, CP43, and CP64 are seen. The D1 protein is not separated from D2.

between synthesis of Chl and Chl *a* apoproteins. The failure to detect the synthesis of the CP64 apoprotein after irradiation of isolated etioplasts (Laing *et al.*, 1988) can be explained with more recent results (Eichacker *et al.*, 1990). Etioplasts lose GGPP and PhyPP during isolation (see Section II,B) and need an exogenous supply of the isoprenoid diphosphate for esterification of Chlide. Esterification is the switch for starting the synthesis of Chl *a* apoproteins. The nature of the substrate for esterification

## 6. Esterification of Chlorophyllide

seems to play only a secondary role: translation can be induced with GGPP as well as with PhyPP (Eichacker, 1991). The esterification reaction itself seems to be essential for translation. Incubation with esterified Chl, the product of the synthetase reaction, does not induce synthesis of Chl $a$ apoproteins (Fig. 5). Therefore, Chl synthetase seems to be involved more closely in the synthesis of Chl $a$ apoproteins than Pchlide-reductase. In a certain range (up to 50% of the endogenous Pchlide content), synthesis of Chl $a$ apoproteins depends stoichiometrically on the amount of newly formed Chl. In this case, it does not matter whether Chlide is added exogenously or formed by phototransformation of Pchlide.

Various mechanisms have been proposed to explain the Chl-dependent accumulation of Chl $a$ apoproteins (Mullet, 1988). Stabilization of apoproteins by newly formed Chl, that is, regulation not via increased rate of synthesis but via decreased rate of proteolysis in illuminated plants, was concluded from pulse-chase experiments (Mullet *et al.*, 1990). A careful investigation using pulses of 5 to 30 min and chases of 0 to 30 min (Eichacker, 1991; Eichacker *et al.*, 1992) revealed that synthesis of Chl did not increase the stability of Chl $a$ apoproteins *in vitro*. Instead, a higher turnover of Chl $a$ apoproteins was found after induction of Chl synthesis than in the Chl-free dark controls. Turnover and synthesis of Chl $a$ apoproteins was increased by Chl formation to about the same extent, so measurable proteolysis of CP47, CP43, and D2 occurred under all conditions (about 20% of synthesis); only CP64 did not show significant degradation. This result clearly eliminates stabilization of apoproteins by Chl as the only regulatory effect, but suggests an additional block at the translational level (see Klein and Mullet, 1986; see also Chapters 8 and 10).

Whether initiation and elongation of translation are involved in regulation was investigated using inhibitors of initiation (Eichacker, 1991; Eichacker *et al.*, 1992). Kasugamycin is known to inhibit only initiation; aurin-tricarboxylic acid inhibits initiation at lower and elongation as well at higher concentrations. Incorporation of [$^{35}$S]methionine into Chl $a$ apoproteins was blocked to about 50% by addition of 20–100 $\mu M$ kasugamycin or 30 $\mu M$ aurin-tricarboxylic acid; higher concentrations (up to 100 $\mu M$) of aurin-tricarboxylic acid inhibit translation more effectively (up to 90%; Eichacker *et al.*, 1992). The results show that initiation (or reinitiation) of translation is possible in the *in vitro* system and that this process is at least in part involved in Chl-induced accumulation of Chl $a$ apoproteins. The results do not exclude the possibility of regulation at the level of translation elongation. A more detailed analysis must show how the Chl synthetase reaction interferes with initiation and elongation of translation.

## V. INFLUENCE OF METAL CHELATORS AND 5-AMINOLEVULINATE ON CHLOROPHYLL SYNTHETASE OF DEVELOPING PLASTIDS

The specific precursor of tetrapyrroles, 5-aminolevulinate (ALA), often has been used in biosynthetic studies of Chl and other tetrapyrroles. Etiolated seedlings accumulate large amounts of Pchlide in the dark after incubation with ALA. The same is true after incubation with several aromatic heterocyclic nitrogenous bases (e.g., 2,2'-dipyridyl) that are metal chelators. In both cases, the rate of greening after transfer of the seedlings into continuous white light is reduced by the pretreatment (Hendry and Stobart, 1978; Rao *et al.*, 1981). Several explanations for chelator effects have been given by several authors. (1) Increased accumulation of Pchlide in the dark could be explained by a decrease in the heme level due to chelation of iron ions and subsequent lack of feedback inhibition by heme on early steps of tetrapyrrole biosynthesis (Duggan and Gassman, 1974; Hendry and Stobart, 1978; Chereskin and Castelfranco, 1982; Kannangara *et al.*, 1988). (2) Chelation of iron ions, which are required by coproporphyrinogen oxidase (Hsu and Miller, 1970), for the last steps of Pchlide formation (Vlcek and Gassman, 1979; Spiller *et al.*, 1982), and for the Shibata shift (Rao *et al.*, 1981), otherwise could lead to the reduced rate of greening in the light. However, inhibition of enzymatic steps before Pchlide by chelation of iron would contradict the increased amounts detected after incubation with metal chelators. (3) According to Rebeiz *et al.* (1984), increased accumulation of Pchlide (and other tetrapyrroles) in the dark could lead to photodynamic damage in the subsequent light period, which would explain effects detected in the dark and in the light.

A careful analysis in greening cress seedlings (Kittsteiner *et al.*, 1991a) revealed that photodynamic damage cannot explain the observed retardation in the light-dependent greening process entirely. These authors offered another explanation instead. Investigation of the last steps of Chl biosynthesis revealed a strong decrease in the activity of Chl synthetase *in vivo* after treatment with metal chelators (Table V). Whereas esterification was nearly completed (90% of Chlide) in control cress seedlings, only 25% of Chlide was esterified within 1 hr in pretreated seedlings. Similar values were found in excised cress cotyledons (Kittsteiner *et al.*, 1991a). Decreased esterification of Chlide also can be seen in young (2-day-old) etiolated oat seedlings but not in older (3-day-old) oat seedlings pretreated with 1 m$M$ 2,2'-dipyridyl (A. Sappok and W. Rüdiger, 1986, unpublished observations). Contrary to the report for greening groundnut leaves (Rao *et al.*, 1981), no decrease in the photoconversion was observed in cress or oat seedlings. When discussing such differences, not only the different

TABLE V

**Photoconversion of Pchlide and Esterificaiton of Chlide in Etiolated Plants**[a]

| Treatment | Chlorophyllide (photoconversion) (%) | Chlorophyll (esterification) (%) |
|---|---|---|
| Water control | 34 | 21 |
| 2,2'-Dipyridyl | 36 | 9 |
| 8-Hydroxyquinoline | 32 | 8 |

[a] Cress seedlings (*Lepidium sativum* L.) were grown in the dark for 96 hr. During the last 17 hr of this period, the plants were incubated with either 2 m$M$ 2,2'-dipyridyl or 3 m$M$ 8-hydroxyquinoline, as indicated. The plants then were flash irradiated, incubated for 1 hr in the dark for esterification of photoconverted Chlide, and extracted for pigment analysis. All values are percentages of Pchlide in the water controls before flash irradiation (7.4 nmol · g fresh wt$^{-1}$). After Kittsteiner (1990).

plant species but also the developmental stages of plants must be considered. Dramatic age-dependent changes in the capacity of the tetrapyrrole pathway have been found in mustard and cress seedlings (Gehring *et al.*, 1977; Oster *et al.*, 1991). Also in 6.5-day-old etiolated wheat seedlings treated with 8-hydroxyquinoline, esterification of Chlide was inhibited more drastically than was the photoconversion of Pchlide to Chlide (Ryberg *et al.*, 1989; M. Ryberg, 1991, personal communication). Lack of esterification means not only lack of esterified Chl but also lack of translation of Chl *a* apoproteins (see Section IV). Esterified Chl is essential also for assembly of the light-harvesting complex of photosystem II, as concluded from *in vitro* reconstitution experiments (see Chapter 10). Pretreatment with metal chelators therefore must result in strongly delayed plastid development (see also Ryberg *et al.*, 1989). The possibility that Chl synthetase needs metal ions for activity was tested and rejected by *in vitro* activity assays: neither the membrane-bound nor the solubilized enzyme was inhibited by 2,2'-dipyridyl or 1,10-phenanthroline in concentrations up to 5 m$M$ (P. Lauterbach and W. Rüdiger, 1992, unpublished results). The metal ion needed for Chl synthetase activity is magnesium; this ion is not complexed by the nitrogen-containing chelators applied in these studies.

Another possibility is inhibition of enzyme biosynthesis by chelators, in accordance with a general inhibition of transcription by the chelators (Kittsteiner *et al.*, 1991b). This possibility also could explain why no chelator effect was observed in older oat seedlings. The preformed enzyme is not inhibited. Since inhibition occurs in cress seedlings even when the enzyme has been synthesized already, we must assume a rapid turnover of the enzyme in etiolated cress seedlings. So far, no independent proof supports or refutes this possibility.

## VI. CONCLUDING REMARKS

Esterification of Chlide plays a particular role among the many steps of Chl biosynthesis. It is the last step of Chl formation in which more polar precursors are converted into an extremely apolar molecule. At least one substrate (GGPP) comes from the stroma site. The product, Chl, must be released either into the lipid phase of the membrane or directly onto hydrophobic binding sites of apoproteins. The process of release has not yet been investigated in detail. Chl synthetase activity is latent in well-preserved PLBs and only appears after disaggregation of PLBs *in vitro*. It is feasible that the activity also rises *in vivo* during disaggregation of PLBs, that is, immediately after irradiation of etiolated seedlings. However, such an increase remains to be shown. The well-known light-dependent accumulation of plastid-encoded Chl *a* apoproteins now can be understood to some extent. The signal transduction chain implies photoconversion of Pchlide bound to Pchlide-reductase and esterification of newly formed Chlide. The latter reaction is coupled to translation of Chl *a* apoproteins. Factors other than Chl synthetase that are necessary for this translational control are still unknown.

Many aspects of thylakoid development have been investigated during greening of etiolated angiosperm seedlings. Also, most experiments on esterification of Chlide, including consequences for translational control, have been performed with etioplast preparations. The basic results and conclusions are likely to be valid for other developmental stages of plants also. Nevertheless, validity of all details still must be verified experimentally for fully green plants before far-reaching generalizations can be made.

### ACKNOWLEDGMENTS

The author thanks L. Eichacker, M. Helfrich, and P. Lauterbach for making unpublished results available for this chapter. The work of the author was supported by the Deutsche Forschungsgemeinschaft, Bonn and the Fonds der Chemischen Industrie, Frankfurt.

# 6. Esterification of Chlorophyllide

## REFERENCES

Akhtar, M., Ajaz, A. A., and Corina, D. L. (1984). The mechanism of the attachment of esterifying alcohol in bacteriochlorophyll *a* biosynthesis. *Biochem. J.* **224**, 187–194.

Apel, K. (1979). Phytochrome-induced appearance of mRNA activity for the apoprotein of the light-harvesting chlorophyll *a/b* of barley (*Hordeum vulgare*). *Eur. J. Biochem.* **97**, 183–188.

Apel, K., and Kloppstech, K. (1980). The effect of light on the biosynthesis of the light-harvesting chlorophyll *a/b* protein. Evidence for the requirement of chlorophyll *a* for the stabilization of the apoprotein. *Planta* **150**, 426–430.

Artus, N. N., Ryberg, M., Lindsten, A., Ryberg, H., and Sundqvist, C. (1992). The Shibata shift and the transformation of etioplasts to chloroplasts in wheat with clomazone (FMC 57020) and amiprophos-methyl (Tokunol M). *Plant Physiol.* **98**, 253–263.

Bennett, J. (1981). Biosynthesis of the light-harvesting chlorophyll *a/b* protein. *Eur. J. Biochem.* **118**, 61–70.

Benz, J. (1980). "Untersuchungen zur Chlorophyllbiosynthese in etiolierten Haferkeimlingen: Veresterung von Chlorophyllid *in Vitro*." Ph.D. Thesis, University of München.

Benz, J., and Rüdiger, W. (1981a). Incorporation of 1-$^{14}$C-isopentenyldiphosphate, geraniol, and farnesol into chlorophyll in plastid membrane fractions of *Avena sativa* L. *Z. Pflanzenphysiol.* **102**, 95–100.

Benz, J., and Rüdiger, W. (1981b). Chlorophyll biosynthesis: Various chlorophyllides as exogenous substrates for chlorophyll synthetase. *Z. Naturforsch.* **36C**, 51–57.

Benz, J., Wolf, C., and Rüdiger, W. (1980). Hydrogenation of geranylgeraniol. *Plant Sci. Lett.* **19**, 225–230.

Benz, J., Hampp, R., and Rüdiger, W. (1981). Chlorophyll biosynthesis by mesophyll protoplasts and plastids from etiolated oat (*Avena sativa* L.) leaves. *Planta* **152**, 54–58.

Benz, J., Fischer, I., and Rüdiger, W. (1983a). Determination of phytyl diphosphate and geranylgeranyl diphosphate in etiolated oat seedlings (*Avena sativa* L.). *Phytochemistry* **22**, 2801–2804.

Benz, J., Haser, A., and Rüdiger, W. (1983b). Changes in the endogenous pools of tetraprenyl diphosphates in etiolated oat seedlings after irradiation. *Z. Pflanzenphysiol.* **111**, 349–356.

Benz, J., Lempert, U., and Rüdiger, W. (1984). Incorporation of phytol precursors into chlorophylls of tobacco cell cultures. *Planta* **162**, 215–219.

Bergweiler, P. (1987). "Charakterisierung von Bau und Funktion der Photosynthesemembranen ausgewählter Pflanzen unter den Extrembedingungen des Hochgebirges." Ph.D. Thesis. University of Köln.

Blank-Huber, M. (1986). "Untersuchungen zur Chlorophyll-Biosynthese Solubilisierung und Eigenschaften der Chlorophyll-Synthetase." Ph.D. Thesis, University of München.

Block, M. A., Joyard, J., and Douce, R. (1980). Site of synthesis of geranylgeraniol derivatives in intact spinach chloroplasts. *Biochim. Biophys. Acta* **631**, 210–219.

Böddi, B., Lindsten, A., Ryberg, M., and Sundqvist C. (1989). On the aggregational states of protochlorophyllide and its protein complexes in wheat etioplasts. *Physiol. Plant.* **76**, 135–143.

Bogorad, L. (1976). Chlorophyll biosynthesis. *In* "Chemistry and Biochemistry of Plant Pigments" (T. W. Goodwin, ed.), 2nd Ed., Vol. 1, pp. 64–148. Academic Press, London.

Buetow, D. E. (1986). Chloroplast development. *In* "Regulation of Chloroplast Differentiation" (G. Akoyunoglou and H. Senger, eds.), Vol. 2, pp. 427–432. A. R. Liss, New York.

Chereskin, B. M., and Castelfranco, P. A. (1982). Effects of iron and oxygen on chlorophyll biosynthesis. II. Observations on the biosynthetic pathway in isolated etiochloroplasts. *Plant Physiol.* **69,** 112–116.
Dehesh, K., van Cleve, B., Ryberg, M., and Apel, K. (1986). Light-induced changes in the distribution of the 36,000 $M_r$ polypeptide of NADPH-protochlorophyllide oxidoreductase within different cellular compartments of barley (*Hordeum vulgare* L.). II. Localization by immunogold labeling in ultrathin sections. *Planta* **169,** 172–183.
Dogbo, O., Bardat, F., and Camara, B. (1984). Terpenoid metabolism in plastids: Activity, localization, and substrate specificity of chlorophyll synthetase in *Capsicum annuum* plastids. *Physiol. Veg.* **22,** 75–83.
Duggan, J., and Gassman, M. (1974). Induction of porphyrin synthesis in etiolated bean leaves by chelators of iron. *Plant Physiol.* **53,** 206–215.
Eichacker, L. (1991). "Regulation der Chl *a*-Apoprotein Synthese in Etioplasten aus Gerste." Ph.D. Thesis, University of München.
Eichacker, L., Soll, J., Lauterbach, P., Rüdiger, W., Klein, R. R., and Mullet, J. E. (1990). *In vitro* synthesis of chlorophyll *a* in the dark triggers accumulation of chlorophyll *a* apoproteins in barley etioplasts. *J. Biol. Chem.* **23,** 13566–13571.
Eichacker, L., Paulsen, H., and Rüdiger, W. (1992). Synthesis of chlorophyll *a* regulates translation of chlorophyll *a* apoproteins P700, CP47, CP43, and D2 in barley etioplasts. *Eur. Biochem. J.* **205,** 17–24.
Emery, V. C., and Akhtar, M. (1987). Mechanistic studies on the phytylation and methylation steps in bacteriochlorophyll *a* biosynthesis: An applicant of the $^{18}$O-induced isotope effect in $^{13}$C NMR. *Biochemistry* **26,** 1200–1208.
Feierabend, J. (1977). Capacity for chlorophyll synthesis in heat-bleached 70S ribosome-deficient rye leaves. *Planta* **135,** 83–88.
Forreiter, C., van Cleve, B., Schmidt, A., and Apel, K. (1991). Evidence for a general light-dependent negative control of NADPH-protochlorophyllide oxidoreductase in angiosperms. *Planta* **183,** 126–132.
Gehring, H., Kasemir, H., and Mohr, H. (1977). The capacity of chlorophyll-*a* biosynthesis in the mustard selling cotyledons as modulated by phytochrome and circadian rhythmicity. *Planta* **133,** 295–302.
Hehlein, C. (1979). "Einfluβ der Anaerobiose auf die Chlorophyllbiosynthese bei ergrünenden Haferkeimlingen." Master's Thesis, University of München.
Hendry, G. A. F., and Stobart, A. K. (1978). Effect of 2,2'-bipyridyl on porphyrin synthesis in etiolated and light-treated barley leaves. *Phytochemistry* **17,** 671–674.
Herrmann, R. G., Westhoff, P., Alt, J., Tittgen, J., and Nelson, N. (1985). Thylakoid membrane proteins and their genes. *In* "Molecular Form and Function of the Plant Genomes" (L. van Vloten-Doting, G. S. P. Groot, and T. C. Hall, eds.), pp. 233–256. Plenum, Amsterdam.
Hess, W. R., Blank-Huber, M., Fieder, B., Börner, T., and Rüdiger, W. (1992). Chlorophyll synthetase and chloroplast tRNA$^{glu}$ are present in heat-bleached, ribosome-deficient plastids. *J. Plant Physiol.* **139,** 427–430.
Hsu, W. P., and Miller, G. W. (1970). Coproporphyrinogenase in tobacco (*Nicotiana tabacum* L.). *Biochem. J.* **117,** 215–220.
Kannangara, C. G., Gough, S. P., Bruyant, P., Hoober, J. K., Kahn, A., and von Wettstein, D. (1988). tRNA$^{glu}$ as a cofactor in δ-aminolevulinate biosynthesis: Steps that regulate chlorophyll synthesis. *Trends Biochem. Sci.* **13,** 139–143.
Keckeis, F.-J. (1983). "Die letzten Schritte der Chlorophyll-Biosynthese im etiolierten Haferkeimling, in Abhängigkeit seines morphologischen Alters." Master's Thesis, University of München.

## 6. Esterification of Chlorophyllide

Kittsteiner, U. (1990). "Der Ergrünungsprozeβ in Gartenkresse unter dem Einfluβ von Metallchelatoren und 5-Aminolävulinat." Ph.D. Thesis, University of München.

Kittsteiner, U., Mostowska, A., and Rüdiger, W. (1991a). The greening process in cress seedlings. I. Pigment accumulation and ultrastructure after application of 5-aminolevulinate and complexing agents. *Physiol. Plant.* **81,** 139–147.

Kittsteiner, U., Brunner, H., and Rüdiger, W. (1991b). The greening process in cress seedlings. II. Complexing agents and 5-aminolevulinate inhibit accumulation of *cab*-mRNA coding for the light-harvesting chlorophyll *a/b* protein. *Physiol. Plant.* **81,** 190–196.

Klein, R. R., and Mullet, J. E. (1986). Regulation of chloroplast-encoded chlorophyll-binding protein translation during higher plant chloroplast biogenesis. *J. Biol. Chem.* **261,** 11138–11145.

Klein, R. R., Gamble, P. E., and Mullet, J. E. (1988). Light-dependent accumulation of radiolabeled plastid-encoded chlorophyll *a*-apoproteins requires chlorophyll *a*. *Plant Physiol.* **88,** 1246–1256.

Kreuz, K., and Kleinig, H. (1981). Chlorophyll synthetase in chlorophyll-free chromoplasts. *Plant Cell Rep.* **1,** 40–42.

Kreuz, K., Dehesh, K., and Apel, K. (1986). The light-dependent accumulation of the P700 chlorophyll *a* protein of the photosystem I reaction center in barley. *Eur. J. Biochem.* **159,** 459–467.

Laing, W., Kreuz, K., and Apel, K. (1988). Light-dependent, but phytochrome-independent, translational control of the accumulation of the P700 chlorophyll-*a* protein of photosystem I in barley (*Hordeum vulgare* L.). *Planta* **176,** 269–276.

Lindsten, A., Welch, C. J., Schoch, S., Ryberg, M., Rüdiger, W., and Sundqvist, C. (1990). Chlorophyll synthetase is latent in well preserved prolamellar bodies of etiolated wheat. *Physiol. Plant* **80,** 277–285.

Lütz, C., Benz, J., and Rüdiger, W. (1981). Esterification of chlorophyllide in prolamellar body (PLB) and prothylakoid (PT) fractions from *Avena sativa* etioplasts. *Z. Naturforsch.* **36C,** 58–61.

Mapleston, R. E., and Griffiths, W. T. (1978). Effect of illumination of etiolated leaves on the redox state of NADPH in the plastids. *FEBS Lett.* **92,** 168–172.

Mullet, J. E. (1988). Chloroplast development and gene expression. *Annu. Rev. Plant Physiol. Plant Mol. Biol.* **39,** 475–502.

Mullet, J. E., Klein, G. P., and Klein, R. R. (1990). Chlorophyll regulates accumulation of the plastid-encoded chlorophyll apoproteins CP43 and D1 by increasing apoprotein stability. *Proc. Natl. Acad. Sci. USA* **87,** 4038–4042.

Oelmüller, R. (1989). Photooxidative destruction of chloroplasts and its effect on nuclear gene expression and extraplastidic enzyme levels. *Photochem. Photobiol.* **49,** 229–239.

Oster, U., Blos, I., and Rüdiger, W. (1991). The greening process in cress seedlings. III. Age-dependent changes in the capacity of the tetrapyrrole pathway. *Z. Naturforsch.* **46C,** 1052–1058.

Rao, S. R., Sainis, J. K., and Sane, P. V. (1981). Inhibition of chlorophyll biosynthesis by $\alpha,\alpha'$-dipyridyl during greening of groundnut leaves. *Phytochemistry* **20,** 2683–2686.

Rebeiz, C. A., Montazer-Zouhoor, A., Hopen, H. J., and Wu, S. M. (1984). Photodynamic herbicides. I. Concept and phenomenology. *Enzyme Microb. Technol.* **6,** 390–401.

Rüdiger, W. (1987). Chlorophyll synthetase and its duplication for regulation of chlorophyll biosynthesis. *In* "Progress in Photosynthetic Research" (J. Biggins, ed.), Vol. 4, pp. 461–467. Nijhoff, Dordrecht, The Netherlands.

Rüdiger, W., and Benz, J. (1979). Influence of aminotriazol on the biosynthesis of chlorophyll and phytol. *Z. Naturforsch.* **34C,** 1055–1057.

Rüdiger, W., Benz, J., Lempert, U., Schoch, S., and Steffens, D. (1976). Inhibition of phytol accumulation with herbicides: Geranylgeraniol and dihydrogeranylgeraniol-containing chlorophyll from wheat seedlings. *Z. Pflanzenphysiol.* **80,** 131–143.
Rüdiger, W. (1987). Chlorophyll synthetase and its duplication for regulation of chlorophyll biosynthesis. *In* "Progress in Photosynthetic Research" (J. Biggins, ed.), Vol. 4, pp. 461–467. Nijhoff, Dordrecht, The Netherlands.
Rüdiger, W., Benz, J., and Guthoff, C. (1980). Detection and partial characterization of activity of chlorophyll synthetase in etioplast membranes. *Eur. J. Biochem.* **190,** 193–200.
Ryberg, M., and Dehesh, K. (1986). Localization of NADPH: protochlorophyllide oxidoreductase in dark-grown wheat (*Triticum aestivum*) by immuno-electron microscopy before and after transformation of the prolamellar bodies. *Physiol. Plant.* **66,** 616–624.
Ryberg, M., Lindsten, A., and Artus, N. (1989). Etioplast to chloroplast development is inhibited by the metal chelator 8-hydroxyquinoline. *Physiol. Plant.* **76,** A138.
Schindler, C. (1983). "Chlorophyllsynthetase in ergrünenden Haferkeimlingen." Master's Thesis, University of München.
Schoch, S. (1978). The esterification of chlorophyllide *a* in greening bean leaves. *Z. Naturforsch.* **33C,** 712–714.
Schoch, S., Lempert, U., and Rüdiger, W. (1977). On the last steps of chlorophyll biosynthesis: Intermediates between chlorophyllide and phytol-containing chlorophyll. *Z. Pflanzenphysiol.* **83,** 427–436.
Schoch, S., Hehlein, C., and Rüdiger, W. (1980). Influence of anaerobiosis on chlorophyll biosynthesis in greening oat seedlings (*Avena sativa* L.). *Plant Physiol.* **66,** 576–579.
Soll, J., Schultz, G., Rüdiger, W., and Benz, J. (1983). Hydrogenation of geranylgeraniol: Two pathways exist in spinach chloroplasts. *Plant Physiol.* **71,** 849–854.
Spiller, S. C., Castelfranco, A. M., and Castelfranco, P. A. (1982). Effects of iron and oxygen on chlorophyll biosynthesis. I. *In vivo* observation on iron- and oxygen-deficient plants. *Plant Physiol.* **69,** 107–111.
Steiger, A., Mitzka-Schnabel, U., Rau, W., Soll, J., and Rüdiger, W. (1985). Inhibition of geranylgeranyl diphosphate synthesis in *in vitro* systems. *Phytochemistry* **24,** 739–743.
Taylor, W. C. (1989). Regulatory interactions between nuclear and plastid genomes. *Annu. Rev. Plant Physiol.* **40,** 211–233.
Vlcek, L. M., and Gassman, M. L. (1979). Reversal of $\alpha,\alpha'$-dipyridyl-induced porphyrin synthesis in etiolated and greening red kidney bean leaves. *Plant Physiol.* **64,** 393–397.
Wilkinson, R. E. (1985). Inhibition of conversion of geranylgeranyl-chlorophyll to phytol-chlorophyll by *S*-ethyl dipropylthiocarbamate (EPTC). *Pestic. Biochem. Physiol.* **23,** 289–293.
Willstätter, R., and Stoll, A. (1913). "Untersuchungen über Chlorophyll." Springer-Verlag, Berlin.
Wolff, J. B., and Price, L. (1957). Terminal steps of chlorophyll *a* biosynthesis in higher plants. *Arch. Bioch. Biophys.* **72,** 293–301.

# 7

# Chloroplast Lipids and the Assembly of Membranes

## EVA SELSTAM AND ANNA WIDELL WIGGE

Department of Plant Physiology
University of Umeå
S-901 87 Umeå, Sweden

I. Introduction
II. Plastid Lipids and Localization of Their Biosynthesis
    A. Acyl Lipids of Chloroplasts and Etioplasts
    B. Phase Behavior of Galactolipids
    C. Biosynthesis and Transfer of Lipids from Envelope to Thylakoid Membranes
    D. Possible Significance of Monogalactosyl Diacylglycerol in Thylakoids
III. Heterogeneity of Lipids in Inner Plastid Membranes
    A. Lateral Asymmetry
    B. Boundary Lipids
    C. Transverse Asymmetry
IV. Specific Interaction between Thylakoid Lipids and Chlorophyll Protein Complexes in Photosystem II
    A. Acylation of the D1 Protein
    B. Lipids in the Photosystem II Reaction Center
    C. Lipids in the Light-Harvesting Complex II
V. Assembly of Prolamellar Body Membranes
    A. Nature of the Regularly Branched Prolamellar Body Membrane
    B. Protein–Protein Interaction of Protochlorophyllide Reductase
    C. Lipid–Protein Interaction in Prolamellar Body Membranes
    D. Phototransformation of Prolamellar Body Membranes
References

## I. INTRODUCTION

All the pigment–protein complexes discussed in this volume are localized to the lipid matrix of the inner membranes of chloroplasts and etioplasts. This chapter discusses the lipids that surround the pigment–

protein complexes, their general role as structural components, and the known specific interactions between lipids and pigment–protein complexes in chloroplasts and etioplasts. Since the lipids constitute the matrix for the membrane proteins, it can be difficult to distinguish their structural and functional aspects; only limited unequivocal evidence for specific roles of plastid lipids has been presented to date. In this presentation, we focus on the significance of the phase structure formed by the membrane lipids. According to the fluid mosaic membrane model, the role of the lipids is to form the membrane barrier; this role is fulfilled by lipids forming a lamellar phase with water.The discovery of reversed hexagonal lipids among the membrane lipids was puzzling at first. However, increased knowledge about the principles of self-assembly of lipids into different phase structures has reached a state at which it is possible to understand the structural role not only of the lamellar but also of the reversed hexagonal lipids in the membrane. When searching for a function of lipids in the membrane not only the role of lipids forming a barrier but also the dynamic functions of the membrane must be considered. The role of thylakoid membrane lipids has been reviewed (Quinn and Williams, 1983, 1985; Gounaris *et al.*, 1986; Murphy, 1986; Sprague, 1987).

## II. PLASTID LIPIDS AND LOCALIZATION OF THEIR BIOSYNTHESIS

### A. Acyl Lipids of Chloroplasts and Etioplasts

All inner membranes of chloroplasts and etioplasts have the same specific lipid class composition (Table I). The relative composition varies somewhat depending on growth conditions and plant species, as has been reviewed extensively elsewhere (Douce and Joyard, 1980; Harwood, 1980; Mudd, 1980). The two plastic-specific galactolipids, monogalactosyl diacylglycerol (MGDG) and digalactosyl diacylglycerol (DGDG), are the dominating lipids. Together, these neutral lipids constitute approximately 80% of the total lipids. In addition to these galactolipids, some negatively charged lipids are found, namely the specific sulfur-containing lipid sulfoquinovosyl diacylglycerol (SQDG; 5–10%) and phosphatidyl glycerol (PG; 5–10%, Table I). A small percentage of phosphatidyl inositol (PI) also has been shown to be endogenous to these membranes. Most lipid analyses of inner plastid membranes show the presence of phosphatidyl choline (PC). Dorne *et al.* (1990) showed that the PC found in isolated spinach thylakoids was present due to contamination of the preparations by the outer envelope membrane. The outer envelope membrane contains about

## 7. Chloroplast Lipids and Membrane Assembly

**TABLE I**

**Lipid Composition of Different Fractions from Wheat (*Triticum aestivum* L.)**

| Membrane fraction[a] | Mol % lipid[b] | | | | | |
|---|---|---|---|---|---|---|
| | MGDG | DGDG | SQDG | PG | PC | PI |
| Thylakoid | 49 | 31 | 8.6 | 9.7 | 1.1 | t[c] |
| Prothylakoid | 45 | 32 | 6.2 | 8.9 | 6.2 | 1.6 |
| Prolamellar body | 51 | 29 | 7.7 | 9.5 | 3.3 | t |

[a] Thylakoids and etioplast membranes were isolated according to Walker (1971) and Selstam and Widell Wigge (1989), respectively. Lipid extraction and quantification according to Selstam and Öquist (1985).

[b] MGDG, monogalactosyl diacylglycerol; DGDG, digalactosyl diacylglycerol; SQDG, sulfoquinovosyl diacylglycerol; PG, phosphatidyl glycerol; PC, phosphatidyl choline; PI, phosphatidyl inositol.

[c] Traces.

20% PC. Highly purified thylakoids from spinach showed no PC, so PC as an endogenous component of the inner plastid membranes must be reconsidered.

The specific lipid composition of plastids in higher plants is seen in thylakoids of cyanobacteria (Murata and Nishida, 1987) and prochlorophyta (Murata and Nishida, 1987; Gombos and Murata, 1991) also. This similarity supports the endosymbiotic theory which claims that the chloroplast is of prokaryotic origin. The high degree of evolutionary conservation of the specific lipid composition suggests that it is essential for the function of the pigment–protein complexes and photosynthetic electron transport. This suggestion is supported further by the finding of MGDG in the photosystem II (PSII) reaction center (Murata *et al.*, 1990; e.g., Section IV,B). Another ecophysiological interpretation of this evolutionary conservation is that the plastid lipid composition might be of importance for optimum plant growth. The galactolipids do not contain phosphorus or nitrogen, which are the most common growth-limiting factors for photosynthetic bacteria as well as for plants.

The fatty acyl group composition of the plastid lipids shows some variation among plastids, growth conditions, and species (Table II; Douce and Joyard, 1980; Harwood, 1980; Quinn and Williams, 1985). Generally, however, the lipids of the inner plastid membranes are very unsaturated. The average number of double bonds of the total thylakoid lipids is between 4 and 5. All double bonds are of *cis*-configuration except in *trans*-hexadecanoic acid ($\Delta$3-16:1t). For MGDG and DGDG, the tri-unsaturated acyl group of linolenic acid (18:3) alone or, in some species, with hexadeca-

TABLE II

Fatty Acid Composition of Thylakoids Isolated from Wheat (*Triticum aestivum* L.) "18.3" Species and Spinach (*Spinacia oleracea* L.) "16.3" Species

| Lipid class[a] | Mol % fatty acid | | | | | | | | |
|---|---|---|---|---|---|---|---|---|---|
| | 16:0 | 16:1 | 16:2 | 16:3 | 18:0 | 18:1 | 18:2 | 18:3 | 18:4 |
| Wheat[b] | | | | | | | | | |
| MGDG | 1 | 1 | — | — | t | t | 4 | 94 | t[c] |
| DGDG | 7 | 1 | — | — | 1 | 1 | 2 | 88 | t |
| SQDG | 26 | 2 | — | — | 2 | 1 | 6 | 63 | — |
| PG | 17 | 33[d] | — | — | 1 | 1 | 4 | 44 | t |
| Spinach | | | | | | | | | |
| MGDG | 1 | 1 | 1 | 22 | — | — | 2 | 73 | t |
| DGDG | 9 | 1 | t | 5 | 1 | 1 | 2 | 81 | t |
| SQDG | 40 | 2 | — | 2 | 1 | 2 | 9 | 44 | — |
| PG | 12 | 36[d] | — | — | 1 | 3 | 7 | 40 | 1 |

[a] MGDG, monogalactosyl diacylglycerol; DGDG, digalactosyl diacylglycerol; SQDG, sulfoquinovosyl diacylglycerol; PG, phosphatidyl glycerol.

[b] Thylakoids were isolated according to Walker (1971) and the analysis of fatty acids was performed according to Selstam and Öquist (1985).

[c] Traces.

[d] *trans*-Hexadecanoic acid.

trienoic acid (16:3) constitutes up to 85–95% of the fatty acids (Table II). The acyl groups of SQDG and PG are somewhat less unsaturated, due to a large proportion of palmitic acid (16:0). The 16:1t acid is only esterified to PG. This unusual fatty acid naturally has received much attention. Most often it has been ascribed a specific role in the oligomerization of the light-harvesting chlorophyll–protein complex of PSII (LHCII), a phenomenon that is discussed further in Section IV,C. The positional distribution of acyl chains in galactolipids and PG is published elsewhere (Heinz, 1977; Nishihara *et al.*, 1980; Murata, 1983). The significance of the unsaturated thylakoid membrane lipids is discussed in other publications as well (Quinn and Williams, 1985; Wada *et al.*, 1990).

## B. Phase Behavior of Galactolipids

The most obvious function of the lipids in a biological membrane is to form the membrane barrier. To achieve this function, only a few membrane lipids are needed. In all biological membranes, however, a great variety of lipids with differing acyl group compositions is present. The importance of this diversity is still an intriguing question. One approach to obtaining a better understanding of the function of different lipids is analysis of the physicochemical properties of the lipids, alone and in combinations. Analysis of the phase structure of membrane lipids has shown that most membranes contain at least one lipid that can form a nonlamellar phase with water. The function of these lipids in the membrane is probably to give a more dynamic bilayer that is able to be reorganized during processes such as cell division, membrane biogenesis, and membrane fusion. These membrane processes play an important role in the maintenance of the bilayer.

The phase behavior of the chloroplast galactolipids has been studied by X-ray diffraction (Shipley et al., 1973), freeze fracture electron microscopy (Quinn and Williams, 1983; Sprague and Staehelin, 1984), and $^2$H NMR spectroscopy (Brentel et al., 1985). In these studies, it has been demonstrated that, with water, MGDG forms a reversed hexagonal phase ($H_{II}$)

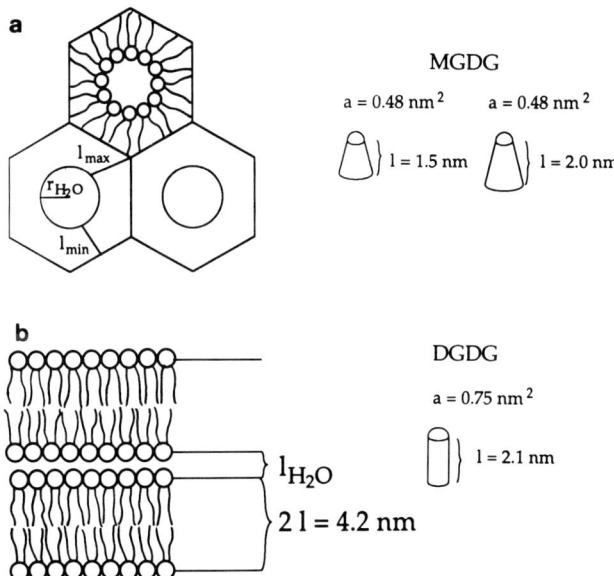

**Fig. 1.** Cross sections of reversed hexagonal ($H_{II}$) and lamellar phase ($L_\alpha$) of plant MGDG and DGDG in water. The dimensions of the molecules and phase structures are average for the structures at maximum water content, calculated from values given by Shipley et al. (1973). In the $L_\alpha$ phase the thickness of the lipid bilayer is constant, whereas in the $H_{II}$ phase, the lengths of the lipids are different at $l_{min}$ and $l_{max}$. (a) MGDG and water in an $H_{II}$ phase, (b) DGDG and water in an $L_\alpha$ phase.

and DGDG a lamellar phase ($L_\alpha$; Fig. 1). SQDG also was demonstrated to form an $L_\alpha$ phase with water (Shipley et al., 1973).

Why membrane lipids form different phase structures with water can be explained in part by the different shapes of the molecules (Israelachvili et al., 1980). The shape of the lipid molecule depends on the size of the polar head group and the length and the unsaturation of the hydrocarbon chains (Rilfors et al., 1984). MGDG, with a small polar head group, is a conically shaped molecule that forms an $H_{II}$ phase structure whereas DGDG, with a larger polar head group, is a more cylindrically shaped molecule that forms an $L_\alpha$ phase structure (Fig. 1). Plant MGDG, with mostly trienoic acyl chains, forms only an $H_{II}$ phase when swelled with water (Brentel et al., 1985) whereas saturated distearoyl-MGDG forms only an $L_\alpha$ phase when dispersed in an excess of water (Quinn and Williams, 1983).

The underlying mechanisms for the formation of $L_\alpha$ and $H_{II}$ phases are more complicated, however. A deeper understanding of these phase structures may give a better background for the interpretation of the structural roles of the thylakoid galactolipids. To develop the model further, a thermodynamic and a material physical approach has been used to explain how different phase structures are formed (Helfrich, 1973; Kirk et al., 1984; Gruner et al., 1985). This model considers only the terms that dominate the free energy in a phase transition from $L_\alpha$ to $H_{II}$ phase, namely, the curvature energy of the lipid monolayer and the stretching energy of the lipid chains (Gruner, 1985; Tate and Gruner, 1987). The lipid monolayer has a spontaneous tendency to curve because the interactions between the polar head groups in the monolayer differ from the interactions between the hydrocarbon chains. To force the curved monolayer into a flat monolayer requires free energy. The formation of the curved monolayer necessarily will lead to a stretching of the lipids away from their equilibrium length, which also requires free energy. Thus, the phase structure formed by a membrane lipid depends on which phase has the lowest free energy.

In mixtures of MGDG and DGDG, $L_\alpha$, $H_{II}$, and a cubic liquid crystalline ($I_2$) phase are formed (Fig. 2; Brental et al., 1985). In mixtures of $^2H_2O$, MGDG, and DGDG where the proportion of MGDG is low, an $L_\alpha$ phase is formed, whereas an $H_{II}$ phase is formed when larger amounts of MGDG are present. The more MGDG that is present in the mixture, the more nonbilayer phases are formed. In the ternary phase diagram, a cubic phase is found in the area between the $L_\alpha$ and $H_{II}$ phase areas (Fig. 2; Selstam et al., 1990). Viewed by freeze fracture electron microscopy, the cubic phase appears as regular arrays of a rounded structure (Sen et al., 1981, 1982a; Sprague and Staehelin, 1984). With different NMR methods and X-ray diffraction, the cubic phase was identified to have a reversed Ia3d (3-armed) symmetry and to be bicontinuous, that is, both water and lipids

# 7. Chloroplast Lipids and Membrane Assembly

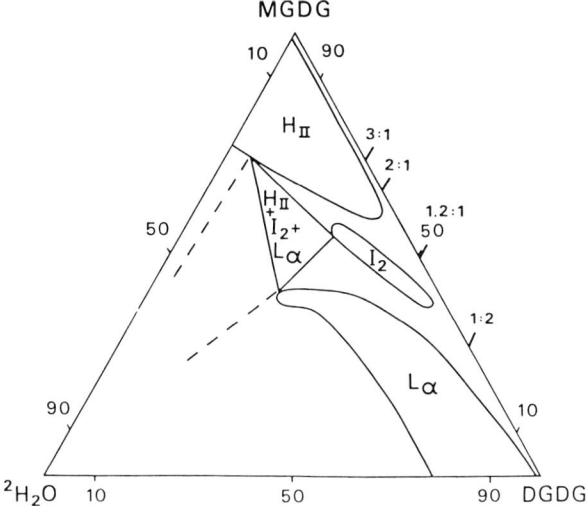

**Fig. 2.** Tentative ternary phase diagram of the $^2$H$_2$O/MGDG/DGDG system. Pure lamellar liquid crystalline (L$_\alpha$) phase and pure reversed hexagonal liquid crystalline (H$_{II}$) phase are found at low and high concentrations of MGDG, respectively. In mixtures of approximately equal amounts of MGDG and DGDG, a pure liquid crystalline cubic phase (I$_2$) is formed. MGDG and DGDG were isolated from wheat leaves. Their acyl chains were those given in Table II. The phase structure was determined as described in Brentel *et al.* (1985). Reprinted with permission from Lindblom and Rilfors (1989).

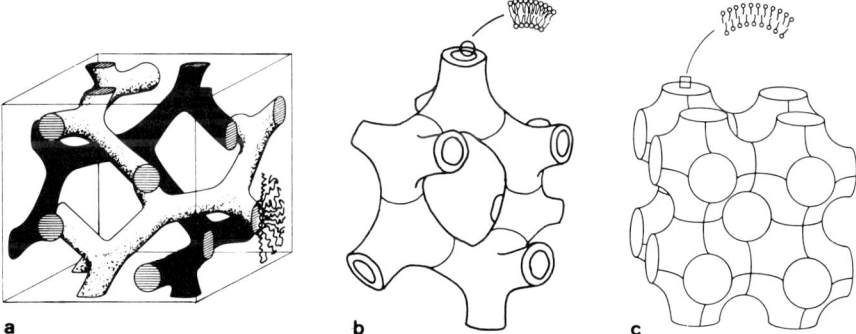

**Fig. 3.** Schematic presentations of bicontinuous cubic phases with three different symmetries: (a) Ia3d, a 3-armed symmetry; (b) Pn3m, a 4-armed symmetry; and (c) Im3m, a 6-armed symmetry. The Ia3d symmetry has been identified in $^2$H$_2$O, MGDG, and DGDG mixtures as well as in $^2$H$_2$O, MGDG, DGDG, SQDG, and PG mixtures. The Pn3m and Im3m symmetries are found in other lipid–water systems and in the PLB membrane. The bicontinuous cubic phases are formed by a regularly branching bilayer, where the water phases on each side of the bilayer form separate branching channels. For clarity, one water channel in the Ia3d symmetry (a) is shown in black. (a) Reprinted with permission from Seddon (1990). (c) Reprinted with permission from Lindblom and Rilfors (1989).

are continuous in the structure (Fig. 3a; Brentel *et al.*, 1985). The water forms two independent and interwoven branched channels. At each branching point, three branches meet. A monolayer of lipids surrounds each water channel. The hydrocarbon chains from one monolayer meet the hydrocarbon chains from the other, thus forming a branched bilayer between the water channels. Since the distance between the water channels could not be constant in such a structure, the bilayer of lipids must have a varying thickness (Anderson *et al.*, 1988).

Description of the cubic phase is difficult. The best description of different bicontinuous cubic phases to date was made using infinite periodic minimal surfaces (see Lindblom and Rilfors, 1989, for further information). In the phase diagram of $^2H_2O$, MGDG, and DGDG, the cubic phase area is located between the $L_\alpha$ and $H_{II}$ phase areas (Fig. 2). This location indicates that the cubic phase is a structure formed when the thermodynamic restrictions for neither the $L_\alpha$ phase nor the $H_{II}$ phase can be fulfilled (see previous text). Using the thermodynamic theory for molecular packing, it is also possible to calculate that, in the bicontinuous cubic phase, the frustration between the spontaneous curvature of the lipid monolayer and the geometrical elasticity of the lipid molecules is at a minimum (Anderson *et al.*, 1988). The bicontinuous cubic phase is, thus, thermodynamically the most stable molecular organization of the substances involved.

Cubic phases also have been found in a "native" mixture of all the chloroplast lipids, namely MGDG, DGDG, SQDG, PG (52 : 26 : 10 : 10 by mol), and $^2H_2O$. In this mixture, a cubic Ia3d phase is formed at water contents of 5–13%. A cubic phase also is formed in equilibrium with an $L_\alpha$ phase at water contents of 20–85% (Selstam *et al.*, 1990). This cubic phase probably is not of Ia3d symmetry, since the structural symmetry is dependent on the water content (Lindblom and Rilfors, 1989). The Ia3d symmetry is formed at low water content and the Pn3m and Im3m symmetries at higher water contents (Hyde *et al.*, 1984; Siegel and Banschbach, 1990). The Pn3m and Im3m symmetries are also bicontinuous cubic phases, but differ from the Ia3d symmetry by having 4-armed and 6-armed structural units (Fig. 3; Lindblom and Rilfors, 1989).

## C. Biosynthesis and Transfer of Lipids from Envelope to Thylakoid Membranes

The biosynthesis of chloroplast lipids is dependent on reactions that are localized in three different compartments of the cell: the chloroplast stroma, the envelope membranes, and the endoplasmic reticulum (ER). Significant progress in understanding lipid biosynthesis and its localization

## 7. Chloroplast Lipids and Membrane Assembly

has been made during the last decade. The reader is directed to various reviews for details on the subject (Stumpf, 1980, 1987; Heemskerk and Wintermans, 1987; Douce and Joyard, 1990; Browse and Somerville, 1991).

The final assembly of MGDG, DGDG, SQDG, and PG is localized to the chloroplast envelope membranes (Heemskerk and Wintermans, 1987; Heemskerk et al., 1990). After assembly, the acyl groups of these lipids are desaturated by four different desaturases that probably are localized in the chloroplast envelopes (Browse and Somerville, 1991). The final characteristic composition (Table II) and distribution of acyl groups at the sn-1 and sn-2 positions of the glycerol backbone are dependent on both the source of the diacylglycerol backbone and the substrate specificity of the desaturases. The diacylglycerol backbone is formed inside the chloroplast by the prokaryotic pathway or is originated from PC in the ER membrane by the eukaryotic pathway. Two very specific desaturases desaturate PG 16:0/18:1 to PG 16:1t/18:1 and MGDG 16:0/18:1 to MGDG 16:1/18:1. The other two less substrate-specific desaturases will desaturate monoenoic acyl groups to dienoic groups and dienoic groups to trienoic groups (Browse and Somerville, 1991).

The mechanisms for the transportation of thylakoid lipids from their site of synthesis in the envelope to their localization in the thylakoid membranes are not established. Three different mechanisms have been proposed: membrane flow of vesicles from the inner envelope membrane to the thylakoid, a lateral diffusion of lipids via membrane bridges between the inner envelope and the thylakoid, or a protein-mediated lipid transfer. These mechanisms also have been proposed for lipid transfer between cytoplasmic membranes (Kader, 1985; Sleight, 1987; Bishop and Bell, 1988).

Several proteins that transfer galactolipids *in vitro* have been isolated from spinach chloroplasts and whole leaves and from castor bean endosperm (Nishida and Yamada, 1985; Watanabe and Yamada, 1986). The proposed function of the galactolipid transfer proteins has not yet been demonstrated *in vivo*. Phospholipid transfer proteins are well known from plant material (Kader, 1985), but chloroplast-specific phospholipid transfer proteins for transfer of PG from the envelope to the thylakoid have not been found (Schwitzguebel and Siegenthaler, 1985). It was found that a lipid transfer protein from spinach was translated like a secretory protein (Bernhard et al., 1991) and that lipid transfer proteins from castor bean seedlings are localized *in vivo* in glyoxysomes and vessel cell walls (Yamada et al., 1990). At present, a proposed role of the lipid transfer proteins could be doubtful (Bernhard et al., 1991; Browse and Somerville, 1991).

Numerous observations of electron micrographs of chloroplasts, etioplasts, and developing plastids suggest that the inner envelope membrane forms invaginations, vesicles, and a peripheral reticulum (Gunning and Steer, 1975; Douce and Joyard, 1979; Wellburn, 1982). One function of these structures could be transportation of membrane lipids from the envelope to the thylakoids. The developing plastids always contain numerous invaginations and continuous membrane contacts from the envelope to the developing thylakoids or prolamellar bodies (Gunning, 1965; Carde et al., 1982; Oross and Possingham, 1991). The importance of vesicle or lateral transportation of lipids in mature chloroplasts has been questioned, since envelope invaginations and membrane connections seldom are observed (Douce and Joyards, 1979; Carde et al., 1982; Oross and Possingham, 1991). In mature chloroplasts, a substantial transfer of lipids by lateral diffusion over membrane bridges from the inner envelope membrane to the thylakoid is less likely. Such bridges would decrease the $\Delta$pH between the chloroplast stroma and the thylakoid lumen (dicussed in Douce and Joyard, 1979; Wellburn, 1982). Observations of the formation of vesicles by chloroplast envelopes in mature leaves of spinach, pea, soybean, and tobacco (Morré et al., 1991b) have shown that numerous vesicles can be observed only when the leaves have been kept at low temperature. Also, vesicles formed at 12°C disappear after some time at 23°C (Morré et al., 1991b), indicating that numerous vesicles are formed in mature chloroplasts also. These vesicles are observed most easily at low temperature when the fusion of the vesicles with the thylakoids is slower than the formation of vesicles. In a cell-free system, the formation of vesicles from the inner envelope was shown to be stimulated by ATP (Morré et al., 1991a). In the same system, ATP stimulated transfer of MGDG from envelope membranes to thylakoids (Morré et al., 1991a).

The presence of a membrane flow of chloroplast lipids by vesicles from the inner envelope membrane to the thylakoid is supported by evidence that these membranes have the same lipid composition (Douce and Joyard, 1990) and that final chloroplast lipid biosynthesis is only localized to the envelope (Heemskerk and Wintermans, 1987; Douce and Joyard, 1990). In addition, carotenoids, plastoquinones, and chlorophyllide synthesized In the envelope could be transported by this mechanism (Douce and Joyard, 1990). Vesicle formation at the chloroplast envelope and vesicle fusion with the thylakoid both demand reorganization of the membrane lipids. In model systems of nonthylakoid lipids, such a reorganization has been shown to involve formation of intermediate phase structures such as lipidic particles or inverted (reversed) micelle intermediates (IMI) by the two approaching bilayers (Fig. 4; Verkleij, 1984; Siegel, 1986; Ellens et al., 1989; Talmon et al., 1990). IMIs form by an exchange of lipids between the monolayers of the approaching bilayers. Fusion of the entire bilayers is completed by a second rearrangement of the lipids into a structure called an interlamellar attachment (ILA). The fusion described is likely to occur

# 7. Chloroplast Lipids and Membrane Assembly

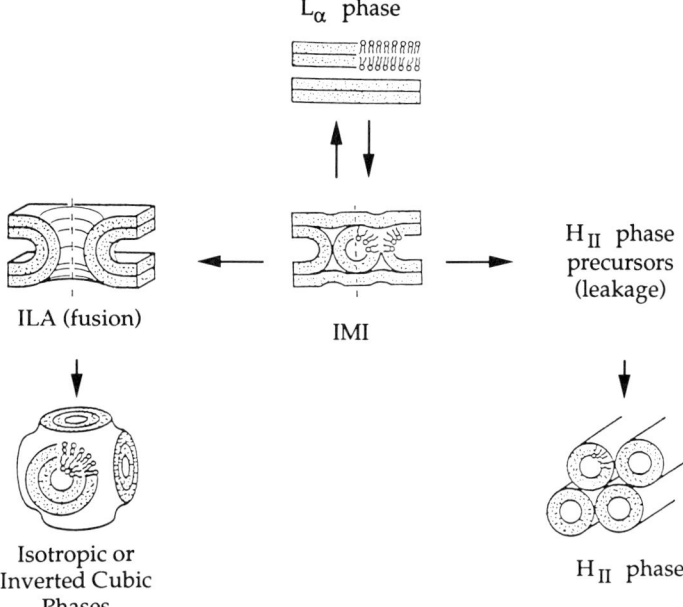

**Fig. 4.** Proposed scheme for fusion of bilayers. During fusion between two bilayers, the approaching bilayers first form inverted (reversed) micellar intermediates (IMI). This structure is short-lived. IMI will revert to the original bilayers, reorganize to an $H_{II}$ phase, or reorganize into interlamellar attachments (ILA). By the formation of ILA, a fusion of the original bilayers is achieved. The ILA structure is longer lived than IMI and is formed as an intermediate between $L_\alpha$ and inversed (reversed) cubic phase structures. Reprinted with permission from Ellens *et al.* (1989).

when the lipid mixture can form reversed cubic phase structures (Siegel, 1986). In model systems, chloroplast lipids form both lipidic particles (Sen *et al.*, 1981, 1982b; Sprague and Staehelin, 1984) and a reversed cubic phase structure (Brentel *et al.*, 1985). The formation of these structures is dependent on the presence of a lipid forming an $H_{II}$ phase with water (Sen *et al.*, 1981; Sprague and Staehelin, 1984; Verkleij, 1984; Brentel *et al.*, 1985; Siegel, 1986). The vesicle transfer of lipids in the chloroplast and etioplast thus could be facilitated by the presence of a large amount of MGDG in the envelope, thylakoid, and prothylakoid membranes.

A vesicle flow also could allow some proteins to be transported from the inner envelope membrane to the thylakoid. In chloroplasts, all chlorophyll *a*/*b*-binding proteins are synthesized in the cytoplasm and imported to the plastid. (For reviews, see Chapters 9 and 10). An import mechanism has been suggested for the LHCP that does not involve any vesicle transportation. During greening of dark-grown *Chlamydomonas*, where the proplastids lack inner membranes, thylakoids bud off from the envelope membrane (Hoober *et al.*, 1991). In this case, both LHC and functional PSII complexes are assembled in the envelope membrane.

### D. Possible Significance of Monogalactosyl Diacylglycerol in Thylakoids

In the thylakoid and the inner etioplast membrane, 50% of the membrane lipids are MGDG (Table I). This high content of lipid that forms only an $H_{II}$ phase with water is probably a major destabilizing factor for these membranes. In native mixtures of MGDG, DGDG, SQDG, and PG, the presence of MGDG results in the formation of $H_{II}$ and cubic phases (e.g., Section II,B). However, the thylakoid normally shows a stable bilayer without cubic or $H_{II}$ structures. An exception is the observation of an $H_{II}$ phase that is formed at high temperature (Gounaris *et al.*, 1984). Also, in the etioplast, a part of the inner membrane forms a cubic structure, the crystalline prolamellar body (PLB) membrane (e.g., Section V,A).

Most biological membranes contain some lipids that will form a nonlamellar ($H_{II}$ or $I_2$) phase with water. A comparison of the lipid compositon of the thylakoid with that of the thoroughly studied plasma membrane in the prokaryotic mycoplasma *Acholeplasma laidlawii* gives some information on the role of MGDG in the thylakoid. The lipid composition of the *A. laidlawii* membrane is dominated by two sugar lipids, monoglucosyl diacylglycerol and diglucosyl diacylglycerol. By inhibition of the endogenous synthesis of fatty acids, *A. laidlawii* can accept exogenously added fatty acids for growth. In experiments in which the fatty acids were varied, with different proportions of 16:0 and 18:1, an increase in the amount of 18:1 incorporated into the membrane lipids was found to be accompanied by a reduction in the ratio of monoglucosyl and diglucosyl diacylglycerol in the membrane (Lindblom *et al.*, 1986). This result can be explained by the molecular shapes of the membrane lipids. The longer and the more unsaturated the fatty acids are, the more cone-shaped the lipids. Hence, less of the especially cone-shaped monoglucosyl diacyglycerol is needed to maintain the membrane structure. The total lipid mixture at each growth condition always formed an $L_\alpha$ phase with water. By elevating the temperature 10–15°C above growth temperature, the lipid mixture started to form nonbilayer phases. This indicates that the lipid composition of the *A. laidlawii* membrane is regulated to be in $L_\alpha$ phase but is always near a phase transition, so local and transient nonbilayer structures can be formed easily in response to some triggering mechanisms (Lindblom *et al.*, 1986).

*Acholeplasma laidlawii* also can adjust the monoglucosyl to diglucosyl diacylglycerol ratio when grown on 18:2, but growth on 18:3 is impossible unless a stabilizing lipid such as cholesterol or chlorophyll is added (Wieslander and Selstam, 1987). The large amount of polyunsaturated MGDG found in the thylakoids require stabilization by some unknown factors to form the thylakoid bilayer. Sterols are not present in thylakoid membranes. Possible stabilizing factors are the large amounts of transmembrane protein complexes; PSI, PSII, LHC, Cyt $b_6$–$f$ and ATP synthase. The transmembrane proteins have a fixed location in the bilayer because of the hydrophobic interaction between the transmembrane spanning part of the protein and the hydrocarbon chains of the lipids. This interaction regulates membrane thickness and could be a geometrical packing constraint that hinders the formation of nonlamellar structures of varying thickness in the membrane monolayer.

From the structural analysis of the LHC and the reaction center of *Rhodopseudomonas sphaeroides* it has been calculated that the hydrophobic regions of both these membrane protein complexes are approximately 4.0 nm wide (Yeates *et al.*, 1987, Kühlbrandt and Wang, 1991). Although the extent of the hydrophobic region is difficult to measure exactly (Rees *et al.*, 1989), the calculated hydrophobic region is close to the thickness of 4.2 nm for the hydrophobic region of DGDG in an $L_\alpha$ phase (Fig. 1; Shipely *et al.*, 1973). Earlier suggestions that large proteins could be surrounded by inverted micelles of MGDG (Murphy, 1982) must be ruled out since inverted micelles inside the bilayer would increase the hydrophobic region in the bilayer by approximately 100%.

In perspective, considering the significance of MGDG, it was thought first that this $H_{II}$ lipid was needed for packing large protein complexes in the bilayer. The high concentration of MGDG in grana membranes was thought to surround the LHC complexes (Murphy, 1982; Quinn and Williams, 1985). A thermodynamic approach to membrane structure, however, implies that many transmembrane spanning proteins form a rigid backbone that keeps the lipids in a bilayer. Large amounts of MGDG then could be present before the lipids in the membrane start to form nonbilayer phases. Consequently, large amounts of MGDG are needed in the membrane to achieve the instability necessary for fusion and vesicle formation.

Presence of MGDG also has been shown to be important for the energy transfer from LHC to the reaction centers of PSI and PSII, and for the activity of ATPase and chlorophyllase (Sieferman-Harms *et al.*, 1982; Terpstra and Lambers, 1983; Pick *et al.*, 1984; Siefermann-Harms *et al.*, 1987). The regulation of membrane enzyme activity by MGDG, DGDG, and other structurally different lipids has been demonstrated for a rat liver enzyme (Jensen and Schutzbach, 1988, 1989).

## III. HETEROGENEITY OF LIPIDS IN INNER PLASTID MEMBRANES

Lateral or transverse heterogeneity of the lipds in the inner plastid membranes could give indirect evidence for specific roles of the different lipids. Thus such heterogeneity is of interest for understanding membrane function.

### A. Lateral Asymmetry

#### 1. Thylakoids

The inner membrane of the chloroplast, the thylakoid, is built from flattened membrane sacs (see Fig. 2, Chapter 2). The membrane has two different regions, the grana lamellae, or appressed region, with stacked membranes and the stroma lamellae, or nonappressed region, with nonstacked membranes. The relative amounts of grana and stroma lamellae are generally about equal but can vary with environmental conditions (Anderson, 1986). The structural differentiation of the thylakoid into two regions is caused by the asymmetrical distribution of the photosynthetic protein complexes in the membrane (Andersson and Anderson, 1980). In fractionation studies it has been shown that mainly $PSI_\beta$ and some $PSII_\beta$ with small light antennae are localized to the stroma larmellae, whereas some $PSI_\alpha$ and mainly $PSII_\alpha$ with large light antennae are localized to the grana lamellae (Andreasson et al., 1988; Albertsson et al., 1990). Also, the margins of the grana stacks have been suggested to form a distinct domain of the thylakoid, containing mainly PSI and ATP synthase (Svensson and Albertsson, 1989)

The pronounced lateral heterogeneity of the thylakoid proteins is not accompanied by any extensive asymmetry among the thylakoid lipids. Studies of the lipid composition in mechanically isolated fractions of appressed and nonappressed thylakoids have revealed only an enrichment of MGDG in the appressed region (Gounaris et al., 1983; Murphy and Woodrow, 1983a,b; Chapman et al., 1984). Also, in some studies, a large difference in the lipid to protein ratios of the two thylakoid regions has been found (Murphy and Woodrow, 1983a,b). An objection that can be directed against the results just mentioned is the method of fractionation used. By this method, only a minor part of the stroma lamellae is obtained. This fraction is extremely enriched in PSI and cannot be regarded as representative of the total nonappressed region (Andreasson et al., 1988). Also, several of the discrepancies observed could be due to lipid degradation in the membrane fractions. Severe degradation and modification of galactolipids has been observed during handling of spinach chloroplasts (Henry et al., 1983).

## 2. Etioplast Inner Membranes

Lateral asymmetries of the protein and the lipid components (Table I) have been demonstrated in the inner plastid membranes of the etioplast. These membranes show a pronounced structural heterogeneity because the regularly branched membrane of the PLB exists in continuum with the flat perforated sacs of the prothylakoids (PTs) (Gunning and Steer 1975; see Fig. 7, Chapter 2). The membrane is dominated by two proteins, NADPH-protochlorophyllide oxidoreductase (Pchlide-reductase; EC 1.3.1.33) and ATP synthase. These proteins are separated: Pchlide-reductase is localized to the PLBs and ATP synthase is localized to the PTs (Wellburn, 1977; Ikeuchi and Murakami, 1983; Ryberg and Dehesh, 1986). The ratio of MGDG to DGDG is 1.8 in the PLB and 1.4 in the flat PTs (Table I; Ryberg et al., 1983; Selstam and Sandelius, 1984; Protoschill-Krebs and Kesselmeier, 1988). The lower MGDG to DGDG ratio of the PT fraction could, however, be due to some contamination by the inner envelope membrane.

In electron micrographs, the inner envelope shows numerous connections with the PTs (Gunning, 1965; Sandelius and Selstam, 1984; Selstam and Sandelius, 1984). The implications of the structural and compositional asymmetry of the inner etioplast membrane are discussed in detail subsequently (e.g., Section V).

## B. Boundary Lipids

Since there is a spatial separation of the different chlorophyll–protein complexes, the presence of specific lipids, the so-called boundary lipids around the complexes, could induce a lateral asymmetry of the thylakoid lipids. The concept of "boundary lipids" was associated initially with the idea of a "long-lived" layer of immobilized lipids surrounding the membrane proteins. Several spectroscopic approaches have been applied to study this phenomenon; in no case have the initially suggested static lipid–protein complexes been found.

Early applications of $^2$H NMR clearly showed that the exchange of lipids between the protein boundary and the bulk of the membrane lipids must be fast, since very little perturbance of the lipid order or the dynamics dependent on the presence of membrane proteins could be seen in the NMR spectra (Gennis, 1989, pp.166–198). More refined instrumentation, combined with advanced line-shape simulations, has shown that changes in NMR spectra can be attributed to lipid protein interactions (Meier et al., 1987). These interactions are similar to the rapid exchange of lipids at the protein surface observed by electron spin resonance (ESR) measurements, but also to so-called "trapped" lipids of protein aggregates, exchanging on a slower time scale (Van Gorkom et al.,1990).

The first experimental evidence for the existence of boundary lipids appeared in 1973 (Jost et al., 1973). ESR measurements on isolated membrane systems into which an extrinsic spin-label was introduced revealed a two-component ESR spectrum. An additional spectral component was observed from lipid–protein mixtures compared with the spectra of pure lipid samples. This component was interpreted to result from a fraction of the lipids that was motionally restricted by the presence of membrane proteins. Later it was shown that the rate of lipid exchange at the protein–lipid interface was almost independent of the lipid/protein ratio, although the fraction of restricted lipids increased with decreasing lipid/protein ratios (Horváth et al., 1988). The calculated rate of lipid exchange between the boundary layer and the bulk phase is about $10^6$–$10^7$ sec.$^{-1}$, which is slow on the ESR time scale but too fast to be measured by $^2$H NMR. The rate of exchange is only about one order of magnitude slower than the free diffusion of lipids in the bulk phase, $10^{-8}$ cm$^2$ sec.$^{-1}$ (Knowles et al., 1979; Gennis, 1989). The 'boundary lipids' thus cannot be regarded as bound to the membrane proteins. Irrespective of whether the motionally restricted lipids are present as a boundary or annulus around the protein or as "trapped" lipids between the proteins of a membrane with a low lipid to protein ratio as described by Silver (1985), the exchange with the bulk lipids is substantial. Different opinions also have been forwarded; no consensus has yet been reached concerning how many layers of lipid around a membrane protein can be affected by the protein (Jost et al., 1973; Knowles et al., 1979). The typical two-component ESR spectra have been derived from measurements on a wide variety of lipid–protein mixtures or biological membranes. However, in some cases the "motionally restricted" component has not been observed (Fretten et al., 1980; Tanaka and Freed, 1985; Arvidson et al., 1989). The reason for this discrepancy is unclear and sheds some doubt on the interpretation of the two-component spectra.

ESR studies of isolated thylakoid membranes using spin-labeled MGDG, PG, and PC have shown typical two-component spectra indicating the presence of motionally restricted lipids (Li et al., 1989, 1990a,b). In most cases, PG showed a specificity in the interaction with the thylakoid membrane proteins (Li et al., 1989, 1990b). This result is in accordance with studies on other membrane systems, in which negatively charged lipids in some cases have shown a specific interaction with membrane proteins

(Marsh, 1987). The observed specificity is generally weak and depends on both lipid and protein structure. The state of hydration seems to be of major importance and the specificity cannot be explained only in terms of electrostatics (Lee, 1987; Marsh, 1987). The low specificity and the rapid exchange in the lipid–protein interactions of the thylakoid membrane, as measured by ESR, do not create a lateral heterogeneity in the thylakoid. This result is also consistent with earlier results obtained with radiolabeled galactolipids for which no specific binding could be observed (Heinz and Siefermann-Harms, 1981). From the results of Li et al. (1989), one can estimate the overall motionally restricted fraction of pea thylakoid lipids to be approximately 40%, assuming that the motion of DGDG, the second major galactolipid, is restricted to the same extent as that of MGDG, and that the motional restriction of the charged SQDG is similar to that of PG. It seems likely that the multipolypeptide chlorophyll–protein complexes contain some trapped lipids, exchanging on a time scale that cannot be distinguished using ESR. Also, these related studies do not exclude the specific interaction of single lipid molecules within the chlorophyll–protein complexes. These issues will be emphasized further in Section IV.

Doubts concerning the current evaluation of the ESR spectra, predicting a two-site model for simulation, have arisen. Meirovich et al. (1984) have shown that an equally good description of experimental ESR spectra can be achieved using a one-site model. In conclusion, current knowledge of whether boundary lipids exist and how they might contribute to the function of the membrane is obscure.

## C. Transverse Asymmetry

Transverse asymmetry of membrane proteins is an obvious consequence of the way proteins are inserted into membranes. A transverse asymmetry of membrane lipids is not as obvious but still would seem both functionally and structurally reasonable. All methods available for determining the lipid composition of one of the bilayer leaflets have methodological problems; hence, the results at hand are often inconsistent. One system in which transverse asymmetry of the lipids has been demonstrated unequivocally by different methods is the mammalian erythrocyte membrane (Gennis, 1989).

Various approaches, including immunology, radiolabeling, and different lipolytic treatments, have been taken to determine whether the inner plastid membrane lipids show any transverse asymmetry. The results from these studies are rather conflicting. Several different transverse asymmetries of the thylakoid lipids have been suggested. Siegenthaler and co-workers have showed an accumulation (60–70%) of the total MGDG in the outer leaflet of the thylakoid in all their lipolytic and radiolabeling studies of different species and different developmental stages (Rawyler and Siegenthaler, 1985; Rawyler et al., 1987; Giroud and Siegenthaler, 1988). This result is in accordance with the radiolabeling results of Sundby and Larsson (1985), whereas Unitt and Harwood (1982) observed the opposite distribution of MGDG (20% outside, 80% inside) after treatment with lipases. Concerning DGDG, Siegenthaler and co-workers found 85–90% of the DGDG in the inner leaflet and Sundby and Larsson (1985) found the distribution of DGDG to be similar to that of MGDG (60% outside, 40% inside). Unitt and Harwood (1982) found 44% of the total DGDG outside and 56% inside. However, using thylakoids and inside-out thylakoid vesicles for lipolytic studies, Siegenthaler et al. (1988) have established the outside/inside distribution of MGDG and DGDG further, to 60%/40% and 20%/80%, respectively. Phospholipase treatments have resulted in determinations of 70–86% of the total PG in the outer leaflet in all studies (Unitt and Harwood, 1985; Siegenthaler and Giroud, 1986). For SQDG, only indirect estimates from lipolytic treatments are at hand, pointing to an asymmetric distribution with 70–90% of the SQDG located in the inner leaflet (Rawyler et al., 1987). Substantial evidence exists for transverse asymmetry of the thylakoid lipids. The accumulation of MGDG and PG and the depletion of DGDG in the outer leaflet seems rather clear, although the function of this asymmetry is still unknown.

How the transverse asymmetry is maintained remains unclear. An active enzymatic process for translocation of aminophospho lipids has been found in some mammalian plasma membranes, but no protein has been isolated. Thus, the process still remains speculative in animal as well as plant systems (Tilley et al., 1986; Zachowski et al., 1986, 1987).

## IV. SPECIFIC INTERACTIONS BETWEEN THYLAKOID LIPIDS AND CHLOROPHYLL PROTEIN COMPLEXES IN PHOTOSYSTEM II

In this section, the acylations of D1 and LHCP as well as the lipids associated with the PSII reaction center are described. Also, different ideas about the role of lipids in the LHCII complex are discussed. Various preparations of chlorophyll–protein complexes have been used to study specific interactions between the complexes and the thylakoid lipids. Often, inconsistent results have been achieved (Krupa 1988), indicating that many preparations contain residual membrane lipids and not only lipids specifically associated with the chlorophyll–protein complexes. Also, the presence of detergents in the isolated chlorophyll–protein complexes makes the identification and quantification of the specific lipids difficult (Murata et al., 1990).

## A. Acylation of the D1 Protein

Post-translational acylation of the D1 protein in the PSII reaction center has been investigated by Mattoo and Edelman (1987). The D1 protein is synthesized and processed in the stroma membrane. After translocation to the grana membrane, the D1 protein is palmitoylated. Acylation is probably specific for palmitic acid; the acid is bound to the protein by ether or amide linkage in the membrane-anchoring part of the protein. The palmitoylation is light regulated and much shorter lived than the protein. The function of the palmitoylation of the D1 protein is not known. Similar palmitoylation characteristics are found in proteins in viruses, bacteria, and animal cells (Schultz et al., 1988; Grand, 1989; Schmidt, 1989; McIlhinney, 1990). Palmitoylation has been found in intrinsic proteins, where acylation occurs near a transmembrane hydrophobic $\alpha$-helix. Acylation enhances the hydrophobicity of the protein and influences the secondary structure. The palmitoylation gives the D1 protein a certain configuration that might be needed for the interaction between the protein subunits in the reaction center. Since the acylation is temporary, it could have a regulatory function in the turnover of the D1 protein.

## B. Lipids in the Photosystem II Reaction Center

In a study of the lipids associated with PSII, the lipid content and composition in different PSII particles were analyzed (Murata et al., 1990). The lipid composition in spinach thylakoids, PSII membranes, PSII core complexes, and PSII reaction centers was compared. The core complex performed both oxygen evolution and charge separation, whereas the reaction center performed only charge separation. One molecule of MGDG per P680 was found in the PSII reaction center. In the PSII core complex, 10 lipid molecules per P680 were found. These were composed of equal amounts of MGDG, DGDG, and PG. Neither PC nor SQDG was found in the PSII core nor PSII reaction center complexes. In an earlier investigation, both MGDG and SQDG were found in the PSII reaction center (Gounaris and Barber, 1985). The discrepancy is probably caused by the detergents present in the lipid extract that must be eliminated before chromatographic separation of the lipids can be performed successfully (Murata et al., 1990). The MGDG found in the PSII reaction center was more saturated than the MGDG found in the bulk of thylakoid lipids (Gounaris and Barber, 1985; Murata et al., 1990). The relative amount of trienoic acyl groups had decreased from 94 to 29%, and the amount of saturated acyl groups had increased from 2 to 50% (Murata et al., 1990).

The preferentially more saturated acyl groups in MGDG of the PSII reaction center indicate that special properties are needed for a proper structure of the lipid–protein complex.

## C. Lipids in the Light-Harvesting Complex II

Palmitic acid can bind to LHCII covalently (Mattoo and Edelman, 1987). This palmitoylation is probably temporary since stoichiometric amounts of palmitic acid are not bound to the complex (Jansson et al., 1990). The function of this palmitoylation is not known. LHCII isolated from spinach is composed of two polypeptides of 25 and 27 kDa. The difference in molecular mass of these polypeptides is not because of palmitoylation but because they are different gene products (Jansson et al., 1990).

Chlorophylls and xanthophylls are associated with the hydrophilic region of the LHC. Chl $a$, Chl $b$, and xanthophylls are needed for the assembly of LHCII monomers (Plumley and Schmidt, 1987; Paulsson et al., 1990; see also Chapter 10). In vivo, LHCII forms an oligomer complex, probably a trimer (Butler and Kühlbrandt, 1988; Thornber et al., 1991). Trimers are formed also when isolated LHCII is crystallized (Kühlbrandt and Wang, 1991). Whether thylakoid lipids are needed for the formation of oligomers or trimers in vivo is unknown, but crystallization of detergent-solubilized LHCII is dependent on the presence of thylakoid lipids (W. Kühlbrandt, 1991, personal communication). In the crystallized LHCII, thylakoid lipids and detergents are probably seen both in the periphery and in the center of the LHCII trimers (Kühlbrandt and Wang, 1991).

LHCII isolated by repeated solubilization and precipitation by KCl and $MgCl_2$ contains 0.58 mol acyl lipids per mol Chl (Krupa, 1988). If 15 chlorophyll molecules are associated to one polypeptide of LHCII (Thornber et al., 1991), the number of lipid molecules per LHCII polypeptide would be 8 or 9. Approximately equal amounts of MGDG, DGDG, and PG and some SQDG are present in the complex. However, the nonthylakoid lipids PC and phosphatidyl ethanolamine (PE; 9–13% of total lipids) were present in these isolated LHCIIs (Krupa et al., 1987), indicating that residual membrane lipids were still present in the isolated LHCIIs.

LHCII but not PSI nor PSII is stable at pH 1.6 (Siefermann-Harms and Ninnemann, 1983). However, addition of a liposome preparation of SQDG (2–30 µg/ml) to an LHCII preparation of 2.5 µg Chl $a$ and $b$ per ml resulted in rapid pheophytin formation (D. Siefermann-Harms and H. Y. Yamamoto, 1982, personal communication). In control experiments in which liposomes of egg lecithin (PC 0–200 µg/ml) were added there was no destruction of chlorophylls, indicating that the chlorophyll molecules are sheltered in a hydrophobic environment by detergents and residual thylakoid lipids in the isolated LHCII and that addition of the acidic SQDG destroys the chlorophyll. The sulfonate group of SQDG gives the lipid the characteristics of a strong acidic detergent. A specific localization of SQDG next to the exposed Chl *a* and *b* therefore could not be expected. Liposomes of the other negatively charged lipid PG do not destroy the chlorophyll in isolated LHCIIs (Rémy *et al.*, 1984).

The enrichment of PG in the detergent-isolated LHCII (Rémy *et al.*, 1982; Krupa *et al.*, 1987) has focused a lot of interest on whether an oligomer formation of LHCII is dependent on PG esterified to 16:1t. Positive correlations between the amount of PG (16:1t) and the resolution of an oligomer of LHCII by PAGE have been described for light- and dark-grown seedlings of spruce, nonhardened and frosthardened rye leaves, cadmium-exposed radish cotyledons, and a 16:1t-less mutant of *Chlamydomonas* (Lemoine *et al.*, 1982; Huner *et al.*, 1987; Krupa, 1987; Garnier *et al.*, 1990). However, the LHCII oligomer also can be found by PAGE when thylakoids from a 16:1t-less mutant of *Arabidopsis* are used, if the NaCl concentration is kept low (McCourt *et al.*, 1985). Thus, the LHCII oligomer formed during PAGE is not specifically dependent on PG 16:1t. Rather, the stability of the detergent-isolated complex in the presence of ions is dependent on PG (16:1t; McCourt *et al.*, 1985). LHCII oligomers also have been shown to reconstitute when LHCII monomers are mixed with liposomes formed by different membrane phospholipids or DGDG (Rémy *et al.*, 1984).

## V. ASSEMBLY OF PROLAMELLAR BODY MEMBRANES

In etioplasts or reetiolated plastids, the presence of large amounts of Pchlide-reductase results in the formation of a PLB membrane. The PLB membrane differs from the thylakoid by its regularly branched structure. Since Pchlide-reductase totally dominates the protein content of this membrane, specific protein–protein and lipid–protein interactions are probably important for the assembly of the PLB membrane.

### A. Nature of the Regularly Branched Prolamellar Body Membrane

The PLB is thought to be a membrane in a cubic phase structure. Evidence for this suggestion can be summarized in three points.

## 1. Structural Evidence

The structure of the PLB has been studied extensively by electron microscopy (Gunning and Steer, 1975; Murakami et al., 1985; Fig. 7, Chapter 2). The PLB is built from a bilayer that is continuous throughout the branching structure (Gunning, 1965). The membrane separates two "water" channels—the stroma of the etioplast and the PLB lumen. Thus, the crystalline structure is formed by two different branched and interwoven water channels separated by a bilayer. This organization is the same as that of the bicontinuous cubic phases of membrane lipids (e.g., Section II,B; Lindblom and Rilfors, 1989). The most common structural unit in the PLB is the tetrahedral or 4-armed unit. Most often six tetrahedral units form a hexagon, and the hexagons are held together in a structure with the same symmetry as the carbon atoms in the diamond crystal (Murakami et al., 1985). Several other structural symmetries are formed by the 4-armed unit also (Gunning and Steer, 1975).

In some plant species, the PLB is formed by a cubic or 6-armed unit (Gunning, 1965). In avocado, this structural symmetry is found in the etioplasts of the edible mesocarp (Fig. 5). In the freeze fracture image of the PLB from avocado, the curving and branching of the 6-armed PLB membrane is clearly visualized. The similarity between the 6-armed PLB membrane and the cubic Im3m (6-armed) structural symmetry is obvious (Fig. 3c; Larsson et al., 1980; Lindstedt and Liljenberg, 1990). The 4-armed PLB membrane of the diamond type (Simpson, 1978) has a Pn3m structural symmetry (Lindstedt and Liljenberg, 1990). The other various types of 4-armed PLB membrane structure have not been identified in model systems of lipids. Various 4-armed and 6-armed PLB structures can be described by the mathematics of infinite periodic minimal surfaces (Andersson et al., 1988; Lindstedt and Liljenberg, 1990). Collectively, the PLB as visualized by electron microscopy has the same structural organization as the cubic phase structure formed by membrane lipids. The reason that a large variety of PLB structures is formed is unknown, but although the different types of PLB structures look very different, the energy needed to change from one cubic symmetry to another is very small (Hyde et al., 1984).

To the structural evidence for the cubic phase nature of the PLB can be added the processes for PLB formation and disappearance. In electron micrographs, we have seen that the PLB is formed from perforated PTs (Lütz, 1981). Perforated PTs also are formed when the PLB disappears on irradiation (Henningsen and Boynton, 1970; Gunning and Steer, 1975; Simpson, 1978). The perforations can be described as holes through two bilayers that form one flat PT sac. The PT membrane is, thus, continuous

# 7. Chloroplast Lipids and Membrane Assembly

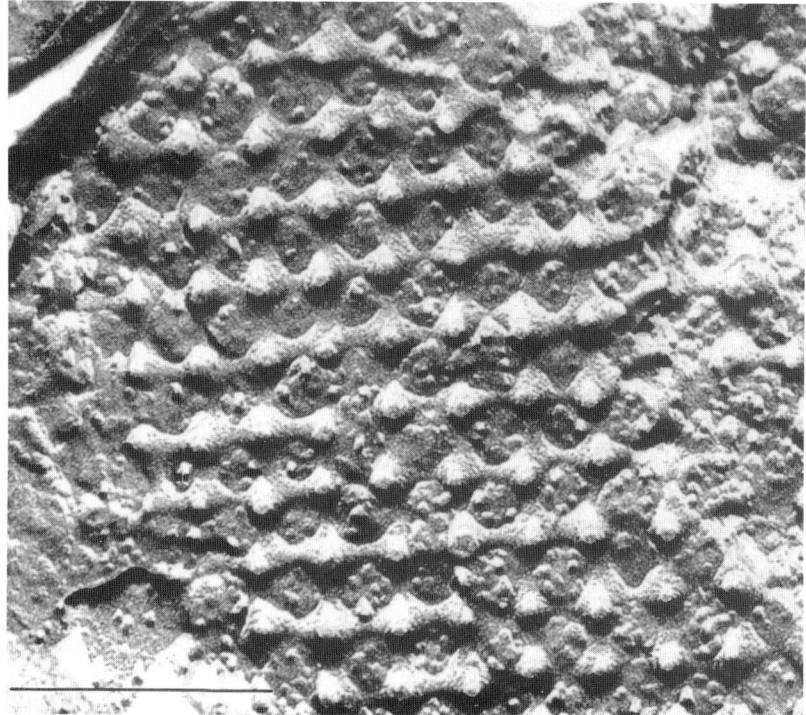

**Fig. 5.** Freeze fracture electron micrograph of a PLB from the mesocarp of avocado (Persea Americana Mill). The crystalline membrane is formed by 6-armed units. The unit length is 50 nm. The surface of the PLB membrane is smooth. No particles similar to freeze fracture particles in thylakoid membranes can be seen. Bar: 0.2 μm. Reprinted courtesy of W. W. Thomson and K. A. Platt-Aloia.

through the holes. These perforations are similar to ILAs (Fig. 4), the intermediate structures formed between the $L_\alpha$ and cubic phases. The PLB is developed from perforated PTs in a manner that is strikingly similar to the proposed model for how approaching bilayers first fuse and then form a cubic phase structure (e.g., Section I,C; Ellens *et al.*, 1989).

## 2. Properties of Plastid Lipids

The lipid composition of the PLB membrane is similar to that of the thylakoid (Table I); the MGDG/DGDG ratio is higher in the PLB than in the PT (e.g., Section III,A). In model systems, the PLB lipids form different nonlamellar phases: $H_{II}$, cubic, and lipidic particles (e.g., Section II,B). The cubic phase so far identified in mixtures of $^2H_2O$, MGDG, and DGDG has an Ia3d symmetry (Brentel *et al.*, 1985). This 3-armed symmetry has not been identified among PLB membrane structures. In these galactolipid mixtures, the Ia3d symmetry is formed at low water content (5–13%). In the "native" mixture of chloroplast lipids, cubic phases are formed both at low and high water contents (Selstam *et al.*, 1990). At higher water contents, the 3-armed cubic symmetry changes into a 4- or 6-armed symmetry (e.g., Section II,B). To date, the 4-armed and 6-armed unit structures of the PLB membranes have not been identified in the model systems of plastid lipids. However, these systems do not include all membrane components. The phase structures of lipids are known to be dependent on pH, cations, temperature, osmotically active substances, and proteins (Ericsson *et al.*, 1983; Quinn and Williams, 1983; Rilfors *et al.*, 1984).

The PLB structure has been shown to be dependent on cations. The PLB membrane has a p*I* of 4.5 and, thus, is negatively charged at physiological pH (Widell Wigge and Selstam, 1990). In the presence of ethylenediaminetetraacetic acid (EDTA), the PLB transforms into a structure similar to perforated thylakoids (Protoschill-Krebs and Kesselmeier, 1988), whereas low amounts (5 m*M*) of $Mg^{2+}$ stabilize the regular structure (Ikeuchi and Murakami, 1983). Nonphysiological amounts of $Ca^{2+}$ or $Mg^{2+}$ ions also distort the regular PLB structure (Grevby *et al.*, 1989; Lachmann and Kesselmeier, 1989; Widell Wigge and Selstam, 1990). One possible explanation is that addition of salt to the membrane leads to screening of the surface charges, according to the Guy–Chapman theory. Screening changes the forces at the membrane surface and affects membrane stability. Neutralization of the negatively charged lipids reduces the surface area of the polar head groups at the hydrocarbon–water interface, which allows the membrane components to pack closer together at the surface while the hydrophobic volume of the membrane remains constant (Rilfors *et al.*, 1984). This rearrangement results in changes in the packing of the membrane components which, in some cases, can trigger phase transitions in pure lipid bilayers. High concentrations of ions also disturb the Pchlide-reductase 650–657 complex (Widell Wigge and Selstam, 1990).

### *3. Prolamellar Body Membrane Function*

The PLB membrane is osmotically inactive and, unlike most other membranes, has no barrier function. The PLB membrane can be ascribed the function of storage site for pigments, lipids, and amino acids (in the form of Pchlide-reductase) that will be used to form thylakoids during greening.

7. Chloroplast Lipids and Membrane Assembly 265

The cubic PLB membrane structure stores large amounts of Pchl(ide), Pchlide-reductase, and membrane lipids in a small volume.

## B. Protein–Protein Interaction of Protochlorophyllide Reductase

Pchlide-reductase is a polypeptide with a molecular mass of 36 kDa (Apel *et al.*, 1980; Oliver and Griffiths, 1980). In dark-grown plants, Pchlide-reductase forms spectroscopically different complexes with Pchl(ide) and NADPH (Oliver and Griffiths, 1982). A high concentration of Pchlide-reductase is found in the PLB membrane, where it forms a complex with characteristic absorption and fluorescence maxima at 650 and 657 nm, respectively (Ryberg and Sundqvist, 1982; Ikeuchi and Murakami, 1983; Selstam *et al.*, 1987). In this Pchlide-reductase complex, several Pchl(ide) molecules interact (Böddi *et al.*, 1989). Since only 2–3 Pchl(ide) molecules are associated with one polypeptide of Pchlide-reductase (Apel *et al.*, 1980), it is possible that Pchlide-reductase 650–657 is an oligomer (Böddi *et al.*, 1989). Cross-linking studies of the Pchlide-reductase also indicate that the pigment–protein complex 650–657 is an oligomer (Sundqvist *et al.*, 1992). The presence of the oligomeric Pchlide-reductase–pigment complex, either in interaction with Pchl(ide) or Chlide, is strongly correlated to the regularly branched PLB membrane (Ryberg and Sundqvist, 1988; Ryberg and Sundqvist, 1991; Sundqvist *et al.*, 1992).

The oligomer of Pchlide-reductase is more fragile than other complexes of membrane proteins. After solubilization with Triton X-100, Pchlide-reductase is isolated in monomer form (Apel *et al.*, 1980; Oliver and Griffiths, 1980) whereas the complexes of, for instance, ATP synthase, LHCII, and PSII reaction centers can be solubilized with Triton X-100 without dissociating into their subunits (Anderson *et al.*, 1978; Murata *et al.*, 1990). The cross-linkers, 1-ethyl-3-[3-(dimethylamino)propyl] carbodiimide and 2-iminothiolane, also link Pchlide-reductase and ATP synthase together (Sundqvist *et al.*, 1992). These observations raise the question of whether the Pchlide-reductase–pigment complexes form a true oligomer or whether they are only located close to each other in the PLB membrane.

## C. Lipid–Protein Interaction in Prolamellar Body Membranes

Isolated PLB membranes contain proportionally more lipids per mg protein than isolated PTs and envelope membranes (Ryberg *et al.*, 1983; Selstam and Sandelius, 1984). The significance of the lipid to protein ratio is, however, difficult to interpret because of the clearly different types of proteins present in these membranes (Selstam and Sandelius, 1984).

The nature of the association of Pchlide-reductase to the membrane is not clear. From the amino acid sequence (Schulz et al., 1989), the hydrophobicity plot for Pchlide-reductase has been calculated and shows that Pchlide-reductase is hydrophilic and has no typical transmembrane spanning parts (Benli et al., 1991; see also Chapter 5). Nonetheless, Pchlide-reductase has amphiphilic characters as judged by TX-114 phase partitioning (Selstam and Widell Wigge, 1989) and probably is not bound electrostatically to the membrane (Grevby et al., 1989; Widell Wigge and Selstam, 1990). High concentrations of the Pchlide-reductase–pigment complex are found in the PLB membrane; since this membrane has the characteristics of a cubic phase structure, it is likely that the nature of the interactions between Pchlide-reductase and the lipids facilitate the formation of the cubic membrane structure. If Pchlide-reductase is not anchored in the PLB membrane with a transmembrane spanning $\alpha$-helix, and no other transmembrane proteins are present, there will be a limited geometrical hindrance for the plastid lipids to curve and form the cubic phase structure (e.g., Section II,B). The lipid composition in the PLB is suitable for packing of the large transmembrane protein complexes in the thylakoid membrane (e.g., Section II,D), but with a small extrinsic protein that lacks a transmembrane spanning $\alpha$-helix it will form a cubic phase. The large production of Pchlide-reductase–pigment complexes results in a phase separation of the membrane into one lamellar and one cubic part. This separation can be compared with the overproduction shown for a transmembrane protein in the ER membrane, which results in the formation of regularly arranged membrane cylinders but no branched cubic structures (Chin et al., 1982).

The crystalline PLB membrane is in direct continuation with the planar PT membrane. The PT membrane is packed densely with ATP synthase. The lateral heterogeneity, with Pchlide-reductase complex in the PLB membrane and ATP synthase in the PT membrane, could be caused by the inability of the large transmembrane ATP synthase complex to participate in the cubic structure, both because of its transmembrane part and because of the large CF1 unit. The interactions between the Pchlide-reductase complexes and between Pchlide-reductase and the lipids that form the cubic PLB structure are thermodynamically more favorable than the free diffusion of ATP synthase into the cubic PLB structure. PLB membranes containing the Pchlide-reductase complex also are formed in equilibrium with green thylakoid membranes, as in dark-grown cotyledons of Scots pine (Selstam and Widell, 1986) and Jeffrey pine (Selstam et al., 1987), and in reetiolated etiochloroplasts (Minkov et al., 1988). Minor amounts of Pchlide-reductase are present in PTs (Shaw et al., 1985; Böddi *et al.*, 1989) and in the outer envelope membrane (Joyard *et al.*, 1990) without the formation of a cubic phase structure.

From the current state of knowledge about how the PLB is formed, the cubic structure is thermodynamically the most favorable organization of the extrinsic Pchlide-reductase complex and the plastid lipids. Whether the formation of a Pchlide-reductase–pigment oligomer complex or the lack of transmembrane-spanning stabilization triggers the phase separation of the lipids is a matter of debate.

## D. Phototransformation of Prolamellar Body Membranes

Irradiation of dark-grown leaves results in the transformation of the PLB into perforated thylakoids (Henningsen and Boynton, 1970; Gunning and Steer, 1975; Simpson, 1978). *In vitro*, the transformation is correlated with a dissociation of the Pchlide-reductase oligomer complex (Böddi *et al.*, 1990; Sundqvist *et al.*, 1992). However, the mechanism for the phototransformation of the PLB membrane is not clear. Several factors could disturb the stability of the cubic phase: (1) degradation of some Pchlide-reductase (Santel and Apel, 1981; Kay and Griffiths, 1983; Häuser *et al.*, 1984, 1987), (2) degradation of MGDG (Selldén and Selstam, 1976), or (3) release of Chl *a* in the membrane (Tanaka and Tsuji, 1983; Wieslander and Selstam, 1987). Once the forces that keep the membrane in the cubic phase have disappeared, the crystallinity disappears and a flat membrane is formed. At the same time, the restricted diffusion of the ATP synthase and Pchlide-reductase complexes ceases, so both complexes will diffuse and mix in the entire etiochloroplast inner membrane (Wellburn, 1977; Ryberg and Dehesh, 1986). Viewed by electron microscopy, the structural change during phototransformation is very drastic. However, the membrane lipids are organized in a bilayer both in the PLB and in the perforated PT membrane. Thus, the change is less drastic than it appears.

## ACKNOWLEDGMENTS

The authors are grateful to Göran Lindblom, Åke Wieslander, and Göran Wikander for stimulating discussions. We gratefully acknowledge Werner Kühlbrandt, Dorothea Siefermann-Harms, Harry Y. Yamamoto, Kathryn A. Platt Aloia, and William W. Thomson for sending us unpublished results and David Siegel, John Sedon, and Leif Rilfors for the permission to use their original illustrations. We also thank Monica Nilsson and Karin Degerman for typing and correcting the manuscript. This work was supported by the Swedish Natural Science Research Council.

## REFERENCES

Albertsson, P.-Å., Andreasson, E., and Svensson, P. (1990). The domain organization of the plant thylakoid membrane. *FEBS Lett.* **273**, 36–40.
Anderson, D. M., Gruner, S. M., and Leibler, S. (1988). Geometrical aspects of the frustration in the cubic phases of lyotrophic liquid crystals. *Proc. Natl. Acad. Sci. USA* **85**, 5364–5368.
Anderson, J. M. (1986). Photoregulation of the composition, function, and structure of thylakoid membranes. *Annu. Rev. Plant Physiol.* **37**, 93–136.
Anderson, J. M., Waldron, J. C., and Thorne, S. W. (1978). Chlorophyll–protein complexes of spinach and barley thylakoids. Spectral characterization of six complexes resolved by an improved electrophoretic procedure. *FEBS Lett.* **92**, 227–233.
Andersson, B., and Anderson, J. M. (1980). Lateral heterogeneity in the distribution of chlorophyll–protein complexes of the thylakoid membranes of spinach chloroplasts. *Biochim. Biophys. Acta* **593**, 427–440.
Andersson, S., Hyde, S. T., Larsson, K., and Lidin, S. (1988). Minimal surfaces and structures: From inorganic and metal crystals to cell membranes and biopolymers. *Chem Rev.* **88**, 221–242.
Andreasson, E., Svensson, P., Weibull, C., and Albertsson, P.-Å. (1988). Separation and characterization of stroma and grana membranes—Evidence for heterogeneity in antenna size of both photosystem I and photosystem II. *Biochim. Biophys. Acta* **936**, 339–350.
Apel, K., Santel, H.-J., Redlinger, T. E., and Falk, H. (1980). The protochlorophyllide holochrome of barley (*Hordeum vulgare* L.): Isolation and characterization of the NADPH:protochlorophyllide oxidoreductase. *Eur. J. Biochem.* **111**, 251–258.
Arvidson, G., Ronquist, G., Wikander, G., and Öjteg, A.-C. (1989). Human prostasome membranes exhibit very high cholesterol/phospholipid ratios yielding high molecular ordering. *Biochim. Biophys. Acta* **984**, 167–173.
Benli, M., Schulz, R., and Apel, K. (1991). Effect of light on the NADPH-protochlorophyllide oxidoreductase of *Arabidopsis thaliana*. *Plant Mol. Biol.* **16**, 615–625.
Bernhard, W. R., Thoma, S., Botella, J., and Somerville, C. R. (1991). Isolation of a cDNA clone for Spinach lipid transfer protein and evidence that the protein is synthesized by the secretory pathway. *Plant Physiol.* **95**, 164–170.
Bishop, W. R., and Bell, R. M. (1988). Assembly of phospholipids into cellular membranes: Biosynthesis, transmembrane movement, and intracellular translocation. *Annu. Rev. Cell Biol.* **4**, 579–610.
Böddi, B., Lindsten, A., Ryberg, M., and Sundqvist, C. (1989). On the aggregational states of protochlorophyllide and its protein complexes in wheat etioplasts. *Physiol. Plant.* **76**, 135–143.
Böddi, B., Lindsten, A., Ryberg, M., and Sundqvist, C. (1990). Phototransformation of aggregated forms of protochlorophyllide in isolated etioplast inner membranes. *Photochem. Photobiol.* **52**, 83–87.
Brentel, I., Selstam, E., and Lindblom, G. (1985). Phase equilibria of mixtures of plant galactolipids. The formation of a bicontinuous cubic phase. *Biochim. Biophys. Acta* **812**, 816–826.
Browse, J., and Somerville, C. (1991). Glycerolipid synthesis: Biochemistry and regulation. *Annu. Rev. Plant Physiol. Plant Mol. Biol.* **42**, 467–506.
Butler, P. J. G., and Kühlbrandt, W. (1988). Determination of the aggregate size in detergent solution of the light-harvesting chlorophyll *a/b*–protein complex from chloroplast membranes. *Proc. Natl. Acad. Sci. USA* **85**, 3797–3801.

Carde, J. P., Joyard, J., and Douce, R. (1982). Electron microscopic studies of envelope membranes from spinach plastids. *Biol. Cell* **44**, 315-324.

Chapman, D. J., De-Felice, J., and Barber, J. (1984). Lipids at sites of quinone and herbicide interaction with the photosystem two pigment-protein complex of chloroplast thylakoids. *In* "Structure, Function, and Metabolism of Plant Lipids" (P.-A. Siegenthaler and W. Eichenberger, eds.), pp. 457-464. Elsevier, Amsterdam.

Chin, D. J., Luskey, K. L., Anderson, R. G. W., Faust, J. R., Goldstein, J. L., and Brown, M. S. (1982). Appearance of crystalloid endoplasmic reticulum in compactin-resistant Chinese hamster cells with a 500-fold increase in 3-hydroxy-3-methylglutaryl-coenzyme A reductase. *Proc. Natl. Acad. Sci. USA* **79**, 1185-1189.

Dorne, A.-J., Joyard, J., and Douce, R. (1990). Do thylakoids really contain phosphatidylcholine? *Proc. Natl. Acad. Sci. USA* **87**, 71-74.

Douce, R., and Joyard, J. (1979). Structure and function of the plastid envelope. *In* "Advances in Botanical Research" (H. W. Woolhouse, ed.), Vol. 7, pp. 1-116. Academic Press, London.

Douce, R., and Joyard, J. (1980). Plant galactolipids. *In* "The Biochemistry of Plants" (P. K. Stumpf, ed.), Vol. 4, pp. 321-362. Academic Press, New York.

Douce, R., and Joyard, J. (1990). Biochemistry and function of the plastid envelope. *Annu. Rev. Cell Biol.* **6**, 173-216.

Ellens, H., Siegel, D. P., Alford, D., Yeagle, P. L., Boni, L., Lis, L. J., Quinn, P. J., and Bentz, J. (1989). Membrane fusion and inverted phases. *Biochemistry* **28**, 3692-3703.

Ericsson, B., Larsson, K., and Fontell, K. (1983). A cubic protein-monoolein-water phase. *Biochim. Biophys. Acta* **729**, 23-27.

Fretten, P., Morris, S. J., Watts, A., and Marsh, D. (1980). Lipid-lipid and lipid-protein interactions in chromaffin granule membranes: A spin label ESR study. *Biochim. Biophys. Acta* **598**, 247-259.

Garnier, J., Wu, B., Maroc, J., Guyon, D., and Trémoliéres, A. (1990). Restoration of both an oligomeric form of the light-harvesting antenna CP II and a fluorescence state II-state I transition by $\Delta^3$-*trans*-hexadecenoic acid-containing phosphatidylglycerol, in cells of a mutant of *Chlamydomonas reinhardtii*. *Biochim. Biophys. Acta* **1020**, 153-162.

Gennis, R. B. (1989). "Biomembranes: Molecular Structure and Function." Springer-Verlag, New York.

Giroud, C., and Siegenthaler, P.-A. (1988). Development of oat prothylakoids into thylakoids during greening does not change transmembrane galactolipid asymmetry but preserves the thylakoid bilayer. *Plant Physiol.* **88**, 412-417.

Gombos, Z., and Murata, N. (1991). Lipids and fatty acids of *Prochlorothrix hollandica*. *Plant Cell Physiol.* **32**, 73-77.

Gounaris, K., and Barber, J. (1985). Isolation and characterisation of a photosystem II reaction centre lipoprotein complex. *FEBS Lett.* **188**, 68-72.

Gounaris, K., Sundby, C., Andersson, B., and Barber, J. (1983). Lateral heterogeneity of polar lipids in the thylakoid membranes of spinach chloroplasts. *FEBS Lett.* **156**, 170-174.

Gounaris, K., Brain, A. R. R., Quinn, P. J., and Williams, W. P. (1984). Structural reorganisation of chloroplast thylakoid membranes in response to heat stress. *Biochim. Biophys. Acta* **766**, 198-208.

Gounaris, K., Barber, J., and Harwood, J. L. (1986). The thylakoid membranes of higher plant chloroplasts. *Biochem. J.* **237**, 313-326.

Grand, R. J. A. (1989). Acylation of viral and eukaryotic proteins. *Biochem. J.* **258**, 625-638.

Grevby, C., Engdahl, S., Ryberg, M., and Sundqvist, C. (1989). Binding properties of NADPH-protochlorophyllide oxidoreductase as revealed by detergent and ion treatments of isolated and immobilized prolamellar bodies. *Physiol. Plant.* **77**, 493-503.

Gruner, S. M. (1985). Intrinsic curvature hypothesis for biomembrane lipid composition: A role for nonbilayer lipids. *Proc. Natl. Acad. Sci. USA* **82**, 3665–3669.
Gruner, S. M., Cullis, P. R., Hope, M. J., and Tilcock, C. P. S. (1985). Lipid polymorphism: The molecular basis of nonbilayer phases. *Annu. Rev. Biophys. Biophys. Chem.* **14**, 211–238.
Gunning, B. E. S. (1965). The greening process in plastids. 1. The structure of the prolamellar body. *Protoplasma* **60**, 111–130.
Gunning, B. E. S., and Steer, M. W. (1975). "Plant Cell Biology: An Ultrastructural Approach." Edward Arnold, London.
Häuser, I., Dehesh, K., and Apel, K. (1984). The proteolytic degradation *in vitro* of the NADPH–protochlorophyllide oxidoreductase of barley (*Hordeum vulgare* L.). *Arch. Biochem. Biophys.* **228**, 577–586.
Häuser, I., Dehesh, K., and Apel, K. (1987). Light-induced changes in the amounts of the 36,000-$M_r$ polypeptide of NADPH–protochlorophyllide oxidoreductase and its mRNA in barley plants grown under a diurnal light/dark cycle. *Planta* **170**, 453–460.
Harwood, J. L. (1980). Plant acyl lipids: Structure, distribution, and analysis. *In* "The Biochemistry of Plants" (P. K. Stumpf, ed.), Vol. 4, pp. 1–55. Academic Press, New York.
Heemskerk, J. W. M., and Wintermans, J. F. G. M. (1987). Role of the chloroplast in the leaf acyl-lipid synthesis. *Physiol. Plant.* **70**, 558–568.
Heemskerk, J. W. M., Storz, T., Schmidt, R. R., and Heinz, E. (1990). Biosynthesis of digalactosyldiacylglycerol in plastids from 16:3 and 18:3 plants. *Plant Physiol.* **93**, 1286–1294.
Heinz, E. (1977). Enzymatic reactions in galactolipid biosynthesis. *In* "Lipids and Lipid Polymers in Higher Plants" (M. Tevini and H. K. Lichtenthaler, eds.), pp. 102–120. Springer-Verlag, Berlin.
Heinz, E., and Siefermann-Harms, D. (1981). Are galactolipids integral components of the chlorophyll–protein complexes in spinach thylakoids? *FEBS Lett.* **124**, 105–111.
Helfrich, W. (1973). Elastic properties of lipid bilayers: Theory and possible experiments. *Z. Naturforsch.* **28 C**, 693–703.
Henningsen, K. W., and Boynton, J. E. (1970). Macromolecular physiology of plastids. VIII. Pigment and membrane formation in plastids of barley greening under low light intensity. *J. Cell Biol.* **44**, 290–304.
Henry, L. E. A., Mikkelsen, J. D., and Møller, B. L. (1983). Pigment and acyl lipid composition of photosystem I and II vesicles and of photosynthetic mutants in barley. *Carlsberg Res. Commun.* **48**, 131–148.
Hoober, J. K., Boyd, C. O., and Paavola, L. G. (1991). Origin of thylakoid membranes in *Chlamydomonas reinhardtii* y-1 at 38°C. *Plant Physiol.* **96**, 1321–1328.
Horváth, L. I., Brophy, P. J., and Marsh, D. (1988). Exchange rates at the lipid–protein interface of myelin proteolipid protein studied by spin-label electron spin resonance. *Biochemistry* **27**, 46-52.
Huner, N. P. A., Krol, M., Williams, J. P., Maissan, E., Low, P. S., Roberts, D., and Thompson J. E. (1987). Low temperature development induces a specific decrease in *trans*-$\Delta^3$-hexadecenoic acid content which influences LHCII organization. *Plant Physiol.* **84**, 12–18.
Hyde, S. T., Andersson, S., Ericsson, B., and Larsson, K. (1984). A cubic structure consisting of a lipid bilayer forming an infinite periodic minimum surface of the gyroid type in the glycerol monooleat–water system. *Z. Kristallogr.* **168**, 213–219.
Ikeuchi, M., and Murakami, S. (1983). Separation and characterization of prolamellar bodies and prothylakoids from squash etioplasts. *Plant Cell Physiol.* **24**, 71–80.

Israelachvili, J. N., Marčelja, S., and Horn, R. G. (1980). Physical principles of membrane organization. *Q. Rev. Biophys.* **13**, 121–200.

Jansson, S., Selstam, E., and Gustafsson, P. (1990). The rapidly phosphorylated 25-kDa polypeptide of the light-harvesting complex of photosystem II is encoded by the type 2 *cab-II* genes. *Biochim. Biophys. Acta* **1019**, 110–114.

Jensen, J. W., and Schutzbach, J. S. (1988). Modulation of dolichyl-phosphomannose synthase activity by changes in the lipid environment of the enzyme. *Biochemistry* **27**, 6315–6320.

Jensen, J. W., and Schutzbach, J. S. (1989). Phospholipase-induced modulation of dolichyl-phosphomannose synthase activity. *Biochemistry* **28**, 851–855.

Jost, P. C., Griffith, O. H., Capaldi, R. A., and Vanderkooi, G. (1973). Evidence for boundary lipid in membranes. *Proc. Natl. Acad. Sci. USA* **70**, 480–484.

Joyard, J., Block, M., Pineau, B., Albrieux, C., and Douce, R. (1990). Envelope membranes from mature spinach chloroplasts contain a NADPH:protochlorophyllide reductase on the cytosolic side of the outer membrane. *J. Biol. Chem.* **265**, 21820–21827.

Kader, J.-C. (1985). Lipid-binding proteins in plants. *Chem. Phys. Lipids* **38**, 51–62.

Kay, S. A., and Griffiths, W. T. (1983). Light-induced breakdown of NADPH–protochlorophyllide oxidoreductase *in vitro*. *Plant Physiol.* **72**, 229–236.

Kirk, G. L., Gruner, S. M., and Stein, D. L. (1984). A thermodynamic model of the lamellar to inverse hexagonal phase transition of lipid membrane–water systems. *Biochemistry* **23**, 1093–1102.

Knowles, P. F., Watts, A., and Marsh, D. (1979). Spin-label studies of lipid immobilization in dimyristoylphosphatidylcholine-substituted cytochrome oxidase. *Biochemistry* **18**, 4480–4487.

Krupa, Z. (1987). Cadmium-induced changes in the composition and structure of the light-harvesting chlorophyll *a/b* protein complex II in radish cotyledons. *Physiol. Plant.* **73**, 518–524.

Krupa, Z. (1988). Acyl lipids in the supramolecular chlorophyll–protein complexes of photosystems—Isolation artifacts or integral components regulating their structure and functions? *Acta Soc. Bot. Pol.* **57**, 401–408.

Krupa, Z., Huner, N. P. A., Williams, J. P., Maissan, E., and James, D. R. (1987). Development at cold-hardening temperatures: The structure and composition of purified rye light harvesting complex II. *Plant Physiol.* **84**, 19–24.

Kühlbrandt, W., and Wang, D. N. (1991). Three-dimensional structure of plant light-harvesting complex determined by electron crystallography. *Nature (London)* **350**, 130–134.

Lachmann, K. U., and Kesselmeier, J. (1989). Influence of divalent cations and chelators on the structure of prolamellar bodies of *Avena sativa*. *Plant Cell Physiol.* **30**, 1081–1088.

Larsson, K., Fontell, K., and Krog, N. (1980). Structural relationships between lamellar cubic and hexagonal phases in monoglyceride–water systems, possibility of cubic structures in biological systems. *Chem. Phys. Lipids* **27**, 321–328.

Lee, A. G. (1987). Interactions of lipids and proteins: Some general principles. *J. Bioenerg. Biomembr.* **19**, 581–603.

Lemoine, Y., Dubacq, J.-P., and Zabulon, G. (1982). Changes in light-harvesting capacities and $\Delta^3$*trans*-hexadecenoic acid content in dark- and light-grown *Picea abies*. *Physiol. Veg.* **20**, 487–503.

Li, G., Knowles, P. F., Murphy, D. J., Nishida, I., and Marsh, D. (1989). Spin-label ESR studies of lipid–protein interactions in thylakoid membranes. *Biochemistry* **28**, 7446–7452.

Li, G., Knowles, P. F., Murphy, D. J., and Marsh, D. (1990a). Lipid–protein interactions in stacked and destacked thylakoid membranes and the influence of phosphorylation and illumination. Spin-label ESR studies. *Biochim. Biophys. Acta* **1024**, 278–284.

Li, G., Knowles, P. F., Murphy D. J., and Marsh, D. (1990b). Lipid–protein interactions in thylakoid membranes of chilling-resistant and -sensitive plants studied by spin label electron spin resonance spectroscopy. *J. Biol. Chem.* **265**, 16867–16872.

Lindblom, G., and Rilfors, L. (1989). Cubic phases and isotropic structures formed by membrane lipids—Possible biological relevance. *Biochim. Biophys. Acta* **988**, 221–256.

Lindblom, G., Brentel, I., Sjölund, M.,Wikander, G., and Wieslander, Å. (1986). Phase equilibria of membrane lipids from *Acholeplasma laidlawii:* Importance of a single lipid forming nonlamellar phases. *Biochemistry* **25**, 7502–7510.

Lindstedt, I., and Liljenberg, C. (1990). On the periodic minimal surface structure of the plant prolamellar body. *Physiol. Plant.* **80**, 1–4.

Lütz, C. (1981) On the significance of prolamellar bodies in membrane development of etioplasts. *Protoplasma* **108**, 99–115.

McCourt, P., Browse, J., Watson, J., Arntzen, C. J., and Somerville, C. R. (1985). Analysis of photosynthetic antenna function in a mutant of *Arabidopsis thaliana* (L.) lacking *trans*-hexadecenoic acid. *Plant Physiol.* **78**, 853–858.

McIlhinney, R. A. J. (1990). The fats of life: The importance and function of protein acylation. *Trends Biochem. Sci.* **15**, 387–391.

Marsh, D. (1987). Selectivity of lipid–protein interactions. *J. Bioenerg. Biomembr.* **19**, 677–689.

Mattoo, A. K., and Edelman, M. (1987). Intramembrane translocation and posttranslational palmitoylation of the chloroplast 32-kDa herbicide-binding protein. *Proc. Natl. Acad. Sci. USA* **84**, 1497–1501.

Meier, P., Sachse, J.-H., Brophy, P. J., Marsh, D., and Kothe, G. (1987). Integral membrane proteins significantly decrease the molecular motion in lipid bilayers: A deuteron NMR relaxation study of membranes containing myelin proteolipid apoprotein. *Proc. Natl. Acad. Sci. USA* **84**, 3704–3708.

Meirovich, E., Nayeem, A., and Freed, J. H. (1984). Analysis of protein–lipid interactions based on model simulations of electron spin resonance spectra. *J. Phys. Chem.* **88**, 3454–3465.

Minkov, I. N., Ryberg, M., and Sundqvist, C. (1988). Properties of reformed prolamellar bodies from illuminated and redarkened etiolated wheat plants. *Physiol. Plant.* **72**, 725–732.

Morré, D. J., Morré, J. T., Morré, S. R., Sundqvist, C., and Sandelius, A. S. (1991a). Chloroplast biogenesis, cell-free transfer of envelope monogalactosyl glycerides to thylakoids. *Biochim. Biophys. Acta* **1070**, 437–445.

Morré, D. J., Selldén, G., Sundqvist, C., and Sandelius, A. S. (1991b). A stromal low temperature compartment derived from the inner membrane of the chloroplast envelope. *Plant Physiol.* **97**, 1558–1564.

Mudd, J. B. (1980). Phospholipid biosynthesis. *In* "The Biochemistry of Plants" (P. K. Stumpf, ed.), Vol. 4, pp. 249–282. Academic Press, New York.

Murakami, S., Yamada, N., Nagano, M., and Osumi, M. (1985). Three-dimensional structure of the prolamellar body in squash etioplasts. *Protoplasma* **128**, 147–156.

Murata, N. (1983). Molecuar species composition of phosphatidylglycerols from chilling-sensitive and chilling-resistant plants. *Plant Cell Physiol.* **24**, 81–86.

Murata, N., and Nishida I. (1987). Lipids of blue-green algae (Cyanobacteria). *In* "The Biochemistry of Plants" (P. K. Stumpf and E. E. Conn, ed.), Vol. 9, pp. 315–347. Academic Press, Orlando, Florida.

Murata, N., Higashi, S.-I., and Fujimura, Y. (1990). Glycerolipids in various preparations of photosystem II from spinach chloroplasts. *Biochem. Biophys. Acta* **1019**, 261–268.

Murphy, D. J. (1982). The importance of non-planar bilayer regions in photosynthetic membranes and their stabilisation by galactolipids. *FEBS Lett.* **150,** 19–26.

Murphy, D. J. (1986). The molecular organisation of the photosynthetic membranes of higher plants. *Biochim. Biophys. Acta* **864,** 33–94.

Murphy, D. J., and Woodrow, I. E. (1983a). The lateral segregation model. A new paradigm for the dynamic role of acyl lipids in the molecular organization of photosynthetic membranes. *In* "Biosynthesis and Function of Plant Lipids" (W. W. Thomson, J. B. Mudd, and M. Gibbs, eds.), pp. 104–125. Waverly Press, Baltimore.

Murphy D. J., and Woodrow, I. E. (1983b). Lateral heterogeneity in the distribution of thylakoid membrane lipid and protein components and its implications for the molecular organisation of photosynthetic membranes. *Biochim. Biophys. Acta* **725,** 104–112.

Nishida, I., and Yamada, M. (1985). Semisynthesis of a spin-labeled monogalactosyldiacylglycerol and its application to the assay for galactolipid-transfer activity in spinach leaves. *Biochim. Biophys. Acta* **813,** 298–306.

Nishihara, M., Yokota, K., and Kito, M. (1980). Lipid molecular species composition of thylakoid membranes. *Biochim. Biophys. Acta* **617,** 12–19.

Oliver, R. P., and Griffiths, W. T. (1980). Identification of the polypeptides of NADPH–protochlorophyllide oxidoreductase. *Biochem. J.* **191,** 277–280.

Oliver, R.P., and Griffiths, W. T. (1982). Pigment–protein complexes of illuminated etiolated leaves. *Plant Physiol.* **70,** 1019–1025.

Oross, J. W., and Possingham, J. V. (1991). Tabular structures in developing plastids of three dicotyledonous species. *Can. J. Bot.* **69,** 136–139.

Paulsen, H., Rümler, U., and Rüdiger, W. (1990). Reconstitution of pigment-containing complexes from light-harvesting chlorophyll $a/b$-binding protein overexpressed in *Escherichia coli*. *Planta* **181,** 204–211.

Pick, U., Gounaris, K., Admon, A., and Barber, J. (1984). Activation of the $CF_0$-$CF_1$, ATP synthase from spinach chloroplasts by chloroplast lipids. *Biochim. Biophys. Acta* **765,** 12–20.

Plumley, F. G., and Schmidt, G. W. (1987). Reconstitution of chlorophyll $a/b$ light-harvesting complexes: Xanthophyll-dependent assembly and energy transfer. *Proc. Natl. Acad. Sci. USA* **84,** 146–150.

Protoschill-Krebs, G., and Kesselmeier, J. (1988). Prolamellar bodies of oat, wheat, and rye: Structure, lipid composition, and adsorption of saponins. *Protoplasma* **146,** 1–9.

Quinn, P. J., and Williams, W. P. (1983). The structural role of lipids in photosynthetic membranes. *Biochim. Biophys. Acta* **737,** 223–266.

Quinn, P. J., and Williams, W. P. (1985). Environmentally induced changes in chloroplast membranes and their effects on photosynthetic function. *In* "Photosynthetic Mechanisms and the Environment" (J. Barber and N. R. Baker, eds.), Vol. 6, pp. 1–47. Elsevier, Amsterdam.

Rawyler, A., and Siegenthaler, P.-A. (1985). Transversal localization of monogalactosyldiacylglycerol and digalactosyldiacylglycerol in spinach thylakoid membranes. *Biochim. Biophys. Acta* **815,** 287–298.

Rawyler, A., Unitt, M. D., Giroud, C., Davies, H., Mayor, J.-P, Harwood, J. L., and Siegenthaler, P.-A. (1987). The transmembrane distribution of galactolipids in chloroplast thylakoids is universal in a wide variety of temperate climate plants. *Photosynth. Res.* **11,** 3–13.

Rees, D. C., Komiya, H., Yeates, T. O., Allen, J. P., and Feher, G. (1989). The bacterial photosynthetic reaction center as a model for membrane proteins. *Annu. Rev. Biochem.* **58,** 607–633.

Rémy, R., Trémolières, A., Duval, J. C., Ambard-Bretteville, F., and Dubacq, J. P. (1982). Study of the supramolecular organization of light-harvesting chlorophyll protein

(LHCP). *FEBS Lett.* **137**, 271–275.
Rémy, R., Trémolières, A., and Ambard-Bretteville, F. (1984). Formation of oligomeric light-harvesting chlorophyll a/b protein by interaction between its monomeric form and liposomes. *Photobiochem. Photobiophys.* **7**, 267–276.
Rilfors, L., Lindblom, G., Wieslander, Å., and Christiansson, A. (1984). Lipid bilayer stability in biological membranes. *In* "Membrane Fluidity" (M. Kates and L. A. Manson, eds.), pp. 205–245. Plenum, New York.
Ryberg, M., and Dehesh, K . (1986). Localization of NADPH–protochlorophyllide oxidoreductase in dark-grown wheat (*Triticum aestivum*) by immunoelectron microscopy before and after transformation of the prolamellar bodies. *Physiol. Plant.* **66**, 616–624.
Ryberg, M., and Sundqvist, C. (1982). Characterization of prolamellar bodies and prothylakoids fractionated from wheat etioplasts. *Physiol. Plant.* **56**, 125–132.
Ryberg, M., and Sundqvist, C. (1988). The regular ultrastructure of isolated prolamellar bodies depends on the presence of membrane-bound NADPH–protochorophyllide oxidoreductase. *Physiol. Plant.* **73**, 218–226.
Ryberg, M., and Sundqvist, C. (1991). Structural and functional significance of pigment–protein complexes of chlorophyll precursors. *In* "Chlorophylls" (H. Scheer, ed.), pp. 587–612. CRC Press, Boca Raton, Florida.
Ryberg, M., Sandelius, A. S., and Selstam, E. (1983). Lipid composition of prolamellar bodies and prothylakoids of wheat etioplasts. *Physiol. Plant.* **57**, 555-560.
Sandelius, A. S., and Selstam, E. (1984). Localization of galactolipid biosynthesis in etioplasts isolated from dark-grown wheat (*Triticum aestivum* L.). *Plant Physiol.* **76**, 1041–1046.
Santel, H.-J., and Apel, K. (1981). The protochlorophyllide holochrome of barley (*Hordeum vulgare* L.). The effect of light on the NADPH:protochlorophyllide oxidoreductase. *Eur. J. Biochem.* **120**, 95–103.
Schmidt, M. F. G. (1989). Fatty acylation of proteins. *Biochim. Biophys. Acta* **988**, 411–426.
Schultz, A. M., Henderson, L. E., and Oroszlan, S. (1988). Fatty acylation of proteins. *Annu. Rev. Cell Biol.* **4**, 611–647.
Schulz, R., Steinmüller, K., Klaas, M., Forreiter, C., Rasmussen, S., Hiller, C., and Apel, K. (1989). Nucleotide sequence of a cDNA coding for the NADPH–protochlorophyllide oxidoreductase (PCR) of barley (*Hordeum vulgare* L.) and its expression in *Escherichia coli*. *Mol. Gen. Genet.* **217**, 355–361.
Schwitzguebel, J. P., and Siegenthaler, P.-A. (1985). Evidence for a lack of phospholipid transfer protein in the stroma of spinach chloroplasts. *Plant Sci.* **40**, 167–171.
Seddon, J. M. (1990). Structure of the inverted hexagonal ($H_{II}$) phase and nonlamellar phase transitions of lipids. *Biochim. Biophys. Acta* **1031**, 1–69.
Selldén, G., and Selstam, E. (1976). Changes in chloroplast lipids during the development of photosynthetic activity in barley etio-chloroplasts. *Physiol. Plant.* **37**, 35–41.
Selstam, E., and Öquist, G. (1985). Effects of frost hardening on the composition of galactolipids and phospholipids occurring during isolation of chloroplast thylakoids from needles of Scots pine. *Plant Sci.* **42**, 41–48.
Selstam, E., and Sandelius, A. S. (1984). A comparison between prolamellar bodies and prothylakoid membranes of etioplasts of dark-grown wheat concerning lipid and polypeptide composition. *Plant Physiol.* **76**, 1036–1040.
Selstam, E., and Widell, A. (1986). Characterization of prolamellar bodies, from dark-grown seedlings of Scots pine, containing light- and NADPH-dependent protochlorophyllide oxidoreductase. *Physiol. Plant.* **67**, 345–352.
Selstam, E., and Widell Wigge, A. (1989). Hydrophobicity of protochlorophyllide oxidoreductase, characterized by means of Triton X-114 partitioning of isolated etioplast mem-

brane fractions. *Physiol. Plant.* **77,** 401–406.
Selstam, E., Widell, A., and Johansson, L. B.-Å. (1987). A comparison of prolamellar bodies from wheat, Scots pine, and Jeffrey pine. Pigment spectra and properties of protochlorophyllide oxidoreductase. *Physiol. Plant.* **70,** 209–214.
Selstam, E., Brentel, I., and Lindblom, G. (1990). Phase structure of lipids of the photosynthetic membrane. *In* "Plant Lipid Biochemistry, Structure, and Utilization" (P. J. Quinn and J. L. Harwood, eds.), pp. 39–46. Portland Press, London.
Sen, A., Williams, W. P., Brain, A. P. R., Dickens, M. J., and Quinn, P. J. (1981). Formation of inverted micelles in dispersions of mixed galactolipids. *Nature (London)* **293,** 488–490.
Sen, A., Brain, A. P. R., Quinn, P. J., and Williams, W. P. (1982a). Formation of inverted lipid micelles in aqueous dispersions of mixed sn-3-galactosyldiacylglycerols induced by heat and ethylene glycol. *Biochim. Biophys. Acta* **686,** 215–224.
Sen, A., Williams, W. P., Brain, A. P. R., and Quinn, P. J. (1982b). Bilayer and nonbilayer transformations in aqueous dispersions of mixed sn-3-galactosyldiacylglycerols isolated from chloroplasts: A freeze-fracture study. *Biochim. Biophys. Acta* **685,** 297–306.
Shaw, P., Henwood, J., Oliver, R., and Griffiths, T. (1985). Immunogold localization of protochlorophyllide oxidoreductase in barley etioplasts. *Eur. J. Cell Biol.* **39,** 50–55.
Shipley, G. G., Green, J. P., and Nichols, B. W. (1973). The phase behaviour of monogalactosyl, digalactosyl, and sulphoquinovosyl diglycerides. *Biochim. Biophys. Acta* **311,** 531–544.
Siefermann-Harms, D., and Ninnemann, H. (1983). Differences in acid stability of the chlorophyll–protein complexes in intact thylakoids. *Photobiochem. Photobiophys.* **6,** 85–91.
Siefermann-Harms, D., Ross, J. W., Kaneshiro, K. H., and Yamamoto, H. Y. (1982). Reconstitution by monogalactosyldiacylglycerol of energy transfer from light-harvesting chlorophyll $a/b$–protein complex to the photosystems in Triton X-100-solubilized thylakoids. *FEBS Lett.* **149,** 191–196.
Siefermann-Harms, D., Ninnemann, H., and Yamamoto, H. Y. (1987). Reassembly of solubilized chlorophyll–protein complexes in proteolipid particles—Comparison of monogalactosyldiacylglycerol and two phospholipids. *Biochim. Biophys. Acta* **892,** 303–313.
Siegel, D. P. (1986). Inverted micellar intermediates and the transitions between lamellar, cubic, and inverted hexagonal lipid phases. II. Implications for membrane–membrane interactions and membrane fusion. *Biophys. J.* **49,** 1171–1183.
Siegel, D. P., and Banschbach, J. L. (1990). Lamellar/inverted cubic ($L_\alpha/Q_{II}$) phase transition in N-methylated dioleoylphosphatidylethanolamine. *Biochemistry* **29,** 5975–5981.
Siegenthaler, P.-A., and Giroud, C. (1986). Transversal distribution of phospholipids in prothylakoid and thylakoid membranes from oat. *FEBS Lett.* **201,** 215–220.
Siegenthaler, P.-A., Sutter, J., and Rawyler, A. (1988). The transmembrane distribution of galactolipids in spinach thylakoid inside-out vesicles is opposite to that found in intact thylakoids. *FEBS Lett.* **228,** 94–98.
Silver, B. L. (1985). "The Physical Chemistry of Membranes: An Introduction to the Structure and Dynamics of Biological Membranes." Allen & Unwin, Winchester.
Simpson, D. J. (1978). Freeze-fracture studies on barley plastid membranes. I. Wild-type etioplast. *Carlsberg Res. Commun.* **43,** 145–170.
Sleight, R. G. (1987). Intracellular lipid transport in eukaryotes. *Annu. Rev. Physiol.* **49,** 193–208.
Sprague, S. G. (1987). Structural and functional conseqences of galactolipids on thylakoid membrane organization. *J. Bioenerg. Biomembr.* **19,** 691–703.

Sprague, S. G., and Staehelin, L. A. (1984). Effects of reconstitution method on the structural organization of isolated chloroplast membrane lipids. *Biochim. Biophys. Acta* **777**, 306–322.
Stumpf, P. K. (1980). "Lipids: Structure and Function," Vol. 4. Academic Press, New York.
Stumpf, P. K. (1987). "Lipids: Structure and Function," Vol. 9. Academic Press, Orlando, Florida.
Sundby, C., and Larsson, C. (1985). Transbilayer organization of the thylakoid galactolipids. *Biochim. Biophys. Acta* **813**, 61–67.
Sundqvist, C., Wiktorsson, B., Bang, Z. L., Böddi, B., and Ryberg, M. (1992). Cross-linking of NADPH–protochlorophyllide oxidoreductase in isolated prolamellar bodies. *In* "Regulation of Chloroplast Biogenesis," (J. H. Argyroudi-Akoyunoglou and H. Senger, eds.), pp. 225–230. Plenum Press, New York.
Svensson, P., and Albertsson, P.-Å. (1989). Preparation of highly enriched photosystem II membrane vesicles by a non-detergent method. *Photosynth. Res.* **20**, 249–259.
Talmon, Y., Burns, J. L., Chestnut, M. H., and Siegel, D. P. (1990). Time-resolved cryotransmission electron microscopy. *J. Elec. Microsc. Tech.* **14**, 6–12.
Tanaka, A., and Tsuji, H. (1983). Formation of chlorophyll-protein complexes in greening cucumber cotyledons in light and then in darkness. *Plant Cell Physiol.* **24**, 101–108.
Tanaka, H., and Freed, J. H. (1985). Electron spin resonance studies of lipid–gramicidin interactions utilizing oriented multibilayers. *J. Phys. Chem.* **89**, 350–360.
Tate, M. W., and Gruner, S. M. (1987). Lipid polymorphism of mixtures of dioleoylphosphatidylethanolamine and saturated and monounsaturated phosphatidylcholines of various chain lengths. *Biochemistry* **26**, 231–236.
Terpstra, W., and Lambers, J. W. J. (1983). Interactions between chlorophyllase, chlorophyll $a$, plant lipids, and $Mg^{2+}$. *Biochim. Biophys. Acta* **746**, 23–31.
Thornber, J. P., Morishige, D. T., Anandan, S., and Peter, G. F. (1991). Chlorophyll–carotenoid proteins of higher plant thylakoids. *In* "Chlorophylls" (H. Scheer, ed.), pp. 549–585. CRC Press, Boca Raton, Florida.
Tilley, L., Cribier, S., Roelofsen, B., Op den Kamp, J. A. F., and Van Deenen, L. L. M. (1986). ATP-dependent translocation of amino phospholipids across the human erythrocyte membrane *FEBS Lett.* **194**, 21–27.
Unitt, M. D., and Harwood, J. L. (1982). Lipid topography of thylakoid membranes. *In* "Biochemistry and Metabolism of Plant Lipids" (J. F. G. M. Wintermans and P. J. C. Kuiper, eds.), pp. 359–362. Elsevier, Amsterdam.
Unitt, M. D., and Harwood, J. L. (1985). Sidedness studies of thylakoid phosphatidylglycerol in higher plants. *Biochem. J.* **228**, 707–711.
Van Gorkom, L. C. M., Horváth, L. I., Hemminga, M. A., Sternberg, B., and Watts, A. (1990). Identification of trapped and boundary lipid binding sites in M13 coat protein/lipid complexes by deuterium NMR spectroscopy. *Biochemistry* **29**, 3828–3834.
Verkleij, A. J. (1984). Lipidic intramembranous particles. *Biochim. Biophys. Acta* **779**, 43–63.
Wada, H., Gombos, Z., and Murata, N. (1990). Enhancement of chilling tolerance of a cyanobacterium by genetic manipulation of fatty acid desaturation. *Nature (London)* **347**, 200–203.
Walker, D. A. (1971). Chloroplasts (and grana): Aqueous (including high carbon fixation ability). *Meth Enzymol.* **23**, 211–220.
Watanabe, S., and Yamada, M. (1986). Purification and characterization of a nonspecific lipid transfer protein from germinated castor bean endosperms which transfers phospholipids and galactolipids. *Biochim. Biophys. Acta* **876**, 116–123.
Wellburn, A. R. (1977). Distribution of chloroplast coupling factor ($CF_1$) particles on plastid membranes during development. *Planta* **135**, 191–198.

Wellburn, A. R. (1982). Bioenergetic and ultrastructural changes associated with chloroplast development. *Int. Rev. Cytol.* **80,** 133–191.
Widell Wigge, A., and Selstam, E. (1990). Effects of salt wash on the structure of the prolamellar body membrane and the membrane binding of NADPH-protochlorophyllide oxidoreductase. *Physiol Plant.* **78,** 315–323.
Wieslander, Å., and Selstam, E. (1987). Acyl-chain-dependent incorporation of chlorophyll and cholesterol in membranes of *Acholeplasma laidlawii*. *Biochim. Biophys. Acta* **901,** 250–254.
Yamada, M., Tsuboi, S., Osafune, T., Suga, T., and Takishma, K. (1990). Multifunctional properties of non-specific lipid transfer protein from higher plants. *In* "Plant Lipid Biochemistry, Structure, and Utilization" (P. J. Quinn and J. L. Harwood, eds.), pp. 278–280. Portland Press, London.
Yeates, T. O., Komiya, H., Rees, D. C., Allen, J. P., and Feher, G. (1987). Structure of the reaction center from *Rhodobacter sphaeroides* R-26: Membrane–protein interactions. *Proc. Natl. Acad. Sci. USA* **84,** 6438–6442.
Zachowski, A., Favre, E., Cribier, S., Herve, P., and Devaux, P. F. (1986). Outside–inside translocation of aminophospholipids in the human erythrocyte membrane is mediated by a specific enzyme. *Biochemistry* **25,** 2585–2590.
Zachowski, A., Hermann, A., Paraf, A., and Devaux, P. F. (1987). Phospholipid outside–inside translocation in lymphocyte plasma membranes is a protein–mediated phenomenon. *Biochem. Biophys. Acta* **897,** 197–200.

# 8

# Regulation, Synthesis, and Integration of Chloroplast- and Nuclear-Encoded Proteins

## WOLFGANG HACHTEL AND ANDREAS FRIEMANN

Botanical Institute
University of Bonn
W-5300 Bonn, Germany

I. Introduction
II. Chloroplast-Encoded Chlorophyll Apoproteins
   A. Genes and Gene Products
   B. Chloroplast Transcription and Regulation of Plastid mRNA Levels
   C. Translational and Post-translational Regulation
   D. Integration into Thylakoid Membrane
III. Nuclear-Encoded Chlorophyll-Binding Proteins
   A. Genes Encoding Light-Harvesting Complex II Apoproteins
   B. Regulation of Expression of *Cab* Genes
   C. Role of Chlorophylls in Post-translational Stability and Turnover of Light-Harvesting Complex II Apoproteins
   D. Integration of Light-Harvesting Complex II Apoproteins into Thylakoid Membrane
   E. Polypeptides of Light-Harvesting Complex I
IV. Coordination of Nuclear and Plastid Gene Expression
   A. Coordinate Synthesis of Chlorophyll Proteins
   B. Nuclear Mutants That Affect Synthesis of Chloroplast-Encoded Chlorophyll Proteins
   C. Regulation of *Cab* Gene Expression by Chloroplast Signal
V. Concluding Remarks
   References

## I. INTRODUCTION

The development of photosynthetically active chloroplasts from progenitor organelles—proplastids and etioplasts—is accomplished by the cooperation of two genetic systems. The chloroplast DNA encodes several, but not all, polypeptides that have important structural and functional roles in the photosynthetic protein complexes. Many of the genes for photosynthetic proteins have a nuclear location. The presence of protein-encoding genes in the chloroplast genome requires the maintenance of a complete apparatus for their expression, that is, transcription and translation systems operational within the organelle.

The complexity of this situation can be appreciated by considering the origin of proteins that make up photosystem II (PSII) and photosystem I (PSI). PSII and PSI seem to have the same basic structure, a reaction center complex surrounded by other pigment–protein complexes that act as antenna systems. In PSII, the heart of the reaction center is made of two similar polypeptides known as D1 and D2. The peripheral pigment proteins fall into two classes: tightly coupled and less tightly coupled. The former contain only chlorophyll (Chl) $a$ and often are called CP43 and CP47. CP43 and CP47 constitute, with the reaction center, a PSII core complex. In higher plants and green algae, the less tightly coupled light-harvesting complexes of PSII (LHCII) contain Chl $b$ as well as Chl $a$. In addition to the pigment–binding proteins, a number of intrinsic and extrinsic membrane proteins are associated with PSII. Formation of the whole PSII complex requires the synthesis of at least eight different nuclear-encoded proteins including the LHCII apoproteins. These proteins are translated in the cytoplasm and imported into the chloroplast. The imported polypeptides are assembled with at least nine different proteins of chloroplast origin, including D1 and D2 and the CP43 and CP47 apoproteins. In PSI, the chloroplast-encoded P700 Chl $a$ apoproteins are involved directly in the core of the reaction center. A nuclear-encoded class of Chl-binding proteins is associated with the PSI antenna.

The partial elucidation of the structure and composition of the photosystems led to an examination of the mechanisms by which these complexes and their components are synthesized. The development of photosynthetically competent chloroplasts presents a challenging opportunity to decipher how plastid and nuclear gene expression is controlled temporally and spatially. Progress in the field of chloroplast gene expression has been rapid, partly because of the complete DNA sequence analysis of the chloroplast genome and the development of chloroplast-derived *in vitro* transcription systems that allow the examination of the mechanisms that control transcription and mRNA levels in the organelle.

The multicopy gene families encoding LHCII apoproteins in higher plants are a paradigm for studies of light-induced nuclear gene expression. Traditionally, this light regulation has been attributed to phytochrome. Furthering our understanding of the molecular biology of light induction necessitates identifying the *cis*-regulatory DNA elements that mediate light-regulated transcription and the *trans*-acting protein factors that interact with these sequences. The development of efficient vector systems for plant cell transformation has provided an experimental methodology by which normal genes and engineered gene constructs can be evaluated in transgenic plants.

In this chapter, we focus on recent approaches and results in the analysis of the expression of photosynthetic genes and the synthesis of the encoded proteins. No attempts at an exhaustive review of all relevant research in this field are being made here. Rather, we emphasize genes encoding Chl-binding proteins and make use of other available information as a background for discussion.

## II. CHLOROPLAST-ENCODED CHLOROPHYLL APOPROTEINS

### A. Genes and Gene Products

Several chlorophyll apoproteins are encoded by chloroplast genes and synthesized by the transcriptional and translational apparatus of the chloroplast. The P700 Chl *a* apoproteins A1 and A2 of the reaction center of PSI are encoded by the genes *psaA* and *psaB*. The molecular masses of these polypeptides, as calculated from their nucleotide sequences, are 82–83 kDa (Fish *et al.*, 1985), that is, considerably larger than the 65–70 kDa determined by SDS polyacrylamide gel electrophoresis (PAGE). The Chl *a*-binding proteins CP47 (CPa-1) and CP43 (CPa-2) of the PSII core complex are encoded by *psbB* and *psbC*, respectively. The polypeptide molecular masses as calculated from migration on SDS–PAGE are 45–51 kDa for CP47 and 40–45 kDa for CP43. However, the nucleotide sequences give molecular masses of 51 kDa and 44 kDa (Westhoff *et al.*, 1983). The D1 and D2 proteins of the reaction center of PSII are encoded by *psbA* and *psbD* (Zurawski *et al.*, 1982; Alt *et al.*, 1984). Since D1 and D2 are not pigment–proteins in a classical sense, results concerning D1 and D2 will be considered only when they contribute especially to the understanding of regulation, synthesis, and the integration of other Chl apoproteins. The nomenclature for genes follows the proposals made by Hallick and Bottomley (1983).

### B. Chloroplast Transcription and Regulation of Plastid mRNA Levels

#### 1. Transcription and RNA Processing

Most plastid genes are organized into polycistronic transcription units. The sequence analysis of the entire tobacco, liverwort, and rice (Hiratsuka *et al.*, 1989) chloroplast genomes, in addition to the partial sequence and mapping data from other plant chloroplast genomes, has revealed that, in land plants, the arrangement of genes in transcription units is highly conserved. The plastid operons are transcribed into polycistronic precursor RNA that, in most cases, is processed into smaller RNAs encoding only one to three cistrons. The mRNA diversity arising from many chloroplast polycistronic operons has been explained by 5' and internal endonucleolytic cleavage events, 3' exonucleolytic processing reactions, and intermediates in splicing.

One of the most intensely studied plastid DNA operons is the *psbB–psbH–petB–petD* gene cluster (Westhoff *et al.*, 1986; Rock *et al.*, 1987; Gamble *et al.*, 1988; Kohchi *et al.*, 1988; Shinozaki *et al.*, 1988; Westhoff and Herrmann, 1988; Woodbury *et al.*, 1988). Northern blot hybridization and reverse transcription analyses have revealed that this gene cluster is transcribed as a polycistronic unit, but is processed rapidly. The RNA processing involves the splicing of the introns in *petB* and *petD* and the rapid processing of the polycistronic precursor RNA into several smaller RNA species that finally represent one or two cistrons. Analysis of immunoselected polysomes from maize (Barkan, 1988) revealed that all transcripts from the *psbB–psbH–petB–petD* operon containing spliced *petB* or *petD* sequences are translated to give these proteins, regardless of upstream or downstream sequences, and that the *psbB* gene is translated from all transcripts encoding it. Thus, intercistronic processing is not a prerequisite for translation of these RNAs, although certain processing steps may enhance translational efficiency. In *Vicia* chloroplasts, monocistronic *psbB* mRNA was found to be the major *psbB*-encoding transcript that is associated with polysomes (Friemann and Hachtel, 1988). Our understanding of the role of plastid RNA processing events in the regulation of plastid gene expression may be enhanced by the study of nuclear mutants such as *hcf*38 in maize, which accumulates aberrant sets of transcripts from the *psbB* operon (Barkan *et al.*, 1986).

The *psbC* gene has been shown to be part of a complex operon comprising *psbK*, *psbI*, *psbD*, *psbC*, *orf*62, and *trnG* in barley (Berends *et al.*, 1987; Berends-Sexton *et al.*, 1990). At least 12 different RNAs are produced from this DNA region, including two transcripts that accumulate in response to blue light (Gamble *et al.*, 1988). *In vitro* capping, transcription, and RNA processing experiments provided evidence that much of the *psbD–psbC* mRNA diversity can be explained by transcription initiation at multiple promoters (Woodbury *et al.*, 1989; Yao *et al.*, 1989; Berends-Sexton *et al.*, 1990).

The *psaA* and *psaB* genes are cotranscribed in spinach (Kirsch *et al.*, 1986), barley (Berends *et al.*, 1987), and *Euglena* (Cushman *et al.*, 1988). Both bicistronic and monocistronic mRNAs derived from the *psaA–psaB* operon were found in thylakoid bound polysomes in *Vicia* (Friemann and Hachtel, 1988). An impressing example of *trans* splicing was found in the green alga *Chlamydomonas reinhardtii*, in which the *psaB* gene is a linear continuous sequence whereas the *psaA* gene consists of three exons that are scattered around half of the circular plastid genome. Exon 2 and exon 3 are transcribed from the same strand, whereas exon 1 is transcribed from the opposite strand (Kück *et al.*, 1987; Choquet *et al.*, 1988). *trans* splicing of the independently transcribed three exons is the most probable mechanism for the formation of a translatable *psaA* mRNA. The analysis of several nuclear and plastid mutants of *C. reinhardtii* that are deficient in PSI activity showed several genes to be involved in *trans* splicing and assembly of *psaA* mRNA (Choquet *et al.*, 1988). Herrin and Schmidt (1988) analyzed two mutants of *C. reinhardtii* that lack mature *psaA* mRNA but accumulate unspliced and partially spliced *psaA* transcripts. These studies indicate that (1) exon 1 of *psaA* is transcribed separately from exon 2 whereas exon 2 and the *psbD* gene are cotranscribed, (2) at least two nuclear gene products are required for exon 1–exon 2 *trans* splicing, and (3) exon 2–exon 3 ligation occurs by *trans* splicing.

## 2. Regulation of Chloroplast Transcription

After illumination, mRNAs from several plastid genes that encode different photosynthetic proteins accumulate during the development of chloroplasts. However, the extent to which these mRNAs increase in the light is variable. A light-dependent change in the rate of transcription, similar to that of nuclear genes (Section III,B,1), also has been proposed as a major control by which light may regulate the expression of plastid-specific genes. It has been shown that the expression of several plastid genes can be adjusted rapidly at the translational level (Section II,C). However, translational control often is superimposed on long-term changes at the RNA level. Plastid transcription activity and RNA levels are high during rapid chloroplast growth in barley; both decline when this phase of chloroplast development is completed (Mullet and Klein, 1987). In spinach, overall plastid transcription activity increases slowly during chloroplast development and decreases slowly in chloroplasts during leaf maturation (Deng and Gruissem, 1987). These results document the importance of transcription activity in the overall determination of plastid RNA levels. The induction of plastid transcription during chloroplast biogenesis could result from increased DNA levels, implicating that chloroplast transcription activity is template limited. In spinach, however, transcription activity per DNA template varies up to 5-fold during chloroplast biogenesis (Deng and Gruissem, 1987). Promoter strength and competition between promoters have been reported to influence the relative transcription rate of chloroplast genes (reviewed by Mullet, 1988). Promoter strength therefore can be considered the first control step that establishes the basic transcription activities of individual transcription units (Deng *et al.*, 1987; Gruissem *et al.*, 1988). Further, the rate of transcription of plastid genes is influenced by changes in plastid RNA polymerase activity, and possibly by DNA conformation (Mullet, 1988).

Divergent overlapping transcription was demonstrated in the *psbB* operon of *Marchantia polymorpha* (Kohchi *et al.*, 1988). An *orf*43 on the opposite DNA strand in the spacer region between the *psbB* and *psbH* genes encodes a transcript that is entirely complementary to a part of the primary transcript of the *psbB* operon. These results suggest premature transcription termination or controlled mRNA processing as a mechanism for gene expression in the *psbB* operon. Since the amount of mRNA corresponding to *orf*43 increases with greening in deetiolating seedlings (Kohchi *et al.*, 1988), such a mechanism could explain the differential expression of the *psbB* gene and the genes *psbH*, *petB*, and *petD* in light- as opposed to dark-grown spinach, as observed by Westhoff *et al.* (1983, 1986).

## 3. Transcriptional and Post-transcriptional Control of Plastid mRNA Levels

The accumulation of transcripts to different steady-state levels during light-induced chloroplast development has been examined in several plant species. Changes in mRNA levels could result from altered transcription of plastid genes (transcriptional control). On the other hand, mRNA turnover rates could be the principal determinants of steady-state transcript levels (post-transcriptional control). The plastid run-on transcription assay can be used to assess the relationship between changes in the transcription activity of individual genes and the accumulation of their RNAs (Deng *et al.*, 1987; Mullet and Klein, 1987; Schrubar *et al.*, 1990). This type of examination for some genes shows parallel changes between RNA levels and gene transcription (Mullet and Klein, 1987). However, such correlation is not always present. For a number of developmental stages of spinach, such analyses revealed that transcription activity and RNA accumulation act independently for most of the genes studied (Deng and Gruissem, 1987; Deng *et al.*, 1987; reviewed in Gruissem *et al.*, 1988; Gruissem, 1989). The transcriptional activities and steady-state transcript levels of the *psaA* and *psaB* genes were analyzed during the cell cycle of *Chlamydomonas*. Because of the similar transcriptional patterns observed, the differential steady-state levels of the transcripts appeared to be regulated post-transcriptionally (Leu *et al.*, 1990). These studies support the hypothesis that differential stability of plastid mRNAs, as influenced by post-transcriptional events, is an important mechanism by which differential accumulation of mRNAs occurs during chloroplast development.

A novel transcriptional regulatory mechanism for a chloroplast operon was proposed by Berends-Sexton *et al.* (1990), who suggested a light-induced switch in promoter utilization that results in maintenance of *psbD–psbC* gene expression in chloroplasts of illuminated barley. Illumination of dark-grown barley causes the decline of 10 *psbD–psbC* RNAs and the accumulation of 2 different *psbD–psbC* RNAs. Capping assays, *in vitro* transcription and RNA processing experiments, and treatment of plants with tagetitoxin (a selective inhibitor of chloroplast transcription) indicate that the light-induced transcripts arise by transcription initiation. Run-on transcription and RNA quantitation experiments provided evidence that both light-induced transcription and RNA stability play roles in the accumulation of the light-induced RNAs.

In seedlings of a $C_4$ plant, *Sorghum bicolor*, accumulation of transcripts of plastid genes including *psbA*, *psbB*, *psbC*, and *psaA* was due to a light-induced increase in plastid transcriptional activity that was paralleled by rising levels of plastid RNA polymerase, not to altered RNA stability (Schrubar *et al.*, 1990). Investigations of differential expression of plastid genes in mesophyll and bundle sheath cells of $C_4$ species revealed that levels of the oligocistronic transcripts derived from the *psbB* and *psbD/psbC* transcription units, with the exception of *psbH*, are reduced selectively in bundle sheath cells whereas RNAs carrying the nonphotosystem II components are present in similar quantities in the two cell types (Westhoff *et al.*, 1991). This result demonstrates that segmental RNAs within a single transcription unit can accumulate to different degrees.

## *4. Differential Stability of Transcripts*

Structure and mode of expression of plastid transcription units suggest that RNA processing and differential stabilities of transcripts are involved in post-transcriptional regulation. Protein coding regions in plastids are generally flanked by inverted repeat (IR) sequences that are located in the 3'-untranslated region of transcription units. These IRs potentially can form stem-loop structures and their 3' ends coincide with the 3' ends of the transcripts. Plastid 3' IRs, however, are ineffective transcription terminators (Stern and Gruissem, 1987; Chen and Orozco, 1988). Instead, IR sequences were shown to serve as efficient RNA processing elements *in vitro* and that the 3' ends of *in vitro*-processed mRNAs are identical or nearly identical to those identified *in vivo*. Further, 3' IR sequences are capable of stabilizing upstream RNA segments *in vitro*, whereas linear mRNA fragments are degraded rapidly, suggesting a potential role in plastid mRNA stabilization (Stern and Gruissem, 1987). Soluble proteins were identified that specifically bind IR-containing plastid mRNAs *in vitro*. These proteins may function in plastid mRNA maturation and modulate the effectiveness of the 3' IRs as stabilizing elements (Stern *et al.*, 1989). Other processing events and other mechanisms such as ribosome protection also may modulate the stability of plastid mRNAs (discussed by Mullet, 1988).

## C. Translational and Post-translational Regulation

### 1. Control of Translation

The notion that translational and post-translational regulation are important in plastid gene expression is evident from the lack of correspondence between plastid mRNA and protein levels in different growth conditions, tissues, and mutants that affect photosynthetic functions.

Increasing evidence suggests that translation of certain plastid mRNAs is controlled directly or indirectly by light. Although most of the soluble proteins and many membrane proteins found in chloroplasts accumulate in dark-grown plants (Klein and Mullet, 1987), plastids of these plants lack Chl and Chl apoproteins (Kreuz *et al.*, 1986; Klein and Mullet, 1987; Sutton *et al.*, 1987; Laing *et al.*, 1988). Nevertheless, the plastid genes encoding Chl apoproteins are transcribed in dark-grown plants (Mullet and Klein, 1987) and Chl apoprotein mRNA accumulates (Herrmann *et al.*, 1985; Kreuz *et al.*, 1986; Berends *et al.*, 1987; Klein and Mullet, 1987). This mRNA is associated with polysomes (Klein *et al.*, 1988b), although radiolabeling studies failed to detect amino acid incorporation into D1, CP43, CP47, and the P700 Chl *a* apoproteins (Klein and Mullet, 1987). Within 5 min of illuminating dark-grown plants, amino acid incorporation into the Chl *a* apoproteins of PSI and PSII can be detected as well, and the Chl apoproteins begin to accumulate without concomitant increase in apoprotein mRNA level and distribution in polysomes (Klein *et al.*, 1988a; Laing *et al.*, 1988). These results can be explained in terms of light-dependent translation in the developing chloroplast or in terms of rapid turnover of proteins in the absence of light. The light-dependent activation of Chl apoprotein accumulation is controlled by protochlorophyllide reductase and requires formation of Chl *a* (Klein *et al.*, 1988a; Laing *et al.*,

## 8. Regulation, Synthesis, and Integration

1988). On the basis of these data, Chl *a* is suggested to activate Chl apoprotein accumulation, either by overcoming a block in translation elongation or by binding to and stabilizing nascent Chl apoproteins. An interesting regulation mechanism has been found in homologous run-off translations using lysed etioplasts of barley. Simultaneous *de novo* synthesis of Chl *a* from exogenously added chlorophyllide *a* and phytylpyrophosphate was necessary and sufficient to trigger accumulation of the plastid-encoded Chl *a* apoproteins (P700 Chl *a* apoprotein, CP47, and CP43) in the dark (Eichacker *et al.*, 1990; see also Chapters 6 and 10).

In *Euglena*, the levels to which individual plastid proteins accumulate in light-grown as opposed to dark-grown cells and during greening is due, at least in part, to their stabilization and is less controlled by mRNA levels. Therefore, a post-translational control operating at the level of turnover of the protein products regulates the expression of a variety of plastid genes including *psaA*, *psbB*, and *psbC* (Buetow *et al.*, 1988; L. S. H. Yi, G. Erdös, and D. E. Buetow, 1989, unpublished observations).

Many thylakoid proteins are synthesized by thylakoid membrane-bound polysomes (Section II,D). In the *psaB* gene, one nucleotide domain has been determined that apparently is involved in stimulation of translation or stabilization of the nascent polypeptide in the presence of thylakoids.Therefore, interaction of thylakoids with sequences that enable membrane protein mRNA to be translated efficiently in the presence of thylakoids is suggested as a mechanism that influences chloroplast gene expression (Leu *et al.*, 1989; A. Michaels, and S. Leu, 1990, unpublished observations).

### 2. Stability and Turnover of Apoproteins

The rapid increase in amino acid incorporation into Chl apoproteins in the absence of changes in mRNA content and distribution in polysomes is consistent with activation of translation by Chl *a*. Rapid stabilization of apoproteins, however, is not ruled out. Chl *a* was shown to activate apoprotein accumulation by stabilizing newly synthesized D1 and CP43 (Mullet *et al.*, 1990). This mechanism is consistent with apoprotein mRNA association with polysomes in dark-grown plants and with pulse–chase assays showing Chl-induced stabilization of newly synthesized CP43. Regulation of apoprotein stability by cofactor binding is not unique to the plastid-encoded Chl apoproteins (Section III,C). In higher plant chloroplasts, the regulatory mechanism described affects the accumulation of a large number of abundant chloroplast proteins, giving it special significance to chloroplast biogenesis.

## D. Integration into Thylakoid Membrane

Chloroplast polysomes have been shown to occur free in the stroma and attached to thylakoids in both algae and higher plants. When thylakoid-bound ribosomes were found to be attached to the membranes by ionic interactions as well as by nascent polypeptide chains, they were expected to be engaged in the synthesis and cotranslational integration of chloroplast-encoded thylakoid proteins in analogy to the well-documented situation at the rough endoplasmic reticulum. At least some of the intrinsic thylakoid proteins encoded by the chloroplast genome were found to be synthesized *in vitro* on thylakoid-bound ribosomes (reviewed by Jagendorf and Michaels, 1990). Among these, the P700 Chl *a* apoprotein of PSI (Minami and Watanabe, 1984; Margulies *et al.*, 1987) and the CP47 and CP43 apoproteins of PSII (Minami *et al.*, 1986) were identified, as was the D2 protein (Herrin *et al.*, 1981; Margulies, 1983) and cytochrome *f* (Willey *et al.*, 1983).

Using DNA probes, the mRNAs of free and thylakoid-bound polysomes from broad bean were analyzed by Northern hybridization (Friemann and Hachtel, 1988). The type of attachment of polysomes to the thylakoids was defined by sensitivity to release by high salt or puromycin. High salt releases electrostatically bound polysomes whereas puromycin is supposed to release polysomes that are bound via a nascent polypeptide chain. Transcripts of the genes *psaA, psaB, psbB, psbC, psbD,* and *petA* (encoding cytochrome *f*) were found predominantly on thylakoid-bound polysomes. Release of *psaA, psaB,* and *psbC* transcripts from the thylakoids required both high salt and puromycin. Transcripts of *psbB, psbD,* and *petA* were released in part under high salt conditions and much more with high salt plus puromycin (Friemann and Hachtel, 1988). In agreement with the results from *in vitro* translation assays, one may conclude that a cotranslational mechanism is involved in the integration of the PSI reaction center polypeptides, the CP47 and CP43 polypeptides, the D2 protein of PSII, and cytochrome *f*. The initial hypothesis that all thylakoid membrane proteins are synthesized by bound ribosomes (see Jagendorf and Michaels, 1990) could not be verified, since transcripts of a number of genes (*psbE, petD, atpA, atpB, atpE,* and *atpH*) were found predominantly on free polysomes (Friemann and Hachtel, 1988), corresponding to a stroma-located translation of the respective mRNAs. *In vitro* synthesis and membrane integration of integral thylakoid proteins encoded in the chloroplast genome was studied in cell-free reconstitution systems (Friemann *et al.* 1992). Heterologous prokaryotic and eukaryotic systems were used for transcription and translation of distinct chloroplast genes. Thylakoid membranes were added to study membrane integration of newly forming polypeptides. The D2 protein of PSII was synthesized *in vitro* from the tobacco *psbD* gene. A large amount of labeled D2 was found incorporated into broad bean thylakoids that were added to the translation assays either before protein synthesis started or after translation was finished. Integration was not affected by adding membrane-free chloroplast lysates. Chloroplast stroma factors, if any are at work *in vivo*, might have been replaced by corresponding factors in the *Escherichia coli* and reticulocyte lysate, respectively, used for *in vitro* translation. Results similar to those with D2 were obtained with *psbB* and *petA* gene products.

## III. NUCLEAR-ENCODED CHLOROPHYLL-BINDING PROTEINS

### A. Genes Encoding Light-Harvesting Complex II Apoproteins

The Chl *a*- and *b*-binding apoproteins of the major light-harvesting complex of PSII (LHCII) are encoded by nuclear genes, termed LHCPII genes or *Cab* genes. Each of the genes characterized to date encodes a cytoplasmically synthesized precursor polypeptide (pLHCPII) that has a transit peptide of 33–35 amino acids attached to the N terminus of the mature polypeptide of approximately 233 amino acids. LHCII in angiosperms and gymnosperms can be classified into two subpopulations, designated type 1 and type 2, by virtue of differences in the amino acid sequences of the proteins (Karlin-Neumann *et al.*, 1985; Smeekens *et al.*, 1986; Stayton *et al.*, 1986; Pichersky *et al.*, 1987; Yamamoto *et al.*, 1988; Jansson *et al.*, 1990; Matsuoka, 1990). The structures of type 1 and type 2 are similar to one another with respect to the internal and C-terminal regions of the mature proteins, but the sequences of the transit peptides and the N-terminal regions of mature proteins differ. In both monocot and dicot plants, genes for type 1 and type 2 LHCPII are present as small multigene families (Cashmore, 1984; Dunsmuir, 1985; Lamppa *et al.*, 1985a; Kohorn *et al.*, 1986; Leutwiler *et al.*, 1986; Castresana *et al.*, 1987; Pichersky *et al.*, 1987). The number of type 2 genes is probably similar to or slightly smaller than that of the type 1 genes (Pichersky *et al.*, 1987; Matsuoka, 1990). The relationship between the different genes and the different mature LHCII polypeptides observed by denaturing and nondenaturing electrophoresis has not been established fully.

## B. Regulation of Expression of *Cab* Genes

Many techniques have been used to probe different steps in the expression of *Cab* genes. Immunoprecipitation of translation products of polyadenylated mRNA gives levels of translatable mRNA for this protein. Northern blots are used to measure the level of a particular message in the total RNA population. *In vitro* nuclear run-off transcription studies with isolated nuclei permit determination of the rate of transcription. In such experiments, isolated nuclei are used to incorporate labeled uridine triphosphate into RNA transcripts initiated *in vivo*. The transcriptional activity of a gene is proportional to the fraction of nuclear RNA that hybridizes with a specific gene clone by filter or gel blot hybridization procedures. Regulation of *Cab* genes has been reported to occur on many different levels of gene expression and is one of the best characterized systems for analyzing light- and developmentally regulated as well as organ- and tissue-specific gene expression in plants (reviewed by Chitnis and Thornber, 1988). Additionally, regulation of *Cab* gene expression by chlorophyll (Section III,C), carotenoids (Section IV,C), cytokinins (Flores and Tobin, 1986), and nitrogen availability (Plumley and Schmidt, 1989) has been reported.

### *1. Light-Inducible Transcription*

LHCPII gene transcription is induced 10- to 30-fold by light. This induction is mediated by phytochrome (Silverthorne and Tobin, 1984; Mösinger *et al.*, 1985; Wehmeyer *et al.*, 1990). Depending on the amount of red light required, phytochrome-responsive genes can be classified as exhibiting low fluence rate (LF) and very low fluence rate (VLF) responses (Kaufmann *et al.*, 1984). Pea *Cab* gene expression occurs in the VLF range whereas the expression of the *RbcS* gene encoding the small subunit of ribulose-bisphosphate carboxylase requires a fluence rate at least four orders of magnitude higher (Kaufmann *et al.*, 1984, 1985). Transcriptional regulation of the *Cab* gene family by blue light was demonstrated in etiolated tomato and tobacco seedlings (Wehmeyer *et al.*, 1990), in pea seedlings grown under dim red light (Marrs and Kaufmann, 1989, 1991), and in light-grown maize plants (Eskins *et al.*, 1989). Blue-light mediated accumulation of *Cab* transcripts was observed in a phytochrome deficient *aurea* mutant of tomato (Oelmüller *et al.*, 1989). These results suggest involvement of a blue-light receptor, in addition to phytochrome, probably operating in concert with phytochrome (Oelmüller and Kendrick, 1991; see also Chapter 3).

Hoober (1988) suggested that light regulation of transcription of plant

… genes might act via chelation of transition metals by the active form of phytochrome. This hypothesis is not supported by the effects of metal complexing agents on the light induction of *Cab* mRNA as observed by Kittsteiner *et al.* (1991). Metal complexing agents that cause accumulation of Chl precursors inhibit accumulation of *Cab* mRNA in garden cress. In *Chlamydomonas*, transcription of *Cab* genes is inhibited by Chl precursors. Metal chelators lead to accumulation of porphyrins in this case and consequently to a decreased level of *Cab* mRNA (Johanningmeier and Howell, 1984; Johanningmeier, 1988). 5-Aminolevulinate, in any case, did not behave as a typical inducer for transcription of *Cab* genes (Kittsteiner *et al.*, 1991) as was proposed by Horwitz *et al.* (1988).

Regulation of expression of type 1 and type 2 LHCPII apoproteins has not been studied separately until recently. In dark-grown rice seedlings, mRNAs for type 1 and type 2 accumulated rapidly and accumulated in a similar manner after illumination of seedlings with white light. The amount of type 1 mRNA was three times larger than that of type 2 mRNA in greening seedlings (Matsuoka, 1990).

## 2. *cis*-Acting Elements and *trans*-Acting Factors

How does the plant cell coordinate the temporal and spatial regulation of differentially expressed structural gene sets? One hypothesis is that genes expressed under the same developmental circumstances share a set of repetitive *cis*-acting control elements that form a regulatory network. *cis* control elements are, in turn, recognized by sequence-specific *trans*-acting factors that regulate gene expression at the transcriptional or post-transcriptional level. With the development of methods for introducing defined DNA segments into the plant cell genome (Zambryski *et al.*, 1983), a rapidly growing catalog of development-specific, organ-specific, and environmentally induced *cis*-control regions has been described in plants. Transgenic plants can be regenerated that differ from untransformed plants only by the presence of the transferred genes of interest. Each step in LHCP gene regulation can be reproduced in transgenic plants: transcription is light-inducible, organ-specific, and phytochrome-mediated (Lamppa *et al.*, 1985b; Nagy *et al.*, 1986; Simpson *et al.*, 1986a); nascent polypeptides are processed correctly and exported to the chloroplast where they are assembled in the thylakoid membrane.

Chimeric genes under the control of 5'-flanking sequences of a *Cab* gene from pea were used first to study expression in transgenic tobacco plants (Simpson *et al.*, 1985, 1986a). A 400-bp fragment is sufficient to direct white-light regulation and organ-specific expression observed for *Cab* genes. The levels of expression are, however, low, indicating that some elements that affect *Cab* transcription quantitatively are located still further upstream. Nagy *et al.* (1986) demonstrated a characteristic VLF phytochrome response of a chimeric gene with approximately 2 kb of 5'-flanking sequences present. Further work has identified a 268-bp fragment, 89–357 bp upstream of the transcription start site, that, when placed in either orientation, can confer enhancement, organ specificity, and phytochrome responsiveness on the constitutive promoter of cauliflower mosaic virus (Nagy *et al.*, 1987). More detailed analyses have been carried out on the number and nature of regulatory elements for light induction of the *RbcS* gene (Kuhlemeier *et al.*, 1988). Similar control sequences were found in the 5'-control region of the *Petunia Cab*22R gene (Gidoni *et al.*, 1989) and the *CabE* gene from *Nicotiana plumbaginifolia* (Castresana *et al.*, 1987, 1988). The expression of this gene is under the control of positive, negative, and light-regulatory promoter elements (Castresana *et al.*, 1988; see Fig. 1). Transcriptional control sequences also regulate light-induced *Cab* gene expression quantitatively. Series of 5'-deletions showed that the truncation of each control region reduced the level of light-induced gene expression by increments (Simpson *et al.*, 1985; Kuhlemeier *et al.*, 1988). A chimeric gene containing two copies of an LHCP control region pro-

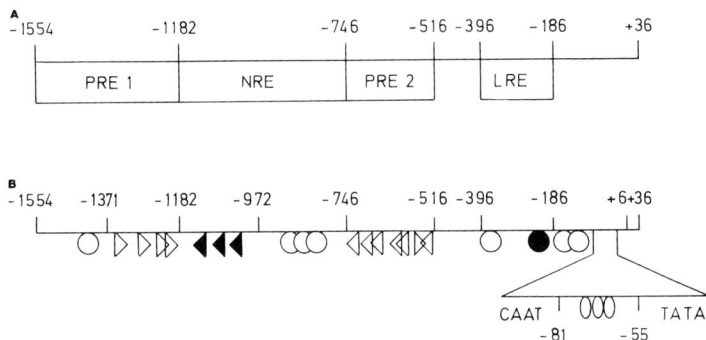

Fig. 1. (A) Different regulatory elements (Castresana *et al.*, 1988) and (B) multiple protein-binding sites (Schindler and Cashmore, 1990) in the *CabE* promoter of *Nicotiana plumbaginifolia*. PRE 1 and PRE 2 are positive regulatory elements that confer maximum levels of photoregulated expression. NRE is a negative regulatory element that reduces the level of gene expression in the light. LRE is a light regulatory element that confers photoregulated expression when fused to a constitutive nopalin synthase promoter. GT-1 is a protein that also interacts with promoters of other light regulated genes. GBF interacts with a sequence homologous to the G-box found in many photoregulated plant promoters. GA-1 binds to the GATA element. GC-1 and AT-1 bind to multiple sites located in GC-rich (from −1371 to −1182 and from −746 to −516) and AT-rich (from −1182 to −972) elements, respectively. Redrawn from Schindler and Cashmore (1990) with permission of Oxford University Press.

duced twice the level of light-induced mRNA of a gene containing just one copy (Simpson et al., 1986a). Moreover, the quantitative effect of these control elements is influenced by the stage of plant development, young leaves being sensitive to control element copy number and mature leaves not (Kuhlemeier et al., 1988).

The transcriptional control of gene expression commonly depends on an interplay between multiple sequence-specific DNA-binding proteins and their cognate promoter elements. Work on light-responsive plant genes has focused on the nuclear proteins that interact with specific elements within the upstream regions of these genes (Lam and Chua, 1990; review by Gilmartin et al., 1990). A tobacco nuclear protein factor, termed activation sequence factor 2 (ASF-2), that binds to the cauliflower mosaic virus 35S promoter was identified by DNase I footprinting. A similarity between the ASF-2 binding site and the paired GATA motifs found in several *Cab* genes was observed (Lam and Chua, 1989a). At least five different nuclear proteins with distinct DNA recognition sites were found, by electrophoretic mobility shift and methylation interference assays, to interact in a specific manner with the *CabE* promoter elements of *N. plumbaginifolia* (Schindler and Cashmore, 1990; see Fig. 1). Both the number of these proteins and the multiplicity of binding sites for the individual factors provide a picture of photoregulated gene expression that is substantially more complex than hitherto described. An interesting feature of the protein factor AT-1 is that binding of AT-1 to an AT-rich negative regulatory element of the *CabE* promoter is modulated by phosphorylation (Datta and Cashmore, 1989).

## 3. Control by Endogenous Circadian Rhythms

Circadian oscillations of *Cab* mRNA levels are documented in green leaves and in fruits (Kloppstech, 1985; Piechulla and Gruissem, 1987; Nagy et al., 1988; Paulsen and Bogorad, 1988; Piechulla, 1988, 1989) and appear to be distributed widely among monocotyledonous and dicotyledonous plant species (Meyer et al., 1989; Stayton et al., 1989; Taylor, 1989a). In continuous darkness, the steady-state *Cab* mRNA level continues to oscillate for 4 days, accompanied by a gradual damping of the amplitude (Piechulla, 1988). *Cab* gene expression is controlled by the light-entrained (endogenous) circadian clock at the transcriptional level (Giuliano et al., 1988; Nagy et al., 1988; Fejes et al., 1990). A chimeric gene comprising the coding region of *Cab* driven by the constitutive 35S promoter is not sensitive to light and circadian rhythm (Nagy et al., 1988). This demonstrates that *Cab* mRNA stability is not regulated by the circadian clock. Fejes *et al.* (1990) have identified a short *Cab* promoter region between -211 and -90 that is responsible for circadian clock-regulated gene expression. The phase of *Cab* gene expression is modulated by the light-to-dark transition (Lam and Chua, 1989b). The regulation of *Cab* expression by an endogenous circadian clock may have some selective advantages, since Chl biosynthesis in higher plants requires light and the apoprotein product of *Cab* is unstable in the absence of Chl *a* and *b*. The regulation by a circadian clock thus will improve the efficiency of the system, since the expression of *Cab* is coordinated and correlated closely with the accumulation of Chl (Lam and Chua, 1989b; Meyer *et al.*, 1989).

## C. Role of Chlorophylls in Post-translational Stability and Turnover of Light-Harvesting Complex II Apoproteins

One possible mechanism for coordinating the two components of the Chl–protein complex is based on the post-translational stabilization of nascent LHCII apoprotein molecules by Chl (Apel and Kloppstech, 1980; Michel *et al.*, 1983; Hoober *et al.*, 1990). In greening higher plant seedlings and in *Chlamydomonas*, Chl *a* synthesis is essential for the stable accumulation of LHCII (Bennett, 1981; Michel *et al.*, 1983; Oelmüller and Schuster, 1987; Klein *et al.*, 1988a). In etiolated seedlings of barley, a red light pulse triggers the appearance mRNA activity for the LHCP; the message is taken up into the polysomes in subsequent darkness and may be translated *in vitro* in a cell-free protein synthesizing system. However, an accumulation of freshly synthesized polypeptide within the plant is not observed (Apel and Kloppstech, 1980). The apparent instability of LHCP might be explained by a deficiency of Chl in red light-treated barley plants, or by the degradation of preexisting Chl in pea seedlings exposed to continuous light for 24 hr and then returned to darkness (Bennett, 1981). Etiolated pea plants do not contain LHCP but do have detectable levels of mRNA encoding it. The accumulation of LHCP in greening pea leaves is not governed primarily by the levels of *Cab* mRNA but by post-translational stablization, in which Chl synthesis is thought to play a necessary role (Bennett *et al.*, 1984).

The rapid coordination between Chl and thylakoid protein components obvious in deetiolating seedlings is not found in fully green leaves of tobacco. In tobacco plants grown under white light, supplementary far-red light reduced the content of Chl *a* and Chl *b* in fully green leaves without changing the Chl *a:b* ratio (Casal *et al.*, 1990). A decrease in Chl *a* content of similar magnitude to that reported to cause the destabilization of LHCP in deetiolating pea (Bennett, 1981) is, however, not correlated

## 8. Regulation, Synthesis, and Integration

with an obvious decrease in the LHCP steady-state level in white-light grown tobacco (Casal et al., 1990). When high-light adapted cells of *Dunaliella tertiolecta* were incubated with gabaculine, which inhibits Chl synthesis, and transferred to low light, the LHCII apoproteins still were synthesized. The $^{35}$S-labeled LHCII apoproteins remained stable after a 24-hr chase (Mortain-Bertrand et al., 1990). These results suggest that synthesis of Chl is not required for stability of the LHCII apoproteins in this alga, although in *D. tertiolectal* as in other plants the synthesis of pigments is coordinated with that of the apoproteins, so normally no excess pigment is synthesized that is not bound to protein and no significant excess protein is synthesized without the simultaneous synthesis of pigments.

The accumulation of LHCII in the thylakoid membrane is correlated with that of Chl *b* (Bennett, 1981; Cuming and Bennett, 1981). Only a small amount of LHCII accumulated in a barley mutant lacking Chl *b* (White and Green, 1987). Very small amounts of Chl *b*, LHCPII, and LHCII were found in tissues illuminated with intermittent light (Slovin and Tobin, 1982). When the duration of the dark period between rounds of intermittent light was reduced, Chl *b* and LHCII apoproteins accumulated (Tzinas et al., 1987). In deetiolating cucumber cotyledons, a small amount of LHCII apoproteins accumulated within the lag phase, that is, before the appearance of Chl *b* (Shimada et al., 1990). The presence of slightly more LHCII apoprotein than would be strictly equivalent, on a stoichiometric basis, to Chl *b* at the early stages of greening suggests the presence of Chl *a*–LHCII apoprotein complexes or LHCII with less Chl *b* than that present in the LHCII of mature chloroplasts. A low level of LHCII apoproteins has been found in thylakoids of a barley mutant that lacks Chl *b*, suggesting the presence of Chl *a*–LHCII apoprotein complexes (Bassi et al., 1985) that may be more stable than apoproteins without Chl.

### D. Integration of Light-Harvesting Complex II Apoproteins into the Thylakoid Membrane

The N-terminal transit peptides of chloroplast protein precursors contain necessary information to target the protein to the chloroplast. In addition, sequences in the mature protein appear to play a role in directing the protein to its final location in the chloroplast (Kohorn et al., 1986). Using a fusion protein, Lamppa (1988) has shown that the mature LHCPII contains the information necessary for insertion into the thylakoid after import. Amino acid charge distribution in the putative membrane-spanning helices is important for the successful accumulation of LHCPII in thylakoids (Kohorn and Tobin, 1987). A region that includes a sequence of 15 amino acids of the carboxy-proximal hydrophobic helix (helix 3) is extremely well conserved in higher plant LHCII apoproteins. An *in vitro*-deletion mutant of LHCPII that lacks these 15 amino acids can be imported but is unable to associate stably with the thylakoid membrane (Kohorn *et al.*, 1986; Kohorn and Tobin, 1989). Although helix 3, when fused to a soluble protein, can target it to the thylakoid, the full integration of helix 3 itself requires the presence of additional regions of LHCPII (Kohorn and Tobin, 1989; Clark *et al.*, 1990). Histidine substitutions have their most prominent effect on the integration of LHCPII into the thylakoid membrane, and also on the assembly of LHCII (Kohorn, 1990). Thus, LHCPII targeting and integration into the thylakoid membranes requires a complex interaction involving a number of different domains in the LHCP polypeptide.

Integration of LHCPII into thylakoids has been reconstituted *in vitro* and shown to require only thylakoids, ATP (Cline, 1986, 1988; Chitnis *et al.*, 1987), and a stromal protein factor (Chitnis *et al.*, 1987; Payan and Cline, 1991). Light can replace ATP partially (Cline, 1988). Both the intact precursor (Chitnis *et al.*, 1987; Cline, 1988) and the mature LHCPII (Viitanen *et al.*, 1988) can serve as substrates in the reconstituted system. A direct demonstration of the sequence of events leading to LHCPII integration is currently unavailable (Chitnis *et al.*, 1988; Reed *et al.*, 1990; see also Chapter 10).

### E. Polypeptides of Light-Harvesting Complex I

A pigment-binding protein of the light-harvesting complex of PSI (LHCI) has been described by Lam *et al.* (1984) and others (Bassi and Simpson, 1987; Nechushtai *et al.*, 1987). These observations were extended by Vainstein *et al.* 1989) to isolate and characterize three different pigment-binding proteins of maize LHCI; each of these pigment–proteins can transfer absorbed energy from its carotenoid and/or Chl *b* components to Chl *a*. A fourth type of LHCPI was described also (Ikeuchi *et al.*, 1991). Studies of the biogenesis of maize PSI during greening of etiolated plants showed that all the core complex polypeptides accumulated to a detectable level prior to the appearance of the LHCI polypeptides (Vainstein *et al.*, 1989). Nuclear genes encoding two different polypeptides purported to be LHCI apoproteins have been cloned (Hoffman *et al.*, 1987; Stayton *et al.*, 1987), but as yet it is impossible to pinpoint unequivocally which of the LHCI polypeptides the cloned genes encode.

# IV. COORDINATION OF NUCLEAR AND PLASTID GENE EXPRESSION

## A. Coordinate Synthesis of Chlorophyll Proteins

Photosystems are composed of different chlorophyll proteins in definite ratios. Therefore, the amount of each chlorophyll protein must be regulated in the formation of functional photosystems. This requirement raises the following questions: Is the synthesis of Chl *a*, Chl *b*, and their apoproteins coordinated? How is the conversion of Chl *a* to Chl *b* regulated to maintain the ratio of Chl *a*/*b* protein complexes to Chl *a* protein complexes? How are Chl *a* and Chl *b* incorporated into the appropriate apoproteins? Shimada *et al.* (1990) examined the changes in levels of LHCII apoproteins by varying the supply of Chl *a* and the levels of Chl *a*-binding apoproteins. Their results can be explained by assuming that the plastid-encoded Chl *a*-binding apoproteins of PSI and PSII have a higher affinity for Chl *a* than do LHCII apoproteins. When the availablity of Chl *a* is limited, these apoproteins compete with one another for Chl *a*, with the result of preferential formation of the Chl *a* protein complexes. When the supply of Chl *a* becomes large enough for saturation of Chl *a*-binding apoproteins, some of the Chl *a* is bound to LHCII apoproteins, either directly or after conversion to Chl *b*.

## B. Nuclear Mutants that Affect Synthesis of Chloroplast-Encoded Chlorophyll Proteins

Examination of nuclear mutants deficient in PSII activity has revealed that nuclear genes play an important role in the synthesis of chloroplast-encoded subunits of PSII and in the assembly of the PSII complex (Kuchka *et al.*, 1988; Gamble and Mullet, 1989). For example, two nuclear mutants of *Chlamydomonas* have been shown to affect the synthesis and/or degration of D2 specifically. These mutants are able to synthesize and integrate the other chloroplast-encoded PSII polypeptides into the thylakoid membrane but these subunits, including the chl *a*-binding gene products of *psbB* and *psbC*, do not accumulate due to increased turnover of D2 protein (Kuchka *et al.*, 1988). Another nuclear mutant of *Chlamydomonas* specifically affects the accumulation of *psbB* mRNA; this mutant also fails to conform stable PSII complexes ( Jensen *et al.*, 1986). Similar findings have been obtained with nuclear photosynthetic mutants of maize (Barkan *et al.*, 1986). In a nuclear mutant of barley that is deficient in PSII activity, the mutation was shown to result in inhibition of translation and decreased stability of the chloroplast encoded D1 and CP47 polypeptides in mature chloroplasts, despite the presence of transcripts for these proteins. Although other chloroplast-encoded PSII polypeptides can be radiolabeled in mutant plastids, they fail to accumulate, whereas the nuclear-encoded polypeptides of the oxygen-evolving complex do (Gamble and Mullet, 1989). Some evidence indicates that a nuclear gene product is required for the production of transcripts that encode D2 and CP43 (P. E. Gamble and J. E. Mullet, 1988, unpublished observations).

## C. Regulation of *Cab* Gene Expression by Chloroplast Signal

A key component of plastid development is the coordination of gene expression between two different genomes. As indicated by a number of facts reviewed by Taylor (1989b), the developmental program controlling the plastid resides in the nucleus. Nevertheless, the state of chloroplast development significantly affects the expression of at least some of the nuclear genes encoding chloroplast proteins. From their studies of pigment-deficient maize mutants, Mayfield and Taylor (1984) found that mutants completely deficient in carotenoids failed to accumulate *Cab* mRNA. Similar results were obtained when carotenoid deficiency was produced with the herbicide norflurazon (Mayfield and Taylor, 1984; Oelmüller and Mohr, 1986). Carotenoid deficiencies cause a wide range of pleiotropic effects. Many of these effects are the result of photooxidative damage to the plastid. Because the extent of photooxidative damage depends on light quantity and quality, many but not all carotenoid-deficient mutants have light-conditional phenotypes. Using this characteristic as an experimental tool, it has been shown that carotenoid deficiency per se has no effect on chloroplast development or gene expression. Rather photooxidation in the chloroplast is responsible for both the block in chloroplast development and the very low level of *Cab* mRNA in the cytosol (Batschauer *et al.*, 1986; Mayfield *et al.*, 1986; Oelmüller and Mohr, 1986; Sagar *et al.*, 1988). The effect of photooxidative damage to the chloroplast seems to be very specific, affecting only chloroplast RNAs and cytosolic mRNAs encoding chloroplast proteins (Mayfield and Taylor, 1987; Burgess and Taylor, 1988).

Evidence for a block of transcription of *Cab* genes caused by photooxidative damage to the chloroplast comes from transcription measurements in isolated nuclei (Batschauer *et al.*, 1986; Burgess and Taylor, 1988) and from experiments with transgenic tobacco plants. The expression of chimeric gene constructions using *Cab* promotors was blocked when transgenic plants were treated with norflurazon to produce carotenoid deficiency but grown under normal light conditions (Simpson *et al.*, 1986b). However, when these promoters were replaced by nopaline synthase promoter, expression was unaffected by norflurazon treatment.

To explain how photooxidative damage to the chloroplast does affect the transcription rate of only a small number of nuclear genes, it was proposed that photooxidation destroys a factor or signal of chloroplast origin that is a necessary component of optimal *Cab* gene transcription (Batschauer *et al.*, 1986; Börner, 1986; Oelmüller and Mohr, 1986; Simpson *et al.*, 1986a; Mayfield and Taylor, 1987; Burgess and Taylor, 1988). This signal has not been identified as a specific molecule. It is most unlikely to be a protein (Oelmüller *et al.*, 1986). Other data indicate that the factor is not dependent on photosynthesis of final steps in chlorophyll biosynthesis (Mayfield and Taylor, 1984), although it is possible that precursors at an earlier step of this pathway are involved (Johanningmeier and Howell, 1987; Johanningmeier, 1988). To date, no evidence exists for a regular exchange of nucleic acids between chloroplasts and cytoplasm, nor for an export of chloroplast-made proteins.

## V. CONCLUDING REMARKS

Coordinate synthesis and assembly of chloroplast pigment–protein complexes is accomplished in part through coactivation of nuclear and plastid gene transcription by cellular or environmental signals. Chloroplast polypeptide accumulation also is regulated by post-transcriptional processes that modify the chloroplast RNA population, as well as at the levels of translation and protein turnover. The expression characteristics of plastid genes may change during the development of chloroplasts or may show organ-specific or species-specific differences, yet it seems clear that, in contrast to the predominant importance of transcriptional control in regulation of bacterial and nuclear genes, post-transcriptional and translational control steps appear to be more significant in regulating chloroplast gene expression in higher plants. The mechanisms of this regulation remain obscure. Post-transcriptional regulation of the expression of plastid genes in illuminated etioplasts is not surprising for several reasons. First, the post-transcriptional adjustment of plastid transcript levels may provide a way to partially uncouple transcription and translation, thus permitting greater control of the production of chloroplast-encoded polypeptides. Second, whereas chloroplast development under normal light–dark regimes is best described as a continuous process, the assembly of the photosynthetic apparatus in greening etiolated seedlings is a step-by-step process: some thylakoid membrane proteins accumulate even in etioplasts whereas others, such as most constituent polypeptides of PSI and PSII reaction centers, are lacking and synthesis and assembly of these polypeptides and their cofactors are coupled. Third, if plastid transcripts are stable relative to cytoplasmic transcripts or if plastid and nuclear gene transcription do not correspond to environmental signals at the same rate, then regulation of protein translation or turnover can help tighten the coupling between nuclear gene expression and production of chloroplast-encoded polypeptides. Fourth, post-transcriptional control could facilitate the adaptation to changes in the doses ratio of nuclear and plastid genes that occur during chloroplast biogenesis as a consequence of an increase in plastid DNA copy number.

Although the available evidence provides strong arguments for transcription as a key control step in the light-regulated expression of nuclear genes for *Cab* proteins, further support could come, for example, from direct *in vitro* transcription studies using purified genes and DNA-dependent transcription extracts. Such extracts have not yet been developed from plant nuclei with full success. Although this deficiency hampers development of this area, the transgenic plant as an experimental system has proven to be a reliable tool for assaying *cis*-regulatory DNA elements. The ongoing characterization of *trans*-acting factors that interact with these regions has served to confirm the expectations derived from mutant promoter analysis in transgenic plants.

Although plastid transformation in higher plants has not yet become a routine task, much progress has been made in the development of plastid *in vitro* systems that are capable of faithful transcription and RNA processing. Present strategies are oriented toward characterization and reconstitution of the various components of such systems, quite analogous to the search for *cis*-acting DNA elements and their cognate *trans*-acting factors presumed to provide key control in nuclear gene expression. The major unresolved problems concern the molecular events of signal transduction, including the question of whether the expression of organelle genes is triggered by light directly or via a signal transmitted by the nucleo-cytoplasmic system. How complex is the network of controls involved in light-regulated gene expression in plants? Light quality and irradiance level interact in the control of *Cab* genes. The presumed control of nuclear *Cab* gene expression by Chl precursors relates to the unresolved problem of the putative chloroplast factor thought to be required for coordination of nuclear and chloroplast gene expression. Identifying the control elements that regulate the expression of light-induced genes under different spectral and developmental conditions at the transcriptional and post-transcriptional level is important to further definition of the regulatory pathways that control light-regulated gene expression in the plant cell.

The discovery of the control of gene expression by endogenous rhythms in mature tissues should lead to new insights into the subtle requirements of the plant to maintain its growth, development, and functions under normal light–dark cycles. Genes that are responsive to circadian rhythms also will prove useful in the molecular analysis of the biological clock itself.

Targeting and integration of LHCPs in thylakoid membranes requires a complex interaction involving a number of different domains of the LHCP and stroma factor(s). Based on current knowledge, it can be assumed that analogous mechanisms are at work for membrane integration of plastid-encoded Chl-binding proteins. The functional roles of polysome binding to thylakoids and of differential affinity of ribosomes to chloroplast mRNA still must be defined.

## REFERENCES

Alt, J., Morris, J., Westhoff, P., and Herrmann, R. G. (1984). Nucleotide sequence of the clustered genes for the 44-kd chlorophyll *a* apoprotein and the "32-kd"-like protein of the photosystem reaction center in the spinach plastid chromosome. *Curr. Genet.* **8**, 597–606.

Apel, K., and Kloppstech, K. (1980). The effect of light on the biosynthesis of the light-harvesting chlorophyll *a/b* protein. *Planta* **150**, 426–430.

Barkan, A. (1988). Proteins encoded by a complex chloroplast transcription unit are each translated from both monocistronic and polycistronic mRNAs. EMBO J. **7**, 2637–2644.

Barkan, A., Miles, D., and Taylor, W. C. (1986). Chloroplast gene expression in nuclear, photosynthetic mutants of maize. *EMBO J.* **5**, 1421–1427.

Bassi, R., and Simpson, D. (1987). Chlorophyll–protein complexes of barley photosystem I. *Eur. J. Biochem.* **163**, 221–230.

Bassi, R. Hinz, U., and Barbato, R. (1985). The role of the light-harvesting complex and photosystem II in thylakoid stacking in the chlorina-f2 barley mutant. *Carlsberg Res. Commun.* **50**, 347–367.

Batschauer, A., Mösinger, E., Kreuz, L., Doerr, I., and Apel, K. (1986). The implication of a plastid-derived factor in the transcriptional control of nuclear genes encoding the light-harvesting chlorophyll *a/b* protein. *Eur. J. Biochem.* **154**, 625–634.

Bennett, J. (1981). Biosynthesis of the light-harvesting chlorophyll *a/b* protein. Polypeptide turnover in darkness. *Eur. J. Biochem.* **118**, 61–70.

Bennett, J., Jenkins, G. I., and Harley, M. R. (1984). Differential regulation of the accumulation of the light-harvesting chlorophyll *a/b*-complex and ribulose bisphosphate carboxylase/oxygenase in greening pea leaves. *J. Cell. Biochem* **25**, 1–13.

Berends, T., Gamble, P. E., and Mullet, J. E. (1987). Characterization of the barley chloroplast transcription units containing *psa*A-*psa*-B and *psb*D-*psb*C. *Nucleic Acids Res.* **15**, 5217–5240.

Berends-Sexton, T., Christopher, D. A., and Mullet, J. E. (1990). Light-induced switch in barley *psb*D-*psb*C promoter utilization: A novel mechanism regulating chloroplast gene expression. *EMBO J.* **9,** 4485–4494.

Börner, T. (1986). Chloroplast control of nuclear gene function. *Endocytobiosis Cell Res.* **3,** 265–274.

Buetow, D. E., Chen, H., Erdos, G., and Yi, L. S. H. (1988). Regulation and expression of the multigene family coding light-harvesting chlorophyll *a*/*b*-binding proteins of photosystem II. *Photosynth. Res.* **18,** 61–97.

Burgess, D. G., and Taylor, W. C. (1988). The chloroplast affects the transcription of a nuclear gene family. *Mol. Gen. Genet.* **214,** 89–96.

Casal, J. C., Whitelam, G. C., and Smith, H. (1990). Phytochrome effects on relationships between chlorophyll and steady-state levels of thylakoid polypeptides in light-grown tobacco. *Plant Physiol.* **94,** 370–374.

Cashmore, A. R. (1984). Structure and expression of a pea nuclear gene encoding chlorophyll *a*/*b*-binding polypeptide. *Proc. Natl. Acad. Sci. USA* **81,** 2960–2964.

Castresana, C., Staneloni, R., Malik, V. S., and Cashmore, A. R. (1987). Molecular characterization of two clusters of genes encoding the type I CAB polypeptides of PSII in *Nicotiana plumbaginifolia*. *Plant Mol. Biol.* **10,** 117–126.

Castresana, C., Garcia-Luque, I., Alonso, E., Malik, V. S., and Cashmore, A. R. (1988). Both positive and negative regulatory elements mediate expression of a photoregulated CAB gene from *Nicotiana plumbaginifolia*. *EMBO J.* **7,** 1929–1936.

Chen, L. -J., and Orozco, E. M., Jr. (1988). Recognition of procaryotic transcription terminators by spinach chloroplast RNA polymerase. *Nucleic Acids Res.* **16,** 8411–8431.

Chitnis, P. R., and Thornber, J. P. (1988). The major light-harvesting complex of photosystem II: Aspects of its molecular and cell biology. *Photosynth. Res.* **16,** 41–63.

Chitnis, P. R., Nechushtai, R., and Thornber, J. P. (1987). Insertion of the precursor of the light-harvesting chlorophyll *a*/*b*-protein into the thylakoids requires the presence of a developmentally regulated stromal factor. *Plant Mol. Biol.* **10,** 3–11.

Chitnis, P. R., Morishige, D. T., Nechushtai, R., and Thornber, P. J. (1988). Assembly of the barley light-harvesting chlorophyll *a*/*b* proteins in barley etiochloroplasts involves processing of the precursors in thylakoids. *Plant Mol. Biol.* **11,** 95–107.

Choquet, Y., Goldschmidt-Clermont, M., Girard-Bascou, J., Kück, U., Bennoun, P., and Rochaix, J. -D. (1988). Mutant phenotypes support a *trans*-splicing mechanism for the expression of the tripartite *psa*A gene in *C. reinhardtii* chloroplast. *Cell* **52,** 903–913.

Clark, S. E., Oblong, J. E., and Lamppa, G. L. (1990). Loss of efficient import and thylakoid insertion due to *N*-and *C*-terminal deletions in the light-harvesting chlorophyll *a*/*b* binding protein. *Plant Cell* **2,** 173–184.

Cline, K. (1986). Import of proteins into chloroplasts: Membrane integration of a thylakoid precursor protein reconstituted in chloroplast lysates. *J. Biol. Chem.* **261,** 14804–14810.

Cline, K. (1988). Light-harvesting chlorophyll a/b protein: Membrane insertion, proteolytic processing, assembly into LHC II, and localization to appressed membranes occurs in chloroplast lysates. *Plant Physiol.* **86,** 1120–1126.

Cuming, A. C., and Bennett, J. (1981). Biosynthesis of the light-harvesting chlorophyll *a*/*b* protein: Control of messenger RNA activity by light. *Eur. J. Biochem.* **118,** 71–80.

Cushman, J. C., Hallick, R. B., and Price, C. A. (1988). The two genes for the P700 chlorophyll *a* apoproteins on the *Euglena gracilis* chloroplast genome contain multiple introns. *Curr. Genet.* **13,** 159–171.

Datta, N., and Cashmore, A. R. (1989). Binding of a pea nuclear protein to promoters of certain photoregulated genes is modulated by phosphorylation. *Plant Cell* **1,** 1069–1077.

Deng, X. -W., and Gruissem, W. (1987). Control of plastid gene expression during development: The limited role of transcriptional regulation. *Cell* **49,** 379–387.

## 8. Regulation, Synthesis, and Integration 303

Deng, X. -W., Stern, D. B., Tonkyn, J. C., and Gruissem, W. (1987). Plastid run-on transcription: Application to determine the transcriptional regulation of plastid genes. *J. Biol. Chem.* **262**, 9641–9648.

Dunsmuir, P. (1985). The petunia chlorophyll *a/b*-binding protein genes: A comparison of *Cab* genes from different gene families. *Nucleic Acids Res.* **13**, 2503–2518.

Eichacker, L. A., Soll, J., Lauterbach, P., Rüdiger, W., Klein, R. R., and Mullet, J. E. (1990). *In vitro* synthesis of chlorophyll *a* in the dark triggers accumulation of chlorophyll *a* apoproteins in barley etioplasts. *J. Biol. Chem.* **265**, 13566–13571.

Eskins, K., Westhoff, P., and Beremand, P. D. (1989). Light quality and irradiance level interaction in the control of expression of light-harvesting complex of photosystem II. *Plant Physiol.* **91**, 163–169.

Fejes, E., Pay, A., Kanevsky, I., Szell, M., Adam, E., Kay, S., and Nagy, F. (1990). A 268-bp upstream sequence mediates the circadian clock-regulated transcription of the wheat *Cab*-1 gene in transgenic plants. *Plant Mol. Biol.* **15**, 921–932.

Fish, L. E., Kück, U., and Bogorad, L. (1985). Two partially homologous adjacent light-inducible maize chloroplast genes encoding polypeptides of the P700 chlorophyll *a*-protein complex of photosystem I. *J. Biol. Chem.* **260**, 1413–1421.

Flores, S., and Tobin, E. M. (1986). Benzyladenine modulation of the expression of two genes for nuclear encoded chloroplast proteins in *Lemna gibba*. *Planta* **186**, 340–349.

Friemann, A., and Hachtel, W. (1988). Chloroplast messenger RNAs of free and thylakoid-bound polysomes from *Vicia faba* L. *Planta* **175**, 50–59.

Friemann, A., Schwarz, H. J., and Hachtel, W. (1992). *In vitro* synthesis and membrane integration of the chloroplast encoded D-2 protein of photosystem II. *In* "Regulation of Chloroplast Biogenesis" (J. H. Argyroudi-Akoyunoglou, ed.), pp. 271–276. Plenum, New York.

Gamble, P. E., and Mullet, J. E. (1989). Translation and stability of proteins encoded by the plastid *psb*A and *psb*B genes are regulated by a nuclear gene during light-induced chloroplast development in barley. *J. Biol. Chem.* **264**, 7236–7243.

Gamble, P. E., Berends-Sexton, T., and Mullet, J. E. (1988). Light-dependent changes in *psb*D and *psb*C transcripts of barley chloroplasts: Accumulation of two transcripts maintains *psb*D and *psb*C translation capability in mature chloroplasts. *EMBO J.* **7**, 1289–1297.

Gidoni, D., Brosio, P., Bond-Nutter, D., Bedbrook, J., and Dunsmuir, P. (1989). Novel *cis*-acting elements in petunia *Cab* gene promoters. *Mol. Gen. Genet.* **215**, 337–344.

Gilmartin, P. M., Sarokin, L., Memelink, J., and Chua, N. -H. (1990). Molecular light switches for plant genes. *Plant Cell* **2**, 369–378.

Giuliano, G., Hoffman, N. E., Skolnik, P. A., and Cashmore, A. R. (1988). A light-entrained circadian clock controls transcription of several plant genes. *EMBO J.* **7**, 3635–3642.

Gruissem, W. (1989). Chloroplast gene expression: How plants turn their plastids on. *Cell* **56**, 161–170.

Gruissem, W., Barkan, A., Deng, X. -W., and Stern, D. (1988). Transcriptional and post-transcriptional control of plastid mRNA in higher plants. *Trends Genet.* **4**, 258–263.

Hallick, R. B., and Bottomley, W. (1983). Proposals for the naming of chloroplast genes. *Plant Mol. Biol. Rep.* **1**, 38–43.

Herrin, D. L., and Schmidt, G. W. (1988). *Trans*-splicing of transcripts for the chloroplast *psa*A1 gene: *In vivo* requirement for nuclear gene products. *J. Biol. Chem.* **263**, 14601–14604.

Herrin, D. L., Hickey, E., and Michaels, A. (1981). Synthesis of a chloroplast membrane polypeptide on thylakoid-bound ribosomes during the cell cycle of *Chlamydomonas reinhardtii*. *Biochim. Biophys. Acta* **655**, 136–145.

Herrmann, R. G., Westhoff, P., Alt, J., Tittgen, J., and Nelson, N. (1985). Thylakoid membrane proteins and their genes. In "Molecular Form and Function of the Plant Genome" (L. van Vloten-Doting, G. S. P. Groot, and T. C. Hall, eds.), pp. 233–256. Plenum, Amsterdam.

Hiratsuka, J., Shimada, H., Whittier, R., Ishibashi, T., Sakamoto, M., Mori, M., Kondo, C., Honji, Y., Sun, C. -R., Meng, B. -Y., Li, Y. -Q., Kanno, A., Nishizawa, Y., Hirai, A., Shinozaki, K., and Sugiura, M. (1989). The complete sequence of the rice (*Oryza sativa*) chloroplast genome: Intermolecular recombination between distinct tRNA genes accounts for a major plastid DNA inversion during the evolution of the cereals. *Mol. Gen. Genet.* **217**, 185–194.

Hoffman, N. E., Pichersky, E., Malik, V. S., Castresana, C., Ko, K., Darr, S. C., and Cashmore, A. R. (1987). A cDNA clone encoding a photosystem I protein with homology to photosystem II chlorophyll *a/b*-binding polypeptides. *Proc. Natl. Acad. Sci. USA* **84**, 8844–8848.

Hoober, J. K. (1988). "Light-derepressible" genes are regulated by metal–protein complexes: A hypothesis. *Carlsberg Res. Commun.* **53**, 27–41.

Hoober, J. K., Maloney, M. A., Asbury, L. R., and Marks, D. B. (1990). Accumulation of chlorophyll *a/b*-binding polypeptides in *Chlamydomonas reinhardii* y−1 in the light or dark at 38°C. *Plant Physiol.* **92**, 419–426.

Horwitz, B. A., Thompson, W. T., and Briggs, W. R. (1988). Phytochrome regulation of greening in *Pisum*: Chlorophyll accumulation and abundance of mRNA for the light-harvesting chlorophyll *a/b* binding proteins. *Plant Physiol.* **86**, 299–305.

Ikeuchi, M., Hirano, A., and Inoue, Y. (1991). Correspondence of apoproteins of light-harvesting chlorophyll *a/b* complexes associated with photosystem I to *cab* genes: Evidence for a novel type IV apoprotein. *Plant Cell Physiol.* **32**, 103–112.

Jagendorf, A. T., and Michaels, A. (1990). Rough thylakoids: Translation on photosynthetic membranes. *Plant Sci.* **71**, 137–145.

Jansson, S., Selstam, E., and Gustafsson, P. (1990). The rapidly phosphorylated 25-kDA polypeptide of the light-harvesting complex of photosystem I is encoded by the type 2 *cab*-II genes. *Biochim. Biophys. Acta* **1019**, 110–114.

Jensen, K. H., Herrin, D. L., Plumley, F. G., and Schmidt, G. W. (1986). Biogenesis of photosystem II complexes: Transcriptional, translational, and posttranslational regulation. *J. Cell Biol.* **103**, 1315–1325.

Johanningmeier, U. (1988). Possible control of transcript levels by chlorophyll precursors in *Chlamydomonas*. *Eur. J. Biochem.* **177**, 417–424.

Johanningmeier, U., and Howell, S. H. (1984). Regulation of light-harvesting chlorophyll-binding protein mRNA accumulation in *Chlamydomonas reinhardtii*. *J. Biol. Chem.* **259**, 13541–13549.

Karlin-Neumann, G. A., Kohorn, B. D., Thornber, J. P., and Tobin, E. M. (1985). A chlorophyll *a/b* protein encoded by a gene containing an intron with characteristics of a transposable element. *J. Mol. Appl. Genet.* **3**, 45–61.

Kaufmann, L. S., Thompson, W. F., and Briggs, W. R. (1984). Different red light requirements for phytochrome-induced accumulation of RNA encoding the small subunit of RuBPCase and that for a chlorophyll *a/b* binding protein. *Science* **226**, 1447–1449.

Kaufmann, L. S., Briggs, W. R., and Thompson, W. F. (1985). Phytochrome control of specific mRNA levels in developing pea buds: The presence of both very low fluence and low fluence responses. *Plant Physiol.* **78**, 388–393.

Kirsch, W., Seyer, P., and Hermann, R. G. (1986). Nucleotide sequence of the clustered genes of two P700 chlorophyll *a* apoproteins of the photosystem I reaction center and the ribosomal protein S14 of the spinach plastid chromosome. *Curr. Genet.* **10**, 843–855.

Kittsteiner, U., Brunner, H., and Rüdiger, W. (1991). The greening process in cress seedlings.

## 8. Regulation, Synthesis, and Integration

II. Complexing agents and 5-aminolevulinate inhibit accumulation of *cab*-mRNA coding for the light-harvesting chlorophyll *a/b* protein. *Physiol. Plant.* **81,** 190–196.

Klein, R. R., and Mullet, J. E. (1987). Control of gene expression during higher plant chloroplast biogenesis. *J. Biol. Chem.* **262,** 4341–4348.

Klein, R. R., Gamble, P. E., and Mullet, J. E. (1988a). Light-dependent accumulation of radiolabelled plastid encoded chlorophyll *a*-apoprotein requires chlorophyll *a*. *Plant Physiol.* **88,** 1246–1256.

Klein, R. R., Mason, H. S., and Mullet, J. E. (1988b). Light-regulated translation of chloroplast proteins. I. Transcripts of *psa*A-*psa*B, *psb*A, and *rbc*L are associated with polysomes in dark-grown and illuminated barley seedlings. *J. Cell Biol.* **106,** 289–301.

Kloppstech, K. (1985). Diurnal and circadian rhythmicity in the expression of light-induced plant nuclear messenger RNAs. *Planta* **165,** 502–506.

Kohchi, T., Yoshida, T., Komano, T., and Ohyama, K. (1988). Divergent mRNA transcription in the chloroplast *psb*B operon. *EMBO J.* **7,** 885–891.

Kohorn, B. D. (1990). Replacement of histidines of light harvesting chlorophyll *a/b* binding protein II disrupts chlorophyll–protein complex assembly. *Plant Physiol.* **93,** 339–342.

Kohorn, B. D., and Tobin, E. M. (1986). Chloroplast import of light-harvesting chlorophyll *a/b*–proteins with different amino termini and transit peptides. *Plant Physiol.* **82,** 1172–1174.

Kohorn, B. D., and Tobin, E. M. (1987). Amino acid charge distribution influences the assembly of apoprotein into light-harvesting complex II. *J. Biol. Chem.* **262,** 12897–12899.

Kohorn, B. D., and Tobin, E. M. (1989). A hydrophobic, carboxy-proximal region of a light-harvesting chlorophyll *a/b* protein is necessary for stable integration into thylakoid membranes. *Plant Cell* **1,** 159–166.

Kohorn, B. D., Harel, E., Chitnis, P. R., Thornber, J. P., and Tobin, E. M. (1986). Functional and mutational analysis of the light-harvesting chlorophyll *a/b* protein of thylakoid membranes. *J. Cell Biol.* **102,** 972–981.

Kreuz, K., Dehesh, K., and Apel, K. (1986). The light-dependent accumulation of the P700 chlorophyll *a* protein of the photosystem I reaction center in barley. *Eur. J. Biochem.* **159,** 459–467.

Kuchka, M. R., Mayfield, S. P., Rochaix, J. -D. (1988). Nuclear mutations specifically affect the synthesis and/or degradation of the chloroplast-encoded D2 polypeptide of photosystem II in *Chlamydomonas reinhardtii*. *EMBO J.* **7,** 319–324.

Kück, U., Choquet, Y., Schneider, M., Dron, M., and Bennoun, P. (1987). Structural and transcription analysis of two homologous genes for the P700 chlorophyll *a*-apoproteins in *Chlamydomonas reinhardtii*: Evidence for *in vivo trans*-splicing. *EMBO J.* **6,** 2185–2195.

Kuhlemeier, C., Cuozzo, M., Green, P., Goyvaerts, E., Ward, K., and Chua, N. -H. (1988). Localization and conditional redundancy of regulatory elements in *rbc*S-3A, a pea gene encoding the small subunit of ribulose-bisphosphate carboxylase. *Proc. Natl. Acad. Sci. USA* **85,** 4662–4666.

Laing, W., Kreuz, K., and Apel, K. (1988). Light-dependent, but phytochrome-independent, translational control of the accumulation of the P700 chlorophyll *a*-protein of photosystem I in barley (*Hordeum vulgare* L.). *Planta* **176,** 269–276.

Lam, E., and Chua, N. -H. (1989a). ASF-2: A factor that binds to the cauliflower mosaic virus 35S promoter and a conserved GATA motif in *Cab* promoters. *Plant Cell* **1,** 1147–1156.

Lam, E., and Chua, N. -H. (1989b). Light to dark transition modulates the phase of antenna chlorophyll protein gene expression. *J. Biol. Chem.* **264,** 20175–20176.

Lam, E., and Chua, N. -H. (1990). GT-1 binding site confers light responsive expression in transgenic tobacco. *Science* **248,** 471–474.

Lam, E., Ortiz, W., and Malkin, R. (1984). Chlorophyll a/b proteins of photosystem I. *FEBS Lett.* **168**, 10–14.

Lamppa, G. K. (1988). The chlorophyll a/b-binding protein inserts into the thylakoids independent of its cognate transit peptide. *J. Biol. Chem.* **263**, 14996–14999.

Lamppa, G. K., Morelli, G., and Chua, N. -H. (1985a). Structure and developmental regulation of a wheat gene encoding the major chlorophyll a/b binding polypeptide. *Mol. Cell Biol.* **5**, 1370–1378.

Lamppa, G., Nagy, G., and Chua, N. -H. (1985b). Light-regulated and organ-specific expression of a wheat *Cab* gene in transgenic tobacco. *Nature (London)* **316**, 750–752.

Leu, S., Herrin, D., and Michaels, A. (1989). Regulation of chloroplast gene expression in *Chlamydomonas reinhardtii*. *In* "Applied Plant Molecular Biology" (G. Galling, ed.), pp. 22–37. Braunschweig, Braunschweig, Germany.

Leu, S., White, D., and Michaels, A. (1990). Cell cycle-dependent transcriptional and post-transcriptional regulation of chloroplast gene expression in *Chlamydomonas reinhardtii*. *Biochim. Biophys. Acta* **1049**, 311–317.

Leutwiler, L. S., Meyerowitz, E. M., and Tobin, E. (1986). Structure and expression of three light-harvesting chlorophyll a/b-binding protein genes in *Arabidopsis thaliana*. *Nucleic Acids Res.* **14**, 4051–4064.

Margulies, M. M. (1983). Synthesis of photosynthetic membrane proteins directed by RNA from rough thylakoids of *Chlamydomonas reinhardtii*. *Eur. J. Biochem.* **137**, 241–248.

Margulies, M. M., Tiffany, H. L., and Hattori, T. (1987). Photosystem I reaction center polypeptides of spinach are synthesized on thylakoid-bound ribosomes. *Arch. Biochem. Biophys.* **254**, 454–461.

Marrs, K. A., and Kaufmann, L. S. (1989). Blue-light regulation of transcription for nuclear genes in pea. *Proc. Natl. Acad. Sci. USA* **86**, 4489–4492.

Marrs, K. A., and Kaufmann, L. S. (1991). Rapid transcriptional regulation of the *Cab* and pEA207 gene families in peas by blue light in the absence of cytoplasmic protein synthesis. *Planta* **183**, 327–333.

Matsuoka, M. (1990). Classification and characterization of cDNA that encodes the light-harvesting chlorophyll a/b binding protein in photosystem II from rice. *Plant Cell Physiol.* **31**, 519–526.

Mayfield, S. P., and Taylor, W. C. (1984). Carotenoid-deficient maize seedlings fail to accumulate light-harvesting chlorophyll a/b binding protein (LHCP) mRNA. *Eur. J. Biochem.* **144**, 79–84.

Mayfield, S. P., and Taylor, W. C. (1987). Chloroplast photooxidation inhibits the expression of a set of nuclear genes. *Mol. Gen. Genet.* **208**, 309–314.

Mayfield, S. P., Nelson, T., and Taylor, W. C. (1986). The fate of chloroplast proteins during photooxidation in carotenoid-deficient maize leaves. *Plant Physiol.* **82**, 30–35.

Meyer, H., Thienel, U., and Piechulla, B. (1989). Molecular characterization of the diurnal/circadian expression of the chlorophyll a/b-binding proteins in leaves of tomato and other dicotyledonous and monocotyledonous plant species. *Planta* **180**, 5–15.

Michel, H., Tellenbach, M., and Boschetti, A. (1983). A chlorophyll b-less mutant of *Chlamydomonas reinhardtii* lacking in the light-harvesting chlorophyll a/b-protein complex but not in its apoproteins. *Biochim. Biophys. Acta* **725**, 417–424.

Minami, E.-I., and Watanabe, A. (1984). Thylakoid membranes, the translational site of chloroplast DNA-regulated thylakoid polypeptides. *Arch. Biochem. Biophys.* **235**, 562–570.

Minami, E.-I., Shinohara, K., Kuwabara, T., and Watanabe, A. (1986). *In vitro* synthesis and assay of photosystem II proteins of spinach chloroplasts. *Arch. Biochem. Biophys.* **244**, 517–527.

## 8. Regulation, Synthesis, and Integration

Mösinger, E., Batschauer, A., Schaefer, E., and Apel, K. (1985). Phytochrome control of *in vitro* transcription of specific genes in isolated nuclei from barley (*Hordeum vulgare*). *Eur. J. Biochem.* **147,** 137–142.

Mortain-Bertrand, A., Bennett, J., and Falkowski, P. G. (1990). Photoregulation of the light harvesting chlorophyll protein complex associated with photosystem II in *Dunaliella tertiolecta*. Evidence that apoprotein abundance but not stability requires chlorophyll synthesis. *Plant Physiol.* **94,** 304–311.

Mullet, J. E. (1988). Chloroplast development and gene expression. *Annu. Rev. Plant Physiol. Plant Mol. Biol.* **39,** 475–502.

Mullet, J. E., and Klein, R. R. (1987). Transcription and RNA stability are important determinants of higher plant chloroplast RNA levels. *EMBO J.* **6,** 1571–1579.

Mullet, J. E., Gamble-Klein, P., and Klein, R. R. (1990). Chlorophyll regulates accumulation of the plastid encoded chlorophyll apoproteins CP43 and D1 by increasing apoprotein stability. *Proc. Natl. Acad. Sci. USA* **87,** 4038–4042.

Nagy, F., Kay, S. A., Boutry, M., Hsu, M.-Y., and Chua, N.-H. (1986). Phytochrome controlled expression of a wheat *Cab* gene in transgenic tobacco seedlings. *EMBO J.* **5,** 1119–1124.

Nagy, F., Boutry, M., Hsu, M.-Y., Wong, M., and Chua, N.-H. (1987). 5' proximal region of the wheat *Cab*-1 gene contains a 268-bp enhancer-like sequence for phytochrome response. *EMBO J.* **9,** 2537–2542.

Nagy, F., Kay, S. A., and Chua, N.-H. (1988). A circadian clock regulates transcription of the wheat *cab*-1 gene. *Genes Dev.* **2,** 376–382.

Nechushtai, R., Peterson, C. C., Peter, G. F., and Thornber, J. P. (1987). Purification and characterization of a light-harvesting chlorophyll-*a/b*–protein of photosystem I of *Lemna gibba*. *Eur. J. Biochem.* **164,** 345–350.

Oelmüller, R., and Kendrick, R. E. (1991). Blue light is required for survival of the tomato phytochrome-deficient *aurea* mutant and the expression of four nuclear genes coding for plastidic proteins. *Plant Mol. Biol.* **16,** 293–299.

Oelmüller, R., and Mohr, H. (1986). Photooxidative destruction of chloroplasts and its consequences for expression of nuclear genes. *Planta* **167,** 106–113.

Oelmüller, R., and Schuster, G. (1987). Inhibition and promotion by light of the accumulation of translatable mRNA of the light-harvesting chlorophyll *a/b*-binding protein of photosystem II. *Planta* **172,** 60–70.

Oelmüller, R., Levitan, I., Bergfeld, R., Rajasekhar, V. K., and Mohr, H. (1986). Expression of nuclear genes as affected by treatments acting on the plastids. *Planta* **168,** 482–492.

Oelmüller, R., Kendrick, R. E., and Briggs, W. R. (1989). Blue-light mediated accumulation of nuclear-encoded transcripts coding for proteins of the thylakoid membrane is absent in the phytochrome-deficient *aurea* mutant of tomato. *Plant Mol. Biol.* **13,** 223–232.

Paulsen, H., and Bogorad, L. (1988). Diurnal and circadian rhythms in the accumulation and synthesis of mRNA of the light-harvesting chlorophyll *a/b*-binding protein in tobacco. *Plant Physiol.* **88,** 1104–1109.

Payan, L. A., and Cline, K. (1991). A stromal protein factor maintains the solubility and insertion competence of an imported thylakoid membrane protein. *J. Cell Biol.* **112,** 603–614.

Pichersky, E., Hoffman, N. E., Malik, V. S., Bernatzky, R., Tanksley, S. D., Szabo, L., and Cashmore, A. R. (1987). The tomato *Cab*-4 and *Cab*-5 genes encode a second type of *CAB* polypeptides localized in photosystem II. *Plant Mol. Biol.* **9,** 109–120.

Piechulla, B. (1988). Plastid and nuclear mRNA fluctuations in tomato leaves—Diurnal and circadian rhythms during extended dark and light periods. *Plant Mol. Biol.* **11,** 345–353.

Piechulla, B. (1989). Changes of the diurnal and circadian (endogenous) mRNA oscillations of the chlorophyll a/b-binding protein in tomato leaves during altered day/night (light/dark) regimes. *Plant Mol. Biol.* **12**, 317–327.

Piechulla, B., and Gruissem, W. (1987). Diurnal mRNA fluctuations of nuclear and plastid genes in developing tomato fruits. *EMBO J.* **6**, 3593–3599.

Plumley, F. G., and Schmidt, G. W. (1989). Nitrogen dependent regulation of photosynthetic gene expression. *Proc. Natl. Acad. Sci. USA* **86**, 2678–2682.

Reed, J. E., Cline, K., Stephens, L. C., Bacot, K. O., and Viitanen, P. V. (1990). Early events in the import/assembly pathway of an integral thylakoid protein. *Eur. J. Biochem.* **194**, 33–42.

Rock, C. D., Barkan, A., and Taylor, W. C. (1987). The maize plastid *psb*B-*psb*H-*pet*B-*pet*D gene cluster: Spliced and unspliced *pet*B and *pet*D RNAs encode alternative products. *Curr. Genet.* **12**, 69–77.

Sagar, A. D., Horwitz, B. A., Elliott, R. C., Thompson, W. F., and Briggs, W. R. (1988). Light effects on several chloroplast components in norflurazon-treated pea seedlings. *Plant Physiol.* **88**, 340–347.

Schindler, U., and Cashmore, A. R. (1990). Photoregulated gene expression may involve ubiquitous DNA binding proteins. *EMBO J.* **9**, 3415–3427.

Schrubar, H., Wanner, G., and Westhoff, P. (1990). Transcriptional control of plastid gene expression in greening *Sorghum* seedlings. *Planta* **183**, 101–111.

Shimada, Y., Tanaka, A., Tanaka, Y., Takabe, T., Takabe, T., and Tsuji, H. (1990). Formation of chlorophyll-protein complexes during greening. 1. Distribution of newly synthesized chlorophyll among apoprotein. *Plant Cell Physiol.* **31**, 639–647.

Shinozaki, K., Hayashida, N., and Sugiura, M. (1988). *Nicotiana* chloroplast genes for components of the photosynthetic apparatus. *Photosynth. Res.* **18**, 7–31.

Silverthorne, J., and Tobin, E. M. (1984). Demonstration of transcriptional regulation of specific genes by phytochrome action. *Proc. Natl. Acad. Sci. USA* **81**, 1112–1116.

Simpson, J., Timko, M. P., Cashmore, A. R., Schell, J., Van Montagu, M., and Herrera-Estrella, L. (1985). Light-inducible and tissue specific expression of a chimeric gene under control of the 5' flanking sequence of a pea chlorophyll a/b binding protein gene. *EMBO J.* **4**, 2723–2729.

Simpson, J., Schell, J., Van Montagu, M., and Herrera-Estrella, L. (1986a). The light-inducible and tissue specific expression of a pea LHCP gene involves an upstream element combining enhancer and silencer-like properties. *Nature (London)* **323**, 551–553.

Simpson, J., Van Montagu, M., and Herrera-Estrella, L. (1986b). Photosynthesis associated gene families: Differences in response to tissue-specific and environmental factors. *Science* **233**, 34–38.

Slovin, J. P., and Tobin, E. M. (1982). Synthesis and turnover of the light-harvesting chlorophyll a/b-protein in *Lemna gibba* grown with intermittent red light: Possible translational control. *Planta* **154**, 465–472.

Smeekens, S., Van Oosten, J., De Groot, M., and Weisbeek, P. (1986). *silene* cDNA clones for a divergent chlorophyll-a/b-binding protein and a small subunit of ribulose bisphosphate carboxylase. *Plant Mol. Biol.* **7**, 433–440.

Stayton, M. M., Black, M., Bedbrook, J., and Dunsmuir, P. (1986). A novel chlorophyll a/b bidning (Cab) protein gene from petunia which encodes the lower molecular weight Cab precursor protein. *Nucleic Acids Res.* **14**, 9781–9796.

Stayton, M. M., Brosio, P., and Dunsmuir, P. (1987). Characterization of a full-length petunia cDNA clone encoding a polypeptide of the light-harvesting complex associated with photosystem I. *Plant Mol. Biol.* **10**, 127–137.

Stayton, M. M., Brosio, P., and Dunsmuir, P. (1989). Photosynthetic genes of *Petunia* (Mitchell) are differentially expressed during the diurnal cycle. *Plant Physiol.* **89**, 776–782.

Stern, D. B., and Gruissem, W. (1987). Control of plastid gene expression: 3' inverted repeats act as mRNA processing and stabilizing elements, but do not terminate transcription. *Cell* **51**, 1145–1157.

Stern, D. B., Jones, H., and Gruissem, W. (1989). Function of plastid mRNA 3' inverted repeats: RNA stabilization and gene-specific protein binding. *J. Biol. Chem.* **264**, 18742–18750.

Sutton, A., Sieburth, L. E., and Bennett, J. (1987). Light dependent accumulation and localization of photosystem II proteins in maize. *Eur. J. Biochem.* **164**, 571–578.

Taylor, W. C. (1989a). Transcriptional regulation by a circadian rhythm. *Plant Cell* **1**, 259–264.

Taylor, W. C. (1989b). Regulatory interactions between nuclear and plastid genomes. *Annu. Rev. Plant Physiol. Plant Mol. Biol.* **40**, 211–233.

Tzinas, A., Argyroudi-Akoyunoglou, J. H., and Akoyunoglou, G. (1987). The effect of dark interval in intermittent light on thylakoid development: Photosynthetic unit formation and light-harvesting protein accumulation. *Photosynth. Res.* **14**, 241–258.

Vainstein, A., Peterson, C. C., and Thornber, J. P. (1989). Light-harvesting pigment–proteins of photosystem I in maize. Subunit composition and biogenesis. *J. Biol. Chem.* **264**, 4058–4063.

Viitanen, P. V., Doran, E. R., and Dunsmuir, P. (1988). What is the role of the transit peptide in the thylakoid integration of the light-harvesting chlorophyll *a/b* protein? *J. Biol. Chem.* **263**, 15000–15007.

Wehmeyer, B. Cashmore, A. R., and Schaefer, E. (1990). Photocontrol of the expression of genes encoding chlorophyll *a/b* binding proteins and small subunit of ribulose-1,5-bisphosphate carboxylase in etiolated seedlings of *Lycopersicon esculentum* (L.) and *Nicotiana tabacum* (L.). *Plant Physiol* **93**, 990–997.

Westhoff, P., and Herrmann, R. G. (1988). Complex RNA maturation in chloroplasts. *Eur. J. Biochem.* **171**, 551–564.

Westhoff, P., Alt, J., and Herrmann, R. G. (1983). Localization of the genes for the two chlorophyll *a*-conjugated polypeptides (mol. wt. 51 and 44 kd) of the photosystem II reaction center on the spinach plastid chromosome. *EMBO J.* **2**, 2229–2237.

Westhoff, P., Farchaus, J. W., and Herrmann, R. G. (1986). The gene for the $M_r$ 10,000 phosphoprotein associated with photosystem II is part of the *psb*B operon of the spinach plastid chromosome. *Curr. Genet.* **11**, 165–169.

Westhoff, P., Offermann-Steinhard, K., Höfer, M., Eskins, E., Oswald, A., and Streubel, M. (1991). Differential accumulation of plastid transcripts encoding photosystem II components in the mesophyll and bundle sheath cells of monocotyledonous NADP malic enzyme-type C4 plants. *Planta* **184**, 377–388.

White, M. J., and Green, B. R. (1987). Polypeptides belonging to each of the three major chlorophyll *a + b* protein complexes are present in a chlorophyll-*b*-less barley mutant. *Eur. J. Biochem.* **165**, 531–535.

Willey, D. L., Huttley, A. K., Phillips, A. L., and Gray, J. C. (1983). Localization of the gene for cytochrome *f* in pea chloroplast DNA. *Mol. Gen. Genet.* **189**, 85–89.

Woodbury, N. W., Roberts, L. L., Palmer, J. D., and Thompson, W. F. (1988). A transcription map of the pea chloroplast genome. *Curr. Genet.* **14**, 75–89.

Woodbury, N. W., Dobres, M., and Thompson, W. F. (1989). The identification and localization of 33 pea chloroplast transcription initiation sites. *Curr. Genet.* **16**, 433–445.

Yamamoto, N., Matsuoka, M., Kano-Murakami, Y., Tanaka, Y., and Ohashi, Y. (1988).

Nucleotide sequence of a full length cDNA clone of light harvesting chlorophyll *a/b* binding protein gene from dark-grown pine (*Pinus tunbergii*) seedling. *Nucleic Acids Res.* **16,** 11829.

Yao, W. B., Meng, B. Y., Tanaka, M., and Sugiura, M. (1989). An additional promotor within the protein-coding region of the *psb*D-*psb*C gene cluster in tobacco chloroplast DNA. *Nucleic Acids Res.* **17,** 9583–9591.

Zambryski, P., Joos, H., Genetello, C., Leemans, J., Van Montagu, M., and Schell, J. (1983). Ti-plasmid vector for the introduction of DNA into plant cells without alteration of their normal regeneration capacity. *EMBO J.* **2,** 2143–2150.

Zurawski, G., Bohnert, H. J., Whitfeld, P. R., and Bottomley, W. (1982). Nucleotide sequence of the gene for the $M_r$ 32,000 thylakoid membrane protein from *Spinacia oleracea* and *Nicotiana debneyi* predicts a totally conserved primary translation product of $M_r$ 38,950. *Proc. Natl. Acad. Sci. USA* **79,** 7699–7703.

# 9

# Import and Routing of Chloroplast Proteins

## DOUWE DE BOER* AND PETER WEISBEEK[†]

* Agrotechnological Research Institute (ATO-DLO)
6700 AA Wageningen, The Netherlands

[†] Department of Molecular Cell Biology
University of Utrecht
3508 TB Utrecht, The Netherlands

I. Introduction
II. Chloroplast Targeting Signals
   A. Transit Peptides
   B. Thylakoid Transfer Domains
III. Chloroplast Envelope Translocation
   A. Cytosolic Factors
   B. Components of Translocation Complex
   C. Energetics of Envelope Translocation
   D. Stromal Processing
   E. Stromal Factors
IV. Intraorganelle Routing
   A. Membrane Insertion
   B. Thylakoid Membrane Translocation
   C. Thylakoid Processing
References

## I. INTRODUCTION

Plant cells, like all other eukaryotic cells, contain various compartments, called organelles, that serve to separate the different metabolic processes. All organelles are enclosed by one or two selectively permeable membranes constituted of lipids and proteins. Most of these proteins are involved in the import and export of a large variety of molecules (Werner-Washburne and Keegstra, 1985).

One family of organelles, the plastids, is unique to plants, whereas other organelles, such as mitochondria and the endoplasmic reticulum, are also present in other eukaryotes. Plastids are surrounded by a two-membrane envelope and contain a limited amount of DNA (Ellis, 1983; Douce and Joyard, 1990). Several types of plastids can be distinguished in the different tissues of a plant (Thomson and Whatley, 1980), but all originate in development from the proplastid. The most well-known type, the chloroplast, is present in green tissue and is the site of photosynthesis. Plastids and mitochondria are supposed to originate from two independent endosymbiotic events (Gray and Doolittle, 1982). Mitochondria display similarities with some species of purple bacteria (Gabellini, 1988) whereas chloroplasts resemble prochloron (Turner *et al.*, 1989) and certain types of cyanobacteria (Gray and Doolittle, 1982).

About 250 different spots have been detected after two-dimensional gel electrophoresis of chloroplast proteins (Dietz and Bogorad, 1987). These proteins are required to perform a large set of different functions, such as photosynthesis, $CO_2$ fixation, biosynthesis of small molecules, transcription, translation, regulation, and transport of metabolites and proteins (Danks *et al.*, 1983; Ellis, 1983; Douce and Joyard, 1990). The coding capacity of the chloroplast genome is not sufficient to encode all the proteins that are present in the organelle. In the course of evolution, most of the original genome was either lost or transferred to the nucleus (Oliver *et al.*, 1989; Baldauf and Palmer, 1990). About 60–70% of the chloroplast proteins in higher plants are now encoded by the nucleus and must be imported into the organelle from the surrounding cytoplasm. This chapter discusses the different parameters that are required for this import process and the routing of proteins within the organelle.

Chloroplasts are subdivided by membranes into three different compartments: the envelope intermembrane space, the stroma, and the thylakoid lumen. Therefore, including the membranes, there are six different locations in which proteins can reside. The two membranes of the envelope are separated from each other by the intermembrane space, but there are specific points, called contact sites, at which the two membranes are fused or held in close proximity (Cline *et al.*, 1985b; Cremers *et al.*, 1988; Pain *et al.*, 1988). Nearly 75 different proteins have been assigned to the chloroplast envelope (Joyard *et al.*, 1982). These proteins mainly function in metabolite and protein transport but also act in many biosynthetic processes (Douce and Joyard, 1990). About 140 different chloroplast proteins are located in the stroma (Ellis, 1981), where most of the chloroplast functions occur. Photosynthetic electron transport and phosphorylation, however, occur in the thylakoid membranes (Hall and Rao, 1987). The proteins associated with the thylakoid membranes are arranged in

## 9. Import and Routing

four complexes: PSI and PSII, their associated light-harvesting complexes, cytochrome $b_6/f$, and the $F_0/F_1$–ATP synthase complex (Staehelin, 1986). The thylakoid membranes are arranged in stacked (grana lamellae) and unstacked (stroma lamellae) regions. The different protein complexes are distributed unevenly over these regions (Staehelin, 1986). Only a few proteins are found in the thylakoid lumen, where the water-splitting reaction of photosynthesis takes place.

Nuclear-encoded chloroplast proteins are synthesized in the cytoplasm with an N-terminal extension that contains the import signals (see Sections II,A and B). In the cytoplasm, the precursor might interact with cytosolic factors (see Section III,A) during or shortly after its synthesis. After binding to receptor proteins in the outer membrane (see Section II,B), translocation across the envelope occurs. ATP is required for binding to the receptor as well as during translocation (see Section III,C). After translocation, the chloroplast import signal is removed by a stromal processing peptidase (see Section III,D) and interaction with stromal factors might occur (see Section III,E). This transport into the stroma is the default route taken when a transit peptide is present in the precursor. Transport is followed by a routing step to one of the other locations in the chloroplast only when additional routing signals are present in the protein. For membrane insertion, signals are usually present in the mature protein (see Section IV,A), whereas for thylakoid membrane translocation, an increased N-terminal extension composed of two domains is present in front of the mature protein (see Sections II,B and IV,B). The two-domain targeting signal is removed from the precursor in two steps (see Sections IV,B and C).

## II. CHLOROPLAST TARGETING SIGNALS

Nuclear-encoded chloroplast proteins are synthesized as higher molecular weight precursor proteins with an N-terminal extension, the chloroplast targeting signal (Dobberstein *et al.*, 1977; Schmidt *et al.*, 1979). This targeting signal is required for chloroplast import and sometimes for routing within the chloroplast. The part of the targeting signal that is required for translocation across the envelope membranes is called a transit peptide (Chua and Schmidt, 1979). This part is removed by a stromal processing peptidase after import into the chloroplast (Robinson and Ellis, 1984). Thylakoid lumen proteins are synthesized with a targeting signal that contains a thylakoid transfer domain behind the transit peptide (Smeekens *et al.*, 1986). This second domain is essential for thylakoid lumen targeting and is removed from the protein by a thylakoidal processing peptidase after translocation across the thylakoid membrane. A list of cloned and sequenced genes for nuclear-encoded chloroplast proteins is available (De Boer and Weisbeek, 1991), but more and more cloned genes are being published. From these sequencing data, several characteristic features of chloroplast proteins can be determined.

## A. Transit Peptides

The presence of an extension in front of a cytoplasmically synthesized chloroplast protein was reported first by Dobberstein *et al*. (1977). They showed that the *Chlamydomonas* small subunit of ribulose 1,5-bisphosphate carboxylase (Rubisco) was synthesized in a wheat germ system as a precursor with a 3.5-kDa extension. The extension was called a transit peptide to distinguish it from signal sequences, which are structurally and functionally different (Chua and Schmidt, 1979).

Although several groups using mutagenesis experiments have tried to divide the transit peptide into functional regions, a clear subdivision was not obtained. Deletions in the N- and C-terminal domains of the transit peptide resulted in reduced binding and import efficiencies (Reiss *et al*., 1987, 1989; Smeekens *et al*., 1989; Hageman *et al*., 1990). Binding usually was abolished and the import efficiencies reduced when deletions were made in the central region (Reiss *et al*., 1989). Mainly stromal processing was influenced when deletions were made in the C terminus of the transit peptide (Wasmann *et al*., 1988; Ostrem *et al*., 1989). Therefore the transit peptide as a whole seems to be inportant for binding and import, whereas the C terminus is also important for processing.

A comparison of the amino acid sequences of cloned nuclear-encoded chloroplast genes shows no real homology blocks among the transit peptides of different proteins (Keegstra *et al*., 1989). However, a conformity in charge distribution is apparent, and there are some preferential positions for certain amino acids (Von Heijne, *et al*., 1989). Apart from the C-terminal region of the transit peptide (see subsequent text), there seem to be no tendencies for any real secondary structure, theoretically. In fact, the structure of a transit peptide seems to be a "perfect random coil" (Von Heijne and Nishikawa, 1991).

Nonetheless, statistical analysis of higher plant transit peptides shows that they have several features in common (Fig. 1A; Von Heijne *et al*., 1989; Gavel and Von Heijne, 1990). First, the amino-terminal 10 amino acids are almost devoid of proline, glycine, and charged amino acids; the N terminus usually starts with the dipeptide methionine/alanine. This dipeptide is supposed to signal the removal of methionine from the precur-

## A

```
        . .      ..+      ..    .+ ..++ . +     +.           +
        MASLATLAAVQPTTLKGLAGSSIAGTKFTSARRQSFKLNNVRSGAIVA*KY
```

## B

```
        +   +-
        ASLKDVGVVVAATAAAGILAGNAMA*
```

**Fig. 1.** (A) The transit peptide of the spinach subunit VI of PSI (Steppuhn et al., 1989) is shown as an example of a transit peptide. The C- and N-terminal 10 amino acids of the transit peptide are underlined. The amino acids around the cleavage site that partly match the consensus cleavage site are shown in bold face. The actual cleavage site is marked with an asterisk. The charges of the amino acids are given above the sequence. Hydroxy amino acids are marked with dots. (B) Thylakoid transfer domain of the plastocyanin protein (Smeekens et al., 1985). The start of this thylakoid transfer domain was determined by microsequencing (C. Robinson, personal communication). The hydrophobic domain is underlined. Amino acid charges and cleavage site are as indicated in A.

sor after translation (Von Heijne et al., 1989). Second, the central region is rich in hydroxy groups and positively charged amino acids and is variable in length. Third, the carboxy-terminal 10 amino acids before the processing site theoretically have a high potential to form amphiphilic $\beta$-strands. Helix-forming amino acids such as leucine and lysine are excluded from the region. Arginine is usually present in the region $-6$ to $-10$ and at position $-2$. Finally, the consensus amino acid sequence (I/V)-X-(A/C)↓A is determined for the processing site, although a lot of deviation occurs (Gavel and Von Heijne, 1990). Transit peptides do not contain the amphiphilic helix structure that is characteristic of mitochondrial targeting peptides and are usually larger (Von Heijne, 1986a). The central hydrophobic domain (h-region) that is characteristic of signal peptides also is not present in transit peptides (Von Heijne, 1985). These differences and the lack of secondary structure might form the basis for the specificity of chloroplast recognition. Transit peptides of the alga *Chlamydomonas* are small and theoretically have a tendency to form an amphiphilic helix structure (Fränzen et al., 1990); consequently, they resemble mitochondrial targeting peptides. Therefore, the signals for discrimination between organelles seem to be even more obscure in algae.

The size of a transit peptide in higher plants (see De Boer and Weisbeek, 1991, for an overview) ranges from 28 amino acids for ADP-glucose phosphorylase (Anderson et al., 1989) to 83 amino acids for glyceraldehyde

3-phosphate dehydrogenase subunit B (Brinkman *et al.*, 1989). These variations in size do not reflect differences among species, but are related to the type of protein. Similar proteins in different species tend to have corresponding transit peptide lengths. Therefore, the length of a transit peptide and most of its primary structure probably depend on the mature protein and only to a lesser extent on the species or the chloroplast import machinery. During the course of evolution, the transit peptide seems to have adjusted to its mature protein. For example, the basic transit peptide of α-glucan phosphorylase compensates for the very acidic mature part of the protein (Nakano *et al.*, 1989).

## B. Thylakoid Transfer Domains

Proteins that must be routed to the thylakoid lumen contain a two-domain targeting signal (Smeekens *et al.*, 1986) of which the first domain is indistinguishable from a normal transit peptide. The import process of a thylakoid lumen protein is divided into two discrete steps. First, the protein is imported into the chloroplast by the default route to the stroma. Then the protein is routed to the lumen. After translocation across the envelope membranes, an intermediate-sized protein is found in the stroma transiently. This intermediate-sized protein still contains the second thylakoid-transfer domain (Smeekens *et al.*, 1986; Scherer and Knauf, 1987; Ko and Cashmore, 1989) that is required for routing. A mature-sized protein is formed by specific processing after translocation into the thylakoid lumen (Hageman *et al.*, 1986; Smeekens *et al.*, 1986).

Thylakoid transfer domains contain all the characteristics of a signal peptide (Fig. 1B; Von Heijne, 1985, 1986b). The N-terminal part is positively charged (n-region); the central part is hydrophobic (h-region); the C-terminal part (c-region) contains the consensus cleavage site A-X-A ↓ (Von Heijne *et al.*, 1989).

Experiments with fusion proteins showed that the presence of a transit peptide is the only requirement for a protein to be imported into the chloroplast, both *in vitro* and *in vivo*. Almost all fusion proteins are imported into chloroplasts when a transit peptide is present at the N terminus (for an overview, see De Boer and Weisbeek, 1991). Also, the first domain of a targeting signal of a thylakoid lumen protein is able to direct a protein into the chloroplast (Scherer and Knauf, 1987; Ko and Cashmore, 1989). However, the thylakoid transfer domain is, in most cases, not able to direct a passenger protein to the thylakoid lumen *in vitro;* an intermediate-sized protein is found arrested in the stroma in these cases (Smeekens *et al.*, 1986, 1987; Scherer and Knauf, 1987; Ko and Cashmore, 1989; De

Boer *et al.*, 1991). A comparison between *in vitro* and *in vivo* routing of fusion proteins with the *Escherichia coli* β-lactamase protein as a passenger showed that routing the thylakoid lumen was hampered only *in vitro* (De Boer *et al.*, 1991). Therefore, all the necessary information for routing is probably present in the thylakoid transfer domain. The cause for abnormal routing behavior *in vitro* might be that *in vitro* import is normally done with chloroplasts that originate from only one developmental stage of peas, whereas *in vivo* import can take place during the complete course of chloroplast development.

## III. CHLOROPLAST ENVELOPE TRANSLOCATION

Although the presence of a transit peptide at the N terminus of a precursor protein is sufficient to signal import into chloroplasts, several other requirements are necessary for the actual process. Cytosolic factors might be needed to assist in unfolding and translocation. Receptor proteins that recognize the transit peptide are expected to be present in the chloroplast envelope. Energy is needed in the form of ATP. After translocation, a stromal peptidase is needed. Stromal factors might be necessary to assist in refolding and assembly.

### A. Cytosolic Factors

Proteins that enter the chloroplast do so post-translationally. Therefore, they must be transported through the cytoplasm to the chloroplast. During this transport process, the protein must be folded or hydrophobic regions in the protein must be shielded to prevent denaturation and aggregation. However, for translocation across the envelope membranes, a partially unfolded state is thought to be necessary. For mitochondrial import and import into the endoplasmic reticulum, most precursors were found to be present in the cytoplasm complexed with a 70-kDa heat-shock protein (hsp70; Chirico *et al.*, 1988; Deshaies *et al.*, 1988b). This protein was proposed to function as an ATP-dependent unfoldase that prevents the precursor from aggregating (Deshaies *et al.*, 1988a; Meyer, 1988). ATP hydrolysis results in release of hsp70 from the precursor, a process that is thought to occur shortly before translocation across the membrane(s). Proteins with similar functions in preventing proteins from aggregation are found in the lumen of the endoplasmic reticulum (BiP; Pelham, 1988), in mitochondria (Craig *et al.*, 1989), in chloroplasts (Marshall *et al.*, 1990), and in prokaryotes (e.g., the dnaK protein in *E. coli;* Lebowitz *et al.*, 1985;

Zylicz et al., 1989). This family of unfoldases has been called molecular chaperones (for a review, see Ellis and Hemmingsen, 1989).

Expression of precursors in E. coli and subsequent purification has made it possible to study the requirements for cytosolic factors of the import of nuclear-encoded proteins into chloroplasts. Import into chloroplasts of the LHCPII precursor was found to require at least two cytosolic factors (Waegemann et al., 1990). Although hsp70 stimulated the import of the E. coli purified protein, no association with a hsp70 protein was observed in a different study (K. Keegstra, 1991, personal communication). Evidence also exists for the presence of additional cytosolic factors apart from the 70-kDa chaperone in mitochondrial import (Murakami and Mori, 1990). This additional factor might act by presenting the protein in an import-competent state, maybe by exposing the transit peptide. For two other chloroplast precursors, ferredoxin and plastocyanin, that were expressed in E. coli, no evidence exists for the requirement of cytosolic factors (Pilon et al., 1990; A. D. de Boer, 1991, unpublished observation). These small proteins may be folded and unfolded easily and therefore do not require chaperones. Alternatively, unfoldases in the chloroplast envelope may act on these proteins just before translocation.

The first evidence that unfolding of precursor proteins is usually necessary for translocation across membranes came from studies with a fusion protein. Import into mitochondria of this fusion protein, a fusion between a mitochondrial targeting peptide and the cytosolic protein dihydrofolate reductase (DHFR), could be blocked by the presence of the substrate analog methotrexate (Eilers and Schatz, 1986). Methotrexate is known to stabilize the structure of DHFR and was found to stabilize the structure of the fusion protein also. A similar observation was made for import into chloroplasts. Import into chloroplasts of the precursor protein of 5-enolpyruvylshikimate 3-phosphate synthase (pEPSPS) was reduced severely when the structure was stabilized by the addition of substrate and a herbicide (Della-Cioppa and Kishore, 1988). These and other data support the view that unfolding is required for translocation of proteins across membranes.

### B. Components of Translocation Complex

The first step in the import of a precursor into the chloroplast is its binding to a proteinaceous receptor in the envelope membrane. The presence of receptor proteins in the chloroplast envelope was recognized several decades ago. Mild protease treatment of intact chloroplasts was found to reduce binding and import of proteins (Chua and Schmidt, 1978; Bohnert et al., 1985).

## 9. Import and Routing

Binding and import could be separated from each other by destroying the ATP generating system of the chloroplast or by lowering the temperature (Cline et al., 1985a; Friedman and Keegstra, 1989). Both treatments resulted in binding of precursors to the envelope but no import. This binding assay made it possible to titrate the number of binding sites (Friedman and Keegstra, 1989). Saturation was reached between 1500 and 3500 molecules bound per chloroplast. Pretreatment of the chloroplasts with a protease reduced the number of binding sites to about 700, whereas the affinity of the remaining binding sites for the precursor was not influenced. However, when chloroplasts were pretreated with the cysteine-specific protein modifying agent $N$-ethylmaleimide (NEM), the number of binding sites was unaltered whereas the affinity for the precursor was reduced (Friedman and Keegstra, 1989).

Import competition studies with the plastocyanin precursor purified after expression in *E. coli* and with other precursors synthesized in a wheat germ system showed that plastocyanin was able to reduce the import of the other precursor proteins significantly (A. D. de Boer, 1991, unpublished observations). Plastocyanin, ferredoxin, and the small subunit of Rubisco seem to make use of the same import pathway. A similar conclusion was reached when competition studies were performed with synthetic transit peptides (Perry et al., 1991). Competition studies for binding show not only that the translocation complex is identical but also that the amount of different receptors is limited. A limited number of different receptors was expected already since chloroplast proteins could be imported into other plastids and vice versa (De Boer, et al., 1988; Klösgen et al., 1989). Specific receptors are unlikely to be present in plastids in tissues in which the corresponding precursor normally is not present. Mitochondria also make use of only a limited number of receptor proteins for the import of proteins (Pfanner and Neupert, 1990). In *Neurospora crassa* mitochondria, two types of receptor proteins have been identified. One receptor, MOM19 (Söllner et al., 1989), is used by most precursor proteins; the second receptor, MOM72 (Söllner et al., 1990), is specific for the ADP/ATP carrier precursor. In yeast mitochondria one receptor, MAS70 (Hines et al., 1990), also is used by most precursor proteins, including the ADP/ATP carrier precursor; the specificity of another supposed receptor, p32 (Murakami et al., 1990; Pain et al., 1990), is not yet known. None of the yeast receptor proteins is found to be essential for viability (Baker and Schatz, 1991).

In addition to receptor proteins, there are proteins needed for the actual translocation of precursor proteins across the envelope. These proteins probably form a channel through which the precursors are transported across the membranes. These channel-forming proteins are expected to be localized in the contact sites between the inner and outer envelope membranes. In yeast mitochondria, one component, ISP42 (Baker *et al.*, 1990), of this translocation complex has been identified; its homolog in *N. crassa* is MOM38 (the general insertion protein, GIP; Kiebler *et al.*, 1990). Another protein in *N. crassa*, MOM22 (Kiebler *et al.*, 1990), is associated with the translocation complex but its function is still unknown.

The translocation complex in chloroplasts is less well characterized; to date only one protein has been identified using anti-idiotype antibodies against a synthetic transit peptide of the small subunit of Rubisco (Pain *et al.*, 1988). The gene for this 30-kDa protein is cloned and sequenced (Schnell *et al.*, 1990). The protein is present in contact sites between the inner and outer chloroplast envelope. Antibodies against the 30-kDa protein inhibit import of precursor proteins into the chloroplast. There are, however, some arguments against a function in protein transport. The same DNA sequence of the 30-kDa protein, as cloned and sequenced by another group, is claimed to be the gene for the phosphate translocator (see also Section IV,A; Willey *et al.*, 1991). More data are required to determine whether this gene encodes the phosphate translocator, a component of the tranlocation complex, or a protein with a dual function.

## C. Energetics of Envelope Translocation

Translocation of proteins across the chloroplast envelope requires the hydrolysis of ATP (Grossman *et al.*, 1980; Flügge and Hinz, 1986; Theg *et al.*, 1989). In the light, the chloroplast is able to provide the necessary ATP by photophosphorylation; in the dark, ATP must be provided externally. Other plastids are not able to photosynthesize and have an absolute requirement for externally added ATP (Boyle *et al.*, 1986; Schindler and Soll, 1986). ATP on the outside of the plastid probably is shuttled in by the ADP/ATP carrier. Experiments with inhibitors of photosynthesis and with a variety of ionophores have shown that ATP is the only requirement for translocation of proteins across the membrane. No membrane potential or proton motive force is required (Flügge and Hinz, 1986; Pain and Blobel, 1987; Theg *et al.*, 1989). The energy requirements are therefore different from those of mitochondria, in which a membrane potential is crucial for translocation of proteins into the organelle (Hartl *et al.*, 1989).

A small amount of ATP is required to bind precursor proteins to the outer membrane (Olsen *et al.* 1989). The amount of ATP is about five times lower than the total amount required for import of proteins into the chloroplast. Phosphorylation of a 51-kDa protein that was found to be correlated with protein import also required only a limited amount of ATP. Since the phosphorylation site could be removed by protease treatment of the chloroplast, the ATP necessary for binding may be required for phosphorylation of this 51-kDa protein (Hinz and Flügge, 1988). This protein therefore is a good candidate for the receptor or for a subunit of the receptor complex. The low amount of ATP probably is utilized in the intermembrane space (Olsen et al., 1989).

To date, the purpose of the bulk of the ATP required is unknown. Perhaps it is necessary for a conformational change of the translocation complex or for release of the precursor from the complex. In mitochondria, ATP is required also, but only outside the organelle for release of the precursor from the chaperone (Eilers and Schatz, 1988). In the chloroplast, this function for the ATP is not supported by the experimental evidence since ATP is thought to be used on the inside of the chloroplast (Keegstra et al., 1987; Theg et al., 1989), although some controversy exists over this claim. However, for most of the energetics studies with pea chloroplasts, small proteins such as the small subunit of Rubisco, ferredoxin, and plastocyanin have been used (Theg et al., 1989); these proteins may not require a chaperone for transport in the cytoplasm (see Section III,A). For example, experiments with ferredoxin and plastocyanin purified after expression in E. coli do not provide any evidence of a chaperone for import into chloroplasts (M. Pilon, 1991, unpublished observations; A. D. de Boer, 1991, unpublished observations). Therefore, ATP may also be required outside the chloroplast, for example, in the import of the LHCPII protein, which has a clear need for cytosolic factors.

### D. Stromal Processing

After translocation of precursor proteins across the envelope membranes, the transit peptide is removed specifically by the stromal processing peptidase (SPP). This stromal protease first was observed by Dobberstein et al. (1977) and subsequently was localized to the chloroplast stroma (Smith and Ellis, 1979). The activity was partially purified and was found to consist of a 180-kDa complex (Robinson and Ellis, 1984). The activity is inhibited only by metal chelators such as ethylenediaminetetraacetic acid (EDTA). ATP is not required for the activity.

Whether the mitochondrial and chloroplast processing proteases are evolutionarily related is not known, but some similarities exist. The enzyme activity responsible for processing mitochondrial precursor proteins also was found to be a metalloprotease. The enzyme complex in mitochondria also has a high molecular weight and is composed of two subunits, MAS1 and MAS2 in yeast (Jensen and Yaffe, 1988; Witte et al., 1988) and

PEP and MPP in *N. crassa* (Hawlitschek *et al.*, 1988). MAS1 and MAS2 were found to be essential for viability in yeast (Baker and Schatz, 1991).

### E. Stromal Factors

After translocation across the chloroplast envelope membranes, proteins usually associate with stromal factors before they assemble into protein complexes or are routed to a different location. The first stromal factor that was identified was a protein complex that binds the large subunit of Rubisco (Barraclough and Ellis, 1980; Roy *et al.*, 1982). Although the large subunit protein is synthesized in the stroma, many nuclear-encoded proteins were found to associate with this 700-kDa complex also after import into the chloroplast (Lubben *et al.*, 1989). The complex was found to consist of 14 subunits, 7 $\alpha$-subunits of 61 kDa and 7 $\beta$-subunits of 60 kDa (Musgrove *et al.*, 1987). The $\alpha$-subunit was cloned from wheat and was found to be highly homologous to the *E. coli* groEL protein (Hemmingsen *et al.*, 1988). The groEL complex consists of 14 identical subunits that are packed in two stacked rings of 7 subunits each (Horwich *et al.*, 1990). The protein binds to newly synthesized proteins and prevents aggregation. Assembly of the bound protein into protein complexes is catalyzed by the groEL complex. The bound protein is released from the groEL complex after ATP hydrolysis. groEL-like proteins are found in prokaryotes, mitochondria (hsp60), and chloroplasts; this subclass of molecular chaperones has been named chaperonins (Hemmingsen *et al.*, 1988).

Association with chaperonins was observed after import of several nuclear-encoded chloroplast proteins (Lubben *et al.*, 1989), such as the small subunit of Rubisco, the $\beta$-subunit of the $F_1$–ATP synthase complex, glutamine synthetase, and LHCPII. Moreover, nonchloroplast proteins such as chloramphenicol acetyltransferase and $\beta$-lactamase also were found to associate with chaperonins after import into the chloroplast. No indication for association with a chaperonin was observed for ferredoxin, superoxide dismutase, or plastocyanin.

In mitochondria, imported proteins also are found to associate with chaperonins (Ostermann *et al.*, 1989). The presence of a functional gene for the 60-kDa chaperonin (hsp60) is in fact found to be essential for viability in yeast (Baker and Schatz, 1991). The chaperonins are not the only factors inside the organelle that assist in protein import. In mitochondria, a 70-kDa heat-shock protein in the matrix was found to interact with translocated precursor proteins immediately after translocation and before their association with hsp60 (Kang *et al.*, 1990). The presence of this organellar chaperone also was found to be essential for viability in yeast

(Baker and Schatz, 1991). hsp70-like proteins also have been detected in the envelope and in the stroma of the chloroplast (Marshall *et al.*, 1990; K. Keegstra, 1991, personal communication); perhaps they perform the same function as in mitochondria.

Several other, mostly unknown, factors were found to be essential for assembly of proteins into protein complexes or into the thylakoid membrane. The assembly of Rubisco is, in addition to the large subunit-binding protein, dependent on a not fully identified stromal factor that seems to be homologous to the bacterial groES protein (Goloubinoff *et al.*, 1989). The LHCPII protein forms a 120-kDa complex with a stromal protein, probably after its association with the chaperonin complex (Payan and Cline, 1991). Insertion into the membrane seems to be dependent on yet another stromal factor (Payan and Cline, 1991, see also Chapter 10).

## IV. INTRAORGANELLE ROUTING

Nonstromal proteins need to contain information for routing to their final location. Routing information is necessary for nuclear-encoded proteins that enter the chloroplast through the default pathway to the stroma, but also for proteins that are encoded by the chloroplast genome. Proteins destined for the thylakoid lumen usually contain a cleavable thylakoid transfer domain in front of the mature protein (see Sections II,B and IV,B). Proteins destined for the thylakoid membrane normally contain information in the mature protein (see Section IV,A). Little is known about inner envelope membrane proteins, but they seem to contain information for membrane insertion in the mature protein (see Section IV,A). No examples are available of proteins that reside in the envelope intermembrane space. Therefore, it is not known whether they are imported into the chloroplast and subsequently back into the intermembrane space, as has been suggested for several mitochondrial intermembrane space proteins (Hartl *et al.*, 1989), or whether they only translocate across the outer membrane. All known nuclear-encoded outer-membrane proteins are synthesized without a transit peptide and seem to enter the membrane directly, without translocation into the stroma (see Section IV,A).

### A. Membrane Insertion

Only a few cloned genes are available that encode chloroplast envelope proteins. Consequently, only limited knowledge about protein insertion into the envelope membranes exists. The genes for a 6.7-kDa (Salomon *et al.*, 1990) and a 14-kDa (Li *et al.*, 1991) outer membrane protein, both with unknown functions, have been cloned and sequenced. These nuclear-encoded outer envelope membrane proteins are synthesized as native proteins without a cleavable targeting signal. ATP is not required for insertion of either protein into the membrane. Pretreatment of the chloroplasts with protease does not affect insertion. Whether cytosolic factors are involved is not known. Integration into the chloroplast outer membrane probably occurs by direct interaction of the hydrophobic core in the proteins with the lipids in the membrane (see Fig. 2). However, interaction with a protease-resistant protein cannot be excluded.

The cryptic genes for two inner envelope membrane proteins, a 37-kDa (Dreses-Werringloer et al., 1991) and a 30-kDa (K. Keegstra, 1991, personal communication) protein, have been cloned and sequenced. Their products have been used to study integration into the membrane. Both proteins are synthesized in the cytoplasm as precursor proteins with a cleavable N-terminal extension. ATP is necessary for arrival at the inner membrane; interaction with the chloroplast envelope does not occur when the chloroplasts are pretreated with protease. Interaction with a receptor protein in the outer envelope is assumed from these data. Whether the N-terminal extension acts as a normal transit peptide by translocating the

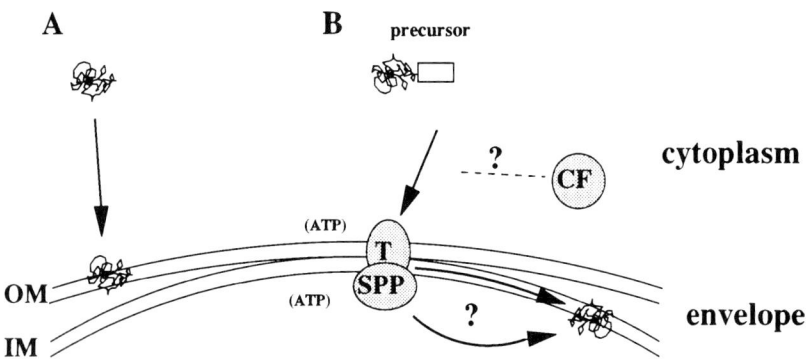

Fig. 2. Model for the insertion of an outer membrane protein (A) and an inner membrane protein (B) into the chloroplast envelope. The outer membrane protein interacts with and integrates into the outer membrane (OM) in the absence of ATP or a proteinaceous receptor. Whether cytosolic factors (CF) are required is not known. The inner membrane protein interacts with a proteinaceous translocation complex (T). Interaction with cytosolic factors is expected to occur. ATP is needed for binding and translocation. Complete translocation into the stroma might occur, followed by insertion into the inner membrane (IM), or the protein might be arrested in the inner membrane during translocation. The transit peptide is removed by stromal processing peptidase (SPP).

protein into the stroma is not known. If this is true, the mature part of the protein must integrate into the inner membrane from the stromal side (see Fig. 2). Whether proteins are involved in this insertion is not clear but, because of the similar lipid composition of the inner envelope membrane and the thylakoid membrane (Douce and Joyard, 1990), it is difficult to imagine proper sorting without them. Alternatively, it is possible that inner envelope membrane proteins are arrested in the membrane during translocation. However, if this is true, it is puzzling why other proteins with a stop-transfer sequence (Lubben et al., 1987) or with membrane-spanning domains (Cheung et al., 1988) are not arrested but are translocated into the stroma instead.

The gene for one other envelope protein is cloned and sequenced. This 30-kDa protein is reported to be the phosphate translocator that is present in the inner envelope membrane (Flügge et al., 1989; Willey et al., 1991), whereas another group claims that it is the import receptor present in the contact sites. More experiments are needed to determine the real function of this protein. The protein is synthesized in the cytoplasm as a precursor protein; therefore, translocation of at least the N-terminal part of the protein is expected for processing. These data do not clarify whether the protein is localized in the inner envelope membrane or in the contact sites. The experimental result that the protein is protease resistant (Flügge et al., 1989) is, however, not expected for the protease-sensitive receptor that is present in the contact sites. Additional data are necessary to clarify this ambiguity.

Proteins that are inserted into the thylakoid membrane contain noncleavable routing information in the protein. The exception of this rule is the chloroplast-encoded protein cytochrome $f$ (Alt and Herrmann, 1984; Willey et al., 1984). The main part of this protein is located at the lumen of the thylakoid membrane; as a consequence, most of the protein must translocate through the membrane. The protein is synthesized in the stroma with a thylakoid transfer domain in front of the protein. A stop-transfer sequence at the C terminus of the protein prevents its complete translocation. The extension is removed from the protein by the thylakoid processing peptidase. All other thylakoid membrane proteins that have been examined enter the membrane without a cleavable sequence. These thylakoid membrane proteins contain one or more membrane-spanning domains that probably also signal membrane insertion. The orientation of the proteins in the membrane is such that most of the positive charges are located at the stromal side of the membrane. Insertion and orientation of proteins in the thylakoid membrane is reviewed more thoroughly elsewhere (De Boer and Weisbeek, 1991, see also Chapter 10).

## B. Thylakoid Membrane Translocation

Translocation of proteins into the thylakoid lumen was studied first with the precursor of plastocyanin (Smeekens et al., 1986). After translocation of the protein into the stroma and partial processing, the intermediate-sized protein was imported into the thylakoid lumen and processed to its mature size. This two-step import pathway (see Fig. 3) also was observed for the 16-, 23-, and 33-kDa subunits of the water splitting complex (James et al., 1989; Ko and Cashmore, 1989).

The first step, the translocation across the envelope membranes, is comparable to import of stromal proteins. The first part of the targeting signal can be exchanged for a normal transit peptide (Ko and Cashmore, 1989; Hageman et al., 1990). The requirements for the second part, translo-

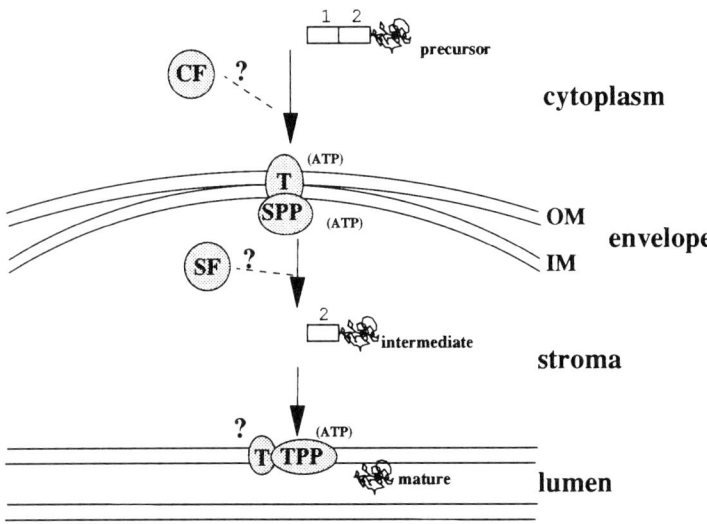

**Fig. 3.** Model for the two-step routing display pathway of a thylakoid lumen protein. The precursor is synthesized with a two-domain targeting signal and is expected to bind cytoplasmic factors (CF) before interacting with a proteinaceous translocation complex (T). After translocation, the first domain of the targeting signal is removed by stromal processing peptidase (SPP). The intermediate-sized protein is supposed to interact with stromal factors (SF). The second domain of the targeting signal is removed by thylakoidal processing peptidase (TPP) after translocation into the lumen. A proteinaceous translocation complex might be present in the thylakoid membrane since the lipid composition of the inner envelope membrane and the thylakoid membrane is very similar. ATP is needed for binding to the translocation complex in the envelope and for translocation across the envelope and the thylakoid membranes.

cation across the thylakoid membrane, were more difficult to determine. Fusions between the two-domain targeting signal and a passenger protein usually were not successful in proper targeting of the passenger *in vitro;* an intermediate-sized protein was found in the stromal fraction instead. This unsuccessful routing of passenger proteins was, however, not observed *in vivo* (see Section II,B; De Boer *et al.*, 1991).

Although an *in vitro* thylakoid translocation system has been developed, it works only with full-length precursor proteins and not with the intermediate-sized protein (Kirwin *et al.*, 1989). There may be a direct coupling of translocation across the envelope and thylakoid membrane. Alternatively, the intermediate-sized protein *in vitro* may adapt a conformation that is not compatible with translocation across the thylakoid membrane. ATP, isolated thylakoids, and the stromal processing peptidase at least were required for efficient thylakoid translocation. Whether stromal factors are required for translocation or whether a proteinaceous translocation complex is present in the thylakoid membranes is not known.

## C. Thylakoid Processing

Thylakoid lumen proteins are processed to their mature sizes by a processing peptidase present in the thylakoid membrane (Hageman *et al.*, 1986). The processing activity has been purified partially and was found to be insensitive to several types of protease inhibitors (Kirwin *et al.*, 1987, 1988). The presence of Triton X-100 was essential during purification to prevent inactivation of the protease. The protease has a low pH optimum of 6.5–7.0; the active site of the protein is present at the lumenal side of the thylakoid membrane. The protein is able to process both the intermediate and the precursor form of thylakoid lumen proteins. The protease is found solely in the stroma lamellae of the thylakoid membrane and is not associated with other protein complexes.

The thylakoidal processing peptidase resembles the leader peptidase of *E. coli* and the eukaryotic signal peptidase in the endoplasmic reticulum with respect to enzymatic activity and membrane orientation (Halpin *et al.*, 1989). The inner membrane protease I of yeast is also related to these three enzymes (Schneider *et al.*, 1991). These relationships imply that the translocation machinery, or at least the processing complex necessary for routing inside organelles, is a remnant of the prokaryotic export device. In cyanobacteria, and probably also in chloroplasts, the situation is a bit more complicated. Proteins are translocated from inside the organelle across both the inner envelope membrane and the thylakoid membrane.

Therefore, we are left to determine whether the supposed translocation/ processing complexes in the two membranes are similar or more likely to be substantially different.

## REFERENCES

Alt, J., and Herrmann, R. G. (1984). Nucleotide sequence of the gene for pre-apocytochrome *f* in the spinach plastid chromosome: *Curr. Genet.* **8,** 551–557.

Anderson, J. M., Hnilo, J., Larson, R., Okita, T. W., Morell, M., and Preiss, J. (1989). The encoded primary sequence of a rice seed ADP-glucose pyrophosphorylase subunit and its homology to the bacterial enzyme. *J. Biol. Chem.* **264,** 12238–12242.

Baker, K. P., and Schatz, G. (1991). Mitochondrial proteins essential for viability mediated protein import into yeast mitochondria. *Nature (London)* **349,** 205–208.

Baker, K. P., Schaniel, A., Vestweber, D., and Schatz, G. (1990). A yeast mitochondrial outer membrane protein essential for protein import and cell viability. *Nature (London)* **348,** 605–609.

Baldauf, S. L., and Palmer, J. D. (1990). Evolutionary transfer of the chloroplast *Tuf*A gene to the nucleus. *Nature (London)* **344,** 262–265.

Barraclough, R., and Ellis, R. J. (1980). Protein synthesis in chloroplasts IX: Assembly of newly synthesized large subunits into ribulose bisphosphate carboxylase in isolated intact pea chloroplasts. *Biochim. Biophys. Acta* **608,** 19–31.

Bohnert, H. J., Michalowski, C., Bevacqua, S., Mucke, E., and Loffelhardt, W. (1985). Cyanelle DNA from *Cyanophora paradoxa* physical mapping and location of protein coding regions. *Mol. Gen. Genet.* **201,** 565–574.

Boyle, S. A., Hemmingsen, S. M., and Dennis, D. T. (1986). Uptake and processing of the precursor to the small subunit of ribulose 1,5-bisphosphate carboxylase by leucoplasts from the endosperm of developing Castor oil seeds. *Plant Physiol.* **81,** 817–822.

Brinkman, H., Cerff, R., Salomon, M., and Soll, J. (1989). Cloning and sequence analysis of cDNAs encoding the cytosolic precursor of subunits GapA and GapB of chloroplast glyceraldehyde-3-phosphate dehydrogenase from pea and spinach. *Plant Mol. Biol.* **13,** 81–94.

Cheung, A. Y., Bogorad, L., Van Montagu, M., and Schell, J. (1988). Relocating a gene for herbicide tolerance: A chloroplast gene is converted into a nuclear gene. *Proc. Natl. Acad. Sci. USA* **85,** 391–395.

Chirico, W. J., Waters, M. G., and Blobel, G. (1988). 70 K heat shock related proteins stimulate protein translocation into microsomes. *Nature (London)* **332,** 805–810.

Chua, N.-H., and Schmidt, G. W. (1978). Post-translational transport into intact chloroplasts of a precursor to the small subunit of ribulose-1,5-bisphosphate carboxylase. *Proc. Natl. Acad. Sci. USA* **75,** 6110–6114.

Chua, N.-H., and Schmidt, G. W. (1979). Transport of proteins into mitochondria and chloroplasts. *J. Cell Biol.* **81,** 461–483.

Cline, K., Werner-Washburne, M., Lubben, T. H., and Keegstra, K. (1985a). Precursors to two nuclear-encoded chloroplast proteins bind to the outer envelope membrane before being imported into chloroplasts. *J. Biol. Chem.* **260,** 3691–3696.

Cline, K., Keegstra, K., and Staehelin, L. A. (1985b). Freeze-fracture electron microscopic analysis of ultrarapidly frozen envelope membranes on intact chloroplasts and after purification. *Protoplasma* **125,** 111–123.

Craig, E. A., Kramer, J., Shilling, J., Werner-Washburne, M., Holmes, S., Kosic-Smither, J., and Nicolet, C. M. (1989). SSC1, an essential member of the *S. cerevisiae hsp*70 multigene family, encodes a mitochondrial protein. *Mol. Cell Biol.* **9**, 3000–3008.

Cremers, A. F. M., Voorhout, W. F., Van der Krift, T. P., Leunissen-Bijvelt, J. J. M., and Verkleij, A. J. (1988). Visualization of contact sites between outer and inner envelope membranes in isolated chloroplasts. *Biochim. Biophys. Acta* **933**, 334–340.

Danks, S. M., Evans, E. H., and Whittaker, P. A. (1983). "Photosynthetic Systems: Structure, Function, and Assembly." Wiley, New York.

De Boer, A. D., and Weisbeek, P. J. (1991). Chloroplast topogenesis: Protein import, sorting, and assembly. *Biochim. Biophys. Acta* **1071**, 221–253.

De Boer, D., Cremers, F., Teertstra, R., Smits, L., Hille, J., Smeekens, S., and Weisbeek, P. (1988). *In vivo* import of plastocyanin and a fusion protein into developmentally different plastids of transgenic plants. *EMBO J.* **7**, 2631–2636.

De Boer, D., Bakker, H., Lever, A., Bouma, T., Salentijn, E., and Weisbeek, P. (1991). Protein targeting towards the thylakoid lumen of chloroplasts: Proper localization of fusion proteins is only observed *in vivo*. *EMBO J.* **10**, 2765–2772.

Della-Cioppa, G., and Kishore, G. M. (1988). Import of a precursor protein into chloroplasts is inhibited by the herbicide glyphosate. *EMBO J.* **7**, 1299–1305.

Deshaies, R. J., Koch, B. D., and Schekman, R. (1988a). The role of stress proteins in membrane biogenesis. *Trends Biochem. Sci.* **13**, 384–388.

Deshaies, R. J., Koch, B. D., Werner-Washburne, M., Craig, E. A., and Schekman, R. (1988b). A subfamily of stress proteins facilitates translocation of secretory and mitochondrial precursor polypeptides. *Nature (London)* **332**, 800–805.

Dietz, K. J., and Bogorad, L. (1987). Plastid development in *Pisum sativum* leaves during greening: A comparison of plastid polypeptide composition and *in organello* translation characteristics. *Plant Physiol.* **85**, 808–815.

Dobberstein, B., Blobel, G., and Chua, N.-H. (1977). *In vitro* synthesis and processing of a putative precursor for the small subunit of ribulose-1,5-bisphosphate carboxylase of *Chlamydomonas reinhardtii*. *Proc. Natl. Acad. Sci. USA* **74**, 1082–1085.

Douce, R., and Joyard, J. (1990). Biochemistry and function of the plastid envelope. *Annu. Rev. Cell Biol.* **6**, 173–216.

Dreses-Werringloer, U., Fisher, K., Wachter, E., Link, T. A., and Flügge, U. I. (1991). cDNA sequence and deduced amino acid sequence of the precursor of the 37-kD inner envelope membrane polypeptide from spinach chloroplasts: Its transit peptide contains an amphiphilic α-helix as the only detectable structural element. *Eur. J. Biochem.* **195**, 361–368.

Eilers, M., and Schatz, G. (1986). Binding of a specific ligand inhibits import of a purified precursor protein into mitochondria. *Nature (London)* **322**, 228–232.

Eilers, M., and Schatz, G. (1988). Protein unfolding and the energetics of protein translocation across biological membranes. *Cell* **52**, 481–483.

Ellis, R. J. (1981). Chloroplast proteins: Synthesis, transport, and assembly. *Annu. Rev. Plant Physiol.* **31**, 111–138.

Ellis, R. J. (1983). Chloroplast protein synthesis: Principles and problems. *Subcell. Biochem.* **9**, 237–261.

Ellis, R. J., and Hemmingsen, S. M. (1989). Molecular chaperones: Proteins essential for the biogenesis of some macromolecular structures. *Trends Biochem. Sci.* **14**, 339–342.

Flügge, U. I., and Hinz, G. (1986). Energy dependence of protein translocation into chloroplasts. *Eur. J. Biochem.* **160**, 563–570.

Flügge, U. I., Fischer, K., Gross, A., Sebald, W., Lottspeich, F., and Eckerskorn, C. (1989). The triose phosphate-3-phosphoglycerate-phosphate translocator from spinach chloroplasts nucleotide sequence of a full-length complementary DNA clone and import of the *in vitro* synthesized precursor protein into chloroplasts. *EMBO J.* **8**, 39–46.

Fránzen, L.-G., Rochaix, J.-D., and Von Heijne, G. (1990). Chloroplast transit peptides from the green alga *Chlamydomonas reinhardtii* share feature with both mitochondrial and higher plant chloroplast presequences. *FEBS Lett.* **260**, 165–168.

Friedman, A. L., and Keegstra, K. (1989). Chloroplast protein import quantitative analysis of precursor binding. *Plant Physiol.* **89**, 993–999.

Gabellini, N. (1988). Organization and structure of the genes for the cytochrome $b/c1$ complex in purple photosynthetic bacteria. A phylogenetic study describing the homology of the $b/c1$ subunits between prokaryotes, mitochondria, and chloroplasts. *J. Bioenerg. Biomembr.* **20**, 59–83.

Gavel, Y., and Von Heijne, G. (1990). A conserved cleavage-site motif in chloroplast transit peptides. *FEBS Lett.* **261**, 455–458.

Goloubinoff, P., Christeller, J. T., Gatenby, A. A., and Lorimer, G. H. (1989). Reconstitution of active dimeric ribulose bisphosphate carboxylase from an unfolded state depends on two chaperonin proteins and Mg-ATP. *Nature (London)* **342**, 884–889.

Gray, M. W., and Doolittle, W. F. (1982). Has the endosymbiont hypothesis been proven? *Microbiol. Rev.* **46**, 1–42.

Grossman, A., Bartlett, S., and Chua, N.-H. (1980). Energy-dependent uptake of cytoplasmically synthesized polypeptides by chloroplasts. *Nature (London)* **285**, 625–628.

Hageman, J., Robinson, C., Smeekens, S., and Weisbeek, P. (1986). A thylakoid processing protease is required for complete maturation of the lumen protein plastocyanin. *Nature (London)* **324**, 567–569.

Hageman, J., Baecke, C., Ebskamp, M., Pilon, R., Smeekens, S., and Weisbeek, P. (1990). Protein import into and sorting inside the chloroplast are independent processes. *Plant Cell* **2**, 479–494.

Hall, D. O., and Rao, K. K. (1987). "Photosynthesis," Edward Arnold, London.

Halpin, C., Elderfield, P. D., James, H. E., Zimmermann, R., Dunba, B., and Robinson, C. (1989). The reaction specificities of the thylakoidal processing peptidase and *Escherichia coli* leader peptidase are identical. *EMBO J.* **8**, 3917–3921.

Hartl, F.-U., Pfanner, N., Nicholson, D. W., and Neupert, W. (1989). Mitochondrial protein import. *Biochim. Biophys. Acta* **988**, 1–45.

Hawlitschek, G., Schneider, H., Schmidt, B., Tropschug, M., Hartl, F.-U., and Neupert, W. (1988). Mitochondrial protein import: Identification of processing peptidase and of PEP, a processing enhancing protein. *Cell* **53**, 795–806.

Hemmingsen, S. M., Woolford, C., Van der Vies, S., Tilly, K., Dennis, D. T., Georgopoulos, C. P., Hendrix, R. W., and Ellis, R. J. (1988). Homologous plant and bacterial proteins chaperone oligomeric protein assembly. *Nature (London)* **333**, 330–334.

Hines, V., Brandt, A., Griffiths, G., Horstmann, H., Brütsch, H., and Schatz, G. (1990). Protein import into yeast mitochondria is accelerated by the outer membrane protein MAS70. *EMBO J.* **9**, 3191–3200.

Hinz, G., and Flügge, U. I. (1988). Phosphorylation of 51-kD envelope membrane polypeptide involved in protein translocation into chloroplasts. *Eur. J. Biochem.* **175**, 649–659.

Horwich, A. L., Neupert, W., and Hartl, F.-U. (1990). Protein-catalysed protein folding. *Trends Biotechnol.* **8**, 126–131.

James, H. E., Bartling, D., Musgrove, J. E., Kirwin, P. M., and Herrmann, R. G. (1989). Transport of proteins into chloroplasts import and maturation of precursors to the 33-kDa, 23-kDa, and 16-kDa proteins of the photosynthetic oxygen-evolving complex. *J. Biol. Chem.* **264**, 19573–19576.

Jensen, R. E., and Yaffe, M. P. (1988). Import of proteins into yeast mitochondria: The

nuclear MAS2 gene encodes a component of the processing protease that is homologous to the MAS1-encoded subunit. *EMBO J.* **7**, 3863–3871.
Joyard, J., Grossman, A., Bartlett, S. G., Douce, R., and Chua, N.-H. (1982). Characterization of envelope membrane polypeptides from spinach chloroplasts. *J. Biol. Chem.* **257**, 1095–1101.
Kang, P.-J., Ostermann, J., Shilling, J., Neupert, W., Craig, E. A., and Pfanner, N. (1990). Requirement for *hsp*70 in the mitchondria matrix for translocation and folding of precursor proteins. *Nature (London)* **348**, 137–143.
Keegstra, K., Bauerle, C., Lubben, T., Smeekens, S., and Weisbeek, P. (1987). Targeting of proteins into chloroplasts *In* "UCLA Symposia on Molecular and Cellular Biology New Series, Plant Gene Systems and Their Biology" (J. L. Key and L. McIntosh, eds.), Vol. 62, pp. 363–370. A. R. Liss, New York.
Keegstra, K., Olsen, L. J., and Theg, S. M. (1989). Chloroplastic precursors and their transport across the envelope membranes. *Annu. Rev. Physiol. Plant Mol. Biol.* **40**, 471–502.
Kiebler, M., Pfaller, R., Söllner, T., Griffiths, G., Horstmann, H., Pfanner, N., and Neupert, W. (1990). Identification of a mitochondrial receptor complex required for recognition and membrane insertion of precursor proteins. *Nature (London)* **348**, 610–616.
Kirwin, P. M., Elderfield, P. D., and Robinson, C. (1987). Transport of proteins into chloroplasts partial purification of a thylakoidal processing peptidase involved in plastocyanin biogenesis. *J. Biol. Chem.* **262**, 16386–16390.
Kirwin, P. M., Elderfield, P. D., Williams, R. S., and Robinson, C. (1988). Transport of proteins into chloroplasts organization and lateral distribution of the plastocyanin processing peptidase in the thylakoid network. *J. Biol. Chem.* **263**, 18128–18132.
Kirwin, P. M., Meadows, J. W., Shackleton, J. B., Musgrove, J. E., Elderfield, P. D., Mould, R., Hay, N. A., and Robinson, C. (1989). ATP dependent import of a lumenal protein by isolated thylakoid vesicles. *EMBO J.* **8**, 2251–2255.
Klösgen, R. B., Saedler, H., and Weil, J.-H. (1989). The amyloplast-targeting transit peptide of the waxy protein of maize also mediates protein transport *in vitro* into chloroplasts. *Mol. Gen. Genet.* **217**, 155–161.
Ko, K., and Cashmore, A. R. (1989). Targeting of proteins to the thylakoid lumen by the bipartite transit peptide of the 33-kd oxygen-evolving protein. *EMBO J.* **8**, 3187–3194.
Lebowitz, J. H., Zylicz, M., Georgopoulos, C., and McMacken, R. (1985). Imitation of DNA replication on single-stranded DNA templates catalyzed by purified replication proteins of bacteriophage lambda and *Escherichia coli. Proc. Natl. Acad. Sci. USA* **82**, 3988–3992.
Li, H. M., Moore, T., and Keegstra, K. (1991). Targeting of proteins to the outer envelope membrane uses a different pathway than transport into chloroplasts. *Plant Cell* **3**, 709–717.
Lubben, T. H., Bansberg, J., and Keegstra, K. (1987). Stop-transfer regions do not halt translocation of proteins into chloroplasts. *Science* **238**, 1112–1114.
Lubben, T. H., Donaldson, G. K., Viitanen, P. V., and Gatenby, A. A. (1989). Several proteins imported into chloroplasts form stable complexes with the GroEL-related chloroplast molecular chaperone. *Plant Cell* **1**, 1223–1230.
Marshall, J., DeRocher, A. E., Keegstra, K., and Vierling, E. (1990). Identification of heat shock *hsp*70 homologues in chloroplasts. *Proc. Natl. Acad. Sci. USA* **87**, 374–378.
Meyer, D. I. (1988). Preprotein conformation: The year's major theme in translocation studies. *Trends Biochem. Sci.* **13**, 471–474.
Murakami, H., Blobel, G., and Pain, D. (1990). Isolation and characterization of the gene for a yeast mitochondrial import receptor. *Nature (London)* **347**, 488–491.

Murakami, K., and Mori, M. (1990). Purified presequence binding factor (PBF) forms an import-competent complex with purified mitochondrial precursor protein. *EMBO J.* **9,** 3201–3206.

Musgrove, J. E., Johnson, R. A., and Ellis, R. J. (1987). Association of the ribulose bisphosphate-carboxylase large-subunit binding protein into dissimilar subunits. *Biochemistry* **163,** 529–534.

Nakano, K., Mori, H., and Fukui, T. (1989). Molecular cloning of complementary DNA encoding potato amyloplast alpha glucan phosphorylase and the structure of its transit peptide. *J. Biochem.* **106,** 691–695.

Oliver, J. L., Martin, A., and Martinez-Zapater, J. M. (1989). Chloroplast genes transferred to the nuclear plant genome have adjusted to nuclear base composition and codon usage. *Nucleic Acids Res.* **18,** 65–73.

Olsen, L. J., Theg, S. M., Selman, B. R., and Keegstra, K. (1989). ATP is required for the binding of precursor proteins to chloroplasts. *J. Biol. Chem.* **264,** 6724–6729.

Ostermann, J., Horwich, A. L., Neupert, W., and Hartl, F.-U. (1989). Protein folding in mitochondria requires complex formation with *hsp*60 and ATP hydrolysis. *Nature (London)* **341,** 125–130.

Ostrem, J. A., Ramage, R. T., Bohnert, H. J., and Wasmann, C. C. (1989). Deletion of the carboxyl-terminal portion of the transit peptide affects processing but not import or assembly of the small subunit of ribulose-1,5-bisphosphate carboxylase. *J. Biol. Chem.* **264,** 3662–3665.

Pain, D., and Blobel, G. (1987). Protein import into chloroplasts requires a chloroplast ATPase. *Proc. Natl. Acad. Sci. USA* **84,** 3288–3292.

Pain, D., Kanwar, Y. S., and Blobel, G. (1988). Identification of a receptor for protein import into chloroplasts and its localization to envelope contact zones. *Nature (London)* **331,** 232–237.

Pain, D., Murakami, H., and Blobel, G. (1990). Identification of a receptor for protein import into mitochondria. *Nature (London)* **347,** 444–449.

Payan, L. A., and Cline, K. (1991). A stromal factor maintains the solubility and insertion competence of an imported thylakoid membrane protein. *J. Cell Biol.* **112,** 603–613.

Pelham, H. (1988). Coming in from the cold. *Nature (London)* **332,** 776–777.

Perry, S. E., Buvinger, W. E., Bennett, J., and Keegstra, K. (1991). Synthetic analogues of a transit peptide inhibit binding or translocation of chloroplast precursor proteins. *J. Biol. Chem.* **266,** 11882–11889.

Pfanner, N., and Neupert, W. (1990). The mitochondrial protein import apparatus. *Annu. Rev. Biochem.* **59,** 331–351.

Pilon, M., De Boer, A. D., Knols, S. L., Koppelman, M. H. G. M., Van der Graaf, R. M., De Kruijff, B., and Weisbeek, P. J. (1990). Expression in *Escherichia coli* and purification of a translocation-competent precursor of the chloroplast protein ferredoxin. *J. Biol. Chem.* **265,** 3358–3361.

Reiss, B., Wasmann, C. C., and Bohnert, H. J. (1987). Regions in the transit peptide of ssu essential for transport into chloroplasts. *Mol. Gen. Genet.* **209,** 116–121.

Reiss, B., Wasmann, C. C., Schell, J., and Bohnert, H. J. (1989). Effect of mutations on the binding and translocation functions of a chloroplast transit peptide. *Proc. Natl. Acad. Sci. USA* **86,** 886–890.

Robinson, C., and Ellis, R. J. (1984). Transport of proteins into chloroplasts partial purification of a chloroplast protease involved in the processing of imported precursor polypeptides. *Eur. J. Biochem.* **142,** 337–342.

Roy, H., Bloom, M., Milos, P., and Monroe, M. (1982). Studies on the assembly of large subunits of ribulose bisphosphate carboxylase in isolated pea chloroplasts. *J. Cell Biol.* **94**, 20–27.

Salomon, M., Fisher, K., Flügge, U. I., and Soll, J. (1990). Sequence analysis and protein import studies of an outer chloroplast envelope polypeptide. *Proc. Natl. Acad. Sci. USA* **87**, 5778–5782.

Scherer, D. E., and Knauf, V. C. (1987). Isolation of a complementary DNA clone for the acyl carrier protein-i of spinach. *Plant Mol. Biol.* **9**, 127–134.

Schindler, C., and Soll, J. (1986). Protein transport in intact, purified pea etioplasts. *Arch. Biochem. Biophys.* **247**, 211–220.

Schmidt, G. W., Devillers-Thiery, A., Desruisseaux, H., Blobel, G., and Chua, N. H. (1979). $NH_2$-terminal amino acid sequences of precursor and mature forms of the ribulose-1,5-bisphosphate carboxylase small subunit from *Chlamydomonas reinhardtii*. *J. Cell Biol.* **83**, 615–622.

Schneider, A. Behrens, M., Scherer, P., Pratje, E., Michaelis, G., and Schatz, G. (1991). Inner membrane protease I, an enzyme mediating intramitochondrial protein sorting in yeast. *EMBO J.* **10**, 247–254.

Schnell, D. J., Blobel, G., and Pain, D. (1990). The chloroplast import receptor is an integral membrane protein of chloroplast envelope contact sites. *J. Cell Biol.* **111**, 1825–1838.

Smeekens, S., De Groot, M., Van Binsbergen, J., and Weisbeek, P. (1985). Sequence of the precursor of the chloroplast thylakoid lumen protein plastocyanin. *Nature (London)* **317**, 456–458.

Smeekens, S. Bauerle, C., Hageman, J., Keegsrta, K., and Weisbeek, P. (1986). The role of the transit peptide in the routing of precursors toward different chloroplast compartments. *Cell* **46**, 365–376.

Smeekens, S., Van Steeg, H., Bauerle, C., Bettenbroek, H., and Keegstra, K. (1987). Import into chloroplasts of a yeast mitochondrial protein directed by ferredoxin and plastocyanin transit peptides. *Plant Mol. Biol.* **9**, 377–388.

Smeekens, S., Geerts, D., Bauerle, C., and Weisbeek, P. (1989). Essential function in chloroplast recognition of the ferredoxin transit peptide processing region. *Mol Gen. Genet.* **216**, 178–182.

Smith, S. M., and Ellis, R. J. (1979). Processing of small subunit precursor of ribulose bisphosphate carboxylase and its assembly into whole enzyme are stromal events. *Nature (London)* **278**, 662–664.

Söllner, T., Griffiths, G., Pfaller, R., Pfanner, N., and Neupert, W. (1989). MOM19, an import receptor for mitochondrial precursor proteins. *Cell* **59**, 1061–1070.

Söllner, T., Pfaller, R., Griffiths, G., Pfanner, N., and Neupert, W. (1990). A mitochondrial import receptor for the ADP/ATP carrier. *Cell* **62**, 107–115.

Staehelin, L. A. (1986). Chloroplast structure and supramolecular organization of photosynthetic membranes. *In* "Encyclopedia of Plant Physiology" (L. A. Staehelin and C. J. Arntzen, eds.), Vol. 19, pp. 1–84. Springer-Verlag, New York.

Steppuhn, J., Hermans, J., Nechushtai, R., Herrmann, G. S., and Hermann, R. G. (1989). Nucleotide sequences of cDNA clones encoding the entire precursor polypeptide for subunit VI and of the plastome-encoded gene for subunit VII of the photosystem reaction center from spinach. *Curr. Genet.* **16**, 99–108.

Theg, S. M., Bauerle, C., Olsen, L. J., Selman, B. R., and Keegstra, K. (1989). Internal ATP is the only energy requirement for the translocation of precursor proteins across chloroplastic membranes. *J. Biol. Chem.* **264**, 6730–6736.

Thomson, W. W., and Whatley, J. M. (1980). Development of non-green plastids. *Annu. Rev. Plant Physiol.* **31**, 375–394.

Turner, S., Burger-Wiersma, T., Giovannoni, S. T., Mur, L. R., and Pace, N. R. (1989). The relationship of a prochlorophyte *Prochlorothrix hollandica* to green chloroplasts. Nature (London) **337**, 380–382.

Von Heijne, G. (1985). Signal sequences: The limits of variation. *J. Miol. Biol.* **184**, 99–105.

Von Heijne, G. (1986a). Mitochondrial targeting sequences may form amphiphilic helices. *EMBO J.* **5**, 1335–1342.

Von Heijne, G. (1986b). Towards a comparative anatomy of *N*-terminal topogenic protein sequences. *J. Mol. Biol.* **189**, 239–242.

Von Heijne, G., and Nishikawa, K. (1991). Chloroplast transit peptides: The perfect random coil? *FEBS Lett.* **278**, 1–3.

Von Heijne, G., Steppuhn, J., and Herrmann, R. G. (1989). Domain structure of mitochondrial and chloroplast targeting peptids. *Eur. J. Biochem.* **180**, 535–545.

Waegemann, K., Paulsen, H., and Soll, J. (1990). Translocation of proteins into isolated chloroplasts requires cytosolic factors to obtain import competence. *FEBS Lett.* **261**, 89–92.

Wasmann, C. C., Reiss, B., and Bohnert, H. J. (1988). Complete processing of a small subunit of ribulose-1,5-bisphosphate carboxylase/oxygenase from pea requires the amino acid sequence Ile-Thr-Ser. *J. Biol. Chem.* **263**, 617–619.

Werner-Washburne, M., and Keegstra, K. (1985). Metabolite transport across the chloroplast envelope. *Oxf. Surv. Plant Mol. Biol.* **2**, 123–145.

Willey, D. L., Auffret, A. D., and Gray, J. C. (1984). Structure and topology of cytochrome *f* in pea chloroplast membranes. *Cell* **36**, 555–562.

Willey, D. L., Fisher, K., Wachter, E., Link, T. A., and Flügge, U. I. (1991). Molecular cloning and structural analysis of the phosphate translocator from pea chloroplasts and its comparison to spinach phosphate translocator. *Planta* **183**, 451–461.

Witte, C., Jensen, R. E., Yaffe, M. P., and Schatz, G. (1988). MAS1, a gene essential for yeast mitochondrial assembly, encodes a subunit of the mitochondrial processing protease. *EMBO J.* **7**, 1439–1447.

Zylicz, M., Ang, D., Liberek, K., and Georgopoulos, C. (1989). Initiation of lambda DNA replication with purified host- and bacteriophage-encoded proteins: The role of the *dna*K, *dna*J and *grp*E heat shock proteins. *EMBO J.* **8**, 1601–1608.

# 10

## Assembly and Reconstitution of Chlorophyll *a/b*-Containing Complexes

**HARALD PAULSEN**

Botanisches Institut III der Universität
W-8000 München 19, Germany

I. Introduction
II. Chlorophyll *a/b*-Containing Complexes and Apoproteins
   A. Antenna of Photosystem I
   B. Antenna of Photosystem II
   C. Structural Comparisons between Chl *a/b*-Containing Complexes
III. Assembly of Light-Harvesting Complex II
   A. Processing
   B. Insertion into Thylakoid Membrane
   C. Assembly with Pigments
   D. Aggregation of Pigment–Protein Complexes
IV. Reconstitutions
   A. Reconstitution of Antenna Complexes *in Vitro*
   B. Synergetic Organization of Chl *a/b*-Containing Complexes
   C. Reconstitution of Antenna Complexes into Membranes
V. Concluding Remarks
   References

## I. INTRODUCTION

All chlorophyll *a/b* (Chl *a/b*)-containing complexes known to date in higher plants and green algae are antenna complexes of the photosynthetic apparatus. These complexes are involved in harvesting light energy and funneling it into the reaction centers of photosystems I and II (PSI and PSII, respectively). The Chl *a/b*-containing complexes comprise the light-harvesting complex of PSI (LHCI) and the major light-harvesting complex of PSII (LHCII), as well as the minor light-harvesting complexes associated with PSII (CP29, CP26, and CP24). All these complexes share close structural similarity. They are intrinsic components of the thylakoid membrane. All apoproteins with known sequences contain three distinct hydrophobic regions that most likely correspond to three membrane-spanning domains. The pigments in these complexes are noncovalently associated with the proteins. They include Chl *a* and Chl *b* in various ratios, xanthophylls, and, in some cases, β-carotene.

Our knowledge about Chl *a/b*-containing complexes is far from well balanced: LHCII and its apoproteins are among the best-studied biomolecules, much less is known about LHCI, and some of the minor light-harvesting complexes have only come to our attention recently. The biochemistry of these pigment–protein complexes has been covered in excellent reviews (Green 1988; Bassi *et al.*, 1990; Thornber *et al.*, 1991). Therefore, I present only a short overview of some recent progress in this field and concentrate on our knowledge about the assembly of Chl *a/b*-containing complexes in the thylakoid membrane.

For the reasons just mentioned, most of the information reviewed here is about LHCII. Because of the functional and structural similarities between LHCII and the other Chl *a/b*-containing complexes, however, it would not be unexpected if the assembly pathways of all these complexes turn out to be similar. Thus, for the time being, LHCII may be viewed as a provisional model for the other Chl *a/b*-containing complexes.

## II. CHLOROPHYLL *a/b*-CONTAINING COMPLEXES AND APOPROTEINS

### A. Antenna of Photosystem I

The number of pigment-containing subunits of LHCI is not clear yet. Two Chl *a/b*-containing complexes have been separated in LHCI preparations from spinach and barley and called LHCIa and LHCIb or LHCI-680 and LHCI-730, respectively, according to their emission wavelengths at 70 K. At least three different polypeptides have been assigned to these complexes, all in the range of 20 to 25 kDa (Lam *et al.*, 1984; Bassi and Simpson, 1987; White and Green, 1987a). On the other hand, only one pigmented complex has been detected in LHCI of *Lemna gibba*, containing an apoprotein of 20 kDa (Nechushtai *et al.*, 1987). A similar situation has been described for *Chlamydomonas reinhardtii;* although several polypeptides of 20–30 kDa have been found in LHCI preparations, only one of 22 kDa was complexed with Chl. Of this polypeptide 10 copies have been estimated per reaction center (Schuster *et al.*, 1988). However, three distinct Chl *a/b* complexes were isolated from maize LHCI, each of which contained a single apoprotein of 24, 21, or 17 kDa (Vainstein et al., 1989b). Moreover, immunological studies on isolated LHCI from maize indicate that a number of different polypeptides are involved in pigment binding (Di Paolo et al., 1990). In another study on LHCI from *Chlamydomonas*, six polypeptides were found, all of them present in pigment–protein complexes (R. Bassi, 1991, personal communication). These apparently contradictory results need clarification.

LHCI contains roughly 100 Chl molecules (Bassi and Simpson, 1987; Bruce and Malkin, 1988), xanthophylls (Vainstein et al., 1989b), and $\beta$-carotene (Nechushtai et al., 1987). Three types of genes for LHCI apoproteins have been identified, most of them in tomato (Hoffman et al., 1987; Stayton et al., 1987; Pichersky et al., 1988, 1989; Pichersky and Green, 1990). By partial sequencing of isolated apoproteins, Ikeuchi et al. (1991) assigned the gene types I, II, and III to 20.5-, 23-, and 25-kDa proteins, respectively, in spinach. A fourth apoprotein of 21 kDa did not fit into any of these types and, thus, represents a fourth type of gene.

In *C. reinhardtii*, the PSI antenna can be isolated as a stable complex, called CP0, of about 200 kDa (Wollman and Bennoun, 1982; Herrin et al., 1987), containing a variety of polypeptides similar to the ones found in higher plant complexes. The isolation of a high molecular weight complex containing LHCI from barley has been described (Anandan and Thornber, 1990). This complex is equivalent to CP0 from *C. reinhardtii*, although somewhat smaller in size.

In *Euglena*, a gene has been isolated that encodes a large precursor protein that is subsequently split into five LHCI apoproteins of 18–19 kDa (Houlné and Schantz, 1988).

## B. Antenna of Photosystem II

The antenna of PSII is more complex than the one of PSI. It contains two proteins that bind Chl *a* but not Chl *b*, the 43- and 47-kDa proteins. In addition, at least four different Chl *a/b*-containing complexes have been identified in PSII preparations: the major PSII antenna complex LHCII and the so-called minor Chl *a/b*–protein complexes, CP29, CP26, and CP24.

### 1. Major Chl a/b-Containing Complex of PSII: LHCII

LHCII is by far the most abundant pigmented complex in higher plants, containing as much as half the protein and pigment in the thylakoid membrane. In addition to its role as a major antenna, LHCII is thought to function in distributing energy between PSI and PSII and in stacking thylakoid membranes to grana (Barber, 1982), although stacked thylakoids have been observed in an *Arabidopsis thaliana* mutant lacking LHCII (Murray and Kohorn, 1991). LHCII itself seems to be a mixture of several pigment–protein complexes. By isoelectric focusing under nondenaturing conditions, LHCII can be separated into as many as six bands (Dainese *et al.*, 1990). On denaturation of the complexes, several (up to six, Bassi *et al.*, 1990) apoproteins can be visualized by polyacrylamide gel electrophoresis, ranging in size from 26 to 30 kDa. The variety of apoproteins in LHCII is likely to be even larger, since comigrating LHCII apoproteins from different LHCII complexes display different epitopes (Dainese *et al.*, 1990).

The variety of LHCII apoproteins may reflect the presence of numerous genes encoding these proteins in all plants studied to date. For a review of LHCPII genes, often termed *Cab*-genes, see Buetow *et al.* (1988) and Chitnis and Thornber (1988). The LHCPII gene family can have 3 (in *Arabidopsis*, Leutwiler *et al.*, 1986) or as many as 17 (in *Petunia*, Stayton *et al.*, 1986) members. The majority of genes, designated type I, encodes closely related polypeptides that are highly conserved among species (Jansson and Gustafson, 1990). Type II genes are somewhat divergent from type I genes (about 80% of amino acids conserved in the mature part of the derived protein sequence). They contain an intron and encode a slightly shorter polypeptide. A third gene type has been added to the classification of LHCPII genes (Morishige and Thornber, 1990; Green *et al.*, 1991). LHCPII type III genes have been found in maize, barley, tomato, and spinach (R. Bassi, 1991, personal communication). Predicted amino acid sequences of all LHCPIIs contain three hydrophobic regions that are thought to correspond to three membrane-spanning domains (Karlin-Neumann *et al.*, 1985).

The presence of a gene family encoding distinct LHCPII species cannot be the only source of the heterogeneity observed among LHCII apoproteins. Precursor proteins translated from a single gene transcript *in vitro* can give rise to several mature protein bands that appear after import into isolated chloroplasts (Kohorn *et al.*, 1986; Lamppa, 1988). This result can only be explained by various translational start or stop sites or by post-translational modification. One such modification is processing, the removal of the transit peptide in the chloroplast. Evidence has been presented that indicates the existence of two processing sites in pLHCPII (Lamppa and Abad, 1987; Abad *et al.*, 1989; see Section III,A). Moreover, LHCPII is phosphorylated reversibly (Bennett 1983) by a membrane-bound kinase (Coughlan and Hind, 1987) whose activity is modulated by the redox state of the photosynthetic electron transport chain (Allen *et al.*, 1981), possibly through the cytochrome $b_6/f$ complex (Gal *et al.*, 1990).

## 10. Assembly of Chlorophyll *a/b*-Containing Complexes

Phosphorylation of LHCII causes destacking of thylakoids and increased interaction of LHCII with PSI (Allen *et al.*, 1981), thus altering the energy distribution between PSI and PSII (state transition). Other modifications of LHCII may occur as well. Palmitoylation of LHCII, although probably to a limited extent, has been detected (Mattoo and Edelman, 1987).

If LHCII is isolated from thylakoid membranes under mild conditions, it is obtained as a trimeric complex that is stable in some detergents (Butler and Kühlbrandt, 1988). These trimers easily form two-dimensional (Wang and Kühlbrandt, 1991) or three-dimensional crystals (Kühlbrandt and Wang, 1991). Analysis of three-dimensional crystals has yielded structural information resolved to 6 Å resolution. The LHCII monomer has three $\alpha$-helical stretches spanning the membrane; two longer ones (about 32 residues) in close contact probably represent the first and third hydrophobic regions in the LHCPII molecule and a shorter one (20 amino acids) represents the central hydrophobic region. Distinct high-density regions in the three-dimensional map have been attributed to 15 Chl molecules (Kühlbrandt and Wang, 1991). The result is consistent with some earlier biochemical measurements (Butler and Kühlbrandt, 1988) but not with others (Mullet 1983; Peter and Thornber, 1988).

Chl *a/b* ratios between 1 and 1.4 have been measured in isolated LHCIIs from higher plants (Thornber *et al.*, 1991). Further pigment components of these LHCIIs are 2–4 xanthophyll molecules per protein monomer, namely lutein, neoxanthin, and violaxanthin (Siefermann-Harms, 1985). In green algae, the pigment composition of LHCII is quite variable. Chl *a/b* ratios reported range from 0.64 in *Acetabularia* (Apel, 1977) to 5 in *Dunaliella tertiolecta* (Sukenik *et al.*, 1988); Chl apoprotein ratios range from about 6 in *D. tertiolecta* (Sukenik *et al.*, 1988) to over 30 in *Mantoniella squamata* (Wilhelm *et al.*, 1990). Moreover, LHCIIs from algae contain a great variety of different carotenoids (Siefermann-Harms, 1985; Brandt and Wilhelm, 1990; Nakayama and Okada, 1990; see also Chapter 14).

### 2. Minor Chl a/b-Containing Complexes of PSII

The minor Chl *a/b* complexes were identified much later than LHCI and LHCII, partially because in nondenaturing "green" gels they are often obscured by more abundant pigmented complexes (Bassi *et al.*, 1990). There may be more of these minor complexes than the three known to date (Irrgang *et al.*, 1990).

CP29 (Machold and Meister, 1979; also referred to as LHCIIa, Thornber *et al.*, 1991) contains one apoprotein (Dainese *et al.*, 1990) or two apoproteins of 26–31 kDa (White and Green, 1987b; Barbato *et al.*, 1989). Estimates of the number of Chl molecules per apoprotein are 4 (Barbato *et al.*, 1989), 8 (Dainese *et al.*, 1990), and 10–12 (Henrysson *et al.*, 1989) with Chl *a*/*b* ratios between 2.8 and 4.5. Also present are lutein and violaxanthin in nearly equal amounts (Thornber *et al.*, 1991). CP29, like the other minor Chl *a*/*b*-containing complexes, is not phosphorylated and is strictly confined to grana thylakoids (Dunahay and Staehelin, 1987). A CP29 gene has been isolated from tomato (Picherski and Green, 1990) that corresponds to a 26-kDa apoprotein (Green and Camm, 1990).

CP26 (Bassi *et al.*, 1987; alternatively named LCHIIc, Thornber *et al.*, 1991) contains one polypeptide (Peter and Thornber, 1988) or two polypeptides of 28–29 kDa (Di Paolo *et al.*, 1990). Reported Chl contents vary between five Chl molecules at a Chl *a*/*b* ratio of 2.7 (Barbato *et al.*, 1989) and nine Chl molecules at a Chl *a*/*b* ratio of 2.2 (Bassi and Dainese, 1990). CP26 contains lutein as the major xanthophyll, along with neoxanthin and some minor amounts of violaxanthin (Thornber *et al.*, 1991).

In CP24 (Dunahay and Staehelin, 1986; called LHCIId by Thornber *et al.*, 1991), two (Dainese *et al.*, 1990) or three (Di Paolo *et al.*, 1990) closely migrating polypeptides of about 20 kDa can be resolved. CP24 is related immunologically to, although distinct from, LHCI-680 (Di Paolo *et al.*, 1990; Morishige *et al.*, 1990). Three Chl *a* and two Chl *b* molecules have been reported per polypeptide (Di Paolo *et al.*, 1990). In addition to Chl, CP24 contains lutein and small amounts of neoxanthin and violaxathin (Thornber *et al.*, 1991). Genes encoding CP24 apoproteins have been identified in tomato (Schwartz and Pichersky, 1990) and spinach (Spangfort *et al.*, 1989).

PSII preparations from *Chlamydomonas reinhardtii* contain polypeptides with characteristics similar to those of CP29, CP26, and CP24 (Bassi and Wollman, 1991).

The organization of LHCII and the minor Chl *a*/*b*-containing complexes into the antenna of PSII is still unclear. A hypothetical model, based on stoichiometries of antenna components in relation to the PSII reaction center and on spectroscopic data, has been proposed (Bassi and Dainese, 1990). In this model, the minor Chl *a*/*b*-containing complexes function as linkers between LHCII and the reaction center of PSII (see also Chapter 13).

## C. Structural Comparisons between Chl *a*/*b*-Containing Complexes

Since all Chl *a*/*b*-containing complexes of the thylakoid membrane are functionally closely related, it should be interesting to compare some of their structural characteristics to find common features that may be related to their function. All antenna proteins are intrinsic membrane proteins with three hydrophobic regions that are likely to span the membrane. A particularly interesting feature is an intramolecular homology of two stretches of 50–60 amino acids within or near the amino-proximal and the carboxy-proximal membrane-spanning regions of LHCPII (Kühlbrandt and Downing, 1989). This homology is shared by other Chl *a/b*-binding proteins; the region of internal homology is also the region of highest conservation among different antenna proteins (Hoffman *et al.*, 1987; Stayton *et al.*, 1987; Pichersky and Green, 1990; Schwartz and Pichersky, 1990). Therefore these regions are likely to play an important role in the antenna function of the respective pigment–protein complexes. Perhaps the two homologous domains in Chl *a/b*-binding proteins have evolved via a gene duplication event. Thus, they could constitute an "urantenna," a modular element of Chl *a/b*-binding proteins with the capacity to bind pigments. However, *in vitro* reconstitution experiments with deletion mutants of LHCPII lacking one of the conserved domains or lacking the membrane-spanning domain between the two conserved regions did not yield stable pigment–protein complexes (see Section IV,B). This result indicates that the conserved domains are not sufficient as protein components for the formation of stable pigment–protein complexes or that they bind pigments cooperatively, thus significantly increasing the stability of the resulting complexes.

The size of the Chl *a/b*-binding proteins identified so far is relatively homogeneous, ranging from 20 to 30 kDa. On the other hand, Chl *a/b* ratios and numbers of Chl molecules per apoprotein seem to vary much more between different Chl *a/b*-containing complexes. Chl *a/b* ratios reported in plants range from close to 1 (LHCII) to 3 or higher (CP29). The estimates of Chl molecules per apoprotein are between 5 (CP24) and 15 (LHCII) In *Mantoniella squamata*, as many as 30 Chl molecules per apoprotein have been measured (Wilhelm *et al.*, 1990). These estimates suggest significant differences concerning pigment–protein interaction in the various complexes.

For a comparison of Chl *a/b*-binding proteins on the basis of deduced amino acid sequences and for resulting considerations on the evolution of these proteins, see Green *et al.* (1991).

## III. ASSEMBLY OF LIGHT HARVESTING COMPLEX II

All Chl *a/b*-binding proteins known to date are encoded in the nucleus. All respective genes analyzed encode proteins that contain an N-terminal transit sequence involved in their transport into the plastid. Therefore all Chl *a/b*-binding proteins are likely to be synthesized and assembled essentially along the pathway that has been outlined for LHCPII. This pathway includes the translation of the mRNA in the cytoplasm into the precursor protein, the transport of the protein over the outer and inner chloroplast envelope membrane, cleavage of the transit sequence to yield the mature protein, insertion into thylakoid membranes, binding of pigments, organization into antenna complexes, and finally the assembly into PSI and PSII (Fig. 1). The transport of proteins into plastids is discussed in Chapter 9. The temporal sequence of the subsequent steps is still unknown. However, the process has been dissected into individual steps that can be reproduced and studied *in vitro*. Information obtained from such experiments is presented next.

## A. Processing

An activity was identified in the soluble fraction of lysed pea and wheat chloroplasts that cleaves a wheat pLHCPII, translated *in vitro*, into a 25-kDa polypeptide (Lamppa and Abad, 1987). A 25-kDa polypeptide also appears next to a 26-kDa polypeptide when the same pLHCPII is imported into isolated pea or wheat chloroplasts. The two LHCPII bands observed are assumed to be created by two different processing sites, a primary and a secondary site that give rise to the 26- and the 25-kDa protein, respectively, implying that processing *in vitro* only yields the product of cleavage at the second site (Abad *et al.*, 1989). The product of *in vitro* processing was shown to lack a basic hexapeptide from the N terminus of LHCPII (Abad *et al.*, 1991). Interestingly, this hexapeptide carries much of the positive charge of the N-terminal LHCPII domain that is exposed to the stroma, and contains a threonine that can be phosphorylated. When either of the two processing sites is blocked by directed mutagenesis of the protein, cleavage of the other site still can occur on import of the mutant precursor into isolated chloroplasts, indicating two independent processing steps (Clark *et al.*, 1989). Further mutations near the processing sites revealed specific sequence motifs required for cleavage at either site. These sequences contain a basic residue in position $-4$ relative to the primary processing site and the residues KAK at the secondary processing site (S. E. Clark and G. K. Lamppa, 1991, personal communication).

These observations provide convincing evidence that differential processing contributes to the heterogeneity of LHCPII sizes observed *in vitro* and, most likely, *in vivo*. However, the functional significance of the distinct processing sites still needs to be defined.

The stroma fraction that processed pLHCPII *in vitro* was estimated to have an average molecular mass of 240 kDa (Abad *et al.*, 1989). This size,

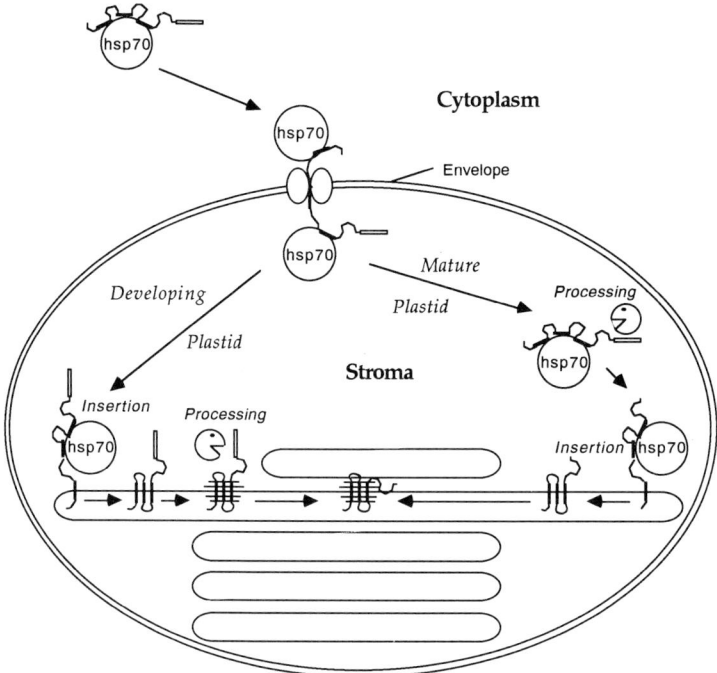

**Fig. 1.** Hypothetical model of the assembly of LHCII. The precursor of the LHCII apoprotein, pLHCPII (the N-terminal transit sequence is indicated by an open bar), in the cytoplasm is maintained in an unfolded import-competent state by hsp70. Probably, pLHCPII interacts with an additional, yet unknown, factor(s) not depicted here (Waegemann *et al.*, 1990). The precursor is imported into the plastid by transiently interacting with a receptor complex located in the envelope (see Chapter 9). Possibly already during the import process, pLHCPII binds to a stromal hsp70 that again keeps it in an unfolded conformation. The order of the following steps could vary depending on the developmental stage of the plastid. In greening plastids, pLHCPII is inserted into stromal lamellae of the thylakoid membrane, migrates to the granal lamellae, and finally is assembled into a stable pigment–protein complex (indicated by hatching) and processed to the mature size. In fully developed chloroplasts, the processing step takes place while pLHCPII is located in the stroma, before insertion into the thylakoid membrane, migration into appressed regions, and formation of a stable pigment–protein complex. It is not known whether pigment–protein interaction aids the insertion of LHCII or pLHCPII into the membrane nor whether the protein is partially pigmented before stable, that is, extractable, LHCII is formed.

as well as some other properties such as pH and temperature optimum, is similar to a partially purified stromal peptidase from pea that cleaves the precursors of ferredoxin and the small subunit (SSU) of ribulose 1,5-bisphosphate carboxylase (Robinson and Ellis, 1984). The presumed pLHCPII-processing enzyme from pea cleaved a number of precursor proteins, including the one of SSU. Therefore, both described processing activities actually may be due to the same enzyme that would mature a broad variety of precursor proteins translocated across the plastid membrane, including the ones of Chl *a/b*-binding proteins, possibly recognizing some common structural features at the region between the transit peptides and the mature part of the proteins (Gavel and von Heijne, 1990).

The presence of a transit peptidase in the stroma does not prove per se that the LHCPII precursor is cleaved to its mature form while in the stroma. The amino terminus of LHCPII in native LHCPII is known to be exposed to the stroma and, thus, would still be accessible for processing. In fact, on importing pLHCPII from *Lemna* or barley into isolated greening barley chloroplasts (Chitnis *et al.*, 1986, 1988), the precursor and varying amounts of the mature protein are found in the LHCII fraction. In a pulse chase experiment with unlabeled pLHCPII, it was shown that labeled pLHCPII already inserted in the thylakoid could be converted to mature-sized LHCPII, demonstrating processing of membrane-bound pLHCPII. Similarly, Vainstein *et al.* (1989a) found pLHCPII as well as LHCPII in LHCII complexes isolated on a nondenaturing gel when they imported pLHCPII from *Lemna* into isolated chloroplasts from maize mesophyll and bundle sheath cells.

The presence of a transit peptide is not required for the insertion of LHCPII into thylakoids (Lamppa, 1988; Viitanen *et al.*, 1988). Moreover, when the insertion of proteins into the thylakoid membrane is inhibited by treating isolated chloroplasts with the uncouplers nigericin and valinomycin, mature LHCPII rather than the precursor accumulates in the stroma (Cline *et al.*, 1989). Reed *et al.* (1990) designed an experimental system in which it is possible to stop rapidly the import of pLHCPII into isolated chloroplasts as well as subsequent steps, using mercuric chloride. In these assays, it was shown that mature LHCPII appears in the stroma immediately after translocation of the precursor across the plastid envelope membrane, well before LHCPII is inserted properly into the thylakoid membrane. These observations favor processing of pLHCPII prior to insertion into the thylakoid.

To reconcile these apparently contradictory results, the developmental stage of the chloroplasts used in these assays may be considered. Insertion of pLHCPII into thylakoids has been observed in greening chloroplasts isolated, after a few hours of illumination, from previously etiolated tissue. After about 24 hr of greening, Chitnis *et al.* (1988) found almost exclusively mature barley LHCPII integrated into the membranes. All evidence for mature LHCPII located in the stroma was obtained from fully green mature chloroplasts. Perhaps both pathways, processing before and after insertion of the protein into thylakoids, are used in a developmentally regulated manner so during greening, when LHCII is assembled at a high rate, most of the pLHCPII is processed after it is inserted into the membrane whereas after greening processing in the stroma prevails.

## B. Insertion into Thylakoid Membrane

### 1. Structural Elements of Protein Required for Insertion

After import into isolated chloroplasts, much of the pLHCPII is found inserted in thylakoid membranes and even assembled with pigments into LHCII (Kohorn and Tobin, 1986). The step of membrane insertion also can be reproduced in chloroplast lysates, independent of import (Cline, 1986; Chitnis et al., 1987). Cline (1988) observed that the inserted protein was assembled into LHCII and localized in grana lamellae. Under specific experimental conditions, up to 50% of the inserted precursor was processed to its mature size during the insertion assay.

The components required for the insertion of LHCPII into thylakoid membranes are Mg-ATP and a stromal activity, the so-called stromal factor. Under some experimental conditions, ATP can be replaced in part by light (Cline, 1988). Membrane uncouplers such as nigericin and valinomycin partially inhibit insertion, suggesting that a transmembrane electrochemical or proton gradient is involved (Cline et al., 1989).

The insertion is not dependent on the presence or absence of a transit peptide in the protein (Lamppa, 1988; Viitanen et al., 1988). Also, a pLHCPII transit peptide linked to neomycin does not direct this marker protein into thylakoid membranes in transgenic plants expressing the chimeric protein (Van den Broeck et al., 1988). Consequently, the information for targeting LHCPII into the thylakoid must reside in the mature protein itself. Kohorn et al. (1986) performed an analysis of the import and subsequent membrane insertion of a number of LHCPII deletion mutants. Mutants carrying deletions in or between the first two hydrophobic regions were still integrated into the thylakoid membrane, albeit with a changed sensitivity to tryptic digestion compared with wild-type LHCPII. Mutants missing the carboxy-proximal hydrophobic region or most of the stretch between the middle and carboxy-proximal hydrophobic regions did not accumulate in the chloroplast at all, possibly because they were not inserted into the membrane and thus were degraded rapidly. Clark et al. (1990) also showed that a pLHCPII deletion mutant missing part of the carboxy-proximal hydrophobic region did not insert into the thylakoid correctly, neither on import into isolated chloroplasts nor in the chloroplast lysate system.

Kohorn and Tobin (1989) tested even more directly whether the carboxy-proximal hydrophobic region of LHCPII is involved in directing the protein into the thylakoid. They showed that SSU, a soluble protein, becomes associated with the thylakoid if it is fused with the third hydrophobic domain of LHCPII but not if it is fused with either of the other two hydrophobic regions. However, the presence of the third hydrophobic domain is not sufficient for proper integration since this domain is still trypsin sensitive after association of the hybrid protein with thylakoid membranes. Consequently, structural elements of other regions of the LHCPII molecule are involved in the membrane insertion process as well.

If single histidine residues located in or near the hydrophobic domains are exchanged for alanine or arginine, the insertion of the mutant LHCPII into the thylakoid is reduced somewhat compared with the wild-type protein (Kohorn 1990). It is unclear, however, whether this effect is caused directly by the amino acids exchanged or indirectly by some overall structural changes caused by the mutations.

Similarly, the distribution of charged amino acids at the ends of N- and C-proximal hydrophobic domains in pLHCPII has been demonstrated to have an impact on the efficiency with which the imported protein is integrated in the thylakoid and assembled into LHCII (Kohorn and Tobin, 1987). Again, it is difficult to distinguish between direct involvement of these changes in the insertion step and destabilization of the complex and some secondary effect via conformational changes of the protein.

## 2. Developmental Stage of Thylakoids

LHCPII precursors are not inserted in prothylakoid membranes isolated from etioplasts, even in the presence of stroma from mature chloroplasts (Chitnis *et al.*, 1987). When thylakoids are isolated from greening chloroplasts after various illumination times, the amount of inserted protein increases during the first 12 hr of greening and then stays at a constant level. Whether this effect is due to increasing amounts of Chl present in the membranes or to some other developmental process is unknown at this point. Chitnis *et al.* (1988) imported pLHCPII into greening chloroplasts from a Chl *b*-less mutant of barley and found as much protein inserted into thylakoids as they did in experiments using wild-type barley, although the Chl *b* content of the mutant was less than 20% of the level in the wild type. This observation, however, does not exclude the possibility that pigments are involved in the insertion of LHCPII or pLHCPII into the membrane, since even the reduced amount of Chl *b* in the barley mutant still could be sufficient for membrane insertion of the small amounts of protein imported into the plastid.

Kohorn and Auchincloss (1991) observed insertion of pLHCPII, with a

# 10. Assembly of Chlorophyll a/b-Containing Complexes

bacterial signal sequence added to its N terminus, into the inner membrane of *Escherichia coli* when this hybrid protein was overexpressed in the bacteria. Pigments, therefore, are not absolutely required for the insertion of LHCPII into membranes, although insertion mediated by a signal sequence may follow a completely different mechanism than integration into thylakoids. The formation of pigment–protein complexes as part of the assembly process of LHCII is discussed further in Section III,C.

### 3. Stromal Factor

The stromal activity required for the insertion of LHCPII into isolated thylakoids (Cline 1986; Chitnis *et al.*, 1987) was soon classified as a soluble protein (Chitnis *et al.*, 1987; Fulsom and Cline, 1988). Isolation of the factor was hampered, at least in part, by the fact that pLHCPII or LHCPII, translated *in vitro* in a wheat germ extract, would integrate into thylakoids to some extent even in the absence of stroma. This problem could be solved by using purified pLHCPII that had been overexpressed in *E. coli*, thus avoiding the wheat germ extract. Using this protein, it was demonstrated that wheat germ extract exhibits the same factor activity in membrane insertion experiments as does stroma extract. Moreover, using overexpressed LHCPII, it could be shown that a member of the family of heat-shock proteins, hsp70, has the same factor activity (S. Yalovsky, H. Paulsen, D. Michaeli, and R. Nechushtai, 1991, unpublished observations). hsp70 was isolated from the stroma fraction; thus, perhaps it represents the stromal factor. Using intrinsic tryptophan fluorescence as a marker of the folding state of pLHCPII, it was shown that hsp70 interacts directly with the precursor protein, slowing down the refolding process of pLHCPII after its complete denaturation at high concentrations of guanidinium chloride. This finding corresponds well to the idea of hsp70 as an "unfoldase" (Rothman, 1989). Some cytosolic heat-shock proteins, often called chaperones (Ellis, 1987), are thought to keep precursor proteins in an unfolded state and, thus, facilitate their import into mitochondria (Pfanner and Neupert, 1990; Pfanner *et al.*, 1991) or chloroplasts (Waegemann *et al.*, 1990). In the mitochondrial matrix, hsp70 is required for the import of precursor proteins. Reinsertion of imported proteins into the inner membrane seems to be dependent on hsp60 (Kang *et al.*, 1990).

"Soluble" LHCPII, accumulated in the stroma when insertion into thylakoids is inhibited, has been characterized as a complex of about 120 kDa, most likely with a stromal protein that does not belong to the hsp60 familiy (Payan and Cline, 1991). Forming this complex keeps LHCPII from forming insoluble aggregates; however, complex formation is not sufficient for the insertion of LHCPII into the thylakoid membrane, which still requires the presence of stroma.

Purified pLHCPII and purified hsp70 facilitate study of the insertion of proteins into thylakoid membranes in a reasonably well-defined experimental system. This assay should make it possible to identify other components necessary for protein insertion into thylakoids, for example, membrane-bound proteins and/or pigments.

## 4. Insertion into Stroma or Grana Lamellae

pLHCPII is inserted predominantly into the stroma lamellae of isolated thylakoids and subsequently migrates to grana lamellae. Only in the grana is pLHCPII found assembled with pigments into LHCII (Yalovsky et al., 1990). On import of pLHCPII into intact chloroplasts, insertion of LHCPII into both stroma and grana lamellae has been observed, followed by migration of LHCPII from nonappressed to stacked membrane regions. This migration was not affected by light or the exchange of N-proximal threonines that are phosphorylated during state I–state II transitions (Kohorn and Yakir, 1990).

These results have been interpreted as contradictory to the hypothesis that links the state I–state II transition to the phosphorylation and subsequent redistribution of LHCII (see Section II,B). According to this hypothesis, the phosphorylation of LHCII under high light conditions triggers the migration of a subpopulation of LHCII from the grana to the stroma lamellae, where it interacts with PSI. After importing pLHCPII into isolated chloroplasts, Kohorn and Yakir (1990) observed migration of the protein in the reverse direction only, namely, from unstacked to stacked thylakoid membranes. Moreover, a migration related to state transition should be influenced by light intensity and the presence or absence of phosphorylation sites, which was not observed in these studies.

However, the data of Yalovsky et al. (1990) show that pLHCPII is not complexed stably with pigments before it has migrated to appressed regions of the thylakoid. Therefore, this migration is likely to be different from the one implied in state transition that causes the redistribution of LHCII, the fully pigmented complex, between grana and stroma lamellae. The phosphorylation as well as the lateral mobility of LHCPII that is not bound into LHCII may be quite different from that of LHCII, the pigment-containing complex.

## C. Assembly with Pigments

Little is known about the mechanism by which pigment–protein complexes of the photosynthetic apparatus are formed. In particular, it is still unclear whether pigment–protein interactions are established before, during, or after insertion of pigment-binding proteins into the membrane. The observation that the insertion of LHCPII into the thylakoid membrane and the assembly of LHCPII into stable pigmented complexes are two separate steps (Kohorn *et al.*, 1986; Yalovsky *et al.*, 1990) favors the possibility that the formation of antenna complexes takes place in the thylakoid membrane. However, since the insertion of LHCPII or pLHCPII seems to be dependent on the greening stage of thylakoid membrane (Chitnis *et al.*, 1987), it is tempting to speculate that pigment–protein interactions play some role in the insertion process, although stable complexes are formed only in a subsequent step.

An unsolved problem in this context is the availability of stoichiometric amounts of pigments and apoproteins for assembly. The coordination between Chl synthesis, which takes place in the plastid, and the synthesis of LHCPII, the most abundant Chl-binding protein, which occurs in the cytoplasm, is an intensively investigated subject (see also Chapter 8). Various mechanisms of mutual regulatory interaction between both synthesis pathways have been suggested. Under some conditions, however, for example, under intermittent illumination, translatable LHCPII mRNA is synthesized although no LHCII is accumulated. It has been suggested that, under such conditions, LHCPII is degraded in the thylakoid membrane because pigments are not available in sufficient amounts to bind to LHCPII and thereby stabilize it (Apel and Kloppstech, 1980; Bennett 1981; Mathis and Burkey, 1989). As an alternative explanation of the effect, post-transcriptional light regulation of pLHCPII synthesis, in particular at the translational level, has been suggested (Slovin and Tobin, 1982). No direct evidence for the degradation of LHCPII in the thylakoid membrane has been presented. Therefore, rapid degradation of soluble pLHCPII (or LHCPII) in the stroma prior to insertion into the thylakoid also must be considered, among other possibilities. In any case, a controlled proteolytic activity would have to be assumed, since the accumulation of LHCPII not stabilized by pigments has been observed under specific conditions (Kohorn *et al.*, 1986; Cline *et al.*, 1989; Yalovsky *et al.*, 1990).

Evidence for a proteolytic degradation of LHCPII unless rescued by pigments also has been presented for *Chlamydomonas reinhardtii* (Hoober *et al.*, 1990). In the green marine alga *Dunaliella tertiolecta*, however, the stability of newly synthesized LHCPII is not affected notably even when Chl synthesis is inhibited by gabaculine (Mortain-Bertrand *et al.*, 1990).

*Dunaliella tertiolecta* has been shown to shift the Chl $a/b$ ratio in LHCII complexes to higher values in response to higher light intensity (Sukenik *et al.*, 1987, 1990). The mechanism is not clear yet. Either the relative amounts of LHCPII species, each binding a different number of Chl $a$ and

Chl*b* molecules, changes (Sukenik *et al.*, 1988) or the pigment composition of individual LHCPII complexes can be adjusted.

The accumulation of LHCPII is correlated with the appearance of Chl *b* (Shimada *et al.*, 1990; Murray and Kohorn, 1991) although this correlation is not a strict one. Low levels of LHCPII accumulate in mutant plants that do not synthesize Chl *b* (Darr *et al.*, 1986; White and Green, 1987b; Mogen *et al.*, 1990) and only moderately reduced levels of LHCII apoproteins are found in a Chl *b*-less mutant of *Chlamydomonas (Michel et al.*, 1983). Based on the correlation between LHCPII and Chl *b* accumulation, a hypothesis describing a stepwise formation of Chl-containing complexes has been proposed. During the early greening phase, at limiting Chl *a* levels, only apoproteins of the PSI reaction center with the highest affinity for Chl *a* are complexed with the pigment. As soon as the amounts of Chl *a* synthesized saturate the available CPI apoprotein, some Chl *a* is bound to antenna proteins; the rest is converted into Chl *b*. In the third step, Chl *b* is incorporated into Chl *a*/*b*-binding proteins, allowing stable antenna complexes to accumulate (Shimada *et al.*, 1990). Another hierarchy has been suggested among Chl *a*/*b*-binding proteins concerning their demand for Chl *b*. CP29 is accumulated at virtually normal levels in a maize mutant in which Chl *b* is limiting. Hence, CP29 ranks higher in this hierarchy than LHCI, which is reduced significantly, and LHCII, which is almost absent (Greene *et al.*, 1988).

Very little is known about how pigments are bound into antenna complexes. In bacteriochlorophyll *a* (BChl *a*)-containing complexes of the bacterial reaction center (Deisenhofer *et al.*, 1985) and in reaction center proteins in plants (Kirsch *et al.*, 1986), Chl molecules are coordinated with histidine residues. Additional coordinations must take place in LHCII since the three histidines present in LHCII are not sufficient to accommodate the 15 Chl molecules most likely bound to the monomeric apoprotein. The same problem occurs in other Chl *a*/*b*-containing complexes. All Chl *a*/*b*-binding proteins for which the sequence is known contain 3–4 histidines. Therefore, it has been suggested that glutamine and asparagine are involved in coordination of chlorophylls also (Wechsler *et al.*, 1985). In fact, several positions in or near hydrophobic domains contain these amino acids and are conserved among species and members of the gene family (Peter and Thornber, 1988).

Even less is known about how carotenoids are assembled into the complex. Some information regarding molecular interactions that stabilize LHCII comes from experiments in which LHCPII–pigment complexes have been reconstituted *in vitro*. These experiments are reviewed in Section IV (See also Chapters 14 and 15).

## D. Aggregation of Pigment–Protein Complexes

In the mature thylakoid, LHCII most likely exists in the form of trimeric complexes that contain three apoproteins and pigments. Evidence for this hypothesis comes from the isolation of stable trimeric complexes from thylakoids solubilized with mild detergents (Butler and Kühlbrandt, 1988). In the presence of stronger detergents such as SDS, the trimeric complexes dissociate into monomers (Ide *et al.*, 1987). Trimers contain lipids, probably as constitutive components, as concluded from analyses of isolated trimers (Krupa, 1988) and from electron crystallographic data on crystallized complexes (Kühlbrandt and Downing, 1989; Kühlbrandt and Wang, 1991). Lipids containing *trans*-C16:1 fatty acids seem to play a role in the oligomerization of LHCII monomers (Huner *et al.*, 1987; Krupa, 1988; Garnier *et al.*, 1990; see also Chapter 7), although normal levels of LHCII trimers have been found in an *Arabidopsis* mutant lacking this particular lipid (McCourt *et al.*,1985.) Whether trimeric complexes of LHCII are formed concomitantly with the assembly of LHCII monomers, form spontaneously from the monomeric complexes, or require the action of an auxiliary factor in the thylakoid membrane is not known. Although an equilibrium between the monomeric and trimeric forms of LHCII in detergent solution has been assumed (Butler and Kühlbrandt, 1988), the formation of trimers from monomers *in vitro* has never been demonstrated.

## IV. RECONSTITUTIONS

### A. Reconstitution of Antenna Complexes *in Vitro*

Clayton and Clayton (1989) demonstrated that, from the B850 antenna of *Rhodobacter sphaeroides* which consists of three BChl *a* molecules and one carotenoid per two apoproteins of 10 and 12 kDa, one BChl *a* could be dissociated and then rebound *in vitro*. Parkes-Loach *et al.* (1988) succeeded in reconstituting the B873 complex of *Rhodospirillum rubrum* from the isolated apoproteins (two polypeptides of about 6 kDa each) and BChl *a*. The reconstituted complex was identical to the isolated B873 complex, according to absorption and CD spectra. The use of BChl *a* analogs in this reconstitution system revealed that the pigment must meet strict structural requirements to be bound into the complex since most of the analogs tested did not form B873 complexes with the proteins (Parkes-Loach *et al.*, 1990).

Surprisingly, antenna complexes from higher plants also can be reconstituted *in vitro*, although their apoproteins are much larger and bind a higher number of pigment molecules than the ones in the bacterial system. Plumley and Schmidt (1987) developed a method for the *in vitro* reconstitution of LHCII by combining heat-denatured or acetone-extracted thylakoids with pigments extracted from thylakoids, using lithium dodecyl sulfate (LDS) as a detergent. They showed that both Chl $a$ and Chl $b$, as well as xanthophylls, are required for the formation of stable complexes. Paulsen *et al.* (1990) and Cammarata *et al.* (1990) extended these experiments using, instead of a protein mixture from thylakoids (containing various LHCPII species), purified LHCPII (or pLHCPII) overexpressed from a pea gene in *Escherichia coli*.

The assumption that the reconstituted pigment–protein complexes structurally resemble LHCII from thylakoid membranes is based on three lines of evidence. First, a number of spectroscopic properties of reconstituted complexes, including absorption, fluorescence emission, and visible CD spectra, as well as energy transfer from Chl $b$ to Chl $a$ are nearly identical to those of LHCII isolated from solubilized thylakoids. Second, the pigment requirement for complex formation is highly specific. Both Chl $a$ and $b$, as well as at least two of three xanthophylls that are found in native LHCII, are necessary for formation of stable complexes. Xanthophylls cannot be replaced with structurally closely related $\beta$-carotene (Plumley and Schmidt, 1987). Finally, the pigment composition of reconstituted complexes is similar to that of the native complex: the Chl $a/b$ ratio is close to 1.1, even if an excess amount of either Chl $a$ or Chl $b$ is present during the reconstitution (Paulsen *et al.*, 1990). The xanthophyll/chlorophyll ratio in the complex is also virtually independent of the relative amounts of pigments in the reconstitution mixture, whereas the violaxanthin/neoxanthin ratio varies within wide margins, indicating that these two xanthophylls can replace one another in the reconstituted complex (H. Paulsen and S. Hobe, 1991, unpublished observations). Moreover, the pigment–protein complexes formed *in vitro* are quite stable. They are separated from free pigment and unbound protein by partially denaturing LDS–polyacrylamide gel electrophoresis. Under the conditions used, of all the pigment–protein complexes solubilized from thylakoids, only CPI, the reaction center of photosystem I, and LHCII stay intact. Collectively, these observations clearly show that the reconstitution of pigmented LHCPII complexes is not due to unspecific association of hydrophobic pigments with a hydrophobic protein but to the highly specific formation of structures that closely resemble those in LHCII.

From the LHCII reconstitution experiments, we can conclude that no particular lipids form constitutive components of LHCII monomers or are required for stabilizing these complexes, since stable LHCPII–pigment complexes can be formed in the absence of lipids (Plumley and Schmidt,

1987). Most likely, lipids are replaced by detergent molecules in the reconstitution process, so reconstituted complexes are embedded in detergent micelles rather than in a membrane. Similar results emerged from neutron diffraction studies on crystallized bacterial photosynthetic reaction centers (Roth *et al.*, 1989). In these studies, the detergent lauryldimethylamine $N$-oxide is organized in rings around the $\alpha$-helical protein domains that are thought to span the photosynthetic membrane.

LHCII complexes also can be reconstituted from apoproteins of *Chlamydomonas*. Steinmetz *et al.* (1990) delipidated thylakoid proteins from a Chl $b$-less mutant and reconstituted these proteins with an extract containing pigments and lipids from wild-type thylakoids. Interestingly, the reverse experiment, that is, reconstitution of wild-type proteins with a pigment extract from the Chl $b$-less mutant, resulted in high yields of a Chl-containing complex with the electrophoretic mobility of LHCII. There are two possible explanations for this apparent contradiction with the results from reconstitution experiments described earlier in which both Chl $a$ and $b$ are required for stable complex formation. (1) In *Chlamydomonas*, LHCPII forms stable pigmented complexes in the absence of Chl $b$. (2) The slightly different experimental conditions used here (reconstitution in a mixture of 0.6% sodium dodecyl sulfate and 0.6% Triton X-100 rather than 2% lithium dodecyl sulfate) preserve a less stable complex that otherwise would have disintegrated and thus escaped detection. The first possibility is consistent with the finding that a Chl $b$-deficient mutant of *Chlamydomonas* accumulated LHCII apoproteins (Michel *et al.*, 1983).

## B. Synergetic Organization of Chl $a/b$-Containing Complexes

In the higher plant system, the stability of reconstituted pigment–LHCPII complexes is dependent on the presence of the complete set of pigments to be bound into the complex; no stable complexes containing a subset of pigments have been observed (Plumley and Schmidt, 1987; Paulsen *et al.*, 1990). This result indicates a highly cooperative binding of pigments into the complex, that is, pigment molecules are involved in more than one stabilizing interaction with the protein or with other pigments. A similarly strict dependence of complex stability on the presence of specific structural elements of the apoprotein has been observed. Mutant LHCPII molecules with one of the membrane-spanning regions (or a part thereof) deleted cannot be reconstituted into stable complexes. On the other hand, hydrophilic stretches on either end of the molecule can be deleted without impairing complex formation. The borders of the protein sections essential for reconstitution are adjacent to the N- and C-proximal membrane-spanning regions. Here, the deletion of a few additional amino acids completely abolishes the ability of the protein to form stable complexes with pigments (H. Paulsen and S. Hobe, 1991, unpublished observations). It can be concluded that essential parts of the protein, including the hydrophobic domains, contribute to complex stability in a cooperative way.

The synergistic organization of pigments and protein in LHCII that was described in the previous section does not support the view of LHCPII as a matrix providing individual and independent binding sites for pigment molecules. Such a view seems particular unlikely considering the fact that, in LHCII, pigments contribute more than one-third of the total molecular mass of the complex. These considerations suggest that pigment–pigment interactions play a major stabilizing role, also. In fact, in the chlorosome antenna of the green bacterium *Chloroflexus aurantiacus*, some of the pigment moieties are thought to be bound solely by pigment–pigment interactions (Blankenship *et al.*, 1988). Indications for chlorosome antennas containing protein-free pigment aggregates have been obtained (Holzwarth *et al.*, 1990).

In higher plants, the Chl/protein molecular ratio in LHCII is relatively highly conserved. However, in the green alga *Mantoniella squamata*, at least 33 Chl molecules and 7 carotenoids are bound to a 23-kDa apoprotein (Wilhelm *et al.*, 1990), making pigments the major component by molecular mass in this pigment–protein complex. *Mantoniella* has been suggested to represent a conserved earlier stage in the evolution of Chl *a*/*b*-containing organisms (Krämer *et al.*, 1988). Perhaps antenna complexes evolved from such complexes with high pigment and relatively low protein content.

In the crystal structure of LHCII, some Chl molecules appear to be too far away from the $\alpha$-helices of the protein for direct interaction between their central $Mg^{2+}$ and a nucleophilic side chain in the protein (Kühlbrandt and Wang, 1991). Structural analysis at higher resolution is necessary to determine whether these Chl molecules are held in place by hydrogen bonds between the tetrapyrrole and the protein or by pigment–pigment binding.

Quite specific interactions between photosynthetic pigments have been observed in aggregates of photosynthetic pigments in the absence of proteins (Abraham *et al.*, 1988; Scherz *et al.*, 1990). These observations raise the possibility that the formation of similar aggregates between pigments in the thylakoid is the initiating step for the assembly of antenna complexes. These aggregates subsequently would be stabilized by the apoproteins. This purely speculative model can be tested. If the formations of pigment aggregates and of pigment–protein interactions truly are sequential steps, it should be possible to reproduce and observe these steps *in vitro*.

## C. Reconstitution of Antenna Complexes into Membranes

Isolated LHCII can be reconstituted relatively easily into liposomes of artificial membranes or into thylakoid membranes. Various techniques have been used, including sonication, freeze–thaw cycles, and removal of detergent.

If LHCII is inserted into liposomes made of phosphatidylcholine (McDonnel and Staehelin, 1980; Mullet and Arntzen, 1980) or diacylglycerols (Ryrie *et al.*, 1980), the liposomes aggregate in the presence of divalent ions. This behavior has been interpreted as mimicking the stacking of thylakoids into grana, suggesting the LHCII is involved in this process (see Section II, B). The efficiency of LHCII insertion and the organization of complexes in the vesicles, as revealed by electron microscopy, varies with the lipid composition of the liposomes (Sprague *et al.*, 1985). When isolated LHCII and oxygen-evolving PSII complexes containing some LHCPII are inserted into phosphatidylcholine liposomes, the two complexes interact with one another, at least photochemically (Murphy *et al.*, 1984). Energy transfer from LHCII to both PSI and PSII reaction centers occurs when these complexes are combined in liposomes of thylakoid lipids (Larkum and Anderson, 1982).

Interaction between LHCII and PSII also can be seen when isolated LHCII is reconstituted into thylakoid membranes isolated from barley that has been grown in intermittent light (Day *et al.*, 1984). Under these culture conditions, PSI and PSII appear in the thylakoids but virtually no LHCII is accumulated. The reconstitution of isolated LHCII in such membranes has been facilitated technically using a dialyzable detergent (Darr and Arntzen, 1986). The reconstituted preparations obtained by this technique reached light-harvesting efficiencies comparable to those measured after 18 hr of continuous illumination of previously intermittent light-treated seedlings. However, the amount of LHCII needed in the reconstituted preparation to obtain this efficiency was three times higher than the amount of LHCII present in the seedlings after 18 hr of greening. One reason for the diminished light-harvesting efficiency of LHCII that has been reconstituted into membranes is likely to be its random orientation in the membrane, compared with the uniformly oriented LHCII in native thylakoids.

No reports have been published that demonstrate the reconstitution into membranes of LHCII complexes that in turn have been reconstituted from the apoprotein and pigments. The interaction of such reconstituted complexes with PSII reaction centers and their light-harvesting efficiency would be an important test of their functional equivalence with native LHCII.

## V. CONCLUDING REMARKS

Some of the information presented in this chapter is summarized in Fig. 1. This sketch oversimplifies the process of LHCII assembly and is hypothetical in many parts. This simplicity means that our knowledge of the assembly of LHCII is still rather limited. Only when we are able to reconstitute every step in the biogenesis of LHCII *in vitro* using isolated and purified material will we be in a position to claim that we know all the components involved in this process and, perhaps, even their functions. Much work remains to be done to reach this goal, and even more in the case of the other Chl *a*/*b*-containing complexes. We can only assume at this point that their assembly is similar to that of LHCII.

Some steps in the assembly of LHCII that can be reproduced *in vitro* include the import of precursor proteins into isolated plastids, the insertion of pLHCPII or LHCPII into the thylakoid membrane, processing, and the formation of stable pigment–protein complexes. Work is needed to identify the components required in these reactions. This information will help identify the specific functions of these components and the structural characteristics that are involved in these functions. The ability to reconstitute Chl *a*/*b*-containing complexes *in vitro* provides a particularly promising tool for understanding on a molecular basis how these complexes work as light antennas in photosynthesis. Achieving this understanding is a particularly challenging goal since the light-harvesting function adds efficiency to the process of photosynthesis.

## ACKNOWLEDGMENTS

I thank R. Bassi and L. Eichacker for critically reading the manuscript, and many colleagues for communicating information prior to publication. The author's work was supported by the Deutsche Forschungsgemeinschaft (SFB 184).

## REFERENCES

Abad, M. S., Clark, S. E., and Lamppa, G. K. (1989). Properties of a chloroplast enzyme that cleaves the chlorophyll *a*/*b* binding protein precursor. *Plant Physiol.* **90,** 117–124.

Abad, M. S., Oblong, J. E., and Lamppa, G. K. (1991). A soluble chloroplast enzyme cleaves preLHCP made in *E. coli* to a mature form lacking a basic *N*-terminal domain. *Plant Physiol.* **96,** 1220–1227.

Abraham, R. J., Goff, D. A., and Smith, K. M. (1988). NMR spectra of porphyrins. 35. An examination of the proposed models of the chlorophyll *a* dimer. *J. Chem. Soc. Perkins Trans.* **1,** 2443–2451.

Allen, J. F., Bennett, J., Steinback, K. E., and Arntzen, C. J. (1981). Chloroplast protein phosphorylation couples plastoquinone redox state to distribution of excitation energy between photosystems. *Nature (London)* **291**, 25–29.

Anandan, S., and Thornber, J. P. (1990). Isolation of the LHCI complex of barley containing multiple pigment-proteins. *In* "Current Research in Photosynthesis" (M. Baltscheffsky, ed.), Vol. 2, pp. 285–288. Kluwer, Dordrecht, The Netherlands.

Apel, K. (1977). Chlorophyll-proteins from *Acetabularia mediterranea*. *Brookhaven Symp. Biol.* **28**, 149–161.

Apel, K., and Kloppstech, K. (1980). The effect of light in the biosynthesis of the light-harvesting chlorophyll *a/b* protein. *Planta* **150**, 426–430.

Barbato, R., Rigoni, F., Giardi, M. T., and Giacometti, G. M. (1989). The minor antenna complexes of an oxygen evolving photosystem II preparation: Purification and Stoichiometry. *FEBS Lett.* **251**, 147–154.

Barber, J. (1982). Influence of surface charges on thylakoid structure and function. *Annu. Rev. Plant Physiol.* **33**, 261–295.

Bassi, R., and Dainese, P. (1990). The role of light-harvesting complex II and of the minor chlorophyll *a/b* proteins in the organization of the photosystem II antenna system. *In* "Current Research in Photosynthesis" (M. Baltscheffsky, ed.), Vol. 2, pp. 209–216. Kluwer, Dordrecht, The Netherlands.

Bassi, R., and Simpson, D. (1987). Chlorophyll–protein complexes of barley photosystem I. *Eur. J. Biochem.* **163**, 221–230.

Bassi, R., and Wollman, A. (1991). The chlorophyll *a/b* proteins of photosystem II in *Chlamydomonas reinhardtii*. Isolation, characterization, and immunological cross-reactivity to higher-plant polypeptides. *Planta* **183**, 423–433.

Bassi, R., Høyer-Hansen, G., Barbato, R., Giacometti, G. M., and Simpson, D. J. (1987). Chlorophyll proteins of the photosystem II antenna system. *J. Biol. Chem.* **262**, 13333–13341.

Bassi, R., Rigoni, F., and Giacometti, G. M. (1990). Chlorophyll binding proteins with antenna function in higher plants and green algae. *Photochem. Photobiol.* **52**, 1187–1206.

Bennett, J. (1981). Biosynthesis of the light-harvesting chlorophyll *a/b* protein. Polypeptide turnover in darkness. *Eur. J. Biochem.* **118**, 61–70.

Bennett, J. (1983). Regulation of photosynthesis by reversible phosphorylation of the light-harvesting chlorophyll *a/b* protein. *Biochem. J.* **212**, 1–13.

Blankenship, R. E., Brune, D. C., and Wittmershaus, B. P. (1988). Chlorosome antennas in green photosynthetic bacteria. *In* "Light Energy Transduction in Photosynthesis" (S. E. Stevens and D. A. Bryant, eds.), pp. 32–46. American Society of Plant Physiologists, Rockville, Maryland.

Brandt, P., and Wilhelm, C. (1990). The light-harvesting system of *Eugenia gracilis* during the cell cycle. *Planta* **180**, 293–296.

Bruce, B. D., and Malkin, R. (1988). Subunit stoichiometry of the chloroplast photosystem I complex. *J. Biol. Chem.* **263**, 7302–7308.

Buetow, D. E., Chen, H., Erdös, G., and Li, L. S. H. (1988). Regulation and expression of the multigene family coding light-harvesting chlorophyll *a/b*-binding proteins of photosystem II. *Photosynth. Res.* **18**, 61–97.

Butler, P. J. G., and Kühlbrandt, W. (1988). Determination of the aggregate size in detergent solution of the light-harvesting chlorophyll *a/b*-protein complex from chloroplast membranes. *Proc. Natl. Acad. Sci. USA* **85**, 3797–3801.

Cammarata, K., Plumley, F. G., and Schmidt, G. W. (1990). Reconstitution of light-harvesting complexes: A single apoprotein binds Chl *a*, Chl *b*, and xanthophylls. *In* "Current

Research in Photosynthesis'' (M. Baltscheffsky, ed.), Vol. 2, pp. 341–344. Kluwer, Dordrecht, The Netherlands.
Chitnis, P. R., and Thornber, J. P. (1988). The major light-harvesting complex of photosystem II: Aspects of its molecular and cell biology. *Photosynth. Res.* **16**, 41–43.
Chitnis, P. R., Harel, E., Kohorn, B. D., Tobin, E., and Thornber, J. P. (1986). Assembly of the precursor and processed light-harvesting chlorophyll *a/b* protein of *Lemna* into the light-harvesting complex II of barley etiochloroplasts. *J. Cell Biol.* **102**, 982–988.
Chitnis, P. R., Nechushtai, R., and Thornber, J. P. (1987). Insertion of the precursor of the light-harvesting chlorophyll *a/b*-protein into the thylakoids requires the presence of a developmentally regulated stromal factor. *Plant Mol. Biol.* **10**, 3–11.
Chitnis, P. R., Morishige, D. T., Nechushtai, R., and Thornber, J. P. (1988). Assembly of the barley light-harvesting chlorophyll *a/b* proteins in barley etiochloroplasts involves processing of the precursor on thylakoids. *Plant Mol. Biol.* **11**, 95–107.
Clark, S. E., Abad, M. S., and Lamppa, G. K. (1989). Mutations at the transit peptide–mature protein junction separate two cleavage events during chloroplast import of the chlorophyll *a/b*-binding protein. *J. Biol. Chem.* **264**, 17544–17550.
Clark, S. E., Oblong, J. E., and Lamppa, G. K. (1990). Loss of efficient import and thylakoid insertion due to N- and C-terminal deletions in the light-harvesting chlorophyll *a/b* binding protein. *Plant Cell* **2**, 173–184.
Clayton, R. K., and Clayton, B. J. (1981). B850 pigment–protein complex of *Rhodopseudomonas sphaeroides*. Extinction coefficients, circular dichroism, and the reversible binding of bacteriochlorophyll. *Proc. Natl. Acad. Sci. USA* **78**, 5583–5587.
Cline, K. (1986). Import of proteins into chloroplasts. Membrane integration of a thylakoid precursor protein reconstituted in chloroplast lysates. *J. Biol. Chem.* **261**, 14804–14810.
Cline, K. (1988). Light-harvesting chlorophyll *a/b* protein. Membrane insertion, proteolytic processing, assembly into LHCII, and localization to appressed membranes occurs in chloroplast lysates. *Plant Physiol.* **86**, 1120–1126.
Cline, K., Fulsom, D. R., and Viitanen, P. V. (1989). An imported thylakoid protein accumulates in the stroma when insertion into thylakoids is inhibited. *J. Biol. Chem.* **264**, 14225–14232.
Coughlan, S., and Hind, G. (1987). Phosphorylation of thylakoid proteins by a purified kinase. *J. Biol. Chem.* **262**, 8402–8408.
Dainese, P., Høyer-Hansen, G., and Bassi, R. (1990). The resolution of chlorophyll *a/b* binding proteins by a preparative method based on flat bed electrofocusing. *Photochem. Photobiol.* **52**, 693–703.
Darr, S. C., and Arntzen, C. J. (1986). Reconstitution of the light harvesting chlorophyll *a/b* pigment–protein complex into developing chloroplast membranes using a dialyzable detergent. *Plant Physiol.* **80**, 931–937.
Darr, S. C., Somerville, S. C., and Arntzen, C. J. (1986). Monoclonal antibodies to the light-harvesting chlorophyll *a/b* protein complex of photosystem II. *J. Cell Biol.* **103**, 733–740.
Day, D. A., Ryrie, I. J., and Fuad, N. (1984). Investigation of the role of the main light-harvesting chlorophyll–protein complex in thylakoid membranes: Reconstitution of depleted membranes from intermittent-light-grown plants with the isolated complex. *J. Cell Biol.* **97**, 163–172.
Deisenhofer, J., Michel, H., and Huber, R. (1985). The structural basis of photosynthetic light reactions in bacteria. *Trends Biochem. Sci.* **10**, 243–248.
Di Paolo, M. L., Dal Belin Peruffo, A., and Bassi, R. (1990). Immunological studies on chlorophyll-*a/b* proteins and their distribution in thylakoid membrane domains. *Planta* **181**, 275–286.

Dunahay, T. G., and Staehelin, L. A. (1986). Isolation and characterization of a new minor chlorophyll $a/b$ protein complex (CP24) of spinach. *Plant Physiol.* **80,** 429–434.

Dunahay, T. G., and Staehelin, L. A. (1987). Immuno-localization of the chlorophyll $a/b$ light-harvesting complex and CP29 under conditions favoring phosphorylation and dephosphorylation of the thylakoid membrane (state I–state II transition). *In* "Progress in Photosynthesis Research" (J. Biggins, ed.), Vol. 2, pp. 701–704. Nijhoff, Dordrecht, The Netherlands.

Ellis, R. J. (1987). Proteins as molecular chaperones. *Nature (London)* **328,** 378–379.

Fulsom, D. R., and Cline, K. (1988). A soluble protein factor is required *in vitro* for membrane insertion of the thylakoid precursor protein, pLHCP. *Plant Physiol.* **88,** 1146–1153.

Gal, A., Hauska, G., Herrmann, R., and Ohad, I. (1990). Interaction between light harvesting chlorophyll-$a/b$ protein (LHCII) kinase and cytochrome $b_6/f$ complex. *J. Biol. Chem.* **265,** 19742–19749.

Garnier, J., Wu, B., Maroc, J., Guyon, D., and Tremolieres, A. (1990). Restoration of an oligomeric form of the light-harvesting antenna CPII and of a fluorescence state II–state I transition by delta$^3$-trans-hexadecanoic acid-containing phosphatidylglycerol, in a mutant of *Chlamydomonas*. *In* "Current Research in Photosynthesis" (M. Baltscheffsky, ed.), Vol. 2, pp. 277–280. Kluwer, Dordrecht, The Netherlands.

Gavel, Y., and von Heijne, G. (1990). A conserved cleavage-site motif in chloroplast transit peptides. *FEBS Lett.* **261,** 455–458.

Green, B. R. (1988). The chlorophyll–protein complexes of higher plant photosynthetic membranes or just what green band is that? *Photosynth. Res.* **15,** 3–22.

Green, B. R., and Camm, E. L. (1990). Relationship of chl $a/b$-binding and related polypeptides in PSII core particles. *In* "Current Research in Photosynthesis" (M. Baltscheffsky, ed.), Vol. 1, pp. 659–663. Kluwer, Dordrecht, The Netherlands.

Green, B. R., Pichersky, E., and Kloppstech, K. (1991). Chlorophyll $a/b$-binding proteins: An extended family. *Trends Biochem. Sci.* **16,** 181–186.

Greene, B. A., Allred, D. R., Morishige, D. T., and Staehelin, L. A. (1988). Hierarchical response of light harvesting chlorophyll–proteins in a light-sensitive chlorophyll $b$-deficient mutant of maize. *Plant Physiol.* **87,** 357–364.

Henrysson, T., Schröder, W. P., Spangfort, M., and Åkerlund, H. E. (1989). Isolation and characterization of the chlorophyll $a/b$ protein complex CP29 from spinach. *Biochim. Biophys. Acta* **977,** 301–308.

Herrin, D. L., Plumley, G. F., Ikeuchi, M., Michaels, A. S., and Schmidt, G. W. (1987). Chlorophyll antenna proteins of photosystem I: Topology, synthesis, and regulation of the 20-kDa subunit of *Chlamydomonas* light-harvesting complex of photosystem I. *Arch. Biochem. Biophys.* **849,** 387–408.

Hoffmann, N. E., Pichersky, E., Malik, V. S., Castresana, C., Ko, K., Darr, S. C., and Cashmore, A. R. (1987). A cDNA clone encoding a photosystem I protein with homology to photosystem II chlorophyll $a/b$-binding polypeptides. *Proc. Natl. Acad. Sci. USA* **84,** 8844–8848.

Holzwarth, A. R., Griebenow, K., and Schaffner, K. (1990). A photosynthetic antenna system which contains a protein-free chromophore aggregate. *Z. Naturforsch.* **45C,** 203–205.

Hoober, J. K., Maloney, M. A., Asbury, L. R., and Marks, D. B. (1990). Accumulation of chlorophyll $a/b$-binding polypeptides in *Chlamydomonas reinhardtii* y-1 in the light or dark at 38°C. *Plant Physiol.* **92,** 419–426.

Houlnè, G., and Schantz, R. (1988). Characterization of cDNA sequences for LHCI apoproteins in *Euglena gracilis:* The mRNA encodes a large precursor containing several consecutive divergent polypeptides. *Mol. Gen. Genet.* **213,** 479-486.

Huner, N. P. A., Krol, M., Williams, J. P., Maissan, E., Low, P. S., Roberts, D., and Thompson, J. E. (1987). Low temperature development induces a specific decrease in trans-$\Delta^3$-hexanoic acid content which influences LHCII organization. *Plant Physiol.* **84,** 12-18.

Ide, J. P., Klug, D. R., Kühlbrandt, W., Giorgi, L. B., and Porter, G. (1987). The state of detergent solubilised light-harvesting chlorophyll-*a/b* protein complex as monitored by picosecond time-resolved fluorescence and circular dichroism. *Biochim. Biophys. Acta* **893,** 349-364.

Ikeuchi, M., Hirano, A., and Inoue, Y. (1991). Correspondence of apoproteins of light-harvesting chlorophyll-*a/b* complexes associated with photosystem-I to *Cab* genes—Evidence for a novel type-IV apoprotein. *Plant Cell Physiol.* **32,** 103-112.

Irrgang, K. D., Bochtel, C., Vater, J., and Renger, G. (1990). A new chlorophyll *a/b*-binding protein in photosystem II from spinach with a $M_r$ of 14 kDa. In "Current Research in Photosynthesis" (M. Baltscheffsky, ed.), Vol. 1, pp. 355-358. Kluwer, Dordrecht, The Netherlands.

Jansson, S., and Gustafson, P. (1990). Type I and type II genes for chlorophyll *a/b*-binding protein in the gymnosperm *Pinus sylvestris* (Scots pine): cDNA cloning and sequence analysis. *Plant Mol. Biol.* **14,** 287-296.

Kang, P. J., Ostermann, J., Shilling, J., Neupert, W., Craig, E. A., and Pfanner, N. (1990). Requirement for *hsp*70 in the mitochondrial matrix for translocation and folding of precursor proteins. *Nature (London)* **348,** 137-143.

Karlin-Neumann, G. A., Kohorn, B. D., Thornber, J. P., and Tobin, E. M. (1985). A chlorophyll *a/b*-binding protein encoded by a gene containing an intron with characteristics of a transposable element. *J. Mol. Appl. Genet.* **3,** 45-61.

Kirsch, W., Seyer, P., and Herrmann, R. G. (1986). Nucleotide sequence of the clustered genes for two P700 chlorophyll *a* apoproteins of the photosystem I reaction center and the ribosomal protein S14 of the spinach plastid chromosome. *Curr. Genet.* **10,** 843-855.

Kohorn, B. D. (1990). Replacement of histidines of light harvesting chlorophyll *a/b* binding protein II disrupts chlorophyll-protein complex assembly. *Plant Physiol.* **93,** 339-342.

Kohorn, B. D., and Auchincloss, A. H. (1991). Integration of a chlorophyll binding protein into *E. coli* membranes in the absence of chlorophyll. *J. Biol. Chem.* **266,** 12048-12052.

Kohorn, B. D., and Tobin, E. M. (1986). Chloroplast import of light-harvesting chlorophyll *a/b*-proteins with different amino termini and transit peptides. *Plant Physiol.* **82,** 1172-1174.

Kohorn, B. D., and Tobin, E. M. (1987). Amino acid charge distribution influences the assembly of apoprotein into light-harvesting complex II. *J. Biol. Chem.* **262,** 12897-12899.

Kohorn, B. D., and Tobin, E. M. (1989). A hydrophobic, carboxy-proximal region of a light-harvesting chlorophyll *a/b* protein is necessary for stable integration into thylakoid membranes. *Plant Cell* **1,** 159-166.

Kohorn, B., and Yakir, D. (1990). Movement of newly imported light-harvesting chlorophyll-binding protein from unstacked to stacked thylakoid membranes is not affected by light treatment or absence of amino-terminal threonines. *J. Biol. Chem.* **2675,** 2118-2123.

Kohorn, B. D., Harel, E., Chitnis, P. R., Thornber, J. P., and Tobin, E. M. (1986). Functional and mutational analysis of the light-harvesting chlorophyll *a/b* protein of thylakoid membranes. *J. Cell Biol.* **102,** 972-981.

Krämer, P., Wilhelm, C., Wild, A. Mörschel, E., and Rhiel, E. (1988). Ultrastructure and

freeze-fracture studies of the thylakoids of *Mantoniella squamata* (Prasinophyceae). *Protoplasma* **147**, 170–177.
Krupa, Z. (1988). Cadmium-induced changes in the composition and structure of the light-harvesting chlorophyll *a*/*b* protein complex II in radish cotyledons. *Physiol. Plant.* **73**, 518–524.
Kühlbrandt, W., and Downing, K. H. (1989). Two-dimensional structure of plant light-harvesting complex at 3.7 Å resolution by electron crystallography. *J. Mol. Biol.* **207**, 823–828.
Kühlbrandt, W., and Wang, D. N. (1991). Three-dimensional structure of plant light-harvesting complex by electron crystallography. *Nature (London)* **350**, 130–134.
Lam, E., Ortiz, W., and Malkin, R. (1984). Chlorophyll *a*/*b* protein of photosystem I. *FEBS Lett.* **168**, 10–14.
Lamppa, G. K. (1988). The chlorophyll *a*/*b*-binding protein inserts into the thylakoids independent of its cognate transit peptide. *J. Biol. Chem.* **263**, 14996–14999.
Lamppa, G. K., and Abad, M. S. (1987). Processing of a wheat light-harvesting chlorophyll *a*/*b* protein by a soluble enzyme from higher plant chloroplasts. *J. Cell. Biol.* **105**, 2641–2648.
Larkum, A. W. D., and Anderson, J. M. (1982). The reconstitution of a photosystem II protein complex, P-700-chlorophyll *a*-protein complex and light-harvesting chlorophyll *a*/*b*-protein. *Biochim. Biophys. Acta* **679**, 410–421.
Leutwiler, L. S., Meyerowitz, E. M., and Tobin, E. M. (1986). Structure and expression of three light-harvesting chlorophyll *a*/*b*-binding protein genes in *Arabidopsis thaliana*. *Nucleic Acids Res.* **14**, 4051–4063.
McCourt, P., Browse, J., Watson, J., Arntzen, C. J., and Somerville, C. R. (1985). Analysis of photosynthetic antenna function in a mutant of *Arabidopsis thaliana* (L.) lacking *trans*-hexadecanoic acid. *Plant Physiol.* **78**, 853–858.
McDonnel, A., and Staehelin, L. A. (1980). Adhesion between liposomes mediated by the chlorophyll *a*/*b* light-harvesting complex isolated from chloroplast membranes. *J. Cell Biol.* **84**, 40–56.
Machold, O., and Meister, A. (1979). Resolution of the light-harvesting chlorophyll *a*/*b*-protein of *Vicia faba* chloroplasts into two different chlorophyll–protein complexes. *Biochim. Biophys. Acta* **546**, 472–480.
Mathis, J. N., and Burkey, K. O. (1989). Light intensity regulates the accumulation of the major light-harvesting chlorophyll–protein in greening seedlings. *Plant Physiol.* **90**, 560–566.
Mattoo, A. K., and Edelman, M. (1987). Intramembrane translocation and posttranslational palmitoylation of the chloroplast 32-kDa herbicide-binding protein. *Proc. Natl. Acad. Sci. USA* **84**, 1497–1501.
Michel, H., Tellenbach, M., and Boschetti, A. (1983). A chlorophyll *b*-less mutant of *Chlamydomonas reinhardtii* lacking in the light-harvesting chlorophyll *a*/*b*-protein complex but not in its apoprotein. *Biochim. Biophys. Acta* **725**, 417–424.
Mogen, K., Eide, J., Duysen, M., and Eskins, K. (1990). Chloramphenicol stimulates the accumulation of light-harvesting chlorophyll *a*/*b*-binding protein by affecting post-transcriptional events in the *Chlorina* CD3 mutant wheat. *Plant Physiol.* **92**, 1233–1240.
Morishige, D. T., and Thornber, J. P. (1990). The major light-harvesting chlorophyll *a*/*b* protein (LHCIIb): The smallest subunit is a novel *cab* gene product. *In* "Current Research in Photosynthesis" (M. Baltscheffsky, ed.), Vol. 2, pp. 261–264. Kluwer, Dordrecht, The Netherlands.
Morishige, D. T., Anandan, S., Jaing, J. T., and Thornber, J. P. (1990). Amino-terminal

sequence of the 21-kDa apoprotein of a minor light-harvesting pigment–protein complex of the photosystem II antenna (LHCIId/CP24). *FEBS Lett.* **264,** 239–242.

Mortain-Bertrand, A., Bennett, J., and Falkowski, P. G. (1990). Photoregulation of the light-harvesting chlorophyll protein complex associated with photosystem-II in *Dunaliella tertiolecta*—Evidence that apoprotein abundance but not stability requires chlorophyll synthesis. *Plant Physiol.* **94,** 304–311.

Mullet, J. E. (1983). The amino acid sequence of the polypeptide segment which regulates membrane adhesion (grana stacking) in chloroplasts. *J. Biol. Chem.* **258,** 9941–9948.

Mullet, J. E., and Arntzen, C. J. (1980). Simulation of grana stacking in a model membrane system. Mediation by a purified light-harvesting pigment–protein complex from chloroplasts. *Biochim. Biophys. Acta* **589,** 100–117.

Murphy, D. J., Crowther, D., and Woodrow, I. E. (1984). Reconstitution of light-harvesting chlorophyll–protein complexes with photosystem II complexes in soybean phosphatidylcholine liposomes: Enhancement of quantum efficiency of sub-saturating light intensities in the reconstituted liposomes. *FEBS Lett.* **165,** 151–155.

Murray, D. L., and Kohorn, B. D. (1991). Chloroplasts of *Arabidopsis thaliana* homozygous for the *CH-1* locus lack chlorophyll-*b*, lack stable LHCPII, and have stacked thylakoids. *Plant Mol. Biol.* **16,** 71–79.

Nakayama, K., and Okada, M. (1990). Purification and characterization of light-harvesting chlorophyll *a*/*b*-protein complexes of photosystem-II from the green alga, *Bryopsis maxima*. *Plant Cell Physiol.* **31,** 253–260.

Nechushtai, R., Peterson, C. C., Peter, G. F., and Thornber, J. P. (1987). Purification and characterization of a light-harvesting chlorophyll *a*/*b*–protein of photosystem I of *Lemna gibba*. *Eur. J. Biochem.* **164,** 345–350.

Parkes-Loach, P. S., Sprinkle, J. R., and Loach, P. A. (1988). Reconstitution of the B873 light-harvesting complex of *Rhodospirillum rubrum* from the separately isolated alpha and beta-polypeptides and bacteriochlorophyll *a*. *Biochemistry* **27,** 2718–2727.

Parkes-Loach, P. S., Michalski, T. J., Bass, W. J., Smith, U., and Loach, P. A. (1990). Probing the bacteriochlorophyll binding site by reconstitution of the light-harvesting complex of *Rhodospirillum rubrum* with bacteriochlorophyll *a* analogues. *Biochemistry* **29,** 2951–2960.

Paulsen, H., Rümler, and Rüdiger, W. (1990). Reconstitution of pigment–containing complexes from light-harvesting chlorophyll *a*/*b*-binding protein overexpressed in *Escherichia coli*. *Planta* **181,** 204–211.

Payan, L. A., and Cline, K. (1991). A stromal protein factor maintains the solubility and insertion competence of an imported thylakoid membrane protein. *J. Cell Biol.* **112,** 603–613.

Peter, G. F., and Thornber, J. P. (1988). The antenna components of photosystem II with emphasis on the major pigment–protein, LHC IIb. *In* "Photosynthetic Light-Harvesting Systems" (H. Scheer and S. Schneider, eds.), pp. 175–187. de Gruyter, Berlin.

Pfanner, N., and Neupert, W. (1990). The mitchondrial protein import apparatus. *Annu. Rev. Biochem.* **59,** 331–353.

Pfanner, N., Söllner, T., and Neupert, W. (1991). Mitochondrial import receptors for precursor proteins. *Trends Biol. Sci.* **16,** 63–67.

Pichersky, E., and Green, B. R. (1990). The extended family of chlorophyll *a*/*b*-binding proteins of PSI and PSII. *In* "Current Research in Photosynthesis" (M. Baltscheffsky, ed.), Vol. 3, pp. 553–556. Kluwer, Dordrecht, The Netherlands.

Pichersky, E., Tanksley, S. D., Piechulla, B., Stayton, M. M., and Dunsmuir, P. (1988). Nucleotide sequence and chromosomal location of *cab* 7, the tomato gene encoding the type II chlorophyll *a*/*b*-binding polypeptide of photosystem I. *Plant Mol. Biol.* **11,** 69–71.

Pichersky, E., Brock, T. G., Nguyen, D., Hoffmann, N. E., Piechulla, B., Tanksley, S. D.,

## 10. Assembly of Chlorophyll a/b-Containing Complexes

and Green, B. R. (1989). A new member of *cab* gene family: Structure, expression and chromosomal location of *Cab*-8, the tomato gene encoding the type III chlorophyll a/b-binding protein of photosystem I. *Plant Mol. Biol.* **12,** 257–270.

Plumley, F. G., and Schmidt, G. W. (1987). Reconstitution of chlorophyll a/b light-harvesting complexes: Xanthophyll-dependent assembly and energy transfer. *Proc. Natl. Acad. Sci. USA* **84,** 146–150.

Reed, J. E., Cline, K., Stephens, L. C., Bacot, K. O., and Viitanen, P. V. (1990). Early events in the import assembly pathway of an integral thylakoid protein. *Eur. J. Biochem.* **194,** 33–42.

Robinson, C., and Ellis, R. J. (1984). Transport of proteins into chloroplasts. Partial purification of a chloroplast protease involved in the processing of imported precursor polypeptides. *Eur. J. Biochem.* **142,** 337–342.

Roth, M., Lewitt-Bentley, A., Michel, H., Deisenhofer, J., Huber, R., and Oesterhelt, D. (1989). Detergent structure in crystals of a bacterial photosynthetic reaction centre. *Nature (London)* **340,** 659–662.

Rothman, J. E. (1989). Polypeptide chain binding proteins: Catalysts of protein folding and related processes in cells. *Cell* **59,** 591–601.

Ryrie, I. J., Anderson, J. M., and Goodchild, D. J. (1980). The role of the light-harvesting chlorophyll a/b-protein complex in chloroplast membrane stacking. *Eur. J. Biochem.* **107,** 345–354.

Scherz, A., Rosenbachbelkin, V., and Fisher, J. R. E. (1990). Distribution and self-organization of photosynthetic pigments in micelles—Implication for the assembly of light-harvesting complexes and reaction centers in the photosynthetic membrane. *Proc. Natl. Acad. Sci. USA* **87,** 5430–5434.

Schuster, G., Nechushtai, R. Ferreira, P. C. G., Thornber, J. P., and Ohad, I. (1988). Structure and biogenesis of *Chlamydomonas reinhardtii* photosystem I. *Eur. J. Biochem.* **177,** 411–416.

Schwartz, E., and Pichersky, E. (1990). Sequence of two tomato nuclear genes encoding chlorophyll a/b-binding proteins of CP24 and a PSII antenna component. *Plant Mol. Biol.* **15,** 157–160.

Shimada, Y., Tanaka, A., Tanaka, Y., Takabe, T., Takabe, T., and Tsuji, H. (1990). Formation of chlorophyll-protein complexes during greening. I. Distribution of newly synthesized chlorophyll among apoproteins. *Plant Cell Physiol.* **31,** 639–647.

Siefermann-Harms, D. (1985). Carotenoids in photosynthesis. I. Location in photosynthetic membranes and light-harvesting function. *Biochim. Biophys. Acta* **811,** 325–355.

Slovin, J. P., and Tobin, E. M. (1982). Synthesis and turnover of the light-harvesting chlorophyll a/b-protein in *Lemna gibba* grown with intermittent red light: Possible translational control. *Planta* **154,** 465–472.

Spangfort, M., Larsson, U., Ljungberg, U., Ryberg, M., Andersson, B., Klein, R., Wedel, N., and Herrmann, R. (1989). The 20-kDa apopolypeptide fo the chlorophyll a/b complex CP 24. *In* "Techniques and New Developments in Photosynthesis Research" (J. Barber and R. Malkin, eds.), pp. 145–148. Plenum Press, New York.

Sprague, S. G., Camm, E. L., Green, B. R., and Staehelin, L. A. (1985). Reconstitution of light-harvesting complexes and photosystem II cores into galactolipid and phospholipid liposomes. *J. Cell Biol.* **100,** 552–557.

Stayton, M. M., Black, M., Bedbrook, J., and Dunsmuir, P. (1986). A novel chorophyll a/b binding (*Cab*) protein from petunia which encodes the lower molecular weight *Cab* precursor protein. *Nucleic Acids Res.* **14,** 9781–9796.

Stayton, M. M., Brosio, P., and Dunmuir, P. (1987). Characterization of a full-length petunia cDNA encoding a polypeptide of the light-harvesting complex associated with photosystem I. *Plant Mol. Biol.* **10,** 127–137.

Steinmetz, D., Damm, I., and Grimme, L. H. (1990). Reconstitution of the light-harvesting chl $a/b$ protein complex of the chl $b$-less mutant $cbn$1-48 of *Chlamydomonas reinhardtii* with a pigment extract derived from wildtype. In "Current Research in Photosynthesis" (M. Baltscheffsky, ed.), Vol. 2, pp. 855–858. Kluwer, Dordrecht, The Netherlands.

Sukenik, A., Wyman, K. D., Bennett, J., and Falkowski, P. G. (1987). A novel mechanism for regulating the excitation of photosystem II in a green alga. *Nature (London)* **327,** 704–707.

Sukenik, A., Bennett, J., and Falkowski, P. (1988). Changes in the abundance of individual apoproteins of light-harvesting chlorophyll $a/b$-protein complexes of photosystem I and II with growth irradiance in the marine chlorophyte *Dunaliella tertiolecta*. *Biochim. Biophys. Acta* **932,** 206–215.

Sukenik, A., Bennett, J., Mortain-Bertrand, A., and Falkowski, P. G. (1990). Adaptation of the photosynthetic apparatus to irradiance in *Dunaliella tertiolecta*. A kinetic study. *Plant Physiol.* **92,** 891–898.

Thornber, J. P., Morishige, D. T., Anandan, S., and Peter, G. (1991). Chlorophyll–carotenoid–proteins of higher plant thylakoids. In "The Chlorophylls" (H. Scheer, ed.), pp. 549–586. CRC Press, Bota Raton, Florida.

Vainstein, A., Ferreira, P., Peterson, C. C., Verbeke, J. A., and Thornber, J. P. (1989a). Expression of the major light-harvesting chlorophyll $a/b$–protein and its import into thylakoids of mesophyll in bundle sheath chloroplasts of maize. *Plant Physiol.* **89,** 602–609.

Vainstein, A., Peterson, C. C., and Thornber, J. P. (1989b). Light-harvesting pigment-proteins of photosystem I in maize. *J. Biol. Chem.* **264,** 4058–4063.

Van den Broeck, G., Van Houten, A., Van Montagu, M., and Herrera-Estrella, L. (1988). The transit peptide of a chlorophyll $a/b$-binding protein is not sufficient to insert neomycin phosphotransferase II in the thylakoid membrane. *Plant Sci.* **58,** 171–176.

Viitanen, P. V., Doran, E. R., and Dunsmuir, P. (1988). What is the role of the transit peptide in thylakoid integration of the light-harvesting chlorophyll $a/b$ protein? *J. Biol. Chem.* **263,** 15000–15007.

Waegemann, K., Paulsen, H., and Soll, J. (1990). Translocation of proteins into isolated chloroplasts requires cytosolic factors to obtain import competence. *FEBS Lett.* **261,** 89–92.

Wang, D. N., and Kühlbrandt, W. (1991). High-resolution electron crystallography of light-harvesting chlorophyll-$a/b$-protein complex in three different media. *J. Mol. Biol.* **217,** 691–699.

Wechsler, T., Suter, F., Fuller, R. C., and Zuber, H. (1985). The complete amino acid sequence of the bacteriochlorophyll $c$ binding polypeptide from chlorosomes of the green photosynthetic bacterium *Chloroflexus auranthiacus*. *FEBS Lett.* **181,** 172–178.

White, M. J., and Green, B. R. (1987a). Antibodies to the photosystem I chlorophyll $a+b$ antenna cross-react with polypeptides of CP29 and LHCII. *Eur. J. Biochem.* **163,** 545–551.

White, M. J., and Green, B. R. (1987b). Polypeptides belonging to each of the three major chlorophyll $a+b$ protein complexes are present in a chlorophyll $b$-less barley mutant. *Eur. J. Biochem.* **165,** 531–535.

Wilhelm, C., Wiedemann, I., and May, M. (1990). Comparative analysis of the composition of two Chl $b$-containing light-harvesting complexes. *Planta* **180,** 456–457.

Wollman, F. A., and Bennoun, P. (1982). A new chlorophyll–protein complex related to photosystem I in *Chlamydomonas reinhardtii*. *Biochim. Biophys. Acta* **680,** 352–360.

Yalovsky, S., Schuster, G., and Nechushtai, R. (1990). The apoprotein precursor of the major light-harvesting complex of photosystem II (LHCIIb) is inserted primarily into stromal lamellae and subsequently migrates to the grana. *Plant Mol. Biol.* **14,** 753–764.

# 11

# Photosynthetic Activities during Early Assembly of Thylakoid Membranes

**FABRICE FRANCK** [1]

Department of Botany B22
University of Liege
Liege, B-4000 Sart-Tilman, Belgium

I. Introduction
II. Time Course of Development for Photosynthetic Activities after Single Turnover of Protochlorophyllide Reductase
   A. $O_2$ Evolution and $CO_2$ Fixation
   B. Photosystem II and Photosystem I Activities
III. Spectral Changes of Chlorophyll(ide) in Relation to Formation of Photosystems
IV. Synthesis of Chlorophyll $a$-Binding Polypeptides Induced by Short Illumination
V. Concluding Remarks
   References

## I. INTRODUCTION

Light is not only used by plants as the energy source for photosynthesis, it is also the most important factor in the regulation of the biogenesis of the photosynthetic apparatus and the adaptation of that apparatus to the environment. The regulatory effect of light is exerted mainly through two photoreceptors, protochlorophyllide (Pchlide) and phytochrome, whose light absorption monitors the amount of pigments and specific proteins involved in photosynthesis. Angiosperms fail to synthesize chlorophyll and chlorophyll-binding proteins when they are grown in darkness. The block in chlorophyll synthesis is at the level of its precursor Pchlide. In plastids of dark-grown leaves, small amounts of this pigment accumulate

---

[1] F. Franck is a Research Associate of the Belgian National Fund of Scientific Research.

in the state of a stable photoactive complex with the enzyme Pchlide-reductase and its cofactor NADPH. In this state, it is reduced into chlorophyllide $a$ (Chlide) within several seconds of exposure to light. This event is sufficient to trigger the formation of small amounts of functional photosynthetic units that will accumulate during prolonged illumination through the repetition of the process. Experiments on short light-pulse induction of the synthesis and assembly of photosynthetic pigment–protein complexes in etiolated leaves represent a particularly favorable approach to studying the basic mechanism of light action on the biogenesis of the photosynthetic apparatus. This chapter combines functional and molecular aspects of this problem.

## II. TIME COURSE OF DEVELOPMENT OF PHOTOSYNTHETIC ACTIVITIES AFTER SINGLE TURNOVER OF PROTOCHLOROPHYLLIDE REDUCTASE

Only a limited number of experiments has been carried out that follow the development of photosynthetic activities in dark-grown leaves after a brief illumination (msec flash or light pulse of several min) that reduces all photoactive Pchlide to Chlide. In such experiments, the amount of synthesized Chl(ide) is equal to (or slightly higher than, due to some photoactive Pchlide regeneration; see Granick and Gassman, 1970) the amount of Pchlide that forms a stable ternary complex with Pchlide-reductase and NADPH (Oliver and Griffiths, 1982). In this section I discuss how the capacity for photosynthesis develops in darkness after the initial photoconversion step.

### A. $O_2$ Evolution and $CO_2$ Fixation

Smith (1954) observed for the first time that $O_2$ evolution occurred in etiolated barley leaves after a short illumination (10 min) followed by a sufficiently long dark period (110 min). The ability to evolve $O_2$ was enhanced greatly if the leaves were illuminated for a second time before measurements were made. From those data, it was deduced that a lag period of 10–20 min precedes the development of $O_2$ evolution ability after the onset of (a first) illumination

A few years after Smith's discovery, Gabrielsen *et al.* (1961) showed that $CO_2$ production is enhanced during a first 2-min illumination of etiolated wheat leaves, due to some photooxidative process or respiration enhancement. After repeated illuminations spaced by 10 min darkness, the rate of $CO_2$ output in the light decreased progressively; a net $CO_2$ uptake was measured at the fifth illumination. The interpretation of these results was that photosynthetic $CO_2$ fixation started at the second illumination and caused the observed decrease in light-induced $CO_2$ output.

Franck and Peltier (1986) measured $O_2$ and $CO_2$ simultaneously in etiolated barley leaves. $O_2$ was measured by mass spectrometry in an argon-enriched atmosphere and $CO_2$ was measured by thermal conductivity in a vessel containing a large amount of leaves. $O_2$ production was calculated by measuring $^{18}O$ and $^{16}O$ simultaneously, according to Peltier and Thibault (1985). The main results are reproduced in Fig. 1. When illuminating the leaves for the first time inside the measuring cell, no significant $O_2$ evolution was detected over 15–20 min. A rapid stimulation of $CO_2$ production occurred and was reversed only partly after illumination ceased. When the leaves were preilluminated for 40 sec and then kept in darkness for 90 min, a progressive induction of $O_2$ evolution and of $CO_2$ uptake was observed during the first 10–15 min of the second illumination. The rate of $O_2$ evolution at this early developmental stage was about 50% higher than in green leaves, on a Chl basis, at saturating light intensities. Induction

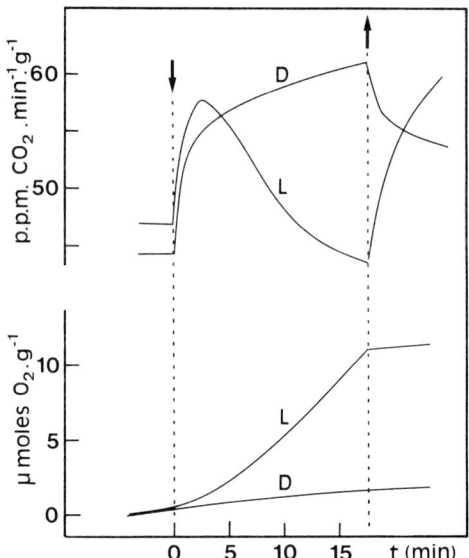

**Fig. 1.** Variations of $O_2$ evolution (bottom) and of first derivative of $CO_2$ concentration (top) at a dark-to-light transition of 6-day-old etiolated barley leaves (D) and of leaves preilluminated for 40 sec and kept in darkness for 90 min thereafter (L). ( ↓ ) Light on; ( ↑ ) light off. White light of 500 $\mu E \cdot m^{-2} \cdot sec^{-1}$.

of low levels of $O_2$ evolution and $CO_2$ fixation could be detected as soon as 20 min after a 40-sec light pulse (Franck and Peltier, 1986).

The experiments just described clearly demonstrate that a complete photosynthetic electron transport chain becomes functional during a second illumination of etiolated leaves that follows the first one by more than 20 min. The fact that $O_2$ evolution appears only progressively during the second illumination suggests that a photoactivation of the water-splitting system takes place in native photosystem II (PSII) that is formed as a result of the initial light pulse. This suggestion is also supported by measurements on isolated plastids (Franck and Schmid, 1984) that show that a second illumination of the intact leaves with continuous light is necessary for isolated plastids to evolve $O_2$ in the light. Photoactivation is known to occur only in intact leaves or intact plastids (Ono and Inoue, 1982).

Any light that invokes the photoreduction of Pchlide at the first light pulse seems to be sufficient to trigger the events that will lead to the further development of photosynthesis. White, blue, red, and even green light pulses of sufficient intensity and duration to reduce all photoactive Pchlide gave maximal rates of $O_2$ evolution after the second illumination (F. Franck, 1986, unpublished observations). No reversion by far-red light was found, indicating the probable absence of direct phytochrome effects on these early steps of greening. Therefore, light initiation of the synthesis and assembly of components of the photosynthetic apparatus seems to occur mainly or only through Pchlide photoreduction. The proposal that Pchlide is the photoreceptor that acts in the induction of Chl-binding polypeptide synthesis (Klein *et al.*, 1988; Eichacker *et al.*, 1990) is consistent with the early development of photosynthetic activity in a phytochrome-independent manner (See Section IV).

## B. Photosystem II and Photosystem I Activities

In which order are different parts of the photosynthetic electron transport chain assembled after the initial photoreduction of Pchlide? It has been reported often that photosystem I (PSI) activity precedes PSII activity under continuous greening. In most cases, measurements were performed on plastid membranes isolated after various lengths of illumination of intact leaves. Since PSII is very unstable at early greening stages, the absence of PSII activity in membranes simply may be due to damaging effects of the isolation procedures. Activities such as 2,6-dichlorophenol-indophenol (DPIP) photoreduction or $O_2$ evolution are suppressed completely after a simple osmotic shock of the plastids during the first hours

## 11. Early Assembly of Thylakoid Membranes

of greening. On the other hand, the interpretation of $O_2$ uptake as an indicator of PSI activity is not always easy because other light-induced $O_2$-consuming processes such as pigment photooxidation take place during early stages of greening (Axelsson, 1976; Redlinger and McDaniel, 1978).

### 1. Photosystem II Activity

PSII activity can be monitored easily using nonintrusive methods. Among these, delayed luminescence and fluorescence variation measurements have been used to trace the formation of PSII after a short light pulse that reduces all Pchlide (Shlyk *et al.* 1985; Franck, 1990). By both methods, PSII-mediated electron transport was detected a very short time after the illumination. Figure 2 represents the development of room temperature delayed luminescence (DL) and 77 K variable fluorescence ($F_v$) of etiolated barley leaves after a msec flash.

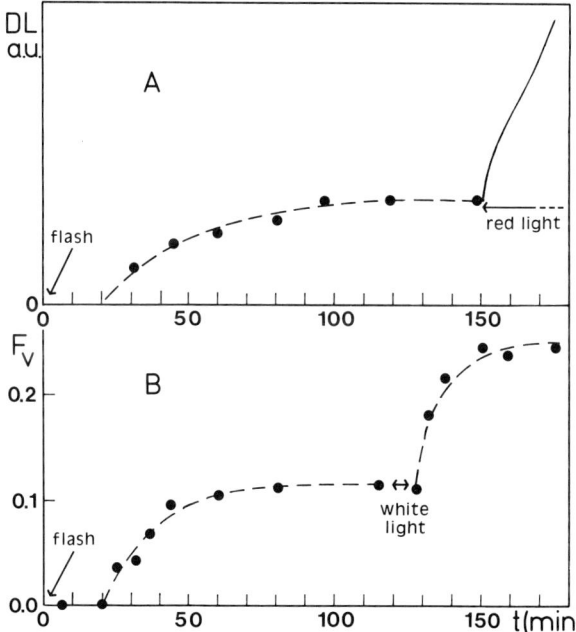

**Fig. 2.** Development of PSII electron transport in intact 5-day-old barley leaves after a 1-msec flash. (A) Room temperature delayed luminescence. Figure courtesy of V. P. Domanskii. (B) 77 K variable fluorescence ($F_V = 1 - F_O/F_M$, where $F_O$ is the constant fluorescence intensity and $F_M$ is the maximum fluorescence intensity at λ660 nm). Also shown are the effects of a second illumination by continuous red or white light, as indicated.

When the integrated DL intensity is measured at various times after one light flash, a lag period of 15–20 min is observed. This lag (which has no relationship to the lag period in Chl accumulation) is followed by a slow increase to a constant value reached after 60–90 min, which reflects the accumulation of PSII core complexes in darkness after the flash (Fig. 2A). Subsequent continuous illumination by red light results in a rapid enhancement of DL because of the formation of additional PSII core complexes as well as the photoactivation of the water-splitting complex in core complexes already present. The involvement of photoactivation is demonstrated by the progressive appearance of the slow component of DL, which was previously shown to increase along with photoactivation in flashed green leaves (Domanskii, 1986; Ichikawa *et al.*, 1975).

Fluorescence variations ($F_v$) in barley leaves frozen in liquid nitrogen at various dark intervals after a msec flash were recorded also (Franck, 1990). The time course of the development of $F_v$ after the flash is very similar to the one of DL (Fig. 2B). A typical lag period of 15–20 min is observed. Both curves reflect the accumulation of photochemically active PSII, whose Chls derive from the Pchlide accumulated during etiolation. Relative $F_v$ values at 77 K depend on the proportion, relative to the total Chl, of Chl molecules that transfer excitation energy to PSII reaction centers. The large increase of $F_v$ after a second illumination (Fig. 2) suggests that the second set of Chl molecules is incorporated into PSII Chl–protein complexes more efficiently than the first set of Chl molecules. Moreover, no lag period of several minutes is observed in $F_v$ development after the second illumination.

## 2. Photosystem I Activity

Precise data on the time course of development of PSI activity after a single turnover of Pchlide-reductase in etiolated leaves is unavailable. Hiller and Boardman (1971) showed that a 3-min illumination of bean leaves is sufficient to induce the ability of PSI to mediate cytochrome photooxidation. Oelze-Karow and Butler (1971) detected cyclic photophosphorylation in intact etiolated bean leaves after only 15 min illumination with white light. Although measurements were not performed at shorter illumination times, extrapolation of their data on phosphorylation rates versus greening time indicated no lag period in the development of this PSI-mediated activity. Primary charge separation in the PSI reaction center (P700 photooxidation) was detected after 1 hr of continuous greening and is likely to appear within the very first minutes of illumination (Baker and Butler, 1976). If the absence of a lag in the formation of

PSI was confirmed, it would indicate some difference in the metabolic processes leading to the formation of the native PSI and PSII complexes. The lag period in PSII formation might be due to some specific delay in the synthesis or assembly of Chl-binding polypeptides of PSII after Pchlide photoreduction. In contrast to most studies on the subject (see Section IV), some authors could detect the P700 apoprotein in unilluminated etiolated leaves (Nechustai and Nelson, 1985; Shlyk *et al.*, 1986).

Oelze-Karow and Butler (1971) showed that the rate of photophosphorylation is not affected by 3-(3,4-dichlorophenyl)-1,1-dimethylurea (DCMU) during the first 30 min of greening in continuous light, indicating that only cyclic photophosphorylation occurs during this period. This observation is consistent with the lag period of several minutes in the formation of PSII, which must cause a delay in the appearance of noncyclic electron transport.

Photoinduced absorbance variation measurements at 340 nm (absorbance maximum of NADPH) by Bertrand *et al.* (1988) strongly suggested that PSI-mediated $NADP^+$ photoreduction occurs less than 10 min after the onset of illumination in bean etioplasts. The absorbance at 340 nm first decreased for 1 min and then increased, provided NADP, FNR, and ferredoxin (Fd) were added in addition to an antioxidation system (glucose oxidase + glucose + catalase). The absorbance increase was reversed in darkness after addition of 3-phosphoglycerate (PGA). These results were interpreted as reflecting an early $NADP^+$ photoreduction. The authors speculated that glucose + glucose oxidase might provide electrons to PSI under their conditions but it is also possible, as suggested earlier by Hiller and Boardman (1971) and by Egneus *et al.* (1972), that endogenous reductants that act as electron donors to PSI are present in etioplasts.

## III. SPECTRAL CHANGES OF CHLOROPHYLL(IDE) IN RELATION TO FORMATION OF PHOTOSYSTEMS

Light absorption *in vivo* by the photoactive $Pchlide_{650-657}$ (the Pchlide-reductase–NADPH complex) initiates a series of spectral shifts forming a cycle during the course of which Chl(ide) is released and photoactive Pchlide is regenerated (Fig. 3; Sironval, 1981). The duration of the shifts increases with age (Sestak, 1984). The half-times indicated here are for 5-day-old barley leaves.

$Chlide_{678-688}$ is formed within 5–10 $\mu$sec of exposure of etiolated leaves to a saturating flash as short as several psec (Franck and Mathis, 1980; Dobek *et al.*, 1981; Inoue *et al.*, 1981). Chlide then undergoes a rapid transformation ($t_{1/2}$ = 4 sec) into $Chlide_{684-696}$ (shift 1A) or, to some extent,

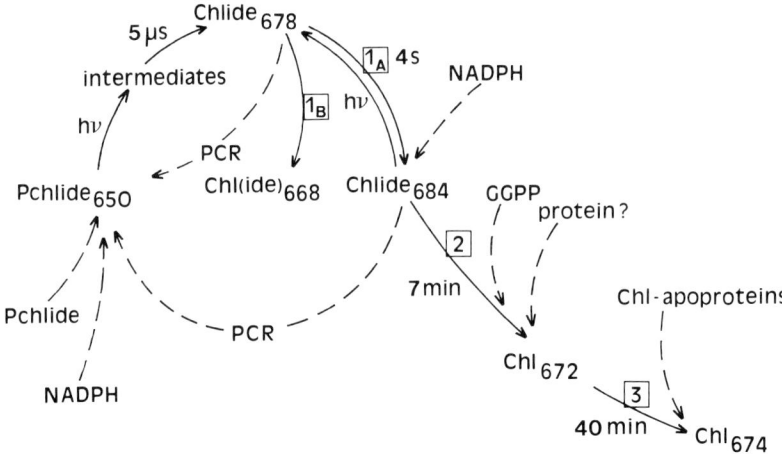

**Fig. 3.** Scheme of the spectral changes during and after Pchlide photoreduction in relation to pigment–protein complex modifications. Pchlide$_X$ and Chlide$_X$ represent protochlorophyllide or chlorophyllide with absorbance maximum at $X$ nm at 77 K.

into Chlide$_{668-675}$ (shift 1B) (Gassman et al., 1968; Bonner, 1969; Mathis and Sauer, 1973). Chlide$_{684-696}$ is then transformed into Chl(ide)$_{672-682}$ (shift 2; $t_{1/2}$ = 7 min). This shift is the so-called Shibata shift (Shibata, 1957; see also Litvin and Krasnovsky, 1957; Thorne, 1971). Chl(ide)$_{672-682}$ is not the final stable product obtained after one flash; it is changed slowly ($t_{1/2}$ = 35 min) into Chl(ide)$_{674-684}$, which represents a mixture of spectrally close Chl forms (Shibata, 1957; Axelsson, 1976; Franck, 1990).

All these spectral changes occur in light as well as in darkness. However, shift 1A is reversed by light absorbed by Chlide$_{684-696}$ in such a way that a Chlide microcycle takes place under continuous light (Bauer and Siegelman, 1972; Franck and Inoue, 1984). According to Oliver and Griffiths (1982) Chlide$_{684-696}$ represents a ternary complex between Pchlide-reductase, Chlide, and NADPH. Franck and Schmid (1985) obtained evidence that NADPH was oxidized or replaced by NADP$^+$ during the light-driven transformation of Chlide$_{684-696}$ to Chlide$_{678-688}$. The occurrence of the Chlide microcycle in light leaves Chlide$_{668-675}$ unaffected and does not influence the course of the Shibata shift significantly (Bauer and Siegelman, 1971; Franck and Inoue, 1984).

The esterification of Chlide *a* to Chl *a* is parallel to, or slightly slower than, shift 2 (Akoyunoglou and Michaelopoulos, 1971; Henningsen and Thorne, 1974). Chl *b* also is synthesized in small amounts after one light pulse, but the rate is slower than that of Chl *a* formation, suggesting transformation of Chl *a* into Chl *b* (Shlyk et al., 1970).

To bind to polypeptides of the photosynthetic light-harvesting and reaction-center complexes, newly formed Chl(ide) molecules must leave the site at which they are formed on Pchlide-reductase. This step certainly is achieved during the Shibata shift (shift 2). In etioplast membranes, addition of exogenous Pchlide after phototransformation induces a Chl(ide) spectral shift identical to the *in vivo* Shibata shift (Oliver and Griffiths, 1982). This result suggests that Chlide molecules are displaced in the presence of excess Pchlide. In intact very young leaves (Akoyunoglou and Siegelman, 1968) or in older leaves treated with 5-aminolevulinic acid (ALA) Brouers and Sironval, 1974), the Shibata shift parallels the regeneration of photoactive Pchlide. These facts indicate the close relationship between the shift and Chlide dissociation from Pchlide-reductase, which allows the binding of new Pchlide molecules to the enzyme. In 5- or 6-day-old barley leaves not fed ALA, the regeneration of photoactive Pchlide is, however, slower than the Shibata shift because of a lag in the biosynthesis of Pchlide pigments that occurs after the shift (Akoyunoglou and Siegelman, 1968; Henningsen and Thorne, 1974). Fradkin and Samoilenko (1985) found that 77 K energy transfer from a 496 nm absorbing carotenoid to Chl(ide) appears gradually during the shift. This result also points out the displacement of Chl(ide) to a microenvironment in which carotenoids are more abundant. The Shibata shift is accompanied by changes at higher organization levels: dissociation of the Pchlide holochrome (probably consisting of aggregated Pchlide-reductase units; Böddi *et al.*, 1989) and loss of the crystalline aspect of the prolamellar body (Henningsen and Boynton, 1969). These processes are reversed as photoactive Pchlide regenerates.

The protein part of Chl(ide)$_{672-682}$ is poorly characterized. Only few reports have been made on analysis of Chl-binding polypeptides at this early developmental step because of the high instability of pigment–protein complexes (Burkey, 1986). Canaani and Sauer (1977) found that the main product of the Shibata shift was a pigment–protein of 16-kDa apparent molecular mass. In green gels of low detergent concentration, a special chlorophyll–protein complex (CPX) appears during the course of the Shibata shift and disappears within 3 hr of illumination in barley (Tanaka and Tsuji, 1985; Ohashi *et al.*, 1989). Chl(ide)$_{672-682}$ is generally considered to consist of a major Chl–protein complex, or "plastic pool" of Chl, that is used for the construction of the photosynthetic apparatus (Shlyk *et al.*, 1986).

A small proportion (approximately 15% in 5-day-old barley) of Chlide pigments formed after one flash leave their formation site before the Shibata shift has even started. This movement can be inferred from the rapid regeneration of photoactive Pchlide to 15% of the initial content that occurs within several seconds after one saturating flash (Granick and Gassman,

1970), during the time that the same proportion of Chlide undergoes shift 1B from Chlide $_{678-688}$ to Chl(ide)$_{668-675}$ (Michel and Sironval, 1977). This proportion decreases with longer etiolation times. In very young leaves, the rapid shift to Chl(ide)$_{668-675}$ is predominant (Klein and Schiff, 1972; Schoefs and Franck, 1991). Starting with Chlide$_{678-688}$, two distinct pools of Chlide are identified on the basis of the time required for their release from Pchlide-reductase. The rapidly released Chlide produces Chl(ide)$_{668-675}$ (shift 1B) whereas the slowly released one produces Chl(ide)$_{672-682}$ (shifts 1A and 2). These pools probably originate from two distinct Pchlide pools located in the prothylakoids (PTs) and the prolamellar bodies (PLBs), since short- and long-wavelength emitting Chlides are produced by purified PTs and PLBs, respectively (Ryberg and Sundqvist, 1988; Lindsten *et al.*, 1990). An alternative, but less plausible, explanation, in view of the distinct properties of PTs and PLBs, is the possible existence of two different photoactive Pchlide species in dynamic equilibrium, as discussed by Kahn and Nielsen (1974) and by Klockare and Virgin (1983).

As seen in Section IIB, PSI activity is likely to start during or before shift 2 (the Shibata shift) whereas PSII activity develops concomitantly with shift 3. However, the correlation between the appearances of particular spectral forms and of PSI or PSII pigments is difficult to establish, because a relatively small proportion of the total amount of pigments becomes part of photosynthetic units of one or the other photosystem and because the spectral properties of the Chl–proteins involved may be different at early developmental stages than in mature green leaves.

Fradkin *et al.* (1982) detected a weak PSI 77 K fluorescence emission at 720 nm at the end of the Shibata shift. Its excitation spectrum showed a band at 683 nm. Small amounts of PSI pigments absorbing at 683 nm are therefore present at this stage in addition to the large Chl(ide)$_{672-682}$ pool. In fact, deconvolution analysis of the 77 K absorbance spectra after shift 2 revealed minor amounts of pigments absorbing in this region (Virgin and French, 1973). PSI units with properties similar to those seen in green leaves thus are assembled within some minutes of Pchlide photoreduction. Their 77 K fluorescence emission increases linearly from the start of illumination (Fradkin and Shlyk, 1978; Shlyk *et al.*, 1986).

The accumulation of PSII units responsible for delayed luminescence and variable fluorescence coincides in time with shift 3. These PSIIs already contain a set of spectrally distinct Chl–proteins, since the action spectrum of DL exhibits bands at 667, 675, and 683 nm (Shlyk *et al.*, 1986), which must be part of the Chl(ide)$_{674-684}$ spectrum. From the early work by Shibata (1957), it is obvious that shift 3 is complex and produces several spectral forms of Chl. The 720-nm fluorescence emission from PSI also increases during the shift (Fradkin *et al.*, 1982; Shlyk *et al.*, 1986).

At this early stage of development, CP1 is the only detectable photosynthetic chlorophyll–protein complex. CPa and light-harvesting complex II (LHCII) of PSII could be detected only after 3–4 hr of continuous illumination in barley, although PSII activity appeared much earlier (Tanaka and Tsuji, 1985; Burkey, 1986). This discrepancy between time of detection of PSII activity and of its Chl–protein complexes is likely to be caused by the difficulty of detecting very small amounts of these complexes and by the lability of Chl at early stages. The native PSII units that appear during shift 3 probably contain the D1/D2 reaction center and the CP47 internal antenna, since this is a minimal composition for variable fluorescence and $O_2$ evolution.

It is not impossible that PSI pigments originate from the rapidly released $Chl(ide)_{668-675}$ whereas PSII pigments originate only from the slowly released $Chlide_{684-696}$. This hypothesis is consistent with an observation by V. P. Domanskii and L. I. Fradkin (personal communication). The authors have examined the dependence of PSII DL development after one flash on the extent of Pchlide photoreduction that was obtained at various flash intensities. A threshold of a photoreduction extent of 15% was found, below which no DL appeared after the flash. The authors suggested that some precursor must accumulate to a certain level to initiate the formation of PSII centers. It is known from other studies that $Chl(ide)_{668-675}$ but not $Chlide_{684-696}$ is formed at a first flash of low intensity (Michel and Sironval, 1977; Böddi et al., 1991). Although experiments combining PSI and PSII activity measurements with Chlide spectral form determination are still to be carried out, separate experiments seem to indicate that only Chlide that undergoes shift 1A is incorporated effectively into PSII units.

## IV. SYNTHESIS Of CHLOROPHYLL a-BINDING POLYPEPTIDES INDUCED BY SHORT ILLUMINATION

A 60-min treatment of intact leaves with chloramphenicol inhibited the development of PSII DL and $F_v$, as well as spectral shift 3 after one flash, almost completely (Franck, 1993). These data support the close relationship between those events and demonstrate that the formation of PSII pigment–protein complexes after a single turnover of Pchlide-reductase depends on a light-induced synthesis of plastid-encoded Chl-binding polypeptides.

Several apoproteins of the PSI and PSII core complexes are absent from or present only in trace amounts in etioplasts (Mullet, 1988). Light exerts a control at the translational or post-translational level, causing a rapid accumulation of these proteins during greening (Boschetti et al., 1990).

The rapid development of PSI and PSII requires biosynthesis of Chl-binding polypeptides shortly after Pchlide has been reduced to Chlide. Eichacker et al. (1990) showed that the CP47 and CP43 polypeptides of PSII and the P700 apoprotein accumulate in plastids when Chl $a$ is synthesized (see Chapter 10). Their results demonstrate that Pchlide-reductase and Chl-synthetase are key enzymes in the light-induced expression of those plastid-encoded polypeptides and clearly identify Pchlide, not phytochrome, as the photoreceptor (see also Laing et al., 1988). Of the PSII reaction center proteins, only D2 is synthesized in dark-grown barley plants (Klein et al., 1988). In spinach, D1 was detected only after 4 hr of continuous illumination (Liveanu et al., 1986). Klein and Mullet (1986) and Klein et al. (1988) found that a short light pulse induced the synthesis of a polypeptide that comigrates with D1 in PAGE, but the peptide was not identified clearly. If D1 is essential for PSII photochemistry, it should begin to be synthesized within 15–20 min of Pchlide photoreduction, since PSII $F_v$ and DL appear at that time in vivo.

The three extrinsic polypeptides of PSII involved in $O_2$ evolution activity are present in etiolated plants (barley, Ryrie et al., 1984; spinach, Liveanu et al., 1986). Two of these proteins, the 16-kDa and the 24-kDa polypeptides, bind to PSII core complexes during the photoactivation of the water-splitting system (Ono et al., 1986), which occurs under continuous light as soon as PSII photochemistry is detected (Section I,B,1). The development of PSII activity seems limited mainly by the synthesis and assembly of the intrinsic PSII polypeptides D1, CP47, and CP43.

The accumulation of nucleus-encoded polypeptides is under the control of phytochrome. The LHCII antenna complex of PSII is detected only after prolonged irradiation with continuous light (Mathis and Burkey, 1987). Therefore, only core complexes of PSII accumulate in single-flash experiments or at the beginning of continuous greening, which explains the small PSII antenna size at early stages (Tsakiris and Akoyunoglou, 1981).

In green leaves, PSI and PSII complexes are associated with enzymes and precursor pigments of the last steps of Chl biosynthesis, forming several multifunctional supramolecular complexes that can be released from plastid inner membranes by mild digitonin treatment and separated by PAGE (Fradkin et al., 1981; Shlyk et al., 1982). These large complexes correspond to particles seen by freeze-fracture electron microscopy of chloroplast membranes (Fradkin et al., 1989). In etioplasts, pigmented supramolecular complexes of a smaller size are obtained (L. I. Fradkin, 1991 personal communication). Fradkin and co-workers speculate that in etioplasts Pchlide-reductase is part of protein complexes that develop into mature multifunctional ones during greening as a result of Chl accumulation and binding of newly synthesized polypeptides. In support of this idea, Pchlide holochrome does not contain only Pchlide-reductase but also polypeptides of the photosynthetic apparatus such as plastocyanin, cytochrome $f$, and FNR (Dujardin et al., 1987).

## V. CONCLUDING REMARKS

This chapter demonstrates that Pchlide photoreduction initiates the rapid formation of PSI and PSII in etiolated leaves. This process, including protein synthesis, pigment translocation, and pigment–protein assembly, proceeds in complete darkness after the initial photoreduction step, which can be achieved by a very short illumination. Although considerable literature exists on the subject, it is not always easy to correlate spectral shifts, development of photosynthetic activities, and protein synthesis in this context. In the future, the various experimental approaches should be combined to improve our knowledge about the molecular mechanism of light-induced formation of the photosynthetic apparatus in higher plants.

## REFERENCES

Akoyunoglou, G., and Michaelopoulos, G. (1971). The relation between the phytylation and the 682→672 nm shift *in vivo* of chlorophyll *a*. *Physiol. Plant.* **25**, 324–329.

Akoyunoglou, G., and Siegelman, H. (1968). Protochlorophyllide resynthesis in dark-grown bean leaves. *Plant Physiol.* **43**, 60–68.

Axelsson, L. (1976). The photostability of different chlorophyll forms in dark-grown leaves of wheat: I. Stability to high intensity red light of forms appearing after photoreduction of protochlorophyllide. *Physiol. Plant.* **38**, 327–332.

Baker, N., and Butler, W. (1976). Development of the primary photochemical apparatus of photosynthesis during greening of etiolated bean leaves. *Plant Physiol.* **58**, 526–529.

Bauer, S., and Siegelman, H. W. (1972). Photoconversion of chlorophyllide 684 to chlorophyllide 678. *FEBS Lett.* **20**, 352–354.

Bertrand, M., Bereza, B., and Dujardin, E. (1988). Evidence for photoreduction of $NADP^+$ in a suspension of lysed plastids from etiolated bean leaves. *Z. Naturforsch.* **43c**, 443–448.

Böddi, B., Lindsten, A., Ryberg, M., and Sundqvist, C. (1989). On the aggregational states of protochlorophyllide and its protein complexes in wheat etioplasts. *Physiol. Plant.* **76**, 135–143.

Böddi, B., Ryberg, M., and Sundqvist, C. (1991). The formation of short-wavelength chlorophyllide form at partial phototransformation of protochlorophyllide in etioplast inner membranes. *Photochem. Photobiol.* **53**, 667–673.

Bonner, B. (1969). A short-lived intermediate form in the *in vivo* conversion of protochlorophyllide 650 to chlorophyllide 684. *Plant Physiol.* **44**, 739–747.

Boschetti, A., Breidenbach, E., and Blätter, R. (1990). Control of protein formation in chloroplasts. *Plant Sci.* **68**, 131–149.

Brouers, M., and Sironval, C. (1974). Evidence for energy transfer from protochlorophyllide to chlorophyllide in leaves treated with δ-aminolevulinic acid. *Plant Sci. Lett.* **2,** 67–72.
Burkey, K. (1986). Chlorophyll–protein complex composition and photochemical activity in developing chloroplasts from greening barley seedlings. *Photosynth. Res.* **10,** 37–49.
Canaani, O., and Sauer, K. (1977). Analysis of the subunit structure of protochlorophyllide holochrome by sodium dodecyl sulfate–polyacrylamide gel electrophoresis. *Plant Physiol.* **60,** 422–429.
Dobek, A., Dujardin, E., Franck, F., Sironval, C., Breton, J., and Roux, E. (1981). The first events of protochlorophyllide photoreduction in etiolated leaves investigated by means of the fluorescence excited by short, 610nm laser flashes at room temperature. *Photobiochem. Photobiophys.* **2,** 35–44.
Domanskii, V. P. (1986). Ph.D. thesis, Institute of Photobiology of the Bielorussian Academy of Science, Minsk.
Dujardin, E., Bertrand, M., Radunz, A., and Schmid, G. H. (1987). Immunological evidence for the presence of proteins of the photosynthetic membrane in etiolated leaves of *Phaseolus vulgaris*. *J. Plant. Physiol.* **128,** 95–107.
Egneus, H., Reftel, S., and Sellden, G. (1972). The appearance and development of photosynthetic activity in etiolated barley leaves and isolated etio-chloroplasts. *Physiol. Plant.* **27,** 48–55.
Eichacker, L. A., Soll, J., Lauterbach, P., Rüdiger, W., Klein, R. R., and Mullet, J. E. (1990). In vitro synthesis of chlorophyll *a* in the dark triggers accumulation of chlorophyll *a* apoproteins in barley etioplasts. *J. Biol. Chem.* **265,** 13566–13571.
Fradkin, L. I., and Samoilenko, S. (1985). The development of energy transfer from carotenoids to chlorophyllide *a* in the course of Shibata shift. *Doclady Acad. Nauk. SSSR* **281,** 1248–1251 (in Russian).
Fradkin, L. I., and Shlyk, A. A. (1978). Spectral research on group localization of protochlorophyllide and chlorophyll molecules in the centres of chlorophyll biosynthesis. *J. Appl. Spectrosc.* **29,** 1029–1039 (in Russian).
Fradkin, L. I., Chkanikova, R. A., and Shlyk, A. A. (1981). Coupling of chlorophyll metabolism with submembrane chloroplast particles, isolated with digitonin and gel electrophoresis. *Plant Physiol.* **67,** 555–559.
Fradkin, L. I., Domanskii, V. P., Samoilenko, A. G., and Shlyk, A. A. (1982). Photosynthetic pigment systems during first hour of de-etiolation of barley leaves. *Doklad. Biophys.* **264,** 93–97.
Fradkin, L. I., Radyuk, M. S., Domanskii, V. P., and Kolyago, V. M. (1989). Structural localization of multifunctional supracomplexes in chloroplast membranes. *Photosynthetica* **23,** 343–350.
Franck, F. (1990). Development of PS II photochemistry after a single white flash in etiolated barley leaves. In "Current Research in Photosynthesis" (M. Baltscheffsky, ed.), Vol. III, pp. 751–754. Kluwer, Dordrecht, The Netherlands.
Franck, F. (1993). On the formation of photosystem II chlorophyll-proteins after a short light flash in etiolated barley leaves. *J. Photochem. Photobiol.*, in press.
Franck, F., and Inoue, Y. (1984). Light-driven reversible transformation of chlorophyllide $P_{696,682}$ into chlorophyllide $P_{688,678}$ in illuminated etiolated bean leaves. *Photobiochem. Photobiophys.* **8,** 85–96.
Franck, F., and Mathis, P. (1980). A short-lived intermediate in the photoenzymatic reduction of protochlorophyllide into chlorophyllide at a physiological temperature. *Photochem. Photobiol.* **32,** 799–803.
Franck, F., and Peltier, G. (1986). Light-induced gas exchanges in shortly illuminated intact etiolated leaves measured by mass-spectrometry. In "Colloque de Photosynthèse" (G. Paillotion, ed.), pp. 128–130. Institut National des Sciences et Techniques Nucléaires,

Saclay, France.
Franck, F., and Schmid, G. H. (1984). Flash pattern of oxygen evolution in greening etioplasts of oat. *Z. Naturforsch.* **39c**, 1091–1096.
Franck, F., and Schmid, G. H. (1985). The role of NADPH in the reversible phototransformation of chlorophyllide $P_{682}$ into chlorophyllide $P_{678}$ in etioplasts of oat. *Z. Naturforsch.* **40c**, 832–838.
Franck, F., Sironval, C., and Schmid, G. H. (1984). Extension of the experiment by J. H. C. Smith on the onset of the oxygen evolution in etiolated bean leaves. In "Protochlorophyllide Reduction and Greening" (M. Brouers and C. Sironval, eds.), pp. 223–235. Nijhoff/Junk Publishers, The Hague, The Netherlands.
Gabrielsen, E. K., Madsen, A., and Vejlby, K. (1961). Induction of photosynthesis in etiolated leaves. *Physiol. Plant.* **14**, 98–110.
Gassman, M., Granick, S., and Mauzerall, D. (1968). A rapid spectral change in etiolated red kidney bean leaves following phototransformation of protochlorophyllide. *Biochem. Biophys. Res. Commun.* **32**, 295–300.
Granick, S., and Gassman, M. (1970). Rapid regeneration of protochlorophyllide 650. *Plant Physiol.* **45**, 201–205.
Henningsen, K. W., and Boynton, J. E. (1969). Macromolecular physiology of plastids: VII. The effect of a brief illumination on plastids of dark-grown barley leaves. *J. Cell Sci.* **5**, 757–793.
Henningsen, K. W., and Thorne, S. W. (1974). Esterification and spectral shifts of chlorophyll(ide) in wild-type and mutant seedlings developed in darkness. *Physiol. Plant.* **30**, 82–89.
Hiller, R. G., and Boardman, N. K. (1971). Light driven redox changes cytochrome $f$ during greening of bean leaves. *Biochim. Biophys. Acta* **253**, 449–458.
Ichikawa, T., Inoue, Y., and Shibata, K. (1975). Delayed light emission and variable fluorescence from intermittently illuminated wheat leaves under continuous illumination related to activation of the latent water-splitting system. *Plant Sci. Lett.* **4**, 369–376.
Inoue, Y., Kobayashi, T., Ogawa, T., and Shibata, K. (1981). A short-lived intermediate in the photoconversion of protochlorophyllide to chlorophyllide *a*. *Plant Cell Physiol.* **22**, 197–204.
Kahn, A., and Nielsen, O. F. (1974). Photoconvertible protochlorophyll(ide) 635–650 *in vivo*: A single species or two species in dynamic equilibrium? *Biochim. Biophys. Acta* **333**, 409–414.
Klein, R. R., and Mullet, J. E. (1986). Regulation of chloroplast-encoded chlorophyll-binding protein translation during higher plant chloroplast biogenesis. *J. Biol. Chem.* **261**, 11138–11145.
Klein, R. R., Gamble, P. E., and Mullet, J. E. (1988). Light-dependent accumulation of radiolabeled plastid-encoded chlorophyll-*a* apoproteins requires chlorophyll *a*: I. Analysis of chlorophyll-deficient mutants and phytochrome involvement. *Plant Physiol.* **88**, 1246–1256.
Klein, S., and Schiff, J. A. (1972). The correlated appearance of prolamellar bodies, protochlorophyll(ide) species and the Shibata shift during development of bean etioplast in the dark. *Plant Physiol* **49**, 619–626.
Klockare, B., and Virgin, H. I. (1983). Chlorophyll(ide) forms after partial phototransformation of protochlorophyll(ide) in etiolated wheat leaves. *Physiol. Plant.* **57**, 28–34.
Laing, W., Kreuz, K., and Apel, K. (1988). Light-dependent, but phytochrome-independent translational control of the accumulation of the $P_{700}$ chlorophyll-*a* protein of photosystem I in barley (*Hordeum vulgare* L.). *Planta* **176**, 269–276.
Lindsten, A., Welch, C. J., Schoch, S., Ryberg, M., Rüdiger, W., and Sundqvist, C. (1990). Chlorophyll synthetase is latent in well preserved prolamellar bodies of etiolated wheat. *Physiol. Plant.* **80**, 277–285.

Litvin, F. F., and Krasnovsky, A. A. (1957). The study of intermediate steps of chlorophyll formation in etiolated leaves by fluorescence spectroscopy. *Dokl. Akad. Nauk. SSSR* **117,** 106–109 (in Russian).

Liveanu, V., Yocum, C. F., and Nelson, N. (1986). Polypeptides of the oxygen-evolving photosystem II complex; Immunological detection and biogenesis. *J. Biol. Chem.* **261,** 5296–5300.

Mathis, J. N., and Burkey, K. O. (1987). Regulation of light-harvesting chlorophyll protein biosynthesis in greening seedlings, a species comparison. *Plant Physiol.* **85,** 971–977.

Mathis, P., and Sauer, K. (1973). Chlorophyll formation in greening bean leaves during the early stages. *Plant Physiol.* **51,** 115–119.

Michel, J.-M., and Sironval, C. (1977). Shifts to $C_{675-670}$ and to $C_{696-684}$ in etiolated leaves illuminated with series of brief flashes. *Plant Cell Physiol.* **18,** 1223–1234.

Mullet, J. E. (1988). Chloroplast development and gene expression. *Annu. Rev. Plant Physiol. Plant Mol. Biol.* **39,** 475–502.

Nechustai, R., and Nelson, N. (1985). Biogenesis of photosystem I reaction center during greening of oat, bean, and spinach leaves. *Plant Mol. Biol.* **4,** 377–384.

Oelze-Karow, H., and Butler, W. L. (1971). The development of photophosphorylation and photosynthesis in greening bean leaves. *Plant Physiol.* **48,** 621–625.

Ohashi, K., Tanaka, A., and Tsuji, H. (1989). Formation of the photosynthetic electron transport system during the early phase of greening in barley leaves. *Plant Physiol.* **91,** 409–414.

Oliver, R. P., and Griffiths, W. T. (1982). Pigment–protein complexes of illuminated etiolated leaves. *Plant Physiol.* **70,** 1019–1025.

Ono, T., and Inoue, Y. (1982). Photoactivation of the water-oxidation system in isolated intact chloroplasts prepared from wheat leaves grown under intermittent flash illumination. *Plant Physiol.* **69,** 1418–1422.

Ono, T., Kajikawa, H., and Inoue, Y. (1986). Changes in protein composition and Mn abundance in photosystem II particles on photoactivation of the latent $O_2$-evolving system in flash-grown leaves. *Plant Physiol.* **80,** 85–90.

Peltier, G., and Thibault, P. (1985). Light-dependent oxygen uptake, glycolate, and ammonia release in L-methionine sulfoximine-treated *Chlamydomonas. Plant Physiol.* **77,** 281–284.

Redlinger, T. E., and McDaniel, R. G. (1978). Comparison of light-dependent oxygen uptake, protochlorophyllide-650 photoconversion, and chlorophyll disappearance in wheat etioplasts. *Plant Physiol.* **61,** 1006–1009.

Ryberg, M., and Sundqvist, C. (1988). The regular ultrastructure of isolated prolamellar bodies depends on the presence of membrane-bound NADPH–protochlorophyllide oxidoreductase. *Physiol. Plant.* **73,** 218–226.

Ryrie, I. J., Young, S., and Andersson, B. (1984). Development of the 33-, 23-, and 16-kDa polypeptides of the photosynthetic oxygen-evolving system during greening. *FEBS Lett.* **177,** 269–273.

Schoefs, B., and Franck, F. (1991). Photosystem II assembly in 2-day-old bean leaves during the first 6 hours of greening. *C. R. Acad. Sci. Paris, Sér. III,* **315,** 441–445.

Sestak, Z. (1984). Effect of leaf age on protochlorophyllide and chlorophyllide (a review). *In*: "Protochlorophyllide Reduction and Greening" (M. Brouers and C. Sironval, eds.), pp. 365–375. Nijhoff/Junk Publishers, The Hague, The Netherlands.

Shibata, K. (1957). Spectroscopic studies on chlorophyll formation in intact leaves. *J. Biochem.* **44,** 147–173.

Shlyk, A. A., Rudoi, A. B., and Vezitskii, A. Y. (1970). Immediate appearance and accumulation of chlorophyll *b* after a short illumination of etiolated maize seedlings. *Photosynthetica* **4,** 68–77.

Shlyk, A. A., Averina, N. G., and Shalygo, N. V. (1982). Metabolism and intermembrane location of magnesium–protoporphyrin IX monomethyl ester in centers of chlorophyll biosynthesis. *Photobiochem. Photobiophys.* **3**, 197–223.

Shlyk, A. A., Fradkin, L. I., Domanskii, V. P., and Samoilenko, A. G. (1985). Chlorophyll rearrangements at the early steps in the formation of the pigment systems. *In* "Proceedings of the 16th FEBS Congress," Part B, pp. 61–66. VNU Science Press.

Shlyk, A. A., Chaika, M. T., Fradkin, L. I., Rudoi, A. B., Averina, N. G., and Savchenko, G. E. (1986). Biogenesis of photosynthetic apparatus in etiolated leaves during greening. *Photobiochem. Photobiophys.* **12**, 87–96.

Sironval, C. (1981). The protochlorophyllide–chlorophyllide cycle as a source of photosynthetically active chlorophylls. *In* "Photosynthesis V. Chloroplast Development" (G. Akoyunoglou, ed.), pp. 3–13. Balaban, Philadelphia.

Smith, J. H. C. (1954). The development of chlorophyll and oxygen-evolving power in etiolated barley leaves when illuminated. *Plant Physiol.* **29**, 143–148.

Tanaka, A., and Tsuji, H. (1985). Appearance of chlorophyll–protein complexes in greening barley seedlings. *Plant Cell Physiol.* **26**, 893–902.

Thorne, S. W. (1971). The greening of etiolated bean leaves. I. The initial photoconversion process. *Biochim. Biophys. Acta* **226**, 113–127.

Tsakiris, S., and Akoyunoglou, G. (1981). Formation and growth of photosystems I and II units in developing thylakoids of *Phaseolus vulgaris*. *In* "Photosynthesis V. Chloroplast Development" (G. Akoyunoglou, ed.), pp. 513–522. Balaban, Philadelphia.

Virgin, H. I., and French, C. S. (1973). The light-induced protochlorophyll–chlorophyll *a*-transformation and the succeeding interconversions of the different forms of chlorophyll. *Physiol. Plant.* **28**, 350–357.

# 12

# Structure, Function, and Assembly of Photosystem I

**BIRGITTE ANDERSEN AND HENRIK VIBE SCHELLER**

Department of Plant Biology
Royal Veterinary and Agricultural University
DK-1871 Frederiksberg C, Copenhagen, Denmark

I. Introduction
II. Structural and Functional Characterization of Photosystem I
   A. Chloroplast-Encoded Polypeptides
   B. Nuclear-Encoded Polypeptides
   C. Stoichiometry of Photosystem I Subunits
   D. Model of Photosystem I Reaction Center Complex
   E. Crystallization of Photosystem I
III. Assembly of Photosystem I
   A. Expression of Photosystem I Genes
   B. Greening and Transformation Experiments
   C. Synthesis of PSI-A and PSI-B Is Coupled to Chlorophyll Biosynthesis
   D. Reconstitution of Photosystem I Complex
   E. Biosynthesis of Electron Acceptors
IV. Concluding Remarks
   References

## I. INTRODUCTION

In plants, the photochemical reactions of photosynthesis proceed in the thylakoid membranes of chloroplasts. Two photosystems, photosystem I (PSI) and photosystem II (PSII), function in series to convert solar energy into chemical energy. Both photosystems are pigment–protein complexes embedded in the thylakoid membrane.

Antenna pigments absorb the light energy and channel the energy to the reaction centers of the photosystems, P680 in PSII and P700 in PSI (Fig.

1). The photooxidation of P680 in PSII creates a strong oxidant that oxidizes water in the thylakoid lumen. The electrons from the oxidized reaction center P680$^+$ are transferred to plastoquinone via several electron acceptors bound to the PSII complex. Photooxidation and electron transfer in PSI result in reduction of soluble ferredoxin, which is used in the reduction of NADP$^+$ in the stroma of chloroplasts. The photochemical reactions of PSII and PSI are connected by the cytochrome $b_6/f$ Rieske Fe–S complex. From reduced plastoquinone, the electrons from photooxidized P680$^+$ are transferred via the cytochrome $b_6/f$ Rieske complex to plastocyanin in the thylakoid lumen. Plastocyanin donates electrons to the photooxidized P700$^+$. Coincident with the electron transport from plastoquinone to plastocyanin, protons are translocated from the stroma to the thylakoid lumen. The resulting proton gradient serves as the energy potential for the production of ATP.

Electron transport through PSI is mediated by the reaction center P700, the primary electron acceptor $A_0$, the secondary electron acceptor $A_1$, and the three iron–sulfur centers X, B, and A. The reduced form of reaction center P700 presumably is composed of a chlorophyll $a$ dimer and the charge in P700$^+$ is localized on one of the molecules in the dimer. Optical and electron paramagnetic resonance (EPR) spectroscopy of the primary

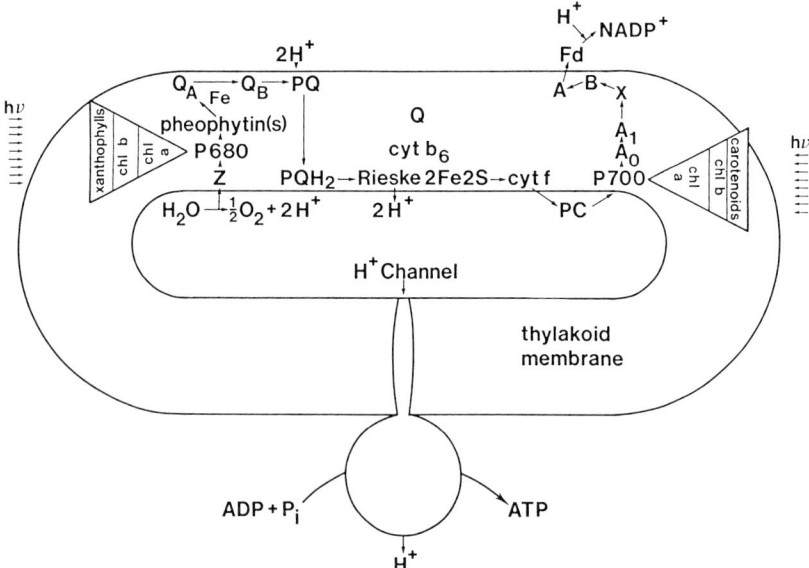

**Fig. 1.** Schematic model of the thylakoid membrane. For explanation, see text.

acceptor $A_0$ indicate that the electron acceptor is a chlorophyll molecule. The chemical nature of the secondary acceptor $A_1$ is not yet resolved, but several groups have suggested that $A_1$ is a phylloquinone molecule (vitamin $K_1$). The three iron–sulfur centers X, A, and B all have been identified as [4Fe–4S] clusters. A detailed description of electron transport in PSI is presented in several reviews (Golbeck, 1987; Lagoutte and Mathis, 1989; Golbeck and Bryant, 1991).

In the thylakoid membrane, the PSI complex is associated with light-harvesting pigment–protein complexes that carry most of the antenna pigments (Bassi and Simpson, 1987). The core complex of PSI is defined as the simplest pigment–protein complex that is able to photoreduce $NADP^+$ in the presence of reduced plastocyanin, ferredoxin, and ferredoxin:$NADP^+$ oxidoreductase (Bengis and Nelson, 1977; Reilly and Nelson, 1988). Twelve different polypeptides have been isolated from the PSI complex of plants, algae, and cyanobacteria. Table I summarizes the nomenclature of the polypeptides, whether the polypeptides are nuclear

**TABLE I**

**Subunits of the PSI Complex**

| Gene[a] | | Polypeptide subunit | | Molecular mass (kDa)[c] | |
|---|---|---|---|---|---|
| | | 1 | 2[b] | Apparent | Calculated |
| psaA | (C) | PSI-A | I | 82 | nd[d] |
| psaB | (C) | PSI-B | I | 82 | nd |
| psaC | (C) | PSI-C | VII, VIb | 9 | 8.8 |
| PsaD | (N) | PSI-D | II | 18 | 17.6 |
| PsaE | (N) | PSI-E | IV | 16 | 10.8 |
| PsaF | (N) | PSI-F | III | 15 | nd |
| PsaG | (N) | PSI-G | — | 9.2 | 10.8 |
| PsaH | (N) | PSI-H | VI | 9.5 | 10.2 |
| psaI | (C) | PSI-I | — | 1.5 | 4.0 |
| psaJ | (C) | PSI-J | — | nd | 4.4 |
| PsaK | (N) | PSI-K | — | 7 | 9.0 |
| PsaL | (N) | PSI-L | V | 14 | 18.0 |

[a] Localization of the genes on the chloroplast (C) or nuclear (N) genomes is indicated.

[b] Nomenclature of Bengis and Nelson (1977).

[c] Molecular mass of the PSI polypeptides from barley. Apparent molecular masses are based on electrophoretic mobilities in SDS–polyacrylamide gels (Høj et al., 1987; Scheller et al., 1989a). Calculated molecular masses are based on amino acid or nucleotide sequencing data (Okkels et al., 1988, 1989, 1991, 1992; Scheller et al., 1989b,c; J. S. Okkels, unpublished results).

[d] Not determined in barley.

or chloroplast encoded, and their molecular masses. The subunit composition of PSI from barley as analyzed by SDS–PAGE is shown in Fig. 2.

The intention of this chapter is to present some of the latest knowledge on the structure and function of PSI polypeptides. The field of studying the assembly of PSI is still in the beginning; the literature concerning this topic is reviewed here.

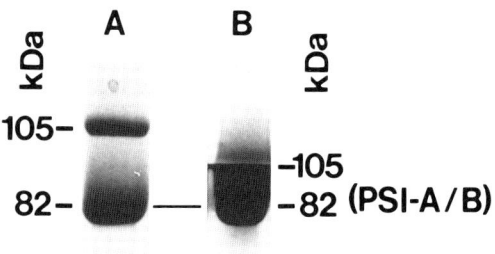

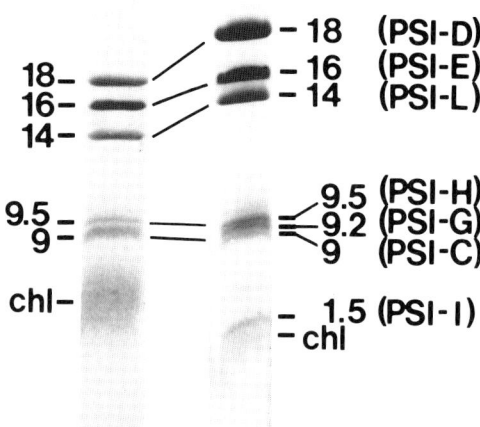

**Fig. 2.** Subunit composition of PSI isolated from barley as analyzed by SDS–PAGE. (A) 8–25% gradient gel. (B) 16.5% Tricine gel.

## II. STRUCTURAL AND FUNCTIONAL CHARACTERIZATION OF PHOTOSYSTEM I

### A. Chloroplast-Encoded Polypeptides

#### 1. PSI-A and PSI-B

The PSI-A and PSI-B polypeptides, presumably in the form of a heterodimer, constitute one of the major chlorophyll proteins in the thylakoids; the dimer is known as P700–chlorophyll *a* protein 1 (CP1). CP1 carries the reaction center P700, the primary electron acceptors $A_0$ and $A_1$, and the iron–sulfur center X. The polypeptides also bind about 90 molecules of chlorophyll *a* and 10–15 molecules of carotene per P700 (Golbeck, 1987; Lagoutte and Mathis, 1989; Scheller and Møller, 1990; Golbeck and Bryant, 1991).

The PSI-A and PSI-B subunits are encoded by the *psaA* and *psaB* genes which, in plants and algae, are located on the chloroplast genome in single copies. The *psaA* and *psaB* genes first were identified and sequenced in maize (Fish *et al.*, 1985) and pea (Lehmbeck *et al.*, 1986) and subsequently have been sequenced in several other species of plants, algae, and cyanobacteria. [See Bryant (1992) for a compilation of sequences.] The *psaA* genes encode polypeptides of 739–751 amino acids and the *psaB* genes encode polypeptides of 733–737 amino acids. The polypeptides are highly conserved among all species analyzed. PSI-A and PSI-B are about 45% identical to each other (Fish and Bogorad, 1986), suggesting that the *psaA* and *psaB* genes are the result of gene duplication. Fish and Bogorad (1986) demonstrated, by amino acid sequencing and immunological methods, that both polypeptides were present in PSI preparations in a ratio of about 1:1. The PSI-A and PSI-B polypeptides are hydrophobic; hydropathy plots of the deduced amino acid sequences indicate the presence of 11 membrane-spanning α-helices in each polypeptide (Fish *et al.*, 1985; Kirsch *et al.*, 1986). As pointed out by Møller *et al.* (1990), the prediction of 11 transmembrane helices should not be accepted readily. However, even the reaction center polypeptide of the phototrophic bacterium *Heliobacillus mobilis* which is similar to PSI-A and PSI-B is predicted to contain 11 membrane-spanning helices (Liebl *et al.*, 1992). The PSI-A and PSI-B polypeptides, in particular the hydrophobic segments, are unusually rich in histidine residues. Histidine residues have been implicated in the binding of chlorophyll molecules (Deisenhofer *et al.*, 1985).

The nature of center X and the number of PSI-A and -B subunits in the

PSI complex have been subjects of discussion. The amount of acid-labile sulfur and nonheme iron bound to the PSI-A and -B subunits has been determined to be about 4 mol acid-labile sulfur (Golbeck and Cornelius, 1986; Høj et al., 1987; Golbeck et al., 1988a) and 4 mol nonheme iron per P700 (Høj and Møller, 1986). This amount can accommodate either one [4Fe-4S] cluster or two [2Fe-2S] clusters. The presence of the iron-sulfur centers A and B has made it difficult to obtain unambiguous results on the nature of center X by EPR and Mössbauer spectroscopy. Binding of both types of iron-sulfur cluster requires four cysteine residues; PSI-A and PSI-B contain three and two conserved cysteine residues, respectively. Using uniformly $^{14}$C-labeled plants (Bruce and Malkin, 1988a, b) or quantitative Coomassie binding studies (Scheller et al., 1989a), it was established that only two 80-kDa subunits are present per P700. Thus, it was concluded that only a single iron-sulfur cluster could be bound. Consequently center X must be a [4Fe-4S] center (Scheller et al., 1989a), although these results conflict with data observed by extended X-ray absorption fine structure (EXAFS) and core extension studies interpreted to indicate that center X was a [2Fe-2S] cluster (Golbeck et al., 1987; McDermott et al., 1988). The isolation of CP1 with intact center X but without centers A and B later permitted an unequivocal identification of center X as a [4Fe-4S] center by Mössbauer (Petrouleas et al., 1989) and EXAFS (McDermott et al., 1989) spectroscopy. Center X is presumed to be bridged between PSI-A and PSI-B, so each subunit contributes two cysteine residues. An examination of the amino acid sequence in the vicinity of the conserved cysteine residues in both PSI-A and PSI-B showed a characteristic distribution of leucine residues, one every seventh amino acid, that locates these residues on one side in an $\alpha$-helix (Kössel et al., 1990; Webber and Malkin, 1990). The hypothesis is that these leucine residues interact like a zipper, thereby ensuring the proper conformation of the polypeptides for the insertion and stabilization of the iron-sulfur cluster. Interestingly, the leucine zipper motif is not present in the reaction center polypeptide of H. mobilis (Liebl et al., 1992).

The binding sites of the reaction center P700 and the early electron acceptors $A_0$ and $A_1$ have not yet been elucidated. Margulies (1991) has shown that the sequence NPFXM, which in purple bacteria and in PSII is involved in the binding of the reaction center, is present in helix X in PSI-B, but not in PSI-A. Search for sequence similarity showed that this short sequence is found only in photosynthetic reaction center polypeptides and a few virus proteins. Margulies (1991) suggests that this sequence in helix X of PSI-B participates in binding P700.

## 2. PSI-C

The electron acceptors A and B were discovered in the early 1970s (Malkin and Bearden, 1971; Bearden and Malkin, 1972; Evans *et al.*, 1972), but identification of the polypeptide carrying the centers took more than 10 years. Høj *et al.* (1987) demonstrated that a 9-kDa polypeptide carries 65–70% of the zero-valence sulfur present in a denatured PSI complex from barley. Zero-valence sulfur is a characteristic of oxidatively denatured iron–sulfur centers. Amino acid sequencing of the 9-kDa polypeptide revealed that it was encoded by a gene of the chloroplast genome designated *psa*C (formerly *frx*A) (Høj *et al.*, 1987). The PSI-C polypeptide has been identified and the amino acid sequence determined or deduced from the gene in a number of plants (Ohyama *et al.*, 1986; Shinozaki *et al.*, 1986; Dunn and Gray, 1988; Oh-oka *et al.*, 1988; Schantz and Bogorad, 1988; Wynn and Malkin, 1988a; Hiratsuka *et al.*, 1989; Scheller *et al.*, 1989b; Steppuhn *et al.*, 1989), algae (Takahashi *et al.*, 1991c), and cyanobacteria (Koike *et al.*, 1989; Bryant *et al.*, 1990; Shimizu, *et al.*, 1990; Mann *et al.*, 1991). The amino acid sequence of the PSI-C polypeptide is highly conserved among species and shows a pattern of cysteine residues that is characteristic of 2[4Fe–4S] proteins. Alignment analysis shows that the structure of PSI-C is similar to the structure of soluble bacterial ferredoxins of the 2[4Fe–4S] type. Compared with the soluble 2[4Fe–4S] proteins, the PSI-C polypeptide contains 14 additional amino acids at the C-terminal end and an insert of 8 amino acids in the middle of the polypeptide (Dunn and Gray, 1988; Oh-oka *et al.*, 1988; Scheller *et al.*, 1989b). The extended C terminus is hydrophilic, whereas the insert shows a higher degree of hydrophobicity. Since the ferredoxins are soluble proteins and the PSI-C polypeptide is membrane bound, the additional amino acids in the insert may be responsible for the binding of PSI-C to the PSI complex. The bacterial 2[4Fe–4S] ferredoxin from *Peptococcus aerogenes* has been crystallized and the structure resolved at the atomic level (Adman *et al.*, 1973). Figure 3 shows the structure of the carbon backbone of ferredoxin from *P. aerogenes* with the additional amino acids in PSI-C inserted in the structure.

Golbeck and co-workers have succeeded in reconstituting the iron–sulfur centers A and B in the PSI-C apoprotein (Parrett *et al.*, 1990; Mehari *et al.*, 1991). Rebinding of the PSI-C polypeptide with the reconstituted iron–sulfur centers to CP1 also regenerated the characteristic EPR signals of centers A and B.

The PSI-C polypeptide is quite hydrophilic and has no membrane-spanning segments (Høj *et al.*, 1987; Dunn and Gray, 1988; Oh-oka *et al.*,

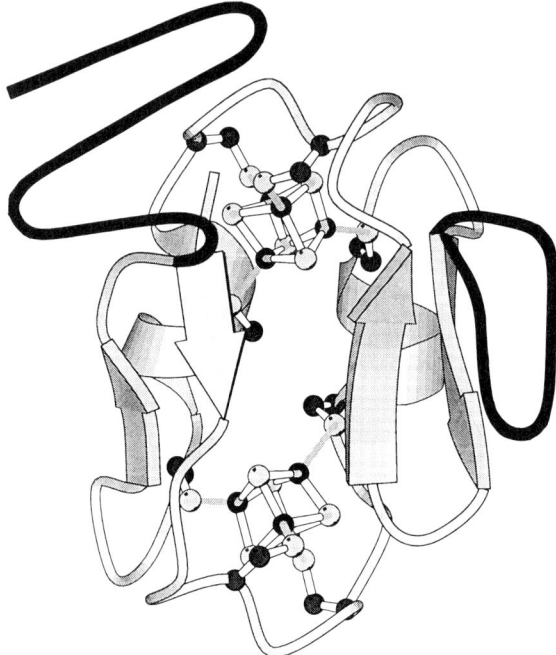

**Fig. 3.** Structure of the carbon backbone of ferredoxin from *Pseudomonas aerogenes* with the additional amino acid residues present in PSI-C inserted (shaded) in the structure. PSI-C contains additional 14 amino acids in the C terminus and an insert of 8 amino acids at position 22 compared with the sequence of ferredoxin from *P. aerogenes* (Adman *et al.*, 1973).

1988). Electron transfer in PSI proceeds from centers B and A directly to soluble ferredoxin located in the stroma. This transfer implies a peripheral location of the PSI-C subunit on the stromal side of the thylakoid membrane. Nevertheless, the PSI-C polypeptide is remarkably resistant to protease treatment of isolated thylakoids (Oh-oka *et al.*, 1989). Crosslinking studies have demonstrated a location of PSI-C, PSI-D, and PSI-E close to one another (Oh-oka *et al.*, 1989). Possibly, the PSI-D and PSI-E subunits shelter the PSI-C subunit from the action of the proteases.

### 3. *PSI-I and PSI-J*

The development of new electrophoretic methods for the separation of low molecular mass polypeptides has revealed the presence of several small polypeptides in the PSI complex. In barley, a small hydrophobic polypeptide migrating with an apparent molecular mass of 1.5 kDa was identified (Scheller et al., 1989c). In spite of N-terminal blockage with a formyl-methionine residue, partial amino acid sequences for this polypeptide were obtained. The gene, *psaI*, was isolated and sequenced from the barley chloroplast genome and the remaining residues deduced. The gene encodes a polypeptide of 36 amino acids with a deduced molecular mass of 4008 Da. Plasma desorption mass spectrometry demonstrated that the polypeptide is not post-translationally processed except for possible conversion of a methionine residue into a methionine sulfone. Subsequently, the presence of the PSI-I polypeptide has been confirmed in PSI from spinach, pea (Ikeuchi et al., 1990), and *Anabaena* sp. (Ikeuchi et al., 1991). The gene has been identified in the chloroplast genome of pea (Smith et al., 1991), rice (Hiratsuka et al., 1989), tobacco (Shinozaki et al., 1986), and liverwort (Ohyama et al., 1986), and in the cyanelle genome of *Cyanophora paradoxa* (Bryant, 1992).

The PSI-I polypeptide is very hydrophobic with a polarity index of 0.28 (Scheller et al., 1989c). The hydropathy plot of the deduced amino acid sequence indicates the presence of one central membrane-spanning α-helix flanked by hydrophilic N and C termini. The PSI-I polypeptide is bound tightly to the heterodimer of PSI-A and PSI-B (Scheller et al., 1989c). Treatment of isolated PSI with chaotropes dissociates many of the smaller polypeptides from the heterodimer, but the PSI-L and PSI-I polypeptides remain associated with the heterodimer.

Koike et al. (1989) reported the presence of an N-terminally blocked 4.1-kDa polypeptide in a PSI preparation from *Synechococcus vulcanus*. The obtained amino acid sequence shared sequence similarity with the deduced amino acid sequence of an open reading frame of the chloroplast genome, now designated the *psaJ* gene. The PSI-J polypeptide subsequently has been identified in PSI complexes from spinach, pea (Ikeuchi et al., 1990), and *Anabaena* sp. (Ikeuchi et al., 1991). The *psaJ* gene is found in the chloroplast genomes of tobacco, liverwort, and rice (Ohyama et al., 1986; Shinozaki et al., 1986; Hiratsuka et al., 1989) as well as in the cyanelle genome of *C. paradoxa* (Bryant, 1992). The PSI-J polypeptide contains a hydrophobic domain that is predicted to form a membrane-spanning α-helix. Nevertheless, the PSI-J subunit appears to be bound relatively loosely since the polypeptide is partly lost by treatments used to separate LHCI and PSI (Ikeuchi et al., 1990).

The PSI-I polypeptide exhibits sequence similarity with helix E of the D2 reaction center polypeptide of PSII (Scheller et al., 1989c). Eleven residues are identical and 11 of the nonidentical residues represent conservative substitutions (Fig. 4). The D1 and D2 polypeptides are known to

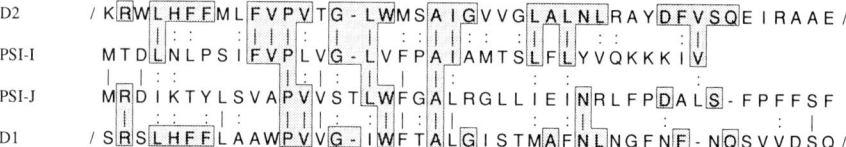

**Fig. 4.** Sequence similarity between PSI-I from barley, PSI-J from rice, and D1 and D2 from barley PSII. The amino acid sequences of PSI-I, D1, and D2 were determined by Scheller *et al.* (1989c), Efimov *et al.* (1988), and Neumann (1988), respectively. The PSI-J sequence was deduced from the nucleotide sequence of the rice chloroplast genome (Hiratsuka *et al.*, 1989). Only part of the D1 and D2 sequence is shown. Residues identical to those in D2 are boxed. Identical residues (|) and conservative substitutions (:) in pairwise comparisons are indicated.

coordinate the binding of the electron transfer components of PSII (Michel and Deisenhofer, 1988). Crystallization and structure determination of the reaction center complex of *Rhodopseudomonas viridis* (Deisenhofer *et al.*, 1985) and *Rhodobacter sphaeroides* (Allen *et al.*, 1987), which contain the reaction center polypeptides L and M that are homologous to D1 and D2 of PSII, elucidated the exact binding of the electron transfer components in these complexes. Helix E participates in the binding of the nonheme iron and the quinones of PSII. Search for proteins with sequence similarity to PSI-I also uncovered several ubiquinone:NADH oxidoreductases. The similarity to helix E of PSII and to ubiquinone:NADH oxidoreductases raises the question of whether PSI-I participates in the binding of some of the primary electron acceptors of PSI.

The similarity between PSI-I and D2 caused Scheller *et al.* (1989c) to suggest the presence of an additional small subunit in PSI homologous to PSI-I. An open reading frame encoding a 34-residue polypeptide with a weak sequence similarity to PSI-I was identified on the chloroplast genome. The 34-residue polypeptide has not been found in PSI, but PSI-J may represent the predicted subunit homologous to PSI-I. Alignment of the PSI-J and PSI-I sequences shows only a very weak similarity (Fig. 4). However, the PSI-J polypeptide has a relatively high sequence similarity to D1 of PSII (Fig. 4). None of the observed sequence similarities are as high as the similarity between the E helices of D1 and D2. However, even the weak similarity between PSI-I and PSI-J is almost as high as the similarity between L and M in the corresponding region. The presence of two subunits in PSI with sequence similarity to D1 and D2 would seem to strengthen the possibility that the PSI-I and PSI-J subunits are involved in the coordination of electron acceptors.

## B. Nuclear-Encoded Polypeptides

### 1. PSI-D and PSI-E

The primary structures of the PSI-D and PSI-E polypeptides have been determined in a number of plants and cyanobacteria by amino acid sequencing and by deduction from the corresponding cDNA sequences. [See Scheller and Møller (1990) and Bryant (1992) for a compilation of sequences.] In plants, the PSI-D and PSI-E polypeptides have molecular masses of about 18 and 11 kDa, respectively, whereas the cyanobacterial polypeptides are somewhat smaller. The PSI-D and PSI-E polypeptides are highly conserved among different species, except for the N-terminal parts which in plants contain about 20 residues more than the cyanobacterial protein. The N-terminal region in plants is rich in proline, alanine, and charged amino acid residues. The PSI-D and PSI-E polypeptides of barley exhibit sequence similarity in the N-terminal region (Scheller *et al.*, 1988). In plants, the molecular mass of the PSI-E polypeptide estimated in SDS-gels is considerably larger than the calculated mass. However, plasma desorption mass spectrometry (Scheller *et al.*, 1990a) demonstrated that the barley PSI-E polypeptide is not post-translationally processed, so the discrepancy between the estimated and calculated molecular mass must be caused by anomalous migration in SDS gels. Both polypeptides are characterized by a high content of basic amino acids. The hydropathy plots indicate the absence of hydrophobic domains likely to form membrane-spanning $\alpha$-helices (Lagoutte, 1988; Münch *et al.*, 1988; Okkels *et al.*, 1988; Franzén *et al.*, 1989a). The peripheral location of the PSI-D and PSI-E polypeptides also is confirmed by dissociation of the polypeptides by treatment with chaotropes (Høj *et al.*, 1987; Oh-oka *et al.*, 1989; Parrett *et al.*, 1989,1990; Li *et al.*, 1991a) or with *n*-butanol (Oh-oka *et al.*, 1988,1989; Koike *et al.*, 1989).

A nuclear-encoded polypeptide is synthesized as a precursor molecule with an N-terminal transit peptide responsible for directing the precursor to the proper cellular compartment. The transit peptides of the PSI-D (Lagoutte, 1988; Münch *et al.*, 1988) and the PSI-E (Münch *et al.*, 1988; Okkels *et al.*, 1988; Franzén *et al.*, 1989a) precursors lack the hydrophobic region characteristic of proteins directed to the thylakoid lumen, indicating a location of the PSI-D and PSI-E polypeptides on the stromal side of the thylakoid membrane. This location is confirmed by the degradation of the polypeptides when intact thylakoids are treated with proteases (Oh-oka *et al.*, 1989).

Chemical cross-linking of PSI isolated from spinach or barley with the water-soluble zero-length cross-linker N-ethyl-3-[3-(dimethylamino) propyl]carbodiimide (EDC) performed in the presence of ferredoxin have shown specific interaction between PSI-D and ferredoxin (Merati and Zanetti, 1987; Zanetti and Merati, 1987; Zilber and Malkin, 1988; Andersen et al., 1990, 1992a). The high content of positively charged residues in PSI-D may enable the negatively charged ferredoxin to react with PSI in spite of the overall negative surface charge of the thylakoid membrane. Other cross-linking studies have demonstrated that the PSI-D polypeptide is located close to the PSI-C, PSI-E, and PSI-H polypeptides (Oh-oka et al., 1989; Andersen et al., 1990,1992a).

Incubation of thylakoids or purified PSI with antibodies against the PSI-D or PSI-E polypeptides inhibits the rate of $NADP^+$ reduction (Andersen et al., 1992a). The accessibility to antibody binding confirms that the PSI-D and -E polypeptides are located externally in the complex. The inhibition shows that both polypeptides are located very near the site of interaction between PSI-C and ferredoxin. Antibodies against PSI-C had no effect on $NADP^+$ reduction, indicating that the PSI-C polypeptide is shielded by other polypeptides, making it inaccessible to antibody binding.

With the purpose of studying the function of the PSI-D subunit, Chitnis et al. (1989b) experimentally inactivated the psaD gene in Synechocystis sp. PCC 6803. This manipulation resulted in a mutant containing a PSI complex devoid of the PSI-D subunit and several other small subunits. The mutant was incapable of photoautotrophic growth but exhibited normal growth rates in the presence of glucose. The purified PSI complex from the mutant was capable of mediating electron transport from plastocyanin to the artificial electron acceptor methylviologen. These results support the experiments that indicated that the PSI-D polypeptide functions as a docking site for ferredoxin.

Chitnis et al. (1989a) also have studied the function of the PSI-E polypeptide by deleting the psaE gene in Synechocystis sp. PCC 6803. The resulting mutant contained a normal PSI complex except for the lack of PSI-E and, possibly, an increased amount of the PSI-D polypeptide. The mutant was capable of photoautotrophic growth at growth rates comparable to the wild type. The increased amount of the PSI-D polypeptide in the mutant could be an indication that the PSI-D polypeptide structurally and functionally can substitute for the PSI-E polypeptide, which further indicates that the PSI-E polypeptide participates in docking ferredoxin to the PSI complex. Results obtained with a PSI complex prepared using mild solubilization of thylakoids indicate a function of PSI-E in binding the ferredoxin:$NADP^+$ oxidoreductase (Andersen et al., 1992b).

The PSI-D and PSI-E polypeptides also may serve to protect the PSI-C polypeptide from the otherwise oxygenic environment. The inaccessibility of the PSI-C polypeptide to proteases and antibodies indicates that the PSI-C polypeptide is well protected from the stroma. This protection may be important since the PSI-C polypeptide is very unstable in the presence of oxygen (Oh-oka *et al.*, 1988).

## 2. PSI-F

The amino acid sequence of the PSI-F polypeptide has been deduced in spinach (Steppuhn *et al.*, 1988), *Chlamydomonas reinhardtii* (Franzén *et al.*, 1989a), *Synechocystis* sp. PCC 6803 (Chitnis *et al.*, 1991), and *Cyanophora paradoxa* (Bryant, 1992). In plants and algae, the *PsaF* gene is localized in the nuclear genome. The mature PSI-F polypeptide from spinach is calculated to have a molecular mass of approximately 17 kDa. The N-terminal sequence of the PSI-F polypeptide has been obtained from pea (Dunn *et al.*, 1988), barley (Anandan *et al.*, 1989; Scheller *et al.*, 1990b), maize (Anandan *et al.*, 1989), *Synechococcus vulcanus* (Koike *et al.*, 1989), and *Synechococcus* sp. PCC 6301 (Li *et al.*, 1991a). The PSI-F polypeptide is rich in charged amino acids, although the hydropathy plot indicates the presence of hydrophobic domains, one of which may be membrane spanning. The experiments on the PSI-F polypeptide have given conflicting results. In barley, the PSI-F polypeptide is bound loosely to the PSI complex and often is lost during purification of PSI (Scheller *et al.*, 1990b). It has been possible to extract the PSI-F polypeptide with *n*-butanol treatment of PSI from *S. vulcanus* (Koike *et al.*, 1989), indicating that PSI-F is an extrinsic polypeptide. In spinach, this treatment does not affect the PSI-F polypeptide, which instead partitions into Triton X-114 micelles (Zilber *et al.*, 1990), a characteristic of hydrophobic polypeptides. In PSI from *Synechococcus* sp. PCC 6301, the polypeptide remains bound to PSI after chaotrope treatment, indicating an intrinsic nature of the polypeptide (Li *et al.*, 1991a). Transit peptides of the PSI-F precursor contain a hydrophobic region typical of transit peptides of proteins located in the thylakoid lumen (Steppuhn *et al.*, 1988; Franzén *et al.*, 1989a). The indication of a luminal localization is in agreement with the presence of a signal peptide in the PSI-F precursor in *C. paradoxa* (Bryant, 1992) and *Synechocystis* sp. PCC 6803 (Chitnis *et al.*, 1991). In 1977, Bengis and Nelson reported that the PSI-F polypeptide (subunit III) has a function in electron donation from plastocyanin to the reaction center P700. A PSI preparation without the PSI-F polypeptide was defective in electron transport from plastocyanin to $NADP^+$, but showed normal electron transport from the artificial electron donor *N*-methylphenozonium 3-sulfonate. Subsequently, studies using chemical cross-linking (Wynn and Malkin, 1988b; Hippler *et al.*, 1989) demonstrated a specific interaction between plastocyanin and the PSI-F polypeptide. Treatment of solubilized barley thylakoids with EDC in the presence of plastocyanin resulted in a cross-linked product with an apparent molecular mass of 26 kDa (Fig. 5). Hippler *et al.* (1989) established that preparations in which plastocyanin had been cross-linked to the PSI-F polypeptide exhibited the same fast flash-induced reduction kinetics of $P700^+$ observed without cross-linking. These results of the cross-linking experiments indicate a function of PSI-F in facilitating the electron transfer between plastocyanin and the PSI complex. The PSI-F polypeptide contains a surplus of basic amino acid residues and may function by enabling negatively charged plastocyanin to react with PSI, in a manner similar to the proposed function of PSI-D with respect to ferredoxin. However, the presence of the PSI-F polypeptide does not seem to be an absolute prerequisite for the interaction between plastocyanin and the PSI complex. Measurement of the rate of $NADP^+$-reduction under

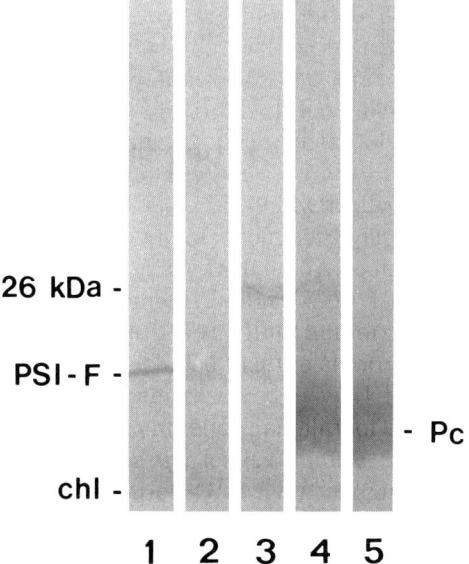

**Fig. 5.** Western blot of thylakoids cross-linked with *N*-ethyl-3-[3-(dimethylamino)propyl]carbodiimide (EDC) in the presence of plastocyanin (Pc). The cross-linking of barley thylakoids solubilized with Triton X-100 was carried out at room temperature for 30 min in 25 m*M* 2-*N*-(morpholino)ethanesulfonic acid (pH 6.3), 5 m*M* $MgCl_2$, 15 m*M* NaCl, 0.5 m*M* Na-ascorbate in the presence of 10 m*M* EDC, 0.25 mg chl/ml, and 12 µ*M* barley Pc. Lane 1, thylakoids; lane 2, thylakoids treated with EDC; lanes 3 and 4, thylakoids treated with EDC in the presence of Pc; lane 5, Pc treated with EDC. Lanes 1–3 were incubated with antibody against the barley PSI-F polypeptide. Lanes 4 and 5 were incubated with antibody against barley Pc.

continuous illumination in PSI preparations containing different amounts of the PSI-F polypeptide showed no correlation between $NADP^+$-reduction activity and amount of PSI-F in the preparations (Scheller et al., 1990b). Chitnis et al. (1991) created a mutant of Synechocystis sp. PCC 6803 with an inactivated psaF gene. The photoautotrophic growth rate of the mutant was comparable to that of the wild type. However, this result does not exclude the possible function of PSI-F in docking plastocyanin, since both plastocyanin and cytochrome $c_{553}$ can act as electron donor to the reaction center in Synechocystis sp. PCC 6803.

## 3. PSI-G and PSI-K

The PSI-D, PSI-E, and PSI-F subunits are peripheral membrane proteins. This location has facilitated their characterization compared with the remaining subunits, which are integral membrane proteins. PSI-G has been characterized by analysis of cDNA clones of spinach (Steppuhn et al., 1989), C. reinhardtii (Franzén et al., 1989b), and barley (Okkels et al., 1992). In addition, N-terminal amino acid sequences are available from pea (Dunn et al., 1988) and cucumber (Iwasaki et al., 1991). The PSI-G subunit has an apparent molecular mass of 9 kDa in barley and a calculated molecular mass of 10.8 kDa deduced from the cDNA clone. In barley, the PSI-G polypeptide was found to be N-terminally blocked to amino acid sequencing (Okkels et al., 1991) but no blocking was observed in the other species.

The PSI-K polypeptide has been characterized by analysis of a cDNA clone in C. reinhardtii (Franzén et al., 1989b) and by N-terminal amino acid sequencing in spinach (Hoshina et al., 1989; Ikeuchi et al., 1990; Wynn and Malkin 1990), pea (Ikeuchi et al., 1990), and cucumber (Iwasaki et al., 1991). The PSI-K polypeptide from C. reinhardtii has an apparent molecular mass of only 3 kDa, although the calculated molecular mass is 8.4 kDa.

Despite the difference in the molecular masses, the PSI-G and PSI-K polypeptides have a pronounced sequence similarity (Fig. 6). The amino acid sequence of a 6.5-kDa polypeptide from S. vulcanus originally identified by Koike et al. (1989; see Bryant, 1992) and the N-terminal sequence of a 6.8-kDa polypeptide from Anabaena sp. (Ikeuchi et al., 1991) are shown in Fig. 6 also. These polypeptides have been claimed to be PSI-K, based on the weak sequence similarity. However, the comparison in Fig. 6 shows that the degree of similarity to PSI-G is about the same. No other sequence information on cyanobacterial PSI-G or PSI-K polypeptides has been published. The transit peptides of the PSI-G and PSI-K polypeptides

```
Barley PSI-G           A L E P S V V I S L S T G L S L V M G R F V F F N F Q R E N V A K Q - - - V P E Q N G K T H F E A G
Spinach PSI-G          E L S P S L V I S L S T G L S L F L G R F V F F N F Q R E N M A K Q - - - V P E Q N G M S H F E A G
C. reinh. PSI-G        A L D P Q I V I S G S T A A F L A I G R F V F L G Y Q R R E A N F D S T V G P K T T G A T Y F D D L
S. vulcanus 6.5 kDa    M V L A T T L P D T T W T P S V G L V V I L S N L F A I A L G R Y A I Q S R G K G P G L - - - - - - - - - - - -
Anabaena 6.8 kDa       A A T T P L E - - W S P T I G L I M V I A N V I A T F G R Q T I /
Cucumber PSI-K         W D Y I G S P T N V I M V I S T I S L M L F A G R F G L A P S A N R /
C. reinh. PSI-K        D G F I G S S T N L I M V A S T I T A T L A A A R F G L A P T V K K N T T A G L K - - - - - - - -

Barley PSI-G           D E R A K E F A G I L K S N D P V G F N L V D V L A W G S I G H I V A Y Y L A T T S N G Y D P P F F G
Spinach PSI-G          D T R A K E Y V S S L K S N D P V G F N I V D V L A W G S I G H I V A Y Y L A T A S N G Y D P S F F
C. reinh. PSI-G        Q - - - K N S T I F A - T N D P A G F N I I D V A G M A L I G H A V G F A V L A I N S L Q G A N L S
S. vulcanus 6.5 kDa    - - - - - - P I A L P A L F E G F G L P E L L A T T S F G H L L A A G V V S G L Q Y A G A L
Anabaena 6.8 kDa
C. reinh. PSI-K        L V D S K N S A G V I - S N D P A G F T I V D V L A M G A A G H G L G V G I V L G L K G I G A L
```

**Fig. 6.** Sequence similarity between the PSI-G and PSI-K subunits. Residues identical to PSI-G from barley are boxed. Data are presented for barley (Okkels *et al.*, 1992), spinach PSI-G (Steppuhn *et al.*, 1988), *Chlamydomonas reinhardtii* (Franzén *et al.*, 1989b), *Synechococcus vulcanus* (Koike *et al.*, 1989; Bryant, 1992), *Anabaena* sp. (Ikeuchi *et al.*, 1991), and cucumber PSI-K (Iwasaki *et al.*, 1991).

share the common features characteristic for proteins directed to a stromal or integral membrane location. Hydropathy analysis of the amino acid sequences of the mature polypeptides showed two hydrophobic regions that may represent membrane-spanning $\alpha$-helices. The structure of the proteins is predicted to be two membrane-inserted helices connected by a hydrophilic segment that most likely protrudes into the stroma (Gavel et al., 1991). The hydrophilic segment between the helices in PSI-G shows a very low degree of conservation among the different species, except for the sequence element SNDPXGFN. This segment is also present in PSI-K of C. reinhardtii but not in the 6.5-kDa polypeptide of S. vulcanus. Thus it seems that the most conserved parts of the polypeptides are remarkably similar in PSI-G and PSI-K. This homology supports the general idea of a pseudosymmetrical reaction center complex analogous to the pseudosymmetrical PSII reaction center complex. The fact that eukaryotic PSI-G and PSI-K are more similar to each other than to the cyanobacterial "PSI-K" polypeptides suggests that a gene duplication took place relatively late in evolution. Most likely, no homolog of the PSI-G polypeptide is present in cyanobacteria. Therefore, either the two-fold symmetry with respect to PSI-G/PSI-K is absent in these organisms or the cyanobacterial PSI contains two copies of the "PSI-K" polypeptide. Unfortunately, no stoichiometric data with respect to the cyanobacterial "PSI-K" polypeptide are available.

What is the function of the PSI-G and PSI-K polypeptides? Studies with spinach PSI have shown that PSI-K is associated more closely with the PSI-A/PSI-B heterodimer than any other small subunit (Hoshina et al., 1989; Wynn and Malkin, 1990). In barley, the PSI-G polypeptide has been shown to be bound tightly to this dimer (Okkels et al., 1991). These observations could indicate an important function of the two subunits in the organization of the PSI core. However, contradictory results were obtained by Ikeuchi et al. (1990), Ikeuchi and Inoue (1991), and Iwasaki et al. (1991), who found that PSI-K and PSI-G were depleted from the PSI complex during isolation. Preliminary studies by Wynn and Malkin (1990) indicate that removal of PSI-K from CP1 in spinach does not affect charge separation under steady-state illumination. Wynn and Malkin pointed out that more detailed kinetic analyses are required to deterimine if PSI-K has a role in PSI photochemistry. An interesting feature of the PSI-G and PSI-K polypeptides is a conserved histidine residue in the second transmembrane helix. Histidine residues have been implicated in the coordination of chlorophyll molecules. Transformation experiments in which *PsaG* or *Psa*K is inactivated or altered may be the best way to obtain information on the function of the two subunits.

## 4. PSI-H

The primary structure of the PSI-H polypeptide is known from sequencing of cDNA clones from barley (Okkels et al., 1989), spinach (Steppuhn et al., 1989), and *C. reinhardtii* (Franzén et al., 1989b). In addition, N-terminal amino acid sequences from pea (Dunn et al., 1988) and cucumber (Iwasaki et al., 1991) are known. A genomic clone for *PsaH* has been characterized in rice (de Pater et al., 1990). This clone is the only genomic clone of a PSI subunit from the nuclear genome of a eukaryote. The PSI-H polypeptide has not been found in cyanobacteria, in spite of much work with this group of organisms. In barley, the PSI-H polypeptide has an apparent molecular mass of 9.5 kDa, which is in good agreement with the calculated molecular mass of 10.2 kDa. The PSI-H polypeptide is bound relatively tightly to the PSI core. Treatment of PSI with chaotropes such as 3.2 $M$ NaSCN does not dissociate the PSI-H polypeptide from the CP1 core, whereas treatment with 8 $M$ urea does (Scheller et al., 1989a; Okkels et al., 1991). The basic structure of the PSI-H polypeptide consists of a very hydrophilic and highly conserved N-terminal domain followed by a hydrophobic domain that may be membrane-spanning and a second hydrophilic domain. The two latter domains show a very low degree of conservation among higher plants and *C. reinhardtii*. In particular, the hydrophobic domain is not conserved. Thus, the most important features of the PSI-H polypeptide are expected to be found in the hydrophilic domains, particularly in the N-terminal domain that is predicted to protrude into the stroma. This pattern is the opposite of that in the PSI-G and PSI-K polypeptide, in which the hydrophobic membrane-spanning segments are highly conserved and the hydrophilic parts are not conserved. The PSI-H polypeptide can be cross-linked efficiently to the PSI-D polypeptide by treatment of PSI with EDC (Andersen et al., 1990, 1992a). Thus, the conserved stromal domain of the PSI-H polypeptide must be in close contact with the PSI-D polypeptide. A possible function of the PSI-H polypeptide could be to constitute an anchor for the PSI-D polypeptide. The PSI-D and PSI-E polypeptides of plants differ from the corresponding cyanobacterial polypeptides by having N-terminal extensions rich in alanine, proline, and charged amino acid residues. The PSI-H polypeptide and the N-terminal domains of the PSI-D and PSI-E polypeptides may constitute a system for binding the peripheral polypeptides to the PSI complex in plants. Indeed, the PSI-D and PSI-E polypeptides of cyanobacteria are dissociated more easily from the rest of the PSI complex by chaotrope treatment than the PSI-D and PSI-E polypeptides of plants (Parrett et al., 1989).

Apart from the absence of PSI-G and PSI-H, the PSI complex of cyanobacteria seems to be remarkably similar to the PSI complex of plants and algae. Therefore, it has been suggested that PSI-H has a function in mediating the binding of LHCI polypeptides (Scheller and Møller, 1990). The different organization of the antenna pigment system in cyanobacteria could explain the absence of PSI-H in these organisms.

## 5. PSI-L

The PSI-L polypeptide has been characterized by analysis of a cDNA clone from barley (Okkels *et al.*, 1991). In addition, N-terminal amino acid sequences are available from *Synechococcus* spp. (Rhiel and Bryant, 1988; Koike *et al.*, 1989; Li *et al.*, 1991a) and cucumber (Iwasaki *et al.*, 1991), and internal amino acid sequence fragments from spinach and *Anabaena* sp. (Ikeuchi and Inoue, 1991). In barley, spinach, and *Anabaena* sp., the PSI-L polypeptide is N-terminally blocked to amino acid sequencing. The apparent molecular mass of the PSI-L polypeptide is 14 kDa in barley, whereas the calculated mass is about 18 kDa. Most likely this discrepancy is created by anomalous migration of the PSI-L polypeptide during SDS–PAGE.

The basic structure of the PSI-L polypeptide is two hydrophobic domains connected with an unusually hydrophilic segment. The hydrophobic domains are predicted to form membrane-spanning helices and the hydrophilic segment is predicted to protrude into the lumen. The hydrophilic segment of PSI-L is characterized by positive charges. Thus, the hydrophilic domain could have a function similar to PSI-F in promoting the reaction between plastocyanin and PSI. The PSI-L polypeptide is bound very tightly to CP1. Chaotropes will not dissociate PSI-L from CP1; even treatment with 2% lithium dodecyl sulfate (LDS) at room temperature for several hours is only moderately efficient. Thus, CP1 preparations with more or less intact electron acceptors most likely also contain PSI-L. The involvement of this polypeptide in the binding of electron acceptors cannot be excluded. The pseudosymmetrical structure of PSI and other reaction center complexes would, however, suggest that a polypeptide homologous to PSI-L should be present if PSI-L were involved in binding electron acceptors. None of the known PSI polypeptides shows homology to PSI-L.

## 6. PSI-M

N-terminal animo acid sequences of a novel 3.5-kDa polypeptide in PSI preparations of *S. vulcanus* and *Synechocystis* sp. PCC 6803 have been obtained (Bryant, 1992). Homologous polypeptides are encoded by the cyanelle genome of *C. paradoxa* and by the chloroplast genome of liverwort, but not by the chloroplast genomes of tobacco and rice (Bryant, 1992). The 3.5-kDa polypeptide tentatively has been designated PSI-M.

## C. Stoichiometry of Photosystem I Subunits

The subunit stoichiometry of PSI has been investigated in a number of species. The most recent studies were carried out using uniformly $^{14}$C-labeled plant material and measuring the radioactivity corresponding to each subunit (Bruce and Malkin, 1988a,b) or analyzing the isolated PSI subunits with respect to amino acid composition and specific Coomassie brilliant blue binding capacity (Scheller *et al.*, 1989a). These studies show the presence of one copy of each subunit per P700. However, the studies were carried out when fewer PSI subunits were known. Thus, the stoichiometry of subunits below 8 kDa has not been studied; the 8- to 10-kDa region needs to be reexamined. In addition, the identity of the various subunits studied by Bruce and Malkin (1988a,b) was not clear at the time of the study. Nevertheless, no data seem to indicate that any subunit is present in more than one copy per P700. Quantitation of the small subunits PSI-J and PSI-M as well as a few additional claimed PSI subunits would be desirable to determine if these components are real constituents of PSI.

## D. Model of Photosystem I Reaction Center Complex

Knowledge about the individual PSI subunits as reviewed earlier forms the basis of the PSI model shown in Fig. 7. The model shows the localization of the various subunits on the stromal and luminal sides of the membrane and indicates which subunits are integral membrane proteins. The pseudosymmetric structure of PSI is emphasized. The PSI-I and PSI-J subunits are shown near the initial electron acceptors because of the possible homology between PSI-I and PSI-J and the electron acceptor binding D2 and D1 proteins of PSII. The model is in agreement with the topological studies of PSI-C, -D, -E, and -H. The ferredoxin:$NADP^+$ oxidoreductase is shown as part of the PSI complex. A number of investigations implicated a special reductase binding protein that is not one of the known PSI subunits (Ceccarelli *et al.*, 1985; Berzborn *et al.*, 1990). However, our investigations of barley PSI show that the reductase is bound to the PSI complex. Cross-linking experiments have shown an interaction between the reductase and the PSI-E polypeptide, as indicated

## 12. Photosystem I

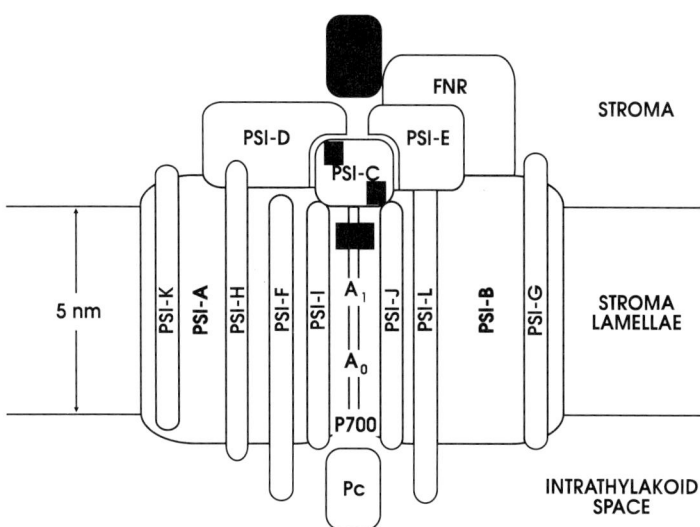

**Fig. 7.** Schematic model of PSI. Adapted from Andersen et al. (1992a).

in the model (Andersen et al., 1992b). The position of the other subunits relative to one another is arbitrary.

### E. Crystallization of Photosystem I

Crystallization and structure determination of the reaction centers from the photosynthetic bacteria *Rhodopseudomonas viridis* (Deisenhofer et al., 1985) and *Rhodobacter sphaeroides* (Allen et al., 1987) have shown that it is possible to crystallize and determine the detailed structure of membrane proteins. Crystals of PSI from the cyanobacteria *Phormidium laminosum* (Ford et al., 1987, 1988), *Mastigocladus laminosus* (Shoham et al., 1990; Almog et al., 1991), and *Synechococcus* sp. (Witt et al., 1987, 1988, 1990) have been reported. The most promising crystals obtained with PSI from *Synechococcus* sp. (Witt et al., 1988,1990) showed diffraction to a resolution of 4 Å. The crystal lattice was characterized as primitive hexagonal with unit cell parameters $a = b = 285$ Å, $c = 167$ Å, $\alpha = \beta = 90°$, $\gamma = 120°$. The space group was determined to be either $P6_3$ or $P6_33$. Witt et al. (1990) are creating heavy atom derivatives of the crystals for evaluation of the phases of the reflections and subsequent structure determination.

Electron microscopy analysis of the PSI complex also has been used for structural information. Electron microscopy of PSI purified from *Synechoccus* sp. (Boekema *et al.*, 1987, 1989) revealed a trimeric structure of the complex with a diameter of 19 nm and a thickness of 6.5 nm. Later the size of the monomeric complex was determined to be $15.3 \times 10.6 \times 6.4$ nm (Rögner *et al.*, 1990). Ford *et al.* (1990) studied PSI from *P. laminosum* and *Synechococcus* sp. OD24 by electron microscopy and digital image processing. The solubilized PSI was reconstituted into ordered two-dimensional arrays with phospholipids. The PSI complexes were determined to be about 12.5 nm in the plane of the membrane and 6.5 nm in height. Surface reliefs reconstructed from freeze-dried metal-shadowed PSI indicate a ridge of about 2.5 nm on one side of the membrane whereas the other side is rather flat. The ridge is implicated to represent the polypeptides located on the stromal side of PSI.

Boekema *et al.* (1990) studied the structure of different types of PSI complexes from spinach by electron microscopy and computer image analysis. The size of the PSI complex with bound light-harvesting complex I was estimated to be $16 \times 12$ nm in the plane of the membrane with a height of 6.8 nm. A shell of eight light-harvesting complex polypeptides is assumed to surround the PSI complex.

## III. ASSEMBLY OF PHOTOSYSTEM I

### A. Expression of Photosystem I Genes

PSI is not detectable in dark-grown angiosperms. However, the genes encoding the PSI subunits are expressed quite strongly in dark-grown plants, as evidenced by Northern blotting (Fig. 8; Klein *et al.*, 1988a,b; Laing *et al.*, 1988; Okkels *et al.*, 1990, 1991; Scheller *et al.*, 1990a; J. S. Okkels and V. S. Nielsen, unpublished results.) Dark expression is found for the chloroplast- as well as the nuclear-encoded genes. On transfer of the dark-grown plant to light, the expression of all the genes shows the same pattern: initially the relative amount of mRNAs for PSI genes decreases but, soon after, it increases to higher levels than in dark-grown plants. The initial decrease probably does not indicate a lower transcription rate for the PSI genes but merely reflects that some non-PSI genes are induced rapidly to high levels, thereby resulting in a lower relative abundance of the PSI mRNAs.

The significant presence of PSI mRNAs in the dark raises the question of why PSI is not present in the dark. The coupling of PSI synthesis with chlorophyll synthesis may answer this question partially (see Section III,C).

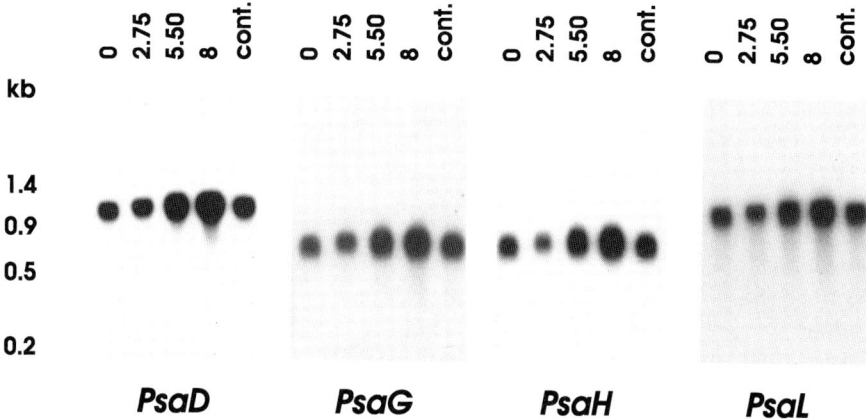

**Fig. 8.** Northern blot analysis of total RNA isolated from 5-day-old barley seedlings and hybridized with barley cDNA clones. The seedlings were grown in the dark and illuminated prior to harvesting for the indicated time periods (hours) or were grown under continuous light for 5 days (control). An equal amount of RNA was loaded in each lane.

## B. Greening and Transformation Experiments

When dark-grown plants are transferred to light, PSI starts to accumulate. Western blotting with antibodies against PSI subunits has been used to study the accumulation. Since the different antibodies may have different sensitivities, these studies must be interpreted with caution. However, the general observation is that all PSI subunits accumulate at about the same time and rate, that is, they can be detected after about 3 hr and their abundance increases in a parallel manner (Vainstein et al., 1989; H. V. Scheller, unpublished results). This observation suggests that the PSI subunits are quickly degraded unless they are incorporated into the PSI complex. Such a scheme has been shown for the cytochrome $b_6/f$ complex (Bruce and Malkin, 1991). In this complex, the Rieske iron–sulfur protein appears to have a central role in the assembly of the complex. The absence of the Reiske iron–sulfur protein leads to the rapid degradation of the other subunits of the complex. It is not clear which of the PSI subunits has a similar key role in the assembly of PSI. Indeed, a protein that is not part of the final complex could be involved in the assembly in a manner similar to the way chloroplast chaperonin is involved in the assembly of ribulose 1,5-bisphosphate carboxylase. Studies with transformants of *Synechocystis* (see Sections II,B,1 and II,B,2) have shown that PSI may be assembled in the absence of the PSI-E or PSI-F subunits. Inactivation of the *psaD* gene resulted in a stable but inactive PSI complex that contained PSI-A, PSI-B, and one additional subunit, possibly PSI-L, whereas the remaining subunits were missing or present in very low amounts. The PSI-D subunit also has been implicated in the binding of PSI-C (see Section III,D). Obviously, the PSI-A and PSI-B subunits are essential for the formation of a PSI complex, since these proteins bind all the pigments and most of the electron transfer components, and must form the framework for binding the remaining subunits. Insertional inactivation of the *psaC* gene in *Chlamydomonas reinhardtii* resulted in a mutant deficient in PSI activity (Takahashi *et al.* 1991c). Neither CP1 nor the smaller polypeptides of PSI accumulate in the thylakoid membrane. Pulse-chase labeling of proteins showed that PSI-A and -B are synthesized normally but turn over rapidly in the mutant cells. This result indicates that the PSI-C polypeptide is required for the stable assembly of PSI in *C. reinhardtii*. A different result was obtained by Mannan *et al.* (1991) by inactivating the *psaC* gene in *Anabaena*. The mutant cyanobacteria contains high amounts of CP1 and shows electron transport from P700 to center X. The different results obtained with inactivation of *psaC* are a reminder that the effects of inactivating *psaD*, *psaE*, and *psaF* in *Synechocystis* should be extrapolated to higher plants with caution.

## C. Synthesis of PSI-A and PSI-B Is Coupled to Chlorophyll Biosynthesis

The PSI-A and PSI-B subunits are not synthesized in the dark, although the mRNAs are abundant and associated with membrane-bound polysomes (Klein and Mullet, 1986, 1987; Klein *et al.*, 1988a,b; Laing *et al.*, 1988). Pulse-labeling experiments clearly showed that the absence of the PSI-A and PSI-B subunits in dark-grown plants is the result of strict translational control and not rapid degradation of the nascent polypeptides (Klein and Mullet, 1986, 1987; Klein *et al.*, 1988a; Laing *et al.*, 1988). Illumination of dark-grown plants for as little as 1 min prior to the isolation of plastids enables the subsequent synthesis of the PSI-A and -B subunits in the isolated plastids (Klein *et al.*, 1988a; Laing *et al.*, 1988). The very fast response and the fact that cycloheximide treatment of the plants prior to illumination does not affect the induction of PSI-A and -B synthesis show that *de novo* synthesis of nuclear-encoded proteins is not involved (Klein and Mullet, 1986; Klein *et al.*, 1988a). The induction of PSI-A and -B synthesis requires red light at a rather high fluency; far-red reversal has not been observed (Klein *et al.*, 1988a; Laing *et al.*, 1988). These observations indicate that the response is mediated by protochlorophyllide rather than by phytochrome. Elegant studies by Mullet and co-workers have shown that the translation of the PSI-A and -B subunits is controlled by the availability of chlorophyll *a*. Thus, a barley mutant that synthesizes chlorophyllide but not chlorophyll *a* is unable to induce the synthesis of the PSI-A and -B subunits (Klein *et al.*, 1988a). However, chlorophyll *a* must be nascent to induce the translation. *In vitro* synthesis of PSI-A and -B can be accomplished in lysed etioplasts provided that chlorophyllide and phytolpyrophosphate is present (Eichacker *et al.*, 1990). In agreement with the mutant studies, chlorophyllide alone is not sufficient (see also Chapter 10). Chlorophyll *a* alone is also not sufficient to induce the translation of the two subunits. New initiation of translation apparently was not occurring in the *in vitro* assays, indicating that chlorophyll *a* somehow releases a translational arrest rather than induces the initiation of translation. The conclusion of these experiments is that the syntheses of chlorophyll *a* and of PSI-A and -B are coupled. This coupling is observed also with some other chlorophyll-proteins of the chloroplast (Mullet *et al.*, 1990). Mullet *et al.* speculated that the coupling serves a dual purpose. By building up a capacity for chlorophyll apoprotein synthesis in the dark, the plant ensures that chlorophyll apoproteins are present to bind the chlorophyll synthesized on illumination. Thereby, accumulation of unbound chlorophyll from which chlorophyll triplets could be generated easily is avoided. Triplets generated in bound chlorophyll can be quenched but triplets in free chlorophyll lead to the generation of toxic superoxide and hydroxyl radicals. On the other hand, the plant cannot solve this problem by accumulating surplus chlorophyll apoproteins, since such a surplus could lead to chlorophyll proteins unsaturated with chlorophyll. Therefore, antennae would be inefficient and easily cause photodamage.

The receptor function of protochlorophyllide for the induction of PSI-A and -B has been established clearly. However, the additional involvement of phytochrome or other receptors cannot be ruled out. A surprising result was obtained by Laing *et al.* (1988), who observed PSI-A and -B synthesis in plastids isolated from dark-grown barley illuminated for only 1 min. However, etioplasts isolated without prior illumination of the plants were unable to synthesize PSI-A or -B in the light. Unfortunately, it was not reported whether the isolated etioplasts actually synthesized chlorophyll *a* during the incubation in light.

## D. Reconstitution of Photosystem I Complex

Reconstitution *in vitro* with isolated PSI subunits is one experimental approach to study the assembly of PSI. Golbeck, Bryant, and co-workers have carried out a number of experiments with *Synechococcus* to determine under what conditions the PSI-C iron–sulfur protein will rebind to the core of PSI (Golbeck *et al.*, 1988b; Parrett *et al.*, 1990; Zhao *et al.*, 1990; Li *et al.*, 1991a,b; Mehari *et al.*, 1991). In these experiments, PSI was denatured with chaotropic agents, which allowed the separation of the peripheral subunits, that is, at least PSI-C, -D, and -E, from the CP1 core. The PSI-I, -J, -K, and -L subunits remain bound to the core after treatment with chaotropic agents (Li *et al.*, 1991a).

Treatment with chaotropic agents caused the loss of centers A and B, but it was possible to avoid the loss of center X by carefully controlling the conditions (Golbeck *et al.*, 1988a; Parrett *et al.*, 1989). P700 and the acceptors $A_0$ and $A_1$ are resistant to chaotropic agents. The core particle produced in this way shows forward electron transfer from P700 to center X and typical back-reaction kinetics. Recombination of the core particle with the peripheral subunits after removal of the chaotropic agents did not cause rebinding of the PSI-C apoprotein containing denatured iron–sulfur clusters (Parrett *et al.*, 1990; Li *et al.*, 1991a). Significantly, PSI-D and -E also were not rebound under these conditions. When the PSI-C polypeptide was isolated from thylakoids in the native form or when the denatured PSI-C polypeptide was renatured chemically prior to or during recombination with the core particle, significant reconstitution was observed. Rebinding of PSI-C was correlated with rebinding of PSI-D and -E. When PSI-C was rebound to the core particle, centers A and B were functional as monitored by spectrophotometrical methods and by EPR. The source of PSI-C was not essential, that is, cyanobacterial core preparations could be reconstituted efficiently with PSI-C from cyanobacteria or from spinach (Mehari *et al.*, 1991). Mehari *et al.* (1991) also showed that a spinach core particle could be reconstituted in the same way as the *Synechococcus* core particle. Since only the native form of PSI-C with intact iron–sulfur centers will bind to CP1, it seems likely that the iron–sulfur centers of the water-soluble PSI-C polypeptide are synthesized in the stroma, thereby making the polypeptide competent for incorporation into PSI. In this context, it would be interesting to know whether PSI-C is synthesized on free or membrane-bound ribosomes, but this issue has not been addressed.

Reconstitution also has been accomplished after denaturation of all three iron–sulfur centers followed by chemical rebuilding of the centers. The time course of reconstitution indicates that center X must be reconstituted prior to the binding of PSI-C (Parrett *et al.*, 1990).

Studies by Zhao *et al.* (1990) and Li *et al.* (1991b) used PSI subunits expressed in and isolated from *Escherichia coli*. These studies showed that binding of PSI-D to the PSI core is a prerequisite for stable binding of PSI-C. Thus, PSI-D must provide a binding site for PSI-C in addition to providing a docking site for ferredoxin. The ability of the reconstituted

PSI to photoreduce NADP$^+$ was not studied. The suggested role of PSI-D is in agreement with the results obtained by insertional inactivation of the *psaD* gene in *Synechocystis*.

## E. Biosynthesis of Electron Acceptors

Assembly of PSI requires the concerted combination of more than 10 different polypeptides. Also, pigment molecules, mainly about 90 chlorophyll *a* molecules, must be incorporated. In addition to these components, the reaction center and the electron acceptors must be synthesized and inserted into their proper location.

The reconstitution experiments with PSI-C indicate that centers A and B are likely to be inserted into the PSI-C polypeptide in the stroma prior to the incorporation of the polypeptide into PSI. The ability of stromal extracts to catalyze the synthesis of the iron–sulfur cluster of ferredoxin has been studied (Takahashi *et al.*, 1991a,b). The iron–sulfur cluster is synthesized in an NADPH- and ATP-requiring process with cysteine as a precursor. Apparently, free sulfide is synthesized in an initial NADPH-requiring step and the sulfide is incorporated efficiently into ferredoxin in a second ATP-requiring step. Although the evidence is circumstantial, the studies suggest that ferredoxin is phosphorylated during the synthesis of the iron–sulfur cluster. Whether the enzyme system responsible for formation of the iron–sulfur cluster in ferredoxin is also responsible for the formation of the iron–sulfur clusters of PSI-C is not known. The very different structure of apoferredoxin and apo-PSI-C, as well as the different structure of the iron–sulfur clusters, suggests that different enzyme systems probably are involved. However, the enzyme system responsible for formation of holo-PSI-C may be of a rather general type, since holo-PSI-C is formed in *E. coli* transformed with the *psaC* gene (Zhao *et al.*, 1990)

The synthesis of the reaction center and of the electron acceptors that are more primary than centers A and B has not been studied. Since PSI-A and PSI-B are synthesized on thylakoid-bound ribosomes and, thus, inserted into the thylakoids during translation, the insertion of the electron acceptors poses a problem. Investigation of the synthesis of D1 and D2 of PSII, which binds a set of electron acceptors similar to PSI-A and PSI-B, has shown that the translation is arrested several times (Mullet *et al.*, 1990; Kim *et al.*, 1991). Pulse-chase studies in isolated chloroplasts suggest that a similar situation exists for PSI-A and PSI-B (Mullet *et al.*, 1986). As described earlier, the chlorophyll molecules bound to PSI-A and PSI-B must be synthesized simultaneously with the polypeptides (see also Chapters 6 and 10). The reaction center and the primary acceptor $A_0$ are chlorophyll *a*-derived molecules. These components are likely to be synthesized simultaneously with the translation of PSI-A and PSI-B as well. The various electron transfer components may be synthesized and inserted during the translational pauses. The pauses also may serve to coordinate the synthesis of the two polypeptides, since they may need to be synthesized at the same rate. The general model is that the synthesis of the highly complex PSI-A and -B heterodimer follows a strict program during which the polypeptides are synthesized simultaneously with antenna pigments and electron transfer components. The synthesis of these components must involve several protein factors in addition to the usual machinery for translation. However, no such factors have been identified. The insertion of electron transfer components during protein synthesis may be likely, but reconstitution experiments have shown that phylloquinone can be inserted into PSI where $A_1$ is absent, thereby reconstituting the ability of PSI to reduce $NADP^+$ (Biggins and Mathis, 1988). Thus, the possibility that $A_1$ is inserted post-translationally should not be excluded. Center X also may be an exception to the general model of cotranslational insertion. Center X is denatured easily by chaotrope treatment of PSI core preparations. The denatured center X can be reconstituted chemically using the same conditions used for reconstitution of centers A and B. Thus, center X seems to be quite accessible; perhaps a stromal enzyme system can synthesize the iron–sulfur cluster and insert it into the PSI-A–PSI-B heterodimer. Again how synthesis of the iron–sulfur cluster is accomplished *in vivo* is not known.

## VI. CONCLUDING REMARKS

Based on the studies reviewed in this chapter, a scheme for the assembly of PSI can be proposed, although the scheme surely will need modifications in light of future experiments. The synthesis of PSI-A and PSI-B is induced when chlorophyll *a* synthesis is induced by light. Simultaneously with the synthesis of the proteins, pigment molecules are inserted. When the leucine zipper motifs are synthesized, the two elongating polypeptides associate and the reaction center and primary acceptors are inserted. Finally, the two polypeptides become covalently linked by the insertion of center X. The newly synthesized CP1 now can bind holo-PSI-C present in the stroma. The binding of PSI-C enables the additional binding of PSI-D and -E, which shield PSI-C from the stroma and anchor PSI-C tightly. In plants, PSI-H is likely to be inserted before PSI-C, -D, and -E. Presumably the other intrinsic subunits are inserted at an early stage also, but it is presently impossible to make plausible suggestions concerning these subunits.

As evident from the information presented, much remains to be learned about the assembly of PSI. Until quite recently, the polypeptide constituents of PSI were not known, which hampered studies on the assembly of the complex. The basic structure of PSI is now reasonably well understood, and investigations of the assembly of the complex can be carried out. Such studies are being carried out actively in a number of laboratories using a variety of different approaches. Therefore, it is likely that the knowledge of the assembly processes will expand significantly in the near future.

## ACKNOWLEDGMENTS

Studies reported from the authors' laboratory were supported by the Center for Plant Biotechnology and by grants from the Danish Agricultural and Veterinary Research Council and the Carlsberg Foundation. B. Andersen was supported by a scholarship from the Nordic Energy Research Programme.

We thank D. A. Bryant, J. H. Golbeck, M. Ikeuchi, J. T. Trost, and R. E. Blankenship for communicating results prior to publication. We also thank Y. Lindquist for making the model of the structure of the PSI-C polypeptide and B. L. Møller and J. S. Okkels for critically reading the manuscript.

## REFERENCES

Adman, E. T., Sieker, L. C., and Jensen, L. H. (1973). The structure of a bacterial ferredoxin. *J. Biol. Chem.* **248**, 1987–3996.

Allen, J. P., Feher, G., Yeates, T. O., Komiya, H., and Rees, D. C. (1987). Structure of the reaction center from *Rhodobacter sphaeroides* R-26: The cofactors. *Proc. Natl. Acad. Sci. USA* **84**, 5730–5734.

Almog, O., Shoham, G., Michaeli, D., and Nechushtai, R. (1991). Monomeric and trimeric forms of photosystem I reaction center of *Mastigocladus laminosus*: Crystallization and preliminary characterization. *Proc. Natl. Acad. Sci. USA* **88**, 5312–5316.

Anandan, S., Vainstein, A., and Thornber, J. P. (1989). Correlation of some published amino acid sequences for photosystem I polypeptides to a 17-kDa LHCI pigment-protein and to subunits III and IV of the core complex. *FEBS Lett.* **256**, 150–154.

Andersen, B., Koch, B., Scheller, H. V., Okkels, J. S., and Møller, B. L. (1990). Nearest neighbor analysis of the photosystem I subunits in barley and their binding of ferredoxin. *In* "Current Research in Photosynthesis" (M. Baltscheffsky, ed.), Vol. II, pp. 671–674. Kluwer, Dordrecht, The Netherlands.

Andersen, B., Koch, B., and Scheller, H. V. (1992a). Structural and functional analysis of the reducing side of photosystem I. *Physiol. Plant.* **84**, 154–161.

Andersen, B., Scheller, H. V., and Møller, B. L. (1992b). The PSI-E subunit of photosystem I binds ferredoxin:NADP$^+$ oxidoreductase. *FEBS Lett.* **311**, 169–173.

Bassi, R., and Simpson, D. (1987). Chlorophyll–protein complexes of barley photosystem I. *Eur. J. Biochem.* **163**, 221–230.

Bearden, A. J., and Malkin, R. (1972). Quantitative EPR studies of the primary reaction of photosystem I in chloroplasts. *Biochim. Biophys. Acta* **283**, 456–468.

Bengis, C., and Nelson, N. (1977). Subunit structure of chloroplast photosystem I reaction center. *J. Biol. Chem.* **252,** 4564–4569.

Berzborn, R. J., Klein-Hitpaβ, L., Otto, J., Schünemann, S., and Oworah-Nkruma, R. (1990). The "additional subunit" $CF_0II$ of the photosynthetic ATP-synthase and the thylakoid polypeptide, binding ferredoxin NADP reductase: Are they different? *Z. Naturforsch.* **45C,** 772–784.

Biggins, J., and Mathis, P. (1988). Functional role of vitamin $K_1$ in photosystem I of the cyanobacterium *Synechocystis* 6803. *Biochemistry* **27,** 1494–1500.

Boekema, E. J., Dekker, J. P., van Heel, M. G., Rögner, M., Saenger, W., Witt, I., and Witt, H. T. (1987). Evidence for a trimeric organization of the photosystem I complex from the termophilic cyanobacterium *Synechococcus* sp. *FEBS Lett.* **217,** 283–286.

Boekema, E. J., Dekker, J. P., Rögner, M., Witt, I., Witt, H. T., and van Heel, M. G. (1989). Refined analysis of the trimeric structure of the isolated photosystem I complex from the termophilic cyanobacterium *Synechococcus* sp. *Biochim. Biophys. Acta* **974,** 81–87.

Boekema, E. J., Wynn, R. M., and Malkin, R. (1990). The structure of spinach photosystem I studied by electron microscopy. *Biochim. Biophys. Acta* **1017,** 49–56.

Bruce, B. D., and Malkin, R. (1988a). Subunit stoichiometry of the chloroplast photosystem I complex. *J. Biol. Chem.* **263,** 7302–7308.

Bruce, B. D., and Malkin, R. (1988b). Structural aspects of photosystem I from *Dunaliella salina*. *Plant Physiol.* **88,** 1201–1206.

Bruce, B. D., and Malkin, R. (1991). Biosynthesis of the chloroplast cytochrome $b_6f$ complex: Studies in a photosynthetic mutant of *Lemna*. *Plant Cell* **3,** 203–212.

Bryant, D. A. (1992). Molecular biology of photosytem I. *In* "Current Topics in Photosynthesis" (J. Barber, ed.), Vol. 11, pp. 501–549. Elsevier, Amsterdam.

Bryant, D. A., Rhiel, E., deLorimier, R., Zhou, J., Stirewalt, V. L., Gasparich, G. E., Dubbs, J. M., and Snyder, W. (1990). Analysis of phycobilisome and photosystem I complexes of cyanobacteria. *In* "Current Research in Photosynthesis" (M. Baltscheffsky, ed.), Vol. II, pp. 1–9. Kluwer, Dordrecht, The Netherlands.

Ceccarelli, E. A., Chan, R. L., and Vallejos, R. H. (1985). Timeric structure and other properties of the chloroplast reductase binding protein. *FEBS Lett.* **190,** 165–168.

Chitnis, P. R., Reilly, P. A., Miedel, M. C., and Nelson, N. (1989a). Structure and targeted mutagenesis of the gene encoding 8-kDa subunit of photosystem I from the cyanobacterium *Synechocystis* sp. PCC 6803. *J. Biol. Chem.* **264,** 18374–18380.

Chitnis, P. R., Reilly, P. A., and Nelson, N. (1989b). Insertional inactivation of the gene encoding subunit II of photosystem I from the cyanobacterium *Synechocystis* sp. PCC 6803. *J. Biol. Chem.* **264,** 18381–18385.

Chitnis, P. R., Purvis, D., and Nelson, N. (1991). Molecular cloning and targeted mutagenesis of gene *psa*F encoding subunit III of photosystem I from the cyanobacterium *Synechocystis* sp. PCC 6803. *J. Biol. Chem.* **266,** 20146–20151.

Deisenhofer, J., Epp, O., Miki, R., Huber, R., and Michel, H. (1985). Structure of the protein subunits in the photosynthetic reaction centre of *Rhodopseudomonas viridis* at 3 Å resolution. *Nature (London)* **318,** 618–624.

de Pater, S., Hensgens, L. A. M., and Schilperoort, R. A. (1990). Structure and expression of a light-inducible shoot-specific rice gene. *Plant Mol. Biol.* **15,** 399–406.

Dunn, P. P. J., and Gray, J. C. (1988). Localization and nucleotide sequence of the gene for the 8-kDa subunit of photosystem I in pea and wheat chloroplast DNA. *Plant Mol. Biol.* **11,** 311–319.

Dunn, P. P. J., Packman, L. C., Pappin, D., and Gray, J. C. (1988). N-terminal amino acid sequence analysis of the subunits of pea photosystem I. *FEBS Lett.* **228,** 157–161.

Efimov, V. A., Andreeva, A. V., Reverdatto, S. V., Jung, R., and Chakhmakcheva, O. G.

(1988). Nucleotide sequence of the barley chloroplast *psb*A gene for the $Q_B$ protein of photosystem II. *Nucleic Acids Res.* **16,** 5685.

Eichacker, L. A., Soll, J., Lauterbach, P., Rüdiger, W., Klein, R. R., and Mullet, J. E. (1990). In vitro synthesis of chlorophyll *a* in the dark triggers accumulation of chlorophyll *a* apoproteins in barley etioplasts. *J. Biol. Chem.* **265,** 13566–13571.

Evans, M. C. W., Telfer, A., and Lord, A. V. (1972). Evidence for the role of a bound ferredoxin as the primary electron acceptor of photosystem I in spinach chloroplasts. *Biochim. Biophys. Acta* **267,** 530–537.

Fish, L. E., and Bogorad, L. (1986). Identification and analysis of the maize P700 chlorophyll *a* apoproteins PSI-A1 and PSI-A2 by high pressure liquid chromatography analysis and partial sequence determination. *J. Biol. Chem.* **261,** 8134–8139.

Fish, L. E., Kück, U., and Bogorad, L. (1985). Two partially homologous adjacent light-inducible maize chloroplast genes encoding polypeptides of the P700 chlorophyll *a*-protein complex of photosystem I. *J. Biol. Chem.* **260,** 1413–1421.

Ford, R. C., Picot, D., and Garavito, R. M. (1987). Crystallization of the photosystem I reaction centre. *EMBO J.* **6,** 1581–1586.

Ford, R. C., Paupit, R. A., and Holzenburg, A. (1988). Structural studies on improved crystals of the photosystem I reaction centre from *Phormidium laminosum. FEBS Lett.* **238,** 385–389.

Ford, R. C., Hefti, A., and Engel, A. (1990). Ordered arrays of the photosystem I reaction centre after reconstitution: Projections and surface reliefs of the complex at 2 nm resolution. *EMBO J.* **9,** 3067–3075.

Franzén, L.-G., Frank, G., Zuber, H., and Rochaix, J.-D. (1989a). Isolation and characterization of cDNA clones encoding the 17.9 and 8.1 kDa subunits of photosystem I from *Chlamydomonas reinhardtii. Plant Mol. Biol.* **12,** 463–474.

Franzén, L.-G., Frank, G., Zuber, H., and Rochaix, J.-D. (1989b). Isolation and characterization of cDNA clones encoding photosystem I subunits with molecular masses 11.0, 10.0, and 8.4 kDa from *Chlamydomonas reinhardtii. Mol. Gen. Genet.* **219,** 137–144.

Gavel, Y., Steppuhn, J., Herrmann, R., and von Heijne, G. (1991). The "positive-inside rule" applies to thylakoid membrane proteins. *FEBS Lett.* **282,** 41–46.

Golbeck, J. H. (1987). Structure, function and organization of the photosystem I reaction center complex. *Biochim. Biophys. Acta* **895,** 167–204.

Golbeck, J. H., and Bryant, D. A. (1991). Photosystem I. In "Current Topics in Bioenergetics" (C. P. Lee, ed.), Vol. 16, pp. 83–177. Academic Press, New York.

Golbeck, J. H., and Cornelius, J. M. (1986). Photosystem I charge separation in the absence of centers A and B. I. Optical characterization of center "$A_2$" and evidence for its association with a 64-kDa peptide. *Biochim. Biophys. Acta* **849,** 16–24.

Golbeck, J. H., McDermott, A. E., Jones, W. K., and Kurtz, D. M. (1987). Evidence for the existence of [2Fe-2S] as well as [4Fe-4S] clusters among $F_A$, $B_B$, and $F_X$. Implications for the structure of the photosystem I reaction center. *Biochim. Biophys. Acta* **891,** 94–98.

Golbeck, J. H., Parrett, K. G., Mehari, T., Jones, K. L., and Brand, J. J. (1988a). Isolation of the intact photosystem I reaction center core containing P700 and iron-sulfur center $F_X$. *FEBS Lett.* **228,** 268–272.

Golbeck, J. H., Mehari, T., Parrett, K., and Ikegami, L. (1988b). Reconstitution of the photosystem I complex from the P700 and $F_X$-containing reaction center core protein and the $F_A/F_B$ polypeptide. *FEBS Lett.* **240,** 9–14.

Hippler, M., Ratajczak, R., and Haenel, W. (1989). Identification of the plastocyanin binding subunit of photosystem I. *FEBS Lett.* **250,** 280–284.

Hiratsuka, J., Shimada, H., Whittier, R., Ishibashi, T., Sakamoto, M., Mori, M., Kondo,

C., Honji, Y., Sun, C.-R., Meng, B.-Y., Li, Y.-Q., Kanno, A., Nishizaea, Y., Hirai, A., Shinozaki, K., and Sugiura, M. (1989). The complete sequence of the rice (*Oryza sativa*) chloroplast genome: Intermolecular recombination between distinct tRNA genes accounts for a major plastic DNA inversion during the evolution of the cereals. *Mol. Gen. Genet.* **217**, 185–194.

Høj, P. B., and Møller, B. L. (1986). The 110-kDa reaction center protein of photosystem I, P700-chlorophyll *a*-protein 1, is an iron-sulfur protein. *J. Biol. Chem.* **261**, 14292–14300.

Høj, P. B., Svendsen, I., Scheller, H. V., and Møller, B. L. (1987). Identification of a chloroplast-encoded 9-kDa polypeptide as a 2[4Fe-4S] protein carrying centers A and B of photosystem I. *J. Biol. Chem.* **262**, 12676–12684.

Hoshina, S., Sue, S., Kunishima, N., Kamide, K., Wada, K., and Itoh, S. (1989). Characterization and N-terminal sequence of a 5-kDa polypeptide in the photosystem I core complex from spinach. *FEBS Lett.* **258**, 305–308.

Ikeuchi, M., and Inoue, Y. (1991). Two new components of 9- and 14-kDa from spinach photosystem I complex. *FEBS Lett.* **280**, 332–334.

Ikeuchi, M., Hirano, A., Hiyama, T., and Inoue, Y. (1990). Polypeptide composition of higher plant photosystem I complex. *FEBS Lett.* **263**, 274–278.

Ikeuchi, M., Nyhus, K. J., Inoue, Y., and Pakrasi, H. B. (1991). Identities of four low-molecular-mass subunits of the photosystem I complex from *Anabaena variabilis* ATCC 29413. *FEBS Lett.* **287**, 5–9.

Iwasaki, Y., Ishikawa, H., Hibino, T., and Takabe, T. (1991). Characterization of genes that encode subunits of cucumber PS I complex by N-terminal sequencing. *Biochim. Biophys. Acta* **1059**, 141–148.

Kim, J., Klein, P. G., and Mullet, J. E. (1991). Ribosomes pause at specific sites during synthesis of membrane-bound chloroplast reaction center protein D1. *J. Biol. Chem.* **266**, 14931–14938.

Kirsch, W., Seyer, P., and Herrmann, R. G. (1986). Nucleotide sequence of the clustered genes for two $P_{700}$ chlorophyll *a* apoproteins of the photosystem I reaction center and the ribosomal protein S14 of the spinach plastid chromosome. *Curr. Genet.* **10**, 843–855.

Klein, R. R., and Mullet, J. E. (1986). Regulation of chloroplast-encoded chlorophyll-binding protein translation during higher plant chloroplast biogenesis. *J. Biol. Chem.* **261**, 11138–11145.

Klein, R. R., and Mullet, J. E. (1987). Control of gene expression during higher plant chloroplast biogenesis. *J. Biol. Chem.* **262**, 4341–4348.

Klein, R. R., Gamble, P. E., and Mullet, J. E. (1988a). Light-dependent accumulation of radiolabeled plastid-encoded chlorophyll *a*-apoproteins requires chlorophyll *a*. *Plant Physiol* **88**, 1246–1256.

Klein, R. R., Mason, H. S., and Mullet, J. E. (1988b). Light-regulated translation of chloroplast proteins I: Transcripts of *psa*A-*psas*B, *psb*A, and *rbc*L are associated with polysomes in dark-grown and illuminated barley seedlings. *J. Cell Biol.* **106**, 289–301.

Kössel, H., Döry, I., Igloi, G., and Maier, R. (1990). A leucine-zipper motif in photosystem I. *Plant Mol. Biol.* **15**, 497–499.

Koike, H., Ikeuchi, M., Hiyama, T., and Inoue, Y. (1989). Identification of the photosystem I components from the cyanobacterium *Synechococcus vulcanus* by N-terminal sequencing. *FEBS Lett.* **253**, 257–263.

Lagoutte, B. (1988). Cloning and sequencing of spinach cDNA clones encoding the 20-kDa PS I polypeptide. *FEBS Lett.* **232**, 275–280.

Lagoutte, B., and Mathis, P. (1989). The photosystem I reaction center: structure and photochemistry. *Photochem. Photobiol.* **49**, 833–844.

Laing, W., Kreuz, K., and Apel, K. (1988). Light-dependent, but phytochrome-independent, translational control of the accumulation of the P700 chlorophyll-*a* protein of photosystem I in barley (*Hordeum vulgare* L.). *Planta* **176**, 269–276.

Lehmbeck, J., Rasmussen, O. F., Bookjans, G. B., Jepsen, B. R., Stummann, B. M., and Henningsen, K. W. (1986). Sequence of two genes in pea chloroplast DNA coding for 84- and 82-kD polypeptides of the photosystem I complex. *Plant Mol. Biol.* **7**, 3–10.

Li, N., Warren, P. V., Golbeck, J. H., Frank, G., Zuber, H., and Bryant, D. A. (1991a). Polypeptide composition of the photosystem I complex and the photosystem I core protein from *Synechococcus* sp. PCC 6301. *Biochim. Biophys. Acta* **1059**, 215–225.

Li, N., Zhao, J., Warren, P. V., Warden, J. T., Bryant, D. A., and Golbeck, J. H. (1991b). PsaD is required for the stable binding of psaC to the photosystem I core protein of *Synechococcus* sp. PCC 6301. *Biochemistry* **30**, 7863–7872.

Liebl, U., Mockenstorm-Wilson, M., Trost, J. T., Brune, D. C., Blankenship, R. E., and Vermaas, W. F. J. (1992). The reaction center core polypepticle in the photosynthetic bacterium *Heliobacillus Mobilis*. In "Proc. IXth International Congress on Photosynthesis" (M. Murata, ed.). Kluwer, Dordrecht, The Netherlands (in press).

McDermott, A. E., Yachandra, V. K., Guiles, R. D., Britt, R. D., Dexheimer, S. L., Sauer, K., and Klein, M. P. (1988). Low potential iron-sulfur centers in photosystem I: An X-ray absorption spectroscopy study. *Biochemistry* **27**, 4013–4020.

McDermott, A. E., Yachandra, V. K., Guiles, R. D., Sauer, K., and Klein, M. P. (1989). EXAFS structural study of $F_X$, the low-potential Fe-S center in photosystem I. *Biochemistry* **28**, 8056–8059.

Malkin, R., and Bearden, A. J. (1971). Primary reactions of photosynthesis: Photoreduction of a bound chloroplast ferredoxin at low temperature as detected by EPR spectroscopy. *Proc. Natl. Acad. Sci. USA* **68**, 16–19.

Mann, K., Schlenkrich, T., Bauer, M., and Huber, R. (1991). The amino-acid sequence of three proteins of photosystem I of the cyanobacterium *Fremyella diplosiphon* (*Calothrix* sp PCC 7601). *Biol. Chem. Hoppe-Seyler* **372**, 519–524.

Mannan, R. M., Whitmarsh, J., Nyman, P., and Pakrasi, H. B. (1991). Directed mutagenesis of an iron sulfur protein of the photosystem I complex in the filamentous cyanobacterium *Anabaena variabilis* ATCC 29413. *Proc. Natl. Acad. Sci. USA* **88**, 10,168–10,172.

Margulies, M. M. (1991). Sequence similarity between photosystem I and II: Identification of a photosytem I reaction center transmembrane helix that is similar to transmembrane helix IV of the D2 subunit of photosystem II and the M subunit of the non-sulfur purple and flexible green bacteria. *Photosynth. Res.* **29**, 133–147.

Mehari, T., Parrett, K. G., Warren, P. V., and Golbeck, J. H. (1991). Reconstitution of the iron-sulfur clusters in the isolated $F_A/F_B$ protein: EPR spectral characterization of same-species and cross-species photosystem I complexes. *Biochim. Biophys. Acta* **1056**, 139–148.

Merati, G., and Zanetti, G. (1987). Chemical cross-linking of ferredoxin to spinach thylakoids. *FEBS Lett.* **215**, 37–40.

Michel, H., and Deisenhofer, J. (1988). Relevance of the photosynthetic reaction center from purple bacteria to the structure of photosystem II. *Biochemistry* **27**, 1–7.

Møller, B. L., Scheller, H. V., Okkels, J. S., Koch, B., Andersen, B., Nielsen, H. L., Olsen, I., Halkier, B. A., and Høj, P. B. (1990). Chloroplast encoded photosystem I polypeptides of barley. In "Current Research in Photosynthesis" (M. Baltscheffsky, ed.), Vol. II, pp. 523–530. Kluwer, Dordrecht, The Netherlands.

Münch, S., Ljungberg, U., Steppuhn, J., Schneiderbauer, A., Nechustai, R., Beyreuter, K., and Herrmann, R. G. (1988). Nucleotide sequences of cDNAs encoding the entire precursor polypeptides of subunits II and III of the photosystem I reaction center from spinach. *Curr. Genet.* **14**, 511–518.

Mullet, J. E., Klein, R. R., and Grossmann, A. (1986). Optimization of protein synthesis in isolated higher plant chloroplasts. Identification of paused translation intermediates. *Eur. J. Biochem.* **155,** 331–338.

Mullet, J. E., Klein, P. G., and Klein, R. R. (1990). Chlorophyll regulates accumulation of the plastid-encoded chlorophyll apoproteins CP43 and D1 by increasing apoprotein stability. *Proc. Natl. Acad. Sci. USA* **87,** 4038–4042.

Neumann, E. M. (1988). Primary structure of barley genes encoding quinone and chlorophyll *a* binding proteins of photosystem II. *Carlsberg Res. Commun.* **53,** 259–275.

Oh-oka, H., Takahashi, Y., Kuriyama, K., Saeki, K., and Matsubara, H. (1988). The protein responsible for center A/B in spinach photosystem I: Isolation with iron-sulfur cluster(s) and complete sequence analysis. *J. Biochem.* **103,** 962–968.

Oh-oka, H., Takahashi, Y., and Matsubara, H. (1989). Topological considerations of the 9-kDa polypeptide which contains centers A and B, associated with the 14- and 19-kDa polypeptides in the photosystem I complex of spinach. *Plant Cell Physiol* **30,** 869–875.

Ohyama, K., Fukuzawa, H., Kohchi, T., Shirai, H., Sano, T., Sano, S., Umesono, K., Shiki, Y., Takeuchi, M., Chang, Z., Aota, S.-i., Inokuchi, H., and Ozeki, H. (1986). Chloroplast gene organization deduced from complete sequence of liverwort *Marchantia polymorpha* chloroplast DNA. *Nature (London)* **322,** 572–574.

Okkels, J. S., Jepsen, L. B., Hønberg, L. S., Lehmbeck, J., Scheller, H. V., Brandt, P. Høyer-Hansen, G., Stummann, B., Henningsen, K. W., von Wettstein, D., and Møller, B. L. (1988). A cDNA clone encoding a 10.8-kDa photosystem I polypeptide of barley. *FEBS Lett.* **237,** 108–112.

Okkels, J. S., Scheller, H. V., Jepsen, L. B., and Møller, B. L. (1989). a cDNA clone encoding the precursor for a 10.2-kDa photosystem I polypeptide of barley. *FEBS Lett.* **250,** 575–579.

Okkels, J. S., Scheller, H. V., Jepsen, L. B., Andersen, B., and Møller, B. L. (1990). Characterization of a cDNA clone for the *Psa*H gene from barley and mRNA level of PS I genes in light-induced barley seedlings. *In* "Current Research in Photosynthesis" (M. Baltscheffsky, ed.), Vol. III, pp. 613–616. Kluwer, Dordrecht, The Netherlands.

Okkels, J. S., Scheller, H. V., Svendsen, I., and Møller, B. L. (1991). Isolation and characterization of a cDNA clone encoding an 18-kDa hydrophobic photosystem I subunit (PSI-L) from barley (*Hordeum vulgare* L.). *J. Biol. Chem.* **266,** 6767–6733.

Okkels, J. S., Nielsen, V. S., Scheller, H. V., and Møller, B. L. (1992). A CDNA clone from barley encoding the precursor for the photosystem I polypeptide PSI-G: Sequence similarity to PSI-K. *Plant Mol. Biol.* **18,** 989–994.

Parrett, K. G., Mehari, T., Warren, P. G., and Golbeck, J. H. (1989). Purification and properties of the intact P700 and $F_X$-containing photosystem I core protein. *Biochim. Biophys. Acta.* **973,** 324–332.

Parrett, K. G., Mehari, T., and Golbeck, J. H. (1990). Resolution and reconstitution of the cyanobacterial photosystem I complex. *Biochim. Biophys. Acta* **1015,** 341–352.

Petrouleas, V., Brand, J. J., Parrett, K. G., and Golbeck, J. H. (1989). Mössbauer analysis of the low-potential iron-sulfur center in photosystem I: Spectroscopic evidence that $F_X$ is a [4Fe-4S] cluster. *Biochemistry* **28,** 8980–8983.

Reilly, P., and Nelson, N. (1988). Photosystem I complex. *Photosynth. Res.* **19,** 73–84.

Rhiel, E., and Bryant, D. A. (1988). Preliminary results concerning the *psa*C, *psa*D, and *psa*F genes and their products in the cyanobacteria *Synechococcus* sp. PCC 7002 and *Nostoc* sp. PCC 8009. *In* "Light-Energy Transduction in Photosynthesis: Higher Plants and Bacterial Models" (S. E. Stevens, Jr., and D. A. Bryant, eds.), pp. 320–323. American Society of Plant Physiologists, Rockville, Maryland.

Rögner, M., Mühlenhoff, U., Boekema, E. J., and Witt, H. T. (1990). Mono-, di-, and trimeric PS I reaction center complexes isolated from the thermophilic cyanobacterium *Synechococcus* sp. *Biochim. Biophys. Acta* **1015**, 415–424.

Schantz, R., and Bogorad, L. (1988). Maize chloroplast genes *ndh*D, *ndh*E, and *psa*C. Sequences, transcripts, and transcript pools. *Plant Mol. Biol.* **11**, 239–247.

Scheller, H. V., and Møller, B. L. (1990). Photosystem I polypeptides. *Physiol. Plant.* **78**, 484–494.

Scheller, H. V., Høj, P. B., Svendsen, I., and Møller, B. L. (1988). Partial amino acid sequences of two nuclear-encoded photosystem I polypeptides from barley. *Biochim. Biophys. Acta* **933**, 501–505.

Scheller, H. V., Svendsen, I., and Møller, B. L. (1989a). Subunit composition of photosystem I and identification of center X as a [4Fe-4S] cluster. *J. Biol. Chem.* **264**, 6929–6934.

Scheller, H. V., Svendsen, I., and Møller, B. L. (1989b). Amino acid sequence of the 9-kDa iron–sulfur protein of photosystem I in barley. *Carlsberg Res. Commun.* **54**, 11–15.

Scheller, H. V., Okkels, J. S., Høj, P. B., Svendsen, I., Roepstorff, P., and Møller, B. L. (1989c). The primary structure of a 4.0-kDa photosystem I polypeptide encoded by the chloroplast *psa*I gene. *J. Biol. Chem.* **264**, 18402–18406.

Scheller, H. V, Okkels, J. S., Roepstorff, P., Jepsen, L. B., and Møller, B. L. (1990a). Characterization of a cDNA clone for the *Psa*E gene from barley and plasma desorption mass spectrometry of the corresponding photosystem I polypeptide PSI-E. *In* "Current Research in Photosynthesis" (M. Baltscheffsky, ed.), Vol. III, pp. 609–612. Kluwer, Dordrecht, The Netherlands.

Scheller, H. V., Andersen, B., Okkels, J. S., Svendsen, I., and Møller, B. L. (1990b). Photosystem I in barley: Subunit PSI-F is not essential for the interaction with plastocyanin. *In* "Current Research in Photosynthesis" (M. Baltscheffsky, ed.), Vol. II, pp. 679–682. Kluwer, Dordrecht, The Netherlands.

Shimizu, T., Hiyama, T., Ikeuchi, M., Koike, H., and Inoue, Y. (1990). Nucleotide sequence of the *psa*C gene of the cyanobacterium *Synechococcus vulcanus*. *Nucleic Acids Res.* **18**, 3644.

Shinozaki, K., Ohme, M., Tanaka, M., Wakasugi, T., Hayashida, N., Matsubayashi, T., Zaita, N., Chunwongse, J., Obokata, J., Yamaguchi-Shinozaki, K., Ohto, C., Torazawa, K., Meng, B. Y., Sugita, M., Deno, H., Kamogashira, T., Yamada, K., Kusuda, J., Takaiwa, F., Kato, A., Tohdoh, N., Shimada, H., and Sugiura, M. (1986). The complete nucleotide sequence of the tobacco chloroplast genome: its gene organization and expression. *EMBO J.* **5**, 2043–2049.

Shoham, G., Michaeli, D., and Nechushtai, R. (1990). Photosystem I reaction center of *Mastigocladus laminosus*: Structural and functional aspects. *In* "Current Research in Photosynthesis" (M. Baltscheffsky, ed.), Vol. II, pp. 555–562. Kluwer, Dordrecht, The Netherlands.

Smith, A. G., Wilson, R. M., Kaethner, T. M., Willey, D. L., and Gray, J. C. (1991). Pea chloroplast genes encoding a 4-kDa polypeptide of photosystem I and a putative enzyme of $C_1$ metabolism. *Curr. Genet.* **19**, 403–410.

Steppuhn, J., Hermans, J., Nechushtai, R., Ljungberg, U., Thümmler, F., Lottspeich, F., and Herrmann, R. G. (1988). Nucleotide sequence of cDNA clones encoding the entire precursor polypeptides for subunits IV and V of the photosystem I reaction center from spinach. *FEBS Lett.* **237**, 218–224.

Steppuhn, J., Hermans, J., Nechustai, R., Herrmann, G. S., and Herrmann, R. G. (1989). Nucleotide sequences of cDNA clones encoding the entire precursor polypeptide for subunit VI and of the plastome-encoded gene for subunit VII of the photosystem I reaction center from spinach. *Curr. Genet.* **16**, 99–108.

Takahashi, Y., Mitsui, A., and Matsubara, H. (1991a). Roles of ATP and NADPH in formation of the Fe–S cluster of spinach ferredoxin. *Plant Physiol.* **95**, 104–110.

Takahashi, Y., Mitsui, A., Fujita, Y., and Matsubara, H. (1991b). Formation of the Fe–S cluster of ferredoxin in lysed spinach chloroplasts. *Plant Physiol* **95**, 97,103.

Takahashi, Y., Goldschmidt-Clermont, M., Soen, S.-Y., Franzén, L. G., and Rochaix, J.-D. (1991c). Directed chloroplast transformation in *Chlamydomonas reinhardtii:* Insertional inactivation of the *psa*C gene encoding the iron sulfur protein destabilizes photosystem I. *EMBO J.* **10**, 2033–2040.

Vainstein, A., Peterson, C. C., and Thornber, J. P. (1989). Light-harvesting pigment–proteins of photosystem I in maize. *J. Biol. Chem.* **264**, 4058–4063.

Webber, A. N., and Malkin, R. (1990). Photosystem I reaction-centre proteins contain leucine zipper motifs—A proposed role in dimer formation. *FEBS Lett.* **264**, 1–4.

Witt, I, Witt, H. T, Gerken, S., Saenger, W., Dekker, J. P., and Rögner, M. (1987). Crystallization of reaction center I of photosynthesis. *FEBS Lett.* **221**, 260–264.

Witt, I., Witt, H. T., Fiore, D. D., Rögner, M., Hinrichs, W., Saenger, W., Granzin, J., Betzel, C., and Dauter, Z. (1988). X-Ray characterization of single crystals of the reaction center I of water splitting photosynthesis. *Ber. Bunsenges. Phys. Chem.* **92**, 1503–1506.

Witt, H. T., Rögner, M., Mühlenhoff, U., Witt, I., Hinrichs, W., Saenger, W., Betzel, C., Dauter, Z., and Boekema, E. J. (1990). On isolated complexes of reaction center I and X-ray characterization of single crystals. *In* "Current Research in Photosynthesis" (M. Baltscheffsky, ed.), Vol. II, pp. 547–554. Kluwer, Dordrecht, The Netherlands.

Wynn, R. M., and Malkin, R. (1988a). Characterization of an isolated chloroplast membrane Fe-S protein and its identification as the photosystem I Fe-$S_A$/Fe-$S_B$ binding protein. *FEBS Lett.* **229**, 293–297.

Wynn, R. M, and Malkin, R. (1988b). Interaction of plastocyanin with photosystem I: A chemical cross-linking study of the polypeptide that binds plastocyanin. *Biochemistry* **27**, 5863–5869.

Wynn, R. M., and Malkin, R. (1990). The photosystem I 5.5-kDa subunit (the *psa*K gene product). *FEBS Lett.* **262**, 45–48.

Zanetti, G., and Merati, G. (1987). Interaction between photosystem I and ferredoxin. Identification by chemical cross-linking of the polypeptide which binds ferredoxin. *Eur. J. Biochem.* **169**, 143–146.

Zhao, J., Warren, P. V., Bryant, D. A., and Golbeck, J. H. (1990). Reconstitution of electron transport in photosystem I with PsaC and PsaD proteins expressed in *Escherichia coli*. *FEBS Lett.* **276**, 175–180.

Zilber, A. L., and Malkin, R. (1988). Ferredoxin cross-links to a 22-kD subunit of photosystem I. *Plant Physiol.* **88**, 810–814.

Zilber, A. L., Wynn, R. M., Webber, A., and Malkin, R. (1990). Organization of PS I subunits in thylakoid membranes. *In* "Current Research in Photosynthesis" (M. Baltscheffsky, ed.), Vol. II, pp. 575–578. Kluwer, Dordrecht, The Netherlands.

# 13

# Function and Organization of Photosystem II

## HANS-ERIK ÅKERLUND

Department of Plant Biochemistry
University of Lund
S-220 07 Lund, Sweden

I. Introduction
II. Function of Photosystem II
    A. Primary Events
    B. Acceptor Side
    C. Donor Side
III. Structural Organization of Photosystem II
    A. Central Core/Reaction Center
    B. Antenna System
    C. Extrinsic Proteins
    D. Miscellaneous Protein Components
IV. Photoinhibition and Turnover of Photosystem II Components
V. Dynamic Changes and Photosystem II Heterogeneity
    A. Phosphorylation
    B. Heat
    C. Heterogeneity
VI. Concluding Remarks
    References

## I. INTRODUCTION

Photosynthetic oxygen evolution basically is a result of light-driven electron transfer from water to plastoquinone. This seemingly simple reaction is catalyzed by photosystem II, a multiprotein complex of fascinating composition. This supracomplex is embedded in the thylakoid membrane and consists of both membrane-spanning proteins (intrinsic) and surface-bound proteins (extrinsic). The complex in higher plants consists of some 20 different proteins and at least 7 different prosthetic groups. It has an estimated molecular mass in the range of 1.0–1.5 MDa, of which 20–30% is constituted by chlorophyll. The complex is chimeric with respect to genetic origin, that is, some of the proteins in the complex are encoded by the chloroplast genome whereas others are encoded by the nuclear genome.

Studies in the fields of molecular biology, biophysics, and biochemistry have provided a dramatic increase in detailed information about the system. Biochemical work has provided us with a number of photosystem II preparations with different levels of structural organization, as well as with pure protein preparations. Molecular biology has provided detailed information on individual protein components. Thus, most of the genes and their protein products have been identified and sequenced (see also Chapters 8 and 9). Biophysics has provided kinetic information and, thus, information on electron transfer events. The work that has made the largest contribution in recent years is undoubtedly the structural visualization by X-ray crystallography of the reaction center of nonsulfur purple bacteria (Michel and Deisenhofer, 1988). The current concept of photosystem II structure and function is based on a fruitful combination of results from these different areas of research. This perspective relies heavily on results obtained from work on photosynthetic microorganisms such as purple bacteria, cyanobacteria, and the green alga *Chlamydomonas*.

This chapter mainly discusses the higher plant photosystem II, and addresses the following questions. What are the main electron transfer events? What are the functional molecules involved? Which proteins are present, how are they arranged, and what is known about their particular function? A number of related reviews on different aspects of photosystem II are recommended (Murphy, 1986; Andersson and Åkerlund, 1987; Glazer and Melis, 1987; Chitnis and Thornber, 1988; Green, 1988; Bassi *et al.*, 1990; Ghanotakis and Yocum, 1990; Hansson and Wydrzynski, 1990).

## II. FUNCTION OF PHOTOSYSTEM II

### A. Primary Events

The absorption of light by the antenna pigments of photosystem II creates excited pigments, mainly chlorophyll *a*, *b*, and carotenoids. The energy of these excited pigments then migrates within the complex array of pigments toward the photosystem II reaction center. This exciton moves from carotenoids and chlorophyll *b* to chlorophyll *a*. The movement is in the direction of decreasing energy as the exciton approaches the reaction center. This organization provides a trap for the exciton. The efficiency of this energy transfer is higher than 90% (van Grondell, 1985). A special pair of chlorophyll *a* molecules, called P680, is located in the reaction center (Diner, 1986). This dimer seems to be oriented parallel to the membrane in photosystem II (Rutherford, 1985), which differs from the perpendicular orientation found in the analogous purple bacteria. When the exciton reaches the reaction center, the special dimer becomes excited; P680 becomes P680*. P680* is a strong reducing agent and rapidly (within a few psec) donates an electron to a pheophytin molecule (Fig. 1). This electron transfer may involve a chlorophyll *a* molecule located between the special pair of chlorophyll *a* molecules (P680) and pheophytin (Rutherford, 1981). The primary charge separation generates pheophytin$^-$ and P680$^+$. P680$^+$ is actually the strongest oxidant generated in living systems. Al-

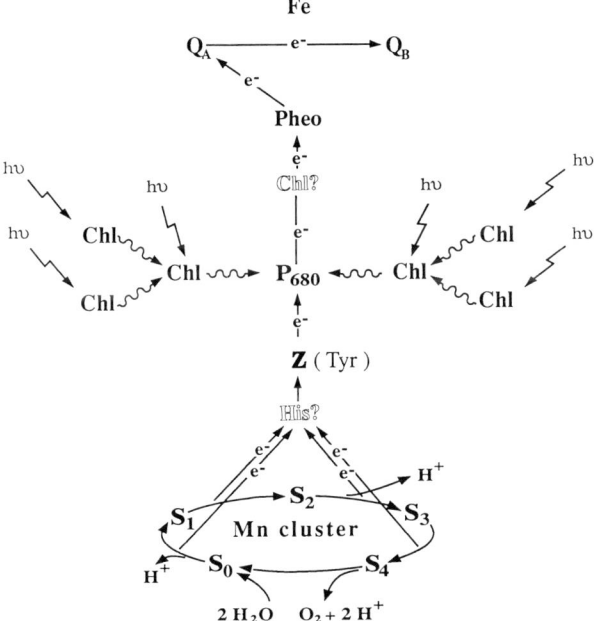

**Fig. 1.** Schematic diagram of electron transfer events and components involved in the reactions of photosystem II. Courtesy of P. O. Arvidsson.

ready at this stage, light energy has been converted into chemical energy. The subsequent steps are "just" for stabilization.

## B. Acceptor Side

The electron on the pheophytin is transferred within a few hundred psec to a permanently bound quinone at the $Q_A$ site, which is reduced to the semiquinone form. The reduction at $Q_A$ has been shown to be electrogenic, which means that this quinone is located on the side of the membrane opposite to pheophytin (Trissl et al., 1987). Although pheophytin is the primary electron acceptor in this system, it is often referred to as the intermediate electron acceptor because of its short lifetime. The $Q_A$ quinone is, for the same reason, called the primary stable acceptor. The electron at $Q_A$ is transferred to a transiently bound plastoquinone molecule at the $Q_B$ site. This plastoquinone molecule is reduced in two successive photoreactions. In the first, $Q_B$ is reduced to a semiquinone anion without uptake of protons from the bulk medium. In the second photoreaction, the reduced quinone at $Q_A$ donates a second electron so the plastoquinone at $Q_B$ is doubly reduced. At this stage, two protons are taken up directly or indirectly from the external medium; plastoquinol is formed. Plastoquinone at the $Q_B$ site is bound only weakly and exchanges with the plastoquinone pool when in the fully oxidized quinone form or in the fully reduced quinol form but is bound strongly and stably when in the semiquinone form (Velthuys, 1981). Herbicides such as 3-(3,4-dichlorophenyl)-1,1-dimethylurea (DCMU) and atrazin act at this $Q_B$ site (Renger, 1986). Bicarbonate has been known for a long time to be an essential cofactor for photosystem II function, and is believed to influence the electron flow between $Q_A$ and $Q_B$.

A nonheme iron is associated closely with the $Q_A$ and $Q_B$ sites and interacts with the quinones, but does not participate directly in the electron transfer reaction. The function of this iron is not well understood. The acceptor sides of photosystem II and purple bacteria appear to be very similar, that is, the iron is positioned between the two quinones (Rutherford, 1987). The major difference seems to be that the fifth ligand to iron is not conserved between purple bacteria and photosystem II. In purple bacteria, a glutamyl residue functions as the fifth ligand. An interesting hypothesis is that bicarbonate functions as the fifth ligand in photosystem II (Michel and Deisenhofer, 1988). This function could explain many of the observed effects of bicarbonate on photosystem II. Another possible role of bicarbonate is protonation of the doubly reduced plastoquinone at the $Q_B$ site (Blubaugh and Govindjee, 1988).

## C. Donor Side

The $P680^+$ generated in the primary charge separation is reduced by a donor, Z, within 50–250 nsec (Gerken et al., 1988; Rutherford, 1989). For some time, this Z-donor was thought to be a specially bound quinone. However, some studies (Barry and Babcock, 1987; Debus et al., 1988a,b; Vermaas et al., 1988), including site-directed mutagenesis, provide strong evidence that this donor is a tyrosine residue on the D1 reaction center protein (see subsequent text). The primary reaction of the reaction center is a one-electron process, whereas the water oxidation is a four-electron process. To couple these two processes, photosystem II has developed a charge accumulation device. The presence of such a device was demonstrated elegantly by $O_2$ flash yield measurements (Joliot et al., 1969). The observed damped oscillation with a period of four was described in terms of an S-state cycle. The index on S represents the number of electrons that have been extracted (Kok et al., 1970). When the state $S_4$ has been reached, oxygen is released and the system returns to $S_0$. The physical counterpart of the S states is the different oxidation states of the manganese cluster (Amesz, 1983; Babcock, 1987). The manganese cluster contains four manganese ions that are involved directly or indirectly in the accumulation of charges. During the charge accumulation process, protons are released (Förster and Junge, 1985) and manganese is oxidized, most likely from III to IV in four successive steps. Finally, oxygen is released. The details of charge accumulation and the mechanism of water oxidation are still under considerable debate.

Several studies suggest that the four manganese ions are not equal but are divided into at least two subpools. The electron flow from the manganese cluster to $P680^+$ has been suggested to involve not only the Z-tyrosine but also a histidine (Kambara and Govindjee, 1985; Boussac et al., 1989; Tamura et al., 1989; Boussac et al., 1990; Ono and Inoue, 1991). This histidine is thought to provide a ligand to manganese also. Chloride has been known for a long time to be an essential cofactor in the oxygen evolving process (Critchley, 1985; Coleman and Govindjee, 1987; Homann, 1987). Its specific function is not yet established, but may involve charge neutralization for stabilization of protein structure or for stabilization of higher oxidation states at the manganese cluster. Further, the requirement for chloride is not strict. Thus, bromide and, to a lesser extent, nitrate, iodine, formate, and bicarbonate ions can be used instead of chloride. Calcium is also an essential cofactor of photosystem II and appears to be required specifically for the electron flow from Z to $P680^+$ (Boussac et al., 1985; Satoh and Katoh, 1985; Ono and Inoue, 1986; Homann, 1987). The nature and site of calcium binding is not yet known.

Further aspects of the $Ca^{2+}$ and $Cl^-$ cofactors will be discussed in relation to the extrinsic proteins of photosystem II.

### III. STRUCTURAL ORGANIZATION OF PHOTOSYSTEM II

The current concept of photosystem II organization relies, to a great extent, on results from the biochemical isolation of protein complexes that retain different photosynthetic activities. Basically, two methods of obtaining pure and highly active preparations of intact photosystem II are available. One is based on mechanical fragmentation of thylakoids followed by aqueous two-phase fractionation (Åkerlund and Andersson, 1983; Andersson et al., 1985; Svensson and Albertsson, 1989). In this type of preparation, photosystem II is still retained in the membrane. Moreover, these photosystem II membranes are shown to be turned inside-out with respect to the original sidedness of the membrane. Generally these preparations contain small amounts of photosystem I and cytochrome $b/f$ complex. The other type of pure and active photosystem II preparation is obtained by detergent solubilization followed by centrifugal fractionation. The original and most applied method is that of Berthold et al. (1981). This method was followed by a number of modified procedures that have been compared with respect to preparation details and properties of obtained fractions (Dunahay et al., 1984). All the preparations completely lack photosystem I and ATP synthase but contain varying amounts of the cytochrome $b/f$ complex. The photosystem II material isolated by the mechanical fragmentation procedure or by the detergent procedure contains intact photosystem II with at least 20 different proteins.

### A. Central Core/Reaction Center

Further treatment of photosystem II particles with detergent yields a number of particle preparations consisting of the central core of photosystem II (Ikeuchi et al., 1985; Tang and Satoh, 1985; Franzén et al., 1986; Ghanotakis et al., 1987). These photosystem II core particles show a drastic reduction in the amount of antenna proteins, especially the chlorophyll $b$-containing ones. They also show variable reduction in the amounts of other proteins. All contain chlorophyll $a$ proteins, CP47 and CP43, two reaction center proteins of 34 (D1) and 32 (D2) kDa, and cytochrome $b$-559, as well as an extrinsic 33-kDa protein. These core complexes appear to represent the structural minimum required for function of oxygen evolution. In addition, several low molecular weight proteins have been found to be associated with the complex (Ljungberg et al., 1986a; Ikeuchi and Inoue, 1988).

Several groups have been able to isolate reaction centers (RCII) of photosystem II (Barber et al., 1987; Nanba and Satoh, 1987; Ghanotakis et al., 1989; Fotinou and Ghanotakis, 1990; Satoh and Nakane, 1990). The principal protein components of this complex are the D1 and D2 proteins, cytochrome b-559 and a 4.8-kDa protein. All these proteins are encoded by the chloroplast genome. All cofactors needed for the primary photoreactions except the quinones appear to be present in these preparations. Thus, the presence of a light-induced spectral change due to pheophytin reduction and the appearance of a spin-polarized triplet at cryogenic temperature established a functional P680 (Takahashi et al., 1987). Seibert et al. (1988) and Wasielewski et al. (1989) measured the kinetics of charge separation. Gounaris et al. (1988) succeeded in restoring the presence of the primary stable acceptor $Q_A$. Mathis et al. (1989) obtained kinetic evidence for the function of Z in isolated photosystem II reaction centers. The presence of P680 and Z was demonstrated also by Nugent et al. (1989). The D1 and D2 proteins generally are believed to carry these functional groups. These results completely rule out CP43 and CP47, which once were thought to harbor the reaction center. Now they are given roles as inner chlorophyll a antennae close to the reaction center.

The first clue to the identification of D1/D2 as the reaction center came from sequence comparisons between the D1/D2 proteins of photosystem II and the L/M proteins of the reaction center of purple bacteria. (Deisenhofer et al., 1985; Michel et al., 1986; Trebst, 1986; Michel and Deisenhofer, 1988). Although a low overall homology was found, specific regions of the proteins displayed higher homology. Among these were the quinone and chlorophyll binding sites. Based on the results described earlier, and in analogy with the reaction center of the purple bacteria (Deisenhofer et al., 1985), the following hypothesis was presented. The D1 and D2 proteins form an intertwined heterodimeric complex. In this complex, P680 is located between the D1 and D2 subunits. The donor Z-tyrosine and the acceptor pheophytin, as well as the $Q_B$ site, all are located in the D1 protein. This complex is called the active branch. The D2 protein carries the $Q_A$ site but also contains a donor, D, and a pheophytin. These latter components, however, are not involved directly in the primary charge separation and constitute the "dead" branch. The donor D is thought to be a tyrosine residue also, located in the D2 protein (Barry and Babcock, 1987; Debus et al., 1988a,b; Vermaas et al., 1988). The role of this donor is not totally clear. However, under some conditions it can donate an electron to P680$^+$. In the oxidized form, it can extract an electron from the manganese cluster slowly in the dark, converting it from state $S_0$ to state $S_1$ (Vermaas et al., 1984; Styring and Rutherford, 1987; Vass and Styring, 1991).

The D1 and D2 proteins are hydrophobic membrane proteins. Based on the primary sequences of the proteins (Trebst, 1986; Michel and Deisenhofer, 1988) and on the use of epitope-specific antibodies on right-side-out and inside-out thylakoids, both proteins show five membrane-spanning helices with the N terminus on the stromal side (Sayre et al., 1986). The D- and Z-tyrosine residues are located close to the lumenal side of the membrane. Amino acids that are thought to be involved in plastoquinone binding and herbicide binding or resistance are located at or close to the stromal side.

Although the reaction center of the purple bacteria and photosystem II show many similarities, especially in the primary charge separation and the reactions on the acceptor side, a number of distinct differences exist. Photosystem II seems to lack the H subunit but has three additional polypeptides, the small and large subunits of cytochrome *b*-559 and the 4.8-kDa product of the *psbI* gene (Nanba and Satoh, 1987; Webber et al., 1989). Photosystem II also appears to have a large number of chlorophyll *a* molecules in the reaction center (Dekker et al., 1989; Kobayashi et al., 1990; see also the comparison made by Rutherford, 1987).

Cytochrome *b*-559 is the most puzzling component of photosystem II. As a redox component, it is expected to be involved in electron transfer reactions. However, the kinetics of the redox changes seen are far too slow to be explained by simple models. Despite this conflict, it is obvious that the cytochrome is of central importance. Thus all organisms that are capable of photosynthetic oxygen evolution, without exceptions, have this cytochrome. The presence of cytochrome *b*-559 in reaction center preparations also suggests a close physical association. The number of cytochrome *b*-559 molecules per photosystem II appears to be two (Murata et al., 1984; Whitmarsh and Ort, 1984; Yamada et al., 1987; Shuvalov et al., 1989) but others suggest it to be one (Ghanotakis et al., 1984a; Miyazaki et al., 1989; Gounaris et al., 1990). Some heterogeneity within cytochrome *b*-559 has been reported also (Barabás and Garab, 1989; Shuvalov et al., 1989). The two subunits of 9 and 4 kDa have a single histidine residue that is suggested to be involved in heme binding (Babcock et al., 1985). Although cytochrome *b*-559 is a small protein, it is still membrane spanning (Tae et al., 1988; Vallon et al., 1989). For further information on the structure and possible function of cytochrome *b*-559, see Cramer et al. (1986).

A fundamental difference between photosystem II and the reaction center of purple bacteria is that the center of purple bacteria does not evolve oxygen. Photosystem II contains a manganese cluster that is required for oxygen evolution. The precise location of the manganese is still not clear, since a functionally active manganese-binding protein has not yet been identified. Almost all the proteins present in photosystem II core particles have been suggested to be the manganese-binding one. However, at present, most attention is focused on the D1/D2 proteins. These proteins have been suggested to be manganese binding (Coleman and Govindjee, 1987; Taylor *et al.*, 1988), a theory that is supported experimentally by mutational studies (Seibert *et al.*, 1989) and photoinhibition studies (Virgin *et al.*, 1988). [See also the interesting modeling work by Svensson *et al.* (1990).] The D1/D2 proteins in the isolated reaction center particles do not bind manganese whereas core particles do bind manganese. Thus removal of CP47/CP43 and the extrinsic 33-kDa protein causes a destabilization of manganese. Therefore, these proteins may be involved in the binding of manganese or may improve the stability of the binding indirectly.

### B. Antenna System

The components of the antenna system of photosystem II can be divided into three groups: (1) the inner chlorophyll *a* antennae CP43 and CP47, (2) the major chlorophyll *a/b*-binding complex LHCII, and (3) the minor chlorophyll *a/b*-binding complexes CP29, CP26, and CP24. These complexes have been given different names by different authors; synonyms are found in the review by Green (1988).

CP43 and CP47 are chloroplast encoded (*psbC* and *B*, respectively) and are exclusively chlorophyll *a* binding (20–25 each). The proteins are associated closely with the reaction center (Green, 1988). CP43 and CP47, in addition to harvesting light, have been suggested to act as anchor proteins for the remaining photosystem II proteins (Callahan *et al.*, 1989). LHCII constitutes 30% of the thylakoid protein and binds about 50% of the chlorophyll found in the thylakoids, with 10–15 chlorophylls per protein and a chlorophyll *a/b* ratio of 1–1.2 (Chitnis and Thornber, 1988). LHCII is very heterogeneous. Thus, SDS gel electrophoresis reveals two apoproteins of 25 and 27 kDa. At least five slightly different (three 27-kDa and two 25-kDa) apoproteins have been found by a combination of isoelectric focusing and SDS gel electrophoresis (Spangfort and Andersson, 1989). This result is in agreement with the finding that the LHCII proteins are encoded by a whole family of nuclear genes (Chitnis and Thornber, 1988).

The minor chlorophyll *a/b*-binding protein complexes CP29, CP26, and CP24 bind about 5–10% of the chlorophyll in photosystem II each. All appear to be encoded by nuclear genes. CP24 was described first by

Dunahay and Staehelin (1986). It has a chlorophyll $a/b$ ratio of 1–1.6 and consists of an apoprotein of 20 kDa that has been N-terminally sequenced (Morishige et al., 1990). CP26 was described first by Bassi et al. (1987) and Dunahay et al. (1987). It has a chlorophyll a/b ratio of 2.0–2.2 and appears to have two apoproteins of 28 and 29 kDa. CP29 was described first by Machold and Meister (1979) and later by Camm and Green (1980). The reported properties of this complex have been quite variable. Thus, the chlorophyll $a/b$ ratio has been reported to be 1.8–4.0 and the complex has been suggested to consist of either one or two apoproteins with molecular masses of 22–31 kDa. CP29 was purified by a nondenaturing procedure and characterized (Henrysson et al., 1989). The isolated complex contained 10–12 chlorophylls per protein with a chlorophyll $a/b$ ratio of 3.0–3.2. When analyzed by denaturing gel electrophoresis, it was shown to have only one apoprotein with a molecular mass of 29–31.5 kDa. A partial amino acid sequence of the protein was obtained also. Pichersky and Green (1990) reported a tomato gene sequence that showed specific homologies with known chlorophyll $a/b$-binding proteins. It is not yet clear to which protein this sequence corresponds. It may correspond to the CP26 of Bassi et al. (1987) or to a hitherto unknown chlorophyll $a/b$-binding protein. However, it does not correspond to CP29 since the partial amino acid sequence of the isolated protein is different (Henrysson et al., 1989). Thus, to avoid confusion, the gene isolated by Pichersky and Green (1990) should not be called the CP29 gene. The role of the minor chlorophyll $a/b$ complexes, apart from light harvesting, is not yet established. However, these proteins should not be neglected simply because they are minor relative to LHCII. Actually, they are as abundant as some of the subforms of LHCII.

A model for the organization of the chlorophyll protein complexes of photosystem II is presented in Fig. 2. This model is based primarily on subfractionation results. The most systematic studies have been done by Bassi and co-workers (Bassi and Dainese, 1990, and references therein; Dainese and Bassi, 1991). From these and other observations, it is clear that CP47 and CP43 are the complexes closest to the reaction center; CP47 interacts directly with the reaction center. The minor complexes are found in the next layer, where CP29 and CP26 appear to be closer to the reaction center than CP24. The most distant antenna is LHCII, which can be divided functionally into an inner and a peripheral LHCII. The inner portion is associated tightly with photosystem II whereas the peripheral one can detach from photosystem II on phosphorylation (see Section V,A). The inner pool contains the 27-kDa apoprotein of LHCII exclusively whereas the peripheral pool contains both the 25- and 27-kDa apoproteins (Spangfort and Andersson, 1989). Further, a trimer organization of LHCII

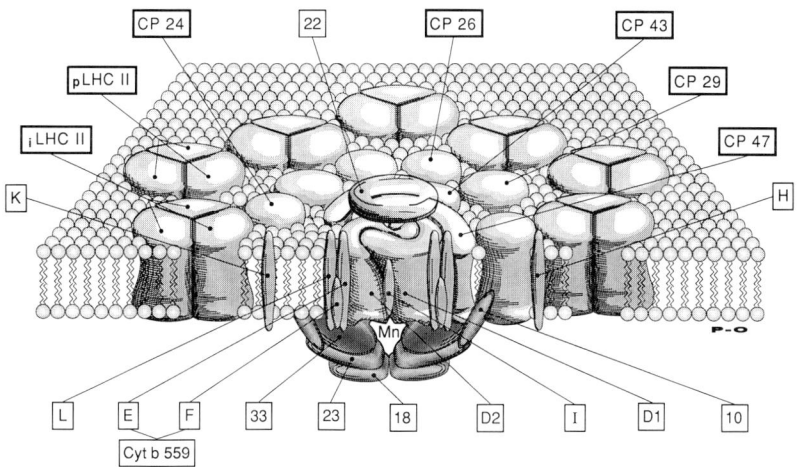

**Fig. 2.** Model of structural organization of protein components of photosystem II. Single letters represent the products of corresponding chloroplast genes. i, inner; p, peripheral. Boldface boxes represent chlorophyll–protein complexes. Courtesy of P. O. Arvidsson.

has been suggested based on mild electrophoresis and crystallization work (Kühlbrandt, 1984). Dainese and Bassi (1991) obtained evidence from cross-linking experiments that the three minor complexes, CP24, CP26, and CP29, might be trimers. As a consequence, LHCII should be organized as a cluster of nine proteins. The organization of the complexes depicted earlier coincides with the observed chlorophyll $b$ content of the complexes and the direction of exciton movement (see Section II). Thus, the chlorophyll $b$ content decreases in the order of LHCII > CP24 > CP26 > CP29 > CP47/CP43.

## C. Extrinsic Proteins

At least four proteins with molecular masses of 10, 18, 23, and 33 kDa can be released specifically from photosystem II particles (Åkerlund and Jansson, 1981; Åkerlund et al., 1982; Murata and Miyao, 1985; Ljungberg et al., 1986b; Andersson and Åkerlund, 1987), with concomitant impairment of oxygen evolution. Thus treatment with $1 M$ NaCl releases the 18- and 23-kDa proteins, with $1 M$ $CaCl_2$ releases the 18-, 23-, and 33-kDa proteins, and with $1 M$ NaCl/0.06% Triton X-100 releases the 18-, 23-, and 10-kDa proteins and a large part of a 22-kDa protein (see Section III,D). All have been purified and characterized. These proteins lack prosthetic groups and, thus, are not likely to be involved directly in either light capture or electron transfer reactions. All have been shown to be located on the lumenal side of the membrane. The 18-, 23-, and 33-kDa proteins are found in equal amounts, but whether there are one or two copies per photosystem II unit is still in question. The model in Fig. 2 assumes the presence of two copies. The 18-, 23-, and 33-kDa proteins are encoded by nuclear genes and synthesized in the cytoplasm with N-terminal targeting sequences that are removed by a membrane-bound peptidase on transfer to the thylakoid lumen (Westhoff *et al.*, 1985; Jansen *et al.*, 1987; Tyagi *et al.*, 1987). Also, the 10-kDa protein appears to be nucleus encoded.

Direct reconstitution experiments have established firmly the involvement of the 18-, 23-, and 33-kDa proteins in photosystem II function. Both the 18- and the 23-kDa proteins are thought to be essential for high affinity binding of the $Cl^-$ and $Ca^{2+}$ required in photosystem II. At high $CaCl_2$ concentrations, oxygen evolution can proceed without the 18- and 23-kDa proteins (Ghanotakis *et al.*, 1984b; Miyao and Murata, 1985), implying that the binding sites for $Cl^-$ and $Ca^{2+}$ are not on the 18- and 23-kDa proteins, but that the binding of these proteins causes a change in another photosystem II component that, in turn, facilitates binding of $Cl^-$ and $Ca^{2+}$. The role of the 18-kDa protein can be seen only at very low $Cl^-$ concentrations and requires the presence of the 23-kDa protein (Imaoka *et al.*, 1984). The binding is electrostatic and requires the N-terminal part (9 kDa) of the 18-kDa protein (Kuwabara *et al.*, 1986). The involvement of the 33-kDa protein also has been demonstrated by reconstitution experiments (Imaoka *et al.*, 1984; Ono and Inoue, 1984; Kuwabara *et al.*, 1985). The 33-kDa protein apparently is bound to components of the photosystem II core, since its presence protects against added reactants (Tae and Cramer, 1989; Hansson and Wydrzynski, 1990). It may even bind directly to the reaction center proteins D1 and D2 (Mei *et al.*, 1989). The 33-kDa protein once was thought to bind manganese. However, this is obviously not true since the protein can be released with manganese still bound to photosystem II. Still, the 33-kDa protein appears to be required for stabilization of manganese binding. Thus, when the 33-kDa protein has been released, manganese binding becomes unstable and manganese gradually releases from the sites in photosystem II. Interestingly, initial removal of manganese results in destabilization of the binding of the 33-kDa protein. These results suggest that the 33-kDa protein provides at least one ligand to manganese (Andersson and Åkerlund, 1987). A 15-kDa fragment of the 33-kDa protein was still active (Völker *et al.*, 1985) whereas removal of only 16 amino acids from the N-terminal part of the protein prohibited rebinding of the protein to photosystem II (Eaton-Rye and Murata, 1989). Two conserved cysteines in the 33-kDa protein have been shown to be important for binding and activation of oxygen evolution (Tanaka and Wada, 1988). Further, the 33-kDa protein is required for binding of the 23-kDa protein and, thus, indirectly for binding of the 18-kDa protein (Andersson *et al.*, 1984). Detailed quantification of the binding properties of the 18-, 23-, and 33-kDa proteins has been performed (Miyao and Murata, 1989).

Several lines of evidence suggest that the 18-, 23-, and 33-kDa proteins are associated closely on the lumenal side of the membrane and form a cavity, or a regulatory cap, around the water-splitting site (Fig. 2). Thus, removal of the 18- and 23-kDa proteins exposes the 33-kDa protein and allows reductants to reach the manganese, which becomes destabilized and rapidly equilibrates with the surrounding medium (Andersson and Åkerlund, 1987).

Electron microscopy studies show that the 18-, 23-, and 33-kDa proteins form tetrameric structures protruding from the membrane surface on the lumenal side of appressed membranes (Simpson and Andersson, 1986; Seibert *et al.*, 1987; Bassi *et al.*, 1989).

The 10-kDa extrinsic protein has been shown by immunoprecipitation to be associated closely with the 23- and 33-kDa proteins (Ljungberg *et al.*, 1984b). The isolated protein shows a relatively high content of hydrophobic amino acids and requires detergents for solubility (Ljungberg *et al.*, 1986b). However, it can be released from inside-out thylakoids by a Tris wash in the absence of detergents (Ljungberg *et al.*, 1984a). The protein is distinct from both cytochrome *b*-559 and the 10-kDa phosphoprotein (Farchaus and Dilley, 1986). The 10-kDa protein found in wheat has been shown to correspond to the derived amino acid sequence of a light-inducible gene ST-LS1 from potato (Eckes *et al.*, 1986). The role of the 10-kDa protein is not clear, but it appears to be required for binding of the 23-kDa protein to photosystem II. Thus, rebinding of the 23-kDa protein and reconstitution of oxygen evolution was prohibited when the 10-kDa protein was removed. However, high concentrations of $CaCl_2$ in the absence of both the 23- and 10-kDa protein still could restore the activity.

## D. Miscellaneous Protein Components

Isolated photosystem II particles contain several protein components of yet unknown function. All may not have a direct function in photosystem II, but it is important to be aware of their presence.

A 22-kDa protein was indicated by immunological nearest-neighbor analysis to interact with the extrinsic 23- and 33-kDa proteins (Ljungberg *et al.*,

1984b). The protein has been purified by ion-exchange chromatography in the presence of detergents (Ljungberg et al., 1986b). No cofactor was found in the isolated protein. The protein can be partly released from photosystem II by treatments with high salt concentrations in the presence of small amounts of detergent. The 22-kDa protein is found in varying amounts in photosystem core particles. Interestingly, the presence of the 22-kDa protein appears to correlate with photosystem II ability to use quinone acceptors more efficiently than ferricyanide (Ghanotakis and Yocum, 1986; Henrysson et al., 1987). This result suggests a function for the 22-kDa protein on the acceptor side of photosystem II as well as a location on top of the D1/D2 complex (Fig. 2), but with a membrane-spanning part that interacts with the 23- and 33-kDa proteins.

Photosystem II has been reported to contain at least one copper ion. It is not yet clear to which protein this copper is bound. Droppa et al. (1987) have shown that copper deficiency results in the disappearance of a 29-kDa protein, whereas Sibbald and Green (1987) found copper to follow LHCII on purification. However, the number of LHCII molecules per photosystem II is much greater than the number of copper atoms, suggesting that copper is only bound to a subform to LHCII or that another protein copurifies with LHCII. Purification results from our group suggest that the copper is bound to a 28-kDa protein (Arvidsson et al., 1992). The 60% pure protein retains 70–80% of the copper found in photosystem II. It appears to be distinct from CP29 and LHCII,but may correspond to the 28-kDa subunit of CP26. Sequencing experiments are now in progress to distinguish these possibilities clearly.

Several low molecular weight proteins are associated with photosystem II also (Ljungberg et al., 1986a; Schröder et al., 1988; Ikeuchi et al., 1989). In addition to the 10-kDa extrinsic protein and the 9- and 4-kDa subunits of cytochrome $b$-559, there are a 10-kDa phosphoprotein and about eight other proteins with molecular masses below 10 kDa. The 10-kDa phosphoprotein is a product of the chloroplast gene *psbH* (Farchaus and Dilley, 1986; Dedner et al., 1988). Although its physiological function is still uncertain, it is the protein that is phosphorylated most strongly on overexcitation of photosystem II. The 4.8-kDa protein product of the *psbI* gene copurifies with the reaction center (see Section III A). A 5-kDa product of the chloroplast *psbL* gene was found by Ikeuchi et al. (1989). Interestingly, the *psbL* gene is located just downstream of *psbE and psbF* genes for cytochrome $b$-559 and is assumed to be cotranscribed with them. All three products are present in core particles, but whether the 5-kDa *psbL* protein is a subunit of cytochrome $b$-559 remains to be established.

## IV. PHOTOINHIBITION AND TURNOVER OF PHOTOSYSTEM II COMPONENTS

Photoinhibition is a problem commonly encountered by plants and involves a decrease of photosynthetic activity induced by high light intensity. The idea of photoinhibition and possible mechanisms for it can be quite different based on whether whole plants or isolated photosynthetic complexes are discussed. However, one major cause for photoinhibition is the damage of the reaction center of photosystem II (Cleland, 1988). The molecular mechanism of this photoinhibition is still controversial. The first step of photoinhibition has been suggested to be damage at the $Q_B$ (Kirilovsky et al., 1988) or $Q_A$ site (Styring et al., 1990) but also at the donor side (Callahan et al., 1986; Jegerschöld et al., 1990). Subsequent to this initial step is a degradation of the reaction center protein D1 (Arntz and Trebst, 1986; Virgin et al., 1988). The degree of photoinhibition can be influenced by several factors such as chilling (Öquist et al., 1987), the presence of oxygen (Nedbal et al., 1990; Ohad et al., 1990), or bicarbonate (Sundby et al., 1989). Whether these effectors increase or decrease photoinhibition is highly controversial (e.g., Jegerschöld and Styring, 1991; Kirilovsky and Étienne, 1991). This controversy may be alleviated by the finding that photoinhibition may not always cause a degradation of the D1 protein and that activity may recover without synthesis of new D1 protein (Hundal et al., 1990a). In living cells, the degraded D1 protein is replaced rapidly with newly synthesized protein. Actually, the D1 protein is the protein in photosystem II with the highest turnover rate (Ellis, 1981; Mattoo et al., 1989). This rapid turnover of the D1 protein alone is somewhat difficult to envision since it is located in the heart of photosystem II, intertwined with the D2 protein, and surrounded by a large number of other proteins (Fig. 2). However, some results suggest that the degradation of the D1 protein results in a total disassembly of the photosystem II complex (Hundal et al., 1990b), which also suggests that the D1 protein controls the assembly of photosystem II proteins.

## V. DYNAMIC CHANGES AND PHOTOSYSTEM II HETEROGENEITY

### A. Phosphorylation

The photosynthetic apparatus is not a fixed system but has an excellent ability to adapt to changes in light conditions and changes in demands within the plant. Long-term adaptations involve changes in the amount of photosystem II relative to photosystem I as well as changes in the size of the light-harvesting antenna of each photosystem (Anderson and Andersson, 1988). Short-term changes, within minutes or seconds, involve the reorganization of protein complexes within the thylakoid membrane as well as the reorganization of whole chloroplasts within the cell. The most studied short-term regulation involves reversible phosphorylation of thylakoid membrane proteins. Thylakoid protein phosphorylation first was characterized by Bennett (1977) and since has been studied intensively (for further references, see Murphy, 1986). The tightly membrane-bound protein kinase is activated by light that preferentially overexcites photosystem II. Activation of the kinase generally is thought to involve reduced plastoquinone or reduced cytochrome $b/f$ complex, but this hypothesis has not been proved. The target proteins for phosphorylation are all components of photosystem II, including the reaction center proteins D1 and D2, the inner chlorophyll $a$ antenna CP43/CP47, the light-harvesting complex LHCII, and a low molecular mass component of 10 kDa. However, neither the minor chlorophyll protein complexes CP29, CP26, and CP24 nor cytochrome $b$-559 become phosphorylated (Dunahay and Staehelin, 1986). The most heavily phosphorylated component is LHCII, which is phosphorylated on a threonine residue located very close to the N terminus of the protein (Chitnis and Thornber, 1988; Michel and Bennett, 1989). Important consequences of protein phosphorylation are that grana partitions are partially destacked and that some of the peripheral pool of LHCII detaches from photosystem II in the partition region and migrates laterally to the stroma-exposed region. This migration clearly reduces the excitation of photosystem II and possibly allows the migrated LHCII to transfer excitation energy to PSI. The movement of LHCII appears to be the molecular explanation for the fluorescence decrease known as the state I–II transition. The destacking and migration of LHCII generally is thought to be caused by the increased number of negative charges introduced by phosphorylation that leads to increased repulsive forces. However, phosphorylation also has been suggested to affect intramolecular forces in the hydrophilic domains of the membrane proteins, causing a conformational change that reduces protein–protein interaction (Allen, 1990). This model also explains how the bacterial photosynthetic system can show state transition changes although it has no grana. The effect of phosphorylation is reversible and is counteracted by the presence of a phosphatase that is always active. The concept of regulation of the excitation energy distribution by protein phosphorylation is very attractive. However, the effect of phosphorylation on photosystem II antenna size is marginal, with a maximum decrease of 20–25%. This result may indicate that phosphorylation is required for fine tuning the system. Also, protein phosphorylation may affect some unknown event. In this context, it is interesting to note the work by Islam (1989a,b), who obtained changes in fluorescence corresponding to state transitions without protein phosphorylation and also obtained phosphorylation without state transitions.

## B. Heat

The organization of the thylakoid membrane in spinach is influenced by temperature (Gounaris *et al.*, 1984; Sundby and Andersson, 1985). Thus, at temperatures above 30°C, part of the photosystem II is disconnected from the peripheral LHCII. Photosystem II migrates out to the nonappressed regions, leaving peripheral LHCII in the appressed region. Although the movement of complexes here opposes that observed after protein phosphorylation, the consequence is essentially the same. The excitation energy reaching photosystem II is decreased. The effect of moderate and short heat treatment is reversible. The temperature-controlled migration of complexes provides a mechanism for protection against overexcitation since plants often are subjected to high light intensities and high temperatures at the same time (see also Chapter 3).

## C. Heterogeneity

Photosystem II is heterogeneous. This characteristic is highly controversial so only a simplified model is given here. For more details, see Black *et al.* (1986). The basic heterogeneities are related to (1) the acceptor side behavior and (2) the antenna size. Most early studies were based on fluorescence measurements, but the results have been supported by other types of measurements. Photosystem II can be divided into at least two subpopulations based on the properties on the acceptor side; B-type ($Q_H$,$Q_1$) and non-B-type ($Q_L$,$Q_2$). B-type centers have a quinone acceptor located on the stromal side of the membrane with a redox potential around 0 mV, and can be reduced by a single flash of light. This type is the normal photosystem II. The non-B-type centers have an acceptor with a redox potential around -250 mV. This acceptor is located on the lumenal side of the membrane and requires repetitive flashing or continuous illumination to be reduced.

Another type of heterogeneity is related to the antenna size of photosystem II. This variation was first demonstrated by detailed quantification of the fluorescence rise after continuous illumination of dark-adapted chloroplasts in the presence of DCMU (Melis and Homann, 1975, 1976). Two kinetic phases were resolved and were explained by two populations of photosystem II, $\alpha$ and $\beta$. The observed amounts of photosystem II$\alpha$ and $\beta$ vary; the interpretations are quite controversial. However, fractionation of the thylakoid membrane revealed that the $\alpha$ and $\beta$ centers exist as separate entities. Thus the $\alpha$ centers with the larger antennae were found in the appressed part of the grana stacks, whereas the $\beta$ centers were found in the stroma lamella region (Anderson and Melis, 1983; Albertsson et al., 1990). In addition to the antenna size differences in the $\alpha$ and $\beta$ centers, there are also differences on the donor side and the acceptor side. A minimal model has been presented by Black et al. (1986). The dominating part of photosystem II is photosystem II$\alpha$, with a large antenna system located in the grana region. Photosystem II$\alpha$ is only of the B-type and is connected to the plastoquinone pool and photosystem I. Photosystem II$\beta$ with its smaller antenna system is located mainly in the stroma region. The photosystem II$\beta$ centers are themselves heterogeneous, consisting of both B-type and non-B-type centers. They do not appear to be connected to photosystem I and also may have differences on the donor side. The role of the $\beta$ centers is still obscure. Some of them apparently result from regulatory processes such as protein phosphorylation, destacking, and heat. Some of them also may represent damaged photosystem II units or units under degradation or assembly.

Several observations suggest that photosystem II can be inactive, both in isolated thylakoids and in leaves (Melis, 1985; Graan and Ort, 1986; Chylla and Whitmarsh, 1989; Cao and Govindjee, 1990). The systems are inactive in the sense that they cannot transfer electrons to the plastoquinone pool rapidly. Chow et al. (1991) obtained results suggesting that inactive centers could be light activated. The mechanism behind this light activation is not clear, but the results may explain some of the contradictory results obtained in other studies.

## VI. CONCLUDING REMARKS

This chapter summarized some of the recent achievements in studying photosystem II. Isolation of active photosystem II particles has helped identify most of the proteins associated with photosystem II. As many as 20 different proteins have been identified. The relative positions of different proteins within the photosystem II complex are gradually appearing. The isolation of active reaction center particles has established unequivocally the role of the D1/D2 proteins in the primary charge separation. Several of the components in electron transfer from water to plastoquinone have been identified. In particular, Z, the immediate donor to P680, has been identified as a tyrosine residue of the D1 protein. The antenna system of photosystem II appears to be more complex than earlier thought. At least seven different chlorophyll proteins have been found. Part of this diversity is required for the regulation of light capture. A number of small proteins with molecular masses less than 10 kDa have been found to be associated with photosystem II. One of these, a 4.8-kDa protein, is associated closely with the reaction center. The role of many of the proteins found in photosystem II is still obscure. They may be important not only for the central electron transfer reactions but also for regulation, protection, and repair. A full understanding of photosystem II function will depend on the elucidation of the contributions of the different subunits. The results obtained so far are encouraging and the future is promising.

## REFERENCES

Åkerlund, H. E., and Andersson, B. (1983). Quantitative separation of spinach thylakoids into photosystem II enriched inside-out vesicles and photosystem I-enriched right-side-out vesicles. *Biochim. Biophys. Acta* **725**, 34–40.
Åkerlund, H. E., and Jansson, C. (1981). Localization of a 34,000 MW polypeptide to the lumenal side of the thylakoid membrane. *FEBS Lett.* **124**, 229–232.
Åkerlund, H. E., Jansson, C., and Andersson, B. (1982). Reconstitution of photosynthetic water splitting in inside-out thylakoid vesicles and identification of a participating polypeptide. *Biochim. Biophys. Acta* **681**, 1–10.
Albertsson, P. Å., Andreasson, E., and Svensson, P. (1990). The domain organization of the plant thylakoid membrane. *FEBS Lett.* **273**, 36–40.
Allen, J. F. (1990). How does protein phosphorylation control protein–protein interactions in the photosynthetic membrane? *In* "Current Research in Photosynthesis" (M. Baltscheffsky, ed.), Vol. II, pp. 915–918. Kluwer, Dordrecht, The Netherlands.
Amesz, J. (1983). The role of manganese in photosynthetic oxygen evolution. *Biochim. Biophys. Acta* **726**, 1–12.
Anderson, J. M., and Andersson, B. (1988). The dynamic photosynthetic membrane and regulation of solar energy conversion. *Trends Biochem. Sci.* **13**, 351–355.
Anderson, J. M., and Melis, A. (1983). Localization of different photosystems in separate regions of chloroplast membranes. *Proc. Natl. Acad. Sci. USA* **80**, 745–749.
Andersson, B., and Åkerlund, H. E. (1987). Proteins of the oxygen-evolving complex. *In* "The Light Reactions" (J. Barber, ed.), pp. 379–420. Elsevier, Amsterdam.
Andersson, B., Larsson, C., Jansson, C., Ljungberg, U., and Åkerlund, H. E. (1984). Immunological studies on the organization of proteins in photosynthetic oxygen evolution. *Biochim. Biophys. Acta* **766**, 21–28.
Andersson, B., Sundby, C., Åkerlund, H. E., and Albertsson, P. Å. (1985). Inside-out thylakoid vesicles—Important tools for the characterization of the photosynthetic membrane. *Physiol. Plant.* **65**, 322–330.
Arntz, B., and Trebst, A. (1986). On the role of the $Q_B$ protein of PS II in photoinhibition. *FEBS Lett.* **194**, 43–49.
Arvidsson, P. O., Bratt, C. E., Andréasson, L. E., and Åkerlund, H. E. (1992). Copper

present in photosystem II is associated with CP26. *In* "Proc. IXth Int. Congr. Photosynth." Nagoya, Japan (in press).

Babcock, G. T. (1987). The photosynthetic oxygen evolving process. *In* "New Comprehensive Biochemistry, Photosynthesis" (J. Amesz, ed.), Vol. 15, pp. 125–158. Elsevier, Amsterdam.

Barabás, K., and Garab, G. (1989). Two populations of the high-potential form of cytochrome $b$-559 in chloroplasts treated with 2-(3-chloro-4-trifluoromethyl)anilino-3,5-dinitrothiophene (Ant 2p). *FEBS Lett.* **248,** 62–66.

Barber, J., Chapman, D. J., and Telfer, A. (1987). Characterization of a PS II reaction center isolated from the chloroplasts of *Pisum sativum*. *FEBS Lett.* **220,** 67–73.

Barry, A. B., and Babcock, G. T. (1987). Tyrosine radicals are involved in the photosynthetic oxygen-evolving system. *Proc. Natl. Acad. Sci. USA* **84,** 7099–7103.

Bassi, R. and Dainese, P. (1990). The role of light harvesting complex II and of the minor chlorophyll $a/b$ proteins in the organization of the photosystem II antenna system. *In* "Current Research in Photosynthesis" (M. Baltscheffsky, ed.), Vol. II, pp. 209–216. Kluwer, Dordrecht, The Netherlands.

Bassi, R., Høyer-Hansen, G., Barbato, R., Giacometti, G. M., and Simpson, D. J. (1987). Chlorophyll-proteins of the photosystem II antenna system. *J. Biol. Chem.* **262,** 13333–13341.

Bassi, R., Magaldi, A. G., Tognon, G., Giacometti, G. M., and Miller, K. R. (1989). Twodemensional crystals of the photosystem II reaction center complex from higher plants. *Eur. J. Cell Biol.* **50,** 84–93.

Bassi, R., Rigioni, F., and Giacometti, G. M. (1990). Chlorophyll binding proteins with antenna function in higher plants and green algae. *Photochem. Photobiol.* **52,** 1187–1206.

Bennett, J. (1977). Phosphorylation of chloroplast membrane polypeptides. *Nature (London)* **269,** 344–346.

Berthold, D. A., Babcock, G. T., and Yocum, C. F. (1981). A highly resolved, oxygenevolving photosystem II preparation from spinach thylakoid membranes. *FEBS Lett.* **134,** 231–234.

Black, M. T., Brearley, T. H., and Horton, P. (1986). Heterogeneity in chloroplast photosystem II. *Photosynth. Res.* **8,** 193–207.

Blubaugh, D. J., and Govindjee (1988). The molecular mechanism of the bicarbonate effect at the plastoquinone reductase site of photosynthesis. *Photosynth. Res.* **19,** 85–128.

Boussac, A., Maison-Peteri, B., Étienne, A. L., and Vernotte, C. (1985). Reactivation of oxygen evolution of NaCl-washed photosystem-II particles by Ca and or the 24-kDa protein. *Biochim. Biophys. Acta* **808,** 231–234.

Boussac, A., Zimmermann, J. L., and Rutherford, A. W. (1989). EPR signals from modified charge accumulation states of the oxygen-evolving enzyme in calcium-deficient photosystem II. *Biochemistry* **28,** 8984–8989.

Boussac, A., Zimmermann, J. L., and Rutherford, A. W. (1990). Histidine oxidation in the oxygen-evolving photosystem II enzyme. *Nature (London)* **347,** 303–306.

Callahan, F. E., Becker, D. W., and Cheniae, G. M. (1986). Studies on the photoreactivation of the water-oxidizing enzyme. 2. Characterization of weak light photoinhibition of PS II and its light-induced recovery. *Plant Physiol.* **82,** 261–269.

Callahan, F. E., Wergin, W. P., Nelson, N., Edelman, M., and Mattoo, A. K. (1989). Distribution of thylakoid proteins between stromal and granal lamellae in *Spirodela*—dual location of photosystem II components. *Plant Physiol.* **91,** 629–635.

Camm, E. L., and Green, B. R. (1980). Fractionation of thylakoid membranes with the nonionic detergent octyl-b-D-glucopyranoside. *Plant Physiol.* **66,** 428–432.

Cao, J., and Govindjee (1990). Chlorophyll $a$ fluorescence transient as an indicator of active

and inactive photosystem II in thylakoid membranes. *Biochim. Biophys. Acta* **1015**, 180–188.
Chitnis, P. R., and Thornber, P. (1988). The major light-harvesting complex of photosystem II: Aspects of its molecular and cell biology. *Photosynth. Res.* **16**, 41–63.
Chow, W. S., Hope, A. B., and Anderson, J. M. (1991). Further studies on quantifying photosystem II *in vivo* by flash-induced oxygen yield from leaf discs. *Aust. J. Plant Physiol.* **18**, 397–410.
Chylla, R. A., and Whitmarsh, J. (1989). Inactive photosystem II complexes in leaves. *Plant Physiol.* **90**, 765–772.
Cleland, R. E. (1988). Molecular events of photoinhibitory inactivation in the reaction center of photosystem II. *Aust. J. Plant Physiol.* **15**, 135–150.
Coleman, W. J., and Govindjee (1987). A model for the mechanism of chloride activation of oxygen evolution in photosystem II. *Photosynth. Res.* **13**, 199–223.
Cramer, W. A., Theg, S. M., and Widger, W. R. (1986). On the structure and function of cytochrome $b$-559. *Photosynth. Res.* **10**, 393–403.
Critchley, C. (1985). The role of chloride in photosystem II. *Biochim. Biophys. Acta* **811**, 33–46.
Dainese, P., and Bassi, R. (1991). Subunit stoichiometry of the chloroplast photosystem II antenna system and aggregation state of the component chlorophyll $a/b$ binding proteins. *J. Biol. Chem.* **266**, 8136–8142.
Debus, R. J., Barry, B. A., Babcock, G. T., and McIntosh, L. (1988a). Site-directed mutagenesis identifies a tyrosine radical involved in the photosynthetic oxygen-evolving system. *Proc. Natl. Acad. Sci. USA* **85**, 427–430.
Debus, R. J., Barry, B. A., Sithole, I., Babcock, G. T., and McIntosh, L. (1988b). Directed mutagenesis indicates that the donor to P680 in photosystem II is tyrosine-161 of the D1 polypeptide. *Biochemistry* **27**, 9071–9074.
Dedner, N., Meyer, H. E., Ashton, C., and Wildner, G. F. (1988). $N$-Terminal sequence analysis of the 8-kDa protein in *Chlamydomonas reinhardtii*. Localization of the phosphothreonine. *FEBS Lett.* **236**, 77–82.
Deisenhofer, J., Epp, O., Miki, K., Huber, R., and Michel, H. (1985). Structure of the protein subunits in the photosynthetic reaction centre of *Rhodopseudomonas viridis* at 3Å resolution. *Nature (London)* **318**, 618–624.
Dekker, J. P., Bowlby, N. R., and Yocum, C. F. (1989). Chlorophyll and cytochrome $b$-559 content of the photochemical reaction center of photosystem II. *FEBS Lett.* **254**, 150–154.
Diner, B. A. (1986). The reaction center of photosystem II. *In* "Encyclopedia of Plant Physiology, New Series, Photosynthesis III" (L. A. Staehelin, and C. J. Arntzen, eds.), Vol. 19, pp. 422–435. Springer-Verlag, Berlin.
Droppa, M., Masojidek, J., Rózsa, Z., Wolak, A., Horváth, L. I., Farkas, T., and Horváth, G. (1987). Characteristics of Cu deficiency-induced inhibition of photosynthetic electron transport in spinach chloroplasts. *Biochim. Biophys. Acta* **891**, 75–84.
Dunahay, T. G., and Staehelin, L. A. (1986). Isolation and chracterization of a new minor chlorophyll $a/b$ protein complex (CP24) from spinach. *Plant Physiol.* **80**, 429–434.
Dunahay, T. G., Staehelin, L. A., Seibert, M., Ogilvie, P. D., and Berg, S. P. (1984). Structural, biochemical, and biophysical characterization of four oxygen-evolving photosystem II preparations from spinach. *Biochim. Biophys. Acta* **764**, 179–193.
Dunahay, T. G., Schuster, G., and Staehelin, L. A. (1987). Phosphorylation of spinach chlorophyll–protein complexes: CPII, but not CP29, CP27, or CP24, is phosphorylated *in vitro*. *FEBS Lett.* **215**, 25–30.
Eaton-Rye, J. J., and Murata, N. (1989). Evidence that the amino-terminus of the 33-

kDa extrinsic protein is required for binding to the photosystem II complex. *Biochim. Biophys. Acta* **977**, 219–226.

Eckes, P. Rosahl, S., Schell, J., and Willmitzer, L. (1986). Isolation and characterization of a light-inducible, organ specific gene from potato and analysis of its expression after tagging and transfer into tobacco and potato shoots. *Mol. Gen. Genet.* **205**, 14–22.

Ellis, J. R. (1981). Chloroplast proteins: Synthesis, transport, and assembly. *Annu. Rev. Plant Physiol.* **32**, 111–137.

Farchaus, J., and Dilley, R. A. (1986). Purification and partial sequence of the $M_r 10,000$ phosphoprotein from spinach thylakoids. *Arch. Biochem. Biophys.* **244**, 94–101.

Förster, V., and Junge, W. (1985). Stoichiometry and kinetics of proton release upon photosynthetic water oxidation. *Photobiochem. Photobiophys.* **41**, 183–190.

Fotinou, C., and Ghanotakis, D. (1990). A preparative method for the isolation of the 43-kDa, 47-kDa and the D1-D2-Cyt $b$559 species directly from thylakoid membranes. *Photosynth. Res.* **25**, 141–146.

Franzén, L. G., Styring, S., Étienne, A. L., Hansson, Ö., and Vernotte, C. (1986). Spectroscopic and functional characterization of a highly oxygen evolving photosystem II reaction center complex from spinach. *Photobiochem. Photobiophys.* **13**, 15–28.

Gerken, S., Brettel, K., Schlodder, E., and Witt, H. T. (1988). Optical characterization of the immediate electron donor to chlorophyll $a$ + II in $O_2$-evolving photosystem II complexes: Tyrosine as possible electron carrier between chlorophyll $a$II and the water-oxidizing manganese complex. *FEBS Lett.* **237**, 69–75.

Ghanotakis, D. F., and Yocum, C. F. (1986). Purification and properties of an oxygen-evolving reaction center complex from photosystem II membranes. *FEBS Lett.* **197**, 244–248.

Ghanotakis, D. F., and Yocum, C. F. (1990). Photosystem II and the oxygen evolving complex. *Annu. Rev. Plant Physiol. Plant Mol. Biol.* **41**, 255–276.

Ghanotakis, D. F., Babcock, G. T., and Yocum, C. F. (1984a). Structural and catalytic properties of the oxygen evolving complex: Correlation of polypeptide and manganese release with the behaviour of $Z^+$ in chloroplasts and a highly resolved preparation of the PSII complex. *Biochim. Biophys. Acta* **765**, 388–398.

Ghanotakis, D. F., Topper, J. N., and Babcock, G. (1984b). Water-soluble 17- and 23-kDa polypeptides restore oxygen evolution activity by creating a high-affinity binding site for $Ca^{2+}$ on the oxidizing side of photosystem II. *FEBS Lett.* **170**, 169–173.

Ghanotakis, D. F., Demetriou, D. M., and Yocum, C. F. (1987). Isolation and characterization of an oxygen evolving photosystem II reaction center core preparation and a 28-kDa Chl-$a$-binding protein. *Biochim. Biophys. Acta* **891**, 15–21.

Ghanotakis, D. F., de Paula, J. C., Demetriou, D. M., Bowlby, N. R., Petersen, J., Babcock, G., and Yocum, C. F. (1989). Isolation and characterization of the 47-kDa and the D1-D2-cytochrome $b$559 complex. *Biochim. Biophys. Acta* **947**, 45–53.

Glazer, A. N., and Melis, A. (1987). Photochemical reaction centers: Structure, organization, and function. *Annu. Rev. Plant Physiol.* **38**, 11–45.

Gounaris, K., Brain, A. R. R., Quinn, P. J., and Williams, W. P. (1984). Structural reorganization of chloroplast thylakoid membranes in response to heat stress. *Biochim. Biophys. Acta* **766**, 198–208.

Gounaris, K., Chapman, D. J., and Barber, J. (1988). Reconstitution of plastoquinone in D1/D2/cytochrome $b$-559 photosystem II reaction centre complex. *FEBS Lett.* **240**, 143–147.

Gounaris, K., Chapman, D. J., Booth, P., Crystall, B., Giorgi, L. B., Klug, D. R., Porter, G., and Barber, J. (1990). Comparison of the D1/D2/cytochrome $b$559 reaction center complex of photosystem two isolated by two different methods. *FEBS Lett.* **265**, 88–92.

Graan, T., and Ort, D. R. (1986). Detection of oxygen-evolving photosystem II centers inactive in plastoquinone reduction. *Biochim. Biophys. Acta* **852**, 320–330.

Green, B. R. (1988). The chlorophyll–protein complexes of higher plant photosynthetic membranes or just what green band is that? *Photosynth. Res.* **15**, 3–32.
Hansson, Ö., and Wydrzynski, T. (1990). Current perception of photosystem II. *Photosynth. Res.* **23**, 131–162.
Henrysson, T., Ljungberg, U., Franzén, L. G., Andersson, B., and Åkerlund, H. E. (1987). Low molecular weight polypeptides in photosystem II and protein depleted acceptor requirement for photosystem II. In "Progress in Photosynthesis Research" (J. Biggens, ed.), Vol. II, pp. 125–128. Nijhoff, Dordrecht, The Netherlands.
Henrysson, T., Schröder, W. P., Spangfort, M., and Åkerlund, H. E. (1989). Isolation and characterization of the chlorophyll $a/b$ protein complex of CP29 from spinach. *Biochim. Biophys. Acta* **977**, 301–308.
Homann, P. H. (1987). The relations between the chloride, calcium, and polypeptide requirements of photosynthetic water oxidation. *J. Bioenerg. Biomembr.* **19**, 105–123.
Hundal, T., Aro, E. M., Carlberg, I., and Andersson, B. (1990a). Restoration of light induced photosystem II inhibition without *de novo* protein synthesis. *FEBS Lett.* **267**, 203–206.
Hundal, T., Virgin, I., Styring, S., and Andersson, B. (1990b). Changes in the organization of photosystem II following light-induced D1-protein degradation. *Biochim. Biophys. Acta* **1017**, 235–241.
Ikeuchi, M., and Inoue, Y. (1988). A new 4.8-kDa polypeptide intrinsic to the PS-II reaction center, as revealed by modified SDS–PAGE with improved resolution of low-molecular weight proteins. *Plant Cell Physiol.* **28**, 1233–1239.
Ikeuchi, M., Yuasa, M., and Inoue, Y. (1985). Simple and discrete isolation of an $O_2$-evolving PSII reaction center complex retaining Mn and the extrinsic 33-kDa protein. *FEBS Lett.* **185**, 316–322.
Ikeuchi, M., Takio, K., and Inoue, Y. (1989). N-Terminal sequencing of photosystem II low-molecular-mass proteins: 5- and 4.1-kDa components of the oxygen evolving complex from higher plants. *FEBS Lett.* **242**, 263–269.
Imaoka, A., Yanagi, M., Akabori, K., and Toyoshima, Y. (1984). Reconstitution of photosynthetic charge accumulation and oxygen evolution in $CaCl_2$-treated PSII particles. *FEBS Lett.* **176**, 341–345.
Islam, K. (1989a). GTP-induced chloroplast membrane protein phosphorylation and photosystem II fluorescence changes: Evidence for multiple protein kinase activities. *Biochim. Biophys. Acta* **974**, 261–266.
Islam, K. (1989b). Thylakoid protein phosphorylation and associated photosystem II fluorescence changes: A study with the ATP analogue adenosine-5-O-thiotriphosphate (ATP-gamma-S). *Biochim. Biophys. Acta* **974**, 267–273.
Jansen, T., Rother, C., Steppuhn, J., Reinke, H., Beyreuther, K., Jannson, C., Andersson, B., and Herrmann, R. G. (1987). Nucleotide sequence of cDNA clones encoding the complete "23 kDa" and "16 kDa" precursor proteins associated with the photosynthetic oxygen-evolving complex from spinach. *FEBS Lett.* **216**, 234–240.
Jegerschöld, C., and Styring, S. (1991). Fast oxygen-independent degradation of the D1 reaction center protein in photosystem II. *FEBS Lett.* **280**, 87–90.
Jegerschöld, C., Virgin, I., and Styring, S. (1990). Light-dependent degradation of the D1 protein in photosystem II is accelerated after inhibition of the water splitting reaction. *Biochemistry* **29**, 6179–6186.
Joliot, P., Barberi, G., and Chabaud, R. (1969). A new model of photochemical centers in system-2. *Photochem. Photobiol.* **10**, 309–329.
Kambara, T., and Govindjee (1985). Molecular mechanism of water oxidation in photosynthesis based on the functioning of manganese in two different environment. *Proc. Natl. Acad. Sci. USA* **82**, 6119–6123.

Kirilovsky, D., and Étienne, A. L. (1991). Protection of reaction center II from photodamage by low temperature and anaerobiosis in spinach chloroplasts. *FEBS Lett.* **279,** 201–204.

Kirilovsky, D., Vernotte, C., Astier, C., and Étienne, A. L. (1988). Reversible and irreversible photoinhibition of herbicide-resistant mutants of *Synechocystis* 6714. *Biochim. Biophys. Acta* **933,** 124–131.

Kobayashi, M., Maeda, H., Watanabe, T., Nakane, H., and Satoh, K. (1990). Chlorophyll $a$ and $\beta$-carotene content in the D1/D2/cytochrome $b$-559 reaction center complex from spinach. *FEBS Lett.* **260,** 138–140.

Kok, B., Forbush, B., and McGloin, M. (1970). Cooperation of charges in photosynthetic oxygen evolution. I. A linear four step mechanism. *Photochem. Photobiol.* **11,** 457–475.

Kühlbrandt, W. (1984). Three-dimensional structure of the light-harvesting chlorophyll $a/b$-protein complex. *Nature (London)* **307,** 478–480.

Kuwabara, T., Miyao, M., Murata, T., and Murata, N. (1985). The function of 33-kDa protein in the photosynthetic oxygen-evolution system studied by reconstitution experiments. *Biochim. Biophys. Acta* **806,** 283–289.

Kuwabara, T., Murata, T., Miyao, M., and Murata, N. (1986). Partial degradation of the 18-kDa protein of the photosynthetic oxygen-evolving complex: A study of binding site. *Biochim. Biophys. Acta* **850,** 146–155.

Ljungberg, U., Åkerlund, H. E., and Andersson, B. (1984a). The release of a 10-kDa polypeptide from everted photosystem II thylakoid membranes by alkaline Tris. *FEBS Lett.* **175,** 255–258.

Ljungberg, U., Åkerlund, H. E., Larsson, C., and Andersson, B. (1984b). Identification of polypeptides associated with the 23- and 33-kDa proteins of photosynthetic oxygen evolution. *Biochim. Biophys. Acta* **767,** 145–152.

Ljungberg, U., Henrysson, T., Rochester, C. P., Åkerlund, H. E., and Andersson, B. (1986a). The presence of low-molecular-weight polypeptides in spinach photosystem II core preparations: Isolation of a 5-kDa hydrophilic polypeptide. *Biochim. Biophys. Acta* **849,** 112–120.

Ljungberg, U., Åkerlund, H. E., and Andersson, B. (1986b). Isolation and characterization of the 10-kDa and 22-kDa polypeptides of higher plant photosystem 2. *Eur. J. Biochem.* **158,** 477–482.

Machold, O., and Meister, A. (1979). Resolution of the light-harvesting chlorophyll $a/b$ protein of *Vicia faba* chloroplasts into two different chlorophyll–protein complexes. *Biochim. Biophys. Acta* **546,** 472–480.

Mathis, P., Satoh, K., and Hansson, Ö. (1989). Kinetic evidence for the function of Z in isolated photosystem II reaction centers. *FEBS Lett.* **251,** 241–244.

Mattoo, A. K., Marder, J. B., and Edelman, M. (1989). Dynamics of the photosystem II reaction center. *Cell* **56,** 241–246.

Mei, R., Green, J. P., Sayre, R. T., and Frasch, W. D. (1989). Manganese-binding proteins of the oxygen-evolving complex. *Biochemistry* **28,** 5560–5567.

Melis, A. (1985). Functional properties of photosystem II$\beta$ in spinach chloroplasts. *Biochim. Biophys. Acta* **808,** 334–342.

Melis, A., and Homann, P. H. (1975). Kinetic analysis of fluorescence induction in 3-(3,4-dichlorophenyl)-1,1-dimethylurea poisoned chloroplasts. *Photochem. Photobiol.* **21,** 431–437.

Melis, A., and Homann, P. H. (1976). Heterogeneity of photochemical centers in system-2 of chloroplasts. *Photochem. Photobiol.* **23,** 343–350.

Michel, H., and Bennett, J. (1989). Use of synthetic peptides to study the subtstrate specificity of a thylakoid protein kinase. *FEBS Lett.* **254,** 165–170.

Michel, H., and Deisenhofer, J. (1988). Relevance of the photosynthetic reaction center from purple bacteria to the structure of photosystem II. *Biochemistry* **27,** 1–7.

Michel, H., Weyer, K. A., Gruenberg, H., Dunger, I., Oesterhelt, D., and Lottspeich, F. (1986). The "light" and "medium" subunits of the photosynthetic reaction center from *Rhodopseudomonas viridis:* Isolation of the genes, nucleotide, and amino acid sequence. *EMBO J.* **5,** 1149–1158.

Miyao, M., and Murata, N. (1985). The Cl$^-$ effect on photosynthetic oxygen evolution: Interaction of Cl$^-$ with 18-kDa, 24-kDa, and 33-kDa proteins. *FEBS Lett.* **180,** 303–308.

Miyao, M., and Murata, N. (1989). The mode of binding of the three extrinsic proteins of 33-kDa, 23-kDa, and 18-kDa in the photosystem II complex of spinach. *Biochim. Biophys. Acta* **977,** 315–321.

Miyazaki, A., Shina, T., Toyoshima, Y., Gounnaris, K., and Barber, J. (1989). Stoichiometry of cytochrome *b*-559 in photosystem II. *Biochim. Biophys. Acta* **975,** 142–147.

Morishige, D. T., Anandan, S., Jaing, J. T., and Thornber, J. P. (1990). Amino-terminal sequence of the 21-kDa apoprotein of a minor light-harvesting pigment–protein complex of the photosystem II antenna (LHC IId/CP24). *FEBS Lett.* **264,** 239–242.

Murata, N., and Miyao, M. (1985). Extrinsic membrane proteins in the photosynthetic oxygen evolving system. *Trends Biochem. Sci.* **10,** 122–124.

Murata, N., Miyao, M., Omata, T., Matsunami, H., and Kuwabara, T. (1984). Stoichiometry of components in the photosynthetic oxygen evolving system of photosystem II particles prepared with Triton X-100 from spinach chloroplasts. *Biochim. Biophys. Acta* **765,** 363–369.

Murphy, J. D. (1986). The molecular organization of the photosynthetic membranes of higher plants. *Biochim. Biophys. Acta* **864,** 33–94.

Nanba, O., and Satoh, K. (1987). Isolation of a photosystem II reaction center consisting of D1 and D2 polypeptides and cytochrome *b*559. *Proc. Natl. Acad. Sci. USA* **84,** 109–112.

Nedbal, L., Masojidek, J., Komeda, J., Prasil, O., and Setlik, I. (1990). Three types of photoactivation. *Photosynth. Res.* **24,** 89–97.

Nugent, J. H. A., Telfer, A., Demetriou, C., and Barber, J. (1989). Electron transfer in the isolated photosystem II reaction centre complex. *FEBS Lett.* **255,** 53–58.

Öquist, G., Greer, D. H., and Ögren, E. (1987). Light stress at low temperature. *In* "Photoinhibition" (D. J. Kyle, C. B. Osmond, and C. J. Arntzen, eds.), pp. 67–87. Elsevier, Amsterdam.

Ohad, I., Adir, N., Koike, H., Kyle, D. J., and Inoue, Y. (1990). Mechanism of photoinhibition *in vivo:* A reversible light-induced conformational change of reaction center II is related to an irreversible modification of the D1 protein. *J. Biol. Chem.* **265,** 1972–1979.

Ono, T. A., and Inoue, Y. (1984). $Ca^{2+}$-dependent restoration of $O_2$-evolving activity in CaCl-washed PSII particles depleted of 33-, 24-, and 16-kDa proteins. *FEBS Lett.* **168,** 281–286.

Ono, T. A., and Inoue, Y. (1986). Effects of removal and reconstitution of the extrinsic 33-, 24-, and 16-kDa proteins on flash yield in photosystem II particles. *Biochim. Biophys. Acta* **850,** 380–389.

Ono, T. A., and Inoue, Y. (1991). Biochemical evidence for histidine oxidation in photosystem II depleted of the Mn-cluster for oxygen evolution. *FEBS Lett.* **278,** 183–186.

Pichersky, E., and Green, B. R. (1990). The extended family of chlorophyll *a*/*b*-binding proteins of PS I and PS II. *In* "Current Research in Photosynthesis" (M. Baltscheffsky, ed.), Vol. III, pp. 553–556. Kluwer, Dordrecht, The Netherlands.

Renger, G. (1986). Herbicide interactions with photosystem II: Recent developments. *Physiol. Veg.* **24,** 509–521.

Rutherford, A. W. (1981). EPR evidence for an acceptor functioning in photosystem II when the pheophytin acceptor is reduced. *Biochem. Biophys. Res. Commun.* **102,** 1065–1070.

Rutherford, A. W. (1985). Orientation of EPR signals arising from components in photosystem II membranes. *Biochim. Biophys. Acta* **807**, 189–201.
Rutherford, A. W. (1987). How close is the analogy between the reaction center of PSII and that of purple bacteria? 2. The electron acceptor side. *In* "Progress in Photosynthesis Research" (J. Biggens, ed.), Vol. I, pp. 227–283. Nijhoff, Dordrecht, The Netherlands.
Rutherford, A. W. (1989). Photosystem II, the water-splitting enzyme. *Trends Biochem. Sci.* **14**, 227–232.
Satoh, K., and Katoh, S. (1985). A functional site of $Ca^{2+}$ in the oxygen-evolving photosystem II preparation from *Synechococcus* sp. *FEBS Lett.* **190**, 199–203.
Satoh, K., and Nakane, H. (1990). Refined purification and characterization of the D1-D2 reaction center of photosystem II. *In* "Current Research in Photosynthesis" (M. Baltscheffsky, ed.), Vol. I, pp. 271–274. Kluwer, Dordrecht, The Netherlands.
Sayre, R. T., Andersson, B., and Bogorad, L. (1986). The topography of membrane protein: The orientation of the 32-kd Qb-binding chloroplast thylakoid membrane protein. *Cell* **47**, 601–608.
Schröder, W. P., Henrysson, T., and Åkerlund, H. E. (1988). Characterization of low molecular mass proteins of photosystem II by *N*-terminal sequencing. *FEBS Lett.* **235**, 289–292.
Seibert, M., DeWit, M., and Staehelin, A. (1987). Structural localization of the $O_2$-evolving apparatus to multimeric (tetrameric) particles on the lumenal surface of freeze-etched photosynthetic membranes. *J. Cell Biol.* **105**, 2257–2265.
Seibert, M., Picorel, R., Rubin, A. B., and Connolly, J. S. (1988). Spectral, photophysical, and stability properties of isolated photosystem II reaction center. *Plant Physiol.* **87**, 303–306.
Seibert, M., Tamura, N., and Inoue, Y. (1989). Lack of photoactivation capacity in *Scenedesmus obliquus* LF-1 results from loss of half the high-affinity manganese-binding site: Relationship to the unprocessed D1 protein. *Biochim. Biophys. Acta* **974**, 185–191.
Shuvalov, V. A., Heber, U., and Schreiber, U. (1989). Low temperature photochemistry and spectral properties of a photosystem 2 reaction center complex containing the proteins D1 and D2 and two hemes of cyt *b*-559. *FEBS Lett.* **258**, 27–31.
Sibbald, P. R., and Green, B. R. (1987). Copper in photosystem II: associated with LHCII. *Photosynth. res.* **14**, 201–209.
Simpson, D. J., and Andersson, B. (1986). Extrinsic polypeptides of the chloroplast oxygen evolving complex constitute the tetrameric ESs particles of higher plant thylakoids. *Carlsberg Res. Commun.* **51**, 467–474.
Spangfort, M., and Andersson, B. (1989). Subpopulations of the main chlorophyll *a*/*b* light-harvesting complex of photosystem II—Isolation and biochemical characterization. *Biochim. Biophys. Acta* **977**, 163–170.
Styring, S., and Rutherford, W. (1987). In the oxygen-evolving complex of photosystem II the $S_0$ state is oxidized to the $S_1$ state by $D^+$ (signal II slow). *Biochemistry* **26**, 2401–2405.
Styring, S., Virgin, I., Ehrenberg, A., and Andersson, B. (1990). Strong light photoinhibition of electron transport in photosystem II: Impairment of the function of the first quinone acceptor, $Q_A$. *Biochim. Biophys. Acta* **1015**, 269–278.
Sundby, C., and Andersson, B. (1985). Temperature-induced reversible migration along the thylakoid membrane of photosystem II regulates its association with LHC-II. *FEBS Lett.* **191**, 24–28.
Sundby, C., Larsson, U. K., and Henrysson, T. (1989). Effects of bicarbonate on thylakoid protein phosphorylation. *Biochim. Biophys. Acta* **975**, 277–282.
Svensson, B., Vass, I., Cedergren, E., and Styring, S. (1990). Structure of donor side components in photosystem II predicted by computer modelling. *EMBO J.* **9**, 2051–2059.

Svensson, P., and Albertsson, P. Å. (1989). Preparation of highly enriched photosystem II membrane vesicles by a non-detergent method. *Photosynth. Res.* **20**, 249–256.

Tae, G. S., and Cramer, W. A. (1989). Lumen-side topography of the alpha-subunit of the chloroplast cytochrome $b$-559. *FEBS Lett.* **259**, 161–164.

Tae, G. S., Black, M. T., Cramer, W. A., Vallon, O., and Bogorad, L. (1988). Thylakoid membrane protein topography: Transmembrane orientation of the chloroplast cytochrome $b559$ $psb$E gene product. *Biochemistry* **27**, 9075–9080.

Takahashi, Y., Hansson, Ö., Mathis, P., and Satoh, K. (1987). Primary radical pair in photosystem II reaction center. *Biochim. Biophys. Acta* **893**, 49–59.

Tamura, N., Ikeuchi, M., and Inoue, Y. (1989). Assignment of histidine residues in D1 protein as possible ligands for functional manganese in photosynthetic water-oxidizing complex. *Biochim. Biophys. Acta* **973**, 281–289.

Tanaka, S., and Wada, K. (1988). The status of cysteine residues in the extrinsic 33-kDa protein of spinach photosystem II complexes. *Photosynth. Res.* **17**, 255–266.

Tang, X. S., and Satoh, K. (1985). The oxygen-evolving photosystem II core complex. *FEBS Lett.* **179**, 60–64.

Taylor, M. A., Packer, J. C. L., and Bowyer, J. R. (1988). Processing of the D1 polypeptide of the photosystem II reaction centre and photoactivation of a low fluorescence mutant (LF-1) of *Scenedesmus obliquus*. *FEBS Lett.* **237**, 229–233.

Trebst, A. (1986). The topography of the plastoquinone and herbicide binding peptides of photosystem II in the thylakoid membrane. *Z. Naturforsch.* **41C**, 240–245.

Trissl, H. W., Breton, J., Deprez, J., and Leibl, W. (1987). Primary electrogenic reactions of photosystem II as probed by the light-gradient method. *Biochim. Biophys. Acta* **893**, 305–309.

Tyagi, A., Hermans, J., Steppuhn, J., Jansson, C., Vater, F., and Herrmann, R. G. (1987). Nucleotide sequence of cDNA clones encoding the complete "33 kDa" precursor protein associated with the photosynthetic oxygen-evolving complex from spinach. *Mol. Gen. Genet.* **207**, 288–293.

Vallon, O., Tae, G. S., Cramer, W. A., Simpson, D., Høyer-Hansen, G., and Bogorad, L. (1989). Visualization of antibody binding to the photosynthetic membrane: The transmembrane orientation of cytochrome $b$-559. *Biochim. Biophys. Acta* **975**, 132–141.

van Grondelle, R. (1985). Excitation energy transfer, trapping, and annihilation in photosynthetic systems. *Biochim. Biophys. Acta* **811**, 147–195.

Vass, I., and Styring, S. (1991). pH-dependent charge equilibria between tyrosine-$D$ and the $S$-states in photosystem II: Estimation of relative midpoint redox potentials. *Biochemistry* **30**, 830–839.

Velthuys, B. (1981). Electron-dependent competition between plastoquinone and inhibitors for binding in photosystem II. *FEBS Lett.* **126**, 277–281.

Vermaas, W. F. J., Renger, G., and Dohnt, G. (1984). The reduction of the oxygen-evolving system in chloroplasts by thylakoid components. *Biochim. Biophys. Acta* **764**, 194–202.

Vermaas, W. F. J., Rutherford, A. W., and Hansson, Ö. (1988) Site directed mutagenesis in photosystem II of the cyanobacterium *Synochocystis* sp. PCC 6803: Donor D is a tyrosine residue in the D2 protein. *Proc. Natl. Acad. Sci. USA* **85**, 8477–8481.

Virgin, I., Styring, S., and Andersson, B. (1988). Photosystem II disorganization and manganese release after photoinhibition of isolated spinach thylakoid membranes. *FEBS Lett.* **233**, 408–412.

Völker, M., Ono, T., Inoue, Y., and Renger, G. (1985). Effect of trypsin on PS-II particles: Correlation between Hill-activity, Mn-abundance, and peptide pattern. *Biochim. Biophys. Acta* **806**, 25–34.

Wasielewski, M. R., Johnson, D. G., Seibert, M., and Govindjee (1989). Determination of the primary charge separation rate in isolated photosystem II reaction centers with 500-fs time resolution. *Proc. Natl. Acad. Sci. USA* **86,** 524–528.

Webber, A. N., Packman, L., Chapman, D. J., Barber, J., and Gray, J. C. (1989). A fifth chloroplast-encoded polypeptide is present in the photosystem II reaction center complex. *FEBS Lett.* **242,** 259–262.

Westhoff, P., Jansson, C., Kleinhitpass, L., Berzborn, R., Larsson, C., and Bartlett, S. G. (1985). Intracellular coding sites of polypeptides associated with photosynthetic oxygen evolution of photosystem II. *Plant Mol. Biol.* **4,** 137–146.

Whitmarsh, J., and Ort, D. R. (1984). Stoichiometries of electron transport complexes in spinach chloroplasts. *Arch. Biochem. Biophys.* **231,** 378–389.

Yamada, Y., Tang, X. S., Itoh, S., and Satoh, K. (1987). Purification and properties of an oxygen evolving photosystem II reaction-center complex from spinach. *Biochim. Biophys. Acta* **891,** 129–137.

# 14

# Carotenoids in Chloroplast Pigment–Protein Complexes

## GEORGE BRITTON

Department of Biochemistry
University of Liverpool
Liverpool L69 3BX, England

 

   I. Introduction
  II. Occurrence and Distribution of Carotenoids
      A. Higher Plants
      B. Algae
 III. Location of Carotenoids in Pigment–Protein Complexes
      A. Photosystem I
      B. Photosystem II
      C. Light-Harvesting Proteins
 IV. Functions of Carotenoids in Pigment–Protein Complexes
      A. Light Harvesting
      B. Photoprotection
  V. Biosynthesis of Carotenoids
      A. General Features
      B. Early Stages and Relation to Other Isoprenoids
      C. Formation of Phytoene
      D. Desaturation
      E. Cyclization
      F. Introduction of Oxygen Functions and Final Modifications
      G. Inhibitors of Carotenoid Biosynthesis: Bleaching Herbicides
      H. Genetics and Organization of Enzymes
 VI. Carotenoid Biosynthesis as Part of Chloroplast Development
      A. Features of Regulation
      B. Carotenoid Transformations in Greening Etiolated Plants
      C. Environmental Factors that Affect Carotenoid Biosynthesis and Incorporation into Thylakoid Pigment–Protein Complexes
      D. Effects of Stress on Chloroplast Carotenoids
VII. Concluding Remarks
     References

## I. INTRODUCTION

"There would be no photosynthesis as we now recognize it were it not for the presence of carotenoids!" (Cogdell and Frank, 1987). The carotenoids play a vital role in photosynthesis. Their biosynthesis is an essential part of chloroplast differentiation and the construction of the functional pigment–protein complexes (PPC). In spite of this, however, the carotenoids generally are given much less attention than the other photosynthetic pigments, the chlorophylls. Even in this book all relevant aspects of carotenoids must be covered in this single chapter. Therefore, some particular aspects necessarily are given only a brief treatment, but it is hoped that this chapter achieves its goal of giving a broad overview of carotenoids in relation to PPCs and chloroplast differentiation, and that it provides sufficient details of the most important characteristics. A book devoted entirely to carotenoids in photosynthesis (Young and Britton, 1993) provides all the details that cannot be incorporated into this chapter.

## II. OCCURRENCE AND DISTRIBUTION OF CAROTENOIDS

### A. Higher Plants

#### 1. Chloroplasts

The carotenoid composition of green leaves of almost all higher plants is strikingly consistent (Goodwin and Britton, 1988). The main chloroplast carotenoids are $\beta$-carotene (25–30% of the total) and the xanthophylls lutein (45%), violaxanthin (10–15%), and neoxanthin (10–15%). In a few members of the Compositae, notably lettuce (*Lactuca sativa*), lactucaxanthin replaces 25–40% of the lutein (Siefermann-Harms et al., 1981). In some cases, $\beta$-carotene is accompanied by its isomer $\alpha$-carotene. The latter usually occurs in only small amounts, but in some cases, for example, cocoa and pine, it can constitute up to half the total carotene. Small amounts of other xanthophylls, for example, lutein 5,6-epoxide, $\alpha$-cryptoxanthin, and $\beta$-cryptoxanthin, usually can be detected also, and zeaxanthin and antheraxanthin, which are components of the xanthophyll cycle, are seen under appropriate conditions (Section IV,B,2). Phytoene and other biosynthetic intermediates are not normally detectable, but can appear in large amounts in plants after treatment with some bleaching herbicides (Section V,G). Chemical structures are given in the Appendix.

In normal functional chloroplasts, the xanthophylls are unesterified but, in senescent autumn leaves of some trees and in some drought-stressed plants such as barley, substantial amounts of acyl esters with long-chain fatty acids accumulate.

In addition to lacking chlorophyll, dark-grown etiolated plants have carotenoid compositions that differ substantially from those of green plants. In particular, little or no carotene is present and neoxanthin levels are very low. Substantial amounts of lutein and violaxanthin are normally present, however, often in conjunction with a high level of antheraxanthin. The total carotenoid concentration can approach 50% of that in the green leaf.

The leaves or needles of many gymnosperms, especially in autumn and winter, can accumulate unusual ketocarotenoids, sometimes in quantities sufficient to impart an orange or red color to the tissues (Czeczuga, 1986). As an example, the *retro*carotenoid rhodoxanthin (6,6'-di*trans*) constitutes a significant proportion of the total carotenoid in the bronze winter foliage of *Thuja occidentalis* var. Rheingold. In some species of cycads (e.g., *Ceratozamia fuscoviridis*), the young leaflets are colored red by large amounts of semi-β-carotenone and related secocarotenoids (Cardini *et al.*, 1987). These pigments, however, are located not in the PPCs of the thylakoids but in oil droplets similar to plastoglobuli. They appear to play no part in photosynthesis.

## 2. Chromoplasts

Chapter 15 of this book discusses the molecular genetics of chromoplast development. A frequent feature of chromoplast development as fruits ripen or flowers open is the synthesis of large amounts of carotenoids, which give rise to the yellow-orange colors of many flowers and orange-red colors of many fruits. These carotenoids are biosynthesized and accumulate in the chromoplasts, frequently in oil droplets or as microcrystalline aggregates (Sitte *et al.*, 1980). They may be associated with protein but are not localized in specific functional PPCs as they are in the chloroplast. The xanthophylls are frequently present as complex mixtures of long-chain acyl esters.

The carotenoids that accumulate in chromoplasts are biosynthesized in the chromoplast. Cell-free preparations of chromoplasts from tomato and *Capsicum* fruits and daffodil flowers have proved extremely useful in studies of the enzymes of carotenoid biosynthesis.

Those fruits that remain green or greenish even when ripe usually contain the normal collection of chloroplast carotenoids, although perhaps in reduced concentration. Fruits that, when ripe, are colored strongly by large amounts of water-soluble red or blue anthocyanins (e.g., raspberry, blackberry) contain little or no carotenoid. In fruits in which the synthesis of large amounts of carotenoids in the chromoplasts is a feature of ripening, several different carotenoid patterns can be recognized:

1. large amounts of the acyclic carotenoid lycopene and its hydroxy derivatives, for example, in tomato (*Lycopersicon esculentum*)
2. large amounts of β-carotene and its hydroxy derivatives (β-cryptoxanthin, zeaxanthin), for example, in peach (*Prunus persica*)
3. collections of 5,6-epoxy- and 5,8-epoxycarotenoids, for example, in star fruit (*Averrhoa carambola*)
4. large amounts of apocarotenoids, for example, in oranges and related *Citrus* species
5. large amounts of one or more specific or unusual carotenoids, for example, capsanthin and capsorubin in red peppers (*Capsicum annuum*)

Distinctive carotenoid patterns also are apparent in many flowers, although, again, many species that are white or colored red or blue by anthocyanins contain little or no carotenoid. The main carotenoid patterns are:

1. substantial amounts of the common carotenes lycopene, α-carotene, or β-carotene and their hydroxy derivatives, for example, lutein (esters) in sunflower (*Helianthus* species)
2. large amounts of epoxycarotenoids, usually 5,6-epoxy- and 5,8-epoxyxanthophylls, often in complex mixtures, for example, violaxanthin in *Viola tricolor*
3. unusual carotenoids of restricted distribution, for example, eschscholtzxanthin in *Eschscholtzia californica*

Carotenoids may be present in any or all anatomical parts of flowers, especially petals, sepals, stamens, and pollen. Details, including extensive tabulated data, of the distribution and accumulation of carotenoids in flowers and fruits are given in books by Goodwin (1980) and by Gross (1987).

## B. Algae

In contrast to higher plants, which all have virtually the same set of chloroplast carotenoids, the algae show great diversity in their pigments. Indeed, carotenoid composition provides an important criterion for algal taxonomy and systematics. The distribution of carotenoids in algae is discussed in great detail by Goodwin (1980) and in a monograph by Rowan (1989). Only a brief outline of the main trends can be given here.

## 14. Carotenoids in Chloroplast Complexes

In general, $\beta$-carotene is the main carotene in algae, although in some cases it is accompanied and may be replaced by $\alpha$-carotene or even $\varepsilon$-carotene. The major variations are seen in the xanthophylls, in which some otherwise unusual structural features are common. In spite of the great structural diversity of algal carotenoids, they are considered to have the same functions in light harvesting and photoprotection as the carotenoids of higher plants.

### *1. Chlorophyta*

*a. Chlorophyceae.* Members of the Chlorophyceae (green algae) generally have carotenoid compositions similar to those of higher plants, with $\beta$-carotene, lutein, violaxanthin, neoxanthin, and, usually, zeaxanthin predominating. Additional carotenoids, particularly loroxanthin and siphonaxanthin, frequently are present also. Some green algae, particularly *Chlorella, Chlamydomonas,* and *Scenedesmus* species, have been used extensively for studies of chloroplast structure and development as well as of photosynthesis.

Under conditions of extreme environmental stress, for example, high light intensity, high temperature, high salt concentration, or mineral deficiency, some microalgae of the Chlorophyceae accumulate large amounts of $\beta$-carotene or its keto derivatives. $\beta$-Carotene concentrations up to 10% of cell dry weight have been reported for *Dunaliella* species. These secondary carotenoids, however, are located in oil droplets or the cell wall and not in the photosynthetic PPCs.

*b. Prasinophyceae.* The Prasinophyceae are similar to the Chlorophyceae, but frequently contain prasinoxanthin instead of siphonaxanthin.

*c. Euglenophyceae.* The Euglenophyceae are members of the Chlorophyta but their carotenoids are more typical of the Chromophyta; the acetylenic xanthophylls diadinoxanthin, diatoxanthin, and heteroxanthin are the major components.

### *2. Chromophyta*

Carotenoid composition of members of the Chromophyta can be very complex. Acetylenic and allenic structures are common and many different individual carotenoids have been identified.

*a. Phaeophyceae.* Fucoxanthin, the main carotenoid of the brown seaweeds, is the most abundant of all naturally occurring carotenoids. Substantial amounts of violaxanthin are also usually present.

*b. Dinophyceae.* The dinoflagellates are characterized by the $C_{37}$ compound peridinin as their main carotenoid, which appears to be restricted to the Dinophyceae. Some species contain the related nor-carotenoid pyrrhoxanthin or, in some cases, fucoxanthin or its derivatives instead of peridinin.

*c. Bacillariophyceae.* Like the brown seaweeds, members of the Bacillariophyceae (diatoms) contain fucoxanthin as their main carotenoid but, unlike the Phaeophyceae, they also have the acetylenic xanthophylls diatoxanthin and diadinoxanthin.

*d. Chrysophyceae, Synurophyceae, and Prymnesiophyceae.* Fucoxanthin and its 19'-esters are the main carotenoids. Acetylenic carotenoids frequently are present also, particularly in the Prymnesiophyceae.

*e. Tribophyceae and Eustigmatophyceae (formerly Xanthophyceae).* The usual main carotenoid is heteroxanthin, sometimes accompanied by neoxanthin and the related vaucheriaxanthin. The Tribophyceae also contain diatoxanthin and diadinoxanthin, but the Eustigmatophyceae may contain zeaxanthin, antheraxanthin, and violaxanthin.

*f. Cryptophyceae.* The major carotenoids present are acetylenic xanthophylls, primarily the bis-acetylenic diol alloxanthin, which is restricted to this class. Carotenoids containing $\varepsilon$-rings are present, but epoxyxanthophylls are claimed to be absent from the Cryptophyceae.

### 3. Rhodophyta

*a. Rhodophyceae.* The red algae contain phycobilins as accessory light-harvesting pigments, but also have a (rather limited) range of carotenoids. Usually, only $\alpha$-carotene and $\beta$-carotene and their hydroxy derivatives are present in appreciable amounts.

### 4. Prokaryotic Algae

Although they do not contain chloroplasts, the prokaryotic blue-green algae or bacteria (cyanobacteria) and the members of the genus *Prochloron* do contain functional PPCs similar to those in the thylakoids of plants and algae in their photosynthetic membranes. The carotenoid biosynthesis capabilities of these prokaryotic species are somewhat limited: only $\beta$-ring compounds can be made and 5,6-epoxyxanthophylls do not occur. The main carotenoids are usually $\beta$-carotene, echinenone, zeaxanthin, caloxanthin, and nostoxanthin, but the glycosides myxoxanthophyll and oscillaxanthin are characteristic of many cyanobacteria.

## III. LOCATION OF CAROTENOIDS IN PIGMENT–PROTEIN COMPLEXES

Although there are reports of the presence of some carotenoids, particularly xanthophylls, in the chloroplast envelope (Jeffrey et al., 1974), the bulk of the chloroplast carotenoid is located, along with chlorophyll, in the functional PPCs in the thylakoids. Comparatively simple mild SDS–PAGE procedures will separate up to seven PPC bands from a solubilized thylakoid preparation readily. Fractions enriched in the core complexes of PSI and PSII, as well as several light-harvesting complexes, are separated easily (Anderson et al., 1978). Depending on the procedure used for solubilization, some pigment is stripped from the PPC and seen as a free pigment zone. Many refined methods have been developed for purifying a particular complex from a particular source. These techniques, coupled with improved high performance liquid chromatography (HPLC) procedures for pigment analysis, have led to many reports of the pigment compositions of isolated PPCs from many plant and algal sources. The nature of the carotenoids present in different organisms may vary, as outlined earlier, but the universal trend appears to be that carotenes are located largely, perhaps exclusively, in the reaction center (RC) core complexes of PSI and PSII, whereas xanthophylls are present in the light-harvesting antennae. The location of the carotenoids in the photosynthetic PPCs of plants and algae has been surveyed in an authoritative article by Siefermann-Harms (1985). Only the main features are outlined here.

### A. Photosystem I

Reaction center I (RCI) proteins have been isolated from cyanobacteria, algae of several classes, and higher plants. The pigment complement usually consists of chlorophyll $a$ and $\beta$-carotene, in association with the primary electron donor P700. $\alpha$-Carotene or $\varepsilon$-carotene may accompany or replace $\beta$-carotene. Quantitative data on the pigment content of RCI are strongly dependent on the solubilization and isolation procedure used, but the average values give about 100 chlorophyll $a$ and 10–15 $\beta$-carotene molecules per 1–2 P700 units (Siefermann-Harms, 1985).

In the PSI supercomplex, RCI is associated with a specific light-harvesting complex, LHCI, that contains chlorophylls $a$ and $b$ and xanthophylls, mainly lutein and violaxanthin (Siefermann-Harms, 1984). Accurate quantitative analyses on isolated LHCI have not been reported.

### B. Photosystem II

The RCII protein is isolated as one of the bands (CPa) in the standard mild SDS–PAGE systems and consists of two subfractions, a core complex (CPa-1 or CP47) that contains phaeophytin also and an internal or firmly associated light-harvesting protein (CPa-2 or CP43) (Green and Camm, 1984), both of which bind chlorophyll $a$ and carotene. Quantitative data suggest that the relative proportion of carotene in RCII preparations is somewhat greater than that in RCI.

Detailed structural studies on RCII of higher plants (Satoh et al., unpublished work, as cited in Mimuro and Katoh, 1991) showed that two molecules of $\beta$-carotene are associated with the core D1–D2 polypeptides, which are similar to the L and M subunits of the bacterial reaction center. One of the $\beta$-carotene molecules, which is in the all-$E$ configuration, was inferred to be close to the accessory chlorophyll in the D2 subunit, whereas the second one may be in the D1 side.

Details of pigment distribution and orientation in the suggested internal light-harvesting protein of PSII are lacking.

### C. Light-Harvesting Proteins

The bulk of the carotenoid in $O_2$-evolving photosynthetic organisms is present in extended light-harvesting systems (light-harvesting chlorophyll $a/b$ (or -$a/c$)–xanthophyll–proteins). These LHCs show enormous variations in the pigments present as well as in their organization and location within, or adjacent to, the membrane. Although structurally diverse, these antenna complexes all function by channeling harvested light energy into PSII. The LHC of higher plants is obtained by SDS–PAGE in three forms that have been designated LHCP1 (trimeric), LHCP2 (dimeric), and LHCP3 (monomeric). LHC represents the major protein of the thylakoid membrane and has been studied intensively. It binds 40–60% of the total chlorophyll $a$ and most of the chlorophyll $b$, as well as the bulk of the xanthophylls, mainly lutein and also neoxanthin, which appears to be bound very tightly. The quantitative pigment composition is not absolutely clear but, in general, there appear to be four chlorophyll $a$, three chlorophyll $b$, and two carotenoid or six chlorophyll $a$, five chlorophyll $b$, and three carotenoid molecules bound to one polypeptide (Ryrie et al., 1980; Braumann et al., 1982).

Many LHCs, some of them water-soluble, have been isolated from a variety of algae. These include peridinin–chlorophyll $a$–protein complexes of dinoflagellates (Song et al., 1976) and fucoxanthin–chlorophyll $a/c$–protein assemblies from the diatom *Phaeodactylum tricornutum* (Friedman and Alberte, 1984) and the brown alga *Dictyota dichotoma* (Katoh et al., 1989), which have been studied in considerable detail. In some of these complexes, the carotenoid spectrum shows a substantial red shift, allowing light absorption up to 540 nm.

Analytical data on algal LHCs have been tabulated by Rowan (1989).

## IV. FUNCTIONS OF CAROTENOIDS IN PIGMENT–PROTEIN COMPLEXES

The physical, chemical, and photochemical properties of carotenoids in nonaqueous organic solutions have been studied extensively. *In vivo*, however, the carotenoids occupy specific sites in the photosynthetic pigment–protein complexes, so their physical properties and photochemical reactivities are modified by and become strongly dependent on the various pigment–pigment and pigment–protein interactions within the complexes. Carotenoids in the PPCs have been shown to have two major functions in photosynthesis, namely, in photoprotection and in light harvesting. The carotenoids that are normally present in the PPCs of plants and algae, and in those of the phototrophic bacteria, that is, molecules that have a chromophore consisting of at least seven conjugated double bonds, have energy properties that make them ideally suited to this dual role (Siefermann-Harms, 1987; Cogdell, 1988).

### A. Light Harvesting

Although chlorophylls (or bacteriochlorophylls) are the major light-harvesting pigments in photosynthetic organisms, an accessory role for carotenoids in light harvesting is well established. The absorption of light by a carotenoid molecule generates an excited singlet state, the energy of which is higher than that of singlet-excited chlorophyll. Therefore, excitation energy can be transferred from carotenoid to chlorophyll, provided the molecules are in close proximity, as they are in the PPCs.

Energy transfer from carotenoids to chlorophyll $a$ in higher plants has been studied mainly by means of action spectra for chlorophyll $a$ fluorescence. By performing fluorescence measurements at low temperature (77 K or below), when fluorescence emission at 735 nm and near 685–695 nm appears to be associated with PSI and PSII, respectively (Strasser and Butler, 1977), it is possible to distinguish between energy transfers to the two photosystems. Carotenoids are involved in the transfer of energy to chlorophyll $a$ of either system in spinach chloroplasts and in isolated PPCs (Kramer *et al.*, 1981). Efficiency of transfer in the major LHC is particularly high; energy transfer from xanthophylls to chlorophyll $a$ in lettuce has been determined to be 100% (Siefermann-Harms and Ninnemann, 1982). Specific structural requirements govern the energy transfer capacity of carotenoids, particularly the integrity of the PPC structures.

Etiolated bean leaves contain high levels of carotenoids but no chlorophylls. Excitation energy absorbed by these existing carotenoids cannot be transferred to newly formed chlorophyll $a$ immediately (Butler, 1961; Goedheer, 1961). After the photoconversion of protochlorophyllide and the formation of chlorophyll $a$, however, the capacity for energy transfer develops during a dark period, as carotenoids and chlorophyll are incorporated into newly synthesized PPCs.

There have been many reports of energy transfer from carotenoid to chlorophyll in algae of most classes, as revealed by excitation spectra for chlorophyll fluorescence and action spectra for photosynthesis. Particularly efficient energy transfer has been demonstrated in isolated PPCs, such as a diadinoxanthin–chlorophyll $a/c_2$–protein complex from *Chroomonas* (Cryptophyceae; Ingram and Hiller, 1983), the water-soluble peridinin–chlorophyll $a$–protein complex of dinoflagellates (Song *et al.*, 1976), the fucoxanthin–chlorophyll $a/c_2$–protein complex of diatoms (Friedman and Alberte, 1984) and brown algae (Barrett and Anderson, 1980), and a siphonaxanthin–chlorophyll $a/b$–protein complex of *Codium* (Chlorophyceae; Anderson, 1985). Some other isolated PPCs, for example, a violaxanthin–chlorophyll $a/c$–protein complex from the brown alga *Acrocarpia* do not exhibit such energy transfer (Barrett and Anderson, 1980).

## 1. Mechanism of Energy Transfer from Carotenoid to Chlorophyll

Studies with model systems have shown that energy transfer from carotenoid to chlorophyll can occur only over very short intermolecular distances (Sineshchekov *et al.*, 1972; Moore *et al.*, 1980). Two possible mechanisms usually are considered, namely, a dipole–dipole resonance interaction (Förster mechanism) and an electron-exchange interaction (Dexter mechanism; Siefermann-Harms, 1985,1987; Cogdell and Frank, 1987; Cogdell, 1988). The low fluorescence yield and extremely short excited singlet state lifetime of carotenoids make a long-range Förster-type process seem unlikely. The actual mechanism of energy transfer has not yet been elucidated. Current ideas, however, are that the energy transfer occurs not from the $^1Bu^*$ state, that is, the excited singlet state into which absorption is symmetry allowed, but from a lower excited singlet state ($^1Ag^*$) into which absorption is symmetry forbidden but transition from the $^1Bu^*$ state is allowed. The energy of the $^1Ag^*$ state allows good overlap with the ground state-to-$Q_X$ transition of chlorophyll $a$ (Thrash et al., 1979). The role of protein in maintaining the carotenoid and chlorophyll in close proximity and in the correct orientation for energy transfer to occur is obviously crucial.

## B. Photoprotection

### 1. Protection against Singlet Oxygen

The photoprotective role of carotenoids is vital. Indeed, the general view is that photosynthesis under oxygenic conditions would be impossible without carotenoids.

When the rate of light absorption by chlorophyll in the PPCs exceeds the rate at which the energy can be used in photosynthesis, a significant proportion of the excited chlorophyll molecules may undergo intersystem crossing to produce the somewhat lower energy but longer lived triplet state, $^3Chl^*$. In the presence of molecular oxygen (itself a triplet state), energy can be transferred from $^3Chl$ to generate singlet oxygen, $^1O_2$, which is highly reactive and a sufficiently strong oxidant to degrade chlorophyll, protein, lipids, and cell structures. Carotenoids afford protection by preventing the formation of $^1O_2$. Triplet–triplet energy transfer from $^3Chl$ to carotenoid generates the triplet state carotenoid, $^3Car^*$, the energy of which is too low to produce $^1O_2$. $^3Car^*$ loses its excitation energy harmlessly by nonradiative processes. Thus $^3Chl$ is removed from the system and the formation of $^1O_2$ is prevented. This quenching process is efficient (around 90%) but some $^1O_2$ still may be produced. If it is, the carotenoid provides a second line of defense because of its ability to accept excitation energy efficently directly from $^1O_2$, again generating harmless $^3Car^*$ (Foote, 1976; Krinsky, 1979). Any carotenoid with a chromophore of at least seven conjugated double bonds has triplet energy that is sufficiently low to be able to quench $^3Chl$ or $^1O_2$, but this property and function are particularly important ones for $\beta$-carotene in the reaction center in plants.

### 2. Protective Role of Xanthophylls and Xanthophyll Cycle

When the transfer of excitation energy from the light-harvesting antennae to the reaction center exceeds the rate of photochemical reaction, photoinhibition, that is, inhibition of photosynthetic activity, occurs (Osmond, 1981). Photoinhibition is a reversible compensatory process and does not cause chlorophyll bleaching or tissue damage directly but, if the conditions are prolonged, bleaching or photodestruction of chloroplast pigments and structural damage to chloroplasts and tissues will follow. Interest has been rekindled in the possibility that carotenoids may have another important protective role by preventing photoinhibition from leading inevitably to photooxidative damage. This protective role involves the xanthophylls and is associated with the xanthophyll cycle, the existence of which has been well known for many years (Yamamoto, 1979). In higher plants, a violaxanthin–zeaxanthin cycle operates. On exposure of the plant to high intensity light, a substantial proportion of the violaxanthin undergoes rapid enzymic deepoxidation via antheraxanthin to zeaxanthin. The reverse process, the reepoxidation of zeaxanthin to violaxanthin, takes place in the dark. At least two pools of violaxanthin exist in the thylakoid; only about 60% of the total is available for conversion into zeaxanthin (Siefermann-Harms, 1984). An interconversion of violaxanthin and zeaxanthin in the chloroplast envelope has been reported also (Jeffrey *et al.*, 1974). The production of zeaxanthin, either rapidly via the cycle or by *de novo* synthesis in response to prolonged stress, has been correlated with the rate of thermal dissipation of excitation energy (as measured by the nonphotochemical quenching of chlorophyll fluorescence) when a plant is subjected to high intensity light (Demmig *et al.*, 1987,1988; Rees *et al.*, 1989).

The results suggest that some xanthophylls but not others (in this case zeaxanthin but not violaxanthin) have the ability to protect the photosynthetic apparatus, especially PSII, by dissipating excitation energy and preventing photochemical reactions and fluorescence emission. The capacity for $\Delta$pH-dependent dissipation of excitation energy (which can account for the dissipation of up to 90% of this energy) is dependent on the level of zeaxanthin in the thylakoid. The mechanism by which zeaxanthin interrupts the chain of events leading from photoinhibition to photooxidative damage is not known.

A violaxanthin–zeaxanthin cycle similar to that in higher plants also has been demonstrated in some green algae (e.g., *Dunaliella* species) that contain these carotenoids; a correlation with chlorophyll fluorescence quenching has been demonstrated. In algae from some other classes, the violaxanthin–zeaxanthin cycle is replaced by a similar cycle in which the participating carotenoids are the acetylenic analogs diatoxanthin and its 5,6-epoxide diadinoxanthin. This cycle has not been studied as intensively but it appears to be similar in its characteristics and functional significance to the cycle in higher plants (Hager, 1980).

## V. BIOSYNTHESIS OF CAROTENOIDS

### A. General Features

The carotenoids are isoprenoid compounds and, as such, are biosynthesized as a branch of the great isoprenoid or terpenoid pathway. The early stages of the pathway, that is, the reactions by which the $C_5$ isoprene unit is constructed and used to build the isoprenoid chain, are common to the biosynthesis of all isoprenoid compounds. The formation of the $C_{40}$ tetraterpene skeleton and all subsequent reactions are specific to carotenoids. Figure 1 shows the main stages of carotenoid biosynthesis. The main features of the pathway are outlined here. Further details can be obtained in specialized review articles by Britton (1976, 1988, 1990a).

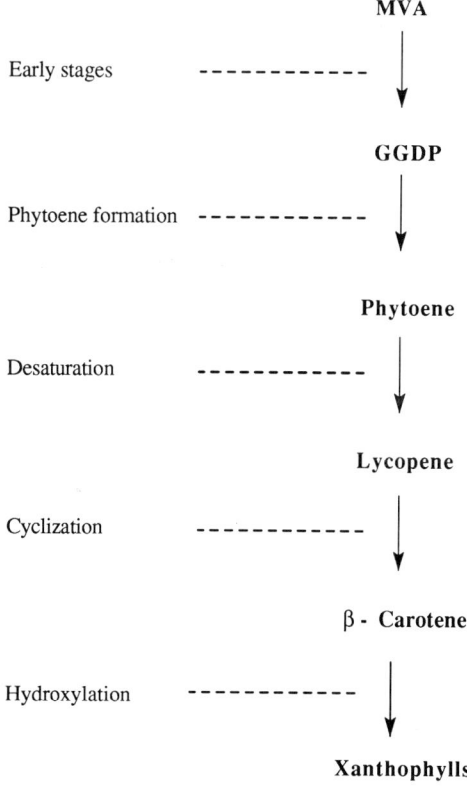

**Fig. 1.** Major stages of carotenoid biosynthesis. MVA, mevalonic acid; GGDP, geranylgeranyl diphosphate.

There has, in fact, been very little work on the pathways and enzymes of carotenoid biosynthesis in chloroplasts. Most of the information outlined here comes from work with chromoplasts or microorganisms, but there is no reason to believe that the main features of the pathway in chloroplasts should differ.

## B. Early Stages and Relation to Other Isoprenoids

The carotenoids, like all other isoprenoid compounds, are biosynthesized from isopentenyl diphosphate (IDP) and its isomer dimethylallyl diphosphate (Fig. 2). Condensation of these two molecules generates the $C_{10}$ intermediate geranyl diphosphate, after which the addition of further $C_5$ units, as IDP, extends the isoprenoid chain via the $C_{15}$ compound farnesyl diphosphate (precursor to squalene, sterols, and triterpenes) to geranylgeranyl diphosphate (GGDP, $C_{20}$), which is a precursor of the phytol side chain of the chlorophylls, phylloquinone, and the tocopherols, as well as of the $C_{40}$ carotenoids. The continued addition of IDP units gives the longer ($C_{45}$) side chain of plastoquinone. These chain extension reactions, and the prenyl transferase enzymes that catalyze them (although not specifically those of the chloroplast), have been well characterized.

There have been conflicting ideas about the ability of chloroplasts to synthesize the basic isoprenoid precursor IDP. For example, it has been reported that the chloroplast is a biosynthetically autonomous organelle that can biosynthesize carotenoids from $CO_2$ via acetyl-CoA and mevalonic acid (MVA); a chloroplast HMG-CoA reductase enzyme (for the conversion of 3-hydroxy-3-methylglutaryl-CoA into MVA) has been reported (Grumbach and Bach, 1979). Other authors, however (e.g., Kreuz and Kleinig, 1984), have concluded that the chloroplast must import IDP. The controversy may have been resolved by work by Heintze *et al.* (1990), which revealed that the biosynthetic capability of chloroplasts changes as the chloroplasts develop or the leaves mature. Thus, in the leaf base segment of seedlings of barley (*Hordeum vulgare*), photosynthetically fixed $CO_2$ was converted into chloroplast isoprenoids including $\beta$-carotene by a pathway, via pyruvate, acetyl-CoA, and MVA, that was located entirely within the chloroplast. (The leaf base contains developing chloroplasts with incomplete thylakoid stacking but a considerable rate of photosynthetic $CO_2$ fixation.) However, in the leaf tip, which contains mature chloroplasts with maximal photosynthetic activity, the chloroplasts were unable to use $CO_2$ for isoprenoid synthesis and had to import extraplastidic IDP. The change was attributed to a drastic decrease in pyruvate decarboxylase–dehydrogenase activity as the chloroplasts developed and matured. These findings are in agreement with earlier reports that very young tissues

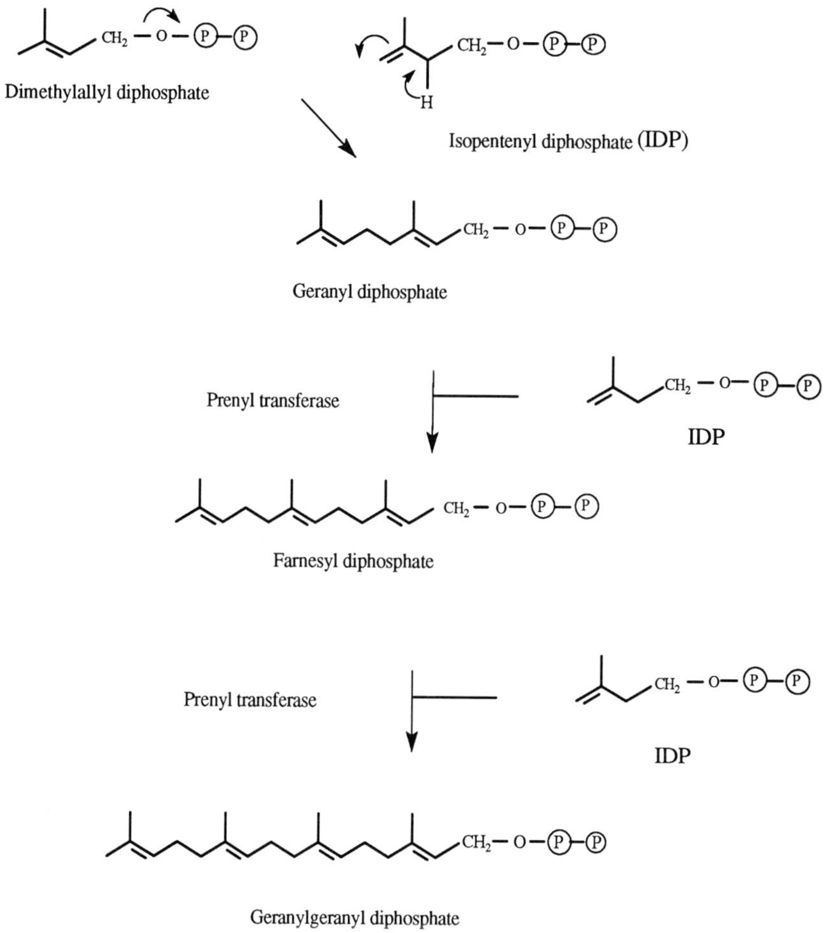

## 14. Carotenoids in Chloroplast Complexes

**Fig. 2.** Early reactions of isoprenoid biosynthesis. Formation of geranylgeranyl diphosphate from dimethylallyl diphosphate.

of spinach and other plants can use photosynthetically fixed $CO_2$ very efficiently for synthesis of chloroplast isoprenoids (Rogers et al., 1966).

## C. Formation of Phytoene

The basic $C_{40}$ carotenoid skeleton is formed by the condensation of two molecules of GGDP. This is a two-step process (Fig. 3) that proceeds via an intermediate, prephytoene diphosphate, that is analogous to presqualene diphosphate, the precursor of squalene and sterols. A single monomeric

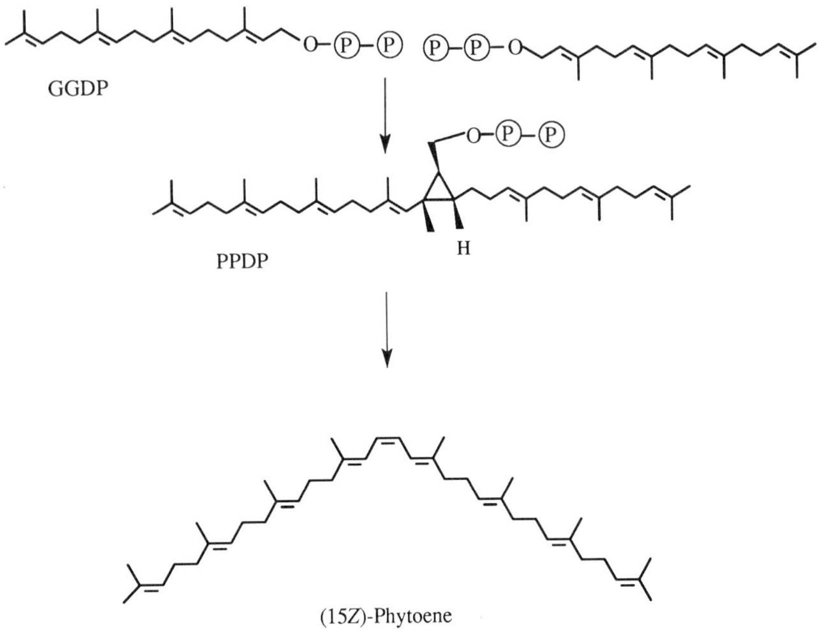

Fig. 3. Formation of phytoene from GGDP via prephytoene diphosphate (PPDP).

47-kDa enzyme, phytoene synthase, catalyzes both stages of the reaction (Dogbo *et al.*, 1988). The phytoene produced in higher plants appears to be the 15Z isomer.

## D. Desaturation

Phytoene, with a chromophore of only three conjugated double bonds (cdb), is colorless but undergoes a series of desaturation reactions, each of which involves the *trans* elimination of hydrogen and introduces a new double bond, to extend the chromophore by two conjugated double bonds (Fig. 4). The intermediates are phytofluene (5 cdb), ζ-carotene (7 cdb), and neurosporene (9 cdb); the normal final product of the desaturation sequence is lycopene (11 cdb).

Working with a preparation of the membrane-bound enzymes obtained by homogenization of daffodil chromoplasts, Beyer and Kleinig (1990) described the desaturation of (15Z)-phytoene to (15Z)-ζ-carotene in the dark, followed by the photoisomerization of (15Z)- to (all-E)-ζ-carotene and the subsequent desaturation of the latter to prolycopene [(7Z,9Z,7'Z,9'Z)-lycopene] in preference to the (all-E) form. In the desatu-

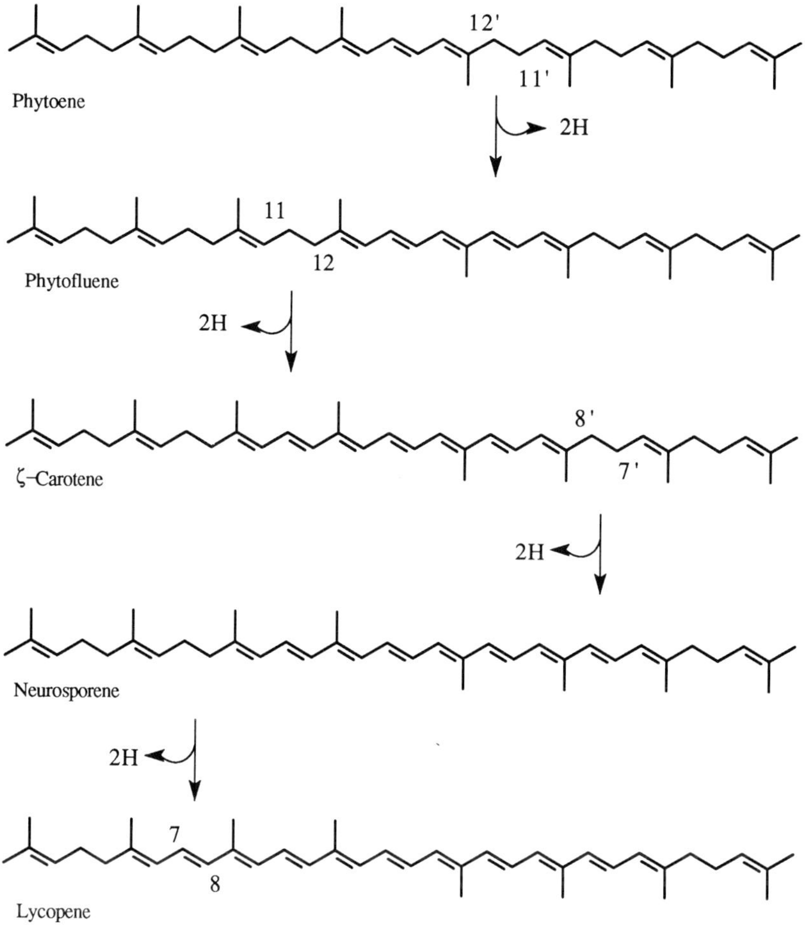

**Fig. 4.** Formation of the colored acyclic intermediate lycopene by desaturation of phytoene.

rations, $O_2$ was an essential cofactor and the involvement of an associated redox system was confirmed.

## E. Cyclization

Cyclization of the carotenoid end groups is now generally accepted to take place at the lycopene level of desaturation. Cyclization is initiated by $H^+$ attack on C-2 of a suitably folded acyclic end group; subsequent alternative loss of a proton from C-6 or C-4 of the proposed carbenium ion intermediate then gives the β or ε ring, respectively (Fig. 5), to produce β- and α-carotene. The two ring types are not interconverted. This mechanism has been confirmed and the stereochemistry of the reactions established by experiments involving labeling with stable isotopes (Britton, 1985, 1990b).

The work of Beyer and Kleinig (1990), cited earlier, showed that cyclization *in vitro* proceeded efficiently only with prolycopene or (7Z,7'Z)-lycopene as substrate and not with (all-*E*)-lycopene, although the cyclic products were the all-*E* isomers. The efficient direct cyclization of (all-*E*)-lycopene has been achieved, however, with other enzyme systems, especially ones from bacteria. The strict requirement of $O_2$ for desaturation shown by Beyer and Kleinig (1990), but of anaerobic conditions for cyclization in the same cell-free system, shows that the microenvironment in which the enzymes operate *in vivo* must be controlled strictly.

### F. Introduction of Oxygen Functions and Final Modifications

Oxygen functions, such as hydroxy groups, normally are introduced in the final steps of carotenoid biosynthesis. Thus, the common 3-hydroxy xanthophylls lutein and zeaxanthin are formed by the stereospecific hydroxylation of the corresponding α- and β-carotenes by mixed-function oxidase enzymes (Fig. 6) (Walton *et al.*, 1969; Milborrow *et al.*, 1982).

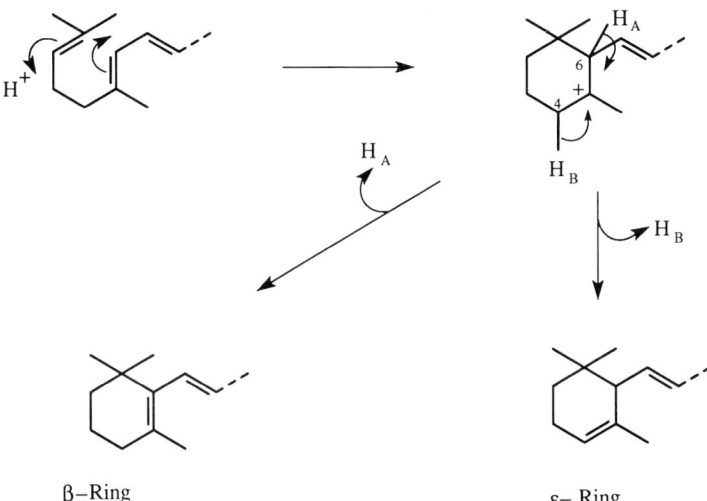

**Fig. 5.** Mechanism of the cyclization reactions by which the β and ε rings are formed.

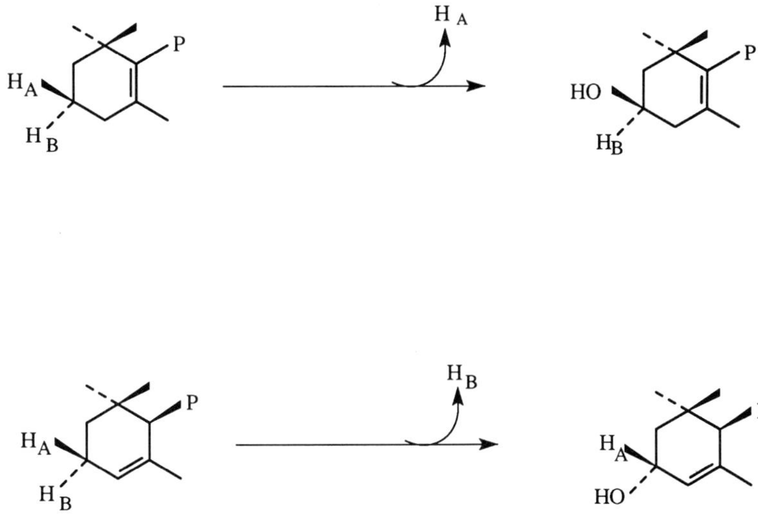

**Fig. 6.** Stereospecific hydroxylation at C-3 in xanthophyll biosynthesis.

Little is known about the other late reactions, especially end-group modifications, by which the many different individual carotenoids that are found in plant and algal chloroplasts are elaborated. Pathways have been proposed for the formation of several different end groups from the 3-hydroxy-5,6-epoxy-$\beta$ end group in violaxanthin (Britton, 1990a), but there is little or no biochemical evidence to support these proposals (Fig. 7).

### G. Inhibitors of Carotenoid Biosynthesis: Bleaching Herbicides

Because carotenoids play a vital role in protecting plants against chlorophyll-sensitized photooxidative damage, any compounds that inhibit the biosynthesis of these carotenoids can be lethal. Many such compounds have been discovered; some have been investigated intensively and have found use as herbicides (Britton, 1979; Ridley, 1982; Britton et al., 1989). The best known of these substances is norflurazon (SAN 9789), which inhibits the desaturation of phytoene and thus prevents the formation of the normal colored carotenoids. Seedlings grown in low light intensity are pale green. The phytoene that accumulates is not able to quench the energy of triplet chlorophyll or singlet oxygen so, in high light intensity, chlorophyll synthesized by the plant immediately acts as a photosensitizer, $^1O_2$ is produced, and the plant suffers lethal damage.

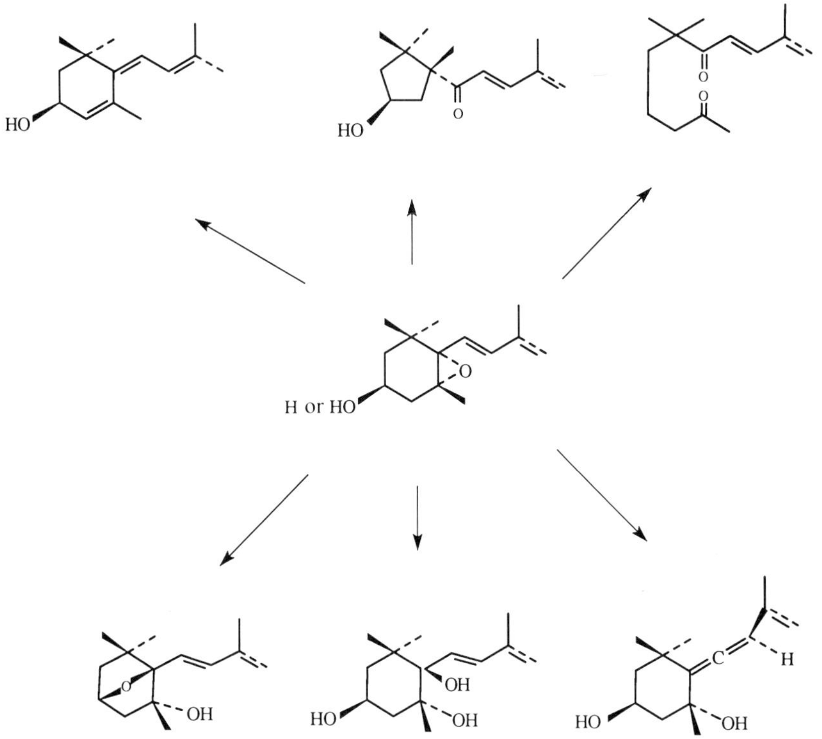

**Fig. 7.** Scheme for the proposed formation of a variety of carotenoid endgroups from a 5,6-epoxy-5,6-dihydro-β end group.

## H. Genetics and Organization of Enzymes

The enzymes responsible for carotenoid biosynthesis in the chloroplast are encoded by nuclear genes. No progress has been made on the isolation and characterization of these genes. Over the past several years, extensive work with phototrophic bacteria, in which the carotenoid biosynthesis genes are located in the photosynthesis gene cluster, has provided information that is likely to be relevant to the organization of the carotenogenic enzymes in the plastids of plants and algae also. From this work (Giuliano et al., 1988; Armstrong et al., 1989) as well as from work on the molecular genetics of carotenogenesis in nonphototrophic bacteria (Ausich et al., 1990; Misawa et al., 1990), a number of conclusions can be drawn (summarized by Britton, 1991). It is clear from the number of genes that have been identified that there is simply one gene for each type of reaction, for example, only one gene for the desaturase enzymes and not a different gene for each step in the desaturation sequence. Similarly, there is only one gene for an end group reaction, not a different gene for the reaction of each end group or for each possible alternative substrate. This result confirms the view that the enzymes of carotenoid biosynthesis recognize a particular end group or structural feature and the remainder of the molecule is unimportant. The efficiency with which biosynthesis occurs and the nature of the eventual product are, however, critically dependent on the organization of the gene products (the enzymes) into a functional membrane-bound or membrane-associated multienzyme complex. Evidence from work with nonphototrophic bacteria (Ausich et al., 1990) suggests that at least one of the carotenogenic enzymes is able to recognize an anchoring site on the membrane and that the multienzyme complex is assembled around this site. Although it seems likely that similar principles apply to the organization and assembly of the carotenogenic enzymes in other systems also, a greater degree of complexity can be expected in chloroplasts, where several different carotenoids are biosynthesized that are destined for different pigment–protein complexes and functions. Clearly, a large amount of detailed and sophisticated molecular genetics work is needed to elucidate these points, but some indications have been obtained from classical labeling studies (Section VI,A).

## VI. CAROTENOID BIOSYNTHESIS AS PART OF CHLOROPLAST DEVELOPMENT

### A. Features of Regulation

Because carotenoid biosynthesis is an essential part of the construction of pigment–protein complexes in the thylakoids, its regulation is linked closely to that of other components such as chlorophylls, proteins, and lipids. Genetic control of carotenogenesis as part of chloroplast development as a whole is obviously important, but other control mechanisms operating on the individual biosynthetic pathways are needed to regulate the supply of the various components, including the carotenoids, as required during the assembly of the PPCs and thylakoids (Britton 1986).

Several different carotenoids have different locations and functions within the PPC. Therefore it is an oversimplification to think of the biosynthesis of total carotenoid in leaf or chloroplast as a single uniform entity. This concept raises several questions about the factors that regulate the biosynthesis of each individual carotenoid and determine the carotenoid composition of each PPC.

Some experimental information relevant to these points has been obtained. When young seedlings that had been grown in the light or in the dark were incubated in the light with [$^3$H]MVA and $CO_2$ simultaneously, different $^3$H:$^{14}$C ratios were observed for β-carotene and lutein, indicating that the time courses for synthesis of these two carotenoids are different (Sergeant and Britton, 1984). Earlier labeling studies had shown that β and ε rings are not interconverted (Williams *et al.*, 1967), so the final stages of lutein and β-carotene synthesis must be independent. The stage at which the two pathways diverge is not known, however. Lutein and β-carotene could be made by two distinct enzyme assemblies. Alternatively, the divergence may occur at a later stage, so different final cyclization and hydroxylation reactions are used after phytoene synthesis and desaturation have taken place on a common enzyme assembly. In the same series of experiments (Sergeant and Britton, 1984), no great differences were seen in either total incorporations or $^3$H:$^{14}$C ratios for any particular carotenoid in different thylakoid fractions. There must be delicate regulatory mechanisms that determine the specific destination and location within the thylakoid PCC of each carotenoid molecule that is made. It may be significant that carotenoid molecules that are labeled with various substrates in a short incubation lose their labeling much more quickly than ones that are labeled in a longer incubation. This suggests that newly formed carotenoid molecules may enter a labile pool from which they are either taken for incorporation into the PPC or degraded rapidly.

The regulation of carotenoid biosynthesis in chloroplasts is very complicated. Several different phases or variations must be considered:

1. synthesis of the bulk of the pigment during the initial construction of the photosynthetic apparatus
2. synthesis that continues in mature chloroplasts as part of turnover
3. in greening etiolated plants, continuing synthesis by the enzymes that were active in the etiolated leaves, especially in the early stages of greening before new enzyme synthesis is established
4. synthesis to modify carotenoid composition in response to or as a consequence of changes in environmental conditions, especially light intensity

Different regulatory mechanisms may apply under these different circumstances. Also, chloroplast differentiation and its associated carotenoid biosynthesis are unlikely to follow the same course universally. Substantial differences can be expected among different plant species and even among different varieties of the same species. Growth conditions and the age of the plant or tissue under investigation are also very important. Cotyledons and true leaves, even of the same plant, may behave quite differently and are not strictly comparable. Finally, the course of chloroplast differentiation and its associated carotenoid biosynthesis in a greening etiolated plant is not the same as in a plant growing under a normal light–dark regime. Extreme caution should be exercised, therefore, when comparing systems, especially when extrapolating from one system to another.

## B. Carotenoid Transformations in Greening Etiolated Plants

Although it is not the normal course of plant growth and chloroplast development (under normal conditions of a day–night light regime, chloroplasts develop direct from proplastids), the greening of etiolated seedlings and the associated etioplast to chloroplast transformation provide an extremely convenient experimental system. Many details of this transformation have been elucidated. It is certainly a convenient system in which to study the incorporation of carotenoids into the developing chloroplasts. Etiolated seedlings generally contain substantial amounts of carotenoids, although the distribution of the individual pigments differs considerably from that in the green plant. In particular, etiolated leaves and etioplasts normally contain little or no carotene and a modified collection of xanthophylls. Lutein and violaxanthin are predominant, but a substantial amount of antheraxanthin is usually present; there is little or no neoxanthin (Bahl, 1977; Barry *et al.*, 1991).

The carotenoids are present in the etioplasts and there are reports that they are located in prothylakoids, prolamellar bodies, and the etioplast envelope (Britton, 1986). The enzymes of carotenoid biosynthesis are also present in the etioplast. Evidence suggests that the phytoene synthase complex is in a soluble form in the stroma, whereas the subsequent enzymes can be found in membrane fractions containing the prothylakoids, prolamellar bodies, and envelope (Lütke-Brinkhaus and Kleinig, 1987).

When dark-grown etiolated plants are illuminated, the prolamellar bodies break down and the prothylakoids grow to form thylakoids and, eventually, grana stacks (Virgin *et al.*, 1963; Henningsen and Boynton, 1970). Over a period of 24–48 hr, chlorophyll is synthesized and the characteristic carotenoid composition of the chloroplast is attained. The stepwise assembly of the PPC during greening has been studied by experiments in which the seedlings are exposed to intermittent flashes of light, but changes in carotenoid composition have not been followed in detail under these conditions. Grumbach (1981), however, reported that the biosynthesis of xanthophylls in barley was inhibited by intermittent light–dark cycles (2 min white light, 98 min dark). This result is consistent with the findings of Akoyunoglou and Argyroudi-Akoyunoglou (1986) that only the core complexes of PSI and PSII are formed under these conditions, not the extended light-harvesting antenna. Experiments involving illumination with either short periods of red light (Kleudgen and Grumbach, 1983) or continuous red light (Oelmüller and Mohr, 1985) led to the conclusion that there are three major controls for the formation and accumulation of carotenoids, namely, a coarse control through phytochrome, a fine tuning correlated with the accumulation of chlorophyll, and the stabilization of the complexes against photodamage. $\beta$-Carotene and neoxanthin levels, however, remained low under conditions of intermittent light–dark cycles. Evidence obtained through the use of the carotenoid biosynthesis inhibitor SAN 9789 to block the formation of colored carotenoids in etiolated seedlings showed that the presence of carotenoids was necessary for the proper development of chloroplasts. In their absence, chlorophyll-sensitized photooxidation damage occurred (Axelsson *et al.*, 1981; Klockare *et al.*, 1981; Ryberg *et al.*, 1981; Sandelius *et al.*, 1981).

A series of experiments from this laboratory compared the effects of different light intensities on carotenoid biosynthesis and changes in carotenoid compositions in different sections of the leaves of greening etiolated barley seedlings (Barry *et al.*, 1991). In general, for normal chloroplast pigment composition to be attained and the same concentration of pigments to be accumulated, it took 24 hr longer in low intensity (1000 lux) than in high intensity light (10,000 lux). Unusual short-lived effects on the concentration of the xanthophyll cycle pigments, violaxanthin and zeaxanthin, were seen, especially a transient (6-hr) high level of zeaxanthin at 10,000 lux and an increase in violaxanthin at 1000 lux. The rate of pigment synthesis was, as expected, greatest in the rapidly growing bottom zone of the shoot, where carotenoid levels increased by around 100% during greening. Virtually no increase was seen in the shoot tip. The levels of $\beta$-carotene and neoxanthin increased severalfold and that of antheraxanthin (and eventually that of zeaxanthin) decreased markedly.

Until recently, it was not known whether the carotenoid molecules that were present in the etioplast could be used and incorporated into the chloroplast and its PPCs. Experiments that employed labeling with deuterium have addressed this question and provided some answers (P. Barry and G. Britton, 1987, unpublished results). These experiments used the novel approach of allowing the etiolated seedlings to become green under watering with $H_2O/D_2O$ (1:1). The carotenoids were isolated from three zones of the shoot (bottom, middle, and tip) and analyzed by mass spectrometry. Under these conditions, existing molecules synthesized in the etiolated seedlings contained no deuterium, whereas molecules synthesized during greening in $D_2O$ showed substantial levels of deuterium incorporation. Patterns characteristic of synthesis from photosynthetically fixed

## 14. Carotenoids in Chloroplast Complexes

$CO_2$ or from existing reduced carbon substrates also could be distinguished. From this work, the following conclusions could be drawn:

1. The xanthophylls lutein and violaxanthin that were present in the etiolated seedlings were retained during greening and were incorporated into the chloroplasts and thylakoids.
2. $\beta$-Carotene was synthesized almost entirely *de novo*, initially from existing reduced carbon substrates and subsequently from $CO_2$, especially in the leaf base region.
3. Although the level of neoxanthin increased approximately 5-fold during greening, no incorporation of deuterium was seen. Neoxanthin therefore must be produced from a pool of carotenoid molecules that was present in the etiolated shoot, the most likely source being antheraxanthin.

### C. Environmental Factors that Affect Carotenoid Biosynthesis and Incorporation into Thylakoid Pigment–Protein Complexes

Many nutritional factors have been shown to influence the carotenoid content of leaves. In general, plants grown under nutritionally favorable conditions have the highest carotenoid contents. The main environmental factor, however, is light. The effects of light are best illustrated in what has become known as the sun–shade response (Björkmann, 1981). Changes in the light environment of a plant can have profound effects on the structure and function of the photosynthetic apparatus; sun-type and shade-type leaves and patterns of pigment organization can be recognized. Under low light conditions, shade-leaf characteristics are developed. There are larger numbers of PSI and PSII units, the PSII:PSI ratio is higher, and the amount of chlorophyll, particularly of chlorophyll *b*, is increased. As the light-harvesting complexes acquire more importance, the ratio of xanthophylls to carotene becomes significantly higher. In contrast, sun-type plants and leaves have a relatively greater amount of the PSI and PSII core complexes, so chlorophyll *b/a* ratios are lower and the ratio of xanthophylls to carotene is also lower (Lichtenthaler *et al.*, 1982, 1983). These differences do not reflect changes in the pigment content and composition of the individual PPCs but result from variations in the relative amounts of the various PPCs (Anderson and Osmond, 1987).

Changes in pigment content and composition when algae are subjected to different light environments are also apparent, but have not been studied widely.

## D. Effects of Stress on Chloroplast Carotenoids

Among the many effects and consequences of environmental stress on plants, changes that affect their pigment systems are easily detected; chlorosis or bleaching is a frequent long-term consequence. The carotenoid compositions of the chloroplast and its PPCs are affected by environmental stress of many kinds, for example, heat, cold, light, drought, atmospheric pollution, mineral deficiency, and chemicals (Young and Britton, 1990). Major changes also show during the course of senescence. Changes that are seen may be of two kinds. In the short term, the stress may elicit a response from the plant, for example, the synthesis of increased amounts of a particular pigment, but in the long term chlorosis and chloroplast degradation are accompanied by severe losses of one or more carotenoids by oxidative destruction.

The most typical short-term effect seen in response to high light intensity and prolonged photoinhibition is a substantial increase in the level of zeaxanthin, initially rapidly via the xanthophyll cycle but then by synthesis *de novo*. Zeaxanthin can reach levels up to 15% of total xanthophyll.

Another change that is sometimes seen is the accumulation of acyl esters of the xanthophylls. It is well known that, in senescent autumn leaves of some deciduous trees, the loss of chlorophyll as the chloroplasts are degraded is accompanied by the formation of large amounts of xanthophyll esters (Goodwin, 1958). A similar phenomenon has been seen in seedlings of some crop plants, especially barley, that suffer severe drought stress. The most common ester formed is that between the major xanthophyll, lutein, and the major thylakoid fatty acid, linolenic acid (Barry *et al.*, 1992). These esters accumulate not in the photosynthetic PPCs but in plastoglobuli (Trevini and Steinmüller, 1985).

In plants that are exposed to an atmosphere containing increased levels of oxygen, to atmospheric pollutants such as ozone or nitrogen oxides, or to some herbicidal chemicals such as DCMU or paraquat that affect photosynthetic electron transport, severe alterations in pigment compositions are seen as a result of oxidative degradation and free radical-mediated processes (Van Hasselt and Van Berlo, 1980; Mikulska *et al.*, 1987; Öquist *et al.*, 1987; Wolfenden *et al.*, 1988; Barry *et al.*, 1990). Reduction in carotenoid levels accompanies chlorophyll degradation but significant variations are seen in the effects on different carotenoids. In many cases, for example, after prolonged photoinhibition or in needles of conifers suffering forest damage, β-carotene is the most susceptible carotenoid to oxidative and enzymic degradation. The appearance of β-carotene-5,6-epoxide is an indication that lipid peroxidation is occurring (Young *et al.*, 1989). However, in other cases, notably exposure to ozone, β-carotene is less susceptible to degradation than the xanthophyll lutein. The significance of these differences is not understood, but they presumably arise because the different factors initiate oxidative processes in different PPCs.

Mineral deficiency also reduces carotenoid contents. The chlorotic effect of iron deficiency is particularly strong. Up to 80% reduction in total pigment may be seen and the xanthophyll:carotene ratio increases (Monge et al., 1987).

## VII. CONCLUDING REMARKS

Although much is known about the carotenoids and their localization within the chloroplast, a great deal is still not known. For example, details of the molecular interactions between the carotenoids and the other components of the pigment–protein complexes, namely, protein, chlorophyll, and lipids, remain to be elucidated. Much progress is being made in localizing carotenoids in the PPCs of phototrophic bacteria now that procedures are available for crystallizing these complexes and for incorporating labeled carotenoid molecules into them. The extension of such studies to the carotenoids in chloroplast PPCs is awaited eagerly.

The overall pathway of carotenoid biosynthesis is well established, although the role of *cis*-(Z) isomers is still not clear and little is known of the enzymes and their organization. The major questions, however, concern the regulation of carotenoid biosynthesis in relation to other features of chloroplast development, for example, what factors determine the individual carotenoid that is made and how this carotenoid is incorporated correctly into its destined PPC. The time is ripe for exciting developments in this area, which surely will be achieved by a combination of traditional biochemical and analytical methods and modern molecular genetics.

## APPENDIX

### Chemical Structures for Some Representative Carotenoids

β-Carotene

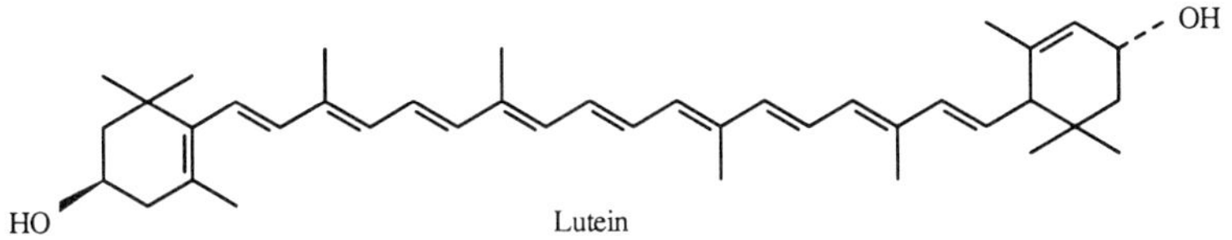

Lutein

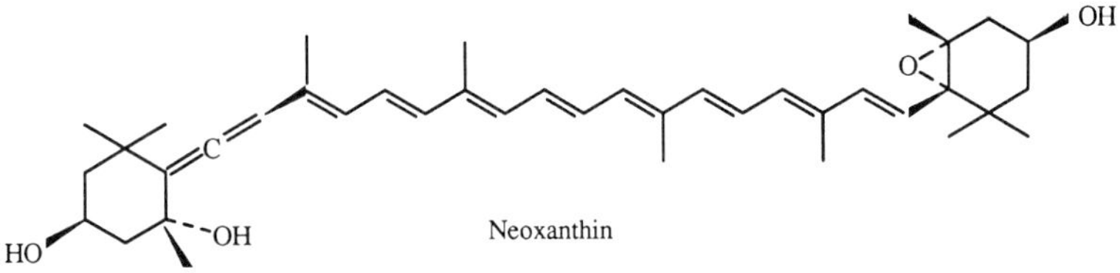

Neoxanthin

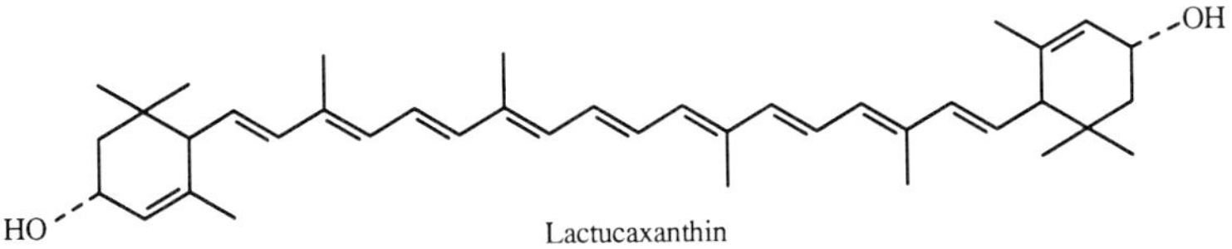

Lactucaxanthin

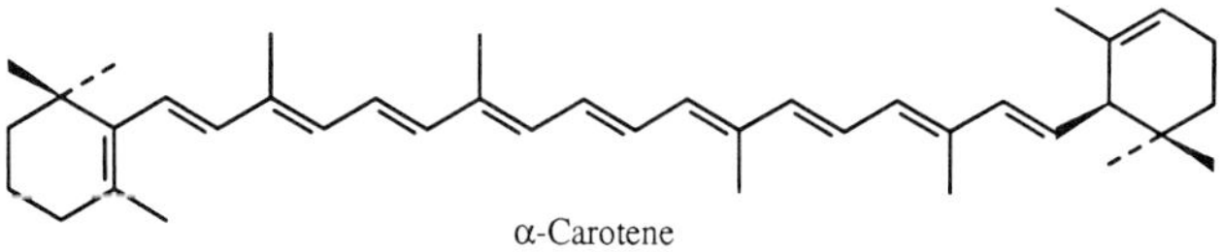

α-Carotene

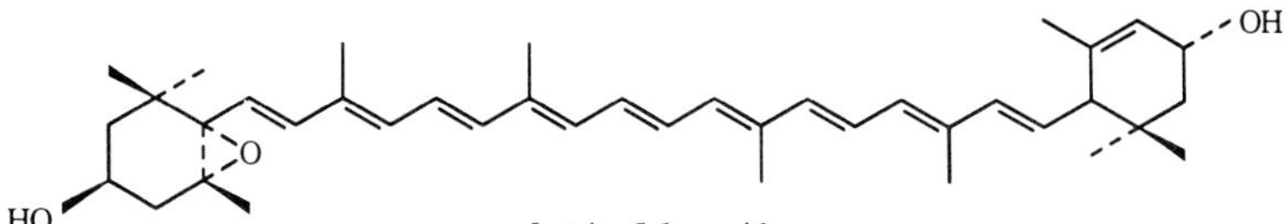

Lutein -5,6-epoxide

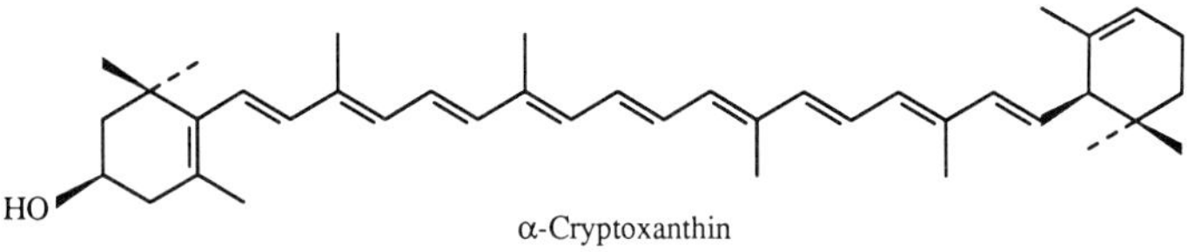

α-Cryptoxanthin

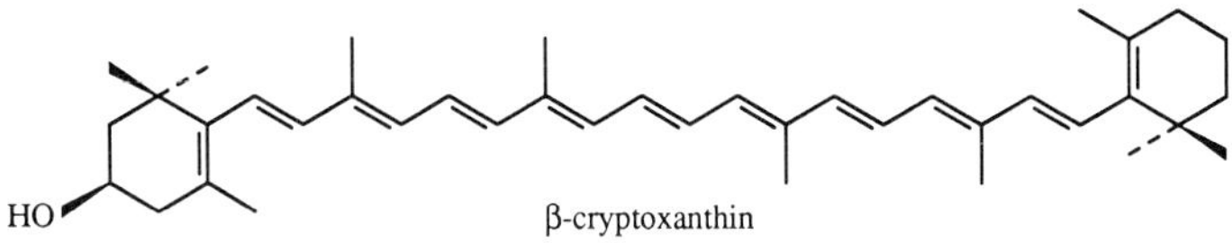

β-cryptoxanthin

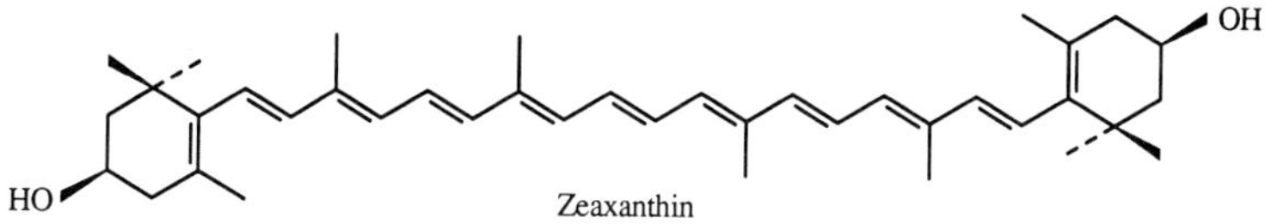

## 14. Carotenoids in Chloroplast Complexes

Zeaxanthin

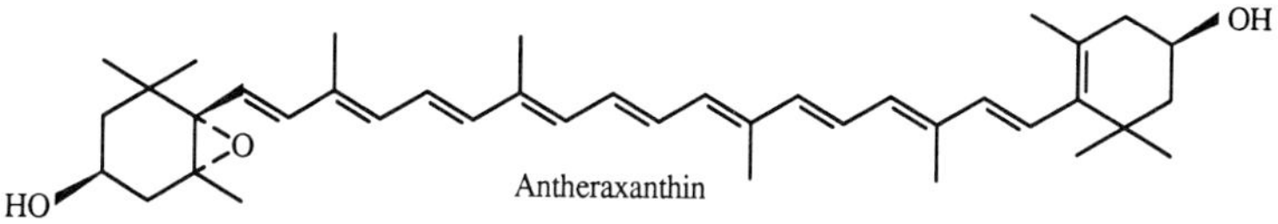

Antheraxanthin

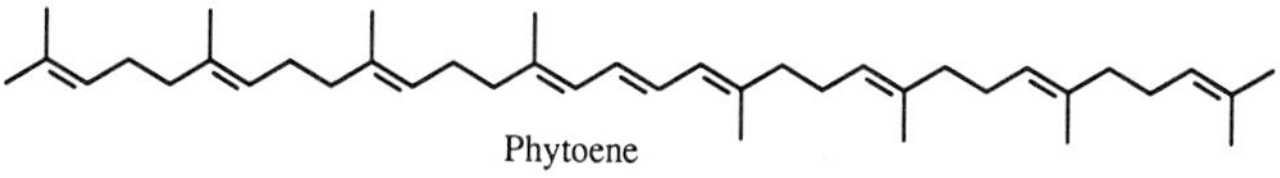

Phytoene

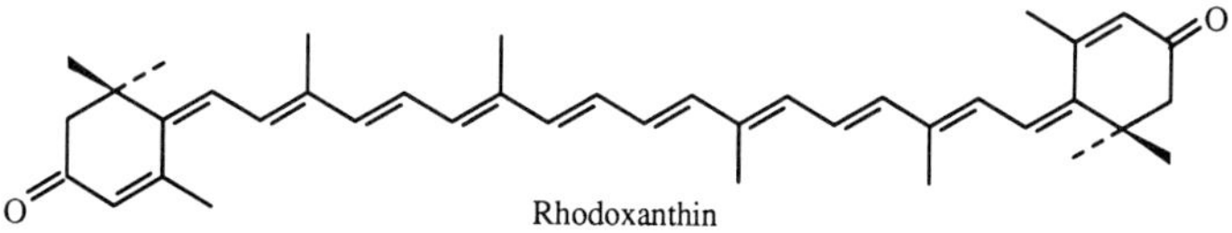

Rhodoxanthin

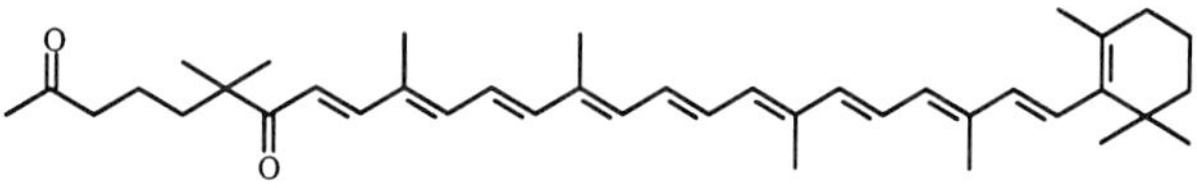

Semi-β-carotenone

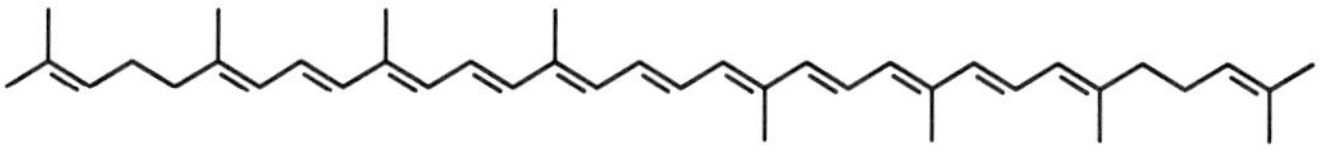

Lycopene

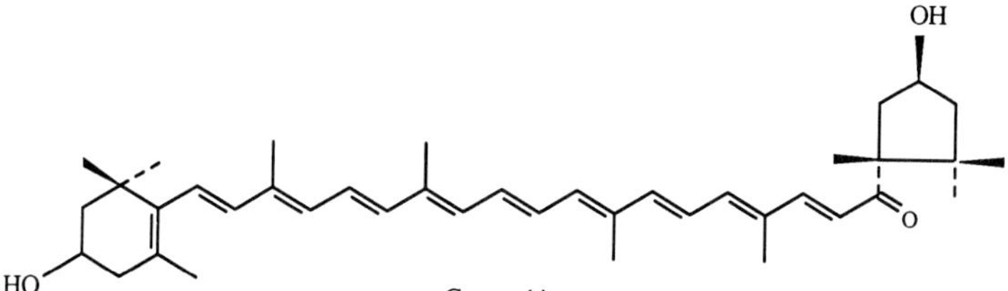

Capsanthin

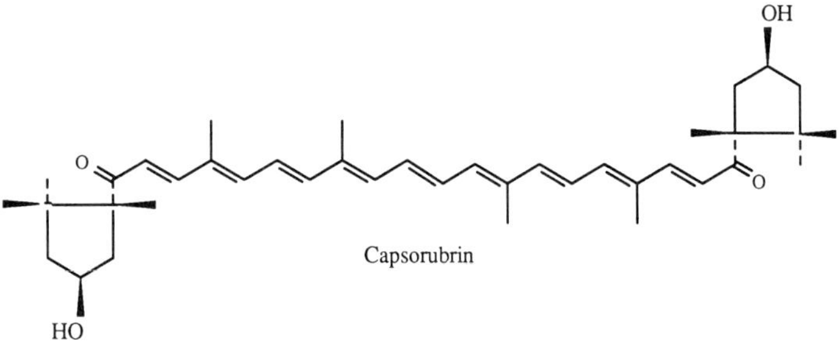

Capsorubrin

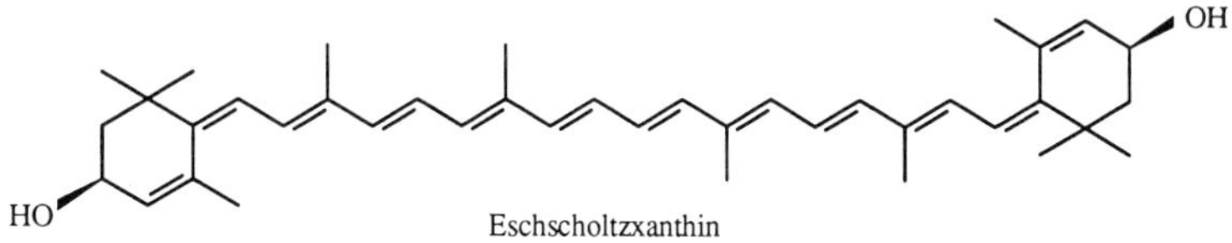

Eschscholtzxanthin

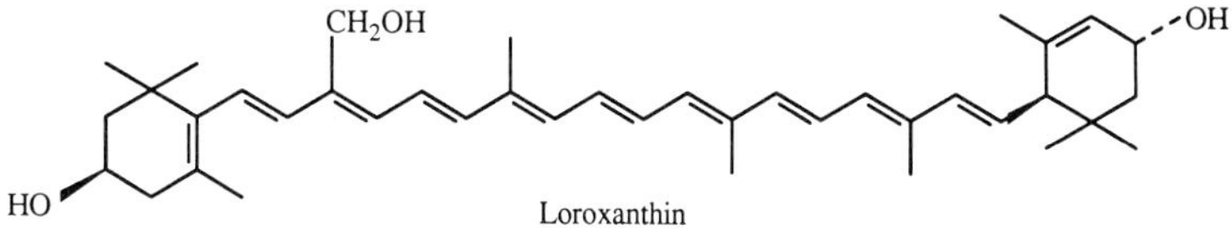

Loroxanthin

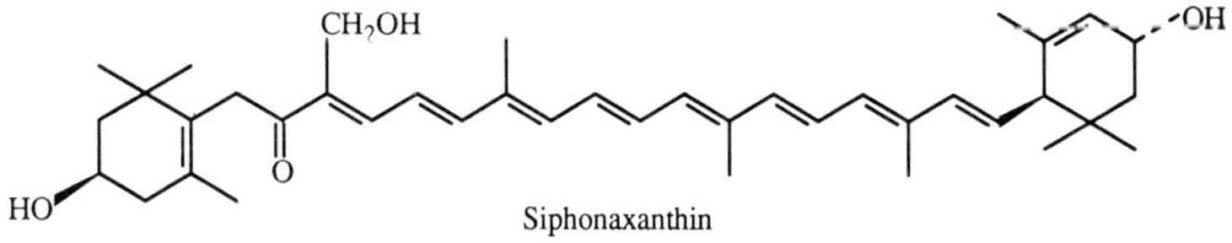

Siphonaxanthin

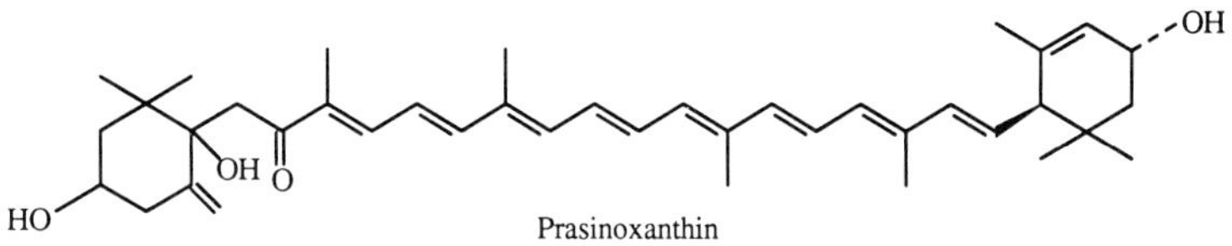

Prasinoxanthin

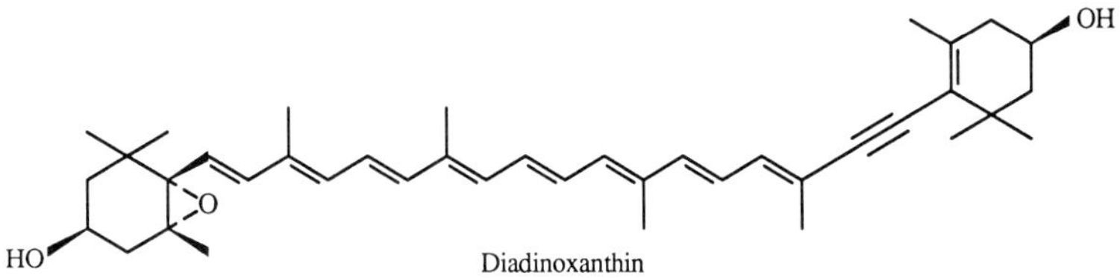

Diadinoxanthin

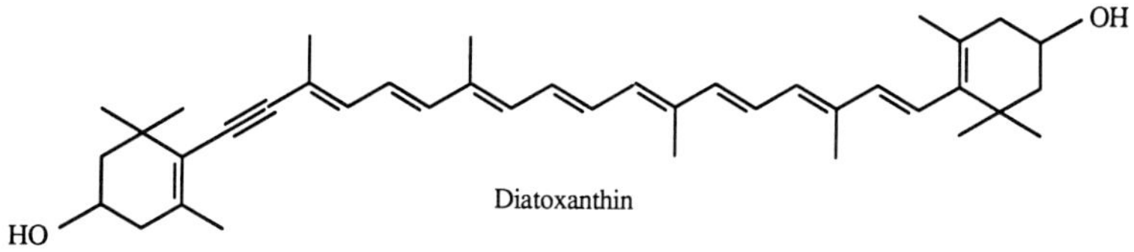

## 14. Carotenoids in Chloroplast Complexes

Diatoxanthin

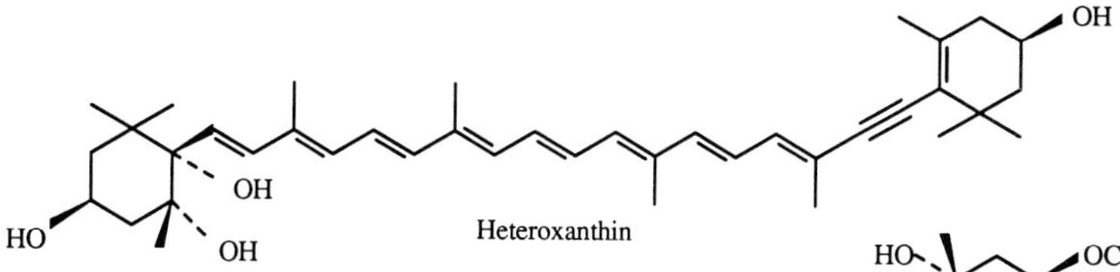

Heteroxanthin

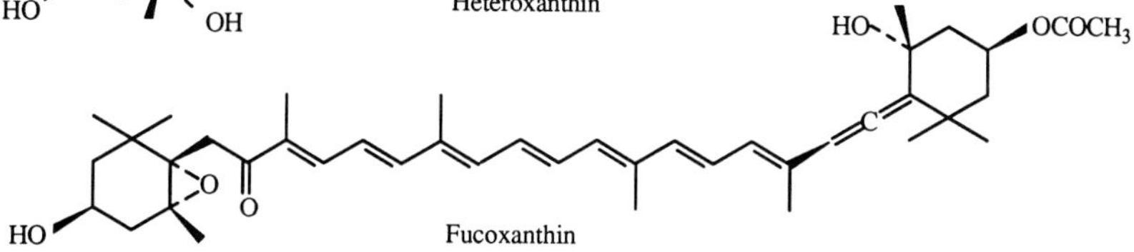

Fucoxanthin

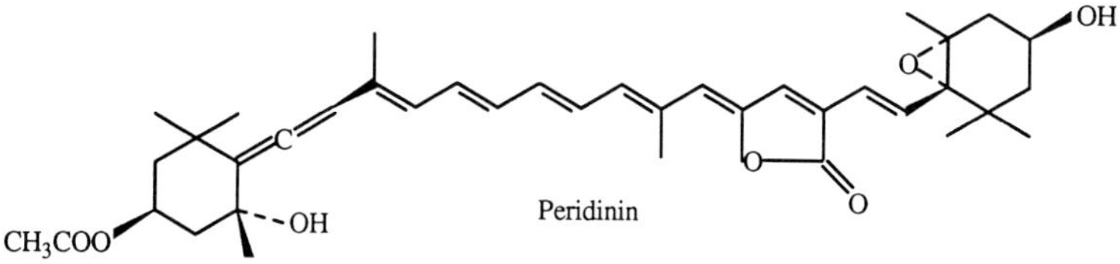

Peridinin

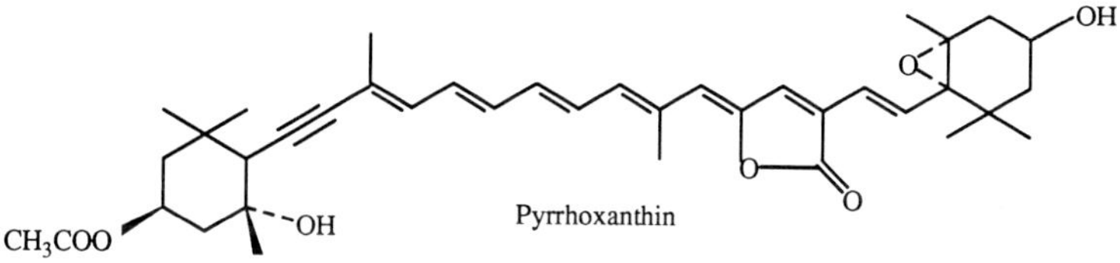

Pyrrhoxanthin

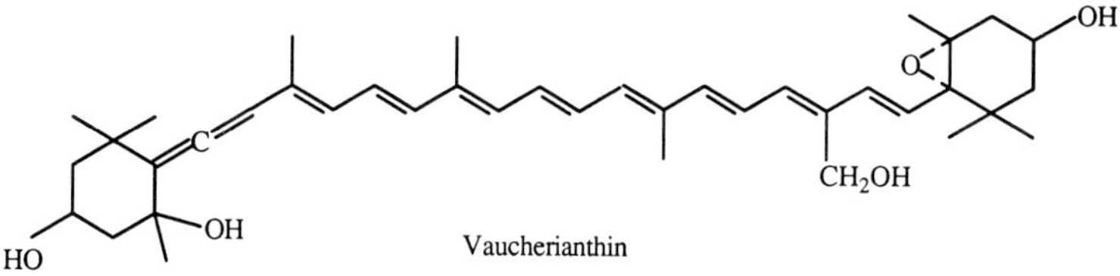

Vaucherianthin

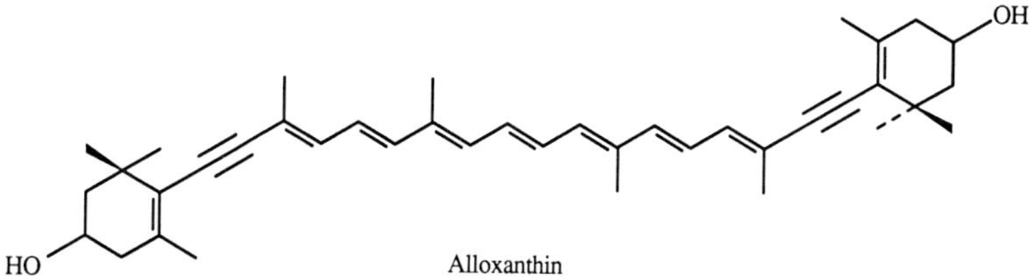

Alloxanthin

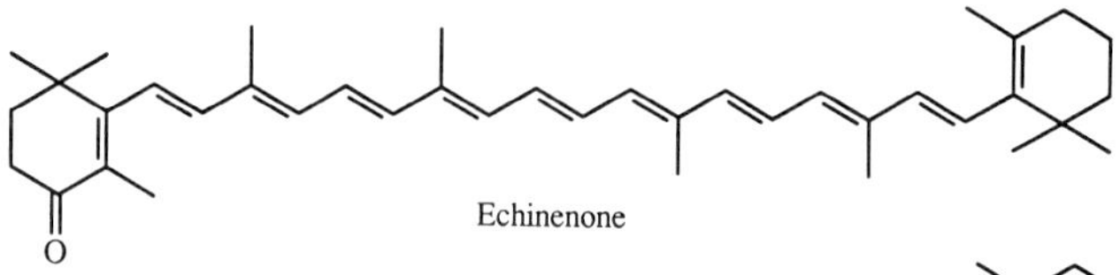

[Structures shown: Echinenone, Caloxanthin, Nostoxanthin, Myxoxanthophyll, Oscillaxanthin, ζ-Carotene, β-Carotene 5,6-epoxide]

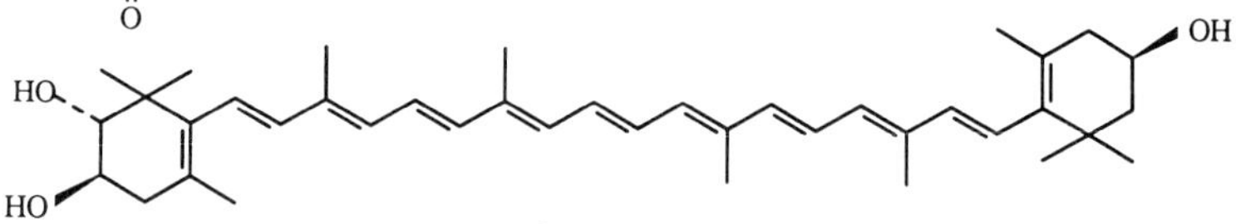

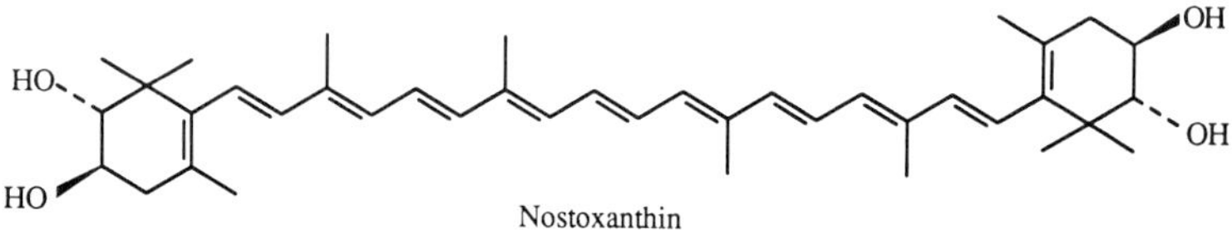

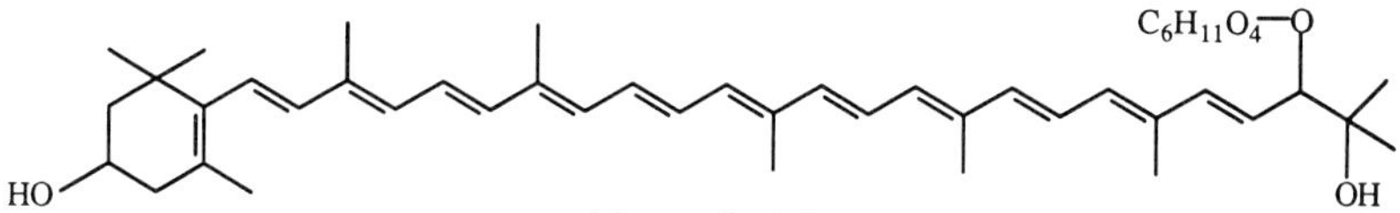

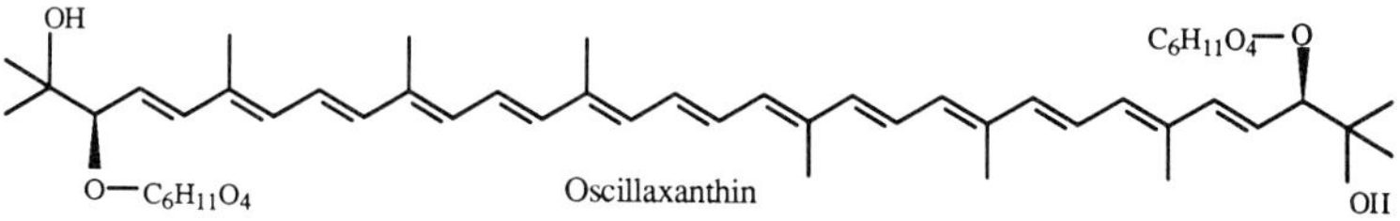

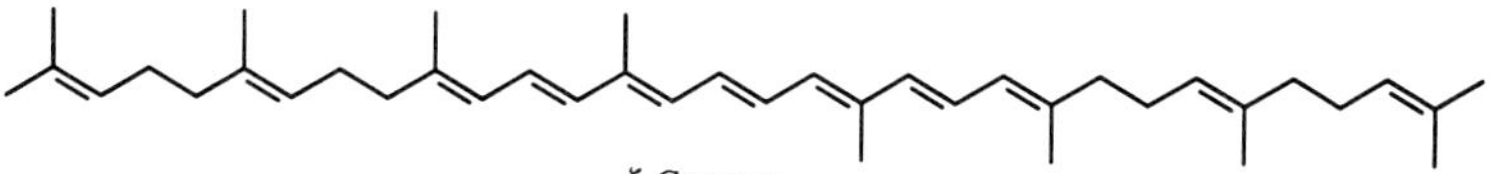

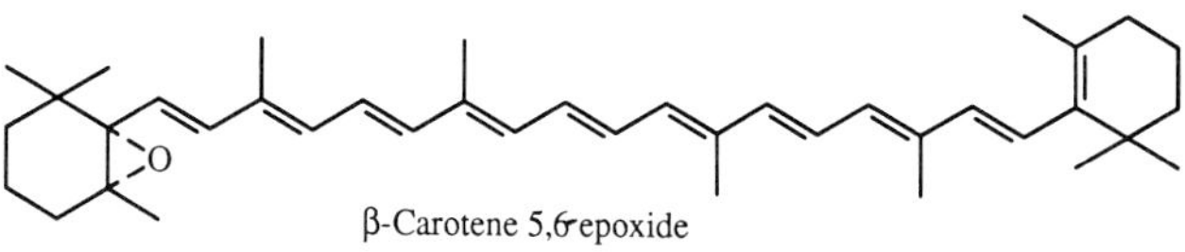

## REFERENCES

Akoyunoglou, G., and Argyroudi-Akoyunoglou, J. (1986). Posttranslational regulation of chloroplast differentiation. In "Regulation of Chloroplast Differentiation" (G. Akoyunoglou and H. Senger, eds.), pp. 571–582. A. R. Liss, New York.

Anderson, J. M. (1985). Chlorophyll–protein complexes of a marine green alga, *Codium* species (Siphonales). *Biochim. Biophys. Acta* **806,** 145–153.

Anderson, J. M., and Osmond, C. B. (1987). Shade–sun responses: Compromises between acclimation and photoinhibition. *In* "Photoinhibition, Topics in Photosynthesis" (D. J. Kyle, C. B. Osmond, and C. J. Arntzen, eds.), Vol. 9, pp. 1–38. Elsevier, Amsterdam.

Anderson, J. M., Waldron, J. C., and Thorne, S. W. (1978). Chlorophyll–protein complexes of spinach and barley thylakoids. *FEBS Lett.* **92,** 227–233.

Armstrong, G. A., Alberti, M., Leach, F., and Hearst, J. E. (1989). Nucleotide sequence, organization, and nature of the protein products of the carotenoid biosynthesis gene cluster of *Rhodobacter capsulatus. Mol. Gen. Genet.* **216,** 254–268.

Ausich, R., Brinkhaus, F., Mukharji, I., Proffitt, J., Yarger, J., Yen, H. C. B., Fink, M., Hardin, L., Hatch, E., Hunsaker, W., Wilber, K., and Wohlfahrt, L. (1990). Isolation of the genes for geranylgeranyl pyrophosphate synthase and phytoene synthase and the use of these genes for phytoene production in bacteria and yeasts. *Abstr. 9th Int. Symp. Carotenoids Kyoto.*, p. 58.

Axelsson, L., Klockare, B., Ryberg, H., and Sandelius, A. S. (1981). The function of carotenoids during chloroplast development. II. Photostability and organization of early forms of chlorophyll(ide). *In* "Photosynthesis, Proceedings of the 5th International Congress, 1980" (G. Akoyunoglou, ed.), Vol. 5, pp. 285–293. Balaban, Philadelphia.

Bahl, J. (1977). Chlorophyll, carotenoid, and lipid content in *Triticum sativa* L. Plastid envelopes, prolamellar bodies, stroma lamellae and grana. *Planta* **136,** 21–24.

Barrett, J., and Anderson, J. M. (1980). The P-700-chlorophyll *a*-protein complex and two major light-harvesting complexes of *Acrocarpia paniculata* and other brown seaweeds. *Biochim. Biophys. Acta* **590,** 309–323.

Barry, P., Young, A. J., and Britton, G. (1990). Photodestruction of pigments in higher plants by herbicide action. I. The effect of DCMU (diuron) on isolated chloroplasts. *J. Exp. Bot.* **41,** 123–129.

Barry, P., Young, A. J., and Britton, G. (1991a). Accumulation of pigments during the greening of etiolated seedlings of *Hordeum vulgare* L. *J. Exp. Bot.* **42,** 229–234.

Barry, P., Evershed, R. P., Young, A., Prescott, M. C., and Britton, G. (1992). Characterisation of carotenoid acyl esters produced in drought-stressed barley seedlings. *Phytochemistry* **31,** 3163–3168.

Beyer, P., and Kleinig, H. (1990). On the desaturation and cyclization reactions of carotenes in chromoplast membranes. *In* "Carotenoids: Chemistry and Biology" ( N. I. Krinsky, M. M. Mathews-Roth, and R. F. Taylor, eds.), pp. 195–206. Plenum, New York.

Björkmann, O. (1981). Responses to different quantum flux densities. *In* "Physiological Plant Ecology. 1. Responses to the Physical Environment" (O. L. Lange, P. S. Nobel, C. B. Osmond, and H. Ziegler, eds.), Encyclopedia of Plant Physiology, Vol. 12A, pp. 109–134. Springer-Verlag, Berlin.

Braumann, T., Weber, G. and Grimme, L. H. (1982). Carotenoid and chlorophyll composition of light-harvesting and reaction centre proteins of the thylakoid membrane. *Photobiochem. Photobiophys.* **4,** 1–8.

Britton, G. (1976). Biosynthesis of carotenoids. *In* "The Chemistry and Biochemistry of Plant Pigments" (T. W. Goodwin, ed.), 2nd Ed., pp. 262–327. Academic Press, London.

Britton, G. (1979). Carotenoid biosynthesis—A target for herbicide activity. *Z. Naturforsch.* **34C,** 979–985.

Britton, G. (1985). Stable isotopes in carotenoid biochemistry. *Pure Appl. Chem.* **57,** 701–708.

Britton, G. (1986). Biosynthesis of chloroplast carotenoids. *In* "Regulation of Chloroplast Differentiation" (G. Akoyunoglou and H. Senger, eds.), pp. 125–134. A. R. Liss, New York.

Britton, G. (1988). Biosynthesis of carotenoids. *In* "Plant Pigments" (T. W. Goodwin, ed.), pp. 133–182. Academic Press, London.

Britton, G. (1990a). Carotenoid biosynthesis—An overview. In "Carotenoids: Chemistry and Biology" (N. I. Krinsky, M. M. Mathews-Roth, and R. F. Taylor, eds.), pp. 167–184. Plenum, New York.
Britton, G. (1990b). The stereochemistry of carotenoid biosynthesis. In "Studies in Natural Products Chemistry" (Atta-ur-Rahman, ed.), Vol. 7, pp. 317–367. Elsevier, Amsterdam.
Britton, G. (1991). The biosynthesis of carotenoids: A progress report. *Pure Appl. Chem.* **63**, 101–108.
Britton, G., Barry, P., and Young, A. J. (1989). Carotenoids and chlorophylls: Herbicidal inhibition of pigment biosynthesis. In "Herbicides and Plant Metabolism" (A. D. Dodge, ed.), pp. 51–72. Cambridge University Press, Cambridge.
Butler, W. L. (1961). Chloroplast development: Energy transfer and structure. *Arch. Biochem. Biophys.* **97**, 287–295.
Cardini, F., Ginaneschi, M., Selva, A., and Chelli, M. (1987). Semi-$\beta$-carotenone from leaves of two cycads. *Phytochemistry* **26**, 2029–2031.
Cogdell, R. J. (1988). The function of pigments in chloroplasts. In "Plant Pigments" (T. W. Goodwin, ed.), pp. 183–230. Academic Press, London.
Cogdell, R. J., and Frank, H. A. (1987). How carotenoids function in photosynthetic bacteria. *Biochim. Biophys. Acta* **895**, 63–79.
Czeczuga, B. (1986). Investigations on carotenoids in Embryophyta. 6. Carotenoids in gymnosperms. *Biochem. Syst. Ecol.* **14**, 13–15.
Demmig, B., Winter, K., Kruger, A., and Czygan, F. -C. (1987). Photoinhibition and zeaxanthin formation in intact leaves. *Plant Physiol.* **84**, 218–224.
Demmig, B., Winter, K., Kruger, A., and Czygan, F. -C. (1988). Zeaxanthin and the heat dissipation of excess light energy in *Nerium oleander* exposed to a combination of high light and water stress. *Plant Physiol.* **87**, 17–24.
Dogbo, O., Laferrière, A., D'Harlingue, A., and Camara, B. (1988). Carotenoid biosynthesis: Isolation and characterization of a bifunctional enzyme catalyzing the synthesis of phytoene. *Proc. Natl. Acad. Sci. USA* **85**, 7054–7058.
Foote, C. S. (1976). Photosensitized oxidation and singlet oxygen: Consequences in biological systems. In "Free Radicals in Biology" (W. A. Pryor, ed.), Vol. II, pp. 85–133. Academic Press, New York.
Friedman, A. L., and Alberte, R. S. (1984). A diatom light-harvesting pigment–protein complex. *Plant Physiol.* **76**, 483–489.
Giuliano, G., Pollock, D., Stapp, M., and Scolnik, P. A. (1988). A genetic–physical map of the *Rhodobacter capsulatus* carotenoid biosynthesis gene cluster. *Mol. Gen. Genet.* **213**, 78–83.
Goedheer, J. C. (1961). Effect of changes in chlorophyll concentration on photosynthetic properties. I. Fluorescence and absorption of greening bean leaves. *Biochim. Biophys. Acta.* **51**, 494–504.
Goodwin, T. W. (1958). Carotenogenesis. XXIV. Changes in carotenoid and chlorophyll pigments in the leaves of deciduous trees during autumn necrosis. *Biochem. J.* **68**, 503–511.
Goodwin, T. W. (1980). "The Biochemistry of the Carotenoids," 2nd Ed., Vol. 1. Chapman and Hall, London.
Goodwin, T. W., and Britton, G. (1988). Distribution and analysis of carotenoids. In "Plant Pigments" (T. W. Goodwin, ed.), pp. 61–132. Academic Press, London.
Green, B. R., and Camm, E. L. (1984). Evidence that CP47 (CPa-1) is the reaction center of photosystem II. In "Advances in Photosynthesis Research" (C. Sybesma, ed.), Vol. II, pp. 95–98. Nijhoff/Junk Publishers, Dordrecht, The Netherlands.

## 14. Carotenoids in Chloroplast Complexes

Gross, J. (1987). "Pigments in Fruits." Academic Press, London.

Grumbach, K. H. (1981). Formation of photosynthetic pigments and quinones and development of photosynthetic activity in barley etioplasts during greening in intermittent and continuous white light. *Physiol. Plant.* **51,** 53–62.

Grumbach, K. H., and Bach, T. J. (1979). The effect of PS-II herbicides, amitrole, and SAN 6706 on the activity of 3-hydroxy-3-methylglutaryl coenzyme A reductase and the incorporation of [2-$^{14}$C]-acetate and DL-[2-$^{3}$H]-mevalonate into chloroplast pigments of radish seedlings. *Z. Naturforsch.* **34C,** 941–943.

Hager, A. (1980). The reversible, light-induced conversions of xanthophylls in the chloroplast. *In* "Pigments in Plants," (F. -C. Czygan, ed.), 2nd Ed., pp. 57–79. Fischer, Stuttgart.

Heintze, A., Gorlach, J., Leuschner, C., Hoppe, P., Hagelstein, P., Schulze-Siebert, D., and Schultz, G. (1990). Plastidic isoprenoid synthesis during chloroplast development: Change from metabolic autonomy to a division-of-labour stage. *Plant Physiol.* **93,** 1121–1127.

Henningsen, K. W., and Boynton, J. E. (1970). Macromolecular physiology of plastids. VIII. Pigment and membrane formation in plastids of barley greening under low light intensity. *J. Cell Biol.* **44,** 290–304.

Ingram, K., and Hiller, R. G. (1983). Isolation and characterization of a major chlorophyll $a/c_2$ light-harvesting protein from a *Chroomonas* species (Cryptophyceae). *Biochim. Biophys. Acta* **722,** 310–319.

Jeffrey, S. W., Douce, R., and Benson, A. A. (1974). Carotenoid transformations in the chloroplast envelope. *Proc. Natl. Acad. Sci. USA* **71,** 807–810.

Katoh, T., Mimuro, M., and Takaichi, S. (1989). Light-harvesting particles isolated from a brown alga *Dictyota dichotoma*: A supramolecular assembly of fucoxanthin–chlorophyll–protein complexes. *Biochim. Biophys. Acta* **976,** 233–240.

Kleudgen, H. K., and Grumbach, K. H. (1983). Differences in phytochrome action on the formation of carotenoids and quinones during chloroplast development in seedlings of barley (*Hordeum vulgare*). *Physiol. Plant.* **57,** 363–366.

Klockare B., Axelsson, L., Ryberg, H., Sandelius, A. S., and Widell, K. O. (1981). The function of carotenoids during chloroplast development. I. Effects of the herbicide SAN-9789 on chlorophyll synthesis and plastid ultrastructure. *In* "Photosynthesis, Proceedings of the 5th International Congress, 1980" (G. Akoyunoglou, ed.), Vol. 5, pp. 277–284. Balaban, Philadelphia.

Kramer, H. J. M., Amesz, J., and Rijgersberg, C. P. (1981). Excitation spectra of chlorophyll fluorescence in spinach and barley chloroplasts at 4K. *Biochim. Biophys. Acta* **637,** 272–277.

Kreuz, K., and Kleinig, H. (1984). Synthesis of prenyllipids in cells of spinach leaf: Compartmentation of enzymes for formation of isopentenyl diphosphate. *Eur. J. Biochem.* **141,** 531–535.

Krinsky, N. I. (1979). Carotenoid protection against oxidation. *Pure Appl. Chem.* **51,** 649–660.

Lichtenthaler, H. K., Ruhn, G., Prenzel, U., Buschmann, C., and Meier, D. (1982). Adaptation of chloroplast-ultrastructure and of chlorophyll–protein levels to high-light and low-light conditions. *Z. Naturforsch.* **37C,** 464–476.

Lichtenthaler, H. K., Burgstahler, R., Buschmann, C., Meier, D., Prenzel, U., and Schonthal, A. (1983). Effect of high light and high light stress on composition, function, and structure of the photosynthetic apparatus. *In* "Effects of Stress on Photosynthesis" (R. Marcelle, H. Clijsters, and M. Van Pouke, eds.), pp. 353–370. Nijhoff, The Hague, The Netherlands.

Lütke-Brinkhaus, F., and Kleinig, H. (1987). Carotenoid and chlorophyll biosynthesis in isolated plastids from mustard seedling cotyledons (*Sinapis alba* L.) during etioplast–chloroplast conversion. *Planta* **170**, 121–129.

Mikulska, M., Siedlecka, M., Poskuta, J. W., and Maleszewski, S. (1987). Effect of long-term action of high oxygen concentration on photosynthetic apparatus in french bean leaves. I. Changes in net photosynthetic rate and in chlorophylls and carotenoids contents. *Photosynthetica* **21**, 175–178.

Milborrow, B. V., Swift, I. E., and Netting, A. G. (1982). Stereochemistry of hydroxylation of the carotenoid lutein in *Calendula officinalis*. *Phytochemistry* **21**, 2853–2857.

Mimuro, M., and Katoh, T. (1991). Carotenoids in photosynthesis: Absorption, transfer and dissipation of light energy. *Pure Appl. Chem.* **63**, 123–130.

Misawa, N., Nakugawa, M., Kobayashi, K., Yamano, S., Izawa, Y., Nakamura, K., and Harashima, K. (1990). Elucidation of the *Erwinia uredovora* carotenoid biosynthetic pathway by functional analysis of gene products expressed in *Eschevichia coli*. *J. Bacteriol.* **172**, 6704–6712.

Monge, E., Val, J., Heras, L., and Abadia, J. (1987). Photosynthetic pigment composition of higher plants grown under iron stress. *In* "Progress in Photosynthesis Research" (J. Biggins, ed.), Vol. 4, pp. 201–204. Nijhoff, Dordrecht, The Netherlands.

Moore, A. L., Dirks, G., Gust, D., and Moore, T. A. (1980). Energy transfer from carotenoid polyenes to porphyrins: A light-harvesting antenna. *Photochem. Photobiol.* **32**, 691–695.

Oelmüller, R., and Mohr, H. (1985). Carotenoid composition in milo (*Sorghum vulgare*) shoots as affected by phytochrome and chlorophyll. *Planta* **164**, 390–395.

Öquist, G., Greer, D. H., and Ogren, E. (1987). Light stress at low temperatures. *In* "Photoinhibition, Topics in Photosynthesis" (D. J. Kyle, C. B. Osmond, and C. J. Arntzen, eds.), Vol. 9, pp. 67–88. Elsevier, Amsterdam.

Osmond, C. B. (1981). Photorespiration and Photoinhibition: Some implications for the energetics of photosynthesis. *Biochim. Biophys. Acta* **639**, 77–98.

Rees, D., Young, A. J., Noctor, G., Britton, G., and Horton, P. (1989). Enhancement of the ΔpH-dependent dissipation of excitation energy in spinach chloroplasts by light activation: Correlation with the synthesis of zeaxanthin. *FEBS Lett.* **256**, 85–90.

Ridley, S. M. (1982). Carotenoids and herbicide action. *In* "Carotenoid Chemistry and Biochemistry" (G. Britton and T. W. Goodwin, eds.), pp. 353–369. Pergamon, Oxford.

Rogers, L. J., Shah, S. P. J., and Goodwin, T. W. (1966). Compartmentation of terpenoid biosynthesis in green plants. *Biochem. J.* **114**, 395–405.

Rowan, K. A. (1989). "Photosynthetic Pigments of Algae." Cambridge University Press, Cambridge.

Ryberg, H., Axelsson, L., Klockare, B., and Sandelius, A. S. (1981). The function of carotenoids during chloroplast development. III. Protection of the prolamellar body and the enzymes for chlorophyll synthesis from photodestruction, sensitized by early forms of chlorophyll. *In* "Photosynthesis, Proceedings of the 5th International Congress, 1980" (G. Akoyunoglou, ed.), Vol. 5, pp. 295–304. Balaban, Philadelphia.

Ryrie, I. J., Anderson, J. M., and Goodchild, D. J. (1980). The role of light-harvesting chlorophyll-*a*/*b*-protein complex in chloroplast membrane stacking. Cation-induced aggregation of reconstituted proteoliposomes. *Eur. J. Biochem.* **107**, 345–354.

Sandelius, A. S., Axelsson, L., Klockare, B., Ryberg, H., and Widell, K. O. (1981). The function of carotenoids during chloroplast development. IV. Protection of galactolipids from photodecomposition, sensitized by early forms of chlorophyll. *In* "Photosynthesis, Proceedings of the 5th International Congress, 1980" (G. Akoyunoglou, ed.), Vol. 5, 305–309. Balaban, Philadelphia.

Sergeant, J. M., and Britton, G. (1984). Observations on the biosynthesis of chloroplast carotenoids. *In* "Advances in Photosynthesis Research" (C. Sybesma, ed.), Vol. IV,

pp. 779–782. Nijhoff/Junk, Dordrecht, The Netherlands.
Siefermann-Harms, D. (1984). Evidence for a heterogeneous organization of violaxanthin in thylakoid membranes. *Photochem. Photobiol.* **40,** 507–512.
Siefermann-Harms, D. (1985). Carotenoids in photosynthesis. I. Location in photosynthetic membranes and light-harvesting function. *Biochim. Biophys. Acta* **811,** 325–355.
Siefermann-Harms, D. (1987). The light-harvesting and protective functions of carotenoids in photosynthetic membranes. *Physiol. Plant.* **69,** 561–568.
Siefermann-Harms, D., and Ninnemann, H. (1982). Pigment organization in the light-harvesting chlorophyll *a*/*b*-protein complex of lettuce chloroplasts: Evidence obtained from protection of the chlorophylls against proton attack, and from energy transfer. *Photochem. Photobiol.* **35,** 719–731.
Siefermann-Harms, D., Hertzberg, S., Borch, G., and Liaaen-Jensen, S. (1981). Carotenoids of higher plants. 13. Lactucaxanthin, an $\varepsilon,\varepsilon$-carotene-3,3'-diol from *Lactuca sativa*. *Phytochemistry* **20,** 85–88.
Sineshchekov, V. A., Litvin, F. F., and Das, M. (1972). Chlorophyll *a* and carotenoid aggregates and energy migration in monolayers and thin films. *Photochem. Photobiol.* **15,** 187–197.
Sitte, P., Falk, H., and Liedvogel, B. (1980). Chromoplasts. In "Pigments in Plants," (F. -C. Czygan, ed.), 2nd Ed., pp. 117–148. Fischer, Stuttgart.
Song, P. -S., Koka, P., Prézelin, B. B., and Haxo, F. T. (1976). Molecular topology of the photosynthetic light-harvesting pigment complex, peridinin–chlorophyll *a*–protein, from marine dinoflagellates. *Biochemistry* **15,** 4422–4427.
Strasser, R. J., and Butler, W. L. (1977). Fluorescence emission spectra of photosystem I, photosystem II, and light-harvesting chlorophyll *a*/*b* complex of higher plants. *Biochim. Biophys. Acta* **462,** 307–313.
Thrash, R. J., Fang, H. L. -B., and Leroi, G. E. (1979). On the role of forbidden low-lying excited states of light-harvesting carotenoids in energy transfer in photosynthesis. *Photochem. Photobiol.* **29,** 1049–1050.
Trevini, M., and Steinmüller, D. (1985). Composition and function of plastoglobuli. II. Lipid composition of leaves and plastoglobuli during beech leaf senescence. *Planta* **163,** 91–96.
Van Hasselt, P. R., and Van Berlo, H. A. C. (1980). Photooxidative damage to the photosynthetic apparatus during chilling. *Physiol. Plant.* **50,** 52–56.
Virgin, H. I., Kahn, A., and Von Wettstein, D. (1963). The physiology of chlorophyll formation in relation to structural changes in the chloroplast. *Photochem. Photobiol.* **2,** 83-91.
Walton, T. J., Britton, G., and Goodwin, T. W. (1969). Biosynthesis of xanthophylls in higher plants: Stereochemistry of hydroxylation at C-3. *Biochem. J.* **112,** 383–385.
Williams, R. J. H., Britton, G., and Goodwin, T. W. (1967). The biosynthesis of cyclic carotenes. *Biochem. J.* **105,** 99–105.
Wolfenden, J., Robinson, D. C., Cape, J. N., Paterson, I. S., Francis, B. J., Mehlhorn, H., and Wellburn, A. R. (1988). Use of carotenoid ratios, ethylene emissions and buffer capacities for the early diagnosis of forest decline. *New Phytol.* **109,** 85–95.
Yamamoto, H. Y. (1979). Biochemistry of the violaxanthin cycle in higher plants. *Pure Appl. Chem.* **51,** 639–648.
Young, A., and Britton, G. (1990). Carotenoids and stress. In "Stress Responses in Plants: Adaptation and Acclimation Mechanisms" (R. Alscher, J. Amther, and J. R. Cumming, eds.), pp. 87–112. Wiley-Liss, New York.
Young, A. J., and Britton, G. (eds.) (1993). "Carotenoids in Photosynthesis." Spinger-Verlag, Stuttgart.
Young, A. J., Barry, P., and Britton, G. (1989). The occurrence of $\beta$-carotene-5,6-epoxide in the photosynthetic apparatus of higher plants. *Z. Naturforsch.* **44c,** 959–965.

# 15

# Molecular Biology of Chromoplast Development

CARL A. PRICE,* MIGUEL CERVANTES-CERVANTES,[†]
NOUREDDINE HADJEB,* LEE A. NEWMAN,*
AND MICHAL OREN-SHAMIR[‡]

*Waksman Institute
Rutgers University
Piscataway, New Jersey

[†]Biochemistry Department
Escuela Nacional de Ciencias Biológicas
Instituto Politechnico Nacional
Mexico City, Mexico

[‡]Department of Plant Genetics
Weizmann Institute
Rehovot 76100, Israel

    I. Introduction
   II. Chromoplasts: Ultrastructure, Chemistry, and Function
       A. Ultrastructure
       B. Chemistry and Function
  III. Genes Affecting Chromoplasts
  IV. Chromoplast DNA
   V. Chromoplast Proteins
       A. Proteins of Chromoplast Membranes
       B. Enzymes of Carotenoid Biosynthesis
  VI. Chromoplasts and Leucoplasts: Importance of Not Being Green
 VII. Conclusions and Prospects
       References

## I. INTRODUCTION

Most of the chapters in this volume discuss chloroplasts, but plastids appear in many additional guises, for example, amyloplasts, chromoplasts, and elaioplasts (Whatley, 1982). This chapter focuses on the molecular biology of chromoplasts which, as the name implies, are heavily pigmented plastids but are usually devoid of chlorophyll. Because the concept of chromoplasts is inseparable from carotenoids, in terms of the large amounts and often unusual species of carotenoids that accumulate in chromoplasts, the molecular biology of chromoplasts also must concern itself with the genes involved in carotenoid biogenesis and with the structures associated with these carotenoids. First we review the development, structure, and composition of chromoplasts to establish the range of physical and chemical structures under genetic control.

Plastids that accumulate large amounts of carotenoids are called chromoplasts (Simpson *et al.*, 1977; Sitte *et al.*, 1980). Plastids appear to be the sole sites of carotenoid synthesis and accumulation in higher plants. Chloroplasts of course contain $\beta$-carotene and various xanthophylls, but the abundance of carotenoids in chromoplasts can be quite prodigious.

Table I shows the total amounts and proportions of carotenoids in fruits of several genotypes of *Capsicum annuum*. Red-fruited peppers contain almost 1 mg carotenoid per gram fresh weight, whereas yellow varieties contain 1/40 as much. Whereas the carotenoids of chloroplasts are principally lutein and $\beta$-carotene, chromoplast carotenoids tend to the exotic; in this example, the keto-carotenoids capsanthin and capsorubin account for half the total.

The mere loss of chlorophyll and consequent unmasking of carotenoids does not make a proper chromoplast; Sitte *et al.* (1980) coined the term "gerontoplast" for the plastids of senescing leaves in which the color is created by residues of the thylakoid and envelope carotenoids. In contrast, chromoplasts are the sites of active carotenoid biosynthesis.

As we shall see in Section III, the absence of chlorophyll is not a prerequisite of chromoplasts. In the final section the similarities between chromoplasts and leucoplasts shall be noted also.

## II. CHROMOPLASTS: ULTRASTRUCTURE, CHEMISTRY, AND FUNCTION

### A. Ultrastructure

All plastids contain carotenoids in their envelope membranes. The accumulation of carotenoids in chromoplasts occurs in various structures that may derive from the envelope but are inside it. The variety of these

**TABLE I**

**Carotenoid Content of Different Genotypes of *Capsicum annuum*[a]**

| | Fruit color and genotype[b] | | | |
|---|---|---|---|---|
| Carotenoid | Red $y^+c^+$ | Yellow-orange $y^-c_1^+$ | Orange $y^+c_1^-$ | Lemon-yellow $y^-c_1^-$ |
|---|---|---|---|---|
| Total carotenoid (μg/g fresh wt) | 869 | 91 | 24.9 | 22.4 |
| Antheraxanthin | 2.0 | 7.0 | 10.6 | 15.9 |
| Auroxanthin | — | 7.8 | — | — |
| Capsanthin | 39.9 | — | 35.4 | — |
| Capsanthin isomer | 3.5 | — | — | — |
| Capsanthin-5,6-epoxide | — | — | 6.1 | — |
| Capsorubin | 11.3 | — | 14.3 | — |
| Capsorubin isomer | 1.4 | — | — | — |
| ζ-Carotene | — | 2.6 | — | — |
| β-Carotene | 23.0 | 15.7 | — | 1.0 |
| α-Carotene | 1.4 | 2.3 | — | — |
| β-Carotene-1,6-epoxide | — | 3.9 | — | — |
| Cryptocapsin | — | — | 8.0 | — |
| Cryptoxanthin | 3.7 | — | — | — |
| Cryptoxanthin-5,6-epoxide | — | 3.9 | — | — |
| Hydroxy-α-carotene | — | 1.6 | 0 | 5.7 |
| Hydroxy-α-carotene-5,6-epoxide | — | 3.2 | — | — |
| Lutein | — | 17.4 | — | 28.1 |
| Luteoxanthin | — | 4.2 | — | — |
| Neoxanthin | 2.5 | 5.8 | 2.7 | 20.0 |
| Neurosoporene | — | 0.5 | — | — |
| Phytofluene | 0.2 | 7.0 | — | — |
| Unidentified | 2.2 | 3.2 | — | — |
| Violaxanthin | 7.1 | 16.3 | 14.9 | 31.3 |
| β-Zeacarotene | — | 0.5 | — | — |
| Zeaxanthin | 2.7 | — | 7.3 | — |

[a] After Simpson *et al.* (1977).
[b] Genotypes based on system of two loci ($y$, $c$) rather than the present system based on three loci ($y$, $c_1$, $c_2$).
[c] Amounts are percentages of total carotenoids.

structures is related to the kinds and amounts of carotenoids present. Sitte *et al.* (1980) distinguish four major types of chromoplasts based on these structures: globulous, tubulous, membranous, and crystalline. Hybrid types can occur also.

The most common and perhaps the most primitive type of chromoplast is globulous (cf. Gross, 1987): the shape is lenticular or spheroidal and the carotenoids in the stroma are contained in plastoglobuli concentrated under the inner membrane (Sitte *et al.*, 1980).

Tubulous chromoplasts are characterized by tube-shaped internal elements that contain carotenoids. Some tubules resemble prolamellar bodies and others plastoglobuli (Spurr and Harris, 1968). Tubules occur in certain genotypes of *C. annuum* (Simpson *et al.*, 1977), but apparently not with specific genes: the critical element appears to be the kind of carotenoid present. Emter *et al.* (1990) associate tubules with apolar carotenoids such as esters of lutein.

Membranous chromoplasts, found most often in flowers, derive from proplastids or chloroplasts (Thomson and Whatley, 1980). They contain spherical carotenoid-containing bodies surrounded by up to 20 concentric membranes (Sitte *et al.*, 1980).

Crystallous chromoplasts contain crystals of pure carotenoids surrounded by a thin membrane. Some crystals remain small and are visible only in electron micrographs, whereas others grow large enough to distort the shape of the chromoplast. Smaller crystals, found primarily in carrot roots, tomatoes, and watermelons, may total over 100 per cell (Sitte *et al.*, 1980; Thomsom and Whatley, 1980).

It should be noted, however, that multiple kinds of structures can exist within the same plant and even within the same plastid. *Capsicum* fruits, for example, contain membranous (Spurr and Harris, 1968) as well as tubulous (Spurr and Harris, 1968) and globulous chromoplasts (Gross, 1987).

## B. Chemistry and Function

The various types of carotenoids in chromoplasts provide another example of nature's capacity to experiment in organic chemistry, especially when the products are not essential to the survival of the plant. The pathway of synthesis of $\beta$-carotene, shown in Figs. 1, 3, and 4 of Chapter 14, is common to all plastids and generally proceeds with increasing desaturation. The enzymes identified in plants to date are listed in Table II. Carotenoids may be acyclic, such as lycopenes, or alicyclic, such as carotenes and most oxygenated carotenoids. The carotenoids that accumulate in chromoplasts can be $\beta$-carotene, intermediates in the pathway, or a variety of elaborated products in which the alicyclic rings are oxygenated as alcohols, epoxides, ketones, or aldehydes (Szabolcs, 1989). Starting with the classical work of Karrer in the 1930s (cf. Karrer and Jucker, 1948), some 29 carotenoids have been isolated from *Lilium tigrinum* flowers alone. In 1980, Goodwin counted 40 authenticated carotenoids in flowers and 70 in fruits.

Combining the numbers of distinct carotenoids and the kinds of struc-

## 15. Chromoplast Development

**TABLE II**

**Enzymes of Carotenoid Biosynthesis in Higher Plants**

| Enzyme | Product | Location | Reference |
|---|---|---|---|
| GGPP synthase (prenyl transferase) | GGPP | Chromoplast stroma | Spurgeon et al. (1984); Dogbo et al. (1988) |
| Isopentenyl pyrophosphate isomerase | Dimethylallyl pyrophosphate | Chromoplast stroma | Spurgeon et al. (1984); Dogbo et al. (1988) |
| Phytoene synthase | Phytoene | Envelope | Lütke-Brinkhaus et al. (1982); Dogbo et al. (1988) |
| Phytoene dehydrogenase | Lycopene | Envelope | Lütke-Brinkhaus et al. (1982) |
| Phytoene desaturase | Various | Chromoplast membrane | Schmidt et al. (1989); Mayer et al. (1990) |

tures with which they are associated, we can begin to estimate the number of genes that may be involved in chromoplast development. In this respect, the categories of carotenoid-containing flowers and fruits are instructive (Table III). In organs containing only small amounts of carotenoids, the species of carotenoids were the same as those in chloroplasts, as if the chlorophyll was simply degraded or never formed. Many flowers and fruits contain abundant lycopene or $\beta$-carotene, representing overproduction of normal intermediates or constituents of the chloroplast. Others contain abundant hydroxy or epoxy derivatives of $\beta$-carotene, such as zeaxanthin and antheraxanthin, which are also common constituents of chloroplasts.

**TABLE III**

**Flowers and Fruits Categorized by Carotenoid Contents**[a]

| Carotenoid content | Flowers | Fruits |
|---|---|---|
| Insignificant amounts | White flowers | White fruits |
| Small amounts of chloroplast carotenoids | [b] | Lutein |
|  |  | Violaxanthine |
|  |  | Neoxanthine |
| Abundant lycopenes and derivatives | Rare | Lycopene |
| Abundant carotenes and derivatives | $\beta$-Carotene | $\beta$-Carotene |
| Abundant epoxides | Antheraxanthine | Zeaxanthine |
| Species-specific carotenoids | Escholtzxanthine | Capsanthine |
|  | Prolycopene | Taraxaxanthine |

[a] Modified from Goodwin (1980).
[b] Except among aroids and orchids, green flowers are relatively rare.

Still others are species-specific carotenoids that result from more elaborate modifications. Since the genes required for the synthesis of $\beta$-carotene and of a number of xanthophylls are common to all plastids, these genes may or may not be specific to chromoplasts. The accumulation of lycopene or of $\beta$-carotene in chromoplasts could be due to the overexpression of ordinary genes by *trans*-acting factors or by the *de novo* expression of special genes that are transcribed only during chromoplast development. The *trans*-acting factors themselves may be development- or cell-specific proteins for which corresponding genes would be required. Genes also would be needed for the protein components of the special structures seen in tubulous and membranous chromoplasts. Since lycopenes and carotenes typically accumulate as droplets or as crystals, no additional proteins would seem to be required for those types of chromoplasts.

The function of brightly colored flowers and fruits is esthetic. The absence of chlorophyll in an otherwise green landscape attracts the attention of animals, such as ourselves, to flowers and fruits, so that pollination is promoted and seeds are dispersed. Surely it is no coincidence that in nature nearly all green flowers are inconspicuously small and wind pollinated; the principal exceptions are among the aroids, which use odor rather than color to attract insects.

## III. GENES AFFECTING CHROMOPLASTS

Chromoplast-specific genes can be separated into two categories (Fig. 1): those whose action is causal in the differentiation of chromoplasts and those that are expressed as a consequence of chromoplast development. We have proposed the symbol *Chr* for genes in both categories (Newman *et al.*, 1989).

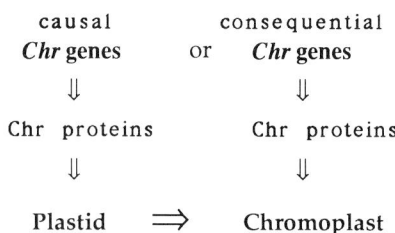

**Fig. 1.** Definition of *Chr* genes. In the class of *Chr* genes as proposed by Newman *et al.* (1989), some act causally to convert a proplastid, leucoplast, or chloroplast into a chromoplast and others are expressed as a consequence of chromoplast development.

## 15. Chromoplast Development

How many chromoplast-specific genes are there? The genetics of fruit color in the Solanaceae provides some clues. Tables IV and V list the genes known to affect the accumulation of carotenoids in *Lycopersicon esculentum* and *C. annuum*. The phenotypes listed refer to the color of the fruit. Whereas color is a valid indicator of genotype against an isogenic background, it has severe limitations. Simpson *et al.* (1977) reported that orange-fruited genotypes included varieties that did and did not contain ketocarotenoids and whose contents of carotenoids differed by a factor of 30. (Interpretation of the data by Simpson *et al.* is suspect because of their use of a genetic model in which carotenoid accumulation in *Capsicum* spp. is determined by two rather than three loci.) There must be additional chromoplast-specific genes for which mutations are not recovered or recognized. The gene $y_1$ in maize, which controls carotene biosynthesis, has been cloned (Buckner *et al.*, 1990), but mutations in $y_1$ affect carotene biosynthesis in plastids generally, not specifically in chromoplasts. To date, all genes known to affect carotenoid accumulation are Mendelian; that is, although the carotenoids are synthesized in the plastid, the plastid genome appears to play a passive role.

Among the genes known to affect the accumulation of carotenoids in chromoplasts, we may find homologs of genes known from bacterial studies to encode the enzymes of carotenoid biosynthesis (Table VI). In fact, Bartley *et al.* (1991) isolated a cDNA from soybean that complements a mutant of *Rhodobacter capsulatus* deficient in *crtI*. The predicted amino acid sequence of the protein, named Pds1 for phytoene desaturase, shows only limited similarity to CrtI or to the related CrtD.

**TABLE IV**

**Genes Known to Affect Carotenoid Accumulation in *Capsicum annuum*[a]**

| $y$ | $c_1$ | $c_2$ | $cl$ | Fruit color |
|---|---|---|---|---|
| + | + | + | − | Red |
| + | − | + | − | Light red |
| + | − | − | − | Orange |
| − | + | + | − | Orange-yellow |
| − | + | − | − | Pale orange-yellow |
| − | − | + | − | Lemon-yellow |
| − | − | − | − | White |
| + | + | + | + | Brown |
| − | + | + | + | Olive-green |

[a] From Smith (1950); Hurtado-Hernandez and Smith (1985).

TABLE V

Genes Known to Affect Carotenoid Accumulation in *Lycopersicon esculentum*[a,b]

| Genes | | | phytoene | phytofluene → ζ-carotene → neurosporene → lycopene → γ-carotene → β-carotene | | | | | δ-carotene → α-carotene | pro-γ-carotene | Total polyenes | Phenotype of fruit color |
|---|---|---|---|---|---|---|---|---|---|---|---|---|
| r | t | at | | | | | | | | | | |
| + | + | | ■ ■ | | | ■ | ■ | ■ | | | 87 | red |
| + | − | | ■ ■ | ■ | | ■ | ■ | ■ | ■ | ■ | 158 | tangerine |
| − | + | | ■ ■ | | | ■ | ■ | ■ | | | 4 | yellow |
| − | − | | ■ ■ | | | ■ | ■ | ■ | ■ | | 34 | yellow-tangerine |
|  |  | − | ■ ■ | | | ■ | ■ | ■ | | | 13 | apricot |
| − | + | − | ■ ○ | | | ■ | ■ | ■ | | | 2 | yellow-apricot |
| + | + | − | ■ ■ | ■ | | ■ | ■ | ■ | ■ | | 29 | tangerine-apricot |
| − | − | − | ■ ■ | | | ■ | ■ | ■ | | | 10 | yellow-tangerine-apricot |

| hp | B | mo_B | Del | | | | | | | | | | |
|---|---|---|---|---|---|---|---|---|---|---|---|---|---|
| − | | | | ■ ■ | | | ■ | ■ | ■ | | | 88 | high pigment |
| | + | | | ■ ■ | | | ■ | ■ | ■ | | | 50 | intermediate β[c] |
| | + | − | | ■ ■ | | | ■ | ■ | ■ | | | 80 | high β[c] |
| | | | + | ■ ■ | ■ | | ■ | ■ | ■ | ■ ■ | ■ | 84 | delta (δ-carotene) |
| − | | | + | ■ ■ | ■ | | ■ | ■ | ■ | ■ ■ | ■ | 55 | high-delta (δ-carotene) |

| gh | nn | | | | | | | | | | | |
|---|---|---|---|---|---|---|---|---|---|---|---|---|
| − | | ■ ■ | | | ○ | | | | | | 295 | ghost |
| | − | ■ ■ | | | ○ | ○ | ○ | | | | | rin |

[a] Reprinted with permission from Goodwin (1981).

[b] Wild-type genotypes are represented as +, homozygous recessives as −. Presence of a polyene is shown for substantial amounts (■) or trace amounts (○). Polyenes in the presumed normal pathway of biosynthesis of β-carotene are shown by dark shading; unusual or aberrant carotenoids are shown by light shading. Gene symbols are *r*, red; *t*, tangerine; *at*, apricot; *hp*, high pigment; *B*, beta; $mo_B$, modifier beta; *Del*, Delta; and *gh*, ghost.

[c] Normal fruits ($B^+B^+mo_Bmo_B$) are red; lycopene constitutes 90% of the pigment. Fruits with the $BBmo_Bmo_B^+$ or $BBmo_Bmo_B$ genotypes are orange and contain 50% lycopene–50% β-carotene, or 10% lycopene–90% β-carotene, respectively.

Single sets of genes may be encoding these enzymes in plants in which overproduction in chromoplasts occurs as the result of one or more *trans*-acting factors that are specific to chromoplasts. A more common strategy for plants, however, would be to have multigene families in which one member is expressed in green leaves and others are expressed in chromoplast-containing tissues.

In ripening fruits of wild-type tomato and pepper, chromoplasts arise

**TABLE VI**

**Genes Encoding Common Pathway of Carotenoid Biosynthesis**[a]

| Gene | Gene product | Function |
|---|---|---|
| crtA | | Oxidation of $\alpha$-carbon to oxo |
| crtB | | Condensation of 2 GGPP to prephytoene pyrophosphate |
| crtC | | Hydroxylation of neurosporene |
| crtD | Desaturase | Desaturation (with crtG) |
| crtE | Dehydrogenase | Prephytoene pyrophosphate dehydrogenation |
| crtF | | O-Methylation |
| crtG | | Desaturation (with crtD) |
| crtI | Desaturase | Desaturation of phytoene to phytofluene |

[a] After Scolnik et al. (1980).

from chloroplasts. Mutations at multiple loci in pepper (Odland and Porter, 1938) and tomato result in a recessive phenotype in which the immature plastids are yellowish white, that is, devoid of chlorophyll (Table VII). These sw loci affect only the immature fruit, so chromoplasts in the mature fruit of $sw^-sw^-$ plants are indistinguishable from isogenic $sw^+$ plants. The idea that yellowish white fruit result from mutations at multiple loci derives from the complicated genetics of the process. A typical set of data is consistent with four loci for sw (cf. Lippert et al., 1966): when one pair of loci is $sw^+$, immature pods are yellow-green; when two pair are $sw^+$, immature pods are cedar green; and four pair $sw^+$, immature pods are very dark green.

Since the leaves of $sw^-sw^-$ plants appear perfectly normal, the inference is that the products of $sw^+$ loci are necessary for normal chloroplast development, but act only in developing fruit tissues.

A single separate locus has an opposite effect: the cl locus in C. annuum

**TABLE VII**

**Genes Affecting Immature Fruit Color in *Capsicum annuum*** [a]

| $SW_1$ | $SW_2$ | Immature fruit color |
|---|---|---|
| + | + | Cedar green |
| − | + | Lettuce green |
| − | − | Sulfury white |

[a] Adapted from Odland and Porter (1938).

and *gf* (greenfruit) in tomato affect the persistence of chlorophyll in ripening fruit. In homozygous recessive plants $cl^-cl^-$, carotenoids accumulate normally, but the fruits retain high levels of chlorophyll and the granal thylakoid system is retained in a rather condensed form (Laborde and Spurr, 1973). The result is a mature fruit of a chocolate-brown color. Protein gels of such plastids show a normal array of chloroplast proteins plus those of wild-type chromoplasts (cf. Fig. 2). The accumulation of chlorophyll in immature $cl^-cl^-$ fruits is dependent on light as it is in leaves, but the loss of chlorophyll in ripening $cl^+$ fruits is also light dependent.

We infer that the product of $cl^+$ is tissue, development, and light specific, acting only in plastids of ripening fruit to cause a disintegration of thylakoid membranes.

## IV. CHROMOPLAST DNA

Until a few years ago, chloroplasts were the only plastids from which DNA had been isolated and characterized. Because of the known interconvertibility among plastids, the untested assumption was that all plastid genomes were identical. From restriction and gene maps of plastid DNA of daffodil (Thompson, 1980), tomato (Iwatsuki *et al.*, 1985; Hunt *et al.*, 1986), and *C. annuum* (Gounaris *et al.*, 1986), the organization of chloroplast and chromoplast DNAs appears to be identical. Kobayashi *et al.* (1990) have assembled overwhelming evidence that at least some chromoplast DNAs are methylated differentially and that methylation corresponds to the down-regulation of specific plastid genes.

With one possible exception (Gounaris and Price, 1987), plastids of higher plants contain no genes that are expressed specifically during chromoplast development or that are involved in carotenoid metabolism. Most genes encoding components of the photosynthetic apparatus are turned off or down, although others continue to be expressed (cf. Piechulla, 1988). The gene encoding geranylgeranyldiphosphate (GGPP) synthase, however, occurs in the plastid DNA of the alga *Cyanophora paradoxa* (Michalowski *et al.*, 1991). Initial indications of the occurrence of *crtA* and *crtE* in the alga subsequently were found to be due to errors in the published sequences (Armstrong *et al.*, 1989) of the carotene gene cluster of *Rhodobacter capsulatus* (Scolnik *et al.*, 1980): part of the published sequence for *crtA* actually encodes BchI, an enzyme involved in the synthesis of bacteriochlorophyll, and the sequence reported for *crtE* actually corresponds to GGPP synthase.

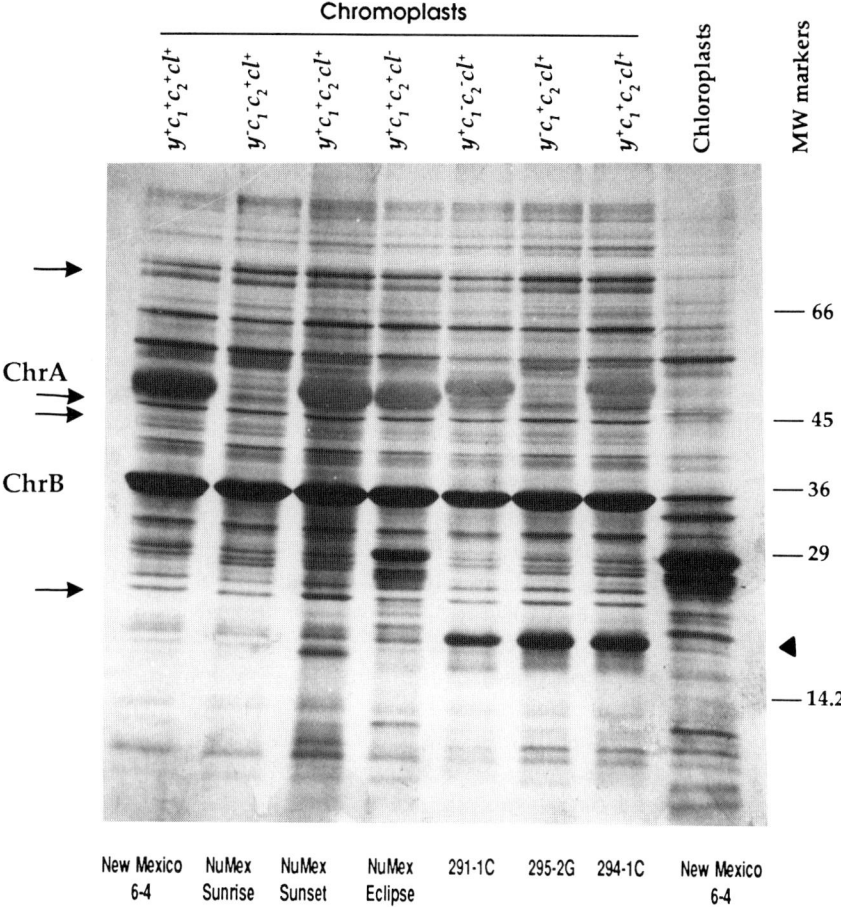

**Fig. 2.** Chromoplast proteins of different genotypes of *Capsicum annuum*. Chromoplasts were isolated from plants that differ in the genes responsible for the color of the ripe fruit. The membranes were recovered, the constituent polypeptides separated by SDS-PAGE, and the gel stained with silver stain. The genotypes of the plants are indicated above each lane. Positions of ChrA (58 kDa) and ChrB (35 kDa) and several minor chromoplast-specific polypeptides are shown by arrows. Note that ChrB occurs in chromoplasts of all genotypes, whereas ChrA occurs only in plants that are $y^+$. A polypeptide of approximately 18 kDa, marked with an arrowhead, appeared in abundance in cultivars 291-1C, 295-2G, and 294-1C, whose fruits are highly pungent ("hot"), but either was absent or occurred in much smaller amounts in the New Mexico series, which are sweet.

## V. CHROMOPLAST PROTEINS

### A. Proteins of Chromoplast Membranes

The differentiation of plastids into chromoplasts is likely to be caused by the action of gene products within the plastid (cf. Fig. 1), but not in most cases by products of plastid genes. Such regulatory proteins are likely to be present in small amounts and none have yet been detected. A consequence of chromoplast differentiation, however, can be the accumulation of structural and enzymatic proteins; the search is for proteins that occur uniquely in chromoplasts.

An essential element in such an approach is the isolation of pure intact chromoplasts. This isolation has been done for *Tropaeolum majus* (Winkenbach *et al.*, 1976), daffodil (cf. Thompson, 1980), *C. annuum* (cf. Camara *et al.*, 1985), and a few other species (cf. Emter *et al.*, 1990). Developing chromoplasts can be isolated from immature tomatoes, but chromoplasts from ripe tomatoes are too fragile.

Winkenbach *et al.* (1976) detected a 30-kDa polypeptide as the dominant protein by SDS–gel electrophoresis in chromoplast membranes (seen as tubules or filaments) from *T. majus*. Emter *et al.* (1990) subsequently reported a correlation, which they took to be causal, between the occurrence of this protein and the formation of tubules. We also employed SDS–gel electrophoresis for the analysis of chromoplasts from ripe fruits of *C. annuum* isolated on silica-sol gradients (Hadjeb *et al.*, 1988; cf. Fig. 2) and isolated two prominent membrane-associated proteins of 35 and 58 kDA, which we named ChrA and ChrB. Indirect analysis of ripening tomato fruits (Bathgate *et al.*, 1985; Jones *et al.*, 1989) has shown proteins of similar molecular mass (Table VIII).

ChrA is the principal membrane protein of pepper chromoplasts, representing half the total protein of the plastid; ChrB is less abundant, representing about 10%. On the basis of western blots (Newman *et al.*, 1989), ChrA begins to accumulate quite late in fruit ripening (Fig. 3) whereas ChrB is visible at the earliest stage (Fig. 4). No cross-reacting proteins are detectable in green-mature fruits or in any other tissue.

ChrA occurs as a carotenoid–protein complex (Cervantes *et al.*, 1990). Because of the nearly exclusive localization of carotenoids in plastids, carotenoproteins are likely to be plastid proteins. Although a few carotenoproteins have been isolated from cyanobacteria (cf. Bullerjahn and Sherman, 1986) and eukaryotic algae (Prézelin, 1987), the only other examples reported in higher plants are in the photosynthetic apparatus (cf. Peter and Thornber, 1991; Thornber *et al.*, 1991) and in isolations from carrot roots (Bryant *et al.*, 1991; Milicua *et al.*, 1991). Emter *et al.* (1990) identi-

**TABLE VIII**

**Chromoplast-Specific Proteins Isolated from Chromoplasts**[a]

| Protein | Function | Source | Reference |
|---|---|---|---|
| 30-kDa protein | Correlates with tubule formation | *Tropaeolum majus* petals | Winkenbach *et al.* (1976); Emter *et al.* (1990) |
| 32.5-kDa protein | ? | *Lycopersicon esculentum* | Bathgate *et al.* (1985) |
| 53-kDa protein | ? | *L. esculentum* | Bathgate *et al.* (1985) |
| ChrA | Complexes with carotenoids | *Capsicum annuum* fruits | Cervantes *et al.* (1990) |
| ChrB | ? | *C. annuum* fruits | Newman *et al.* (1989) |
| GGPP synthase | Carotenoid biosynthesis | *C. annuum* fruits | Dogbo and Camara (1987) |
| Phytoenesynthase | | *C. annuum* fruits | Dogbo *et al.* (1988) |

[a] Evidence that a protein occurs specifically in a chromoplast may be direct, for example, the detection of that protein in isolated chromoplasts, or indirect, for example, increased abundance of that plastid protein during chromoplast development.

fied a 30-kDa protein whose presence correlates with tubules in chromoplasts of *T. majus* and *Impatiens noli-tangere.*

When chromoplast proteins from various genotypes of *C. annuum* that differed with respect to $y$, $c_1$, $c_2$, $sw$, and $cl$ were analyzed, ChrA was found to correlate with $y^+$ (Fig. 2; Table IX). ChrB was present in chromoplasts from all genotypes. No polypeptides were detected that specifically correlate with $c_1$, $c_2$, $sw$, or $cl$.

## B. Enzymes of Carotenoid Biosynthesis

An additional category of chromoplast proteins is that of enzymes involved in carotenoid biosynthesis (cf. Beyer *et al.*, 1989; Camara *et al.*, 1989; Table II). A chromoplast-specific cDNA encoding GGPP synthase has been isolated from *C. annuum* (Schantz *et al.*, 1991; Kuntz *et al.*, 1991); the enzyme previously had been localized in chromoplasts (Dogbo and Camara, 1987). GGPP synthase also must occur in other plastids and has been isolated from etioplasts of *Sinapis alba* (Laferriére and Beyer, 1991). The enhanced accumulation of transcripts in chromoplasts could be either because of the development-specific activation of a common gene or because of the occurrence or a multigene family in which some members are expressed only during fruit ripening.

Both the inner and outer envelope of plastids are bright yellow and

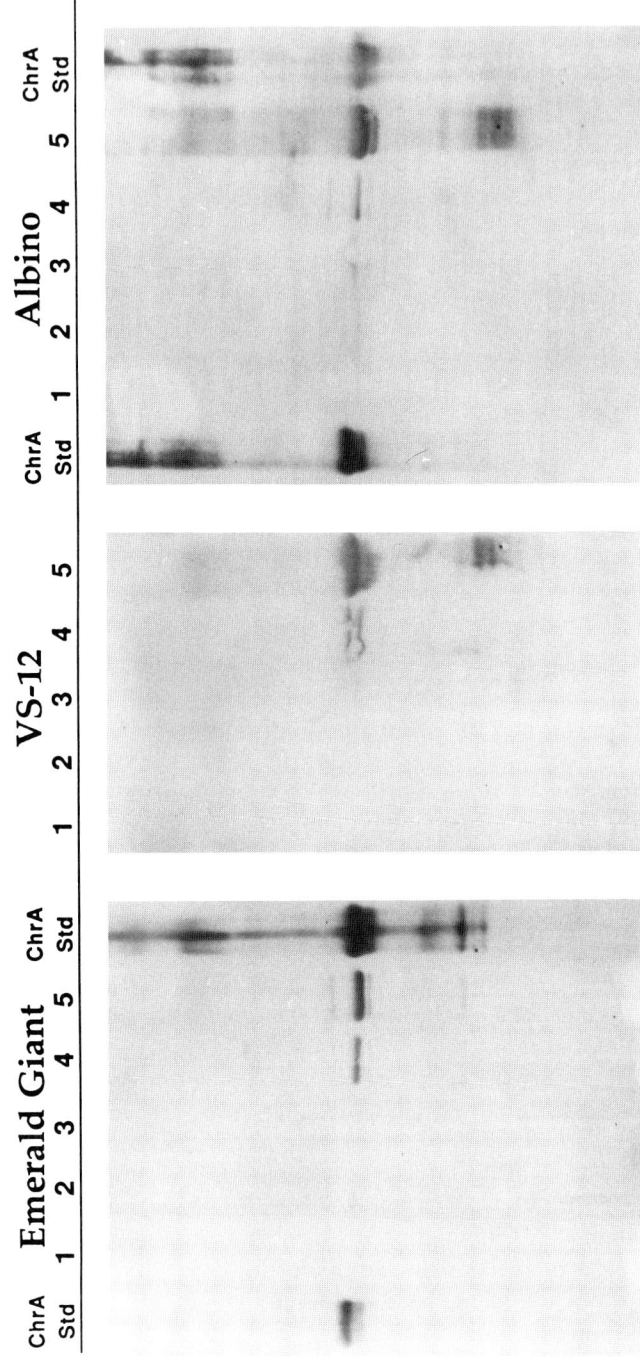

**Fig. 3.** Accumulation of ChrA during fruit ripening in *Capsicum annuum*. Reprinted with permission from Newman *et al.* (1989).

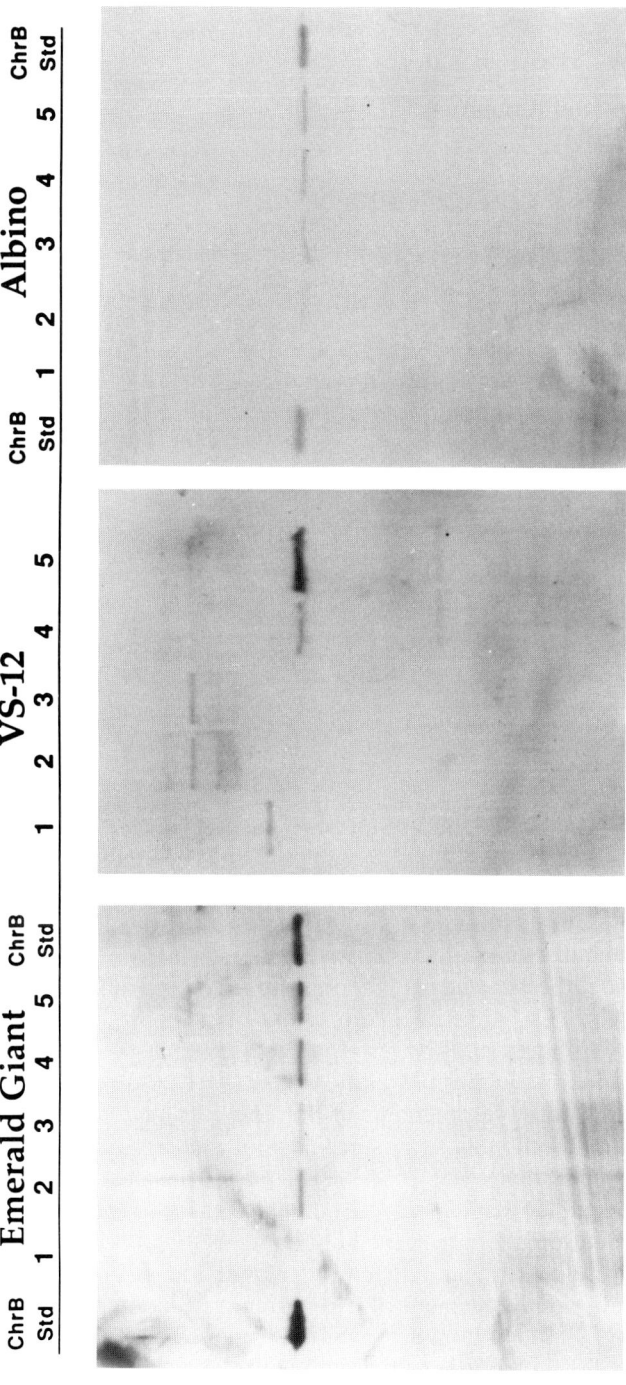

**Fig. 4.** Accumulation of ChrB during fruit ripening in *Capsicum annuun*. Reprinted with permission from Newman *et al.* (1989).

**TABLE IX**

Correlation of Occurrence of ChrA with $y^{+a,b}$

| Cultivar | Color of mature fruit | Genotype | | | |
|---|---|---|---|---|---|
| | | $y$ | $c_1$ | $c_2$ | $cl$ |
| NuMex Sunrise | Yellow | − | − | + | + |
| NuMex Sunset | Orange | + | + | − | + |
| NuMex Eclipse | Brown | + | + | + | − |
| New Mexico 6-4 | Red | + | + | + | + |
| 291-1C | Pale orange | + | − | − | + |
| 294-1C | Orange | + | + | − | + |
| 295-2G | Yellow | − | + | − | + |

[a] Reprinted with permission from Oren-Shamir et al. (1992).
[b] ChrA occurs only in $y^+$ genotypes.

contain carotenoids, principally violaxanthin (Douce and Joyard, 1990). A carotenoid–protein complex was found in the envelope (Markwell et al., 1992).

## VI. CHROMOPLASTS AND LEUCOPLASTS: IMPORTANCE OF NOT BEING GREEN

As noted in Section II,B, chromoplasts play an important role in pollination and seed dispersal. Carotenoids share the esthetic function with anthocyanins, which are located not in plastids but in the central vacuole. Plastids from tissues whose color is due solely to anthocyanins, for example, blue petals, are called leucoplasts (leuco = white) because they are devoid of pigment. Since the persistence of chlorophyll in the presence of carotenoids or anthocyanins results in a loss of brilliance (a muddying of the color), the elimination of chlorophyll in chromoplasts and leucoplasts is essential to their role in reproduction.

What are the possible mechanisms for the loss of chlorophyll from leucoplasts and chromoplasts? First, the loss of chlorophyll must be under strict tissue-specific control. Clp, an ATP-dependent protease system originally described in *Escherichia coli,* has the potential for effecting dissolution of thylakoid membranes in a highly controlled manner. Studies conducted *in organello* (Liu and Jagendorf, 1984; Malek et al., 1984) are consistent with the involvement of Clp in the destruction of excess polypeptides of multisubunit complexes of chloroplasts. The Clp protease is composed of ClpP, a 20-kDa plastid-encoded catalytic subunit (Gottesman *et al.*, 1990) and either ClpA or ClpB, two closely related 50-kDa subunits that are ATP dependent and encoded in the nucleus. The molecular genetics of the Clp system was worked out in *E. coli* (cf. Gottesman *et al.*, 1990). An ORF corresponding to *clpP* is very highly conserved in all plastid genomes examined (Gottesman *et al.*, 1990; Gray *et al.*, 1990). In tomato, *clpB* is represented by a two-member gene family (Gottesman *et al.*, 1990). *clpA* has been isolated from pea (K. Keegstra, 1992 personal communication).

We can imagine a model in which thylakoids and chlorophyll could be degraded by Clp proteases in a tissue- and development-specific manner: ClpP and low concentrations of ClpA are present in normal chloroplasts; the resulting proteolytic activity is responsible for the slow turnover of chloroplast protein. In tissues slated to become nongreen, *clpB* is expressed, resulting in a much more active protease that destroys thylakoid membranes or prevents them from forming. In an alternative model, the more virulent protease is generated from qualitatively different members of *clpA* or *clpB* gene families whose expression is tissue or development specific; that is, one *clpA* might be expressed only in flowers and fruits and another only in senescing leaves.

## VII. CONCLUSIONS AND PROSPECTS

Suppose someone set about to isolate all *Chr* genes. How many would there be in a single plant? In all plants? If we start with a single plant whose chromoplasts accumulate a carotenoid that is a normal component of the photosynthetic apparatus, for example, $\beta$-carotene, or an intermediate, such as lycopene, two *Chr* genes would seem to suffice to convert a proplastid or chloroplast into such a chromoplast. One gene could prevent development of the photosynthetic apparatus or degrade it. A second gene could cause the overproduction of a single carotenoid. The notion of a few chromoplast-specific genes per plant does not, therefore, seem out of the question.

Taking at face value Goodwin's (1980) count of 70 distinct carotenoids in fruits (plus additional flower-specific carotenoids), we might expect to discover several hundred chromoplast-specific genes. The biosynthesis of carotenoids appears, however, to involve type reactions—epoxidation, oxidation, *cis–trans* isomerization—so a relatively few enzymes acting in concert could produce a large array of different chemical species.

The most direct evidence for determining the roles of *Chr* genes will come from complementation assays. The chromoplasts of genotypes that are deficient in chromoplast development would be changed to wild-type chromoplasts by transformation with the isolated genes.

## ACKNOWLEDGMENTS

This study was supported by grants from the New Jersey Commission on Science and Technology, the DNA Plant Technology Corporation, and the Charles and Johanna Busch Memorial Fund. Cervantes-Cervantes received fellowship support from CONACYT, Mexico; N. Hadjeb received fellowship support from the government of Algeria; and L. Newman is a Busch Predoctoral Fellow.

## REFERENCES

Armstrong, G. A., Alberti, M., Leach, F., and Hearst, J. E. (1989). Nucleotide sequence, organization, and nature of the protein products of the carotenoid biosynthesis gene cluster of *Rhodobacter capsulatus*. *Mol. Gen. Genet.* **216,** 254–268.

Bartley, G. D., Viitanen, P. V., Pecker, I., Chamovitz, D., Hirschberg, J., and Scolnik, P. A. (1991). Molecular cloning and expression in a photosynthetic bacteria of a soybean cDNA coding for phytoene desaturase, an enzyme of the carotenoid biosynthesis pathway. *Proc. Natl. Acad. Sci. USA* **88,** 6532–6536.

Bathgate, B., Purton, M. E., Grierson, D., and Goodenough, P. W. (1985). Plastic changes during the conversion of chloroplasts to chromoplasts in ripening tomatoes. *Planta* **165,** 197–204.

Beyer, P. (1989). Carotene biosynthesis in daffodil chromoplasts. In "Physiology, Biochemistry, and Genetics of Nongreen Plastids" (J. Shannon, C. Boyer, and R. C. Hardison, eds.), pp. 157–170. American Society of Plant Physiologist, Rockville, Maryland.

Bryant, J. D., McCord, J. D., and Erdman, J. W., Jr. (1991). Isolation and characterization of carotenoproteins in carrot chromoplasts. *FASEB J.* **5(5),** A1323.

Buckner, B., Kelson, T., and Robertson, D. S. (1990). Cloning of the $y_1$ locus of maize, a gene involved in the biosynthesis of carotenoids. *Plant Cell* **2,** 867–876.

Bullerjahn, G. S., and Sherman, L. A. (1986). Identification of a carotenoid-binding protein in the cytoplasmic membrane from the heterotrophic cyanobacterium *Synechocystis* sp. strain PCC6714. *J. Bacteriol.* **167,** 396–399.

Camara, B., Dogbo, O., d'Harlingue, A., Kleinig, H., and Moneger, R. (1985). Metabolism of plastids terpenoids: Lycopene cyclization by *Capsicum* chromoplast membranes. *Biochim. Biophys. Acta* **836,** 262–266.

Camara, B., Bousquet, J., Cheniclet, C., Carde, J.-P., Kuntz, M., Evrad, J.-L., and Weil, J.-H. (1989). Enzymology of isoprenoid biosynthesis and expression of plastid and nuclear genes during chromoplast differentiation in pepper fruits (*Capsicum annuum*). In "Physiology, Biochemistry, and Genetics of Nongreen Plastids" (J. Shannon, C. Boyer, and R. C. Hardison, eds.), pp. 141–156. American Society of Plant Physiologists, Rockville, Maryland.

Cervantes, M., Hadjeb, N., Newman, L. A., and Price, C. A. (1990). ChrA is a carotenoid-binding protein in chromoplasts of *Capsicum annuum*. *Plant Physiol.* **92,** 1241–1243.

Dogbo, O., and Camara, B. (1987). Purification of isopentenyl pyrophosphate isomerase and geranylgeranyl pyrophosphate synthase from *Capsium* chromoplasts by affinity chromatography. *Biochim. Biophys. Acta* **920,** 140–148.

Dogbo, O., Laferriére, A., d'Harlingue, A., Camara, B. (1988). Carotenoid biosynthesis: Isolation and characterization of a bifunctional enzyme catalyzing the synthesis of phytoene. *Proc. Natl. Acad. Sci. USA* **85,** 7054–7058.

## 15. Chromoplast Development

Douce, R., and Joyard, J. (1990). Biochemistry and function of the plastid envelope. *Annu. Rev. Cell Biol.* **6**, 173–216.
Emter, O., Falk, H., and Sitte, P. (1990). Specific carotenoids and proteins as prerequisites for chromoplast tubule formation. *Protoplasma* **157**, 128–135.
Goodwin, T. W. (1980). "Biochemistry of the Carotenoids," 2d Ed., Vol. 1. Chapman and Hall, London.
Gottesman, S., Squires, C., Pichersky, E., Carrington, M., Hobbs, M., Mattick, J. S., Dalrymple, B., Kuramitsu, H., Shiroza, T., Foster, T., Clark, W. P., Ross, B., Squires, C. L., and Maurizi, M. R. (1990). Conservation of the regulatory subunit for the Clp ATP-dependent protease in prokaryotes and eukaryotes. *Proc. Natl. Acad. Sci. USA* **87**, 3513–3517.
Gounaris, I., and Price, C. A. (1987). Plastid transcripts in chloroplast and chromoplasts of *Capsicum annuum*. *Curr. Genet.* **12**, 219–224.
Gounaris, I., Michalowski, C. B., Bohnert, H. J., and Price, C. A. (1986). Restriction and gene maps of plastid DNA from *Capsicum annuum*: Comparison of chloroplast and chromoplast DNA. *Curr. Genet.* **11**, 7–16.
Gray, J. C., Hird, S. M., and Dyer, T. A. (1990). Nucleotide sequence of a wheat chloroplast gene encoding the proteolytic subunit of an ATP-dependent protease. *Plant Mol. Biol.* **15**, 947–950.
Gross, J. (1987). "Pigments in Fruits, A Series of Monographs." Academic Press, San Diego, California.
Hadjeb, N., Gounaris, I., and Price, C. A. (1988). Chromoplast-specific proteins in *Capsicum annuum*. *Plant Physiol.* **88**, 42–45.
Hunt, C. M., Hardison, R. C., and Boyer, C. D. (1986). Restriction enzyme analysis of tomato chloroplast and chromoplast DNA. *Plant Physiol.* **82**, 1145–1147.
Hurtado-Hernández, H., and Smith, P. G. (1985). Inheritance of mature fruit color in *Capsicum annuum* L. *J. Hered.* **86**, 211–213.
Iwatsuki, N., Hirai, A., and Asahi, T. (1985). A comparison of tomato fruit chloroplast and chromoplast DNAs as analyzed with restriction endonucleases. *Plant Cell Physiol.* **26**, 599–602.
Jones, R. B., Wardley, T. M., and Dalling, M. (1989). Molecular changes involved in the ripening of tomato fruit. *J. Plant Physiol.* **134**, 284–289.
Karrer, P., and Jucker, E. (1948). "Carotenoide." Birkhauser, Basel.
Kobayashi, H., Ngernprasirtsiri, J., and Akazawa, T. (1990). Transcriptional regulation and DNA methylation in plastids during transitional conversion of chloroplasts to chromoplasts. *EMBO J.* **9**, 307–313.
Kuntz, M., Römer, S., Suire, C., Hugueney, P., Weil, J. H., Schantz, R., and Camara, B. (1992). Identification of a cDNA for the plastid-located geranylgeranyl pyrophosphate synthase from *Capsicum annuum*: Correlative increase in enzyme activity and transcript level during fruit ripening. *Plant. J.* **2**(1), 25–34.
Laborde, J. A., and Spurr, A. R. (1973). Chromoplast ultrastructure as affected by genes controlling grana retention and carotenoids in fruits of *Capsicum annuum*. *Am. J. Bot.* **60**, 736–744.
Laferriere, A., and Beyer, P. (1991). Purification of geranylgeranyl diphosphate synthase from *Sinapis alba* etioplasts. *Biochim. Biophys. Acta* **1077**, 167–172.
Lers, A., Levy, H., and Zamir, A. (1991). Co-regulation of an *elip*like gene and β-carotene biosynthesis in the alga *Dunaliella bardawil*. *J. Biol. Chem.* **266**(21), 13698–13705.
Lippert, L. F., Smith, P. G., and Bergh, B. O. (1966). Cytogenetics of the vegetable crops. Garden pepper, *Capsicum* sp. *Bot. Rev.* **32**, 24–55.

Liu, X.-Q., and Jagendorf, A. T. (1984). ATP-dependent proteolysis in pea chloroplasts. *FEBS Lett.* **166,** 248–252.
Lütke-Brinkhaus, F., Liedvogel, B., Kreuz, K., and Kleinig, H. (1982). Phytoene synthase and phytoene dehydrogenase associated with envelope membranes from spinach chloroplasts. *Planta* **156,** 176–180.
Malek, K., Bogorad, L., Ayers, A., and Goldberg, A. L. (1984). Newly synthesized proteins are degraded by an ATP-stimulated proteolytic process in isolated pea chloroplasts. *FEBS Lett.* **166,** 253–257.
Markwell, J., Bruce, B. D., and Keegstra, K. (1992). Isolation of a carotenoid-containing sub-membrane particle from the chloroplastic outer membrane of pea (*Pisum sativum*). *J. Biol. Chem.* **267,** 13933–13937.
Mayer, M. P., Beyer, P., and Kleinig, H. (1990). Quinone compounds are able to replace molecular oxygen as terminal electron acceptor in phytoene desaturation in chromoplasts of *Narcissus pseudonarcissus*. L. *Eur. J. Biochem.* **191,** 359–364.
Michalowski, C. B., Löffelhardt, W., and Bohnert, H. J. (1991). An ORF 323 with homology to *crtE*, specifying prephytoene pyrophosphate dehydrogenase, is encoded by cyanelle DNA in the eukaryotic alga *Cyanophora paradoxa*. *J. Biol. Chem* **266,** 11866–11870.
Milicua, J. C. G., Juarros, J. L., Delasrivas, J., Ibarrondo, J., and Gömez, R. (1991). Isolation of a yellow carotenoprotein from carrot. *Phytochemistry* **30,** 1535–1537.
Newman, L. A., Hadjeb, N., and Price, C. A. (1989). Synthesis of two chromoplast-specific proteins during fruit development in *Capsicum annuum*. *Plant Physiol.* **88,** 42–45.
Odland, M. L., and Porter, A. M. (1938). Inheritance of immature fruit color of peppers. *Proc. Am. Soc. Hort. Sci* **36,** 647–657.
Oren-Shamir, M., Hadjeb, N., Newman, L. A., and Price, C. A. (1992). Occurrence of the chromoplast protein ChrA correlates with a fruit-color gene in *Capsicum annuum*. *Plant Mol. Biol.* (in press).
Peter, G. F., and Thornber, J. P. (1991). Biochemical composition and organization of higher plant photosystem II light-harvesting pigment–proteins. *J. Biol. Chem.* **266,** 16745–16754.
Piechulla, B. (1988). Differential expression of nuclear- and organelle-encoded genes during tomato fruit development. *Planta* **17,** 505–512.
Prézelin, B. (1987). Photosynthetic physiology of dinoflagellates. *In* "The Biology of Dinoflagellates" (F. J. R. Taylor, ed.), pp. 174–223. Blackwell Scientific, Oxford.
Price, C. A., Hadjeb, N., and Newman, L. A. (1989). A search for chromoplast-specific genes and proteins in *Capsicum annuum*. *In* "Physiology, Biochemistry, and Genetics of Nongreen Plastids" (J. Shannon, C. Boyer, and R. C. Hardison, eds.), pp. 215–226. American Society of Plant Physiologists, Rockville, Maryland.
Schantz, R., Römer, S., Weil, J.-H., Kuntz, M., and Camara, B. (1991). Characterization of cDNAs encoding a prenyltransferase and fruit-ripening related proteins in *Capsicum annuum*. Third Int. Cong. Plant Mol. Biol. Abst. 697.
Schmidt, A., Sandmann, G., Armstrong, G. A., Hearst, J. E., and Böger, P. (1989). Immunological detection of phytoene desaturase in algae and higher plants using an antiserum raised against a bacterial fusion-gene construct. *Eur. J. Biochem.* **184,** 375–378.
Scolnik, P. A., Walker, M. A., and Marrs, B. L. (1980). Biosynthesis of carotenoids derived from neurosporene in *Rhodopseudomonas capsulata*. *J. Biol. Chem.* **255,** 2427–2432.
Simpson, D. J., Bauar, M. R., and Lee, T. H. (1977). Chromoplast ultrastructure of *Capsicum* carotenoid mutants. I. Ultrastructure and carotenoid composition of a new mutant. *Z. Pflanzenphysiol.* **83,** 293–308.
Sitte, P., Falk, H., and Liedvogel, B. (1980). Chromoplasts. *In* "Pigments in Plants" (F. C. Czygan, ed.), pp. 117–148. Fischer, Stuttgart.

Smith, P. G. (1950). Inheritance of brown and green mature fruit color in peppers. *J. Hered.* **41,** 138–140.

Spurgeon, S. L., Sathyamoorthy, N., and Porter, J. W. (1984). Isopentenyl pyrophosphate isomerase and prenyltransferase from tomato fruit plastids. *Arch. Biochem. Biophys.* **230,** 446–454.

Spurr, A. R., and Harris, W. M. (1968). Ultrastructure of chloroplasts and chromoplasts in *Capsicum annuum.* I. Thylakoid membrane changes during fruit ripening. *Amer. J. Bot.* **55(10),** 1210–1224.

Szabolcs, J. (1989). Plant carotenoids. *In* "Carotenoids: Chemistry and Biology" (N. I. Krinsky, M. M. Mathews-Roth, and R. F. Taylor, eds.), pp. 39–58. Plenum, New York.

Thompson, J. A. (1980). Apparent identity of chromoplast and chloroplast DNA in the daffodil, *Narcissus pseudonarcissus. Z. Naturforsch.* **35,** 1101–1103.

Thomson, W. W., and Whatley, J. M. (1980). Development of nongreen plastids. *Annu. Rev. Plant Physiol.* **31,** 375–394.

Thornber, J. P., Morishige, D. T., Anandan, S., and Peter G. F. (1991). Chlorophyll-carotenoid proteins of higher plant thylakoids. *In* "Chlorophylls" (H. Scheer, ed.), pp. 549–585. CRC Press, Boca Raton, Florida.

Whatley, J. M. (1982). Ultrastructure of plastid inheritance. Green algae to angiosperms. *Biol. Rev.* **57,** 527–570.

Winkenbach, F., Falk, H., Liedvogel, B., and Sitte, P. (1976). Chromoplasts of *Tropaeolum majus* L.: Isolation and characterization of lipoprotein elements. *Planta* **128,** 23–28.

# Index

## A

Accessory pigment, 14, 455
*Acetabularia*, 339
AcetylCoA, 460
*Acholeplasma laidlawii*, 252
*Acrocarpia*, 456
S-Adenosyl-L-methionine (SAM), 119, 131, 133, 138
S-Adenosyl-L-methionine:magnesium protoporphyrin IX methyltransferase, 126, 129–131
ADP-glucose phosphorylase, transit peptide, 315
Affinity chromatography, 131, 132, 138, 144, 150, 155
Aggregates, of photosynthetic pigments, 354
ALA, 96, 234
   biosynthesis, 99–104, 195
   influence on protochlorophyllide/ protochlorophyll ratio, 193
ALA dehydratase, 98, 104–105
ALA synthase, 96–99
Alloxanthin, 452, 477
Amino acid sequence, *see* specific proteins
4-Amino-5-hexynoic acid, 104
5-Aminolevulinic acid, *see* ALA
Aminomethylbilane, 109
3-Amino-1,2,4-triazol, 230
Amiprophos-methyl, 223
Amyloplast, 33, 49, 486
*Anabaena*, 151, 185, 391, 401
*Anacystis nidulans*, 185, 195
Anaerobiosis, influence on esterification of chlorophyllide, 230

Antenna complex, 335–337, 351–355, 384–385, 427–429, *see also* Light-harvesting complex I; Light-harvesting complex II
   formation, 349
   reconstitution *in vitro*, 351–355
Antheraxanthin, 448, 475, 489
Anthocyanin, 449
Apocarotenoid, 450
*Arabidopsis*, 261, 351
*Arabidopsis thaliana*, 153, 182, 199, 338
*Arum*, 98
ATP, role of
   in chlorophyll biosynthesis, 101, 127
   in plastid protein
     membrane insertion, 324
     import, 317
     translocation, 320
   in protochlorophyllide reduction, 195
ATP synthase, 37, 255, 266, 288, 322
*atpA*, plastid gene, 288
*atpB*, plastid gene, 288
*atpE*, plastid gene, 288
*atpH*, plastid gene, 288
Atrazin, 422
Aurin-tricarboxylic acid, 233
*Avena*, *see* Oat
*Averrhos carambola*, 450
Avocado, 262

## B

Bacteriochlorophyll, 92, 97, 148, 153
Barley, 130, 136, 142, 149, 182, 187, 193, 196, 199, 282, 297, 336, 344, 376, 395, 397, 400, 401
   chlorophyll b less mutant, 80, 346

Bean, 29, 32, 149, 188, 192, 193, 226
Blackberry, 449
Blue-light receptor, 64, 79, 231, 290
*Bougainvillea glabra*, 153
Boundary lipids, 255

## C

*Cab*, nuclear genes, 289–294, 298–299, 338
  expression
    circadian rhytm-regulation of mRNA levels, 293–294
    *cis*-acting control elements, 291–293
    light inducible transcription, 290–293
    regulation by chloroplast signal, 298–299
    *trans*-acting factors, 291–293
Calcium
  effects on prolamellar body structure, 264
  requirement in photosystem II, 423, 430
Caloxanthin, 452, 478
Calvin cycle, 13
Capsanthin, 475, 489
*Capsicum annuum*, 225, 450, 487, 494, 497
Capsorubin, 450, 476
*Ceratozamia fuscoviridis*, 449
Carbon dioxide
  assimilation, 12
  fixation, 366
$N$-(m-Carboxyphenyl)pyridoxamine 5'-phosphate, 103
Carboxysomes, 44
Carotene, 28, 74
  $\alpha$-carotene, 448, 474, 492
  $\beta$-carotene, 66–69, 337, 448, 468, 472, 473, 489, 492
  $\beta$-carotene 5,6-epoxide, 478
  $\gamma$-carotene, 492
  $\delta$-carotene, 492
  $\epsilon$-carotene, 478, 492
Carotenoid, 447–478, 485–501
  biosynthesis, 459, 460, 468
    control of accumulation, 470,
    cyclization, 463
    enzymes, 466–467, 489, 497
    genes, 466, 490–495
    hydroxylation, 464
    inhibitors, 465
    under intermittent light, 469
    oxygenation, 464, 488
  composition, 47
    in *Capsicum annuum* genotypes, 487
    in dark-grown plants, 449, 469
    in developing chloroplasts, 65–68, 469
    developmental regulation, 450, 501
    in pigment–protein complexes, 260, 339–340, 352, 354, 453, 454
    in plant kingdom, 41–43, 448–453
  deficiency, 298
  function, 74, 421, 455
  stress effects, 472
*Ceratodon purpureus*, 82
Channel-forming proteins, 319
Chaperone, 318
Chaperonins, *see* Stromal factor
*Chlamydomonas*, 42, 100, 184, 261, 291, 314, 451
  mutant
    photosystem II-deficient, 297
    chlorophyll b less, 350, 353
    protochlorophyllide reductase-deficient, 152
    temperature-sensitive, 196
*Chlamydomonas reinhardtii*, 185, 283, 336, 340, 349, 395
*Chlorella*, 12, 99, 107, 184, 451
  chlorophyll biosynthesis mutant, 135, 138, 139, 146
*Chlorella fusca*, 185
*Chlorella luteo-viridis*, 184
*Chlorella pyrenoidosa*, 185
*Chlorella regularis*, 185, 193
*Chlorella variegata*, 184
*Chlorella vulgaris*, 185
Chloride, cofactor in oxygen evolution, 423, 430
*Chlorobium*, 102, 129
*Chloroflexus aurantiacus*, 354
$p$-Chloromercuribenzene sulfonate, 127
Chlorophyceae
  chloroplast ultrastructure, 41–42
  pigment composition, 41, 451
Chlorophyll
  a/b ratio, 70, 294, 339–341, 349, 352, 427–429
  accumulation, 65, 82, 370
    light dependence, 6, 70, 81
    in pigment mutants, 66, 76, 185
    temperature dependence, 82

# Index

apoproteins, 231, 281, 336–341, 427–429
  coordination with chlorophyll biosynthesis, 232, 297, 350
  esterification of chlorophyllide, 232
  light regulation, 79, 286, 375
  nuclear genes, 289
  ratio, 339
  stability, 233, 294, 407
  turnover, 287, 294
biosynthesis, 66
  in angiosperms contra gymnosperms, 182
  coordination to protein, 232, 297, 350
  correlation to synthesis of PSI-A, PSI-B, 406
  early stages, 97–123
  influence of light, 6, 10, 73
  later stages, 124–157
  light-independent, 184
  rate, 73
degradation, 472
function, 366–371, 383–386, 421
in pigment–protein complexes, 67, 387, 335–341, 375–376, 407, 427–428
Chlorophyll b
  accumulation of LHCII, 65, 295, 350, 429
  biosynthesis, 9, 154
  rate of, 73
  function, 383–386, 421
  grana formation, 82
  mutants, 346, 350
  in pigment–protein complexes, 67
Chlorophyll–protein complexes, 8, 335–336, 339, 341, 350, see also specific complexes
Chlorophyll synthetase, 9, 154–155, 219–236, see also Chlorophyllide, esterification
  activity, 220
  effect of light, 224
  localization, 155, 223–225
  substrate specificity, 221–223
  temperature dependence, 226–229
Chlorophyll-binding polypeptides, see Chlorophyll apoproteins
Chlorophyllase, 154, 220
Chlorophyllide, 124
  biosynthesis, 45, 154, see also Protochlorophyllide reduction
  conversion to protochlorophyllide, 151
  esterification, 9, 45, 154–156, 219–235, 287, 372
  impact on translation of chloroplast proteins, 206, 231, 286, 344, 376, 407
  microcycle, 372
  protochlorophyllide reductase complexes, 223
  spectral forms, 149, 191
Chlorophyllide b, 154, 223
Chloroplast development, 32, 48, 63–64, see also Plastid development
  carotenoids, 74, 467
  in chlorophyll mutants, 73
  influence
    of light, 68, 72–77, 284
    of temperature, 81
  tissue specificity, 78
Chloroplast DNA, 29, 280, 283–284, 404, see also Chloroplast genes
  light regulation of transcription, 285
  transcription units, 281
Chloroplast-encoded polypeptides, 288, 387–392, see also Plastid-encoded polypeptides
Chloroplast envelope translocation, 317–323, see also Import, of plastid proteins
Chloroplast envelope, see Envelope membranes
Chloroplast genes, 281, 385, 387, 409, 427, 432, see also specific genes
Chloroplast membranes
  inner membranes, see Thylakoid; Thylakoid membrane
  outer membranes, see Envelope membranes
Chloroplast movement, 3
Chloroplast proteins, 286, 387, see also specific proteins
  import, 313–322
  intraorganellar routing, 323–327
  light regulation, 231
  localization, 37–38
  pigment complexes, 67, 231, 288, 336, 339–340, 375, 385–388, 406, 408, 424, 427–429, 454
  targeting signal, 313–317
Chloroplast RNA, 29, 283–287, 404

Chloroplast transcription, 281–285, *see also* Transcription
Chloroplast ultrastructure, *see also* Thylakoid, ultrastructure
  algae, 40–44
  seed plants, 4, 35–40, 70, 71
Chlorosis, 472
Chlorosome, 354
*Chromatium*, 119
*Chromonas*, 456
Chromophyta, carotenoid composition, 451
Chromoplast
  development, 449, 500–501
  genes, 490–495
  membranes, 496
  pigments, 489
  proteins, 495–500
  ultrastructure, 48–49, 486–488
Chrysophyceae
  chloroplast ultrastructure, 41
  pigment composition, 41, 452
Circadian rhytm-regulated gene expression, 293–294
*cis*-acting elements, 192, 291
*Citrus*, 450
Clomazone, 223
*clpA*, nuclear gene, 501
*clpB*, nuclear gene, 501
*clpP*, plastid gene, 501
Cocoa, 448
*Codium*, 456
Coenzyme A, 128, 138
Copper, in photosystem II, 432
Coprogen III oxidase, *see* Coproporphyrinogen III oxidase
Coprogen III, *see* Coproporphyrinogen III
Coproporphyrinogen III, 96, 118–120
Coproporphyrinogen III oxidase, 97, 118–122, 234
Corn, *see* Maize
Coupling factor 1, 5, 70, *see also* ATP-synthase
CP1, chlorophyll–protein complex, 37, 67, 375, 387, 408
CP24, chlorophyll–protein complex, 67, 336, 340, 427–429
CP26, chlorophyll–protein complex, 67, 336, 427–429

CP29, chlorophyll–protein complex, 37, 67, 336, 339, 427–429
CP43, chlorophyll–protein complex, 37, 231, 288, 375, 424, 427, 454
CP47, chlorophyll–protein complex, 37, 231, 288, 375, 424, 427, 454
CP64, chlorophyll–protein complex, 231
CPa, *see* CP43, chlorophyll–protein complex
CPa-1, *see* CP47, chlorophyll–protein complex
CPa-2, *see* CP43, chlorophyll–protein complex
Crassulacean acid metabolism, 31
Cress, 188, 234
Cross-linking, 265, 390, 393–395, 409, 429
*crtA*, gene, in carotenoid biosynthesis, 493
*crtB*, gene, in carotenoid biosynthesis, 493
*crtC*, gene, in carotenoid biosynthesis, 493
*crtD*, gene, in carotenoid biosynthesis, 491
*crtE*, gene, in carotenoid biosynthesis, 493
*crtF*, gene, in carotenoid biosynthesis, 493
*crtG*, gene, in carotenoid biosynthesis, 493
*crtI*, gene, in carotenoid biosynthesis, 491, 493
Cryptophyceae
  chloroplast ultrastructure, 41
  pigment composition, 41, 452
$\alpha$-Cryptoxanthin, 448, 474
$\beta$-Cryptoxanthin, 448, 474
Crystalloid inclusions, 50
Cucumber, 130, 188, 193, 295, 397, 400
*Cucumis sativus*, *see* Cucumber
*Cucurbita pepo*, 193
*Cyanaphora paradoxa*, 391, 395, 494
*Cyanidium*, 103, 184
*Cyanidium caldarium*, 97
Cyanophyta
  lipids, 242
  pigments, 41, 453
  thylakoid organization, 41
Cycloheximide, 406
Cytochrome
  photooxidation, 370
  prosthetic group, 92, 96
Cytochrome $b$-559, 37, 67, 78, 424, 426
Cytochrome $f$, 288, 377
  localization, 37, 325
  membrane insertion, 325

# Index

Cytoplasmic factor, in chloroplast protein import, 318, 326

## D

D1 protein, 37, 67, 79, 231, 258–259, 375, 423–426, 433
D2 protein, 37, 67, 79, 231, 288, 375, 424–426, 433
Daffodil, 225, 462, 494
DCMU, 422, 472
Delayed luminescence, 369
Desaturase, 248, 489, 493
Desaturation
  of carotenoids, 462
  of lipids, 248
Desferal mesylate, 138
*Desulfovibrio gigas*, 123, 124
Diacylglycerols, 355
Diadinoxanthin, 451, 452, 476
4′,6-Diamidino-2-phenylindole (DAPI), 29
4,5-Diaminovaleric acid, 104
Diatoxanthin, 451, 452, 458, 477
*Dictyota dichotoma*, 455
Digalactosyl diacylglycerol, 28, 242
Digitonin, 376
Dinophyceae
  chloroplast ultrastructure, 41
  pigment composition, 41, 452
4,5-Dioxovaleric acid, 104
2,2-Dipyridyl, 133, 234
Dipyrrylmethane cofactor, 112
Divinyl form
  chlorophyll a, 139–148, 154
  chlorophyll b, 139–148, 154
  chlorophyllide, 139–148, 154
  protochlorophyll, 135
  protochlorophyllide, 124, 134, 137, 139–148, 151, 153
DNA, 29, 279–299, 404, 494
DPIP, 368
*Dunaliella tertiolecta*, 295, 339, 349

## E

Echinenone, 452, 478
EDTA, 126, 128, 264
EGTA, 126
Elaioplast, 50, 486
Electron acceptor, 387, 392, 409, 410
Electron donor, artificial, 395
Electron microscopy, 263, 404, *see also* Immunoelectron microscopy
Electron transfer, 396, 419
Endoplasmic reticulum, lipid biosynthesis, 248
Energy transfer, 421, 455–456
Envelope membranes
  biosynthesis
    of carotenoids, 458, 489
    of chlorophyll, 127, 138
  evagination, 27
  invagination, 27, 34, 249
  lipids, 27
  lipid transfer, 248
  pigments, 28
  polypeptides, 27, 312, 323–325
  receptor protein, 318–320, 324
  structural features, 26
  translocation complex, 318–320
  vesicle formation, 250
Eoplast, 32, *see also* Proplastid
*Ephedra distachya*, 183
Epoxidation, 458
Epoxycarotenoids, 450
Equisetaceae, 184
*Eschscholtzia californica*, 450
Eschscholtzxanthin, 476, 489
ESR, 255
Esterification, of chlorophyllide, *see* Chlorophyllide, esterification
S-Ethyl dipropylthiocarbamate, 230
N-Ethylmaleimide, 319
Etiochloroplast, 71, 153, *see also* Etioplast
Etioplast, 4, 44–47, 149
  conversion to chloroplast, 68–75, 80, 180, 231, 299, 469
  pigments, 68
  proteins, 38, 155, 221, 375
  ultrastructure, 46, 155
*Euglena*, 30, 78, 92, 102, 107, 108, 117, 131, 184, 282, 287, 337
*Euglena gracilis*, 38, 98, 99, 129, 193
Euglenophyceae
  chloroplast ultrastructure, 41
  pigment composition, 41, 451
Eustigmatophyceae
  chloroplast ultrastructure, 41
  pigment composition, 41, 452
EXAFS spectroscopy, 388

## F

Farnesol, 221
Farnesyl diphosphate, 155, 460
Fatty acids, $\beta$-oxidation, 133
Ferredoxin, 318, 319, 371, 385, 390
Ferredoxin:NADP$^+$ oxidoreductase, 402
Ferricyanide, 432
Ferrochelatase, 97, 98, 122, 126
Flavin, 124, 151
Flavoprotein, 151
Frosthardening, 261
Fucoxanthin, 42, 451, 477

## G

Gabaculine, 97, 103, 295, 349
Galactolipids, 242, 246
Gene, see also specific genes
  regulation, 279–301
    by light, 79, 231, 181–190, 196–197, 406
Geraniol, 221
Geranyl diphosphate, 460
Geranylgeraniol, 68, 221
Geranylgeranyl chlorophyll, 69, 155
Geranylgeranyl chlorophyllide, 220, 225, 229
Geranylgeranyl diphosphate, 9, 154, 155, 220, 226, 229, 460
Geranylgeranyl diphosphate synthase, 489, 494
Gerontoplast, 48, 486
*Ginkgo biloba*, 183
$\alpha$-Glucan phosphorylase, 316
Glucose oxidase, 371
$\beta$-Glucosidase, 31
Glutamate 1-semialdehyde aminotransferase, 102, 103
Glutamate, 97, 98
Glutamine, 350
Glutamine synthetase, association with chaperonins, 322
Glutamyl-tRNA, 101, 102
  reductase, 102
  synthetase, 100
Glutathione, 127
Glyceraldehyde 3-phosphate dehydrogenase, transit peptide, 315
Glycine, 94
*Gnetum ula*, 183

Grana, 3, 34, see also Thylakoid
  formation, 35, 47, 70, 73
  organization, 34, 41, 436
Greening, 469, see also Etioplast, conversion to chloroplast
GroEL-like proteins, 322

## H

Habituation, to light, 77
Harderoporphyrinogen, 119
Heat-shock, 47
  protein, 82, 317, 318, 323, 343, 347
*Hedera*, 50
*Helianthus*, 450
Heme, 94, 96, 98, 426
Herbicide, 465, 472
Heterogeneity
  LHCII apoproteins, 289, 338
  membrane lipids, 253–258
  photosystem II, 435–436
Heteroxanthin, 451, 452, 477
*trans*-Hexadecanoic acid, 82, 243
High-performance liquid chromatography, see HPLC
Hill reaction, 12
Histidine, role in chlorophyll binding, 350
*Hordeum vulgare*, see Barley
HPLC, 69, 127, 137, 144, 145, 148, 453
Hsp 70, see Heat-shock, protein
Hydrogenase
  activity, 231
  hydrogenation of geranylgeranyl derivatives, 229–231
2-Hydroxy-3-aminotetrahydropyran-1-one, 102
Hydroxyl radicals, 407
Hydroxymethylbilane, 109
8-Hydroxyquinoline, 138, 235

## I

Immunoelectron microscopy, 31, 36
  localization of plastid proteins, 36–39, 192
*Impatiens noli-tangere*, 497
Import, of plastid proteins, 311–328, 342–344
Integration, of thylakoid proteins, 288, 295–297, 323–325, 343–348
Interlamellar attachment, 250

# Index

Intraorganellar routing, 323–328
Intrathylakoidal space, *see* Thylakoid lumen
Inverted micelle intermediates, 250
Inverted repeat sequences, 285
Iron, 119, 234
Iron–sulfur center, 387
*Isoetes*, 184
Isoharderoporphyrinogen, 119
Isopentenyl diphosphate, 221, 460
Isopentenyl pyrophosphate isomerase, 489
Isoprene, 459
Isoprenoid kinase, 221

## K

Kasugamycin, 233

## L

*Lactuca sativa*, 448
Lactucaxanthin, 488, 474
*Larix decidua*, 183
Lateral diffusion, of lipids, 250
*Lemna*, 344
*Lemna gibba*, 336
Leukoplast, 49
Levulinic acid, 105
LHCI, *see* Light-harvesting complex I
LHCII, *see* Light-harvesting complex II
LHCIIa, *see* CP29, chlorophyll–protein complex
LHCIIc, *see* CP26, chlorophyll–protein complex
LHCIId, *see* CP24, chlorophyll–protein complex
LHCPI, *see* Light-harvesting complex I, polypeptides
LHCPII, *see* Light-harvesting complex II, polypeptides
Light-harvesting complex I, 67–69, 335–337, 384–385
  localization, 37
  pigments, 337, 454
  polypeptides, 296, 336
Light-harvesting complex II, 67–69
  assembly, 341–344
  carotenoids, 339, 454
  in chlorophyll b less mutant, 80, 346, 350
  chlorophylls, 70, 294, 354, 375, 428
  degradation, 349
  heterogeneity, 289, 338
  lipids, 260, 351–352
  localization, 37–39
  mutant, 205, 338
  oligomerization, 81, 339, 351
  polypeptides, 427–429
    import, 318, 322–323
    integration, 295, 343–347
    phosphorylation, 338, 348, 428, 434
    processing, 342–345
  reconstitution, 351–355
  regulation
    by light, 78–81, 290, 376
    by temperature, 81–83
  stabilization, 68, 73, 80, 294, 349, 354
  transcription, 80, 205, 289–293
Light regulation
  chlorophyll synthetase activity, 224
  chloroplast development, 68, 72–77, 284, 469
  by irradiance, 70, 349
  by quality of light, 75–77, 406, 470
  in mutants, 73, 76
  formation
    of carotenoids, 469–471
    of chlorophyll, 5, 10, 66
  gene expression
    *Cab* genes, 290–293
    nuclear genes, 78–79, 181–190, 231, 290–293
    plastid genes, 78, 283–286, 376
    protochlorophyllide reductase mRNA, 187
  grana formation, 35, 47
  LHCPII, 79–81, 290
  morphogenesis, 180
  pigment deficiency, 82
    temperature correlation, in mutants, 82
*Lilium tigrinum*, 488
Linolenic acid, 243, 472
Lipid
  biosynthesis, 248
  cubic phase, 248, 263
  desaturation, 249
  hexagonal phase, 245
  lamellar phase, 246
  lateral assymmetry, 254
  diffusion, 250
  monolayer curvature, 246–247
  peroxidation, 472

protein interaction, 256, 258–259, 265
transfer, 248
transverse asymmetry, 257
unsaturation, 243
Liposomes, 260, 355
*Lolium temulentum*, 187, 188
Loroxanthin, 451, 476
Lumen, *see* Thylakoid lumen
Lutein, 67, 68, 74, 340, 448, 450, 468, 473, 474, 489
Lutein epoxide, 68, 69, 448, 474
Lycopene, 450, 475, 492
*Lycopersicum esculentum*, 450, 491, 497

## M

Magnesium, 105, 128, 223
  effects on prolamellar body structure, 264
Magnesium 2,4-divinylpheoporphyrin $a_5$ monomethyl ester, 134
Magnesium 4-ethyl-(4-desvinyl)-protoporphyrin IX monomethylester:NAD+ oxidoreductase, 141
Magnesium branch, in porphyrin biosynthesis, 94, 99, 124
Magnesium chelatase, 126, 127
Magnesium protoporphyrin IX monomethyl ester, 124, 126, 129, 137, 138, 141, 146
Magnesium protoporphyrin IX, 92, 124, 127, 129, 131, 141, 146
Maize, 130, 154, 187, 297, 387, 395
Manganese, 16, 423
*Mantoniella squamata*, 339, 354
*Marchantia polymorpha*, 153, 284
Mass spectrometry, 137
*Mastigocladus laminosus*, 403
*Melilotus alba*, 82
Membrane protein
  hydrophobicity, 203–204, 253, 387, 391, 393, 399–401
  integration, 288, 295, 323–325, 343–347
  transverse asymmetry, 257
Membrane structure, role of lipids, 252–253
Mercuric chloride, 344
Metal chelator, 133, 234
*Metasequoia glyptostroboides*, 183
N-Methylphenozonium 3-sulfonate, 395
Microtubule-like structures, 29, 34

Microtubules, 33
Mitochondria, 94–97, 122
Monogalactosyl diacylglycerol, 28, 242
Monovinyl- and divinyl pathways, in chlorophyll biosynthesis, 139–148
Mössbauer spectroscopy, 388
Mustard, *see Sinapis*
*Myrothamnus*, 40
Myxoxanthophyll, 453, 478

## N

NADH, 102, 119, 127, 138, 141
NADPH
  in chlorophyll biosynthesis, 102, 119, 138
  in chlorophyllide esterification, 155
  in photosynthesis, 13, 394
  in protochlorophyllide reduction, 8, 144, 148–149, 151, 185, 194
Near-UV/blue-light photoreceptor, 180
Neoxanthin, 28, 65, 68, 69, 74, 340, 448, 474, 489
Neurosporene, 462, 492
*Nicotiana plumbaginifolia*, 292
Nigericin, 344
Nitrogen oxide, 472
NMR, 99, 109, 112, 137, 151
Noncyclic electron transport, 371
Norflurazon, 298, 465
Nostoxanthin, 453, 478
Nuclear gene
  inactivation, 394
  light-harvesting complexes, 289–294, 338
  light regulation, 78
  photosystem I, 385, 394–400, 405–406
Nuclear-encoded polypeptides, 289–294, 338, 385
Nucleoids, 28
Nucleomorph, 42

## O

Oat, 149, 187, 188, 192, 199, 226, 234
*Ochromonas*, 42
Orf43, in greening seedlings, 284
Orf62, part of plastid operon, 282
*Oryza*, *see* Rice
Oscillaxanthin, 453, 478
Osmiophilic globuli, *see* Plastoglobuli

# Index

Oxygen
  evolution, 3, 12, 14, 366, 376
  requirement in chlorophyll biosynthesis, 118, 122
Ozone, 472

## P

P680, in electron transport, 421, 425
P700, binding site, 388, 396, *see also* CP1, chlorophyll–protein complex
Palmitic acid, 243
Paraquat, 472
Pea, 107, 130, 182, 188, 342, 387, 391, 395, 397, 400
Pepper, 493, 496
Perforated membranes, *see* Prothylakoid
Peridinin, 452, 456, 477
Periodic minimal surface, 248
*pet*A, plastid gene, 288, 289
*pet*B, plastid gene, 282, 284
*pet*D, plastid gene, 282, 284, 288
*Petunia*, 292
*Phaeodactylum tricornutum*, 455
Phaeophyceae
  chloroplast ultrastructure, 41, 43
  pigment composition, 41, 451
Phase partitioning, 266
*Phaseolus vulgaris*, *see* Bean
1,10-Phenanthroline, 235
*N*-Phenylmaleimide, 150
Pheophorbide geranylgeranyl ester, 221
Pheophytin, 15, 260, 421
*Phormidium laminosum*, 403
Phosphate translocator, in plastid envelope, 27, 320, 325
Phosphatidyl choline, 355
Phosphatidyl ethanolamine, 260
Phosphatidyl glycerol, 242
Phosphatidyl inositol, 242
Phosphoglyceraldehyde, 371
Phospholipase, 258
Phospholipid, transfer proteins, 249
Phosphorylation, of proteins, 338, 348, 428, 433–434
Photoactivation, of water-splitting system, 368
Photoinhibition, 458, 472
Photooxidative damage, 234, 298, 458, 470

Photophosphorylation, 13, 370
Photoprotection, 457
Photoreceptor, 64, 75, 79, 180, 376
Photosensitizer, 465
Photosynthesis, 11, 14, 366–371, 384, 421–423
Photosystem I (PSI), 383–411
  antenna, 280, 336, *see also* Light-harvesting complex I
  assembly, 288, 404
  carotenoids, 453
  crystallization, 403
  degradation, 405
  development, 370, 375
  electron microscopy, 404
  expression of genes, 404
  function, 82, 384
  localization, 37
  polypeptides, 280–281, 288, 385–401, *see also* specific proteins
  reaction center, 67, 70, 78, 288, 396, 402, 453
  reconstitution, 404, 408
Photosystem II (PSII), 419–437
  acceptor side, 422
  antenna system, 340, 427, *see also* Light-harvesting complex II
  assembly, 297, 337–351
  carotenoids, 340, 454
  development of activity, 297, 369
  donor side, 423
  electron microscopy, 431
  function, 82, 384, 420–424
  heat effects, 435
  heterogeneity, 435, 436
  phosphorylation 433
  photoinhibition, 433
  polypeptides, 280–281, 297, 424–432, *see also* specific proteins
  reaction center, 67, 70, 78, 370, 424
  structural organization, 424–432
  subpopulations, 435
  turnover, 433
Phriaporphyrinogen, 117
Phycobilin, 42
Phycobiliprotein, 92
Phycobilisome, 42
Phycoerythrin, 38, 40
Phylloquinone, 410, 460

Phytochrome, 92, 376, 406
  control of carotenoid accumulation, 470
  fluence responses, 75, 77, 80
  effects on greening, 368
  regulation of transcription, 79, 231, 290, 294
Phytoene, 461, 475, 492
  dehydrogenase, 489
  desaturase, 489
  synthase, 489
Phytoferritin, 31
Phytofluene, 462, 492
Phytol, 9, 154, 460
Phytylation, see Chlorophyllide, esterification
Phytylchlorophyllide, 220, 225
Phytyl diphosphate, 154, 155, 407
Picea, 184
Pigment–protein complexes, 67, 335–336, 354, 453–454
  assembly, 70, 348–355
  discovery, 15
  in LHCII, 70, 294, 354, 375, 428
  protochlorophyllide-containing, 8, 45, 149, 190–192, 194, 223, 373
Pine, see Pinus
Ping-pong mechanism, 132
Pinus, 188, 448
Pinus jeffreyi, 184
Pinus pinea, 187
Pisum, see Pea
Plastid, see also specific plastids
  envelope, 26–28
  lipids, 27, 31, 242
  pigments, 41, 65–66, 451
  size, 34
  tissue specificity, 40, 50, 448
  ultrastructure, 26, 28–32
Plastid development, 32–35
  light dependence, 68–81
  metal chelator effects, 234–236
  regulatory proteins, 496
  temperature dependence, 47, 81–83
Plastid DNA, 28, 281–283
Plastid-encoded polypeptides
  accumulation, 287
  in photosystem I, 385, 387–392
  in photosystem II, 424–429
  stabilization, 233, 287
  transcription, 281–286
  translation, 233, 286–289

Plastid factor, 80, 205
Plastid gene
  encoding polypeptides
    of cyt $b/f$ complex, 282
    of photosystem I, 385, 406
    of photosystem II, 282, 426–427, 432
  expression
    in nuclear mutants deficient in PSII, 297
    post-transcriptional control, 284–285
  regulation
    of transcription, 283–284
    of translation, 286
Plastid mRNA
  levels, 283
  processing, 281
  stability, 284–286
  *trans* splicing, 283
Plastocyanin
  plastid import, 318, 319, 326
  thylakoid transfer domain, 315
Plastoglobuli, 4, 31, 48, 49, 449, 472
Plastoquinone, 15, 419, 422
Polysome, 287, 406
Porphobilinogen, 92, 96, 104, 106
*Porphyridium*, 42, 43
*Porphyridium cruentum*, 30
*Posidonia australis*, 182
Prasinophyceae, 451
Prasinoxanthin, 451, 476
Prenyl transferase, 221, 460, 489
Prephytoen diphosphate, 461
Processing
  of plastid mRNA, 281–282
  of polypeptides, 313, 321, 324–327, 342–345
*Prochloron*, 452
Prochlorophyta
  chloroplast ultrastructure, 41
  lipids, 242
  pigments, 41, 452
Prolamellar body
  in developing chloroplasts, 48, 70, 71, 183, 184, 186
  in etioplasts, 44–47, 68
  protochlorophyllide reductase, 45, 70, 149, 225, 255
  transformation, 4, 45, 68, 183, 202, 225, 267
  ultrastructure, 46, 261–265
Prolycopene, 489
Promotor, of *cabE*, 292

Proplastid, 32, 469, 501
*Protaminobacter ruber*, 99
Protease, 390, 500, 501
Protein import, *see also* Import, of plastid proteins
  insertion, translocation complex, 324
  membrane targeting signal, 316
  translocation, into thylakoid lumen, 316, 326–327
Protein phosphorylation, *see* Phosphorylation, of proteins
Proteinoplast, 49
Prothylakoid, 45, 155, 186, 255, 262, 374
Proto IX, *see* Protoporphyrin IX
Protochlorophyll, 6, 9, 49, 68, 69, 192–194
Protochlorophyllide
  in ALA-treated leaves, 8, 195, 234
  biosynthesis, 124, 149–153
  HPLC separation, 69
  inhibitor of tRNA$^{glu}$-ligase, 195
  monovinyl/divinyl-forms, 135, 191, 194
  as photoreceptor, 64, 68, 231, 406
  spectral forms, 149, 190–191, 195, 373–374
Protochlorophyllide holochrome, 7, 149, 377, *see also* Protochlorophyllide reductase
Protochlorophyllide reductase, 148–153, 179–205, *see also* Protochlorophyllide reduction
  active site, 150, 152, 203
  cDNA, 193, 196–202
  genes, 148, 153, 196
  isoelectric species, 182
  isolation, 150, 181
  light independent enzyme, 151, 153, 182
  localization, 38, 45, 150, 181, 191–192, 200, 225, 255
  membrane association, 203, 265–267
  pigment complexes, 8, 45, 149, 190–192, 194, 223, 373
  precursor polypeptide, 182, 196
  primary structure, 196–202, 204
  regulation of
    enzyme activity, 187–190, 194–196, 200–201
    protein content, 187–190
    transcription, 181–190, 196–197
    translation, 187, 206
  substrate specificity, 142, 193
  translocation, 45, 192, 225

Protochlorophyllide reduction, 7, 150–153, 182–185, 190–191, 195, 230, *see also* Protochlorophyllide reductase
  influence of temperature, 195–196
  light independent, 151, 153, 182, 185
  prolamellar body transformation, 45, 180, 184, 267
  synthesis of plastid proteins, 180, 286
Protogen IX, *see* Protoporphyrinogen IX
Protogen IX oxidase, *see* Protoporphyrinogen IX oxidase
Protoheme, 92, 96, 97
Protoporphyrin IX, 94, 97, 117, 122, 126, 141, 146
Protoporphyrinogen IX, 97, 98, 118, 122, 141, 142
Protoporphyrinogen IX oxidase, 97, 98, 120, 122–124
*Prunus persica*, 450
Prymnesiophyceae
  chloroplast ultrastructure, 41
  pigment composition, 41, 452
*psaA*, plastid gene, 281–285, 287–288, 385, 387
*psaB*, plastid gene, 281–282, 284, 288, 385, 387
*psaC*, plastid gene, 385, 389, 406, 409
*PsaD*, nuclear gene, 385, 394, 405–406
*PsaE*, nuclear gene, 385, 394, 406
*PsaF*, nuclear gene, 385, 395, 397, 406
*PsaG*, nuclear gene, 385, 399, 405
*PsaH*, nuclear gene, 385, 400, 405
*psaI*, plastid gene, 385, 391
*psaJ*, plastid gene, 385, 391
*PsaK*, nuclear gene, 385, 399
*PsaL*, nuclear gene, 385, 405
*psbA*, plastid gene, 281, 285
*psbB*, plastid gene, 281–282, 284–285, 287–289, 297, 427
*psbC*, plastid gene, 281–282, 285, 287–288, 297, 427
*psbD*, plastid gene, 281–283, 285, 288–289
*psbE*, plastid gene, 288, 432
*psbF*, plastid gene, 432
*psbH*, plastid gene, 282, 284–285, 432
*psbI*, plastid gene, 282, 426, 432
*psbK*, plastid gene, 282
*psbL*, plastid gene, 432
PSI-A, protein, 385–388, 406
PSI-B, protein, 385–388, 406

PSI-C, protein, 385, 389, 408
PSI-D, protein, 385, 393
  PSI-D-less mutant, 394
PSI-E, protein, 385, 393, 405
PSI-F, protein, 385, 393, 395, 396, 405
PSI-G, protein, 385, 397–399
PSI-H, protein, 385, 400
PSI-I, protein, 385, 391–393
PSI-J, protein, 385, 391–393
PSI-K, protein, 385, 397–399
PSI-L, protein, 385, 401
PSI-M, protein, 401
Pyrenoid, 43
Pyridoxal 5'-phosphate, 103
Pyridoxamine 5'-phosphate, 103
Pyrroxanthin, 452, 477

**Q**

Quinone, 122, 422, 435

**R**

Radish, 261
Raspberry, 449
Reaction center I, see Photosystem I, reaction center
Reaction center II, see Photosystem II, reaction center
Receptor, in envelope, see Envelope, receptor protein
Reconstitution, of pigment–protein complexes, 341, 351–355
  of photosystem I, 404, 408
Retrocarotenoid, 449
*Rhizobium japonicum*, 119
*Rhodobacter capsulatus*, 491, 494
*Rhodobacter sphaeroides*, 97, 104, 105, 107, 118, 119, 122, 123, 126, 129, 130, 134, 146, 148, 351, 392, 403
Rhodophyceae
  chloroplast ultrastructure, 41, 42,
  pigment composition, 41, 452
*Rhodopseudomonas palustris*, 108, 117
*Rhodopseudomonas spheroides*, see *Rhodobacter sphaeroides*
*Rhodopseudomonas viridis*, 392, 403
*Rhodospirillum rubrum*, 351
Rhodoxanthin, 449, 475

Ribosomes, 29, 288, 406
Ribulose 1,5-bisphosphate carboxylase, see Rubisco
Rice, 187, 188, 391, 400
Rieske iron–sulfur protein, stabilizing function, 405
RNA, 29, 279–299, 342, 404
Rubisco
  large subunit, 70
  localization, 30, 43
  small subunit
    precursor polypeptide, 314
    import into plastid, 319
Rye, 82, 149, 188, 261

**S**

*Scenedesmus obliquus*, 77, 99, 184, 193, 194, 195, 451
  mutant, 155, 185, 193, 196
Schiff base, 105
*Secale*, see Rye
Secocarotenoid, 449
*Selaginella*, 184
Semi-$\beta$-carotenone, 449, 475
Semiquinone, 422
Senescence, 472
Shemin pathway, 94
Shibata shift, 45, 190–191, 223, 227, 372
Signal peptide, of thylakoid protein, 316
*Sinapis*, 188, 338
*Sinapis alba*, 77, 80
Singlet oxygen, 457
Siphonoxanthin, 451, 476
Siroheme, 92, 107
*Solanum*, 50
*Sorghum*, 78
*Sorghum bicolor*, 187, 188, 285
Soybean, 130
Spinach, 78, 192, 224, 282, 284, 315, 336, 376, 395, 397, 400, 401
*Spirogyra*, 2
Squash, 149, 188, 192
Starch, 32, 49
State transitions, 348, 434
Sterols, 252
Stress effects, on chloroplast carotenoids, 472
Stroma, 3, 30

# Index

Stroma centers, 31
Stromal factor
  in assembly of Rubisco, 322, 405
  in protein translocation, 322, 326, 345, 347
Stromal inclusions, 30
Stromal processing peptidase, 313, 321, 324, 343
Succinyl-CoA, 94
Sulfoquinovosyl diacylglycerol, 242
Sunflower, 188
Superoxide, 407
*Synechococcus*, 102, 151
*Synechococcus vulcanus*, 391, 395
*Synechocystis*, 100, 102

## T

Tagetitoxin, 285
Taraxaxanthin, 489
Tetraterpen, 459
*Thuja occidentalis*, 449
Thylakoid, *see also* Grana
  development, 45, 68–77, 219
  formation, 34, 68
  stabilization, 352
  ultrastructure, 4, 35, 36, 42–44, 48, 71
Thylakoid lumen, 35
  proteins, 313, 395
Thylakoid membrane
  dissolution, 500–501
  integration of proteins, 288, 295–297, 323–325, 343–348
  lipids, 243, 244, 246, 254, 257
  model, 384, 429
Thylakoid protein, 72, 76, 251, 288
  processing peptidase, 313, 325–327
  transfer domain, 315, 316
Thylakoid-bound ribosomes, 287, 409
Tobacco, 294
Tocopherols, 460
Tomato, 188, 428, 450, 488, 491, 494, 497
*Tradescantia albiflora*, 182
*trans*-acting factors, 291
Transcription
  regulation
    nuclear genes, 187–188, 290–297, 404
    plastid genes, 79, 283–285, 404, 490–494
  units, 281
Transfer proteins, of lipids, 249

Transit peptidase, 343–344
Transit peptide, 313–316, 395
Translation, of chloroplast proteins, 78–79, 286–287
  initiation and elongation, 233
  involvement of cytoplasmic factor, 231
Translocation, of plastid proteins, 317–323, 351
  across envelope, 317–321
  translocation complex, 318–320, 324
*Triticum, see* Wheat
Triton X-100, 150, 429
trnG, plastid gene, 282
*Tropaeolum majus*, 496

## U

Ubiquinone:NADH oxidoreductase, 392
Unfoldase, in import of plastid proteins, 317–318
Urogen III, *see* Uroporphyrinogen III
Urogen III synthase, *see* Uroporphyrinogen III synthase
Uroporphyrinogen III, 92, 96, 106, 107
Uroporphyrinogen III cosynthase, 108
Uroporphyrinogen III decarboxylase, 99, 117
Uroporphyrinogen III synthase, 109, 112, 114

## V

Valinomycin, 344
Vaucheriaxanthin, 452, 477
Vesicles, in thylakoid formation, 34, 250
*Vicia*, 282
4-Vinyl reductase, 141, 144
*Viola tricolor*, 450
Violaxanthin, 28, 65, 68, 69, 74, 340, 448, 458, 489
Vitamin $B_{12}$, 92, 107

## W

Water-splitting system, 78, 368
Watermelon, 488
*Welwitschia mirabilis*, 183
Wheat, 130, 132, 149, 155, 182, 187, 188, 192, 235, 342

## X

Xanthophyll, 28, 65, 67, 337, 339, 352, 457–458
Xanthophyll cycle, 457–458

## Z

*Zea, see* Maize
Zeaxanthin, 42, 448, 458, 475, 489
Zinc, 104, 223
Zinc chelatase, 126
*Zostera capricorni*, 182

ISBN 0-12-676960-5